Technology Strategy and Innovation Management

Strategic Management

Series Editor: Stephen B. Tallman
E. Claiborne Robins Distinguished Professor, Robins School of Business, University of Richmond, USA

1. Cooperative Strategies: Alliance Governance
   Jeffrey J. Reuer, Shivaram Devarakonda and Elko Klijn

2. Strategy Process
   Paul Olk

3. Corporate Strategy
   Margarethe Wiersema and Joseph Beck

4. Technology Strategy and Innovation Management
   Michael J. Leiblein and Arvids A. Ziedonis

Future titles will include:

Competitive Strategy
Catherine A. Maritan and Margaret A. Peteraf

International and Global Strategy
Torben Pederson

Wherever possible, the articles in these volumes have been reproduced as originally published using facsimile reproduction, inclusive of footnotes and pagination to facilitate ease of reference.

For a list of all Edward Elgar published titles visit our website at
www.e-elgar.com

Technology Strategy and Innovation Management

Edited by

Michael J. Leiblein

Associate Professor of Business and Corporate Strategy,
Fisher College of Business, Ohio State University, USA

and

Arvids A. Ziedonis

Assistant Professor of Management,
Charles H. Lundquist College of Business, University of Oregon, USA

STRATEGIC MANAGEMENT

An Elgar Research Collection
Cheltenham, UK • Northampton, MA, USA

Published by
Edward Elgar Publishing Limited
The Lypiatts
15 Lansdown Road
Cheltenham
Glos GL50 2JA
UK

Edward Elgar Publishing, Inc.
William Pratt House
9 Dewey Court
Northampton
Massachusetts 01060
USA

A catalogue record for this book is available from the British Library

Library of Congress Control Number: 2011922867

MIX
Paper from responsible sources
FSC® C018575
FSC www.fsc.org

ISBN 978 1 84844 435 5

Printed and bound by MPG Books Group, UK

Contents

Acknowledgements ix
Introduction Michael J. Leiblein and Arvids A. Ziedonis xi

PART I TYPES OF INNOVATION

1. Michael L. Tushman and Philip Anderson (1986), 'Technological Discontinuities and Organizational Environments', *Administrative Science Quarterly*, **31** (3), September, 439–65 3
2. Rebecca M. Henderson and Kim B. Clark (1990), 'Architectural Innovation: The Reconfiguration of Existing Product Technologies and the Failure of Established Firms', *Administrative Science Quarterly*, **35** (1), March, 9–30 30
3. Clayton M. Christensen and Joseph L. Bower (1996), 'Customer Power, Strategic Investment, and the Failure of Leading Firms', *Strategic Management Journal*, **17** (3), 197–218 52

PART II CAPTURING VALUE FROM INNOVATION

4. David J. Teece (1986), 'Profiting from Technological Innovation: Implications for Integration, Collaboration, Licensing and Public Policy', *Research Policy*, **15**, 285–305 77
5. Richard C. Levin, Alvin K. Klevorick, Richard R. Nelson and Sidney G. Winter (1987), 'Appropriating the Returns from Industrial Research and Development', *Brookings Papers on Economic Activity*, **3**, 783–831 98
6. Sidney G. Winter (2000), 'Appropriating the Gains from Innovation', in George S. Day, Paul J.H. Schoemaker and Robert E. Gunther (eds), *Wharton on Managing Emerging Technologies*, Chapter 11, New York, NY: John Wiley & Sons, Inc., 242–65, 431–4 147
7. Joshua S. Gans and Scott Stern (2003), 'The Product Market and the Market for "Ideas": Commercialization Strategies for Technology Entrepreneurs', *Research Policy*, **32** (2), February, 333–50 175
8. Ashish Arora and Marco Ceccagnoli (2006), 'Patent Protection, Complementary Assets, and Firms' Incentives for Technology Licensing', *Management Science*, **52** (2), February, 293–308 193
9. Mary Tripsas (1997), 'Unraveling the Process of Creative Destruction: Complementary Assets and Incumbent Survival in the Typesetter Industry', *Strategic Management Journal*, **18** Special Issue, 119–42 209

PART III DELIVERING INNOVATIVE VALUE THROUGH RESOURCE ALLOCATION AND ORGANIZATION ACTIVITY

10. Wesley M. Cohen and Daniel A. Levinthal (1990), 'Absorptive Capacity: A New Perspective on Learning and Innovation', *Administrative Science Quarterly*, **35** (1), March, 128–52 235
11. Lee Fleming and Olav Sorenson (2004), 'Science as a Map in Technological Search', *Strategic Management Journal*, **25**, 909–28 260
12. Jack A. Nickerson and Todd R. Zenger (2004), 'A Knowledge-Based Theory of the Firm – The Problem-Solving Perspective', *Organization Science*, **15** (6), November–December, 617–32 280
13. Iain M. Cockburn, Rebecca M. Henderson and Scott Stern (2000), 'Untangling the Origins of Competitive Advantage', *Strategic Management Journal*, **21** (10/11), October–November, 1123–45 296

PART IV DELIVERING INNOVATIVE VALUE THROUGH INTER-ORGANIZATIONAL DRIVERS

14. Gary P. Pisano (1990), 'The R&D Boundaries of the Firm: An Empirical Analysis', *Administrative Science Quarterly*, **35** (1), March, 153–76 321
15. Walter W. Powell, Kenneth W. Koput and Laurel Smith-Doerr (1996), 'Interorganizational Collaboration and the Locus of Innovation: Networks of Learning in Biotechnology', *Administrative Science Quarterly*, **41** (1), March, 116–45 345
16. David C. Mowery, Joanne E. Oxley and Brian S. Silverman (1996), 'Strategic Alliances and Interfirm Knowledge Transfer', *Strategic Management Journal*, **17** Special Issue, 77–91 375
17. Michael J. Leiblein, Jeffrey J. Reuer and Frédéric Dalsace (2002), 'Do Make or Buy Decisions Matter? The Influence of Organizational Governance on Technological Performance', *Strategic Management Journal*, **23** (9), September, 817–33 390
18. Joanne E. Oxley and Rachelle C. Sampson (2004), 'The Scope and Governance of International R&D Alliances', *Strategic Management Journal*, **25** (8/9), 723–49 407
19. Gautam Ahuja and Riitta Katila (2001), 'Technological Acquisitions and the Innovation Performance of Acquiring Firms: A Longitudinal Study', *Strategic Management Journal*, **22** (3), 197–220 434

PART V REAL OPTIONS

20. Rita Gunther McGrath (1997), 'A Real Options Logic for Initiating Technology Positioning Investments', *Academy of Management Review*, **22** (4), October, 974–96 461
21. Bruce Kogut and Nalin Kulatilaka (2001), 'Capabilities as Real Options', *Organization Science*, **12** (6), November–December, 744–58 484

22. Timothy B. Folta and Jonathan P. O'Brien (2004), 'Entry in the Presence of Dueling Options', *Strategic Management Journal*, **25** (2), 121–38 499
23. Michael J. Leiblein and Arvids A. Ziedonis (2007), 'Deferral and Growth Options Under Sequential Innovation', *Advances in Strategic Management*, **24**, 225–45 517
24. Arvids A. Ziedonis (2007), 'Real Options in Technology Licensing', *Management Science*, **53** (10), October, 1618–33 538
25. Ron Adner and Daniel A. Levinthal (2004), 'What Is *Not* A Real Option: Considering Boundaries for the Application of Real Options to Business Strategy', *Academy of Management Review*, **29** (1), 74–85 554

Acknowledgements

The editors and publishers wish to thank the authors and the following publishers who have kindly given permission for the use of copyright material.

Academy of Management Review via Copyright Clearance Center for articles: Rita Gunther McGrath (1997), 'A Real Options Logic for Initiating Technology Positioning Investments', *Academy of Management Review*, **22** (4), October, 974–96; Ron Adner and Daniel A. Levinthal (2004), 'What Is Not A Real Option: Considering Boundaries for the Application of Real Options to Business Strategy', *Academy of Management Review*, **29** (1), 74–85.

Administrative Science Quarterly for articles: Michael L. Tushman and Philip Anderson (1986), 'Technological Discontinuities and Organizational Environments', *Administrative Science Quarterly*, **31** (3), September, 439–65; Rebecca M. Henderson and Kim B. Clark (1990), 'Architectural Innovation: The Reconfiguration of Existing Product Technologies and the Failure of Established Firms', *Administrative Science Quarterly*, **35** (1), March, 9–30; Wesley M. Cohen and Daniel A. Levinthal (1990), 'Absorptive Capacity: A New Perspective on Learning and Innovation', *Administrative Science Quarterly*, **35** (1), March, 128–52; Gary P. Pisano (1990), 'The R&D Boundaries of the Firm: An Empirical Analysis', *Administrative Science Quarterly*, **35** (1), March, 153–76; Walter W. Powell, Kenneth W. Koput and Laurel Smith-Doerr (1996), 'Interorganizational Collaboration and the Locus of Innovation: Networks of Learning in Biotechnology', *Administrative Science Quarterly*, **41** (1), March, 116–45.

Brookings Institution for article: Richard C. Levin, Alvin K. Klevorick, Richard R. Nelson and Sidney G. Winter (1987), 'Appropriating the Returns from Industrial Research and Development', *Brookings Papers on Economic Activity*, **3**, 783–831.

Elsevier for articles: David J. Teece (1986), 'Profiting from Technological Innovation: Implications for Integration, Collaboration, Licensing, and Public Policy', *Research Policy*, **15**, 285–305; Joshua S. Gans and Scott Stern (2003), 'The Product Market and the Market for "Ideas": Commercialization Strategies for Technology Entrepreneurs', *Research Policy*, **32** (2), February, 333–50.

Emerald Group Publishing Ltd for article: Michael J. Leiblein and Arvids A. Ziedonis (2007), 'Deferral and Growth Options Under Sequential Innovation', *Advances in Strategic Management*, **24**, 225–45.

INFORMS (Institute for Operations Research and the Management Sciences) for articles: Bruce Kogut and Nalin Kulatilaka (2001), 'Capabilities as Real Options', *Organization Science*, **12** (6), November–December, 744–58; Jack A. Nickerson and Todd R. Zenger (2004), 'A

Knowledge-Based Theory of the Firm – The Problem-Solving Perspective', *Organization Science*, **15** (6), November–December, 617–32; Ashish Arora and Marco Ceccagnoli (2006), 'Patent Protection, Complementary Assets, and Firms' Incentives for Technology Licensing', *Management Science*, **52** (2), February, 293–308; Arvids A. Ziedonis (2007), 'Real Options in Technology Licensing', *Management Science*, **53** (10), October, 1618–33.

John Wiley and Sons, Inc., for excerpt: Sidney G. Winter (2000), 'Appropriating the Gains from Innovation', in George S. Day, Paul J.H. Schoemaker and Robert E. Gunther (eds), *Wharton on Managing Emerging Technologies*, Chapter 11, 242–65.

John Wiley and Sons Ltd for articles: David C. Mowery, Joanne E. Oxley and Brian S. Silverman (1996), 'Strategic Alliances and Interfirm Knowledge Transfer', *Strategic Management Journal*, **17** Special Issue, 77–91; Clayton M. Christensen and Joseph L. Bower (1996), 'Customer Power, Strategic Investment, and the Failure of Leading Firms', *Strategic Management Journal*, **17** (3), 197–218; Mary Tripsas (1997), 'Unraveling the Process of Creative Destruction: Complementary Assets and Incumbent Survival in the Typesetter Industry', *Strategic Management Journal*, **18** Special Issue, 119–42; Iain M. Cockburn, Rebecca M. Henderson and Scott Stern (2000), 'Untangling the Origins of Competitive Advantage', *Strategic Management Journal*, **21** (10/11), October–November, 1123–45; Gautam Ahuja and Riitta Katilla (2001), 'Technological Acquisitions and the Innovation Performance of Acquiring Firms: A Longitudinal Study', *Strategic Management Journal*, **22** (3), 197–220; Michael J. Leiblein, Jeffrey J. Reuer and Frédéric Dalsace (2002), 'Do Make or Buy Decisions Matter? The Influence of Organizational Governance on Technological Performance', *Strategic Management Journal*, **23** (9), September, 817–33; Timothy B. Folta and Jonathan P. O'Brien (2004), 'Entry in the Presence of Dueling Options', *Strategic Management Journal*, **25** (2), 121–38; Joanne E. Oxley and Rachelle C. Sampson (2004), 'The Scope and Governance of International R&D Alliances', *Strategic Management Journal*, **25** (8/9), 723–49; Lee Fleming and Olav Sorenson (2004), 'Science as a Map in Technological Search', *Strategic Management Journal*, **25** (8/9), 909–28.

Every effort has been made to trace all the copyright holders but if any have been inadvertently overlooked the publishers will be pleased to make the necessary arrangement at the first opportunity.

In addition the publishers wish to thank the Library at the University of Warwick for their assistance in obtaining these articles.

Introduction

Michael J. Leiblein and Arvids A. Ziedonis

Technology Strategy and Innovation Management provides a strategic perspective on management in complex, knowledge-intensive, and dynamic environments. These settings pose unique challenges to the successful identification and pursuit of competitive advantage compared with those chronicled in traditional strategic management frameworks. Although tradeoffs between different resource allocation policies and forms of organization remain critical, settings characterized by innovative activity require new tools to address issues such as (a) how to evaluate highly uncertain investment proposals involving new (often unknown) products and services, (b) whether and how to capture value from assets in 'product' or 'knowledge' markets, (c) whether and when to shift from 'old' to 'new' sources of advantage, and (d) how to design organizations that assemble and organize resources in a manner that facilitates exploration of new opportunities and exploitation of existing advantages.

In this volume we have assembled a collection of writings that consider how different forms of innovation create value, how business organizations capture the value generated by their innovations, and how firms may allocate resources and organize their activities to deliver value through innovation. We conclude with suggestions of future research opportunities that could identify how different investments and organizational forms affect the creation, capture, and delivery of value through varying types of innovation.

The volume is organized into five parts. The five sections that follow each examine a particular conceptual element of a robust technology strategy. We then conclude by offering an organizing framework for future technology strategy research. Although we have endeavored to spotlight some of the central and most influential writings within each category as well as several pieces that we believe contribute new perspectives, the breadth and complexity of technology strategy and innovation management as a field prevents this from being a truly comprehensive volume. We begin our compendium in Part I by first asking the fundamental question, 'what is innovation?' The literature has identified several distinct varieties, each posing unique organizational and strategic challenges. The articles in Part II explore the factors that determine who captures the value from an innovation. Parts III and IV delve more deeply inside the firm to understand what makes a firm innovative, first from a resource allocation perspective followed by consideration of organizational issues more broadly. Part IV, in particular, expands this intra-firm focus by investigating firm boundaries and firm interaction. Finally, the papers in Part V examine an emerging perspective in technology strategy, 'real options.'

I Types of Innovation

Ever since Schumpeter (1934) first advanced the notion of innovation as 'creative destruction,' or the dynamic process by which technological and organizational transformations cause upheaval of existing firms and industries, innovation scholars have attempted to classify innovation into typologies and distinguish among their effects on firm performance. The three articles in Part I each present a different perspective regarding innovation and its affect on business organizations. A recurring theme in these papers is the tension between incumbents and entrants in the commercialization of new technology.

Chapter 1, by Michael Tushman and Philip Anderson, is concerned with technological evolution and the major shifts or discontinuities that often occur as a dominant technology's development unfolds. Building on Abernathy and Utterback's (1978) product-process innovation model,[1] Tushman and Anderson consider an innovation as an event occurring within a technological trajectory. Such a trajectory is characterized as technological progress along a path of incremental improvement interspersed by infrequent technological breakthroughs. Tushman and Anderson observe that these breakthroughs or 'discontinuous' innovations may be of one of two classes. Breakthroughs that render previous technology obsolete, but build upon incumbent firms' core strengths and capabilities are defined as 'competence-enhancing.' The skills and capabilities that firms had developed in producing the earlier technological generation remain critical to participating in the displacing technology. In contrast, a second class of innovation not only displaces the earlier technology, but also renders obsolete existing know-how and technological assets. This innovation is termed 'competence-destroying.'

The distinction between competence-enhancing and competence-destroying innovation suggests the types of firms that are likely to succeed in each of these settings.[2] Incumbent firms benefit from competence-enhancing innovation as their existing skills and capabilities retain their value in the new technological generation. Incumbents are disadvantaged when a competence-destroying innovation occurs, however. Thus, new entry and displacement of leading incumbents are often associated with competence-destroying innovations. Tushman and Anderson find empirical support for their characterization of the evolution of technology and competitive interaction in such disparate industries as airlines, cement, and minicomputers. As we'll see from our other papers in this part, competence-destroying innovations are not the only class of innovations which incumbents must beware.

In their seminal paper, Rebecca Henderson and Kim Clark (1990; Chapter 2) take issue with the simple depictions of innovation as incremental or radical in economic or competence terms. Henderson and Clark examine a class of seemingly incremental innovations, termed 'architectural,' that results not in the reinforcement of the incumbent's position, but its displacement. As does radical innovation, architectural innovation greatly diminishes the value of an incumbent's existing capabilities. Unlike radical innovation, however, architectural innovation destroys the usefulness of knowledge regarding the interaction of components of a technological system rather than the knowledge of the components themselves. Because such knowledge often defines the patterns of interaction between units within an organizational structure, technological changes in the way components interact may go unrecognized within the organization. Thus a seemingly minor change may have profound effects on an organization's ability to respond. The implication is that successful adjustment to architectural innovations will require substantial organization changes to support modified communication channels and information filters.

In Chapter 3 'Customer Power, Strategic Investment, and the Failure of Leading Firms,' Clay Christensen and Joseph Bower agree with Henderson and Clark that the very organizational structure that may lead a firm to success may also prevent it from detecting and responding to certain innovations. Whereas Henderson and Clark emphasize that subtle changes in the way product components interact can lead to the downfall of industry leaders, Bower and Christensen draw attention to the difficulty existing organizations designed to meet the needs of existing customers effectively can have in dealing with 'disruptive' innovations that initially do not meet the need of those critical customers. Bower and Christensen distinguish between innovations that sustain the improvement trajectory of an existing technology from those that redefine that trajectory. Incumbents are well positioned to benefit from the former – these innovations often benefit the very customers these firms are organized to serve and leverage existing resource allocation policies and information channels. In contrast, disruptive innovations, while improving performance on one or more dimensions, actually fall below current standards on the dimension existing customers value most and often call for substantial changes in resource allocation policy. In this situation, even if incumbents are fully capable of understanding and even producing a disruptive innovation, their need to maintain close relationships with existing critical customers and challenges to existing policies and organization structures often prevents them from doing so. Disruptive technologies often gain traction by appealing to different customers or addressing needs of an emerging market. This in itself is not a danger to incumbents if the situation were static. The performance trajectory of disruptive technologies on technological dimensions that matter most to the established market, however, often exceeds the trajectory of the existing technology. By the time absolute performance of the new technology exceeds that of the existing technology, it is too late for the incumbents and they are often displaced.

II Capturing Value from Innovation

The papers in Part II address a central issue within technology strategy, how to capture the value of innovation. The first article in this group, Chapter 4 by David Teece, is one of the most influential and highly cited papers in the innovation literature. His pioneering 'Profiting from Technological Innovation: Implications for Integration, Collaboration, Licensing and Public Policy' has defined a stream of work for the past two decades, so much that the 20th anniversary of the publishing of this article was celebrated by a special issue in the journal *Research Policy* (**35** (8), October 2006). This article separates the act of innovation from the factors that influence whether the innovator can realize the returns from that innovation or whether returns will likely accrue to others. Teece's seminal insight is that in environments in which intellectual property rights are weak (i.e., offer little protection from imitation or substitution by rival innovations) returns often are realized not by the innovator, but by the holders of 'complementary assets,' such as superior manufacturing, distribution, or marketing positions. Considering the interaction of the 'appropriability regime' and the complementary asset positions of the innovator and rivals, Teece predicts who will most likely profit from the innovation. Moreover, he offers detailed prescriptions to innovators on firm boundary strategies (whether to integrate or utilize markets for technology such as licensing) depending on the strength of intellectual property protection surrounding the innovation.

In the year following the appearance of Teece's landmark article, contemporaries Richard Levin, Alvin Klevorick, Richard Nelson, and Sidney Winter published in the *Brookings Papers on Economic Activity*, the widely known and influential 'Yale Survey' of R&D managers on the means by which their corporations appropriate returns from R&D (Chapter 5). This survey of R&D managers from large R&D performing corporations found that industries vary considerably in the extent to which patent and non-patent (e.g., secrecy or lead time) mechanisms afford protection to innovators in capturing value from R&D. Survey responses from this study have been used by many scholars to construct measures of overall industry appropriability conditions. The Yale survey results were updated in 1994 by the 'Carnegie Mellon survey' (Cohen, Nelson, and Walsh, 2000).

Sidney Winter, one of the collaborators in the original Yale Survey, contributes in Chapter 6 a piece directed towards a managerial audience that links many of the ideas in the Teece (1986) article and the findings of Yale and Carnegie Mellon surveys. Winter thoroughly dissects the various value appropriation mechanisms to innovation and offers managers detailed strategic guidelines intended to help them understand the appropriability environment they find themselves in. Managers with unquestioning belief in the efficacy of patent protection will find their confidence shaken and their understanding of value appropriation much more nuanced after reading the Winter article.

The final two articles in this part also build on the 'shoulders of giants' above who pioneered much of our understanding on the capturing value of innovation. Joshua Gans and Scott Stern (Chapter 7) extend Teece (1986) by contributing a framework intended to explicitly guide managers in understanding the appropriability and strategic conditions they encounter in attempting to commercialize a new innovation. They identify four combinations of complementary asset importance and excludability conditions to offer detailed prescriptions to entrepreneurs and innovators on whether to 'go it alone' and challenge incumbents, or to profit from innovation indirectly through the 'market for ideas'.

Ashish Arora and Marco Ceccagnoli (2006; Chapter 8) provide a recent careful cross-industry study of the extent to which the strength of intellectual property protection influences firms' boundary choices. Building on the central ideas outlined in the above articles and based on R&D managers' survey responses in the aforementioned Carnegie Mellon survey, these scholars find that when firms lack the complementary assets needed to commercialize a new technology, they are more likely to license that technology as the effectiveness of patents in protecting the technology increases. Firms that possess the requisite complementary assets, however, are found to be less likely to license a technology than to pursue commercialization internally, although they are more likely to patent.

Our final article in Part II synthesizes heretofore largely independent streams of academic research on the various types of innovation (highlighted in Part I) and the mechanisms to capture value from these innovations. In 'Unraveling the Process of Creative Destruction' (Chapter 9) Mary Tripsas asks why incumbent firms succeed or fail in the face of radical technological change. Specifically, Tripsas examines: (1) how economically radical innovation affects investment incentives of entrants and incumbents, (2) how competence-destroying innovation affects the utility of incumbents' skill base and resource allocation policies, and (3) how the effect of an innovation on the value of complementary assets affects the ability of entrants and incumbents to appropriate value from an innovation. Through a detailed account of the evolution of the typesetting industry, Tripsas emphasizes the effect of the need for new technical skills and new architectural knowledge on incumbent firms' performance in the face of innovative change.

The ideas and findings presented by Arora and Ceccagnoli and the other articles included in Part II suggest that the strength of intellectual property protection and location of complementary assets needed in successful commercialization are key determinants of the vertical structure of firms in innovative industries, as well as the extent to which markets for ideas function within those industries. And, by reminding the reader of the need to broadly consider how different types of innovation affect competition, the Tripsas article provides an important bridge to the topic of Part III – how to deliver value through innovation.

III Delivering Innovative Value Through Resource Allocation and Organization Activity

The central premise common to the papers in Part III is that a firm's ability to recognize, assimilate, and apply information to commercial ends is critical to its capability to deliver innovative value. Wesley Cohen and Daniel Levinthal's seminal article entitled 'Absorptive Capacity: A New Perspective on Learning and Innovation' (Chapter 10), is one of the most influential works within this literature (see also Cohen and Levinthal, 1989). Cohen and Levinthal reason that the basic skills, shared language, and cognitive structures that underlie an organization's existing knowledge base dictate its ability to assimilate and utilize new knowledge. By redirecting attention from the immediate consequences of R&D investment to the innovation process itself, these scholars suggest that the cumulative nature of prior knowledge has important implications for the development of absorptive capacity and the innovative performance of organizations. Further, the observation that R&D creates a capacity to assimilate and exploit new knowledge provides a explanation for why some firms may invest in basic research even when the preponderance of their findings spill out into the public domain. Cohen and Levinthal propose that business firms are often motivated to conduct research not only to solve specific problems but also to develop general skills and understanding that would permit them to exploit external scientific and technological knowledge, a theme picked up by our next article.

Lee Fleming and Olav Sorenson examine the association between scientific research and innovation in their paper 'Science as a Map in Technological Search' (Chapter 11). Consistent with work that describes the various impediments associated with different forms of innovation, these scholars propose that the nature of scientific research itself alters the form of innovative search. In exchange for up-front investment in learning (absorptive capacity), science provides insights and heuristics that allow researchers to (1) conduct comparatively inexpensive offline experimentation, (2) quickly winnow unproductive directions, and (3) identify global optima. While local search processes such as trial and error learning are associated with simple innovations involving independent components, Fleming and Sorenson suggest that a scientific approach is strongly linked to innovation problems that involve multiple, tightly coupled bits of knowledge.

Jack Nickerson and Todd Zenger's Chapter 12 entitled 'A Knowledge-Based Theory of the Firm – The Problem-Solving Perspective' extends Fleming and Sorenson's approach by suggesting ways managers may craft governance forms to influence the innovative search process. Nickerson and Zenger maintain that if problems systematically differ in complexity and if different governance forms are comparatively superior to others in facilitating particular search processes, then a key managerial task is to select the governance device that aligns a particular problem type with a particular search process. Emphasizing distinctions in the allocation of decision rights, the intensity of incentives to motivate search, and the availability of communication channels to support knowledge transfer, these scholars suggest that the selection of market, authority-based hierarchy, or consensus-based hierarchy differs in their support of various forms of search. Market organization is seen as a low-cost and efficient means of supporting directional search, authority-based hierarchy is an efficient means of directing heuristic search in moderately decomposable environments, and consensus-based hierarchy fashions an efficient means to stimulate heuristic search in non-decomposable settings.

In their joint article titled 'Untangling the Origins of Competitive Advantage' (Chapter 13), Ian Cockburn, Rebecca Henderson, and Scott Stern discuss the importance of distinguishing between the influence of initial conditions and directed managerial action on competitive advantage. They exploit the recent diffusion of science-based discovery processes through the pharmaceutical industry to provide evidence suggesting that initial conditions and purposeful strategic adjustment are jointly associated with leading positions. Their work calls attention to opportunities for future research that explore how organizational endowments and *ex ante* organizational decisions influence the manner by which particular firms are able to create and sustain advantage. Future studies could focus on the effect on competitive advantage by decision making on resource allocation policies, use of particular managerial tools, scope of networks, diversification, outsourcing versus relational contracting, or off-shoring and multi-national organization choices, for example.

IV Delivering Innovative Value Through Inter-Organizational Drivers

The papers in Part IV examine factors that determine the boundaries of the firm (i.e., the determinants of which activities are more effectively performed internally and which activities are better accomplished through contracts or alliances).

Gary Pisano's 1990 article, 'The R&D Boundaries of the Firm: An Empirical Analysis' (Chapter 14), is one of the first to apply transaction-cost economics (TCE) reasoning to firm boundary choices within the pharmaceutical industry, and research and development procurement decisions more generally. Pisano highlights two forms of transaction costs as being particularly critical in the choice of whether to conduct R&D in-house or to contract with outside entities (in this context, specialized biotechnology firms). These are bargaining hazards arising from specialized R&D activities and the degree of difficulty that the firm would have in defining and appropriating the value of intellectual property arising from R&D outcomes. The specialized knowledge required for successful research and development of an innovation often reduces the number of potential partners or suppliers in the innovation value chain to a small number. A lack of alternative suppliers with the requisite specialized knowledge can render the downstream firm especially vulnerable to possible opportunistic behavior by the supplier. Pisano finds that in such situations the procuring firm is more likely to conduct R&D internally than through external agreements. Likewise, as the number of rivals grows, firms are more likely to stay out of the market for specialized R&D services to avoid the greater opportunities the supplier would have to leak strategic information to rivals.

Walter Powell, Kenneth Koput, and Laurel Smith-Doerr also investigate firm relationships within the biotechnology and pharmaceutical industries, but come to a very different conclusion than Pisano regarding the primary drivers of firm boundaries and organization. In their 1996 *Administrative Science Quarterly* piece (Chapter 15) these scholars emphasize that in an emerging science-based industry such as biotechnology, knowledge creation is widely dispersed. In such an environment individual firms rarely possess the complete range of scientific capabilities necessary to stay competitive and current. Research breakthroughs and technological development therefore are often products of wide collaborations involving biotech firms, university scientists, and pharmaceutical firms. The central challenge for firms in such an environment is to actively participate in and successfully manage these collaborations. Powell, Koput, and Smith-Doerr argue that the firms that are most centrally located within knowledge networks are those that are best positioned to access the needed capabilities and thus enjoy the greatest prospects for growth.

Whereas Powell et al. highlight the importance of knowledge transfer across firm boundaries, David Mowery, Joanne Oxley, and Brian Silverman look more closely at how it occurs. In their influential article 'Strategic Alliances and Interfirm Knowledge Transfer' (Chapter 16), they assess the effects of cross-firm knowledge transfer on capability development. One novel feature of the paper is their use of firm patent holdings to measure technological capabilities and their transfer. The paper provides modest support of Dan Levinthal and Wes Cohen's notion of 'absorptive capacity' facilitating external knowledge acquisition by the firm (Chapter 10).

While Oxley and Sampson (2004; Chapter 18) concur that alliances can effectively facilitate exchange of technological knowledge, they argue that such organizational arrangements can also leave the alliance partners vulnerable to misappropriation of proprietary knowledge not intended for exchange. They emphasize that the scope of the alliance may be used as a mechanism to limit misappropriation. While narrowing the scope of an alliance may reduce the amount of knowledge that can be transferred across firm boundaries, it can also serve to protect technological assets. Oxley and Sampson find that among R&D alliances involving electronics and telecommunications firms, firms that form partnerships with direct competitors are likely to do so in a limited fashion. Such alliances tend to be restricted to R&D activities rather than encompassing more commercial marketing activities. Equity-based alliances are found to be broader in scope in general, consistent with prior research on alliance governance choice.

Two papers highlighted in this section investigate performance implications of firm boundary decisions. Michael Leiblein, Jeffrey Reuer, and Frédéric Dalsace (Chapter 17) explore outsourcing alliances and vertical integration choices in an industry characterized by innovative activity, semiconductor manufacturing. Like Pisano, these scholars incorporate the transaction cost economics framework into their analysis, and largely confirm his pharmaceutical industry predictions in the semiconductors. Continuing where Pisano left off, however, they examine the performance consequences of these organizational boundary decisions. One of their most important findings is that firms that chose integration when TCE would predict they should enjoyed higher technological performance than they would if they chose the counterfactual, and vice versa. By documenting the costs of transactional misalignment, Leiblein, Reuer, and Dalsace thus provide an important contribution to the empirical transaction cost economic literature.

As do Mowery, Oxley and Silverman, Ahuja and Katila (2001; Chapter 19) measure firm technological capabilities and knowledge bases with patent-based measures, but apply these measures to understanding acquisition performance. A unique feature of this paper is that the authors disaggregate acquisitions into those that are motivated by access to technology and those that are non-technological. While non-technological acquisitions are found to have little effect on post-acquisition innovation output in their sample of chemicals industry firms, acquired technological knowledge bases do influence innovative performance. Similar to the mixed results reported in Mowery, Oxley, and Silverman regarding the influence of absorptive capacity on a firm's ability to integrate knowledge from an alliance partner, Ahuja and Katila find that absorptive capacity is most valuable for acquiring firms whose knowledge bases moderately overlap with that of the acquired firm. Firms whose knowledge bases are either too similar or too dissimilar suffer from reduced post-acquisition innovative performance compared to the moderate firms.

V Real Options

The final six papers in our volume address the 'real options' perspective on technology strategy. The emphasis of this approach on providing a rational method to make sequential decisions under uncertainty suggests that it may offer a logic well-suited to the uncertainty and dynamism typical in technologically intensive settings.

Rita McGrath's paper titled 'A Real Options Logic for Initiating Technology Positioning Investments' (Chapter 20) provides one of the first applications of real options thinking to technology strategy. The paper conceives of technology strategy in terms of a sequence of investments in technology development, technology commercialization, and the right to extract returns from the technology. McGrath treats early investments in technology development or commercialization as options with a claim on future (uncertain) returns from the technology (e.g., through licensing or product-market competition). A key point in the paper is that firms may create competitive advantage by proactively initiating 'strategic actions' that influence the level and structure of uncertainty in subsequent revenue streams. McGrath describes a number of such actions including consumer education, standard setting, and the development of appropriability regimes. She contends that real options logic suggests how actions that alter the structure of demand uncertainty affect the magnitude and timing of potential revenue streams and may offer an under-appreciated means to improve returns.

Bruce Kogut and Nalin Kulatilaka's article entitled 'Capabilities as Real Options' (Chapter 21) further advances our thinking by providing the conceptual apparatus to understand strategic capabilities as real options. The paper employs concepts from strategic management regarding the importance of scarce factors of production, concepts from strategy and organizational ecology regarding irreversibility and inertia, and concepts from complex systems regarding landscape ruggedness and option values to formally illustrate how capabilities and organizations act to derive potential future value.

Kogut and Kulatilaka's provocative paper contributes to the innovation literature in a number of ways. The authors suggest that capabilities act as options to change positions in a competitive landscape, identify the coupling of bundles of assets as a primary source of irreversibility, and describe how option theory may help to evaluate investment in a capability to change a market position of uncertain and changing value. Further, by formally modeling the benefits of exploiting current capabilities and maintaining or switching (at a cost) these capabilities in the future, the paper creates opportunities to carefully explore how firms dynamically choose an optimal set of capabilities over time.

In Chapter 22, titled 'Entry in the Presence of Dueling Options,' Tim Folta and Jonathan O'Brien empirically explore the association between uncertainty and entry or investment in a new business segment. In contrast to prior work that emphasizes the negative ramifications of uncertainty on investment, Folta and O'Brien frame the question of entry in terms of deferral and growth option value. Arguing that growth options are more sensitive to uncertainty than deferral options, they propose that uncertainty has a non-monotonic effect on entry decisions. In their pooled cross-sectional analysis of firms in the Compustat database, they find that uncertainty negatively affects the probability of entry throughout most of the range of measured uncertainty but the effect reverses under very high levels of uncertainty (above the 93rd percentile in their study). In addition to providing one of the first empirical tests of growth and deferral option logic, Folta and O'Brien contribute to the innovation literature by suggesting opportunities to consider questions involving the likelihood and timing of investment in real option terms.

Michael Leiblein and Arvids Ziedonis (Chapter 23) also discriminate between growth and deferral options in their exploration of investments in environments characterized by continual technical progress. They present a model that considers decisions to invest in existing or expected future technologies and the timing of those decisions. Leiblein and Ziedonis employ real option logic to support their claim that environmental-, firm-, and technological-characteristics dictate an optimal investment timing strategy. They describe how growth and deferral option value differ on the object of uncertainty, the mechanisms used to secure a claim on the opportunity to act flexibly in the future, and the expected arrival time of the future opportunity. Leveraging these differences, they suggest how certain characteristics affect the optimal investment strategy. Their paper contributes by outlining a general theoretical framework that helps to distinguish, *ex ante*, between different sources of option value and with which to analyze the factors that influence whether and when firms should adopt a given technology. It also suggests the opportunity for future work that could marry real option logic with theories of competitive advantage based on the timing (early or late) or nature (reversible or irreversible) of investment.

In Chapter 24, titled 'Real Options and Technology Licensing,' Arvids Ziedonis examines the use of options contracts by firms acquiring rights to commercialize university technologies to conduct one of the most careful empirical tests of real options logic published to date. Whereas data constraints have prohibited much of the prior empirical options literature from examining factors that influence the decision to invest in an option or have been forced to rely on proxies of 'virtual' or 'implied' options, this study overcomes these limitations by leveraging a unique dataset which includes observations on decisions by firms whether or not to purchase option contracts in the licensing of technologies arising from university research.

In addition to offering a test of the predicted associations between uncertainty and the propensity to invest in options, the paper offers insights regarding the association between firm-specific knowledge, the propensity to invest in options contracts, and the propensity to exercise these options. In so doing, the study suggests that option value can be firm-specific and that the decision to license a technology is as at least partially conditioned by prior decisions to invest in option agreements.

Whereas the previous articles all assume that real option logic may generate new insights, in 'What is Not a Real Option' (Chapter 25) Ron Adner and Dan Levinthal remind us that many of the assumptions underlying real options logic may be violated in organizational settings. In particular, Adner and Levinthal examine two boundary conditions associated with the application of real options logic to business and technology strategy. Noting that the benefits of flexibility of real options logic stem from the possibility of abandoning investment, they point out that the open-ended nature of search and the lack of explicit and exogenous expiration dates typical of many investments designed to stimulate innovation violate core tenets of the real options logic. Moreover, they propose that other factors such as bounded rationality, escalating commitment, and overconfidence may produce substantial costs in the application of the real options approach. Given the impossibility of proving failure and the absence of formal expiration dates for an option investment, a firm's internal selection regime dictates its ability to manage options. A well-managed real options strategy must necessarily guard against the natural momentum that could come with this approach.

Adner and Levinthal follow by questioning about the distinctiveness of the real options tool compared to the broader set of sequential decision making approaches such as 'slack search.' After comparing real option heuristics with other sequential search processes, they conclude that when target markets and technical agendas are flexible, activities may be more accurately described in terms of path-dependent processes such as slack search.

In the end, we believe the appropriate metric for a management theory is whether or not it is fruitful. As Adner and Levinthal point out, the benefits associated with the precision of a real options approach must be balanced against the boundaries of these initiatives.

Concluding Remarks

The articles selected for this volume illustrate how uncertainty, complexity, and dynamism in innovative settings impose tradeoffs across firms' resource allocation and organization policy decisions. It is our contention that technology strategy and innovation management's emphasis on these tradeoffs provides a number of unique insights and contributions to the broader field of strategic management.

Our compendium begins by reviewing in Part I several categories of innovation identified by leading academic scholars. The precision offered by these definitions sets the stage for delineating the strategic and organizational challenges posed by innovative activity, challenges which are tackled by the remaining articles in the volume. Part II provides a foundation with which to consider mechanisms such as patents, trade secrets, and lead time and their effect on the distribution of profits. Parts III and IV consider associations between intra- and inter-organizational forms and innovation. The volume concludes with a set of articles employing real options logic as a means to understand and manage uncertainty in Part V.

A fuller and more explorative description of our thinking regarding technology strategy and innovation management is summarized in Figure 1. Following an organizing structure commonly used in the fields of competitive and technology strategy, the outer ring of this figure is composed of three elements signifying the creation, capture, and delivery of value through innovation. Creating value involves identifying strategies that increase consumer willingness to pay and/or reduce supplier opportunity cost. Capturing value implies bargaining with consumers and suppliers using mechanisms that distribute portions of the total value created to members of the ecosystem. Delivering value requires the allocation of resources and design of organizational systems that facilitate the coupling and recombination of appropriate knowledge chunks. The inner ring in Figure 1 portrays smaller circles each depicting topical

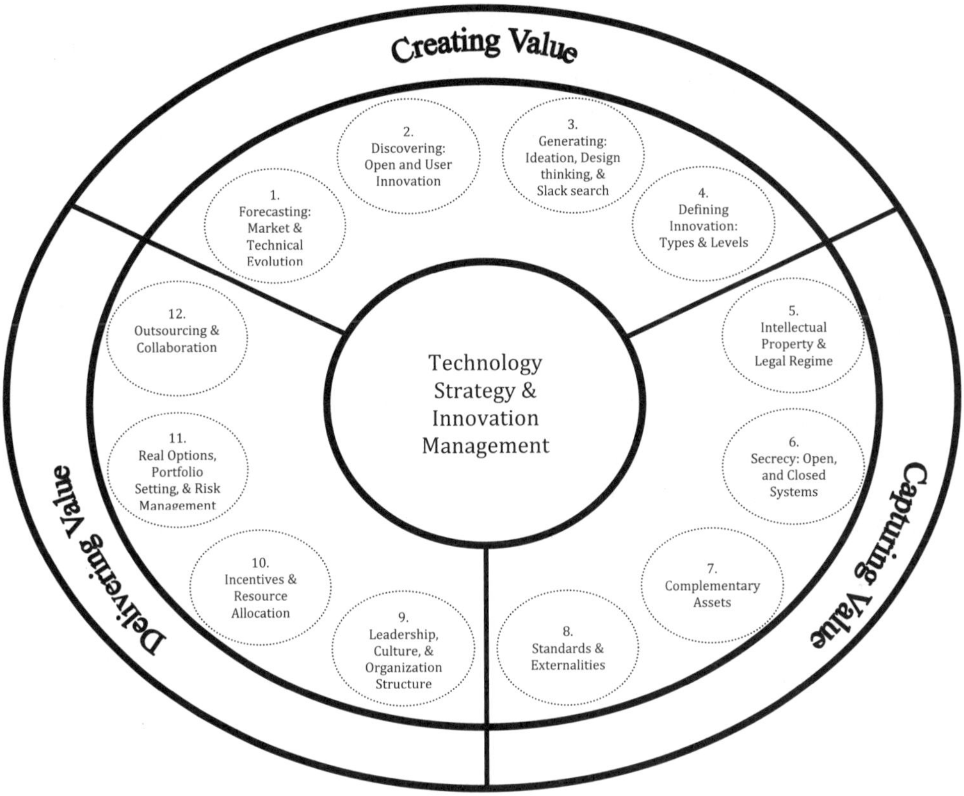

Figure 1 An overview of topics in technology strategy and innovation management

areas within each of the three broader topics. In sum, Figure 1 captures much of the understanding of innovation by academics and practitioners, and points to areas in need of additional research.

The first topic, creating value, captures how organizations harness the creative horsepower inside and outside the organization to increase top line revenue growth and reduce supplier opportunity cost. While our volume emphasizes different types of innovations of interest to managers, other work in technology strategy describes the processes that lead to innovation. Creating value through innovation also involves developing an understanding of the types of products that consumers will enthusiastically reward and an understanding of the improvements in resources and activities required to design, manufacture, sell and distribute these products. Figure 1 depicts processes used to forecast, discover, or generate insights associated with the creation of value through innovation such as trend- and scenario-analysis (e.g., Foster, 1988; Schwartz, 1992), open innovation (e.g., Chesbrough, 2003, 2006a, 2006b; Raymond, 1999) or user innovation (e.g., von Hippel, 1986, 1988), and ideation and design thinking.

Despite frequent discussion, less is known about the causes and consequences of these basic approaches. Future research may productively examine associations between particular environmental characteristics, specific innovative processes, and the generation of various types of innovation. For instance, while analytical forecasting or real option techniques may be valuable in generating innovation in settings where there are high degrees of demand or technological volatility, ideation programs and 'design thinking' may be more useful when there are high degrees of ambiguity regarding consumer needs. Although the ability to develop deep pools of knowledge and support rich communication protocols may favor formal R&D labs for innovations involving poorly structured problems with high degrees of product- or technological complexity, better structured problems involving modular knowledge chunks may benefit from 'user' and 'open innovation' programs that provide access to a broader knowledge base through markets or social structures. In sum, we hope to stimulate additional work regarding the creation of value through innovation by illustrating major topical areas associated with the creation of value through innovation.

The second major component represents questions pertaining to traditional questions in technology strategy and innovation management regarding the mechanisms organizations use to capture value from their inventions. This section focuses on whether and how intellectual property management practices such as patents, secrecy, complementary assets, and lead time affect a firm's ability to manage, protect, and leverage knowledge to capture value. While a great deal of academic research has described these value capture mechanisms and their use in particular settings (e.g., Levin, Klevorick, Nelson and Winter, 1987; Teece, 1986 (Chapter 4); Winter, 2000 (Chapter 6)), opportunities remain to develop and test a robust theory suggesting whether and when each of these mechanisms is likely to be most effective. One promising alternative is to develop theory regarding the industry-, firm-, and technology-level conditions under which each of these mechanisms is most effective. Among other issues, such a theory may indicate conditions under which technology strategies stressing open innovation or secrecy are likely to be successful. Ultimately, an effective knowledge sharing and protection strategy should optimize revenue derived from a firm's patent portfolio (and IP licensing).

The final major element portrayed in Figure 1 summarizes how organizations may allocate resources and manage R&D operations to deliver value. This portion represents how various leadership roles, incentive designs, research portfolios and organization structures affect the level and type of innovation activity and the pursuit of subsequent competitive advantage. Building on the insights of Cohen and Levinthal's (Cohen and Levinthal, 1989, 1990 (Chapter 10)) notion of absorptive capacity, recognition that knowledge required for innovation is often 'sticky' (von Hippel, 1998), and the work on problemistic search (Fleming and Sorenson, 2004 (Chapter 11)) and organization (Nickerson and Zenger, 2004 (Chapter 12)) presented in this volume, we anticipate future opportunities may exist in clarifying and testing whether and how the economic organization of an ecosystem affects innovation activity and the development of an innovation trajectory. It may also be possible to link the insights from research emphasizing linkages between organization and search with real options logic reviewed in this volume suggesting how investment affects capability development (Kogut and Kulatilaka, 2001 (Chapter 21)), the efficacy of early and late mover advantages (Leiblein and Ziedonis, 2007 (Chapter 23)) and the pursuit of competitive advantage (McGrath, 1997 (Chapter 20)). Ultimately, this work should determine which components to source, how to identify and certify appropriate sourcing partners, and whether and at what level to collaborate and share technical, operations, and consumer knowledge.

Overall, this volume provides an overview of influential thinking regarding the means by which technology strategy may be used to dynamically align a firm's goals, investments, resource base, and organization to pursue new sources of competitive advantage and economic value. Ultimately, future research in this area should lead to the development of superior business organizations that outperform their rivals in terms of metrics such as overall margin growth, revenue from new products, and balanced short- and long-term financial performance.

Notes

1. Abernathy and Utterback (1978) suggest that product and process innovation, the competitive environment and organizational structure are all closely linked. They propose that innovations proceed through different phases, an initial phase in which firms compete through product design attempting to establish a standard or 'dominant design,' and a mature phase in which the basic parameters of a new technology become established and further innovative activity is focused on improving the basic technology and reducing costs. According to the Abernathy and Utterback model, the transition from pre-dominant design to a standard is associated with a 'shake-out' period in which firms whose products do not fit within the dominant standard often exit. The remaining firms tend to undergo organizational changes designed to increase competitiveness within an increasingly mature market.
2. In contrast to research that emphasizes how the expected economic benefits of an innovation affect a firm's *incentive* to innovate (e.g., Gilbert and Newberry, 1982; Reinganum, 1983, 1985), Tushman and Anderson (Chapter 1) explain how systemic differences in competencies across entrants and incumbents may affect their comparative *ability* to initiate certain types of innovations.

References

Abernathy, W.J. and Utterback, J.M. (1978), 'Patterns of innovation in technology', *Technology Review*, **80** (7), 40–47.

Chesbrough, H. (2003), *Open Innovation: The New Imperative for Creating and Profiting from Technology*, Boston, MA: Harvard Business School Press.

Chesbrough, H. (2006a), *Open Business Models: How to Thrive in the New Innovation Landscape*, Boston, MA: Harvard Business School Press.

Chesbrough, H. (2006b), *Open Innovation: Researching a New Paradigm*, Oxford: Oxford University Press.

Cohen, W. and Levinthal, D. (1989), 'Innovation and learning: The two faces of R&D', *Economic Journal*, **99** (397), 569–96.

Cohen, W., Nelson, R. and Walsh, J. (2000), Protecting their intellectual assets: appropriability conditions and why U.S. manufacturing firms patent (or not). NBER working paper no. W7552.

Foster, Richard N. (1988), *The Attacker's Advantage*, New York, NY: Simon and Schuster.

Gilbert, Richard J. and Newbery, D.M.G. (1982), 'Preemptive patenting and the persistance of monopoly', *American Economic Review*, **72**, 514–26.

Raymond, Eric S. (1999), *The Cathedral & the Bazaar*, Sebastopol, CA: O'Reilly.

Reinganum, Jennifer F. (1983), 'Uncertain Innovation and the Persistence of Monopoly', *American Economic Review*, **73**, 741–48.

Reinganum, Jennifer F. (1985), 'Innovation and Industry Evolution', *Quarterly Journal of Economics*, **100** (1), 81–99.

Schumpeter, Joseph A. (1934), *Theory of Economic Development*, Cambridge, MA: Harvard University Press.

Schwartz, Peter (1992), *The Art of the Long View: Scenario Planning – Protecting Your Company Against an Uncertain Future*, New York, NY: Random House Business Books.

von Hippel, Eric (1986), 'Lead Users: A Source of Novel Product Concepts', *Management Science*, **32** (7), (July), 791–805.
von Hippel, Eric (1988), *The Sources of Innovation*, New York: Oxford University Press.
von Hippel, Eric (1998), 'Economics of Product Development by Users: The Impact of "Sticky" Local Information', *Management Science*, **44**, (5), (May), 629–44.

Part I
Types of Innovation

[1]

Technological Discontinuities and Organizational Environments

Michael L. Tushman
Columbia University
Philip Anderson
Cornell University

This paper focuses on patterns of technological change and on the impact of technological breakthroughs on environmental conditions. Using data from the minicomputer, cement, and airline industries from their births through 1980, we demonstrate that technology evolves through periods of incremental change punctuated by technological breakthroughs that either enhance or destroy the competence of firms in an industry. These breakthroughs, or technological discontinuities, significantly increase both environmental uncertainty and munificence. The study shows that while competence-destroying discontinuities are initiated by new firms and are associated with increased environmental turbulence, competence-enhancing discontinuities are initiated by existing firms and are associated with decreased environmental turbulence. These effects decrease over successive discontinuities. Those firms that initiate major technological changes grow more rapidly than other firms.•

Since Barnard's (1938) and Selznick's (1949) seminal work, one of the richest streams of research in organizational theory has centered on organization-environment relations (see Starbuck, 1983, for a review). Recent work on organizational life cycles (Miller and Friesen, 1984; Tushman and Romanelli, 1985), organizational adaptation (Aldrich and Auster, 1986), population dynamics (Freeman, 1982), executive succession (Carroll, 1984), and strategy (e.g., Harrigan, 1983) hinges on environment-organization linkages. Environments pose constraints and opportunities for organizational action (Hrebiniak and Joyce, 1985).

If organizational outcomes are critically influenced by the context within which they occur, then better understanding of organizational dynamics requires that we more fully understand determinants of environmental change. While there has been substantial research on environmental conditions and organizational relations (see review in Downey and Ireland, 1979), relatively little research has examined how competitive environments change over time. While it is agreed that environmental conditions are shaped by competitive, legal, political, and technological factors (e.g., Starbuck, 1983; Romanelli and Tushman, 1986), and the interplay between them (Horwitch, 1982; Noble, 1984), there is little data on how these factors change over time or how they affect environmental conditions.

This paper focuses on technology as a central force in shaping environmental conditions. As technological factors shape appropriate organizational forms (McKelvey, 1982), fundamental technological change affects the rise and fall of populations within organizational communities (Astley, 1985). Basic technological innovation affects not only a given population, but also those populations within technologically interdependent communities. For example, major changes in semiconductor technology affected semiconductor firms as well as computer and automotive firms. Technology is, then, an important source of environmental variation and hence a critical factor affecting population dynamics.

This paper specifically investigates patterns of technological change and their impact on environmental conditions. Building on a considerable body of research on technological change, we argue and empirically demonstrate that patterned changes

0001-8392/86/3103-0439/$1.00.

•
This research was generously supported by funds from the Strategy Research Center at Columbia University, the Center for Entrepreneurial Studies at New York University, and the Center for Innovation Management Studies at Lehigh University. We would like to thank Graham Astley, Ellen Auster, Robert Drazin, Kathryn Harrigan, and anonymous *ASQ* reviewers for their helpful comments.

in technology dramatically affect environmental conditions. There exist measurable patterns of technological change that generate consistent patterns of environmental change over time across three diverse industries. While technology is but one force driving the course of environmental evolution, it is a key building block to better understand how environments and ultimately organizations evolve over time.

TECHNOLOGY AND TECHNOLOGICAL DISCONTINUITIES

Technology can be defined as those tools, devices, and knowledge that mediate between inputs and outputs (process technology) and/or that create new products or services (product technology) (Rosenberg, 1972). Technological change has an unequivocal impact on economic growth (Solow, 1957; Klein, 1984) and on the development of industries (Lawrence and Dyer, 1983). The impact of technology and technological change on environmental conditions is, however, less clear.

For over thirty years, technology and workflows have been central topics in organizational theory (e.g., Gerwin, 1981). Most studies of technology in organizational theory, however, have been either cross sectional in design (e.g., Woodward, 1965), have taken place in technologically stable settings (e.g., public and not-for-profit settings), or simply have treated technology as a constant (Astley, 1985). Since technology has been taken as a given, there has been a conspicuous lack of clarity concerning how and why technologies change and how technological change affects environmental and/or organizational evolution. An exception is the work of Brittain and Freeman (1980).

There is a substantial literature on technological evolution and change (e.g., Mensch, 1979; Sahal, 1981; Dutton and Thomas, 1985). Some suggest that technological change is inherently a chance or spontaneous event driven by technological genius, as did Taton (1958) in his discussion of penicillin and radioactivity, and Schumpeter (1961). Others, like Gilfillan (1935), who described the multiple independent discoveries of sail for ships, suggest that technological change is a function of historical necessity; still others view technological progress as a function of economic demand and growth (Schmookler, 1966; Merton, 1968). An analysis of many different technologies over years of evolution strongly indicates that none of these perspectives alone captures the complexity of technological change. Technology seems to evolve in response to the interplay of history, individuals, and market demand. Technological change is a function of both variety and chance as well as structure and patterns (Morison, 1966; Sahal, 1981).

Case studies across a range of industries indicate that technological progress constitutes an evolutionary system punctuated by discontinuous change. Major product breakthroughs (e.g., jets or xerography) or process technological breakthroughs (e.g., float glass) are relatively rare and tend to be driven by individual genius (e.g., C. Carlson and xerography; A. Pilkington and float glass). These relatively rare discontinuities trigger a period of technological ferment. As a new product class opens (or following substitution of one product or process for a previous one), the rate of product variation is substantial as alternative product forms compete for dominance. An example is the competition between electric, wood, and internal combustion engines in automobiles or the competition between incompatible videocassette or microcomputer technologies. This technological experimentation and competition persists within a product class until a dominant design emerges as a synthesis of a number of proven concepts (Utterback and Abernathy, 1975; Abernathy, 1978).

A dominant design reflects the emergence of product-class standards and ends the period of technological ferment. Alternative designs are largely crowded out of the product class, and technological development focuses on elaborating a widely accepted product or process; the dominant design becomes a guidepost for further product or process change (Sahal, 1981; Abernathy and Clark, 1985). Dominant designs and associated shifts in product or process change have been found across industries. The Model T, the DC-3, the Fordson tractor, the Smith Model 5 typewriter and the PDP-11 minicomputer were dominant designs that dramatically shaped the evolution of their respective product classes.

Once a dominant design emerges, technological progress is driven by numerous incremental, improvement innovations (Myers and Marquis, 1969; Dutton and Thomas, 1985). For example, while the basic technology underlying xerography has not changed since Carlson's Model 914, the cumulative effect of numerous incremental changes on this dominant design has dramatically improved the speed, quality, and cost per unit of reprographic products (Dessauer, 1975). A similar effect was documented by Yin and Dutton (1986), who described the enormous performance benefits of incremental process improvement in oil refining.

Incremental technological progress, unlike the initial breakthrough, occurs through the interaction of many organizations stimulated by the prospect of economic returns. This is evident in Hollander's (1965) discussion of rayon, Tilton's (1971) study of semiconductors, and Rosenbloom and Abernathy's (1982) study of VCR technology. These incremental technological improvements enhance and extend the underlying technology and thus reinforce an established technical order.

Technological change is a bit-by-bit, cumulative process until it is punctuated by a major advance. Such discontinuities offer sharp price-performance improvements over existing technologies. Major technological innovations represent technical advance so significant that no increase in scale, efficiency, or design can make older technologies competitive with the new technology (Mensch, 1979; Sahal, 1981). Product discontinuities are reflected in the emergence of new product classes (e.g., airlines, automobiles, plain-paper copiers), in product substitution (e.g., transistors vs. vacuum tubes; diesel vs. steam locomotives), or in fundamental product improvements (e.g., jets vs. turbojets; LSI vs. VSLI semiconductor technology). Process discontinuities are reflected either in process substitution (e.g., mechanical ice making vs. natural ice harvesting; thermal vs. catalytic cracking in crude oil refining; artificial vs. natural gems) or in process innovations that result in radical improvements in industry-specific dimensions of merit (e.g., Dundee kiln in cement; Lubbers machinery in glass).

Technological Discontinuities

These major technological shifts can be classified as *competence-destroying* or *competence-enhancing* (see also Abernathy and Clark, 1985), because they either destroy or enhance the competence of existing firms in an industry. The former require new skills, abilities, and knowledge in both the development and production of the product. The hallmark of competence-destroying discontinuities is that mastery of the new technology fundamentally alters the set of relevant competences within a product class. For example, the knowledge and skills required to make glass using the float-glass method are quite different from those required to master other glass-making technologies. Diesel locomotives required new skills and knowledge that steam-engine manufacturers did not typically possess. Similarly, automatically controlled machine tools required wholesale changes in engineering, mechanical, and data-processing skills. These new technical and engineering requirements were well beyond and qualitatively different from those skills necessary to manufacture conventional paper-punched machine tools (Noble, 1984).

A competence-destroying product discontinuity either creates a new product class (e.g., xerography or automobiles) or substitutes for an existing product (e.g., diesel vs. steam locomotive; transistors vs. vacuum tubes). Competence-destroying process discontinuities represent a new way of making a given product. For example, the float-glass process in glass manufacture substituted for continuous grinding and polishing; mechanical ice making substituted for natural ice harvesting; planar processes substituted for the single-wafer process in semiconductors. In each case, the product remained essentially unchanged while the process by which it was made was fundamentally altered. Competence-destroying process breakthroughs may involve combining previously discrete steps into a more continuous flow (e.g., float glass) or may involve a completely different process (e.g., man-made gems).

Competence-destroying discontinuities are so fundamentally different from previously dominant technologies that the skills and knowledge base required to operate the core technology shift. Such major changes in skills, distinctive competence, and production processes are associated with major changes in the distribution of power and control within firms and industries (Chandler, 1977; Barley, 1986). For example, the ascendance of automatically controlled machine tooling increased the power of industrial engineers within the machine-tool industry (Noble, 1984), while the diffusion of high-volume production processes led to the rise of professional managers within more formally structured organizations (Chandler, 1977).

Competence-enhancing discontinuities are order-of-magnitude improvements in price/performance that build on existing know-how within a product class. Such innovations substitute for older technologies, yet do not render obsolete skills required to master the old technologies. Competence-enhancing product discontinuities represent an order-of-magnitude improvement over prior products yet build on existing know-how. For example, IBM's 360 series was a major improvement in price, performance, and features over prior models yet was developed through the synthesis of familiar technologies (Pugh, 1984). Similarly, the introduction of fan jets or of the screw propeller dramatically improved the speed of jets and ocean-going steamships, and aircraft producers and boatyards were able to take advantage of existing knowledge and skills and rapidly absorb these complementary technologies (Davies, 1972; Headrick, 1981).

Competence-enhancing process discontinuities are process innovations that result in an order-of-magnitude increase in the efficiency of producing a given product. For example, the Edison kiln was a major process innovation in cement manufacture that permitted enormous increases in production capacity yet built on existing skills in the cement industry (Lesley, 1924). Similarly, major process advances in semiconductor integration, strip steel, and glass production eliminated barriers to future growth in their product classes. These advances built on existing knowledge and skills and provided the core for subsequent incremental improvements (Dutton and Thomas, 1985).

Table 1 gives a typology of technological changes with examples of competence-destroying and competence-enhancing product and process technologies.

Table 1

A Typology of Product and Process Technological Changes

Technological Changes: Competence-Destroying	Competence-Enhancing
Product	
New Product Class:	*Major Product Improvements:*
Airlines (1924)	Jet ⟶ turbofan
Cement (1872)	LSI ⟶ VSLI semiconductors
Plain-paper copying (1959)	Mechanical ⟶ electric typewriters
	Continuous aim cannons
	Nonreturnable ⟶ returnable bottles
	Thin-walled iron cylinder block engine
Product Substitution:	
Vacuum tubes ⟶ transistors	
Steam ⟶ diesel locomotives	*Incremental Product Changes*
Piston ⟶ jet engines	
Records ⟶ compact disks	*Dominant Designs:**
Punched paper ⟶ automatic control machine tooling	PDP-11, VHS technology
Discrete ⟶ integrated circuits	IBM 360, DC-3
Open ⟶ closed steel auto bodies	Numerical control machine tools
Process	
Process Substitution:	*Major Process Improvements:*
Natural ⟶ mechanical ice	Edison kiln
Natural ⟶ industrial gems	Resistive metal deposition (semiconductors)
Open hearth ⟶ basic oxygen furnace	Gob feeder (glass containers)
Individual wafer ⟶ planar process	Catalytic cracking ⟶ catalytic reforming
Continuous grinding ⟶ float glass	
Thermal cracking ⟶ catalytic cracking	*Incremental Process Improvements:*
Vertical ⟶ rotary kiln	Learning by doing; numerous process improvements
Blown ⟶ drawn window glass	

*Some dominant designs are incremental improvements (e.g., PDP-11), while others are major improvements (e.g., DC-3, IBM 360).

Both technological discontinuities and dominant designs are only known in retrospect — technological superiority is no guarantee of success. The dominance of a substitute product (e.g., Wankel engines, supersonic jets, or bubble memory), substitute process (e.g., continuous casting), or a dominant design (e.g., VHS vs. beta videocasette systems) is a function of technological, market, legal, and social factors that cannot be fully known in advance. For example, the choice by vacuum tube makers such as RCA, GE, and Philco to concentrate on a dominant design for electron tubes in the early transistor days turned out, in retrospect, to have been an error (Tilton, 1971). Similarly, choices of standard record speeds, widths of railroad track, automatically controlled machine tool technologies or automated office equipment standards are often less a function of technical merit than of market or political power (Noble, 1984).

A number of product-class case studies indicate that technology progresses in stages through relatively long periods of incremental, competence-enhancing change elaborating a particular dominant design. These periods of increasing consolidation and learning-by-doing may be punctuated by competence-destroying technological discontinuities (i.e., product or process substitution) or by further competence-enhancing technological advance (e.g., revitalizing a given product or process with complementary technologies). Technological discontinuities trigger a period of technological ferment culminating in a dominant design and, in turn, leading to the next period of incremental, competence-enhancing, technological change. Thus, we hypothesize:

Hypothesis 1: Technological change within a product class will be characterized by long periods of incremental change punctuated by discontinuities.

Hypothesis 1a: Technological discontinuities are either competence enhancing (build on existing skills and know-how) or competence destroying (require fundamentally new skills and competences).

Competence-destroying and competence-enhancing discontinuities dramatically alter previously attainable price/performance relationships within a product class. Both create technological uncertainty as firms struggle to master an untested and incompletely understood product or process. Existing firms within an industry are in the best position to initiate and exploit new possibilities opened up by a discontinuity if it builds on competence they already possess. Competence-enhancing discontinuities tend to consolidate industry leadership; the rich are likely to get richer.

Competence-destroying discontinuities, in contrast, disrupt industry structure (Mensch, 1979). Skills that brought product-class leaders to preeminence are rendered largely obsolete; new firms founded to exploit the new technology will gain market share at the expense of organizations that, bound by traditions, sunk costs, and internal political constraints, remain committed to outmoded technology (Tilton, 1971; Hannan and Freeman, 1977). We thus hypothesize:

Hypothesis 2: The locus of innovation will differ for competence-destroying and competence-enhancing technological changes. Competence-destroying discontinuities will be initiated by new entrants, while competence-enhancing discontinuities will be initiated by existing firms.

TECHNOLOGICAL DISCONTINUITIES AND ORGANIZATIONAL ENVIRONMENTS

To determine the extent to which technological discontinuities affect environmental conditions, we build on Dess and Beard's (1984) review of environmental dimensions and examine two critical characteristics of organizational environments: uncertainty and munificence. Uncertainty refers to the extent to which future states of the environment can be anticipated or accurately predicted (Pfeffer and Salancik, 1978). Munificence refers to the extent to which an environment can support growth. Environments with greater munificence impose fewer constraints on organizations than those environments with resource constraints.

Both competence-enhancing and competence-destroying technological discontinuities generate uncertainty as firms struggle to master an incompletely understood product or process. Technological breakthroughs trigger a period of technological ferment as new technologies are tried, established price-performance ratios are upset, and new markets open. During these periods of technological upheaval, it becomes substantially more difficult to forecast demand and prices. Technological discontinuities, then, will be associated with increases in environmental uncertainty:

Hypothesis 3: Competitive uncertainty will be higher after a technological discontinuity than before the discontinuity.

Technological discontinuities drive sharp decreases in price-performance or input-output ratios. These factors, in turn, fuel demand in a product class. The role of technological progress in stimulating demand is well documented (e.g., Solow, 1957; Mensch, 1979). As both competence-enhancing and competence-destroying discontinuities reflect major price-performance improvements, both will be associated with increased demand and environmental munificence:

Hypothesis 4: Environmental munificence will be higher after a technological discontinuity than before the discontinuity.

Environments can also be described in terms of different competitive conditions (Scherer, 1980). Important dimensions of competitive conditions include entry-exit patterns and degree of order within a product class. Orderliness within a product class can be assessed by interfirm sales variability. Those environments with substantial net entry and substantial interfirm sales variability will be very different competitive arenas than those environments in which exits dominate and there is minimal interfirm sales variability.

Competence-destroying technological discontinuities have quite different effects on competitive conditions than competence-enhancing discontinuities. Competence-enhancing advances permit existing firms to exploit their competence and expertise and thereby gain competitive advantage over smaller or newer firms. Competence-enhancing discontinuities consolidate leadership in a product class; the rich get richer as liabilities of newness plague new entrants. These order-creating breakthroughs increase barriers to entry and minimum scale requirements. These processes will be reflected in relatively fewer entries relative to exits and a decrease in interfirm sales variability — those remaining firms will share more equally in product-class sales growth.

Competence-destroying discontinuities break the existing order. Barriers to entry are lowered; new firms enter previously impenetrable markets by exploiting the new technology (Astley, 1985; Abernathy and Clark, 1985). These discontinuities favor new entrants at the expense of entrenched defenders. New entrants take advantage of fundamentally different skills and expertise and gain sales at the expense of formerly dominant firms burdened with the legacy (i.e., skills, abilities, and expertise) of prior technologies and ways of operating (Astley, 1985; Tushman and Romanelli, 1985). Competence-destroying discontinuities will be associated with increased entry-to-exit ratios and an increase in interfirm sales variability:

Hypothesis 5: Competence-enhancing discontinuities will be associated with decreased entry-to-exit ratios and decreased interfirm sales variability. These patterns will be reversed for competence-destroying discontinuities.

If competence-destroying discontinuities do not emerge to alter a product class, successive competence-enhancing discontinuities will result in increased environmental orderliness and consolidation. Each competence-enhancing breakthrough builds on prior advances and further raises barriers to entry and minimum scale requirements. As product classes mature, the underlying resource base becomes ever more limited by physical and resource constraints. Successive competence-enhancing discontinuities will have smaller impacts on uncertainty and munificence as successive advances further exploit a limited technology and market-resource base:

Hypothesis 6: Successive competence-enhancing discontinuities will be associated with smaller increases in uncertainty and munificence.

Environmental changes induced by a technological discontinuity present a unique opportunity or threat for individual organizations (Tushman and Romanelli, 1985). Technological discontinuities alter the competitive environment and reward those innovative firms that are first to recognize and exploit technological opportunities. The superiority of a new technology presents organizations with a stark choice: adapt or face decline. Those firms that are among the first to adopt the new product or process proceed down the learning curve ahead of those that follow. The benefits of volume and experience provide early movers with a competitive edge not easily erased (Porter, 1985; MacMillan and McCaffrey, 1984). Therefore, we hypothesize:

Hypothesis 7: Those organizations that initiate major technological innovations will have higher growth rates than other firms in the product class.

RESEARCH DESIGN AND MEASURES

Three product classes were selected for study: domestic scheduled passenger airline transport, Portland cement manufacture, and minicomputer manufacture (excluding firms that merely add peripherals and/or software to another firm's minicomputer and resell the system). These three product classes represent assembled products, nonassembled products, and services; this product-class diversity increases the generalizability of our results. These industries were also selected because most participants historically had been undiversified, so environmental conditions outside the industry had little effect on these firms. Data on each product class was gathered from the year of the niche's inception (1872 for cement, 1924 for airlines, and 1956 for minicomputers) through 1980.

The three populations studied included all U.S. firms that produced cement, flew airplane passengers, or produced minicomputers. These industries were chosen partly because archival sources exist permitting a complete census of population members over time. Two outstanding books (Lesley, 1924; Davies, 1972) chronicle the history of the cement and airline industries and include meticulously researched profiles of early entrants into those product classes. In the airline industry, the Civil Aeronautics Board (CAB) lists of entries and exits after 1938 are definitive, due to licensing requirements. In cement, the very high degree of agreement among two trade journals and two industry directories from 1900 on suggests substantially all firms that ever produced cement are included. Similarly, in minicomputers, the very high degree of agreement among trade journals, an exhaustive annual industry directory in *Computers and Automation*, and International Data Corporation (IDC) product listings indicates that virtually all firms that ever produced a minicomputer are included. All sources included very small firms that survived only briefly; any firms that might have been overlooked in this study have never received published mention in three industries thoroughly covered by numerous archival sources.

Technological change. A thorough review of books and trade publications permitted the identification of price-performance changes and key technological events within the three product classes. Technological change was measured by examining key performance parameters for all new kilns, airplanes, or minicomputers introduced in each year of the industry's existence. For cement and airlines, percentage improvement in the state of the art was calculated by dividing the seat-mile-per-year or barrel-per-day capacity of the most capable plane or largest kiln in existence in a given year by the same capacity figure for the most capable plane or largest kiln in existence the previous year. This review of new equipment also permitted the identification of initiators and early adopters of significant innovations. Technological discontinuities were relatively easy to identify because a few innovations so markedly advanced the state of the art that they clearly stand out from less dramatic improvements.

The key performance parameter in cement production is kiln capacity in barrels of cement per day. For every new kiln, this capacity is reported by the manufacturer and is widely published in trade journals and industry directories. For airlines, the key economic factor is the number of passenger-seat-miles per year a plane can fly, calculated by multiplying the number of seats normally in an aircraft model by the number of miles per year it can fly at normal operating speeds for the average number of flight hours per year it proved able to log. These figures are reported in Davies (1972) for all aircraft models flown by U.S. airliners. In minicomputers, a key performance parameter is the amount of time required for the central processing unit to complete one cycle; this is the primary determinant of computer speed and throughput capability. Both *Computers and Automation*, a leading trade journal and industry directory, and the International Data Corporation (IDC), a leading computer-industry research firm, report cycle time for all minicomputers.

Uncertainty. Uncertainty is typically measured as a function of variance measures (Dess and Beard, 1984). Because environmental uncertainty refers to the extent to which future states of the environment cannot be predicted accurately, we measured uncertainty in terms of forecasting error — the ability of industry analysts to predict industry outcomes. Published forecasts for every SIC code are collected and indexed in *Predicasts Forecasts*. For each of the three niches, published one-year demand growth forecasts were collected and compared to actual historical results. Forecast error is defined as

$$\frac{(\ |\text{Forecast demand growth} - \text{Actual demand growth}| \times 100)}{(\text{Actual demand growth})}$$

To measure environmental uncertainty, the mean forecast error for the five-year period before each technological discontinuity was compared to the mean forecast error for the five-year period following the discontinuity. The choice of five-year periods is arbitrary. Major technological changes do not have an overnight impact; it takes several years for their effect on uncertainty and munificence to appear. Yet in the longer run, extraneous events create demand fluctuations whose noise can drown out the patterns generated by major technological advances. Since the industries selected included discontinuities seven and ten years apart, five years was selected as the maximum practicable period of observation that would not create serious overlap problems between the era following one discontinuity and the era preceding another.

Munificence. Munificence was measured in terms of demand, the basic resource available to niche participants. Annual sales growth in units was obtained from the CAB and Bureau of Mines for the airline and cement niches, respectively. Minicomputer sales data were obtained from the International Data Corporation and from *Computers and Automation*. Since sales figures grow as a result of both inflation and growth in the economy as a whole, these factors were eliminated by dividing demand figures by an index of real GNP growth. Mean demand growth was calculated for five-year periods before and after each technological discontinuity.

Two possible objections may be raised to comparing the means of five-year periods preceding and following a discontinuity. First, if there is a strong upward trend in the time series, then for practically any year chosen, demand in the five succeeding years will be significantly higher than demand in the five preceding years. If this is so, there is nothing special about the eras surrounding a technological discontinuity. On the other hand, it may be that the findings are very sensitive to the exact year chosen to mark the discontinuity. If results are significant comparing, for example, 1960–1964 with 1965–1969, but not significant if the comparison is between 1959–1963 and 1964–1968, or between 1961–1965 and 1966–1970, then the finding is not robust.

Accordingly, the difference-of-means test was performed for every possible combination of two adjacent five-year periods for each industry. In each industry, it was found that eras of significant before and after demand shift are rare. Sixteen of 96 possible comparisons were significant at the .05 level in the cement industry (17 percent), 17 of 45 possible comparisons of airline demand (38 percent), and 2 of 7 possible comparisons of minicomputer demand (28 percent). This suggests that technological discontinuities are not the only events that seem to be associated with sharp increases in demand. However, neither do such shifts occur frequently or at random. In each case, a difference of one year either way in identifying the discontinuity would have made no difference; the demand shift is not particularly sensitive to the specific year chosen as the discontinuity.

Table 2

Summary of Variables, Measures, and Data Sources

Variable	Industry	Measure	Data Source	*N*	Range	Mean	SD
Technological change	Cement	% improvement in barrel/day production capacity of largest kiln.	Published specifications of new kilns in *Rock Products*.	90	0–320%		
	Airlines	% improvement in seat-miles per year capacity of most capable plane flown.	Davies (1972).	54	0–248%		
	Minicomputers	Central processor unit speed.	Published specifications in *Computers and Automation*.	24	.2–9000		
Locus of innovation	Cement	Proportion of new firms among earliest to adopt an innovation.	Reports on new kilns in *Rock Products* and trade directories.	4	.1–1.0		
	Airlines		Davies (1972), CAB annual studies of airplane purchases.	4	0–.9		
	Minicomputers		Published specifications in *Computers and Automation*.	3	0–.5		
Uncertainty	Cement	Mean percentage error of one-year demand growth forecasts	*Predicasts Forecasts*.	28	5.2–266.9	52.0	61.6
	Airlines			88	.1–381.4	59.2	58.4
	Minicomputers			36	3.5–811	138.1	167.1
Munificence	Cement	Annual cement consumption (tons).	U.S. Bureau of Mines.	101	8–85513	30296	27103
	Airlines	Annual passenger-seat-miles (mil.).	Civil Aeronautics Board.	52	.1–156.6	34.6	46.1
	Minicomputers	Annual minicomputer sales (000 units)	International Data Corporation.	16	.1–168	47.3	49.6
Entries	Cement	Number of firms producing for first time (mean, range and SD are entries per year, *N* is number of entries).	*Cement Industry Trade Directory*; *Rock Products*.	281	0–24	2.8	4.2
	Airlines		Davies (1972); CAB annual reports.	147	1–33	11.3	9.8
	Minicomputers		*Computers and Automation*; International Data Corporation.	173	3–30	10.8	7.3
Exits	Cement	Number of firms acquired or no longer producing (mean, SD and range are exits per year, *N* is number of exits).	*Cement Industry Trade Directory, Rock Products*.	218	0–23	2.2	3.9
	Airlines		Davies (1972); CAB annual reports.	126	1–28	9.7	8.2
	Minicomputers		*Computers and Automation*; International Data Corporation.	82	0–14	5.1	4.2
Interfirm sales variance	Airlines	Unweighted variance in five-year sales growth percentage among all firms in the industry.	Same as munificence measure.	4	2.0–13.4	5.5	4.6
	Minicomputers		Same as munificence measure.	4	2.6–21.3	11.2	8.6
Firm growth rate	Airlines	Firm sales at end of five-year era divided by sales at beginning of five-year era.	CAB annual reports.	46	−269–346	61.4	79.9
	Minicomputers		International Data Corporation.	67	−96–11900	635.6	1561.6

Technological Discontinuities

At a few comparatively rare periods in the history of an industry, then, one can locate a demand breakpoint, an era of two or three years during which average demand for the five years following any of these critical years significantly exceeds the average demand in the five years preceding the chosen year. Some of these critical eras are not associated with technological discontinuities. Without exception, every technological discontinuity is associated with such a demand shift.

Entry and exit. Entry and exit data were gathered from industry directories and books chronicling the histories of each product class. An entry was recorded in the year when a firm first began cement production, an airline flew its first passenger-mile, or a firm produced its first minicomputer. An exit was recorded when a firm ceased producing cement, flying passengers, or producing at least one minicomputer. Bankruptcy was recorded as an exit only if production ceased. An exit was recorded whenever a firm was acquired; an entry was recorded only if the acquiring firm did not already produce cement, fly passengers, or produce minicomputers. An entrant was classified as new if the company sold no products prior to its entry into the industry or as an existing firm if it sold at least one product before entering the industry. Entry and exit statistics are not calculated for the airline industry from 1938 through 1979, because entries were forbidden by the CAB, and exits depended more on regulatory action than on market forces. Table 2 provides measures, data sources, and summary data for each variable.

Early adopters. To test hypothesis 7, that those firms initiating technological discontinuities would have higher growth rates than other firms in the product class, we examined the growth rates of the first four adopters. Data were available for airlines after 1955 and for minicomputers. The number of early adopters chosen was arbitrary. Four were selected to provide a group large enough for a mean to be meaningful, yet small enough to argue reasonably that the firms considered were quicker to adopt the innovation than the rest of the industry.

RESULTS

Hypothesis 1 suggested that technological evolution would be characterized by periods of incremental change punctuated by either competence-destroying or competence-enhancing discontinuities. Hypothesis 2 argued that competence-destroying advances would be initiated by new entrants, while competence-enhancing advances would be initiated by existing firms. Table 3 summarizes the key technological discontinuities for each niche, while Figures 1a–1c provide more detailed data on key performance dimensions over time.

The cement, airline, and minicomputer niches opened in 1872, 1924, and 1956, respectively. After the three niche openings, there were six competence-enhancing technological discontinuities and two competence-destroying discontinuities (see Table 3). Each discontinuity had a marked effect on a key measure of cost or performance, far greater than the impact of other, more incremental technological events.[1]

Figure 1a documents the three significant technological changes that have punctuated the history of the Portland cement industry. Portland cement, invented in Europe, was first

1 Other industries may not exhibit such marked differences and eventually a coefficient of technological progress could be developed to help distinguish incremental from discontinuous change; one approach might be to pool annual percentage improvements and select those more than two standard deviations above the mean.

Table 3

Significant Technological Discontinuities

Industry	Year	Event	Importance	Type of discontinuity	Locus of Innovation: New firms	Locus of Innovation: Existing firms	Probability
Cement	1872	First production of Portland cement in the United States.	Discovery of proper raw materials and importation of knowledge opens new industry.	Niche opening	10 of 10	1 of 10	
	1896	Patent for process burning powdered coal as fuel.	Permits economical use of efficient rotary kilns.	Competence-destroying	4 of 5	1 of 5	.333
	1909	Edison patents long kiln (150 ft.).	Higher output with less cost.	Competence-enhancing	1 of 6	5 of 6	.001•
	1966	Dundee Cement installs huge kiln, far larger than any previous.	Use of process control permits operation of very efficient kilns.	Competence-enhancing	1 of 8	7 of 8	.000•
Airlines	1924	First airline.	Mail contracts make transport feasible.	Niche opening	9 of 10	1 of 10	
	1936	DC3 airplane.	First large and fast enough to carry passengers economically.	Competence-enhancing	0 of 4	4 of 4	.005•
	1959	First jet airplane in commercial use.	Speed changes economics of flying.	Competence-enhancing	0 of 4	4 of 4	.005•
	1969	Widebody jets debut.	Much greater capacity and efficiency.	Competence-enhancing	0 of 4	4 of 4	.005•
Minicomputer manufacture	1956	Burroughs E-101.	First computer under $50,000.	Niche opening	1 of 8	7 of 8	
	1965	Digital Equipment Corp. PDP-8.	First integrated-circuit minicomputer.	Competence-destroying	3 of 6	3 of 6	.019•
	1971	Data General Supernova SC.	Semiconductor memory much faster than core.	Competence-enhancing	0 of 7	7 of 7	.533

•$p < .01$.

Note: Fisher's exact test compares the pool of firms that are among the first to enter the niche with the pool of firms that introduce or are among the first to adopt a major technological innovation. The null hypothesis is that the proportion of new firms is the same in each sample; probability is the probability of obtaining the observed proportions if the null hypothesis is correct.

made in this country about 1872, but early attempts to compete with established European brands were largely failures. Two events effectively established the domestic industry. The development of the rotary kiln made the manufacture of large volumes of cement with little labor practicable, and the invention in 1896 of a method for creating a continuous flame fed by powdered coal meant that a high-quality, uniform cement could be made without expensive hand-stoking.

During the following decade, rotary kilns 60 feet in length were standard. In 1909, Thomas Edison patented a technique for making kilns over 150 feet in length, enormously increasing the production capacity of a kiln, and the industry rapidly adopted the new "long kiln." Subsequent progress, though, was gradual; kiln capacity increased greatly over a period of decades, but in a series of incremental advances. In 1960, the industry began experimenting with computerized control of kilns. The introduction of computers permitted the construction of huge kilns, much larger than any that had preceded them. The experimental models of the early 1960s culminated in the enormous Dundee kiln in 1967; previously kilns of such capacity could not have been used because their huge size and weight made them impossible to regulate.

Technological Discontinuities

The revolution that brought powdered coal and rotary kilns to the industry was competence-destroying, rendering almost completely obsolete the know-how required to operate wood-fired vertical kilns. A totally new set of competences was required to make cement, and most vertical kiln operators went out of business. The Edison and Dundee kilns were competence-enhancing innovations; each markedly extended the capability of coal-fired rotary kiln technology. A large investment in new kilns and process-control equipment was required, but existing cement-making techniques were not made obsolete, and the leading firms in the industry proved most able to make the necessary capital expenditures.

Figure 1a. Barrels-per-day production capacity of the largest U.S. cement kiln, 1890–1980.

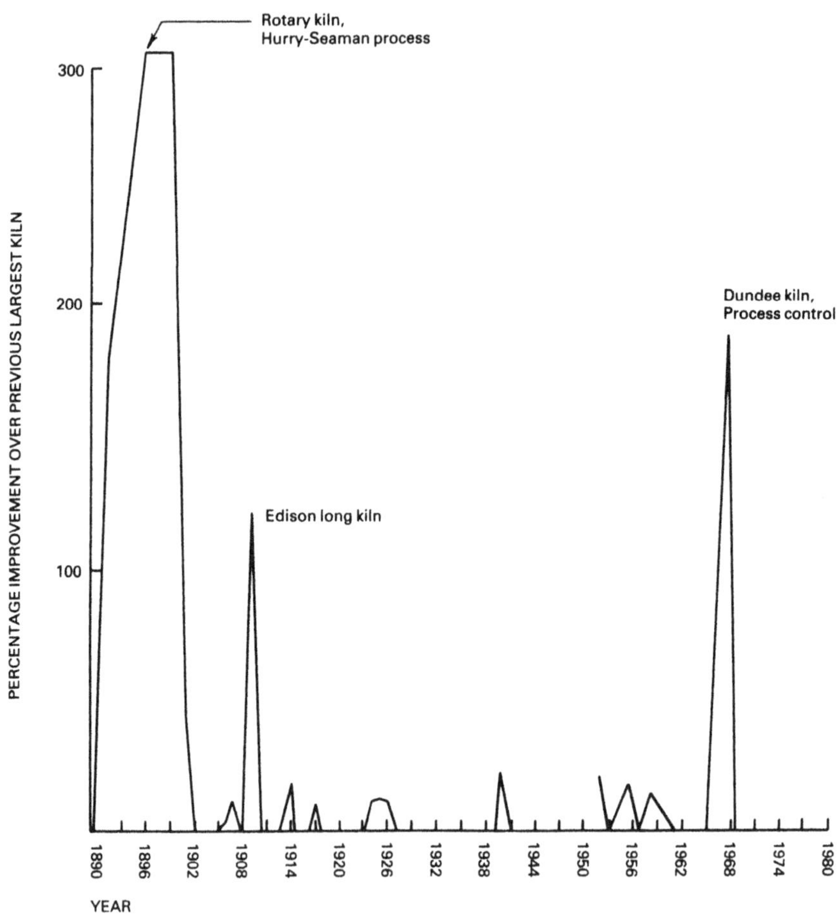

New developments in aircraft construction have been the major technological breakthroughs that have affected the economics of the airline industry, as illustrated in Figure 1b. Numerous flimsy, slow aircraft were flown until the early 1930s, when a flurry of development produced the Boeing 247, Douglas DC-2, and Douglas DC-3 in a span of three years, each a significant improvement on its immediate predecessor. The DC-3, which incorporated some 25 major improvements in aircraft design (Davies, 1972), superseded all other models to become so dominant that by the outbreak of World War II, 80 percent of U.S. airliners in service were DC-3s. Further aircraft improvements were incremental until 1959, when the debut of jet aircraft, with their considerably greater speed and size, again changed the economics of the airline industry. The final breakthrough event was the introduction in 1969 of the Boeing 747, beginning an era dominated by widebody jets.

All three of these major advances were competence enhancing from the perspective of the air carriers (though not from the perspective of aircraft manufacturers). Each advance generated significant economies of scale; airlines could carry many more passengers with each plane than was possible before. Though new skills were required to fly and maintain the new machines, airlines were able to build on their existing competences and take advantage of increased scale economies permitted with new aircraft.

In contrast to cement and airlines, in the minicomputer industry established firms built the first inexpensive computers (usually as an extension of their accounting machine lines). These early minicomputers were based on vacuum tubes and/or transistor technology. The first transistor minicomputer was far faster than its vacuum-tube predecessors, but transistor architecture was replaced by integrated circuitry within two years and thus never diffused widely. Sales were meager until integrated-

Figure 1b. Seat-miles-per-year capacity of the most capable plane flown by U.S. airlines, 1930–1978.

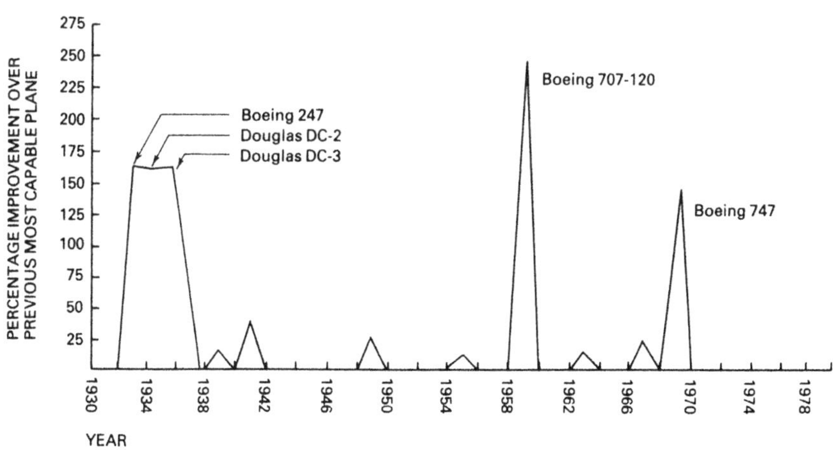

Technological Discontinuities

circuit minicomputers were introduced by a combination of new and older firms. Figure 1c depicts the enormous impact of transistors, immediately followed by integrated circuitry, on computer performance. Integrated circuitry increased minicomputer speed more than 100 times between 1963 and 1965, while size and assembly complexity also decreased substantially. Integrated circuits permitted the construction of compact machines at a greatly reduced cost by eliminating most of the wiring associated with transistors. Integrated-circuit technology was competence-destroying, since expertise in designing, programming, and assembling transistor-based computers was not especially transferable to the design and manufacture of integrated-circuit machines (Fishman, 1981).

The introduction of semiconductor memory in 1971 caused another abrupt performance improvement (see Figure 1c) but did not challenge the fundamental competence of existing minicomputer firms; most companies were able to offer customers versions of their existing models equipped with either magnetic core or semiconductor memory. The effect of semiconductor memory was to increase order in the product class as existing firms were able easily to incorporate this innovation into their existing expertise. For memory manufacturers, however, semiconductor memory was a competence-destroying discontinuity.

Figure 1c. Central-processor-unit cycle time of the fastest minicomputer in production, 1956–1980.

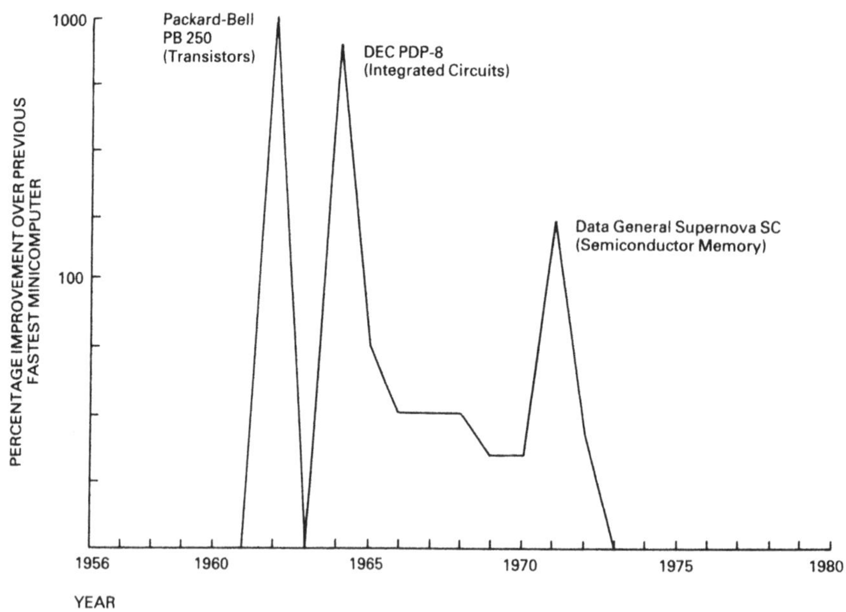

Note: The vertical scale is logarithmic, because the impact of transistors and integrated circuitry on processor speed was so great.

These patterns of incremental technological progress punctuated by discontinuities strongly support hypothesis 1. As suggested in hypothesis 2, the locus of technological innovation for competence-enhancing breakthroughs significantly differs from that of competence-destroying discontinuities. The first cement and airline firms were overwhelmingly new start-ups, not existing companies entering a new industry (Table 3). No product classes existed in 1872 or 1924 whose competences were transferable to cement manufacture or flying airplanes. In contrast, early minicomputers were made by existing accounting machine and electronics manufacturers, who found their existing know-how was readily transferable to the first small, crude computers. New industries can be started either by new organizations or by established ones from other industries; a key variable seems to be whether analogous product classes with transferable competences exist when a new product class emerges.

Patterns in the locus of innovation for discontinuities subsequent to product-class openings are remarkably consistent. The two competence-destroying discontinuities were largely pioneered by new firms (i.e., 7 of 11), while the six competence-enhancing discontinuities were almost exclusively introduced by established industry members (i.e., 35 of 37 firms were existing firms; Fisher's exact test; $p = .0002$). Across these three industries, competence-destroying breakthroughs are significantly more likely to be initiated by new firms, while competence-enhancing breakthroughs are significantly more likely to be initiated by existing firms. Similarly, within each industry, Fisher's exact tests indicate that the proportion of new firms that initiate competence-destroying discontinuities is significantly greater than the proportion of new firms initiating competence-enhancing discontinuities (see last column in Table 3).

Hypothesis 3 suggested that environmental uncertainty would be significantly higher after a technological discontinuity than before it. Since the forecasts we used to test this hypothesis are not available before 1950, only four of the eight technological discontinuities could be tested. In three of the four cases examined, mean forecast error after the discontinuity was significantly higher ($p < .05$) than before the discontinuity (see

Table 4

Forecast Error over Time*

Industry	Era	Mean forecast error	$t(1)$	D.f.	$t(2)$	D.f.
Airlines	1955–1959 1960–1964	16.15% 77.81%	1.78*	18		
Airlines	1965–1969 1970–1974	18.52% 49.13%	4.35**	66	1.91*	54
Cement	1963–1967 1968–1972	38.31% 80.26%	1.85*	26		
Minicomputers	1967–1971 1972–1976	146.31% 136.12%	−.14	34		

*$p < .05$; **$p < .01$.

* $t(1)$ compares mean forecast error of the first period to the mean forecast error of the second period; $t(2)$ compares 1960–1964 with 1970–1974.

Table 4). Except for the period following the introduction of semiconductor memory in minicomputers, the ability of experienced industry observers to predict demand one year in advance was significantly poorer following technological disruption than before.[2] In the semiconductor case, forecast errors were very high both before and after the discontinuity.

Hypothesis 4 suggested that environmental munificence would be higher after a technological discontinuity than before it. The results in Table 5 strongly support the hypothesis. In every case, demand growth following the discontinuity was significantly higher than it was immediately prior to the discontinuity. Further, these demand data indicate the enormous impact of initial discontinuities on product-class demand. Initial discontinuities were associated with, on average, a 529-percent increase in product-class demand. Subsequent discontinuities spark smaller (though still relatively large) increases in demand (226 percent, on average). Technological discontinuities were, then, associated with significantly higher demand after each discontinuity; this effect, though significant in each case, was smaller over successive discontinuities (except for minicomputers, where demand increased substantially after both technological discontinuities).

Table 5

Demand before and after Technological Discontinuity

Industry	Era	Mean annual demand	t*
Cement	1892–1896	168	–3.16••
	1897–1901	1249	
	1905–1909	9271	–6.35••
	1910–1914	15612	
	1963–1967	63348	–2.16•
	1968–1972	77122	
Airlines	1932–1936	2326	–3.01••
	1937–1941	8019	
	1955–1959	244625	–3.68••
	1960–1964	355678	
	1965–1969	742838	–4.42••
	1970–1974	1165943	
Minicomputers	1960–1964	435	–1.96•
	1965–1969	2181	
	1967–1971	7274	–4.60•
	1972–1976	47149	

•$p < .05$; ••$p < .01$.
* t-statistic compares mean demand of first period with mean demand of second period. In each case, there are 8 degrees of freedom.

Hypothesis 5 argued that competence-enhancing discontinuities would be associated with decreased entry-to-exit ratios and decreased interfirm sales variability. Opposite effects were hypothesized for competence-destroying discontinuities. Entry-to-exit ratios were calculated for five years before and after each discontinuity (except for the 1938–1979 period in airlines). Results in Table 6 are partially supportive of hypothesis 5. The ratio of entries to exits was higher in each of the five years before a competence-enhancing discontinuity than during the five subsequent years. None of the differences is statistically significant, though pre-discontinuity entry-to-exit

2
Since data on published forecasts, annual growth in demand, and entry and exit data are available for the three populations, sampling error is not an issue; one could simply report the differences between populations. However, the critical question here is whether consistent differences between pre- and post-discontinuity environments can be discerned. The significance tests show that the probability is small that chance processes could have produced the reported differences between pre- and post-discontinuity eras (Blalock, 1979: 241).

ratios range from 1.15 to over 7 times greater than post-discontinuity entry-to-exit ratios. Entry-to-exit ratios prevailing before a discontinuity are markedly shifted in favor of exits following competence-enhancing discontinuities.

It was expected that entry-to-exit ratios would rise following the two competence-destroying discontinuities; the opposite was observed. Entry-to-exit ratios were quite high following these competence-destroying innovations but were smaller than the extremely large entry-to-exit ratios prevailing just before the discontinuity. Many firms entered and few departed the cement and minicomputer niches in the 1892–1896 and 1960–1964 periods, respectively. Both of these eras were themselves periods of technological ferment in emerging product classes – rotary kilns began to replace vertical kilns in the early 1890s, and transistors began to replace vacuum tubes in the early 1960s. It may be that the rush of new firms to enter emerging product classes confounds the effects of competence-destroying discontinuities.

Table 6

Entry-to-Exit Ratio before and after Discontinuity

Industry	Era	Entry-to-exit ratio*	Discontinuity type
Cement	1872–1896	3.25	Niche opening
	1892–1896	46.00	Competence-destroying
	1897–1901	12.00	
	1905–1909	1.489	Competence-enhancing
	1910–1914	.814	
	1963–1967	1.250	Competence-enhancing
	1968–1972	.160	
Airlines	1913–1930	1.730	Niche opening
	1930–1934	.820	Competence-enhancing†
	1935–1939	.714	
Minicomputers	1956–1960	Not finite‡	Niche opening
	1960–1964	5.500	Competence-destroying
	1965–1969	2.917	
	1967–1971	4.933	Competence-enhancing
	1972–1976	2.708	

*The difference between the pre-discontinuity entry-to-exit ratios and the corresponding post-discontinuity entry-to-exit ratios, while consistent, do not reach statistical significance, due to the large variance between individual years.
†Airline data for subsequent periods were not reported, because entry and exit were regulated.
‡Six entries, no exits.

Entry-to-exit patterns are consistent across these three divergent industries. Entries dominate exits early on, reflecting the rush of new entrants. After competence-enhancing discontinuities in cement and airlines, exits dominate entries, reflecting industry consolidation. In minicomputers, while entry-to-exit ratios decrease over time, entries dominate exits throughout this 20-year period.

Hypothesis 5 also suggested that competence-enhancing discontinuities would decrease interfirm sales variability as those remaining firms adopt industry standards in both products and processes. Small firms drop out of the industry, entry barriers are raised, and firms exploiting similar existing competences experience relatively similar outcomes. Following competence-destroying discontinuities, though, we expected marked variability in sales growth as firms compete with each other on fundamentally different bases; some firms' sales grow explosively while others experience dramatic sales decline.

The results in Table 7 for airlines and minicomputers support this prediction. In minicomputers, integrated circuits triggered explosive growth in the product class and increased interfirm sales variability. Following the other three competence-enhancing discontinuities, though, interfirm sales variability decreased significantly; niche occupants experienced similar results as they built on their existing competences to exploit demand growth.

Table 7

Interfirm Sales Variability before and after Discontinuity

Industry	Era	Discontinuity type	Interfirm variance	F*	D.f.
Airlines	1955–1959	Competence-enhancing	79.24	2.726•	12,12
	1960–1964		29.07		
	1965–1969	Competence-enhancing	103.63	4.096••	11,11
	1970–1974		25.30		
Minicomputers	1960–1964	Competence-destroying	5599.32	−17.480••	8,11
	1965–1969		97873.25		
	1967–1971	Competence-enhancing	86.26	9.960••	9,35
	1972–1976		8.65		

•$p < .05$; ••$p < .01$.

*The F-statistic compares the ratio of interfirm sales variance before the discontinuity to interfirm sales variance after the discontinuity.

Hypothesis 6 suggested that successive competence-enhancing discontinuities would be associated with relatively smaller effects on uncertainty and munificence. Because forecast data are not available before 1950, this hypothesis could only be partially tested in the case of uncertainty. As predicted, the mean forecast error in airlines for the 1960–1964 period is higher than that for the 1970–1974 period ($t = 1.91$; $p < .05$; see Table 4). Hypothesis 6 receives stronger support with respect to munificence. In cement and airlines, mean growth rates in demand are smaller for each successive competence-enhancing discontinuity. These differences are significant for two of the three comparisons (see Table 8). These data suggest that as technology matures, successive competence-enhancing discontinuities increase both uncertainty and munificence, but not as much as those discontinuities that preceded them in establishing the product class. These data, as well as those entry-to-exit data in Table 6, suggest that successive competence-enhancing advances result in increased product-class maturity, reflected in decreased uncertainty, decreased demand growth-rates, and increased product-class consolidation.

Hypothesis 7 argued that those firms initiating technological discontinuities would have higher growth rates than other firms in the product class. Table 9 compares five-year growth rates for the four early adopters to all other firms before and after

Table 8

Demand Patterns Following Successive Competence-Enhancing Discontinuities

Industry	Era	Mean growth*	$t(1)$	D.f.
Cement	1910–1914	48.3%	12.03**	8
	1968–1972	8.4%		
Airlines	1937–1941	161.5%		
	1960–1964	33.4%	30.79**	8
	1970–1974	32.3%	.75	8

**$p < .01$.

*Mean growth is the average annual *percentage* gain in sales for the industry (in contrast to Table 5, which measures demand in units). The t-statistic compares consecutive post-discontinuity periods; e.g., a comparison of mean percentage growth for 1910–1914 with mean percentage growth for 1968–1972 yields a t-statistic of 12.03, failing to support the null hypothesis that there is no difference in percentage growth rates between successive post-discontinuity eras.

technological discontinuities. As hypothesized in each of the four comparisons, early adopters experienced more growth than other firms. Early adopters had significantly higher five-year growth rates than other firms in the airline industry. For jets, early adopters had growth rates similar to others before the discontinuity, while for widebody jets, the early adopters had higher sales growth before and after the discontinuity (see Table 9). In minicomputers, early adopters had annual percentage growth rates that were 105 percentage points higher, on average, than other firms. Technological discontinuities are, then, sources of opportunities (or threats) for firms. While dominant technologies cannot be known in advance, those firms that recognize and quickly adopt a technological breakthrough grow more rapidly than others.

Table 9

Relative Sales Growth of First Four Adopters of a Major Innovation

Industry	Innovation*	Era	Mean sales growth first 4 adopters	Growth all others	t†	D.f.
Airlines	Jet aircraft	1955–1959	38.1%	22.2%	1.268	10
		1960–1964	44.3%	12.3%	2.121**	10
	Widebody jets	1965–1969	101.1%	19.2%	2.487*	9
		1970–1974	16.1%	1.0%	2.642*	9
Minicomputer	Integrated circuits	1960–1964	Not available (new firms)			
		1965–1969	339.2%	179.6%	.44	10
	Semiconductor memory	1967–1971	Not available (new firms)			
		1972–1976	238.0%	188.4%	.14	34

*$p < .05$; **$p < .06$.

*The first four adopters in each case are: *Jet aircraft:* American, TWA, United, Eastern; *Widebody jet:* American, TWA, Continental, United; *Integrated circuits:* Digital Equipment, Computer Control Co., Scientific Data Systems, Systems Engineering Laboratories; *Semiconductors:* Data General, Digital Computer Controls, Interdata, Microdata.

†The t-test compares the mean annual percentage growth rates of the four firms who first introduced or adopted each innovation with the mean annual percentage growth rates of all other firms in the industry. Two periods do not yield interpretable statistics because annual growth for new firms cannot be calculated when the base year contains zero sales.

DISCUSSION

The purpose of this paper has been to explore technological evolution and to investigate its impact on environmental conditions. A better understanding of technological evolution may increase our understanding of a range of phenomena at the population (e.g., structural evolution, population dynamics, strategic groups) as well as the organizational levels of analysis (e.g., organizational adaptation, executive succession patterns, executive demographics, and political dynamics) (Astley, 1985; Tushman and Romanelli, 1985).

Longitudinal data across three diverse industries indicate that technology evolves through relatively long periods of incremental change punctuated by relatively rare innovations that radically improve the state of the art. Such discontinuities occurred only eight times in the 190 total years observed across three industries. Yet in each product class, these technological shifts stand out clearly and significantly altered competitive environments.

The effect of major technological change on the two fundamental dimensions of uncertainty and munificence is unambiguous. Environmental conditions following a discontinuity are sharply different from those that prevailed before the technical breakthrough: the advance makes available new resources to fuel growth within the niche and renders observers far less able to predict the extent of future resource availability. Major technical change opens new worlds for a product class but requires niche occupants to deal with a considerable amount of ambiguity and uncertainty as they struggle to comprehend and master both the new technology and the new competitive environment.

It is also clear that technological discontinuities are not all alike. Competence-enhancing discontinuities significantly advance the state of the art yet build on, or permit the transfer of, existing know-how and knowledge. Competence-destroying discontinuities, on the other hand, significantly advance the technological frontier, but with a knowledge, skill, and competence base that is inconsistent with prior know-how. While competence-enhancing discontinuities build on existing experience, competence-destroying discontinuities require fundamentally new skills and technological competence.

The locus of innovation and the environmental consequences of competence-destroying versus competence-enhancing discontinuities are quite different. Competence-enhancing breakthroughs are overwhelmingly initiated by existing, successful firms. Competence-enhancing discontinuities result in greater product-class consolidation, reflected in relatively smaller entry-to-exit ratios and decreased interfirm sales variability. As competence-enhancing discontinuities build on existing know-how, it appears that the rich get richer, while new firms face liabilities of newness (Stinchcombe, 1965). Product-class conditions become ever more consolidated over successive order-creating discontinuities.

Competence-destroying discontinuities are more rare than competence-enhancing technological advances. Competence-destroying breakthroughs are watershed events in the life of a product class; they open up new branches in the course of industrial evolution (Astley, 1985). These discontinuities are initiated by new firms and open up the product class to waves of new entrants unconstrained by prior technologies and organizational inertia. While liabilities of newness plague new firms confronting competence-enhancing breakthroughs, liabilities of age and tradition constrain existing, successful firms in the face of competence-destroying discontinuities. Although the data were limited, competence-destroying discontinuities seem to break the grip of established firms in a product class. Interfirm sales variability jumped after integrated circuits were introduced in minicomputers, as new firms and established firms pursued different strategies, with markedly different results. Similarly in cement, new firms initiated rotary kilns and went on to dominate the industry.

These patterns are seen most vividly in minicomputer manufacture. The first inexpensive computers were built by established office-equipment firms (e.g., Monroe), electronics firms (e.g., Packard-Bell), and computer firms (e.g., Burroughs). This new product class continued unchanged until the advent of integrated circuits. Without exception, established firms floundered in the face of a technology based on active components. Integrated circuits rendered obsolete much of the engineering knowledge embodied in the first minicomputers. Office-equipment and the existing computer firms were unable to produce a successful model embodying semiconductor technology. Only the few firms explicitly founded to make minicomputers (e.g., DEC) were able to make the transition. By 1965, almost every firm that produced early minicomputers had exited the product class.

Technological discontinuities, whether competence-destroying or competence-enhancing, appear to afford a rare opportunity for competitive advantage for firms willing to risk early adoption. In all four cases, early adopters of major innovations had greater five-year growth rates than the rest of the product class. While these data are not unequivocal, firms that recognize and seize opportunities presented by major advances gain first-mover advantages. Those firms that do not adopt the innovation early or, worse, increase investment in obsolete technology, risk failing, because product-class conditions change so dramatically after the discontinuity.

Technological advance seems to be an important determinant of market as well as intraorganizational power. Competence-enhancing discontinuities are order creating in that they build on existing product-class know-how. These breakthroughs increase the market power of existing firms as barriers to entry are raised and dependence on buyers and suppliers decreases in the face of larger and more dominant producers. Competence-destroying technological advances, on the other hand, destroy order in a product class. These discontinuities create fundamental technological uncertainty as incompatible technologies compete for dominance. New firms, unconstrained by prior competence and history, take advantage of technological opportunities and the lethargy of organizations burdened with the consequences of prior success. Given the enormous impact of technological advance on product-class order, future research could explore the politics of technological change as interest groups attempt to shape technological progress to suit their own competences (e.g., Noble, 1984).

Within the firm, technological discontinuities affect the distribution of power and, in turn, decision-making processes. Those who control technological advances (whether competence destroying or enhancing) will gain power at others' expense (e.g.,

Morison, 1966; Pettigrew, 1973). Because technological dominance is rarely known in advance, the control of technological assumptions and the locus of technological decisions will be an important arena for intraorganizational political processes. Shaping technological advance may be a critical organizational issue, since technology affects both intra- and interorganizational bases of power.

Because technology affects organizational adaptation, organizations may be able to use investment in R&D and technological innovation to shape environmental conditions in their favor. While technological dominance cannot be predicted at the outset (e.g., Wankel engines, bubble memory), organizations that create technological variation, or are able to adopt technological change quickly, maximize their probability of being able to move with a changing technological frontier. Organizations that do not contribute to or keep up with multiple technological bases may lose their ability to be aware of and deal with technological evolution (Dutton and Thomas, 1985).

The patterns of technological change are similar across these three diverse industries. It appears that new product classes are associated with a wave of new entrants, relatively few exits, and substantial technological experimentation. Competence-destroying discontinuities occurred early in both cement and minicomputer manufacture. After competence-destroying breakthroughs, successive competence-enhancing discontinuities resulted in an ever more consolidated and mature product class. While we have no data, subsequent competence-destroying discontinuities may, in turn, break up a mature product class and restart the product class's evolutionary clock (e.g., microcomputers vs. minicomputers or compact disks vs. records).

Competence-destroying discontinuities initiate a period of technological ferment, as alternative technologies compete for dominance. This period of technological competition lasts until a dominant design emerges as a synthesis of prior technological experimentation (e.g., Dundee kiln, DC-3, PDP-11). Dominant designs reflect a consolidation of industry standards. These designs crowd out alternative designs and become guideposts for incremental product as well as major process change (Utterback and Abernathy, 1975). Thus, quite apart from major technological advance, the establishment of a dominant design may also be an important lever in shaping environmental conditions and organizational fate.

CONCLUSION

While these data indicate that technological discontinuities exist and that these discontinuities have important effects on environmental conditions, the data are not conclusive. Though the data are consistent across three diverse industries, the number of cases is relatively small, and some of the tests were limited by data availability. Future research needs to focus more closely on patterns of technological change. If technology is an important determinant of competitive conditions, we need to know more about differences between competence-destroying and competence-enhancing technological advances, what distinguishes between incremental improvements and dramatic advances, what are dominant designs and how they occur, and what are the impacts of competence-destroying advances in mature product classes.

Both product and process innovation are important. While the data here are only suggestive, it may be that different kinds of innovation are relatively more important in different product classes. For nonassembled products (e.g., cement, glass, oil), major process innovations may be relatively more important than product innovations. For assembled products (e.g., mini-computers, VCRs, scientific instruments), major product improvements or substitutions may be relatively more important than process innovations. Future research might explore the differential importance of major product and process innovations by different product-class type.

The effects of nontechnological discontinuities must also be examined to understand more fully how competitive environments change. Technological change does not occur in a vacuum. If frequently sparks a response from the legal, political, or social environments. For example, bioengineering, automatic control machinery, nuclear power, and supersonic transportation each has been directly affected by a complex set of interactions among technological and political, social, and legal considerations (e.g., Horwitch, 1982; Astley and Fombrun, 1983; Noble, 1984). Further, periods of incremental technological change and standardization may become turbulent for non-technological reasons (e.g., airline deregulation or the outlawing of basing-point pricing in cement). More complete analyses of the technology-environment linkages must also take into account the linkages between technological change and these other important social, political, and legal forces.

Technological change clearly affects organizational environments. Beyond exploring more deeply the nature of technological change, future research could also explore the linkage between technological evolution and population phenomena, such as structural evolution, mortality rates, or strategic groups, as well as organizational issues, such as adaptation, succession, and political processes. These results suggest that technology is not a static environmental resource. Rather, technology advances through the competition between alternative technologies promoted by rivalrous organizations. At the organization level, technological action, such as investment in R&D and internal venturing, may be a powerful lever in directly shaping environmental conditions and, in turn, organizational adaptation.

REFERENCES

Abernathy, William
1978 The Productivity Dilemma. Baltimore, MD: Johns Hopkins University Press.

Abernathy, William, and Kim B. Clark
1985 "Innovation: Mapping the winds of creative destruction." Research Policy, 14: 3–22.

Aldrich, Howard, and Ellen R. Auster
1986 "Even dwarfs started small: Liabilities of age and size and their strategic implications." In L. L. Cummings and Barry M. Staw (eds.), Research in Organizational Behavior, 8: 165–198. Greenwich, CT: JAI Press.

Astley, W. Graham
1985 "The two ecologies: Population and community perspectives on organizational evolution." Administrative Science Quarterly, 30: 224–241.

Astley, W. Graham, and Charles Fombrun
1983 "Technological innovation and industrial structure: The case of telecommunications." In Robert Lamb (ed.), Advances in Strategic Management, 1:205–229. Greenwich, CT: JAI Press.

Barley, Stephen R.
1986 "Technology as an occasion for structuring: Evidence from observations of CT scanners and the social order of radiology departments." Administrative Science Quarterly, 31: 78–108.

Technological Discontinuities

Barnard, Chester
1938 The Functions of the Executive. Cambridge, MA: Harvard University Press.

Blalock, Hubert
1979 Social Statistics, rev. 2d ed. New York: McGraw-Hill.

Brittain, Jack, and John Freeman
1980 "Organizational proliferation and density-dependent selection." In John R. Kimberly and Robert Miles (eds.), The Organizational Life Cycle: 291–338. San Francisco: Jossey-Bass.

Carroll, Glenn R.
1984 "Dynamics of publisher succession in newspaper organizations." Administrative Science Quarterly, 29: 93–113.

Chandler, Alfred D., Jr.
1977 The Visible Hand: The Managerial Revolution in American Business. Cambridge, MA: Belknap Press.

Davies, R. E. G.
1972 Airlines of the United States Since 1914. London: Putnam.

Dess, Gregory G., and Donald W. Beard
1984 "Dimensions of organizational task environments." Administrative Science Quarterly, 29: 52–73.

Dessauer, John H.
1975 My Years with Xerox. New York: Manor Books.

Downey, H. Kirk, and R. Duane Ireland
1979 "Quantitative versus qualitative: Environmental assessment in organizational studies." Administrative Science Quarterly, 24: 630–637.

Dutton, John, and Annie Thomas
1985 "Relating technological change and learning by doing." In Richard D. Rosenbloom (ed.), Research on Technological Innovation, Management, and Policy, 2: 187–224. Greenwich, CT: JAI Press.

Fishman, Katherine
1981 The Computer Establishment. New York: Harper & Row.

Freeman, John
1982 "Organizational life cycles and natural selection processes." In Barry M. Staw and L. L. Cummings (eds.), Research in Organizational Behavior, 4: 1–32. Greenwich, CT: JAI Press.

Gerwin, Donald
1981 "Relationships between structure and technology." In Paul Nystrom and William Starbuck (eds.), Handbook of Organizational Design, 2: 3–31. New York: Oxford University Press.

Gilfillan, S. Colum
1935 Inventing the Ship. Chicago: Follett.

Hannan, Michael, and John Freeman
1977 "The population ecology of organizations." American Journal of Sociology, 82: 929–964.

Harrigan, Kathryn
1983 Strategies for Vertical Integration. Lexington, MA: D. C. Heath, Lexington Books.

Headrick, Daniel
1981 The Tools of Empire. Oxford: Oxford University Press.

Hollander, Samuel
1965 The Sources of Increased Efficiency. Cambridge, MA: MIT Press.

Horwitch, Mel
1982 Clipped Wings: A Study of the Supersonic Transport. Cambridge, MA: MIT Press.

Hrebiniak, Lawrence G., and William F. Joyce
1985 "Organizational adaptation: Strategic choice and environmental determinism." Administrative Science Quarterly, 30: 336–349.

Klein, Burton
1984 Wages and Business Cycles: A Dynamic Theory. New York: Pergamon Press.

Lawrence, Paul, and Davis Dyer
1983 Renewing American Industry. New York: Free Press.

Lesley, Robert
1924 A History of the United States Portland Cement Industry. Chicago: Portland Cement Association.

MacMillan, Ian, and M. L. McCaffrey
1984 "Strategy for financial services: Cashing in on competitive inertia." Journal of Business Strategy, 4: 58–73.

McKelvey, Bill
1982 Organizational Systematics — Taxonomy, Evolution, Classification. Berkeley, CA: University of California Press.

Mensch, Gerhard
1979 Stalemate in Technology: Innovations Overcome the Depression. Cambridge, MA: Ballinger.

Merton, Robert
1968 Social Theory and Social Structure. New York: Free Press.

Miller, Danny, and Peter Friesen
1984 Organizations: A Quantum View. Englewood Cliffs, NJ: Prentice-Hall.

Morison, Elting E.
1966 Men, Machines, and Modern Times. Cambridge, MA: MIT Press.

Myers, Sumner, and Donald G. Marquis
1969 Successful Industrial Innovations. Washington, DC: National Science Foundation.

Noble, David
1984 Forces of Production: A Social History of Industrial Automation. New York: Knopf.

Pettigrew, Andrew
1973 Politics of Organizational Decision-Making. London: Tavistock.

Pfeffer, Jeffrey, and Gerald Salancik
1978 The External Control of Organizations. New York: Harper & Row.

Porter, Michael
1985 Competitive Advantage: Creating and Sustaining Superior Performance. New York: Free Press.

Pugh, Emerson W.
1984 Memories That Shaped an Industry: Decisions Leading to the IBM System 360. Cambridge, MA: MIT Press.

Romanelli, Elaine, and Michael L. Tushman
1986 "Inertia, environments and strategic choice: A quasi-experimental design for comparative-longitudinal research." Management Science, 32: 608–621.

Rosenberg, Nathan
1972 Technology and American Economic Growth. Armonk, NY: M.E. Sharpe.

Rosenbloom, Robert, and William Abernathy
1982 "The climate for innovation in industry: The role of management attitudes and practices in consumer electronics." Research Policy, 11: 209–225.

Sahal, Devendra
1981 Patterns of Technological Innovation. Reading, MA: Addison-Wesley.

Scherer, Frederick
1980 Industrial Market Structure and Economic Performance. Boston: Houghton Mifflin.

Schmookler, Jacob
1966 Invention and Economic Growth. Cambridge, MA: Harvard University Press.

Schumpeter, Josef
1961 History of Economic Analysis. New York: Oxford University Press.

Selznick, Philip
1949 TVA and the Grass Roots: A Study of Politics and Organization. Berkeley, CA: University of California Press.

Solow, Robert M.
1957 "Technical change and the aggregate production function." Review of Economics and Statistics, 39: 312–320.

Starbuck, William
1983 "Organizations and their environments." In Marvin D. Dunnette (ed.), Handbook of Organizational and Industrial Psychology: 1069–1123. New York: Wiley.

Stinchcombe, Arthur
1965 "Social structure and organization." In James G. March (ed.), Handbook of Organizations: 142–193. Chicago: Rand McNally.

Taton, René
1958 Reason and Chance in Scientific Discovery. New York: Philosophical Library.

Tilton, John W.
1971 International Diffusion of Technology: The Case of Semiconductors. Washington, DC: Brookings Institution.

Tushman, Michael, and Elaine Romanelli
1985 "Organizational evolution: A metamorphosis model of convergence and reorientation." In L. L. Cummings and Barry M. Staw (eds.), Research in Organizational Behavior, 7: 171–222. Greenwich, CT: JAI Press.

Utterback, James, and William Abernathy
1975 "A dynamic model of process and product innovation." Omega, 33: 639–656.

Woodward, Joan
1965 Industrial Organization: Theory and Practice. Oxford: Oxford University Press.

Yin, Zun-Sheng, and John M. Dutton
1986 "Systems learning and technological change: The evolution of 20th century U.S. domestic petroleum-refining processes." Working paper, New York University Graduate School of Business Administration.

[2]

Architectural Innovation: The Reconfiguration of Existing Product Technologies and the Failure of Established Firms

Rebecca M. Henderson
Massachusetts Institute of Technology
Kim B. Clark
Harvard University

This paper demonstrates that the traditional categorization of innovation as either incremental or radical is incomplete and potentially misleading and does not account for the sometimes disastrous effects on industry incumbents of seemingly minor improvements in technological products. We examine such innovations more closely and, distinguishing between the components of a product and the ways they are integrated into the system that is the product "architecture," define them as innovations that change the architecture of a product without changing its components. We show that architectural innovations destroy the usefulness of the architectural knowledge of established firms, and that since architectural knowledge tends to become embedded in the structure and information-processing procedures of established organizations, this destruction is difficult for firms to recognize and hard to correct. Architectural innovation therefore presents established organizations with subtle challenges that may have significant competitive implications. We illustrate the concept's explanatory force through an empirical study of the semiconductor photolithographic alignment equipment industry, which has experienced a number of architectural innovations.•

The distinction between refining and improving an existing design and introducing a new concept that departs in a significant way from past practice is one of the central notions in the existing literature on technical innovation (Mansfield, 1968; Moch and Morse, 1977; Freeman, 1982). Incremental innovation introduces relatively minor changes to the existing product, exploits the potential of the established design, and often reinforces the dominance of established firms (Nelson and Winter, 1982; Ettlie, Bridges, and O'Keefe, 1984; Dewar and Dutton, 1986; Tushman and Anderson, 1986). Although it draws from no dramatically new science, it often calls for considerable skill and ingenuity and, over time, has very significant economic consequences (Hollander, 1965). Radical innovation, in contrast, is based on a different set of engineering and scientific principles and often opens up whole new markets and potential applications (Dess and Beard, 1984; Ettlie, Bridges, and O'Keefe, 1984; Dewar and Dutton, 1986). Radical innovation often creates great difficulties for established firms (Cooper and Schendel, 1976; Daft, 1982; Rothwell, 1986; Tushman and Anderson, 1986) and can be the basis for the successful entry of new firms or even the redefinition of an industry.

Radical and incremental innovations have such different competitive consequences because they require quite different organizational capabilities. Organizational capabilities are difficult to create and costly to adjust (Nelson and Winter, 1982; Hannan and Freeman, 1984). Incremental innovation reinforces the capabilities of established organizations, while radical innovation forces them to ask a new set of questions, to draw on new technical and commercial skills, and to employ new problem-solving approaches (Burns and Stalker, 1966; Hage, 1980; Ettlie, Bridges, and O'Keefe, 1984; Tushman and Anderson, 1986).

0001-8392/90/3501-0009/$1.00.

•
This research was supported by the Division of Research, Harvard Business School. Their support is gratefully acknowledged. We would like to thank Dataquest and VLSI Research Inc for generous permission to use their published data, the staffs at Canon, GCA, Nikon, Perkin Elmer and Ultratech, and all those individuals involved with photolithographic alignment technology who gave so generously of their time. We would also like to thank the editors of this journal and three anonymous reviewers who gave us many helpful comments. Any errors or omissions remain entirely our responsibility.

The distinction between radical and incremental innovation has produced important insights, but it is fundamentally incomplete. There is growing evidence that there are numerous technical innovations that involve apparently modest changes to the existing technology but that have quite dramatic competitive consequences (Clark, 1987). The case of Xerox and small copiers and the case of RCA and the American radio receiver market are two examples.

Xerox, the pioneer of plain-paper copiers, was confronted in the mid-1970s with competitors offering copiers that were much smaller and more reliable than the traditional product. The new products required little new scientific or engineering knowledge, but despite the fact that Xerox had invented the core technologies and had enormous experience in the industry, it took the company almost eight years of missteps and false starts to introduce a competitive product into the market. In that time Xerox lost half of its market share and suffered serious financial problems (Clark, 1987).

In the mid-1950s engineers at RCA's corporate research and development center developed a prototype of a portable, transistorized radio receiver. The new product used technology in which RCA was accomplished (transistors, radio circuits, speakers, tuning devices), but RCA saw little reason to pursue such an apparently inferior technology. In contrast, Sony, a small, relatively new company, used the small transistorized radio to gain entry into the U.S. market. Even after Sony's success was apparent, RCA remained a follower in the market as Sony introduced successive models with improved sound quality and FM capability. The irony of the situation was not lost on the R&D engineers: for many years Sony's radios were produced with technology licensed from RCA, yet RCA had great difficulty matching Sony's product in the marketplace (Clark, 1987).

Existing models that rely on the simple distinction between radical and incremental innovation provide little insight into the reasons why such apparently minor or straightforward innovations should have such consequences. In this paper, we develop and apply a model that grew out of research in the automotive, machine tool, and ceramics industries that helps to explain how minor innovations can have great competitive consequences.

CONCEPTUAL FRAMEWORK

Component and Architectural Knowledge

In this paper, we focus on the problem of product development, taking as the unit of analysis a manufactured product sold to an end user and designed, engineered, and manufactured by a single product-development organization. We define innovations that change the way in which the components of a product are linked together, while leaving the core design concepts (and thus the basic knowledge underlying the components) untouched, as "architectural" innovation.[1] This is the kind of innovation that confronted Xerox and RCA. It destroys the usefulness of a firm's architectural knowledge but preserves the usefulness of its knowledge about the product's components.

1 In earlier drafts of this paper we referred to this type of innovation as "generational." We are indebted to Professor Michael Tushman for his suggestion of the term architectural.

Architectural Innovation

This distinction between the product as a whole—the system—and the product in its parts—the components—has a long history in the design literature (Marples, 1961; Alexander, 1964). For example, a room fan's major components include the blade, the motor that drives it, the blade guard, the control system, and the mechanical housing. The overall architecture of the product lays out how the components will work together. Taken together, a fan's architecture and its components create a system for moving air in a room.

A component is defined here as a physically distinct portion of the product that embodies a core design concept (Clark, 1985) and performs a well-defined function. In the fan, a particular motor is a component of the design that delivers power to turn the fan. There are several design concepts one could use to deliver power. The choice of one of them—the decision to use an electric motor, for example, establishes a core concept of the design. The actual component—the electric motor—is then a physical implementation of this design concept.

The distinction between the product as a system and the product as a set of components underscores the idea that successful product development requires two types of knowledge. First, it requires component knowledge, or knowledge about each of the core design concepts and the way in which they are implemented in a particular component. Second, it requires architectural knowledge or knowledge about the ways in which the components are integrated and linked together into a coherent whole. The distinction between architectural and component knowledge, or between the components themselves and the links between them, is a source of insight into the ways in which innovations differ from each other.

Types of Technological Change

The notion that there are different kinds of innovation, with different competitive effects, has been an important theme in the literature on technological innovation since Schumpeter (1942). Following Schumpeter's emphasis on creative destruction, the literature has characterized different kinds of innovations in terms of their impact on the established capabilities of the firm. This idea is used in Figure 1, which classifies innovations along two dimensions. The horizontal dimension captures an innovation's impact on components, while the vertical captures its impact on the linkages between components.[2] There are, of course, other ways to characterize different kinds of innovation. But given the focus here on innovation and the development of new products, the framework outlined in Figure 1 is useful because it focuses on the impact of an innovation on the usefulness of the existing architectural and component knowledge of the firm.

Framed in this way, radical and incremental innovation are extreme points along both dimensions. Radical innovation establishes a new dominant design and, hence, a new set of core design concepts embodied in components that are linked together in a new architecture. Incremental innovation refines and extends an established design. Improvement occurs in individual components, but the underlying core design concepts, and the links between them, remain the same.

2
We are indebted to one of the anonymous *ASQ* reviewers for the suggestion that we use this matrix.

Figure 1. A framework for defining innovation.

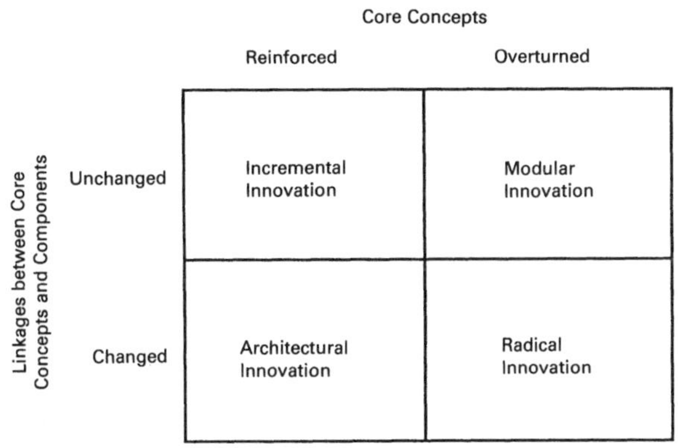

Figure 1 shows two further types of innovation: innovation that changes only the core design concepts of a technology and innovation that changes only the relationships between them. The former is a modular innovation, such as the replacement of analog with digital telephones. To the degree that one can simply replace an analog dialing device with a digital one, it is an innovation that changes a core design concept without changing the product's architecture. Our concern, however, is with the last type of innovation shown in the matrix: innovation that changes a product's architecture but leaves the components, and the core design concepts that they embody, unchanged.

The essence of an architectural innovation is the reconfiguration of an established system to link together existing components in a new way. This does not mean that the components themselves are untouched by architectural innovation. Architectural innovation is often triggered by a change in a component—perhaps size or some other subsidiary parameter of its design—that creates new interactions and new linkages with other components in the established product. The important point is that the core design concept behind each component—and the associated scientific and engineering knowledge—remain the same.

We can illustrate the application of this framework with the example of the room air fan. If the established technology is that of large, electrically powered fans, mounted in the ceiling, with the motor hidden from view and insulated to dampen the noise, improvements in blade design or in the power of the motor would be incremental innovations. A move to central air conditioning would be a radical innovation. New components associated with compressors, refrigerants, and their associated controls would add whole new technical disciplines and new interrelationships. For the maker of large, ceiling-mounted room fans, however, the introduction of a portable fan would be an architectural innovation. While the primary components would be largely the same (e.g., blade, motor, control system), the architecture of the product would be quite different. There would be significant changes in the interactions between components. The smaller size and the co-location of the motor and the blade in the room would focus attention on new types of interaction between the motor size, the blade dimensions, and the amount of air that the fan could circulate, while shrinking the size of the apparatus would probably introduce new interactions between the performance of the blade and the weight of the housing.

The distinctions between radical, incremental, and architectural innovations are matters of degree. The intention here is not to defend the boundaries of a particular definition, particularly since there are several other dimensions on which it may be useful to define radical and incremental innovation. The use of the term architectural innovation is designed to draw attention to innovations that use many existing core design concepts in a new architecture and that therefore have a more significant impact on the relationships between components than on the technologies of the components themselves. The matrix in Figure 1 is designed to suggest that a given innovation may be less radical or more architectural, not to suggest that the world can be neatly divided into four quadrants.

These distinctions are important because they give us insight into why established firms often have a surprising degree of difficulty in adapting to architectural innovation. Incremental innovation tends to reinforce the competitive positions of established firms, since it builds on their core competencies (Abernathy and Clark, 1985) or is "competence enhancing" (Tushman and Anderson, 1986). In the terms of the framework developed here, it builds on the existing architectural and component knowledge of an organization. In contrast, radical innovation creates unmistakable challenges for established firms, since it destroys the usefulness of their existing capabilities. In our terms, it destroys the usefulness of both architectural and component knowledge (Cooper and Schendel, 1976; Daft, 1982; Tushman and Anderson, 1986).

Architectural innovation presents established firms with a more subtle challenge. Much of what the firm knows is useful and needs to be applied in the new product, but some of what it knows is not only not useful but may actually handicap the firm. Recognizing what is useful and what is not, and acquiring and applying new knowledge when necessary, may be quite difficult for an established firm because of the way knowledge—particularly architectural knowledge—is organized and managed.

The Evolution of Component and Architectural Knowledge

Two concepts are important to understanding the ways in which component and architectural knowledge are managed inside an organization. The first is that of a dominant design. Work by Abernathy and Utterback (1978), Rosenberg (1982), Clark (1985), and Sahal (1986) and evidence from studies of several industries show that product technologies do not emerge fully developed at the outset of their commercial lives (Mansfield, 1977). Technical evolution is usually characterized by periods of great experimentation followed by the acceptance of a dominant design. The second concept is that organizations build knowledge and capability around the recurrent tasks that they perform (Cyert and March, 1963; Nelson and

Winter, 1982). Thus one cannot understand the development of an organization's innovative capability or of its knowledge without understanding the way in which they are shaped by the organization's experience with an evolving technology.

The emergence of a new technology is usually a period of considerable confusion. There is little agreement about what the major subsystems of the product should be or how they should be put together. There is a great deal of experimentation (Burns and Stalker, 1966; Clark, 1985). For example, in the early days of the automobile industry, cars were built with gasoline, electric, or steam engines, with steering wheels or tillers, and with wooden or metal bodies (Abernathy, 1978).

These periods of experimentation are brought to an end by the emergence of a dominant design (Abernathy and Utterback, 1978; Sahal, 1986). A dominant design is characterized both by a set of core design concepts that correspond to the major functions performed by the product (Marples, 1961; Alexander, 1964; Clark, 1985) and that are embodied in components and by a product architecture that defines the ways in which these components are integrated (Clark, 1985; Sahal, 1986). It is equivalent to the general acceptance of a particular product architecture and is characteristic of technical evolution in a very wide range of industries (Clark, 1985). A dominant design often emerges in response to the opportunity to obtain economies of scale or to take advantage of externalities (David, 1985; Arthur, 1988). For example, the dominant design for the car encompassed not only the fact that it used a gasoline engine to provide motive force but also that it was connected to the wheels through a transmission and a drive train and was mounted on a frame rather than on the axles. A dominant design incorporates a range of basic choices about the design that are not revisited in every subsequent design. Once the dominant automobile design had been accepted, engineers did not reevaluate the decision to use a gasoline engine each time they developed a new design. Once any dominant design is established, the initial set of components is refined and elaborated, and progress takes the shape of improvements in the components within the framework of a stable architecture.

This evolutionary process has profound implications for the types of knowledge that an organization developing a new product requires, since an organization's knowledge and its information-processing capabilities are shaped by the nature of the tasks and the competitive environment that it faces (Lawrence and Lorsch, 1967; Galbraith, 1973).[3]

In the early stages of a technology's history, before the emergence of a dominant design, organizations competing to design successful products experiment with many different technologies. Since success in the market turns on the synthesis of unfamiliar technologies in creative new designs, organizations must actively develop both knowledge about alternate components and knowledge of how these components can be integrated. With the emergence of a dominant design, which signals the general acceptance of a single architecture, firms cease to invest in learning about alternative configurations of the established set of components. New component knowledge becomes more valuable to a firm than

3
For simplicity, we will assume here that organizations can be assumed to act as boundedly rational entities, in the tradition of Arrow (1974) and Nelson and Winter (1982).

new architectural knowledge because competition between designs revolves around refinements in particular components. Successful organizations therefore switch their limited attention from learning a little about many different possible designs to learning a great deal about the dominant design. Once gasoline-powered cars had emerged as the technology of choice, competitive pressures in the industry strongly encouraged organizations to learn more about gasoline-fired engines. Pursuing refinements in steam- or electric-powered cars became much less attractive. The focus of active problem solving becomes the elaboration and refinement of knowledge about existing components within a framework of stable architectural knowledge (Dosi, 1982; Clark, 1985).

Since in an industry characterized by a dominant design, architectural knowledge is stable, it tends to become embedded in the practices and procedures of the organization. Several authors have noted the importance of various institutional devices like frameworks and routines in completing recurring tasks in an organization (Galbraith, 1973; Nelson and Winter, 1982; Daft and Weick, 1984). The focus in this paper, however, is on the role of communication channels, information filters, and problem-solving strategies in managing architectural knowledge.

Channels, filters, and strategies. An organization's communication channels, both those that are implicit in its formal organization (A reports to B) and those that are informal ("I always call Fred because he knows about X"), develop around those interactions within the organization that are critical to its task (Galbraith, 1973; Arrow, 1974). These are also the interactions that are critical to effective design. They are the relationships around which the organization builds architectural knowledge. Thus an organization's communication channels will come to embody its architectural knowledge of the linkages between components that are critical to effective design. For example, as a dominant design for room fans emerges, an effective organization in the industry will organize itself around its conception of the product's primary components, since these are the key subtasks of the organization's design problem (Mintzberg, 1979; von Hippel, 1990). The organization may create a fan-blade group, a motor group, and so on. The communication channels that are created between these groups will reflect the organization's knowledge of the critical interactions between them. The fact that those working on the motor and the fan blade report to the same supervisor and meet weekly is an embodiment of the organization's architectural knowledge about the relationship between the motor and the fan blade.

The information filters of an organization also embody its architectural knowledge. An organization is constantly barraged with information. As the task that it faces stabilizes and becomes less ambiguous, the organization develops filters that allow it to identify immediately what is most crucial in its information stream (Arrow, 1974; Daft and Weick, 1984). The emergence of a dominant design and its gradual elaboration molds the organization's filters so that they come to embody parts of its knowledge of the key relationships between the components of the technology. For instance, the relationships between the designers of motors and controllers for a room fan are likely to change over time as they are able to express the nature of the critical interaction between the motor and the controller in an increasingly precise way that allows them to ignore irrelevant information. The controller designers may discover that they need to know a great deal about the torque and power of the motor but almost nothing about the materials from which it is made. They will create information filters that reflect this knowledge.

As a product evolves, information filters and communication channels develop and help engineers to work efficiently, but the evolution of the product also means that engineers face recurring kinds of problems. Over time, engineers acquire a store of knowledge about solutions to the specific kinds of problems that have arisen in previous projects. When confronted with such a problem, the engineer does not reexamine all possible alternatives but, rather, focuses first on those that he or she has found to be helpful in solving previous problems. In effect, an organization's problem-solving strategies summarize what it has learned about fruitful ways to solve problems in its immediate environment (March and Simon, 1958; Lyles and Mitroff, 1980; Nelson and Winter, 1982). Designers may use strategies of this sort in solving problems within components, but problem-solving strategies also reflect architectural knowledge, since they are likely to express part of an organization's knowledge about the component linkages that are crucial to the solution of routine problems. An organization designing fans might learn over time that the most effective way to design a quieter fan is to focus on the interactions between the motor and the housing.

The strategies designers use, their channels for communication, and their information filters emerge in an organization to help it cope with complexity. They are efficient precisely because they do not have to be actively created each time a need for them arises. Further, as they become familiar and effective, using them becomes natural. Like riding a bicycle, using a strategy, working in a channel, or employing a filter does not require detailed analysis and conscious, deliberate execution. Thus the operation of channels, filters, and strategies may become implicit in the organization.

Since architectural knowledge is stable once a dominant design has been accepted, it can be encoded in these forms and thus becomes implicit. Organizations that are actively engaged in incremental innovation, which occurs within the context of stable architectural knowledge, are thus likely to manage much of their architectural knowledge implicitly by embedding it in their communication channels, information filters, and problem-solving strategies. Component knowledge, in contrast, is more likely to be managed explicitly because it is a constant source of incremental innovation.

Problems Created by Architectural Innovation

Differences in the way in which architectural and component knowledge are managed within an experienced organization give us insight into why architectural innovation often creates problems for established firms. These problems have two sources. First, established organizations require significant time (and resources) to identify a particular innovation as architectural, since architectural innovation can often initially be accommodated within old frameworks. Radical innovation tends to be obviously radical—the need for new modes of learning and new skills becomes quickly apparent. But information that might warn the organization that a particular innovation is architectural may be screened out by the information filters and communication channels that embody old architectural knowledge. Since radical innovation changes the core design concepts of the product, it is immediately obvious that knowledge about how the old components interact with each other is obsolete. The introduction of new linkages, however, is much harder to spot. Since the core concepts of the design remain untouched, the organization may mistakenly believe that it understands the new technology. In the case of the fan company, the motor and the fan-blade designers will continue to talk to each other. The fact that they may be talking about the wrong things may only become apparent after there are significant failures or unexpected problems with the design.

The development of the jet aircraft industry provides an example of the impact of unexpected architectural innovation. The jet engine initially appeared to have important but straightforward implications for airframe technology. Established firms in the industry understood that they would need to develop jet engine expertise but failed to understand the ways in which its introduction would change the interactions between the engine and the rest of the plane in complex and subtle ways (Miller and Sawyers, 1968; Gardiner, 1986). This failure was one of the factors that led to Boeing's rise to leadership in the industry.

This effect is analogous to the tendency of individuals to continue to rely on beliefs about the world that a rational evaluation of new information should lead them to discard (Kahneman, Slovic, and Tversky, 1982). Researchers have commented extensively on the ways in which organizations facing threats may continue to rely on their old frameworks —or in our terms on their old architectural knowledge—and hence misunderstand the nature of a threat. They shoehorn the bad news, or the unexpected new information, back into the patterns with which they are familiar (Lyles and Mitroff, 1980; Dutton and Jackson, 1987; Jackson and Dutton, 1988).

Once an organization has recognized the nature of an architectural innovation, it faces a second major source of problems: the need to build and to apply new architectural knowledge effectively. Simply recognizing that a new technology is architectural in character does not give an established organization the architectural knowledge that it needs. It must first switch to a new mode of learning and then invest time and resources in learning about the new architecture (Louis and Sutton, 1989). It is handicapped in its attempts to do this, both by the difficulty all organizations experience in switching from one mode of learning to another and by the fact that it must build new architectural knowledge in a context in which some of its old architectural knowledge may be relevant.

An established organization setting out to build new architectural knowledge must change its orientation from one of refinement within a stable architecture to one of active search for new solutions within a constantly changing context. As long as the dominant design remains stable, an organization can segment and specialize its knowledge and rely on standard operating procedures to design and develop products. Architectural innovation, in contrast, places a premium on exploration in design and the assimilation of new knowledge. Many organizations encounter difficulties in their attempts to make this type of transition (Argyris and Schön, 1978; Weick, 1979; Hedberg, 1981; Louis and Sutton, 1989). New entrants, with smaller commitments to older ways of learning about the environment and organizing their knowledge, often find it easier to build the organizational flexibility that abandoning old architectural knowledge and building new requires.

Once an organization has succeeded in reorientating itself, the building of new architectural knowledge still takes time and resources. This learning may be quite subtle and difficult. New entrants to the industry must also build the architectural knowledge necessary to exploit an architectural innovation, but since they have no existing assets, they can optimize their organization and information-processing structures to exploit the potential of a new design. Established firms are faced with an awkward problem. Because their architectural knowledge is embedded in channels, filters, and strategies, the discovery process and the process of creating new information (and rooting out the old) usually takes time. The organization may be tempted to modify the channels, filters, and strategies that already exist rather than to incur the significant fixed costs and considerable organizational friction required to build new sets from scratch (Arrow, 1974). But it may be difficult to identify precisely which filters, channels, and problem-solving strategies need to be modified, and the attempt to build a new product with old (albeit modified) organizational tools can create significant problems.

The problems created by an architectural innovation are evident in the introduction of high-strength-low-alloy (HSLA) steel in automobile bodies in the 1970s. The new materials allowed body panels to be thinner and lighter but opened up a whole new set of interactions that were not contained in existing channels and strategies. One automaker's body-engineering group, using traditional methods, designed an HSLA hood for the engine compartment. The hoods, however, resonated and oscillated with engine vibrations during testing. On further investigation, it became apparent that the traditional methods for designing hoods worked just fine with traditional materials, although no one knew quite why. The knowledge embedded in established problem-solving strategies and communication channels was sufficient to achieve effective designs with established materials, but the new material created new interactions and required the engineers to build new knowledge about them.

Architectural innovation may thus have very significant competitive implications. Established organizations may invest heavily in the new innovation, interpreting it as an incremental extension of the existing technology or underestimating its impact on their embedded architectural knowledge. But new entrants to the industry may exploit its potential much more effectively, since they are not handicapped by a legacy of embedded and partially irrelevant architectural knowledge.

We explore the validity of our framework through a brief summary of the competitive and technical history of the semiconductor photolithographic alignment equipment industry. Photolithographic aligners are sophisticated pieces of capital equipment used in the manufacture of integrated circuits. Their performance has improved dramatically over the last twenty-five years, and although the core technologies have changed only marginally since the technique was first invented, the industry has been characterized by great turbulence. Changes in market leadership have been frequent, the entry of new firms has occurred throughout the industry's history, and incumbents have often suffered sharp declines in market share following the introduction of equipment incorporating seemingly minor innovation. We believe that these events are explained by the intrusion of architectural innovation into the industry, and we use three episodes in the industry's history—particularly Canon's introduction of the proximity aligner and Kasper's response to it—to illustrate this idea in detail.

INNOVATION IN PHOTOLITHOGRAPHIC ALIGNMENT EQUIPMENT

Data

The data were collected during a two-year, field-based study of the photolithographic alignment equipment industry. The study was initially designed to serve as an exploration of the validity of the concept of architectural innovation, a concept originally developed by one of the authors during the course of his experience with the automobile and ceramics industry (Clark, 1987).

The core of the data is a panel data set consisting of research and development costs and sales revenue by product for every product development project conducted between 1962, when work on the first commercial product began, and 1986. This data is supplemented by a detailed managerial and technical history of each project. The data were collected through research in both primary and secondary sources. The secondary sources, including trade journals, scientific journals, and consulting reports, were used to identify the companies that had been active in the industry and the products that they had introduced and to build up a preliminary picture of the industry's technical history.

Data were then collected about each product-development project by contacting directly at least one of the members of the product-development team and requesting an interview. Interviews were conducted over a fourteen-month period, from March 1987 to May 1988. During the course of the research, over a hundred people were interviewed. As far as possible, the interviewees included the senior design engineer for each project and a senior marketing executive from each firm. Other industry observers and participants, including chief executives, university scientists, skilled design engineers, and service managers were also interviewed. Interview data were supplemented whenever possible through the use of internal firm records. The majority of the interviews were semistructured and lasted about two hours. Respondents were asked to describe the technical, commercial, and managerial history of the product-development projects with which they were familiar and to discuss the technical and commercial success of the products that grew out of them.

In order to validate the data that were collected during this process, a brief history of product development for each equipment vendor was circulated to all the individuals who had been interviewed and to others who knew a firm's history well, and the accuracy of this account was discussed over the telephone in supplementary interviews. The same validation procedure was followed in the construction of the technical history of the industry. A technical history was constructed using interview data, published product literature, and the scientific press. This history was circulated to key individuals who had a detailed knowledge of the technical history of the industry, who corrected it as appropriate.

We chose to study the semiconductor photolithographic alignment equipment industry for two reasons. The first is that it is very different from the industries in which our framework was first formulated, since it is characterized by much smaller firms and a much faster rate of technological innovation. The second is that it provides several examples of the impact of architectural innovation on the competitive position of established firms. Photolithographic equipment has been shaken by four waves of architectural innovation, each of which resulted in a new entrant capturing the leadership of the industry. In order to ground the discussion of architectural innovation we provide a brief description of photolithographic technology.

The Technology

Photolithographic aligners are used to manufacture solid-state semiconductor devices. The production of semiconductors requires the transfer of small, intricate patterns to the surface of a wafer of semiconductor material such as silicon, and this process of transfer is known as lithography. The surface of the wafer is coated with a light-sensitive chemical, or "resist." The pattern that is to be transferred to the wafer surface is drawn onto a mask and the mask is used to block light as it falls onto the resist, so that only those portions of the resist defined by the mask are exposed to light. The light chemically transforms the resist so that it can be stripped away. The resulting pattern is then used as the basis for either the deposition of material onto the wafer surface or for the etching of the existing material on the surface of the wafer. The process may be repeated as many as twenty times during the manufacture of a semiconductor device, and each layer must be located precisely with respect to the previous layer (Watts and Einspruch, 1987). Figure 2 gives a very simplified representation of this complex process.

A photolithographic aligner is used to position the mask relative to the wafer, to hold the two in place during exposure, and to expose the resist. Figure 3 shows a schematic diagram of a contact aligner, the first generation of alignment equipment developed. Improvement in alignment technology has meant improvement in minimum feature size, the size of the smallest pattern that can be produced on the wafer surface, yield, the percentage of wafers successfully processed, and

Figure 2. Schematic representation of the lithographic process.

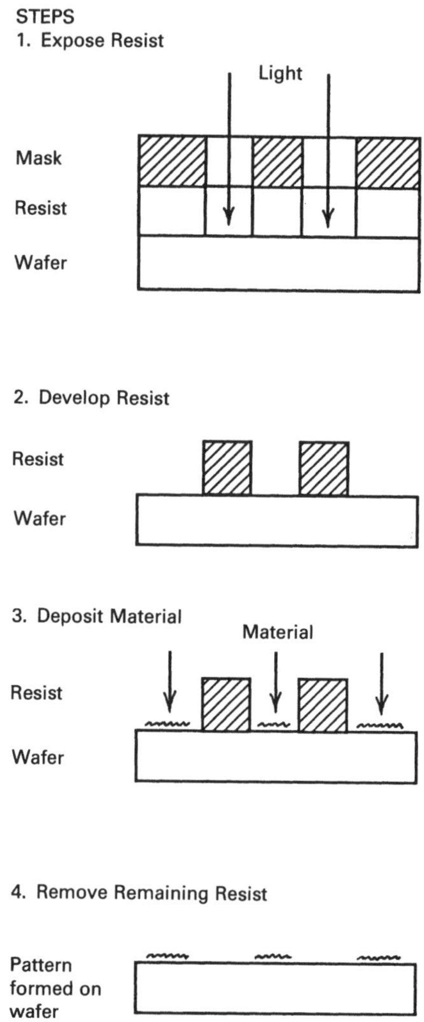

throughput, the number of wafers the aligner can handle in a given time.

Contact aligners were the first photolithographic aligners to be used commercially. They use the mask's shadow to transfer the mask pattern to the wafer surface. The mask and the wafer are held in contact with each other, and light shining through the gaps in the mask falls onto the wafer surface. Contact aligners are simple and quick to use, but the need to bring the mask and the wafer into direct contact can damage the mask or contaminate the wafer. The first proximity aligner was introduced in 1973 to solve these problems.

Figure 3. Schematic diagram of a contact aligner.

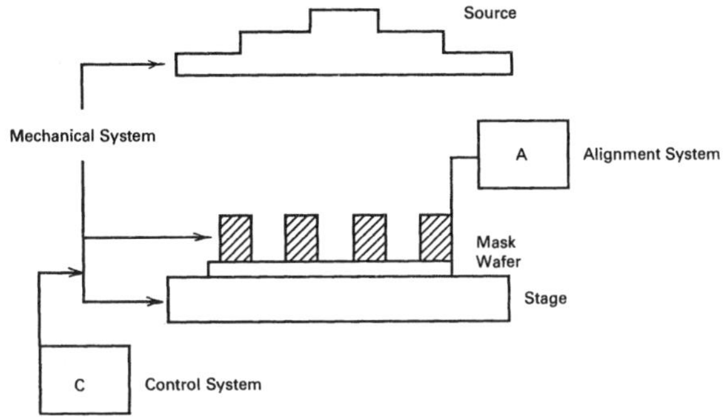

In a proximity aligner the mask is held a small distance away from (in proximity to) the wafer surface, as shown in the simplified drawing in Figure 4. The separation of the mask and the wafer means that they are less likely to be damaged during exposure, but since the mask and wafer are separated from each other, light coming through the mask spreads out before it reaches the resist, and the mask's shadow is less well defined than it is in the case of a contact aligner. As a result, users switching to proximity aligners traded off some minimum feature size capability for increased yield.

The basic set of core design concepts that underlie optical photolithography—the use of a visible light source to transmit the image of the mask to the wafer, a lens or other device to focus the image of the mask on the wafer, an alignment system that uses visible light, and a mechanical system that holds the mask and the wafer in place—have remained unchanged since the technology was first developed, although aligner performance has improved dramatically. The minimum-feature-size capability of the first aligners was about fifteen to twenty microns. Modern aligners are sometimes specified to have minimum feature sizes of less than half a micron.

Figure 4. Schematic diagram of a proximity aligner.

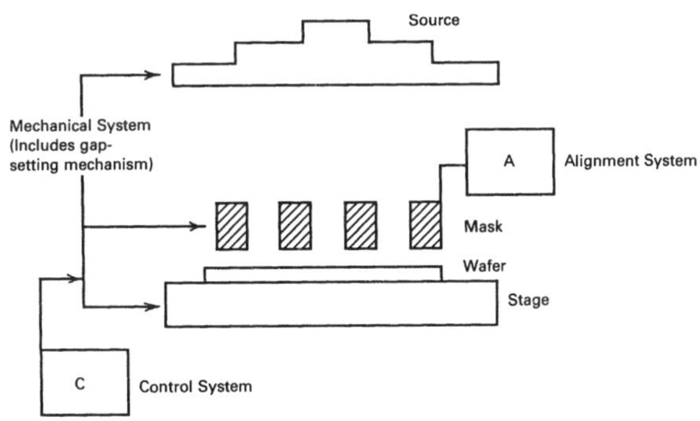

Architectural Innovation

Radical alternatives, making use of quite different core concepts, have been explored in the laboratory but have yet to be widely introduced into full-scale production. Aligners using x-rays and ion beams as sources have been developed, as have direct-write electron beam aligners, in which a focused beam of electrons is used to write directly on the wafer (Chang et al., 1977; Brown, Venkatesan, and Wagner, 1981; Burggraaf, 1983). These technologies are clearly radical. They rely not only on quite different core concepts for the source, but they also use quite different mask, alignment, and lens technologies.

A constant stream of incremental innovation has been critical to optical photolithography's continuing success. The technology of each component has been significantly improved. Modern light sources are significantly more powerful and more uniform, and modern alignment systems are much more accurate. In addition, the technology has seen four waves of architectural innovation: the move from contact to proximity alignment, from proximity to scanning projection alignment, and from scanners to first- and then second-generation "steppers." Table 1 summarizes the changes in the technology introduced by each generation. In each case the core technologies of optical lithography remained largely untouched, and much of the technical knowledge gained in building a previous generation could be transferred to the next. Yet, in each case, the industry leader was unable to make the transition.

Table 1

A Summary of Architectural Innovation in Photolithographic Alignment Technology

Equipment	Major Changes: Technology	Major Changes: Critical relationships between components
Proximity aligner	Mask and wafer separated during exposure.	Accuracy and stability of gap is a function of links between gap-setting mechanism and other components.
Scanning projection	Image of mask projected onto wafer by scanning reflective optics.	Interactions beween lens and other components is critical to successful performance.
First-generation stepper	Image of mask projected through refractive lens. Image "stepped" across wafer.	Relationship between lens field size and source energy becomes significant determinant of throughput. Depth of focus characteristics—driven by relationship between source wavelength and lens numerical aperture—become critical. Interactions between stage and alignment system are critical.
Second-generation stepper	Introduction of "site-by-site" alignment, larger 5× lenses.	Throughput now driven by calibration and stepper stability. Relationship between lens and mechanical system becomes crucial means of controlling distortion.

Source: Field interviews, internal firm records (Henderson, 1988).

Table 2 shows share of deflated cumulative sales, 1962–1986, by generation of equipment for the leading firms. The first commercially successful aligner was introduced by Kulicke and Soffa in 1965. They were extremely successful and held nearly 100 percent of the (very small) market for the next nine years, but by 1974 Cobilt and Kasper had replaced them. In 1974 Perkin-Elmer entered the market with the

Table 2

Share of Deflated Cumulative Sales (%) 1962–1986, by Generation, for the Leading Optical Photolithographic Alignment Equipment Manufacturers*

	Alignment Equipment				
Firm	Contact	Proximity	Scanners	Step and repeat (1)	Step and repeat (2)
Cobilt	44		<1		
Kasper	17	8		7	
Canon		67	21	9	
Perkin-Elmer			78	10	<1
GCA				55	12
Nikon					70
Total	61	75	99+	81	82+

* This measure is distorted by the fact that all of these products are still being sold. For second-generation step and repeat aligners this problem is particularly severe, since in 1986 this equipment was still in the early stages of its life cycle. Source: Internal firm records, Dataquest, VLSI Research Inc.

scanning projection aligner and rapidly became the largest firm in the industry. GCA, in turn, replaced Perkin-Elmer through its introduction of the stepper, only to be supplanted by Nikon, which introduced the second-generation stepper.

In nearly every case, the established firm invested heavily in the next generation of equipment, only to meet with very little success. Our analysis of the industry's history suggests that a reliance on architectural knowledge derived from experience with the previous generation blinded the incumbent firms to critical aspects of the new technology. They thus underestimated its potential or built equipment that was markedly inferior to the equipment introduced by entrants.

The Kasper Saga

The case of Kasper Instruments and its response to Canon's introduction of the proximity printer illustrates some of the problems encountered by established firms. Kasper Instruments was founded in 1968 and by 1973 was a small but profitable firm supplying approximately half of the market for contact aligners. In 1973 Kasper introduced the first contact aligner to be equipped with proximity capability. Although nearly half of all the aligners that the firm sold from 1974 onward had this capability, Kasper aligners were only rarely used in proximity mode, and sales declined steadily until the company left the industry in 1981. The widespread use of proximity aligners only occurred with the introduction and general adoption of Canon's proximity aligner in the late 1970s.

The introduction of the proximity aligner is clearly not a radical advance. The conceptual change involved was minor, and most proximity aligners can also be used as contact aligners. However, in a proximity aligner, a quite different set of relationships between components is critical to successful performance. The introduction of the proximity aligner was thus an architectural innovation. In particular, in a proximity aligner, the relationships between the gap-setting mechanism and the other components of the aligner are significantly different.

Architectural Innovation

In both contact and proximity aligners, the mask and the wafer surface must be parallel to each other during exposure if the quality of the final image on the wafer is to be adequate. This is relatively straightforward in a contact aligner, since the mask and the wafer are in direct contact with each other during exposure. The gap-setting mechanism is used only to separate the mask and the wafer during alignment. Its stability and accuracy have very little impact on the aligner's performance. In a proximity aligner, however, the accuracy and precision of the gap-setting mechanism are critical to the aligner's performance. The gap between the mask and the wafer must be precise and consistent across the mask and wafer surfaces if the aligner is to perform well. Thus, the gap-setting mechanism must locate the mask at exactly the right point above the wafer by dead reckoning and must then ensure that the mask is held exactly parallel to the wafer. Since the accuracy and stability of the mechanism is as much a function of the way in which it is integrated with the other components as it is of its own design, the relationships between the gap-setting mechanism and the other components of the aligner must change if the aligner is to perform well. Thus, the successful design of a proximity aligner requires both the acquisition of some new component knowledge—how to build a more accurate and more stable gap-setting mechanism—and the acquisition of new architectural knowledge.

Kasper's failure to understand the challenge posed by the proximity aligner is especially puzzling given its established position in the market and its depth of experience in photolithography. There were several highly skilled and imaginative designers at Kasper during the early 1970s. The group designed a steady stream of contact aligners, each incorporating significant incremental improvements. From 1968 to 1973, the minimum-feature-size capability of its contact aligners improved from fifteen to five microns.

But Kasper's very success in designing contact aligners was a major contributor to its inability to design a proximity aligner that could perform as successfully as Canon's. Canon's aligner was superficially very similar to Kasper's. It incorporated the same components and performed the same functions, but it performed them much more effectively because it incorporated a much more sophisticated understanding of the technical interrelationships that are fundamental to successful proximity alignment. Kasper failed to develop the particular component knowledge that would have enabled it to match Canon's design. More importantly, the architectural knowledge that Kasper had developed through its experience with the contact aligner had the effect of focusing its attention away from the new problems whose solution was critical to the design of a successful proximity aligner.

Kasper conceived of the proximity aligner as a modified contact aligner. Like the incremental improvements to the contact aligner before it, design of the proximity aligner was managed as a routine extension to the product line. The gap-setting mechanism that was used in the contact aligner to align the mask and wafer with each other was slightly modified, and the new aligner was offered on the market. As a result, Kasper's proximity aligner did not perform well. The gap-setting mechanism was not sufficiently accurate or stable to ensure adequate performance, and the aligner was rarely used in its proximity mode. Kasper's failure to understand the obsolescence of its architectural knowledge is demonstrated graphically by two incidents.

The first is the firm's interpretation of early complaints about the accuracy of its gap-setting mechanism. In proximity alignment, misalignment of the mask and the wafer can be caused both by inaccuracies or instability in the gap-setting mechanism and by distortions introduced during processing. Kasper attributed many of the problems that users of its proximity equipment were experiencing to processing error, since it believed that processing error had been the primary source of problems with its contact aligner. The firm "knew" that its gap-setting mechanism was entirely adequate, and, as a result, devoted very little time to improving its performance. In retrospect, this may seem like a wanton misuse of information, but it represented no more than a continued reliance on an information filter that had served the firm well historically.

The second illustration is provided by Kasper's response to Canon's initial introduction of a proximity aligner. The Canon aligner was evaluated by a team at Kasper and pronounced to be a copy of a Kasper machine. Kasper evaluated it against the criteria that it used for evaluating its own aligners—criteria that had been developed during its experience with contact aligners. The technical features that made Canon's aligner a significant advance, particularly the redesigned gap mechanism, were not observed because they were not considered important. The Canon aligner was pronounced to be "merely a copy" of the Kasper aligner.

Kasper's subsequent commercial failure was triggered by several factors. The company had problems designing an automatic alignment system of sufficient accuracy and in managing a high-volume manufacturing facility. It also suffered through several rapid changes of top management during the late 1970s. But the obsolescence of architectural knowledge brought about by the introduction of architectural innovation was a critical factor in its decline.

Kasper's failure stemmed primarily from failures of recognition: the knowledge that it had developed through its experience with the contact aligner made it difficult for the company to understand the ways in which Canon's proximity aligner was superior to its own. Similar problems with recognition show up in all four episodes of architectural innovation in the industry's history. The case of Perkin-Elmer and stepper technology is a case in point. By the late 1970s Perkin-Elmer had achieved market leadership with its scanning projection aligners, but the company failed to maintain that leadership when stepper technology came to dominate the industry in the early 1980s. When evaluating the two technologies, Perkin-Elmer engineers accurately forecast the progress of individual components in the two systems but failed to see how new interactions in component development—including better resist systems and improvements in lens design—would give stepper technology a decisive advantage.

Architectural Innovation

GCA, the company that took leadership from Perkin-Elmer, was itself supplanted by Nikon, which introduced a second-generation stepper. Part of the problem for GCA was recognition, but much of its failure to master the new stepper technology lay in problems in implementation. Echoing Kasper, GCA first pronounced the Nikon stepper a "copy" of the GCA design. Even after GCA had fully recognized the threat posed by the second-generation stepper, its historical experience handicapped the company in its attempts to develop a competitive machine. GCA's engineers were organized by component, and cross-department communication channels were all structured around the architecture of the first-generation system. While GCA engineers were able to push the limits of the component technology, they had great difficulty understanding what Nikon had done to achieve its superior performance.

Nikon had changed aspects of the design—particularly the ways in which the optical system was integrated with the rest of the aligner—of which GCA's engineers had only limited understanding. Moreover, because these changes dealt with component interactions, there were few engineers responsible for developing this understanding. As a result, GCA's second-generation machines did not deliver the kind of performance that the market demanded. Like Kasper and Perkin-Elmer before them, GCA's sales languished and they lost market leadership. In all three cases, other factors also played a role in the firm's dramatic loss of market share, but a failure to respond effectively to architectural innovation was of critical importance.

DISCUSSION AND CONCLUSIONS

We have assumed that organizations are boundedly rational and, hence, that their knowledge and information-processing structure come to mirror the internal structure of the product they are designing. This is clearly an approximation. It would be interesting to explore the ways in which the formulation of architectural and component knowledge are affected by factors such as the firm's history and culture. Similarly, we have assumed that architectural knowledge embedded in routines and channels becomes inert and hard to change. Future research designed to investigate information filters, problem-solving strategies and communication channels in more detail could explore the extent to which this can be avoided.

The ideas developed here could also be linked to those of authors such as Abernathy and Clark (1985), who have drawn a distinction between innovation that challenges the technical capabilities of an organization and innovation that challenges the organization's knowledge of the market and of customer needs. Research could also examine the extent to which these insights are applicable to problems of process innovation and process development.

The empirical side of this paper could also be developed. While the idea of architectural innovation provides intriguing insights into the evolution of semiconductor photolithographic alignment equipment, further research could explore the extent to which it is a useful tool for understanding the impact of innovation in other industries.

The concept of architectural innovation and the related concepts of component and architectural knowledge have a number of important implications. These ideas not only give us a richer characterization of different types of innovation, but they open up new areas in understanding the connections between innovation and organizational capability. The paper suggests, for example, that we need to deepen our understanding of the traditional distinction between innovation that enhances and innovation that destroys competence within the firm, since the essence of architectural innovation is that it both enhances and destroys competence, often in subtle ways.

An architectural innovation's effect depends in a direct way on the nature of organizational learning. This paper not only underscores the role of organizational learning in innovation but suggests a new perspective on the problem. Given the evolutionary character of development and the prevalence of dominant designs, there appears to be a tendency for active learning among engineers to focus on improvements in performance within a stable product architecture. In this context, learning means learning about components and the core concepts that underlie them. Given the way knowledge tends to be organized within the firm, learning about changes in the architecture of the product is unlikely to occur naturally. Learning about changes in architecture—about new interactions across components (and often across functional boundaries)—may therefore require explicit management and attention. But it may also be that learning about new architectures requires a different kind of organization and people with different skills. An organization that is structured to learn quickly and effectively about new component technology may be ineffective in learning about changes in product architecture. What drives effective learning about new architectures and how learning about components may be related to it are issues worth much further research.

These ideas also provide an intriguing perspective from which to understand the current fashion for cross-functional teams and more open organizational environments. These mechanisms may be responses to a perception of the danger of allowing architectural knowledge to become embedded within tacit or informal linkages.

To the degree that other tasks performed by organizations can also be described as a series of interlinked components within a relatively stable framework, the idea of architectural innovation yields insights into problems that reach beyond product development and design. To the degree that manufacturing, marketing, and finance rely on communication channels, information filters, and problem-solving strategies to integrate their work together, architectural innovation at the firm level may also be a significant issue.

Finally, an understanding of architectural innovation would be useful to discussions of the effect of technology on competitive strategy. Since architectural innovation has the potential to offer firms the opportunity to gain significant advantage over well-entrenched, dominant firms, we might expect less-entrenched competitor firms to search actively for opportunities to introduce changes in product architecture in an industry. The evidence developed here and in other studies suggests that architectural innovation is quite prevalent. As an interpretive lens, architectural innovation may therefore prove quite useful in understanding technically based rivalry in a variety of industries.

REFERENCES

Abernathy, William J.
1978 The Productivity Dilemma: Roadblock to Innovation in the Automobile Industry. Baltimore: Johns Hopkins University Press.

Abernathy, William J., and Kim Clark
1985 "Innovation: Mapping the winds of creative destruction." Research Policy, 14: 3–22.

Abernathy, William J., and James Utterback
1978 "Patterns of industrial innovation." Technology Review, June–July: 40–47.

Alexander, Christopher
1964 Notes on the Synthesis of Form. Cambridge, MA: Harvard University Press.

Argyris, Chris, and Donald Schön
1978 Organizational Learning. Reading, MA: Addison-Wesley.

Arrow, Kenneth
1974 The Limits of Organization. New York: Norton.

Arthur, Brian
1988 "Competing technologies: An overview." In Giovanni Dosi et al. (eds.), Technical Change and Economic Theory: 590–607. New York: Columbia University Press.

Brown, William L., T. Venkatesan, and A. Wagner
1981 "Ion beam lithography." Solid State Technology, August: 60–67.

Burggraaf, Pieter
1983 "X-Ray lithography: Optical's heir." Semiconductor International, September: 60–67.

Burns, Tom, and George Stalker
1966 The Management of Innovation. London: Tavistock.

Chang, T. H. P., M. Hatzakis, A. D. Wilson, and A. N. Broers
1977 "Electron-beam lithography draws a finer line." Electronics, May: 89–98.

Clark, Kim B.
1985 "The interaction of design hierarchies and market concepts in technological evolution." Research Policy, 14: 235–251.
1987 "Managing technology in international competition: The case of product development in response to foreign entry." In Michael Spence and Heather Hazard (eds.), International Competitiveness: 27–74. Cambridge, MA: Ballinger.

Cooper, Arnold C., and Dan Schendel
1976 "Strategic response to technological threats." Business Horizons, 19: 61–69.

Cyert, Richard M., and James G. March
1963 A Behavioral Theory of the Firm. Englewood Cliffs, NJ: Prentice-Hall.

Daft, Richard L.
1982 "Bureaucratic versus nonbureaucratic structure and the process of innovation and change." In Samuel B. Bacharach (ed.), Research in the Sociology of Organizations, 1:129–166. Greenwich, CT: JAI Press.

Daft, Richard L., and Karl E. Weick
1984 "Towards a model of organizations as interpretation systems." Academy of Management Review, 9: 284–295.

David, Paul A.
1985 "Clio and the economics of QWERTY." American Economic Review, 75: 332–337.

Dess, Gregory G., and Donald Beard
1984 "Dimensions of organizational task environments." Administrative Science Quarterly, 29: 52–73.

Dewar, Robert D., and Jane E. Dutton
1986 "The adoption of radical and incremental innovations: An empirical analysis." Management Science, 32: 1422–1433.

Dosi, Giovanni
1982 "Technological paradigms and technological trajectories: A suggested interpretation of the determinants and directions of technical change." Research Policy, 11: 147–162.

Dutton, Jane E., and Susan E. Jackson
1987 "Categorizing strategic issues: Links to organizational action." Academy of Management Review, 12: 76–90.

Ettlie, John E., William P. Bridges, and Robert D. O'Keefe
1984 "Organizational strategy and structural differences for radical vs. incremental innovation." Management Science, 30: 682–695.

Freeman, Christopher
1982 The Economics of Industrial Innovation, 2d ed. Cambridge, MA: MIT Press.

Galbraith, Jay
1973 Designing Complex Organizations. Reading, MA: Addison-Wesley.

Gardiner, J. P.
1986 "Design trajectories for airplanes and automobiles during the past fifty years." In Christopher Freeman (ed.), Design, Innovation and Long Cycles in Economic Development: 121–141. London: Francis Pinter.

Hage, Jerald
1980 Theories of Organization. New York: Wiley Interscience.

Hannan, Michael T., and John Freeman
1984 "Structural inertia and organizational change." American Sociological Review, 49: 149–164.

Hedberg, Bo L. T.
1981 "How organizations learn and unlearn." In P. C. Nystrom and W. H. Starbuck (eds.), Handbook of Organizational Design, 1: 3–27. New York: Oxford University Press.

Henderson, Rebecca M.
1988 "The failure of established firms in the face of technical change: A study of photolithographic alignment equipment." Unpublished Ph.D. dissertation, Harvard University.

Hollander, Samuel
1965 The Sources of Increased Efficiency: A Study of Du Pont Rayon Plants. Cambridge, MA: MIT Press.

Jackson, Susan E., and Jane E. Dutton
1988 "Discerning threats and opportunities." Administrative Science Quarterly, 33: 370–387.

Kahneman, David, Paul Slovic, and Amos Tversky
1982 Judgement under Uncertainty: Heuristics and Biases. Cambridge: Cambridge University Press.

Lawrence, Paul R., and Jay W. Lorsch
1967 Organization and Environment: Managing Differentiation and Integration. Homewood, IL: Irwin.

Louis, Meryl R., and Robert I. Sutton
1989 "Switching cognitive gears: From habits of mind to active thinking." Working Paper, School of Industrial Engineering, Stanford University.

Lyles, Majorie A., and Ian I. Mitroff
1980 "Organizational problem formulation: An empirical study." Administrative Science Quarterly, 25: 102–119.

Mansfield, Edwin
1968 Industrial Research and Technical Innovation. New York: Norton.
1977 The Production and Application of New Industrial Technology. New York: Norton.

March, James G., and Herbert A. Simon
1958 Organizations. New York: Wiley.

Marples, David L.
1961 "The decisions of engineering design." IEEE Transactions on Engineering Management, EM.8 (June): 55–71.

Miller, Ronald, and David Sawyers
1968 The Technical Development of Modern Aviation. New York: Praeger.

Mintzberg, Henry
1979 The Structuring of Organizations. Englewood Cliffs, NJ: Prentice-Hall.

Moch, Michael, and Edward V. Morse
1977 "Size, centralization and organizational adoption of innovations." American Sociological Review, 42: 716–725.

Nelson, Richard, and Sidney Winter
1982 An Evolutionary Theory of Economic Change. Cambridge, MA: Harvard University Press.

Rosenberg, Nathan
1982 Inside the Black Box: Technology and Economics. Cambridge: Cambridge University Press.

Rothwell, Roy
1986 "The role of small firms in the emergence of new technologies." In Christopher Freeman (ed.), Design, Innovation and Long Cycles in Economic Development: 231–248. London: Francis Pinter.

Sahal, Devendra
1986 "Technological guideposts and innovation avenues." Research Policy, 14: 61–82.

Schumpeter, Joseph A.
1942 Capitalism, Socialism and Democracy. Cambridge, MA: Harvard University Press.

Tushman, Michael L., and Philip Anderson
1986 "Technological discontinuities and organizational environments." Administrative Science Quarterly, 31: 439–465.

von Hippel, Eric
1990 "Task partitioning: An innovation process variable." Research Policy (in press).

Watts, Roderick K., and Norman G. Einspruch (eds.)
1987 Lithography for VLSI, VLSI Electronics—Microstructure Science. New York: Academic Press.

Weick, Karl E.
1979 "Cognitive processes in organizations." In B. M. Staw and L. L. Cummings (eds.), Research in Organizational Behavior, 1: 41–47. Greenwich, CT: JAI Press.

[3]

Strategic Management Journal, Vol. 17, 197–218 (1996)

CUSTOMER POWER, STRATEGIC INVESTMENT, AND THE FAILURE OF LEADING FIRMS

CLAYTON M. CHRISTENSEN and JOSEPH L. BOWER
Graduate School of Business Administration, Harvard University, Boston, Massachusetts, U.S.A.

Why might firms be regarded as astutely managed at one point, yet subsequently lose their positions of industry leadership when faced with technological change? We present a model, grounded in a study of the world disk drive industry, that charts the process through which the demands of a firm's customers shape the allocation of resources in technological innovation—a model that links theories of resource dependence and resource allocation. We show that established firms led the industry in developing technologies of every sort—even radical ones—whenever the technologies addressed existing customers' needs. The same firms failed to develop simpler technologies that initially were only useful in emerging markets, because impetus coalesces behind, and resources are allocated to, programs targeting powerful customers. Projects targeted at technologies for which no customers yet exist languish for lack of impetus and resources. Because the rate of technical progress can exceed the performance demanded in a market, technologies which initially can only be used in emerging markets later can invade mainstream ones, carrying entrant firms to victory over established companies.

Students of management have marveled at how hard it is for firms to repeat their success when technology or markets change, for good reason: there are lots of examples. For instance, no leading computer manufacturer has been able to replicate its initial success when subsequent architectural technologies and their corresponding markets emerged. IBM created and continues to dominate the mainframe segment, but it missed by many years the emergence of the minicomputer architecture and market. The minicomputer was developed, and its market applications exploited, by firms such as Digital Equipment and Data General. While very successful in their initial markets, the minicomputer makers largely missed the advent of the desktop computer: a market which was created by entrants such as Apple, Commodore and Tandy, and only later by IBM. The engineering workstation leaders were Apollo and Sun Microsystems, both entrants to the industry. The pioneers of the portable computing market—Compaq, Zenith, Toshiba and Sharp— were not the leaders in the desktop segment.

And yet even as these firms were missing this sequence of opportunities, they were *very* aggressively and successfully leading their industries in developing and adopting many strategically important and technologically sophisticated technologies. IBM's leadership across generations of multi-chip IC packaging, and Sun Microsystems' embrace of RISC microprocessor technology, are two instances. There are many other examples, discussed below, of firms that aggressively stayed at the forefront of technology development for extended periods, but whose industry leadership was later shaken by shifting technologies and markets.

The failure of leading firms can sometimes be ascribed to managerial myopia or organizational lethargy, or to insufficient resources or expertise. For example, cotton-spinners simply lacked the human, financial and technological resources to

Key words: innovation; resource allocation; strategy change; technological change; failure

CCC 0143–2095/96/030197–22

Received 20 September 1993
Final revision received 26 May 1995

compete when DuPont brought synthetic fibers into the apparel industry. But in many instances, the firms that missed important innovations suffered none of these problems. They had their competitive antennae up; aggressively invested in new products and technologies; and listened astutely to their customers. Yet they still lost their positions of leadership. This paper examines why and under what circumstances financially strong, customer-sensitive, technologically deep and rationally managed organizations may fail to adopt critical new technologies or enter important markets—failures to innovate which have led to the decline of once-great firms.

Our conclusion is that a primary reason why such firms lose their positions of industry leadership when faced with certain types of technological change has little to do with technology itself—with its degree of newness or difficulty, relative to the skills and experience of the firm. Rather, they fail because they listen too carefully to their customers— and customers place stringent limits on the strategies firms can and cannot pursue.

The term 'technology', as used in this paper, means the processes by which an organization transforms labor, capital, materials, and information into products or services. All firms have technologies. A retailer such as Sears employs a particular 'technology' to procure, present, sell, and deliver products to its customers, while a discount warehouse retailer such as the Price Club employs a different 'technology'. Hence, our concept of technology extends beyond the engineering and manufacturing functions of the firm, encompassing a range of business processes. The term 'innovation' herein refers to a change in technology.

A fundamental premise of this paper is that patterns of resource allocation heavily influence the types of innovations at which leading firms will succeed or fail. In every organization, ideas emerge daily about new ways of doing things—new products, new applications for products, new technical approaches, and new customers—in a manner chronicled by Bower (1970) and Burgelman (1983a, 1983b). Most proposals to innovate require human and financial resources. The patterns of innovation evidenced in a company will therefore mirror to a considerable degree the patterns in how its resources are allocated to, and withheld from, competing proposals to innovate.

We observe that because effective resource allocation is market-driven, the resource allocation procedures in successful organizations provide impetus for innovations known to be demanded by current customers in existing markets. We find that established firms in a wide range of industries have tended to lead in developing and adopting such innovations. Conversely, we find that firms possessing the capacity and capability to innovate may fail when the innovation does *not* address the foreseeable needs of their current customers. When the initial price/performance characteristics of emerging technologies render them competitive only in emerging market segments, and not with current customers, resource allocation mechanisms typically deny resources to such technologies. Our research suggests that the inability of some successful firms to allocate sufficient resources to technologies that initially cannot find application in mainstream markets, but later invade them, lies at the root of the failure of many once-successful firms.

EARLIER VIEWS OF FACTORS INFLUENCING PATTERNS OF RESOURCE ALLOCATION IN THE INNOVATION PROCESS

Our research links two historically independent streams of research, both of which have contributed significantly to our understanding of innovation. The first stream is what Pfeffer and Salancik (1978) call *resource dependence*: an approach which essentially looks *outside* the firm for explanations of the patterns through which firms allocate resources to innovative activities. Scholars in this tradition contend that firms' strategic options are constrained because managerial discretion is largely a myth. In order to ensure the survival of their organizations, managers lack the power to do anything other than to allocate resources to innovative programs that are required of the firm by external customers and investors: the entities that provide the resources the firm needs to survive. Support for this view comes from the work of historians of technological innovation such as Cooper and Schendel (1976) and Foster (1986). The firms they studied generally responded to the emergence of competitively threatening technologies by intensifying their investments to improve the conventional technologies used by their current customers—which provided the resources the firms needed to survive over the short term.

The second stream of ideas, originally taught by Bower (1970) and amplified by Burgelman (1983a, 1983b), describes the resource allocation process internal to the firm. These scholars suggest that most strategic proposals—to add capacity or develop new products or processes—take their fundamental shape at lower levels of hierarchical organizations. Bower observed that the allocation of funding amongst projects is substantially shaped by the extent to which managers at middle levels of the organization decide to support, or lend *impetus*, to some proposals and to withhold it from others. Bower also observed that risk management and career management were closely linked in the resource allocation process. Because the career costs to aspiring managers of having backed an ultimately unsuccessful project can be severe, their tendency was to back those projects where the demand for the product was assured.

Our study links these two streams by showing how the impetus that drives patterns of resource allocation (and hence innovation) within firms does not stem from autonomous decisions of risk-conscious managers. Rather, whether sufficient impetus coalesces behind a proposed innovation is largely determined by the presence or absence of current customers who can capably articulate a need for the innovation in question. There seems to be a powerful linkage from: (1) the expectations and needs of a firm's most powerful customers for product improvements; to (2) the types of innovative proposals which are given or denied impetus within the firm and which therefore are allocated the resources necessary to develop the requisite technological capabilities; to (3) the markets toward which firms will and will not target these innovations; which in turn leads to (4) the firms' ultimate commercial success or failure with the new technology.

A primary conclusion of this paper is that when significant customers demand it, sufficient impetus may develop so that large, bureaucratic firms can embark upon and successfully execute technologically difficult innovations—even those that require very different competencies than they initially possessed.[1] Conversely, we find that when a proposed innovation addresses the needs of small customers in remote or emerging markets that do not supply a significant share of the resources a firm currently needs for growth and survival, firms will find it difficult to succeed even at innovations that are technologically straightforward. This is because the requisite impetus does not develop, and the proposed innovations are starved of resources.

Our findings build upon the work of earlier scholars who have addressed the question of why leading firms may fail when faced with technological change. Cooper and Schendel (1976) found that new technologies often are initially deployed in new markets, and that these were generally brought into industries by entering firms. They observed that established firms confronted with new technology often intensified investment in traditional technical approaches, and that those that did make initial resource commitments to a new technology rarely maintained adequate resource commitments. Foster (1986) noted that at points when new technologies enter an industry, entrants seem to enjoy an 'attacker's advantage' over incumbent firms. Henderson and Clark (1990) posited that entrant firms enjoyed a particular advantage over incumbents in architectural technology change.

We hope to add additional precision and insight to the work of these pioneering scholars, by stating more precisely the specific sorts of technological innovations that are likely initially to be deployed in new applications, and the sorts that are likely to be used in mainstream markets from the beginning; and to define the types of innovation in which we expect attackers to enjoy an advantage, and the instances in which we expect incumbents to hold the upper hand. By presenting a model of the processes by which resource commitments are made, we hope partially to explain a puzzle posed but not resolved by each of these authors: *why* have incumbent firms generally intensified their commitments to conventional technology, while starving efforts to commercialize new technologies—even while the new technology was gaining ground in the market?

[1] Evidence supporting this conclusion is provided below. In making this statement, we contest the conclusions of scholars such as Tushman and Anderson (1986), who have argued that incumbent firms are most threatened by attacking entrants when the innovation in question destroys, or does not build upon, the competence of the firm. We observe that established firms, though often at great cost, have led their industries in developing critical competence-destroying technologies, when the new technology was needed to meet existing customers' demands.

Finally, by examining why established firms do these things, we hope to provide insights for how managers can more successfully address different types of technological change.

RESEARCH METHODS

Three very different classes of data were used in this study, to establish solid construct validity (Yin, 1989). The first was a data base of the detailed product and performance specifications for every disk drive model announced by every firm participating in the world industry between 1975 and 1990—over 1400 product models in all. These data came from *Disk/Trend Report*, the leading market research publication in the disk drive industry, and from product specification sheets obtained from the manufacturers themselves. The tables and other summary statistics reported in this paper were calculated from this data base, unless otherwise noted. This data set is not a statistical sample, but constitutes a complete census of companies and products for the world industry during the period studied.

The second type of information employed in the study relates to the strategies pursued, and the commercial success and failure, of each of the companies that announced the development of a rigid disk drive between 1976 and 1990. *Disk/Trend* reported each firm's rigid disk drive sales in each of these years, by product category and by market segment. Each monthly issue between 1976 and 1990 of *Electronic Business* magazine, the most prominent trade publication covering the magnetic recording industry, was examined for information about disk drive manufacturers, their strategies and products. We used this information to verify the completeness of the *Disk/Trend* data,[2] and to write a history of the disk drive industry describing the strategies and fortunes of firms in the industry (Christensen, 1993).

The third type of information employed in this study came from over 70 personal, unstructured interviews conducted with executives who are or have been associated with 21 disk drive manufacturing companies. Those interviewed included founders; chief executives; vice presidents of sales and marketing, engineering and finance; and engineering, marketing and managerial members of pivotal product development project teams. The firms whose executives were interviewed together account for over 80 percent of the disk drives produced in the world since the industry's inception. Data from these interviews were used to reconstruct, as accurately as possible, the decision-making processes associated with key innovations in each company's history. Wherever possible, accounts of the same decision were obtained from multiple sources, including former employees, to minimize problems with *post hoc* rationalization. Multiple employees were interviewed in 16 of the 21 companies.

The *Disk/Trend* data enabled us to measure the impact that each new component and architectural technology had on disk drive performance. Furthermore, it was possible to identify which firms were the first to develop and adopt each new technology, and to trace the patterns of diffusion of each new technology through the world industry over time, amongst different types of firms. When analysis of the *Disk/Trend* data indicated a particular entrant or established firm had prominently led or lagged behind the industry in a particular innovation, we could determine the impact of that leadership or followership on the subsequent sales and market shares, by product-market segment, for each company.

Analysis of these data essentially enabled us to develop a theory of *what* will happen when different types of technological change occur—whether we would expect entrant and established firms to take leadership in their development. We then used our interview data to write case histories of key decisions in six companies to understand *why* those patterns of leadership and followership in technology development occur. These case studies covered entrant and established firms, over an extended period of time in which each of them made decisions to invest, or delay investing, in a variety of new technologies. These cases were selected in what Yin (1989) calls a multi-case, nested experimental design, so that through pattern-matching across cases, the external validity of the study's conclusions could be established.[3]

---

[2] *Disk/Trend Report* identified 133 firms that participated in the disk drive industry in the period studied. The search of *Electronic Business* magazine yielded information on one additional firm, Peach Tree Technology, that never generated revenues and somehow had escaped detection by the *Disk/Trend* editors.

[3] Table 3 (which refers to Yin, 1989: 35–37) describes this pattern-matching.

We studied the disk drive industry because its history is one of rapid change in technology and market structure. The world rigid disk drive market grew at a 27 percent annual rate to over $13 billion between 1975 and 1990. Of the 17 firms in the OEM industry in 1976, only one was still in operation in 1990. Over 130 firms entered the industry during this period, and more than 100 of them failed. The cost per megabyte (MB) of the average drive in constant 1990 dollars fell from $560 in 1976 to $5 in 1990. The physical size of a 100 MB drive shrank from 5400 to 8 cubic inches over the same period. During this time, six architecturally distinct product generations emerged, and a new company rose to become market leader in four of these six generations. A description of disk drive technology that may be helpful for some readers is provided in Appendix 1.

TYPOLOGIES OF TECHNOLOGICAL CHANGE

Earlier scholars of technology change have argued that incumbent firms may stumble when technological change destroys the value of established technological competencies (Tushman and Anderson, 1986), or when new architectural technologies emerge (Henderson and Clark, 1990). For present purposes, however, we have found it useful to distinguish between those innovations that *sustained* the industry's rate of improvement in product performance (total capacity and recording density were the two most common measures), and those innovations that *disrupted* or redefined that performance trajectory (Dosi, 1982). The following two sections illustrate these concepts by describing prominent examples of trajectory-sustaining and trajectory-disrupting technological changes in the industry's history. The subsequent sections then describe the role these innovations played in the industry's development; the processes through which incumbent and entrant firms responded to these different types of technological change; and the consequent successes and failures these firms experienced.

Sustaining technological changes

In the disk drive industry's history, most of the changes in competent technology, and two of the six changes in architectural technology, sustained or reinforced established trajectories of product performance improvement. Two examples of such technology change are shown in Figure 1. The left-most graph compares the average recording density of drives that employed conventional particulate oxide disk technology and ferrite head technology, vs. the average density of drives that employed new-technology thin film heads and disks, that were introduced in each of the years between 1976 and 1990. The improvements in the conventional approach are the result of consistent incremental advances such as grinding the ferrite heads to finer, more precise dimensions; and using smaller and more finely dispersed oxide particles on the disk's surface. Note that the improvement in areal density obtainable with ferrite/oxide technology began to level off in the period's later years—suggesting a maturing technology S-curve (Foster, 1986). Note how thin film head and disk technologies emerged to sustain the rate of performance improvement at its historical pace of 35 percent between 1984 and 1990.

The right-most graph in Figure 1 describes a sustaining technological change of a very different character: an innovation in product architecture. In this case, the 14-inch Winchester drive substituted for removable disk packs, which had been the dominant design between 1962 and 1978. Just as in the thin film-for-ferrite/oxide substitution, the impact of Winchester technology was to sustain the historically established rate of performance improvement. Other important innovations, such as embedded servo systems, RLL & PRML recording codes, higher RPM motors and embedded SCSI, SMD, ESDI and AT interfaces, also helped manufacturers sustain the rate of historical performance improvement that their customers had come to expect.[4] Hereafter in this

[4] The examples of technology change presented in Figures 1 and 2 in this paper introduce some ambiguity to the unqualified term 'discontinuity', as it has been used by Dosi (1982), Tushman and Anderson (1986), and others. The innovations in head and disk technology described in the left graph of Figure 1 represent *positive discontinuities* in an established technological trajectory, while the development of trajectory-disrupting technologies charted in Figure 2 represent *negative* discontinuities. As will be shown below, established firms seemed quite capable of leading the industry over positive discontinuities. The negative ones were the points at which established firms generally lost their positions of industry leadership.

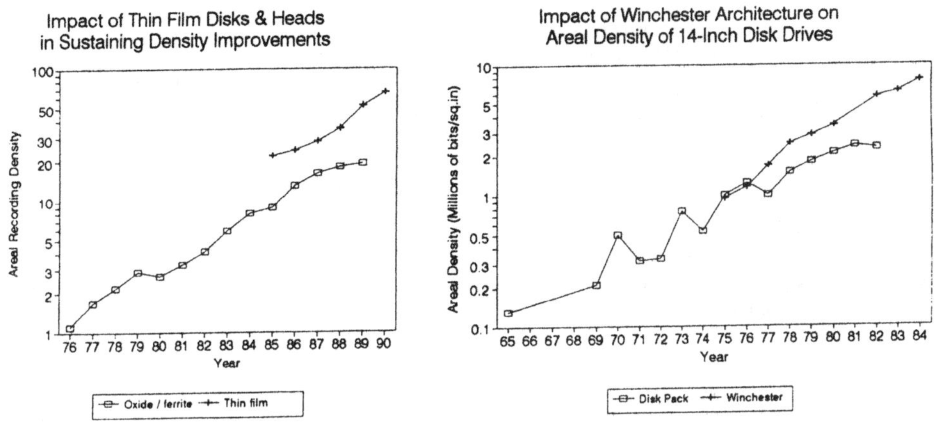

Figure 1. Examples of sustaining technological change in componentry (left) and product architecture (right). Reprinted with permission from *Business History Review*, 1993, **67**, p. 557.

paper, technological changes that have such a sustaining impact on an established trajectory of performance improvement are called *sustaining technologies*.

Disruptive technological changes

Most technological change in the industry's history consisted of sustaining innovations of the sort described above. In contrast, there were just a few trajectory-disrupting changes. The most important of these from a historical viewpoint were the architectural innovations that carried the industry from 14-inch diameter disks to diameters of 8, 5.25 and then 3.5 inches. The ways in which these innovations were disruptive are illustrated in Table 1. Set in 1981, this table compares the attributes of a typical 5.25-inch drive—a new architecture that had been in the market for less than a year at that time—with those of a typical 8-inch drive, which by that time had become the standard drive used by minicomputer manufacturers. Note that along the dimensions of performance which were important to established minicomputer manufacturers—capacity, cost per megabyte, and access time—the 8-inch product was vastly superior. The 5.25-inch architecture did not address the needs of minicomputer manufacturers, as they perceived their needs at that time. On the other hand, the 5.25-inch architecture *did* possess attributes that appealed to the desktop personal computer market segment that was just emerging in 1980–82. It was small and lightweight—important features for this application. And it was priced at around $2000, which means it could economically be incorporated in desktop machines. Hereafter in this paper, technologies such as this, which disrupt an established trajectory of performance improvement, or redefine what performance means, are called *disruptive technologies*.

In general, sustaining technological changes appealed to established customers in existing, mainstream markets. They provided these customers with more of what they had come to

Table 1. The disruptive impact on performance improvement of the 5.25-inch, vs. the 8-inch architecture

Attribute	8-inch drives	5.25-inch drives
Capacity (megabytes)	**60**	10
Volume (cubic inches)	566	*150*
Weight (pounds)	21	*6*
Access time (ms)	**30**	160
Cost per megabyte	**$50**	$200
Total unit cost	$3000	*$2000*

Key: Attributes valued highly in the minicomputer market in 1981 are presented in **boldface**.
Attributes valued in the emerging desktop computing market in 1981 are shown in *italics*.
Source: Analysis of Disk/Trend Report data; from Christensen (1992a: 90)

expect. In contrast, disruptive technologies rarely could initially be employed in established markets. They tended instead to be valued in remote or emerging markets. This tendency consistently appears not just in disk drives, but across a range of industries (Rosenbloom and Christensen, 1995).

THE IMPACT OF SUSTAINING AND DISRUPTIVE TECHNOLOGIES ON INDUSTRY STRUCTURE

The history of sustaining and disruptive technological change in the disk drive industry is summarized in Figure 2. It begins in 1974, the year after IBM's first Winchester architecture model was introduced to challenge the dominant disk pack architectural design. Almost all drives then were sold to makers of mainframe computers. Note that in 1974 the median-priced mainframe computer was equipped with about 130 MB of hard disk capacity. The typical hard disk storage capacity supplied with the median-priced mainframe increased about 17 percent per year, so that by 1990 the typical mainframe was equipped with 1300 MB of hard disk capacity. This growth in the use of hard disk memory per computer is mapped by the solid line emanating from point A in Figure 2. This trajectory was driven by user learning and software developments in the applications in which mainframes were used (Christensen and Rosenbloom, 1995).

The dashed line originating at point A measures the increase in the average capacity of 14-inch drives over the same period. Note that although the capacity of the average 14-inch drive was equal to the capacity shipped with the typical mainframe in 1974, the rate of increase in capacity provided within the 14-inch architecture exceeded the rate of increase in capacity demanded in the mainframe market—carrying this architecture toward high-end mainframes, scientific computers, and supercomputers. Furthermore, note how the new 14-inch Winchester architecture sustained the capacity trajectory that had been established in the earlier removable disk pack architecture. Appendix 2 describes how these trajectories were calculated.

The solid trajectories emanating from points B, C and D represent the average hard disk capacity *demanded* by computer buyers in each market segment, over time.[5] The dashed lines emanating from points B, C, and D in Figure 2 measure trends in the average capacity that disk drive manufacturers were able to *provide* with each successive disk drive architecture. Note that with the exception of the 14-inch Winchester architecture, the maximum capacity initially available in each of these architectures was substantially *less* than the capacity required for the typical computer in the established market—these were *disruptive* innovations. As a consequence, the 8, 5.25 and 3.5-inch designs initially were rejected by the leading, established computer manufacturers, and were deployed instead in emerging market applications for disk drives: minicomputers, desktop PCs and portable PCs, respectively. Note, however, that once these disruptive architectures became established in their new markets, the accumulation of hundreds of sustaining innovations pushed each architecture's performance ahead along very steep, and roughly parallel, trajectories.[6]

Note that the trajectory of improvement that the technology was able to *provide* within each architecture was nearly *double* the slope of the increase in capacity *demanded* in each market. As we will see, this disparity between what the technology could provide and what the market demanded seems to have been the primary source of leadership instability in the disk drive industry.

LEADERS IN SUSTAINING AND DISRUPTIVE TECHNOLOGICAL INNOVATIONS

To better understand why leading firms might successfully pioneer in the development and adoption of many new and difficult technologies, and yet lose their positions of industry leadership by failing to implement others, we compared the innovative behavior of *established* firms with that of *entrant* firms, with respect to each of the

[5] These trajectories represent the disk capacity *demanded* in each market because in each instance, greater disk capacity could have been supplied to users by the computer manufacturers, had the market demanded additional capacity at the cost for which it could be purchased at the time.

[6] The parallel impact of sustaining innovations across these architectural generations results from the fact that the same sustaining technologies, in the form of componentry, were available simultaneously to manufacturers of each generation of disk drives (Christensen, 1992b).

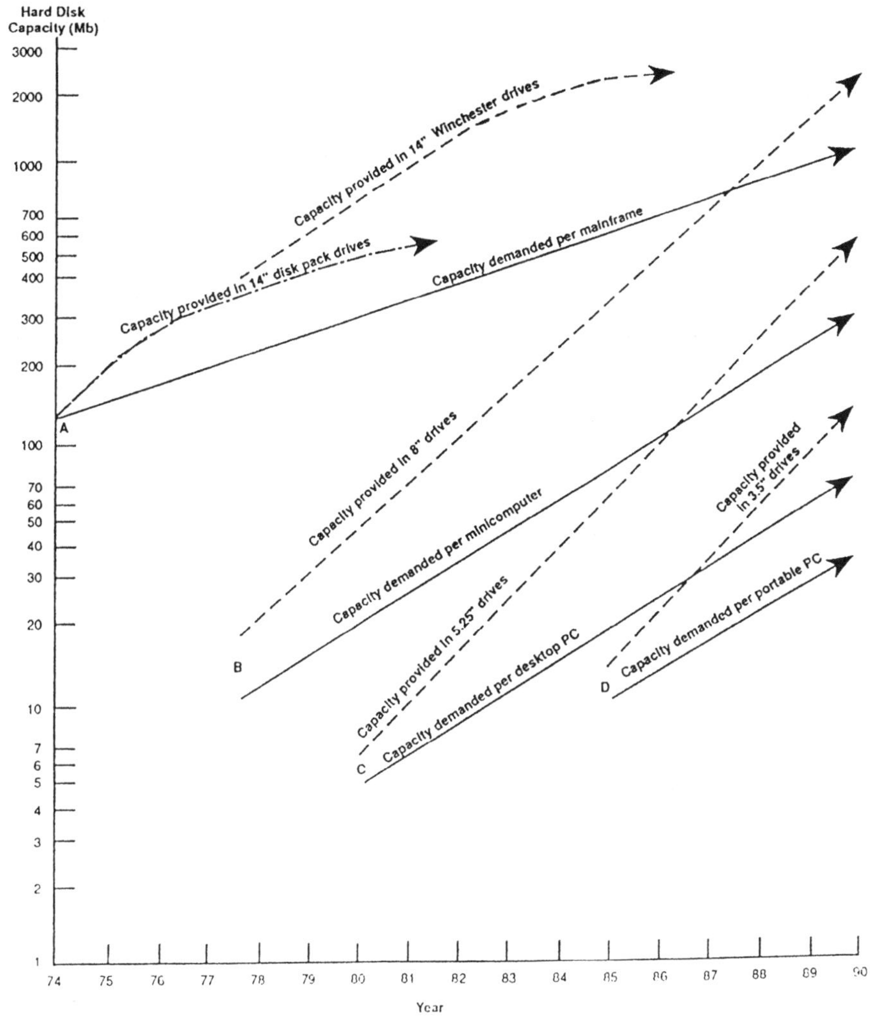

Figure 2. Patterns of entry and improvement in disruptive disk drive technologies. Reprinted with permission from *Business History Review*, 1993, **67**, p. 559.

sustaining and disruptive technological innovations in the history of the disk drive industry. Building upon the approach employed by Henderson and Clark (1990), established firms were defined as firms that had previously manufactured drives which employed an older, established technology, whereas entrant firms were those whose initial product upon entry into the industry employed the new component or architectural technology being analyzed. This approach was used because of this study's longitudinal character, looking at the performance of incumbents and entrants across a sequence of innovations.

In spite of the wide variety in the magnitudes and types of sustaining technological changes in the industry's history, the firms that led in their development and adoption were the industry's leading, established firms. Table 2(a) depicts this leadership pattern for three representative sustaining technologies. In thin-film head technology, it was Burroughs (1976), IBM (1979), and other established firms that first successfully incorporated thin-film heads in disk drives. In the 1981–86 period, when over 60 firms entered the rigid disk drive industry, only five of them (all commercial failures) attempted to do so using thin-film heads as a source of performance advantage in their initial products. All other entrant firms—even aggressively performance-oriented firms such as Maxtor and Conner Peripherals—found it preferable to cut their teeth on ferrite heads in the entry products, before tackling thin-film technology in subsequent generations.

Note the similar pattern in the development and adoption of RLL codes—a much simpler development than thin-film head technology—which consumed at most a few million dollars per firm. RLL enabled a 30 percent density improvement, and therefore represented the type of inexpensive path to performance improvement that ought to be attractive to entrant firms. But in 1985, 11 of the 13 firms which introduced new models employing RLL technology were established firms, meaning that they had previously offered models based on MFM technology. Only two were entrants, meaning that their initial products employed RLL codes. Table 2(a) also notes that six of the first seven firms to introduce Winchester architecture drives were established makers of drives employing the prior disk pack architecture.[7]

The history of literally every other sustaining innovation—such as embedded servo systems, zone-specific recording densities, higher RPM motors and the 2.5-inch Winchester architecture—reveals a similar pattern: the established firms led in the adoption of sustaining technology be it in componentry or architecture. Entrant firms followed. In other words, the failure of leading firms to stay atop the disk drive industry generally was not because they could not keep pace with the industry's movement along the dashed-line technological trajectories mapped in Figure 2. The leading incumbent firms effectively *led* the industry along those trajectories even though many of these were competency-destroying progressions in terms of technologies, skills and manufacturing assets required (Tushman and Anderson, 1986).

In contrast, the firms that led the industry in introducing *disruptive* architectural technologies—in the moves to points B, C and D in Figure 2—tended overwhelmingly to be *entrant*, rather than established firms. This is illustrated in Table 2(b). It shows, for example, that in 1978 an entrant offered the industry's first 8-inch drive. By the end of the second year of that architecture's life (1979), six firms were offering 8-inch drives; two-thirds of them were entrants. Likewise, by the end of the second year of the 5.25-inch generation's life, eight of the 10 firms offering 5.25-inch drives were entrants. Entrants similarly dominated the early population of firms offering 3.5-inch drives. In each of these generations, between half and two-thirds of the established manufacturers of the prior architectural generation *never* introduced a model in the new architecture. And those established drivemakers that did design and manufacture new architecture models did so with an average two-year lag behind the pioneering entrant firms. In this fast-paced industry, such slow response often proved fatal.

These patterns of leadership and followership in sustaining and disruptive technologies are reflected in the commercial success and failure of disk drive manufacturers. The ability of established firms to lead the industry in the sustaining innovations that powered the steep technological trajectories in Figure 2 often were technologically difficult, risky and expensive. Yet in the history of this industry, there is no evidence that the firms that led in sustaining innovations gained market share by virtue of such technology leadership (Christensen, 1992b). This leadership enabled them to maintain their competitiveness only within specific technological trajectories. On the other hand, entrant firms' leadership advantages in disruptive innovations enabled them not only to capture new markets as they emerged, but (because the trajectories of technological progress were steeper than the trajectories of performance demanded) to invade and capture established markets as well.

Hence, all but one of the makers of 14-inch drives were driven from the mainframe computer market by entrant firms that got their start making

[7] Note that the statistics shown in Table 2 are not a sample—they represent the entire population of firms in each of the years shown offering models incorporating the technologies in question. For that reason, tests of statistical significance are not relevant in this case.

Table 2. Trends in technology leadership and followership in sustaining vs. disruptive technologies

(a) Numbers of established and entrant firms introducing models employing selected trajectory-sustaining technologies

		1974	1975	1976	1977	1978	1979	1980	1981	1982	1983	1984	1985	1986	1987	1988
Thin-film heads	Entrants								1		1	2	1		1	4
	Established			1			1	1	3	5	6	8	12	15	17	22
RLL codes	Entrants											1	2	3	6	8
	Established											4	11	20	25	26
Winchester architecture	Entrants				1	4	9									
	Established	1		3	3	7	11									

(b) Numbers of established and entrant firms introducing models based upon disruptive architectural technologies

		1974	1975	1976	1977	1978	1979	1980	1981	1982	1983	1984	1985	1986	1987	1988
8-inch	Entrants					1	4	6	8							
	Established					0	2	5	5							
5.25-inch	Entrants							1	8	8	13					
	Established							1	2	8	11					
3.5-inch	Entrants											1	2	3	4	
	Established											0	1	1	4	

Note: Data are presented in these tables only for those years in which the new technologies were gaining widespread acceptance, to illustrate tendencies in technology leadership and followership. Once the technologies had become broadly accepted, the numbers of firms introducing models using them are no longer reported. Twelve years are covered in the thin-film head category because it took that long for thin film heads to become broadly used in the marketplace. Only 5 years of history are reported for RLL codes because by 1988 the vast majority of established *and* entrant firms had adopted RLL codes. Four years of data are shown for new architectures, because any established firms that had not launched the new architecture within 4 years of its initial appearance in the market had been driven from the industry.

8-inch drives for minicomputers. The 8-inch drivemakers, in turn, were driven from the minicomputer market, and eventually the mainframe market, by firms which led in producing 5.25-inch drives for desktop computers. And the leading makers of 5.25-inch drives were driven from desktop and minicomputer applications by makers of 3.5-inch drives, as mapped in Figure 2.

We began this paper by posing a puzzle: why it was that firms which at one point could be esteemed as aggressive, innovative, customer-sensitive organizations could ignore or attend belatedly to technological innovations with enormous strategic importance. In the context of the preceding analysis of the disk drive industry, this question can be sharpened considerably. The established firms were, in fact, aggressive, innovative, and customer-sensitive in their approaches to sustaining innovations of every sort. But why was it that established firms could not lead their industry in disruptive architectural innovations? For it is only in these innovations that attackers demonstrated an advantage. And unfortunately for the leading established firms, this advantage enabled attacking entrant firms to topple the incumbent industry leaders each time a disruptive technology emerged.[8]

To understand why disruptive technological change was so consistently vexing to incumbent firms, we personally interviewed managers who played key roles in the industry's leading firms, as incumbents or entrants, when each of these disruptive technologies emerged. Our objective in these interviews was to reconstruct, as accurately and from as many points of view as possible, the forces that influenced these firms' decision-making processes relating to the development and commercialization of disruptive architectural technologies. We found the experiences of the firms, and the forces influencing their decisions, to be

8 We believe this insight—that attacking firms have an advantage in disruptive innovations but not in sustaining ones—clarifies but is not in conflict with Foster's (1986) assertions about the attacker's advantage. The historical examples Foster uses to substantiate his theory generally seem to have been disruptive innovations.

remarkably similar. In each instance, when confronted with disruptive technology change, developing the requisite *technology* was never a problem: prototypes of the new drives often had been developed before management was asked to make a decision. It was in the process of allocating scarce resources amongst competing product and technology development proposals, however, that disruptive projects got stalled. Programs addressing the needs of the firms' most powerful customers almost *always* pre-empted resources from the disruptive technologies, whose markets tended to be small and where customers' needs were poorly defined.

In the following section we have synthesized the data from case studies of the six firms we studied in particular depth, into a *six*-step model that describes the factors that influenced how resources were allocated across competing proposals to develop new sustaining vs. disruptive technology in these firms. The struggle of Seagate Technology, the industry's dominant maker of 5.25-inch drives, to successfully commercialize the disruptive 3.5-inch drive, is recounted here to illustrate each of the steps in the model. Short excerpts from a fuller report of other case histories (Christensen, 1992a) are also presented to illustrate what happened in specific companies at each point in the process. Table 3 describes how the findings from each of the case studies support, or do not support, the principal propositions in the model. In Yin's (1989) terms, the high degree of literal and theoretical replication shown in Table 3, and the extent of 'pattern matching' across case studies where more than one firm encountered the same technological change, lend high degrees of reliability and external validity to the model.[9]

[9] For readers who are unfamiliar with the work of scholars such as Yin (1989) and Campbell and Stanley (1966) on research methodology, a *literal* replication of a model occurs when an outcome happens as the model would predict. A *theoretical* replication of the model occurs when a different outcome happens than what would have been predicted by the model, but where this outcome can be explained by elements in the model. In the instance here, the success of entrants and the failure of established forms at points of disruptive technology change are directly predicted by the model, and would be classed as literal replications. Instances where an established firm succeeded in the face of disruptive technological change because it acted in a way that dealt with the factors in the model that typically precipitated failure, would be classed as *theoretical* replications of the model. Several of these instances occurred in the industry's history, as explained later in this paper.

A MODEL OF THE RESOURCE-ALLOCATION PROCESS IN ESTABLISHED FIRMS FACED WITH DISRUPTIVE CHANGE

1. Although entrants were the leaders in *commercializing* disruptive technology, it did not start out that way: the first engineers to develop the disruptive architectures generally did so while employed by a leading established firm, using bootlegged resources. Their work was rarely initiated by senior management. While architecturally innovative, these designs almost always employed off-the-shelf components. For example, engineers at Seagate Technology, the leading 5.25-inch drive maker, were the second in the industry to develop working prototype 3.5-inch models, in 1985. They made over 80 prototype models before the issue of formal project approval was raised with senior management. The same thing happened earlier at Control Data, the dominant 14-inch drivemaker. Its engineers had designed working 8-inch drives internally, nearly 2 years before they appeared in the market.
2. The marketing organization then used its habitual procedure for testing the market appeal of new drives, by showing prototypes to lead customers of the existing product line, asking them to evaluate the new models.[10] Again drawing on the Seagate case, marketers tested the new 3.5-inch drives with IBM and other makers of XT and AT-class desktop personal computers—even though the drives, as shown in Figure 2 above, had significantly less capacity than in the mainstream desktop market demanded.

[10] This is consistent with Burgelman's observation that one of the greatest difficulties encountered by corporate entrepreneurs was finding the right 'beta test sites', where products could be interactively developed and refined with customers. Generally, the entre to the customer was provided by the salesman who sold the firm's established product lines. This helped the firm develop new products for established markets, but did not help it identify new applications for its new technology (Burgelman and Sayles, 1986: 76–80). Professor Rebecca Henderson pointed out to us that this tendency always to take new technologies to mainstream customers reflects a rather narrow *marketing* competence—that although these issues tend to be framed by many scholars as issues of technological competence, a firm's disabilities in finding new markets for new technologies may be its most serious innovative handicap.

Table 3. Support of key elements of model found in each of six in-depth case studies

Companies Studied:	Prototypes of disruptive architecture drive developed internally, well before widespread industry adoption (model step 1)	Marketers show early prototypes to lead customers of prior architecture; they reject product; marketing issues pessimistic forecast (model step 2)	Project to commercialize disruptive product is shelved; company aggressively pursues sustaining innovations (model step 3)	New firms are established to commercialize disruptive architecture; they find new markets, where product's attributes are valued (model step 4)	Entrant firms which initially sold product only in new market improve performance faster than initial market requires, enabling them to attack established markets (model step 5)	In response to entrants' attack, established firms belatedly introduce disruptive product. Sales are largely to existing customers, cannibalizing sales of prior architecture products. (model step 6)
Quantum Corp.	L	L	L, T	L, T	L	L, T
Conner Peripherals	L		L	L	L	
Miniscribe		L		L	L	L
Seagate Technology	L	L	L	L	L	L
Micropolis	T	L	L,T	L,T	L	T
Control Data	L	L	L,T	L,T	L	L,T

Note: An 'L' in the matrix indicates that this step was a clear, explicit element in that firm's case history—in Yin's (1989) terms, a 'literal replication'. Where 'T' is shown, the firm avoided the fate described in the model by explicitly recognizing the factors in the model, and dealing with them in the manner described in the final section of this paper. These constitute what Yin calls 'theoretical replications' of the model. Where no 'L' or 'T' is shown, that step was not a clear or prominent part of the firm's encounter with the disruptive technology being studied. Some firms studied confronted only one disruptive architecture. Miniscribe, for example, started making 5.25-inch drives generally in the pattern indicated by our model; and was subsequently driven from the industry. Other firms, such as Quantum and Control Data, confronted a series of disruptive innovations, and dealt with some of them differently than they did with others, as described in the last section of the paper. In such instances, an 'L' and a 'T' are entered in the matrix. As Yin points out, when multiple case studies are used to support a multi-element model, as in this study, each cell in a matrix such as this constitutes an independent 'observation'. Hence, the model is supported in 32 of the 36 observations.

These customers showed little interest in the disruptive drives, because they did not address their need for higher performance within the established architectural framework. As Figure 2 shows, the established customers needed new drives that would take them *along* their existing performance trajectory. As a consequence, the marketing managers were unwilling to support the disruptive technology and offered pessimistic sales forecasts.

Generally, because the disruptive drives were targeted at emerging markets, initial forecasts of sales were small. In addition, because such products were simpler and offered lower performance, forecast profit margins were also lower than established firms had come to require. Financial analysts in established firms, therefore, joined their marketing colleagues in opposing the disruptive programs. As a result, in the ensuing allocation process resources were explicitly withdrawn, and the disruptive projects were slowly starved.

For example, when Seagate's main customer, IBM's PC division, rejected Seagate's 3.5-inch prototypes for insufficient capacity, sales forecasts were cut and senior managers shelved the program—just as 3.5-inch drives were becoming firmly established in laptops. 'We needed a new model,' recalled a former Seagate manager, 'which could become the next ST412 (a very successful product generating $300 million sales annually in the desktop market that was near the end of its life cycle). Our forecasts for the 3.5-inch drive were under $50 million because the laptop market was just emerging—and the 3.5-inch product just didn't fit the bill.' And earlier, when engineers at Control Data, the leading 14-inch drive maker, developed its initial 8-inch drives, its customers were looking for an average of 300 MB per computer, whereas CDC's earliest 8-inch drives offered less than 60 MB. The 8-inch project was given low priority, and engineers assigned to its development kept getting pulled off to work on problems with 14-inch drives being designed for more important customers. Similar problems plagued the belated launches of Quantum's and Micropolis's 5.25-inch products.

3. In response to the needs of current customers, the marketing managers threw impetus behind alternative *sustaining* projects, such as incorporating better heads or developing new recording codes. These would give their cus-

tomers what they wanted, could be targeted at large markets, and generate the sales and profits required to maintain growth. Although they generally involved greater development expense, such sustaining investments appeared *far* less risky than investments in the disruptive technology, because the customers were there. The rationality of Seagate's decision to shelve the 3.5-inch drive in 1985–86, for example, is stark. Its view downmarket (in terms of Figure 2) was at a $50 million total market forecast for 3.5-inch drives in 1987. What gross margins it could achieve in that market were uncertain, but its manufacturing executives predicted that costs per megabyte in 3.5-inch drives would be much higher than in 5.25-inch products. Seagate's view upmarket was quite different. Volumes in 5.25-inch drives with capacities of 60–100 MB were forecast to be $500 million in size by 1987. And companies serving the 60–100 MB market were earning gross margins of 35–40 percent, whereas Seagate's margins in its high-volume 20 MB drives were between 25 and 30 percent. It simply did not make sense for Seagate to put resources behind the 3.5-inch drive, when competing proposals to move upmarket to develop its ST251 line of drives were also actively being evaluated.

After Seagate executives shelved the 3.5-inch project, it began introducing new 5.25-inch models at a dramatically accelerating rate. In the years 1985, 1986 and 1987, the numbers of new models it introduced each year as a percentage of the total number of its models on the market in the prior year were 57, 78, and 115 percent, respectively. And during the same period, Seagate incorporated complex and sophisticated new component technologies such as thin-film disks, voice coil actuators, RLL codes, and embedded, SCSI interfaces. In each of our other case studies as well, the established firms introduced new models in their established architectures employing an array of new component technologies at an accelerating rate, after the new architectures began to be sold. The clear motivation of the established firms in doing this was to win the competitive wars against each other, rather than to prepare for an attack by entrants from below.

4. New companies, usually including members of the frustrated engineering teams from established firms, were formed to exploit the disruptive product architecture. For example, the founders of the leading 3.5-inch drivemaker, Conner Peripherals, were disaffected employees from Seagate and Miniscribe, the two largest 5.25-inch manufacturers. The founders of 8-inch drive maker Micropolis came from Pertec, a 14-inch manufacturer; and the founders of Shugart and Quantum defected from Memorex.[11] The start-ups were as unsuccessful as their former employers in interesting established computer makers in the disruptive architecture. Consequently, they had to find *new* customers. The applications that emerged in this very uncertain, probing process were the minicomputer, the desktop personal computer, and the laptop (see Figure 2). These are obvious markets for hard drives in retrospect. But at the time, whether these would become significant markets for disk drives was highly uncertain. Micropolis was founded before the market for desk-side minicomputers and word processors, in which its products came to be used, emerged. Seagate was founded 2 years before IBM introduced its PC, when personal computers were simple toys for hobbyists. And Conner Peripherals got its start before Compaq knew the portable computer market had potential. The founders of these firms sold their products without a clear marketing strategy, essentially to whomever would buy them. Out of what was largely a trial-and-error approach to the market, the ultimately dominant applications for their products emerged.
5. Once the start-ups had found an operating base in new markets, they found that by adopting sustaining improvements in new component technologies,[12] they could increase the capacity of their drives at a faster rate than was required by their new market. As shown in Figure 2, they blazed trajectories of 50% annual improvement, fixing their sights on the large, established computer markets immedi-

[11] Ultimately, nearly all North American manufacturers of disk drives can trace their founders' genealogy to IBM's San Jose division, which developed and manufactured its magnetic recording products (Christensen, 1993).

[12] In general, these component technologies were developed within the largest of the established firms that dominated the markets above these entrants, in terms of the technology and market trajectories mapped in Figure 2.

ately above them on the performance scale. As noted above, the established firms' views downmarket, and the entrant firms' views upmarket, were asymmetrical. In contrast to the unattractive margins and market size the established firms saw when eyeing the new markets for simpler drives as they were emerging, the entrants tended to view the potential volumes and margins in the upscale, high-performance markets above them as highly attractive. Customers in these established markets eventually embraced the new architectures they had rejected earlier, because once their needs for capacity and speed were met, the new drives' smaller size and architectural simplicity made them cheaper, faster, and more reliable than the older architectures. For example, Seagate, which started in the desktop personal computer market, subsequently invaded and came to dominate the minicomputer, engineering workstation, and mainframe computer markets for disk drives. Seagate, in turn, was driven from the desktop personal computer market for disk drives by Conner and Quantum, the pioneering manufacturers of 3.5-inch drives.

6. When the smaller models began to invade established market segments, the drivemakers that had initially controlled those markets took their prototypes off the shelf (where they had been put in step #3), and defensively introduced them to defend their customer base in their own market.[13] By this time, of course, the new architecture had shed its disruptive character, and had become fully performance-competitive with the larger drives in the established markets. Although some established manufacturers were able to defend their market positions through belated introduction of the new architecture, many found that the entrant firms had developed insurmountable advantages in manufacturing cost and design experience, and they eventually withdrew from the market. For those established manufacturers that did succeed in introducing the new architectures, survival was the only reward. None of the firms we studied was ever able to win a significant share of the new market whose emergence had been enabled by the new architecture; the new drives simply cannibalized sales of older, larger-architecture products with existing customers. For example, as of 1991 almost none of Seagate's 3.5-inch drives had been sold to portable/laptop manufacturers: its 3.5-inch customers still were desktop computer manufacturers, and many of its 3.5-inch drives continued to be shipped with frames permitting them to be mounted in XT and AT-class computers that had been designed to accommodate 5.25-inch drives. Control Data, the 14-inch leader, never captured even a 1 percent share of the minicomputer market. It introduced its 8-inch drives nearly 3 years after the pioneering start-ups did, and nearly all of its drives were sold to its existing mainframe customers. Miniscribe, Quantum and Micropolis all had the same cannibalistic experience when they belatedly introduced disruptive-technology drives. They failed to capture a significant share of the new market, and at best succeeded in defending a portion of their prior business.

There are curious asymmetries in the *ex post* risks and rewards associated with sustaining and disruptive innovations. Many of the sustaining innovations (such as thin-film heads, thin film disks, and the 14-inch Winchester architecture) were *extremely* expensive and risky from a *technological* point of view. Yet because they addressed well-understood needs of known customers, perceived market risk was low; impetus coalesced; and resources were allocated with only prudent hesitation. Yet, although these innovations clearly helped the innovators retain their customers, there is no evidence from the industry's history that any firm was able to gain observable market share by virtue of such technology leadership.[14]

On the other hand, disruptive innovations were technologically straightforward: several established firms had already developed them by the time formal resource allocation decisions were

[13] Note that at this point, because the disruptive innovation invading below had become fully performance-competitive with the established technology, the innovation had essentially acquired the character of a sustaining innovation—it gave customers what they needed.

[14] Christensen (1992b) shows that there was no discernible first-mover advantage associated with trajectory-sustaining innovations, to firms in the disk drive industry. In contrast, there were *very* powerful first-mover advantages to leaders in trajectory disruptive innovations that fostered the creation of new markets.

made. But these were viewed as extremely risky, because the markets were not 'there'. The most successful of the entrants that accepted the risks of creating new markets for disruptive innovations generated billions in revenues upon foundations of architectural technology that cost at most a few million dollars to put into place.

We argue that although differences in luck, resource endowments, managerial competence, and bureaucratic agility matter, the patterns of technology leadership displayed by established and entrant firms in the disk drive industry accurately reflect differences in the fully informed, rational *ex ante* perceptions of risks and rewards held by managers in the two types of firms. In each of the companies studied, a key task of senior managers was to decide which of the many product and technology development programs continually being proposed to them should receive a formal allocation of resources. The criteria used in these decisions were essentially the total return perceived in each project, adjusted by the perceived riskiness of the project, as these data were presented to them by mid-level managers. Projects targeted at the known needs of big customers in established markets consistently won the rational debates over resource allocation. Sophisticated systems for planning and compensation ensured that this would be the case.[15]

The contrast between the innovative behavior of some *individuals* in the firm, vs. the manner in which the firm's *processes* allocated resources across competing projects, is an important feature of this model.[16] In the cases studied, the pioneering engineers in established firms that developed disruptive-architecture drives were innovative not just in technology, but in their view of the market. They intuitively perceived opportunities for a very different disk drive. But organizational processes allocated resources based on rational assessments of data about returns and risks. Information provided by innovating engineers was at best hypothetical: without existing customers, they could only guess at the size of the market, the profitability of products, and required product performance. In contrast, current customers could articulate features, performance, and quantities they would purchase with *much* less ambiguity. Because of these differences in information clarity, firms were led toward particular sorts of innovations—many of which were extremely challenging and risky—and away from others. In the firms studied here, the issue does not seem so much to be innovativeness *per se*, as it is what *type* of innovation the firms' processes could facilitate.

In light of this research, the popular slogan, 'Stay close to your customers' (which is supported by the research of von Hippel, 1988, and others), appears not always to be robust advice. One instead might expect customers to lead their suppliers toward sustaining innovations, and to provide no leadership—or even to explicitly *mis*lead—in instances of disruptive technology change. Henderson (1993) saw similar potential danger for being held captive by customers in her study of photolithographic aligner equipment manufacturers.

We close our discussion of the model with a final note. Neglect of disruptive technologies proved damaging to established drivemakers because the trajectory of performance improvement that the technology *provided* was steeper than the improvement trajectory *demanded* in individual markets (see Figure 2.) The mismatch in these trajectories provided pathways for the firms that entered new markets eventually to become performance-competitive in established markets as well. If the trajectories were parallel, we would expect disruptive technologies to be deployed in new markets and to stay there; each successive market would constitute a relatively stable niche market out of which technologies and firms would not migrate.

---

[15] It is interesting that 20 years after Bower's (1970) study of resource allocation, we see in leading-edge systems for planning and compensation the same bias against risk taking. Morris and Ferguson's description of how IBM allowed Microsoft to gain control of PC operating system standards is centered on the role of mainframe producers in IBM's resource allocation process. In a 1990 interview with one of the authors, one of the most successful innovators in IBM history recounted how time and again he was forced to battle the controlling influence of middle-management's commitment to serve commercial mainframe customers.

[16] We are indebted to Professor Robert Burgelman for his comments on this issue. He has also noted, given the sequence of events we observed—where engineers inside the established firms began pursuing the disruptive product opportunity before the start-up entrants did—that timing matters a lot. It may be that when individuals in the established firms were pressing their ideas internally, they were too far ahead of the market. In the year or two that it took them to leave their employers, create new firms, and create new products, the nascent markets may have become more ready to accept the new drives.

THE LINKAGE BETWEEN MODELS OF RESOURCE DEPENDENCE AND RESOURCE ALLOCATION

We mentioned at the outset that a contribution of this paper is that it establishes a linkage between the school of thought known as *resource dependence* (Pfeffer and Salancik, 1978) and the models of the resource allocation process proposed by Bower (1970) and Burgelman (1983a, 1983b). Our findings support many of the conclusions of the resource dependence theorists, who contend that a firm's scope for strategic change is strongly bounded by the interests of external entities (customers, in this study) who provide the resources the firm needs to survive. We show that the mechanism through which customers wield this power is the process in which impetus coalesces behind investments in sustaining technologies, directing resources to innovations that address current customers' needs.

But although our findings lend support to the theory of resource dependence, they decidedly do not support a contention that managers are powerless to change the strategies of their companies in directions that are inconsistent with the needs of their customers as resource providers (Pfeffer and Salancik, 1978: 263–265).[17] The evidence from this study is that managers can, in fact, change strategy—but that they can successfully do so only if their actions are consistent with, rather than in counteraction to, the principle of resource dependence. In the disk drive industry's history, three established firms achieved a measure of commercial success in disruptive technologies. Two did so by spinning out organizations that were completely independent, in terms of customer relationships, from the mainstream groups. The third launched the disruptive technology with extreme managerial effort, from within the mainstream organization. This paper closes by summarizing these case histories and their implications for theory.

[17] In Chapter 10 of Pfeffer and Salancik's (1978) book, for example, they assert that the manager's most valuable role is symbolic, and they cite a hypothetical example. When external forces induce hard times in a company, managers can usefully be fired—not because bringing in a new manager will make any difference to the performance of the organization, but because of the symbolic content of that action. It creates the *feeling* in the organization that something is being done to address this problem, even though it will have no effect. The evidence from these case studies does not support this assertion about the ability of managers to change the course of their organizations. *As long as managers act in a manner consistent with the forces of resource dependence*, it appears that they can, indeed, wield significant power.

Distinct organizational units for small drives at Control Data

Control Data (CDC) was the dominant manufacturer of 14-inch disk pack and Winchester drives sold into the OEM market between 1975 and 1982: its market share fluctuated between 55 and 62 per cent. When the 8-inch architecture emerged in the late 1970s, CDC missed it by 3 years. It never captured more than 3–4 percent of the 8-inch market, and those 8-inch drives that it did sell, were sold almost exclusively to its established customer base of mainframe computer manufacturers. The reason given by those interviewed in this study was that engineers and marketers kept getting pulled off the 8-inch program to resolve problems in the launch of next-generation 14-inch products for CDC's mainstream customers.

CDC also launched its first 5.25-inch model 2 years after Seagate's pioneering product appeared in 1980. This time, however, CDC located its 5.25-inch effort in Oklahoma City—according to one manager, 'not to escape CDC's Minneapolis engineering culture, but to isolate the (5.25-inch product) group from the company's mainstream customers. We needed an organization that could get excited about a $50,000 order. In Minneapolis (which derived nearly $1 billion from the sale of 14-inch drives in the mainframe market) you needed a million-dollar order just to turn anyone's head.' Although it was late and never reascended to its position of dominance, CDC's foray into 5.25-inch drives was profitable, and at times it commanded a 20 percent share of higher-capacity 5.25-inch drives.

Having learned from its experience in Oklahoma City, when CDC decided to attack the 3.5-inch market it set up yet another organization in Simi Valley, California. This group shipped its first products in mid-1988, about 18 months behind Conner Peripherals, and enjoyed modest commercial success. The creation of these stand-alone organizations was CDC's way of handling the 'strategic forcing' and 'strategic context determination' challenges described by Burgelman (1983b, 1984).

Quantum Corporation and the 3.5-inch Hardcard

Quantum Corporation, a leading maker of 8-inch drives sold in the minicomputer market, introduced its first 5.25-inch product 3 years after those drives had first appeared in the market. As the 5.25-inch pioneers began to invade the minicomputer market from below, for all of the reasons described above, Quantum launched a 5.25-inch product and was temporarily successful in defending some of its existing customers by selling its 5.25-inch drive to them. But it never sold a single drive into the desktop PC market, and its overall sales began to sag. In 1984 a group of Quantum engineers saw a market for a thin 3.5-inch drive plugged into an expansion slot in IBM XT- and AT-class desktop computers—drives that would be sold to end-users, rather than OEM computer manufacturers. Quantum financed and retained 80 percent ownership of this spin-off venture, called Plus Development Corporation, and set the company up in different facilities. Plus was extremely successful. As sales of Quantum's line of 8-inch drives began to evaporate in the mid-1980s, they were offset by Plus's growing 'Hardcard' revenues. By 1987, sales of 8 and 5.25-inch products had largely evaporated. Quantum purchased the 20 percent of Plus it did not own; essentially closed down the old corporation, and installed Plus's executives in Quantum's most senior positions. They then reconfigured Plus's 3.5-inch products to appeal to desktop computer makers such as Apple, just as the capacity vector for 3.5-inch drives was invading the desktop, as shown in Figure 2. By 1994 the new Quantum had become the largest unit-volume producer of disk drives in the world. Quantum's spin-out of the Hardcard effort and its subsequent strategic reorientation appears to be an example of the processes of strategy change described in Burgelman (1991).

Micropolis: Transition through managerial force

Managers at Micropolis Corporation, also an 8-inch drivemaker, employed a very different approach in which senior management initiated a disruptive program within the mainstream organization that made 8-inch drives. As early as 1982, Micropolis' founder and CEO, Stuart Mabon, intuitively saw the trends mapped in Figure 2 and decided the firm needed to become primarily a maker of 5.25-inch drives. While initially hoping to keep adequate resources focused on the 8-inch line that Micropolis could straddle both markets,[18] he assigned the company's premier engineers to the 5.25-inch program. Mabon recalls that it took '100% of his time and energy for 18 months' to keep adequate resources focused on the 5.25-inch program, because the organization's own mechanisms allocated resources to where the customers were: 8-inch drives. By 1984 Micropolis had failed to keep pace with competition in the minicomputer market for disk drives, and withdrew its remaining 8-inch models. With Herculean effort, however, it did succeed in its 5.25-inch programs. Figure 3 shows why this was necessary: in the transition, Micropolis assumed a position on a very different technological trajectory (Dosi, 1982). In the process it had to walk away from every one of its major customers, and replace the lost revenues with sales of the new product line to an entirely different group of desktop computer makers. Mabon remembers the experience as the most exhausting of his life. Micropolis aborted a 1989 attempt to launch its first 3.5-inch drive, and as of 1992 the company still had not introduced a 3.5-inch product.

Table 4 arrays the experiences of the six companies we studied in depth, as they addressed disruptive technologies from within their mainstream organization, and through independent organizations. Companies are classed as having been successful in this table if their market share in the new market enabled by the disruptive disk drive technology was at least 25% of its percentage share in the prior, established market in which it was dominant. Hence, Control Data, whose share of the 14-inch mainframe computer disk drive market often exceeded 60 percent, was classed as a

[18] The failure of Micropolis to maintain simultaneous competitive commitments to its established technology while adequately nurturing the 5.25-inch technology is consistent with the technological histories recounted in Utterback (1994). Utterback found historically that firms that attempted to develop radically new technology almost always tried simultaneously to maintain their commitments to the old; and that they almost always failed.

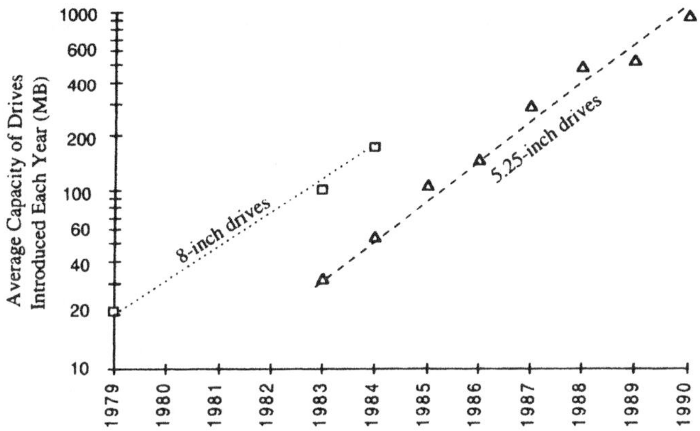

Figure 3. The disruptive impact of 5.25-inch drives on the market position of Micropolis Corp.

failure in its attempt to sell 8-inch drives, because its share of minicomputer disk drives never exceeded 3 percent. Its share of 5.25-inch drives sold to the desktop workstation market, however, reached 20 percent, and it was therefore classed as a success in that effort. An organization was defined as being independent from the mainstream if it was geographically separated; was held accountable for full profit and loss; and included within it all of the functional units of a typical company: sales and marketing, manufacturing, finance, human resources, engineering, and so on.

In addition to the six firms studied in depth, Table 4 lists other firms, shown in *italic type*, whose histories were researched through public sources and a more limited number of personal interviews. The 'L' and 'T' shown next to each company in the table, as in Table 3, denotes whether that firm's experience lends literal or theoretical support (Yin, 1989) to the proposition that managers can effect a strategy change despite resource dependence, by creating independent organizations that depend exclusively upon resources in the targeted market. Micropolis' transition from 8 to 5.25-inch drives is classed as a theoretical replication, because of the enormous managerial effort that was required to counteract the force of resource dependence in that transition.[19] Note that in every instance except Micropolis' 5.25-inch entry, firms that *fought* the forces of resource dependence by attempting to commercialize disruptive technology from within their mainstream organizations failed, as measured by *Disk/Trend* data. And the firms that *accounted for* the forces of resource dependence by spinning out independent organizations succeeded.

Note in Table 4 that there do not seem to be strong firm or managerial effects, compared to the organizational effect. Control Data, Quantum, and Micropolis encountered multiple disruptive technologies; and *the same general managers sat atop these organizations across each of these transitions*. What seems to have distinguished these firms' successful from failed attempts to commercialize these disruptive technologies was not the talent of the managers *per se*, but whether the managers created organizationally distinct units to accomplish the task—where the forces of resource dependence could work in their favor, rather than against them. The successful cases cited here are the only ones in the industry's history in which a leading incumbent stayed atop its market when faced with disruptive technological change—and as a result, the number of data points in the top half of the matrix is limited. But these

[19] The success or failure of these other firms at each point of disruptive technology change was unambiguously determinable from *Disk/Trend Report* data. Similarly, whether these firms managed the launch of disruptive technology products from within their mainstream organization, or through an organizationally separate unit, was a matter of public record and general industry knowledge. Hence, there were no subjective judgments involved in constructing Table 4.

Table 4. The success and failure of companies addressing disruptive technologies through mainstream vs. independent organizations

Succeeded	Control Data 5.25-inch (L) Control Data 3.5-inch (L) Quantum 3.5-inch (L) *Maxtor 3.5-inch (L)*	Micropolis 5.25-inch (T)
Failed		Control Data 8-inch (L) Quantum 5.25-inch (L) Miniscribe 3.5-inch (L) Seagate 3.5-inch (L) Micropolis 3.5-inch (L) *Memorex 8-inch (L)* *Memorex 5.25-inch (L)* *Priam 5.25-inch (L)* *Century Data 8-inch (L)* *Ampex 8-inch (L)* *Ampex 5.25-inch (L)*
	Commercialized from within an independent organization.	Commercialized from within the mainstream organization.

findings do suggest that, while the forces of resource dependence act as strong constraints on managerial discretion, managers can in fact manipulate those constraints effectively in order to achieve strategic change.

CONCLUSIONS

This study highlights an important issue for managers and scholars who strive to understand the reasons why strong, capably managed firms stumble when faced with particular types of technological change. While many scholars see the issue primarily as an issue of *technological competence*, we assert that at a deeper level it may an issue of *investment*. We have observed that when competence was lacking, but impetus from customers to develop that competence was sufficiently strong, established firms successfully led their industries in developing the competencies required for sustaining technological change. Importantly, because sustaining technologies address the interests of established firms' existing customers, we saw that technological change could be achieved without strategy change.

Conversely, when technological competence existed, but impetus from customers was lacking, we saw consistently that firms were unable to commercialize what they already could do. This is because disruptive technologies initially tend to be saleable only in different markets whose economic and financial characteristics render them unattractive to established firms. Addressing these technologies therefore requires a change in strategy in order to attack a very different market. In the end, it appears that although the stumbles of these established firms are *associated* with technological change, the key issue appears to be firms' disabilities in changing strategy, not technology.

Our model is not presented as the path every firm follows when faced with disruptive technology. We believe, however, that it may contribute several insights for scholars interested in the factors that affect strategic change in firms. First, it notes that the allocation of resources to some product development and commercialization programs, and the denial of resources to others, is a key event or decision in the implementation of strategy. The model highlights the process by which impetus and consequent resources may be denied to technological opportunities that do not contribute to the needs of prominent customers. These findings suggest a causal relationship might exist between resource allocation processes, as modeled by Bower (1970) and Burgelman (1983a, 1983b), and the phenomenon of resource dependence (Pfeffer and Salancik, 1978). Our findings suggest that despite the powerful forces of resource dependence, however, managers can, in fact, wield considerable power, and wield it effectively, in changing the strategic course of their firms in directions other than those in which its resource providers are pulling it. By understanding the processes that link customer needs, impetus, and resource allocation, managers can align efforts to commercialize disruptive technology (which entails a change in strategy) with the forces of resource dependence. This involves managing disruptive technology in a manner that is out of the organizational and strategic context of mainstream organizations—where of necessity, incentives and resource allocation processes are designed to nourish sustaining innovations that address current customers' needs. In this way, the model and these case studies illustrate the mechanisms through which autonomous and induced strategic behavior (Burgelman, 1983a) can affect, or fail to affect, a company's course.

Much additional research must be done. Efforts to explore the external validity and usefulness of the model through studies of sustaining and disruptive technological change in other industries has begun (Rosenbloom and Christensen, 1995), but much more is required. In addition, we hope that future researchers can develop clearer models for managerial action and strategic change in the face of disruptive technology change that are consistent with the principles of resource dependence and the processes of resource allocation.

ACKNOWLEDGEMENTS

We gratefully acknowledge the financial support of the Harvard Business School Division of Research in conducting the research for this paper, and thank the editors of *Disk/Trend Report* for sharing their industry data with us. We are indebted to Professors Robert Burgelman of Stanford University, Rebecca Henderson of the Massachusetts Institute of Technology, David Garvin and several of our other colleagues at the Harvard Business School, as well as the anonymous referees, for invaluable suggestions for improving earlier versions of this paper. Any remaining deficiencies are our sole responsibility.

REFERENCES

Bower, J. (1970). *Managing the Resource Allocation Process*. Irwin, Homewood, IL.

Burgelman, R. (1983a). 'A model of the interaction of strategic behavior, corporate context, and the concept of strategy', *Academy of Management Review*, **3**(1), pp. 61–69.

Burgelman, R. (1983b). 'A process model of internal corporate venturing in the diversified major firm', *Administrative Science Quarterly*, **28**, pp. 223–244.

Burgelman, R. (1984). 'Designs for corporate entrepreneurship in established firms', *California Management Review*, **26**, Spring, pp. 154–166.

Burgelman, R. (1991). 'Intraorganizational ecology of strategy-making and organizational adaptation: Theory and field research', *Organization Science*, **2**, pp. 239–262.

Burgelman, R. and L. Sayles (1986). *Inside Corporate Innovation*. Free Press, New York.

Campbell, D. T. and J. C. Stanley (1966). *Experimental and Quasi-Experimental Designs for Research*. Houghton Mifflin, Boston, MA.

Christensen, C. M. (1992a). 'The Innovator's challenge: Understanding the influence of market demand on processes of technology development in the rigid disk drive industry'. Unpublished DBA dissertation. Graduate School of Business Administration, Harvard University.

Christensen, C. M. (1992b). 'Exploring the limits of the technology S-curve', *Production and Operations Management*, **1**, pp. 334–366.

Christensen, C. M. (1993). 'The rigid disk drive industry: A history of commercial and technological turbulence', *Business History Review*, **67**, pp. 531–588.

Christensen, C. M. and R. S. Rosenbloom (1995). 'Explaining the attacker's advantage: Technological paradigms, organizational dynamics, and the value network', *Research Policy*, **24**, pp. 233–257.

Cooper, A. and D. Schendel (February 1976). 'Strategic responses to technological threats', *Business Horizons*, **19**, pp. 61–69.

Data Sources: The Comprehensive Guide to the Information Processing Industry (annual). Ziff-Davis Publishing, New York.

Disk/Trend Report (annual). Disk/Trend, Inc., Mountain View, CA.

Dosi, G. (1982). 'Technological paradigms and technological trajectories', *Research Policy*, **11**, pp. 147–162.

Foster, R. J. (1986). *Innovation: The Attacker's Advantage*. Summit Books, New York.

Henderson, R. M. (1993). 'Keeping too close to your customers', working paper, Sloan School of Management, Massachusetts Institute of Technology.

Henderson, R. M. and K. B. Clark (1990). 'Architectural innovation: The reconfiguration of existing systems and the failure of established firms', *Administrative Science Quarterly*, **35**, pp. 9–30.

Pfeffer, J. and G. R. Salancik (1978). *The External Control of Organizations: A Resource Dependence Perspective*. Harper & Row, New York.

Rosenbloom, R. S. and C. M. Christensen (1995). 'Technological discontinuities, organizational capabilities, and strategic commitments', *Industrial and Corporate Change*, **4**, pp. 655–685.

Tushman, M. L. and P. Anderson (1986). 'Technological discontinuities and organizational environments', *Administrative Science Quarterly*, **31**, pp. 439–465.

Utterback, J. (1994). *Mastering the Dynamics of Innovation*. Harvard Business School Press, Boston, MA.

von Hippel, E. (1988). *The Sources of Innovation*. Oxford University Press, New York.

Yin, R. K. (1989). *Case Study Research: Design and Methods*. Sage, Newbury Park, CA.

APPENDIX 1: A BRIEF PRIMER ON HOW DISK DRIVES WORK

Rigid disk drives are comprised of one or more rotating disks—polished aluminum platters coated with magnetic material—mounted on a central spindle. Data are recorded and read on concentric tracks on the surfaces of these disks. Read/write heads—one each for the top and bottom surfaces of each disk on the spindle—are aerodynamically designed to fly a few millionths of an inch over the surface of the disk. They generally rest on the disk's surface when the drive is at rest; 'take off' as the drive begins to spin; and 'land' again when the disks stop. The heads are positioned over the proper track on the disk by an actuator motor, which moves the heads across the tracks in a fashion similar to the arm on a phonograph. The head is essentially a tiny electromagnet which, when current flows in one direction, orients the polarity of the magnetic domain on the disk's surface immediately beneath it. When the direction of current through the electromagnet reverses, its polarity changes. This induces an opposite switch of the polarity of the adjacent domain on the disk's surface as the disk spins beneath the head. In this manner, data are written in binary code on the disk. To read data, changes in magnetic field on the disk as it spins beneath the head are used to induce changes in the direction of current—essentially the reverse process of writing. Disk drives also include electronic circuitry enabling computers to control and communicate with the drive.

As in other magnetic recording products, *areal recording density* (measured in megabits per square inch of disk surface area, or mbpsi) was the pervasive measure of product performance in the disk drive industry. Historically, areal density in the industry has increased at a steady 35 percent annual rate. A drive's total capacity is the product of the available square inches on the top and bottom surfaces of the disks mounted on the spindle of the drive, multiplied by its areal recording density. Historically, the capacity of drives in a given product architecture has increased at about 50 percent annually. The difference between the 35 percent increase in areal density and the 50 percent increase in total capacity has come from mechanical engineering innovations, which enable manufacturers to squeeze additional disks and heads into a given size of drive.

APPENDIX 2: CALCULATION OF THE TRAJECTORIES MAPPED IN FIGURE 2

The trajectories mapped in Figure 2 were calculated as follows. Data on the capacity provided with computers in the mainframe, minicomputer, desktop personal computer, and portable computer classes were obtained from *Data Sources*, an annual publication that lists the technical specifications of all computer models available from each computer manufacturer. Where particular models were available with different features and configurations, the manufacturer provided *Data Sources* with a 'typical' system configuration, with defined RAM capacity, performance specifications of peripheral equipment (including disk drives), list price, and year of introduction. In instances where a given computer model was offered for sale over a sequence of years, the hard disk capacity provided in the typical configuration generally increased. *Data Sources* divides computers into mainframe, mini/midrange, desktop personal, portable and laptop, and notebook computers. For each class of computers, all models available for sale in each year were ranked by price, and the hard disk capacity provided with the median-priced model was identified, for each year. The best-fit line through the resultant time series for each class of computer is plotted as the solid lines in Figure 2. These single solid lines are drawn in Figure 2 for expository simplification, to indicate the trend in typical machines. In reality, of course, there is a wide band around these lines. The leading and trailing edges of performance—the highest and lowest capacities offered with the most and least expensive computers— were substantially higher and lower, respectively, than the typical values mapped in Figure 2.

The dotted lines in Figure 2 represent the best-fit line through the unweighted average capacity of all disk drives introduced for sale in each given architecture, for each year. These data were taken from *Disk/Trend Report*. Again, for expository simplification, only this average line is shown. There was a wide band of capacities introduced for sale in each year, so that the highest-capacity drive introduced in each year was substantially above the average shown. Stated in another way, a distinction must be made between the full range of products available for purchase, and those in typical systems of use. The upper and lower bands around the median and average trajectories in Figure 2 are generally parallel to the lines shown.

Because higher-capacity drives were available than the capacities offered with the median-priced systems, we state in the text that the solid-line trajectories in Figure 2 represent the capacities 'demanded' in each market. In other words, the capacity per machine was not constrained by technological availability. Rather, it represents a *choice* for hard disk capacity, made by computer users, given the prevailing cost.

Part II
Capturing Value From Innovation

[4]

Profiting from technological innovation: Implications for integration, collaboration, licensing and public policy

David J. TEECE *

School of Business Administration, University of California, Berkeley, CA 94720, U.S.A.

Final version received June 1986

This paper attempts to explain why innovating firms often fail to obtain significant economic returns from an innovation, while customers, imitators and other industry participants benefit. Business strategy – particularly as it relates to the firm's decision to integrate and collaborate – is shown to be an important factor. The paper demonstrates that when imitation is easy, markets don't work well, and the profits from innovation may accrue to the owners of certain complementary assets, rather than to the developers of the intellectual property. This speaks to the need, in certain cases, for the innovating firm to establish a prior position in these complementary assets. The paper also indicates that innovators with new products and processes which provide value to consumers may sometimes be so ill positioned in the market that they necessarily will fail. The analysis provides a theoretical foundation for the proposition that manufacturing often matters, particularly to innovating nations. Innovating firms without the requisite manufacturing and related capacities may die, even though they are the best at innovation. Implications for trade policy and domestic economic policy are examined.

* I thank Raphael Amit, Harvey Brooks, Chris Chapin, Therese Flaherty, Richard Gilbert, Heather Haveman, Mel Horwitch, David Hulbert, Carl Jacobsen, Michael Porter, Gary Pisano, Richard Rumelt, Raymond Vernon and Sidney Winter for helpful discussions relating to the subject matter of this paper. Three anonymous referees also provided valuable criticisms. I gratefully acknowledge the financial support of the National Science Foundation under grant no. SRS-8410556 to the Center for Research in Management, University of California Berkeley. Earlier versions of this paper were presented at a National Academy of Engineering Symposium titled "World Technologies and National Sovereignty," February 1986, and at a conference on innovation at the University of Venice, March 1986.

Research Policy 15 (1986) 285–305
North-Holland

1. Introduction

It is quite common for innovators – those firms which are first to commercialize a new product or process in the market – to lament the fact that competitors/imitators have profited more from the innovation than the firm first to commercialize it! Since it is often held that being first to market is a source of strategic advantage, the clear existence and persistence of this phenomenon may appear perplexing if not troubling. The aim of this article is to explain why a fast second or even a slow third might outperform the innovator. The message is particularly pertinent to those science and engineering driven companies that harbor the mistaken illusion that developing new products which meet customer needs will ensure fabulous success. It may possibly do so for the product, but not for the innovator.

In this paper, a framework is offered which identifies the factors which determine who wins from innovation: the firm which is first to market, follower firms, or firms that have related capabilities that the innovator needs. The follower firms may or may not be imitators in the narrow sense of the term, although they sometimes are. The framework appears to have utility for explaining the share of the profits from innovation accruing to the innovator compared to its followers and suppliers (see fig. 1), as well as for explaining a variety of interfirm activities such as joint ventures, coproduction agreements, cross distribution arrangements, and technology licensing. Implications for strategic management, public policy, and international trade and investment are then discussed.

0048-7333/86/$3.50

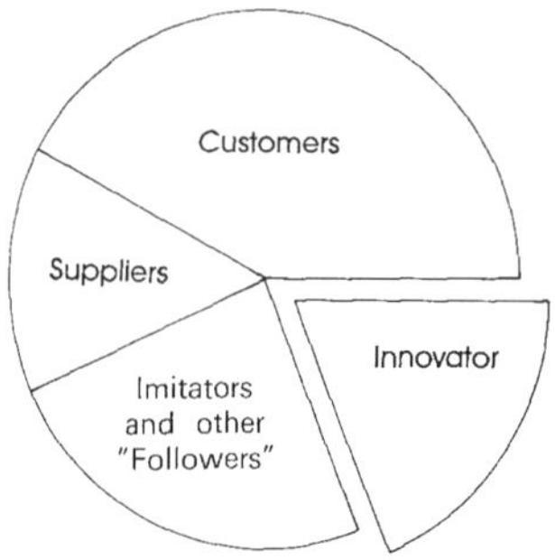

What determines the share of profits captured by the innovator?

Fig. 1. Explaining the distribution of the profits from innovation.

2. The phenomenon

Figure 2 presents a simplified taxonomy of the possible outcomes from innovation. Quadrant 1 represents positive outcomes for the innovator. A first-to-market advantage is translated into a sustained competitive advantage which either creates a new earnings stream or enhances an existing one. Quadrant 4 and its corollary quadrant 2 are the ones which are the focus of this paper.

The EMI CAT scanner is a classic case of the phenomenon to be investigated. [1] By the early 1970s, the UK firm Electrical Musical Industries (EMI) Ltd. was in a variety of product lines including phonographic records, movies, and advanced electronics. EMI had developed high resolution TVs in the 1930s, pioneered airborne radar during World War II, and developed the UK's first all solid-state computers in 1952.

In the late 1960s Godfrey Houndsfield, an EMI senior research engineer engaged in pattern recognition research which resulted in his displaying a scan of a pig's brain. Subsequent clinical work established that computerized axial tomography (CAT) was viable for generating cross-sectional "views" of the human body, the greatest advance in radiology since the discovery of X rays in 1895.

While EMI was initially successful with its CAT scanner, within 6 years of its introduction into the US in 1973 the company had lost market leadership, and by the eighth year had dropped out of the CT scanner business. Other companies successfully dominated the market, though they were late entrants, and are still profiting in the business today.

Other examples include RC Cola, a small beverage company that was the first to introduce cola in a can, and the first to introduce diet cola. Both Coca Cola and Pepsi followed almost immediately and deprived RC of any significant advantage from its innovation. Bowmar, which introduced the pocket calculator, was not able to withstand competition from Texas Instruments, Hewlett Packard and others, and went out of business. Xerox failed to succeed with its entry into the office computer business, even though Apple succeeded with the MacIntosh which contained many of Xerox's key product ideas, such as the mouse and icons. The de Havilland Comet saga has some of the same features. The Comet I jet was introduced into the commercial airline business 2 years or so before Boeing introduced the 707, but de Havilland failed to capitalize on its substantial early advantage. MITS introduced the first personal computer, the Altair, experienced a burst of sales, then slid quietly into oblivion.

If there are innovators who lose there must be followers/imitators who win. A classic example is IBM with its PC, a great success since the time it was introduced in 1981. Neither the architecture nor components embedded in the IBM PC were considered advanced when introduced; nor was the way the technology was packaged a significant departure from then-current practice. Yet the IBM PC was fabulously successful and established MS-DOS as the leading operating system for 16-bit PCs. By the end of 1984, IBM has shipped over 500 000 PCs, and many considered that it had irreversibly eclipsed Apple in the PC industry.

3. Profiting from innovation: Basic building blocks

In order to develop a coherent framework within which to explain the distribution of outcomes illustrated in fig. 2, three fundamental building blocks must first be put in place: the appropriability regime, complementary assets, and the dominant design paradigm.

[1] The EMI story is summarized in Michael Martin, *Managing Technological Innovation and Entrepreneurship*, (Reston Publishing Company, Reston, VA, 1984).

	Innovator	Follower-Imitator
Win	1 • Pilkington (Float Glass) • G.D. Searle (NutraSweet) • Dupont (Teflon)	2 • IBM (Personal Computer) • Matsushita (VHS video recorders) • Seiko (quartz watch)
Lose	4 • RC Cola (diet cola) • EMI (scanner) • Bowmar (pocket calculator) • Xerox (office computer) • DeHavilland (Comet)	3 • Kodak (instant photography) • Northrup (F20) • DEC (personal computer)

Fig. 2. Taxonomy of outcomes from the innovation process.

3.1. Regimes of appropriability

A regime of appropriability refers to the environmental factors, excluding firm and market structure, that govern an innovator's ability to capture the profits generated by an innovation. The most important dimensions of such a regime are the nature of the technology, and the efficacy of legal mechanisms of protection (fig. 3).

It has long been known that patents do not work in practice as they do in theory. Rarely, if ever, do patents confer perfect appropriability, although they do afford considerable protection on new chemical products and rather simple mechanical inventions. Many patents can be "invented around" at modest costs. They are especially ineffective at protecting process innovations. Often patents provide little protection because the legal requirements for upholding their validity or for proving their infringement are high.

In some industries, particularly where the innovation is embedded in processes, trade secrets are a viable alternative to patents. Trade secret protection is possible, however, only if a firm can put its product before the public and still keep the underlying technology secret. Usually only chemical formulas and industrial-commercial processes (e.g., cosmetics and recipes) can be protected as trade secrets after they're "out".

The degree to which knowledge is tacit or codified also affects ease of imitation. Codified knowledge is easier to transmit and receive, and is more exposed to industrial espionage and the like. Tacit knowledge by definition is difficult to articulate, and so transfer is hard unless those who possess the know how in question can demonstrate it to others (Teece [9]). Survey research indicates that methods of appropriability vary markedly across industries, and probably within industries as well (Levin et al. [5]).

The property rights environment within which a firm operates can thus be classified according to the nature of the technology and the efficacy of the legal system to assign and protect intellectual property. While a gross simplification, a dichotomy can be drawn between environments in which the appropriability regime is "tight" (technology is relatively easy to protect) and "weak" (technology is almost impossible to protect). Examples of the former include the formula for Coca Cola syrup; an example of the latter would be the Simplex algorithm in linear programming.

3.2. The dominant design paradigm

It is commonly recognized that there are two stages in the evolutionary development of a given branch of a science: the preparadigmatic stage when there is no single generally accepted conceptual treatment of the phenomenon in a field of study, and the paradigmatic stage which begins when a body of theory appears to have passed the canons of scientific acceptability. The emergence of a dominant paradigm signals scientific maturity and the acceptance of agreed upon "standards" by which what has been referred to as "normal" scientific research can proceed. These "standards" remain in force unless or until the paradigm is overturned. Revolutionary science is what overturns normal science, as when the Copernicus's theories of astronomy overturned Ptolemy's in the seventeenth century.

Abernathy and Utterback [1] and Dosi [3] have provided a treatment of the technological evolution of an industry which appears to parallel

• Legal instruments	• Nature of technology
- Patents	- Product
- Copyrights	- Process
- Trade secrets	- Tacit
	- Codified

Fig. 3. Appropriability regime: Key dimensions.

Kuhnian notions of scientific evolution. [2] In the early stages of industry development, product designs are fluid, manufacturing processes are loosely and adaptively organized, and generalized capital is used in production. Competition amongst firms manifests itself in competition amongst designs, which are markedly different from each other. This might be called the preparadigmatic stage of an industry.

At some point in time, and after considerable trial and error in the marketplace, one design or a narrow class of designs begins to emerge as the more promising. Such a design must be able to meet a whole set of user needs in a relatively complete fashion. The Model T Ford, the IBM 360, and the Douglas DC-3 are examples of dominant designs in the automobile, computer, and aircraft industry respectively.

Once a dominant design emerges, competition shifts to price and away from design. Competitive success then shifts to a whole new set of variables. Scale and learning become much more important, and specialized capital gets deployed as incumbent's seek to lower unit costs through exploiting economies of scale and learning. Reduced uncertainty over product design provides an opportunity to amortize specialized long-lived investments.

Innovation is not necessarily halted once the dominant design emerges; as Clarke [2] points out, it can occur lower down in the design hierarchy. For instance, a "v" cylinder configuration emerged in automobile engine blocks during the 1930s with the emergence of the Ford V-8 engine. Niches were quickly found for it. Moreover, once the product design stabilizes, there is likely to be a surge of process innovation as producers attempt to lower production costs for the new product (see fig. 4).

The Abernathy–Utterback framework does not characterize all industries. It seems more suited to mass markets where consumer tastes are relatively homogeneous. It would appear to be less characteristic of small niche markets where the absence of scale and learning economies attaches much less of a penalty to multiple designs. In these instances, generalized equipment will be employed in production.

[2] See Kuhn [4].

The existence of a dominant design watershed is of great significance to the distribution of profits between innovator and follower. The innovator may have been responsible for the fundamental scientific breakthroughs as well as the basic design of the new product. However, if imitation is relatively easy, imitators may enter the fray, modifying the product in important ways, yet relying on the fundamental designs pioneered by the innovator. When the game of musical chairs stops, and a dominant design emerges, the innovator might well end up positioned disadvantageously relative to a follower. Hence, when imitation is possible and occurs coupled with design modification before the emergence of a dominant design, followers have a good chance of having their modified product annointed as the industry standard, often to the great disadvantage of the innovator.

3.3. Complementary assets

Let the unit of analysis be an innovation. An innovation consists of certain technical knowledge about how to do things better than the existing state of the art. Assume that the know-how in question is partly codified and partly tacit. In order for such know-how to generate profits, it must be sold or utilized in some fashion in the market.

In almost all cases, the successful commercialization of an innovation requires that the know-how in question be utilized in conjunction with other capabilities or assets. Services such as marketing, competitive manufacturing, and after-sales support are almost always needed. These services are often obtained from complementary assets which are specialized. For example, the commercialization of a new drug is likely to require the dissemination of information over a specialized information channel. In some cases, as when the innovation is systemic, the complementary assets may be other parts of a system. For instance; computer hardware typically requires specialized software, both for the operating system, as well as for applications. Even when an innovation is autonomous, as with plug compatible components, certain complementary capabilities or assets will be needed for successful commercialization. Figure 5 summarizes this schematically.

Whether the assets required for least cost production and distribution are specialized to the

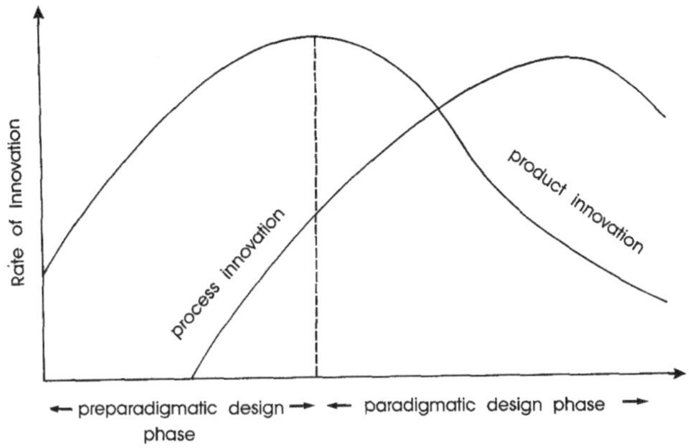

Fig. 4. Innovation over the product/industry life cycle.

innovation turns out to be important in the development presented below. Accordingly, the nature of complementary assets are explained in some detail. Figure 6 differentiates between complementary assets which are generic, specialized, and cospecialized.

Generic assets are general purpose assets which do not need to be tailored to the innovation in question. Specialized assets are those where there is unilateral dependence between the innovation and the complementary asset. Cospecialized assets are those for which there is a bilateral dependence. For instance, specialized repair facilities were needed to support the introduction of the rotary engine by Mazda. These assets are cospecialized because of the mutual dependence of the innovation on the repair facility. Containerization similarly required the deployment of some cospecialized assets in ocean shipping and terminals. However, the dependence of trucking on containerized shipping was less than that of containerized shipping on trucking, as trucks can convert from containers to flat beds at low cost. An example of a generic asset would be the manufacturing facilities needed to make running shoes. Generalized

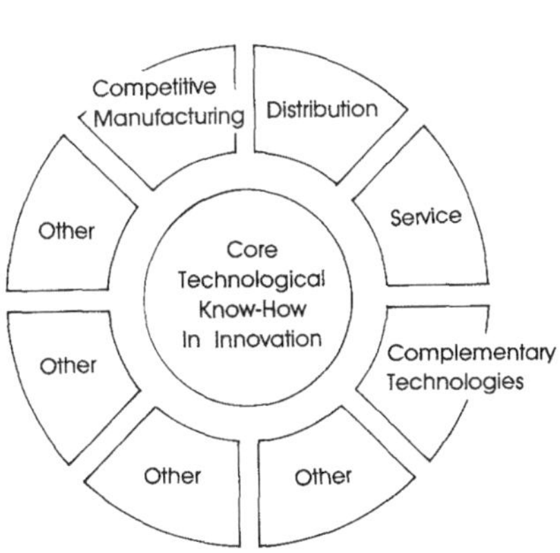

Fig. 5. Complementary assets needed to commercialize an innovation.

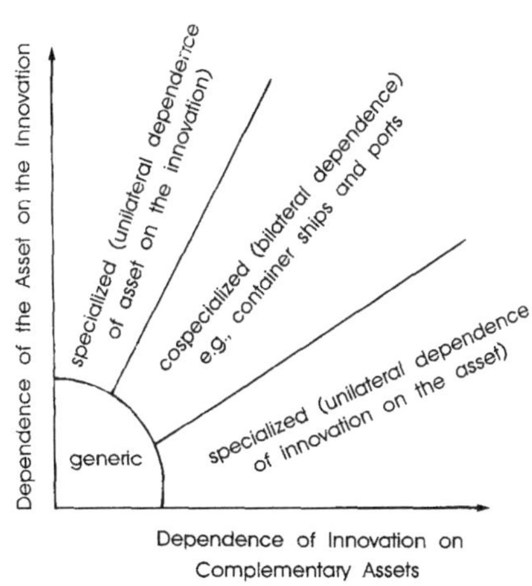

Fig. 6. Complementary assets: Generic, specialized, and cospecialized.

equipment can be employed in the main, exceptions being the molds for the soles.

4. Implications for profitability

These three concepts can now be related in a way which will shed light on the imitation process, and the distribution of profits between innovator and follower. We begin by examining tight appropriability regimes.

4.1. Tight appropriability regimes

In those few instances where the innovator has an iron clad patent or copyright protection, or where the nature of the product is such that trade secrets effectively deny imitators access to the relevant knowledge, the innovator is almost assured of translating its innovation into market value for some period of time. Even if the innovator does not possess the desirable endowment of complementary costs, iron clad protection of intellectual property will afford the innovator the time to access these assets. If these assets are generic, contractual relation may well suffice, and the innovator may simply license its technology. Specialized R&D firms are viable in such an environment. Universal Oil Products, an R&D firm developing refining processes for the petroleum industry was one such case in point. If, however, the complementary assets are specialized or cospecialized, contractual relationships are exposed to hazards, because one or both parties will have to commit capital to certain irreversible investments which will be valueless if the relationship between innovator and licensee breaks down. Accordingly, the innovator may find it prudent to expand its boundaries by integrating into specialized and cospecialized assets. Fortunately, the factors which make for difficult imitation will enable the innovator to build or acquire those complementary assets without competing with innovators for their control.

Competition from imitators is muted in this type of regime, which sometimes characterizes the petrochemical industry. In this industry, the protection offered by patents is fairly easily enforced. One factor assisting the licensee in this regard is that most petrochemical processes are designed around a specific variety of catalysts which can be kept proprietory. An agreement not to analyze the catalyst can be extracted from licensees, affording extra protection. However, even if such requirements are violated by licensees, the innovator is still well positioned, as the most important properties of a catalyst are related to its physical structure, and the process for generating this structure cannot be deduced from structural analysis alone. Every reaction technology a company acquires is thus accompanied by an ongoing dependence on the innovating company for the catalyst appropriate to the plant design. Failure to comply with various elements of the licensing contract can thus result in a cutoff in the supply of the catalyst, and possibly facility closure.

Similarly, if the innovator comes to market in the preparadigmatic phase with a sound product concept but the wrong design, a tight appropriability regime will afford the innovator the time needed to perform the trials needed to get the design right. As discussed earlier, the best initial design concepts often turn out to be hopelessly wrong, but if the innovator possesses an impenetrable thicket of patents, or has technology which is simply difficult to copy, then the market may well afford the innovator the necessary time to ascertain the right design before being eclipsed by imitators.

4.2. Weak appropriability

Tight appropriability is the exception rather than the rule. Accordingly, innovators must turn to business strategy if they are to keep imitators/followers at bay. The nature of the competitive process will vary according to whether the industry is in the paradigmatic or preparadigmatic phase.

4.2.1. Preparadigmatic phase

In the preparadigmatic phase, the innovator must be careful to let the basic design "float" until sufficient evidence has accumulated that a design has been delivered which is likely to become the industry standard. In some industries there may be little opportunity for product modification. In microelectronics, for example, designs become locked in when the circuitry is chosen. Product modification is limited to "debugging" and software modification. An innovator must begin the design process anew if the product doesn't fit the market well. In some respects, however, selecting designs is dictated by the need to meet certain compatibility standards so that new hardware can interface with existing applications software. In one sense, therefore, the design issue for the microprocessor industry today is relatively straightforward: deliver greater power and speed while meeting the the computer industry standards of the existing software base. However, from time to time windows of opportunity emerge for the introduction of entirely new families of microprocessors which will define a new industry and software standard. In these instances, basic design parameters are less well defined, and can be permitted to "float" until market acceptance is apparent.

The early history of the automobile industry exemplifies exceedingly well the importance for subsequent success of selecting the right design in the preparadigmatic stages. None of the early producers of steam cars survived the early shakeout when the closed body internal combustion engine automobile emerged as the dominant design. The steam car, nevertheless, had numerous early virtues, such as reliability, which the internal combustion engine autos could not deliver.

The British fiasco with the Comet I is also instructive. De Havilland had picked an early design with both technical and commercial flaws. By moving into production, significant irreversibilities and loss of reputation hobbled de Havilland to such a degree that it was unable to convert to the Boeing design which subsequently emerged as dominant. It wasn't even able to occupy second place, which went instead to Douglas.

As a general principle, it appears that innovators in weak appropriability regimes need to be intimately coupled to the market so that user needs can fully impact designs. When multiple parallel and sequential prototyping is feasible, it has clear advantages. Generally such an approach is simply prohibitively costly. When development costs for a large commercial aircraft exceed one billion dollars, variations on a theme are all that is possible.

Hence, the probability that an innovator – defined here as a firm that is first to commercialize a new product design concept – will enter the paradigmatic phase possessing the dominant design is problematic. The probabilities will be higher the lower the relative cost of prototyping, and the more tightly coupled the firm is to the market. The later is a function of organizational design, and can be influenced by managerial choices. The former is embedded in the technology, and cannot be influenced, except in minor ways, by managerial decisions. Hence, in industries with large developmental and prototyping costs – and hence significant irreversibilities – and where innovation of the product concept is easy, then one would expect that the probability that the innovator would emerge as the winner or amongst the winners at the end of the preparadigmatic stage is low.

4.2.2. *Paradigmatic stage*

In the preparadigmatic phase, complementary assets do not loom large. Rivalry is focused on trying to identify the design which will be dominant. Production volumes are low, and there is little to be gained in deploying specialized assets, as scale economies are unavailable, and price is not a principal competitive factor. However, as the leading design or designs begin to be revealed by the market, volumes increase and opportunities for economies of scale will induce firms to begin gearing up for mass production by acquiring specialized tooling and equipment, and possibly specialized distribution as well. Since these investments involve significant irreversibilities, producers are likely to proceed with caution. Islands of specialized capital will begin to appear in an industry, which otherwise features a sea of general purpose manufacturing equipment.

However, as the terms of competition begin to change, and prices become increasingly unimportant, access to complementary assets becomes absolutely critical. Since the core technology is easy to imitate, by assumption, commercial success swings upon the terms and conditions upon which the required complementary assets can be accessed.

It is at this point that specialized and cospecialized assets become critically important. Generalized equipment and skills, almost by definition, are always available in an industry, and even if they are not, they do not involve significant irreversibilities. Accordingly, firms have easy access to this type of capital, and even if there is insufficient capacity available in the relevant assets, it can easily be put in place as it involves few risks. Specialized assets, on the other hand, involve significant irreversibilities and cannot be easily accessed by contract, as the risks are significant for the party making the dedicated investment. The firms which control the cospecialized assets, such as distribution channels, specialized manufacturing capacity, etc. are clearly advantageously positioned relative to an innovator. Indeed, in rare instances where incumbent firms possess an airtight monopoly over specialized assets, and the innovator is in a regime of weak appropriability, all of the profits to the innovation could conceivably innure to the firms possessing the specialized assets who should be able to get the upper hand.

Even when the innovator is not confronted by situations where competitors or potential competitors control key assets, the innovator may still be disadvantaged. For instance, the technology embedded in cardiac pacemakers was easy to imitate, and so competitive outcomes quickly came to be determined by who had easiest access to the complementary assets, in this case specialized marketing. A similar situation has recently arisen in the United States with respect to personal computers. As an industry participant recently observed: "There are a huge numbers of computer manufacturers, companies that make peripherals (e.g. printers, hard disk drives, floppy disk drives), and software companies. They are all trying to get marketing distributors because they cannot afford to call on all of the US companies directly. They need to go through retail distribution channels, such as Businessland, in order to reach the marketplace. The problem today, however, is that many of these companies are not able to get shelf space and thus are having a very difficult time marketing their products. The point of distribution is where the profit and the power are in the marketplace today". (Norman [8, p.438])

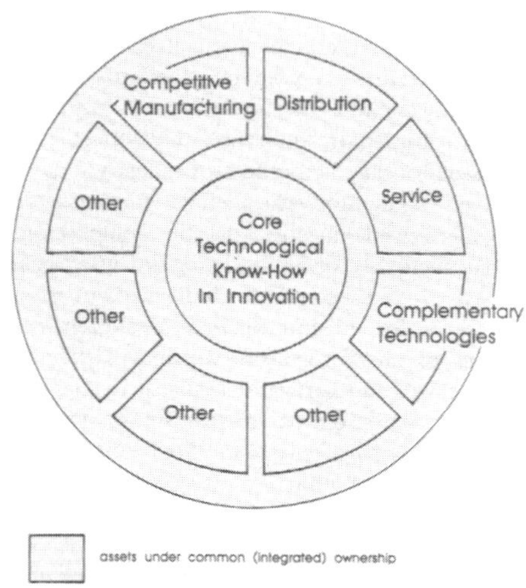

Fig. 7. Complementary assets internalized for innovation: Hypothetical case #1 (innovator integrated into all complementary assets).

There are a myriad of possible channels which could be employed. At one extreme the innovator could integrate into all of the necessary comple-

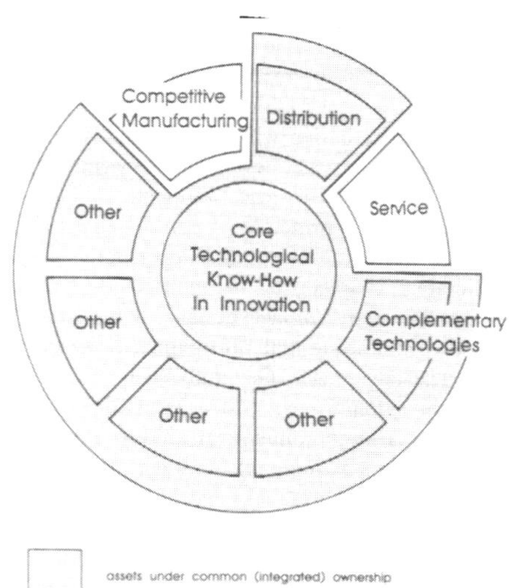

Fig. 8. Complementary assets internalized for innovation: Hypothetical case #2 (innovator subcontracts for manufacturing and service).

5. Channel strategy issues

The above analysis indicates how access to complementary assets, such as manufacturing and distribution, on competitive teams is critical if the innovator is to avoid handling over the lion's share of the profits to imitators, and/or to the owners of the complementary assets that are specialized or cospecialized to the innovation. It is now necessary to delve deeper into the appropriate control structure that the innovator ideally ought to establish over these critical assets.

mentary assets, as illustrated in fig. 7, or just a few of them, as illustrated in fig. 8. Complete integration (fig. 7) is likely to be unnecessary as well as prohibitively expensive. It is well to recognize that the variety of assets and competences which need to be accessed is likely to be quite large, even for only modestly complex technologies. To produce a personal computer, for instance, a company needs access to expertise in semiconductor technology, display technology, disk drive technology, networking technology, keyboard technology, and several others. No company can keep pace in all of these areas by itself.

At the other extreme, the innovator could attempt to access these assets through straightforward contractual relationships (e.g. component supply contracts, fabrication contracts, service contracts, etc.). In many instances such contracts may suffice, although it sometimes exposes the innovator to various hazards and dependencies that it may well wish to avoid. In between the fully integrated and full contractual extremes, there are a myriad of intermediate forms and channels available. An analysis of the properties of the two extreme forms is presented below. A brief synopsis of mixed modes then follows.

5.1. Contractual modes

The advantages of a contractual solution – whereby the innovator signs a contract, such as a license, with independent suppliers, manufacturers or distributors – are obvious. The innovator will not have to make the upfront capital expenditures needed to build or buy the assets in question. This reduces risks as well as cash requirements.

Contracting rather than integrating is likely to be the optimal strategy when the innovators appropriability regime is tight and the complementary assets are available in competitive supply (i.e. there is adequate capacity and a choice of sources).

Both conditions apply in petrochemicals for instance, so an innovator doesn't need to be integrated to be a successful. Consider, first, the appropriability regime. As discussed earlier, the protection offered by patents is fairly easily enforced, particularly for process technology, in the petrochemical industry. Given the advantageous feedstock prices available in hydrocarbon rich petrochemical exporters, and the appropriability regime characteristic of this industry, there is no incentive or advantage in owning the complementary assets (production facilities) as they are not typically highly specialized to the innovation. Union Carbide appears to realize this, and has recently adjusted its strategy accordingly. Essentially, Carbide is placing its existing technology into a new subsidiary, Engineering and Hydrocarbons Service. The company is engaging in licensing and offers engineering, construction, and management services to customers who want to take their feedstocks and integrate them forward into petrochemicals. But Carbide itself appears to be backing away from an integration strategy.

Chemical and petrochemical product innovations are not quite so easy to protect, which should raise new challenges to innovating firms in the developed nations as they attempt to shift out of commodity petrochemicals. There are already numerous examples of new products that made it to the marketplace, filled a customer need, but never generated competitive returns to the innovator because of imitation. For example, in the 1960s Dow decided to start manufacturing rigid polyurethene foam. However, it was imitated very quickly by numerous small firms which had lower costs. [3] The absence of low cost manufacturing capability left Dow vulnerable.

Contractual relationships can bring added credibility to the innovator, especially if the innovator is relatively unknown when the contractual partner is established and viable. Indeed, arms-length contracting which embodies more than a simple buy-sell agreement is becoming so common, and is so multifaceted, that the term strategic partnering has been devised to describe it. Even large companies such as IBM are now engaging in it. For IBM, partnering buys access to new technologies enabling the company to "learn things we couldn't have learned without many years of trial and error." [4] IBM's arrangement with Microsoft to use the latter's MS-DOS operating system software on the IBM PC facilitated the timely introduction of IBM's personal computer into the market.

[3] Executive V.P. Union Carbide, Robert D. Kennedy, quoted in *Chemical Week*, Nov. 16, 1983, p. 48.

[4] Comment attributed to Peter Olson III, IBM's director of business development, as reported in The Strategy Behind IBM's Strategic Alliances, *Electronic Business*, October 1 (1985) 126.

Smaller less integrated companies are often eager to sign on with established companies because of the name recognition and reputation spillovers. For instance Cipher Data Products, Inc. contracted with IBM to develop a low-priced version of IBM's 3480 0.5 inch streaming cartridge drive, which is likely to become the industry standard. As Cipher management points out, "one of the biggest advantages to dealing with IBM is that, once you've created a product that meets the high quality standards necessary to sell into the IBM world, you can sell into any arena." [5] Similarly, IBM's contract with Microsoft "meant instant credibility" to Microsoft (McKenna, 1985, p. 94).

It is most important to recognize, however, that strategic (contractual) partnering, which is currently very fashionable, is exposed to certain hazards, particularly for the innovator, when the innovator is trying to use contracts to access specialized capabilities. First, it may be difficult to induce suppliers to make costly irreversible commitments which depend for their success on the success of the innovation. To expect suppliers, manufacturers, and distributors to do so is to invite them to take risks along with the innovator. The problem which this poses for the innovator is similar to the problems associated with attracting venture capital. The innovator must persuade its prospective partner that the risk is a good one. The situation is one open to opportunistic abuses on both sides. The innovator has incentives to overstate the value of the innovation, while the supplier has incentives to "run with the technology" should the innovation be a success.

Instances of both parties making irreversible capital commitments nevertheless exist. Apple's Laserwriter – a high resolution laser printer which allows PC users to produce near typeset quality text and art department graphics – is a case in point. Apple persuaded Canon to participate in the development of the Laserwriter by providing subsystems from its copiers – but only after Apple contracted to pay for a certain number of copier engines and cases. In short, Apple accepted a good deal of the financial risk in order to induce Canon to assist in the development and production of the Laserwriter. The arrangement appears to have been prudent, yet there were clearly hazards for both sides. It is difficult to write, execute, and enforce complex development contracts, particularly when the design of the new product is still "floating." Apple was exposed to the risk that its co-innovator Canon would fail to deliver, and Canon was exposed to the risk that the Apple design and marketing effort would not succeed. Still, Apple's alternatives may have been rather limited, inasmuch as it didn't command the requisite technology to "go it alone."

In short, the current euphoria over "strategic partnering" may be partially misplaced. The advantages are being stressed (for example, McKenna [6]) without a balanced presentation of costs and risks. Briefly, there is the risk that the partner won't perform according to the innovator's perception of what the contract requires; there is the added danger that the partner may imitate the innovator's technology and attempt to compete with the innovator. This latter possibility is particularly acute if the provider of the complementary asset is uniquely situated with respect to the complementary asset in question and has the capacity to imitate the technology, which the innovator is unable to protect. The innovator will then find that it has created a competitor who is better positioned than the innovator to take advantage of the market opportunity at hand. *Business Week* has expressed concerns along these lines in its discussion of the "Hollow Corporation." [6]

It is important to bear in mind, however, that contractual or partnering strategies in certain cases are ideal. If the innovator's technology is well protected, and if what the partner has to provide is a "generic" capacity available from many potential partners, then the innovator will be able to maintain the upper hand while avoiding the costs of duplicating downstream capacity. Even if the partner fails to perform, adequate alternatives exist (by assumption, the partners' capacities are commonly available) so the innovator's efforts to successfully commercialize its technology ought to proceed profitably.

[5] Comment attributed to Norman Farquhar, Cipher's vice president for strategic development, as reported in *Electronic Business*, October 1 (1985) 128.

[6] See *Business Week*, March 3 (1986) 57–59. *Business Week* uses the term to describe a corporation which lacks in-house manufacturing capability.

5.2. Integration modes

Integration, which by definition involves ownership, is distinguished from pure contractual modes in that it typically facilitates incentive alignment and control. If an innovator owns rather than rents the complementary assets needed to commercialize, then it is in a position to capture spillover benefits stemming from increased demand for the complementary assets caused by the innovation.

Indeed, an innovator might be in the position, at least before its innovation is announced, to buy up capacity in the complementary assets, possibly to its great subsequent advantage. If futures markets exist, simply taking forward positions in the complementary assets may suffice to capture much of the spillovers.

Even after the innovation is announced, the innovator might still be able to build or buy complementary capacities at competitive prices if the innovation has iron clad legal protection (i.e. if the innovation is in a tight appropriability regime). However, if the innovation is not tightly protected and once "out" is easy to imitate, then securing control of complementary capacities is likely to be the key success factor, particularly if those capacities are in fixed supply – so called "bottlenecks." Distribution and specialized manufacturing competences often become bottlenecks.

As a practical matter, however, an innovator may not have the time to acquire or build the complementary assets that ideally it would like to control. This is particularly true when imitation is easy, so that timing becomes critical. Additionally, the innovator may simply not have the financial resources to proceed. The implications of timing and cash constraints are summarized in fig. 9.

Accordingly, in weak appropriability regimes innovators need to rank complementary assets as to their importance. If the complementary assets are critical, ownership is warranted, although if the firm is cash constrained a minority position may well represent a sensible tradeoff.

Needless to say, when imitation is easy, strategic moves to build or buy complementary assets which are specialized must occur with due reference to the moves of competitors. There is no point moving to build a specialized asset, for instance, if one's imitators can do it faster and cheaper.

It is hopefully self evident that if the innovator is already a large enterprise with many of the relevant complementary assets under its control, integration is not likely to be the issue that it might otherwise be, as the innovating firm will already control many of the relevant specialized and cospecialized assets. However, in industries experiencing rapid technological change, technologies advance so rapidly that it is unlikely that a single company has the full range of expertise needed to bring advanced products to market in a timely and cost effective fashion. Hence, the integration issue is not just a small firm issue.

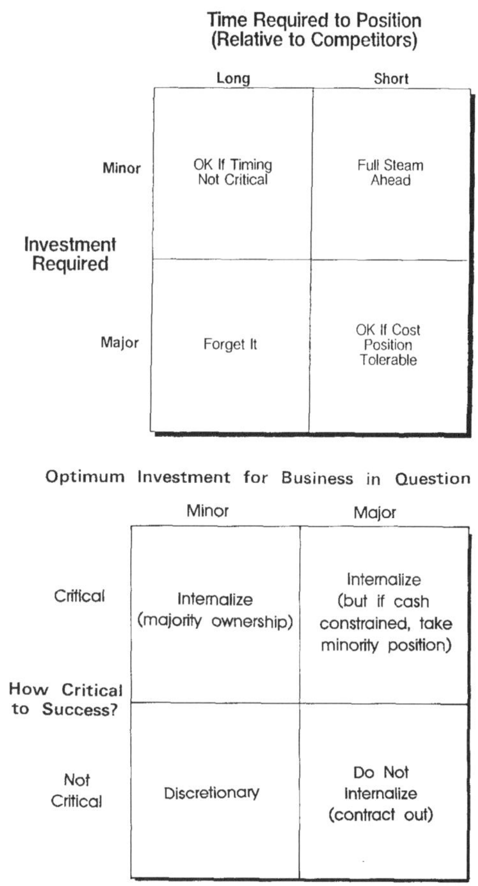

Fig. 9. Specialized complementary assets and weak appropriability: Integration calculus.

5.3. Integration versus contract strategies: An analytic summary

Figure 10 summarizes some of the relevant considerations in the form of a decision flow chart. It indicates that a profit seeking innovator, confronted by weak intellectual property protection and the need to access specialized complementary assets and/or capabilities, is forced to expand its activities through integration if it is to prevail over imitators. Put differently, innovators who develop new products that possess poor intellectual property protection but which requires specialized complementary capacities are more likely to parlay their technology into a commercial advantage, rather than see it prevail in the hands of imitators.

Figure 10 makes it apparent that the difficult strategic decisions arise in situations where the appropriability regime is weak and where specialized assets are critical to profitable commercialization. These situations, which in reality are very common, require that a fine-grained competitor analysis be part of the innovator's strategic assessment of its opportunities and threats. This is carried a step further in fig. 11, which looks only at situations where commercialization requires certain specialized capabilities. It indicates the appropriate strategies for the innovators and predicts the outcomes to be expected for the various players.

Three classes of players are of interst: innovators, imitators, and the owners of cospecialized assets (e.g. distributors). All three can potentially benefit or lose from the innovation process. The latter can potentially benefit from the additional business which the innovation may direct in the asset owners direction. Should the asset turn out

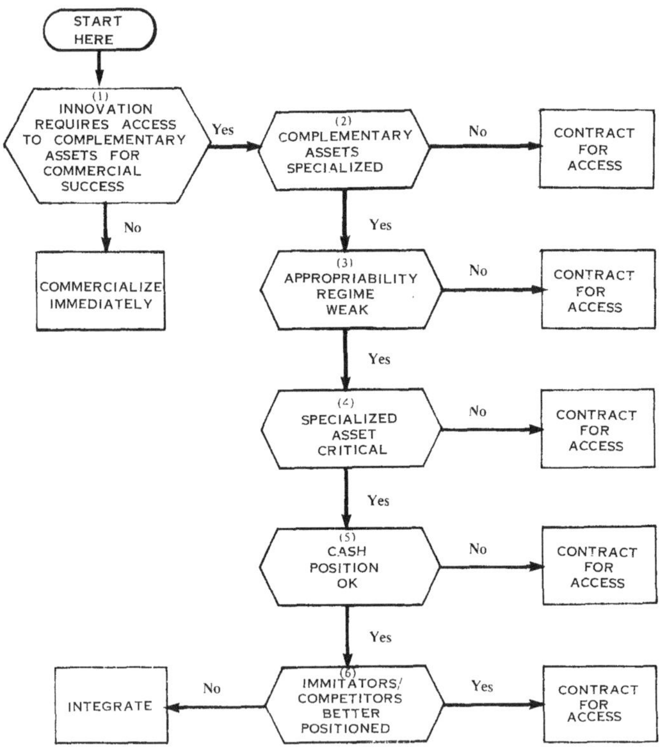

Fig. 10. Flow chart for integration versus contract decision.

to be a bottleneck with respect to commercializing the innovation, the owner of the bottleneck facilities is obviously in a position to extract profits from the innovator and/or imitators.

The vertical axis in fig. 11 measures how those who possess the technology (the innovator or possibly its imitators) are positioned vis à vis those firms that possess required specialized assets. The horizontal axis measures the "tightness" of the appropriability regime, tight regimes being evidence by iron clad legal protection coupled with technology that is simply difficult to copy; weak regimes offer little in the way of legal protection and the essence of the technology, once released, is transparent to the imitator. Weak regimes are further subdivided according to how the innovator and imitators are positioned vis à vis each other. This is likely to be a function of factors such as lead time and prior positioning in the requisite complementary assets.

Figure 11 makes it apparent that even when firms pursue the optimal strategy, other industry participants may take the jackpot. This possibility is unlikely when the intellectual property in question is tightly protected. The only serious threat to the innovator is where a specialized complementary asset is completely "locked up," a possibility recognized in cell 4. This can rarely be done without the cooperation of government. But it frequently occurs, as when a foreign government closes off access to a foreign market, forcing the innovators to license to foreign firms, but with the government effectively cartelizing the potential licensees. With weak intellectual property protection, however, it is quite clear that the innovator will often loose out to imitators and/or asset holders, even when the innovator is pursuing the appropriate strategy (cell 6). Clearly, incorrect strategies can compound problems. For instance, if innovators integrate when they should contract,

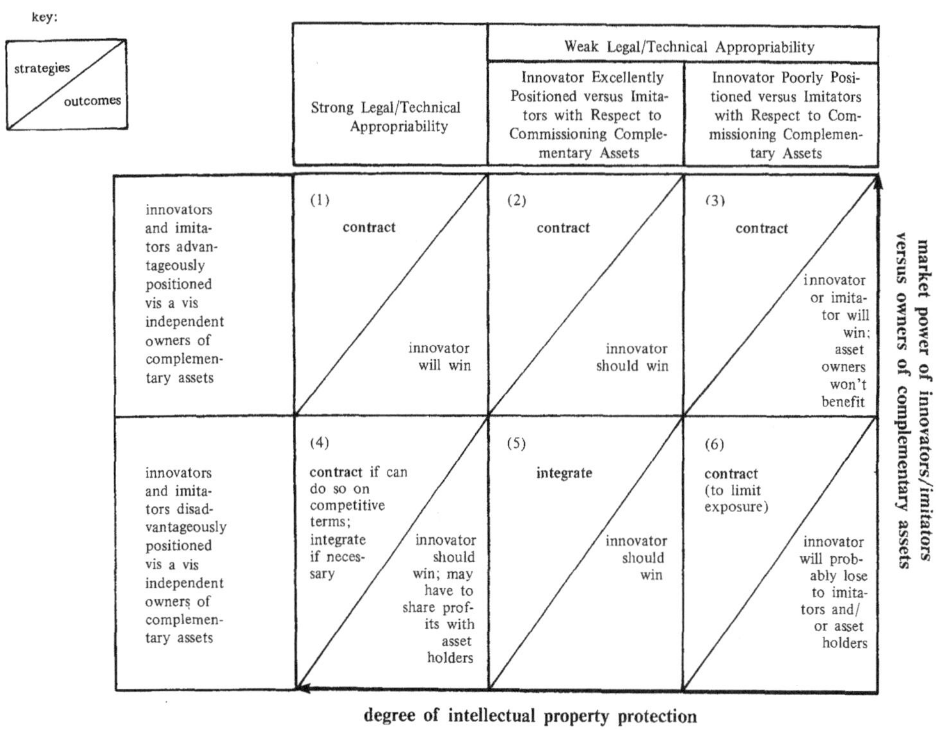

Fig. 11. Contract and integration strategies and outcomes for innovators: Specialized asset case.

a heavy commitment of resources will be incurred for little if any strategic benefit, thereby exposing the innovator to even greater losses than would otherwise be the case. On the other hand, if an innovator tries to contract for the supply of a critical capability when it should build the capability itself, it may well find it has nutured an imitator better able to serve the market than the innovator itself.

5.4. Mixed modes

The real world rarely provides extreme or pure cases. Decisions to integrate or• license involve tradeoffs, compromises, and mixed approaches. It is not surprising therefore that the real world is characterized by mixed modes of organization, involving judicious blends of integration and contracting. Sometimes mixed modes represent transitional phases. For instance, because of the convergence of computer and telecommunication technology, firms in each industry are discovering that they often lack the requisite technical capabilities in the other. Since the technological interdependence of the two requires collaboration amongst those who design different parts of the system, intense cross-boundary coordination and information flows are required. When separate enterprises are involved, agreement must be reached on complex protocol issues amongst parties who see their interests differently. Contractual difficulties can be anticipated since the selection of common technical protocols amongst the parties will often be followed by transaction-specific investments in hardware and software. There is little doubt that this was the motivation behind IBM's purchase of 15 percent of PBX manufacturer Rolm in 1983, a position that was expanded to 100 percent in 1984. IBM's stake in Intel, which began with a 12 percent purchase in 1982, is most probably not a transitional phase leading to 100 percent purchase, because both companies realized that the two corporate cultures are not very compatible, and IBM may not be as impressed with Intel's technology as it once was.

5.5. The CAT scanner, the IBM PC, and NutraSweet: Insights from the framework

EMI's failure to reap significant returns from the CAT scanner can be explained in large measure by reference to the concepts developed above. The scanner which EMI developed was of a technical sophistication much higher than would normally be found in a hospital, requiring a high level of training, support, and servicing. EMI had none of these capabilities, could not easily contract for them, and was slow to realize their importance. It most probably could have formed a partnership with a company like Siemens to access the requisite capabilities. Its failure to do so was a strategic error compounded by the very limited intellectual property protection which the law afforded the scanner. Although subsequent court decisions have upheld some of EMI's patent claims, once the product was in the market it could be reverse engineered and its essential features copied. Two competitors, GE and Technicare, already possessed the complementary capabilities that the scanner required, and they were also technologically capable. In addition, both were experienced marketers of medical equipment, and had reputations for quality, reliability and service. GE and Technicare were thus able to commit their R&D resources to developing a competitive scanner, borrowing ideas from EMI's scanner, which they undoubtedly had access to through cooperative hospitals, and improving on it where they could while they rushed to market. GE began taking orders in 1976 and soon after made inroads on EMI. In 1977 concern for rising health care costs caused the Carter Administration to introduce "certificate of need' regulation, which required HEW's approval on expenditures on big ticket items like CAT scanners. This severely cut the size of the available market.

By 1978 EMI had lost market share leadership to Technicare, which was in turn quickly overtaken by GE. In October 1979, Godfrey Houndsfield of EMI shared the Nobel prize for invention of the CT scanner. Despite this honor, and the public recognition of its role in bringing this medical breathrough to the world, the collapse of its scanner business forced EMI in the same year into the arms of a rescuer, Thorn Electrical Industries, Ltd. GE subsequently acquired what was EMI's scanner business from Thorn for what amounted to a pittance. [7] Though royalties continued to flow to EMI, the company had failed to capture the

[7] See GE Gobbles a Rival in CT Scanners, *Business Week*, May 19, 1980, issue no. 2637.

lion's share of the profits generated by the innovation it had pioneered and successfully commercialized.

If EMI illustrates how a company with outstanding technology and an excellent product can fail to profit from innovation while the imitators succeeded, the story of the IBM PC indicates how a new product representing a very modest technological advance can yield remarkable returns to the developer.

The IBM PC, introduced in 1981, was a success despite the fact that the architecture was ordinary and the components standard. Philip Estridge's design team in Boca Raton, Florida, decided to use existing technology to produce a solid, reliable micro rather than state of the art. With a one-year mandate to develop a PC, Estridge's team could do little else.

However, the IBM PC did use what at the time was a new 16-bit microprocessor (the Intel 8088) and a new disk operating system (DOS) adapted for IBM by Microsoft. Other than the microprocessor and the operating system, the IBM PC incorporated existing micro "standards" and used off-the-shelf parts from outside vendors. IBM did write its own BIOS (Basic Input/Output System) which is embedded in ROM, but this was a relatively straightforward programming exercise.

The key to the PC's success was not the technology. It was the set of complementary assets which IBM either had or quickly assembled around the PC. In order to expand the market for PCs, there was a clear need for an expandable, flexible microcomputer system with extensive applications software. IBM could have based its PC system on its own patented hardware and copyrighted software. Such an approach would cause complementary products to be cospecialized, forcing IBM to develop peripherals and a comprehensive library of software in a very short time. Instead, IBM adopted what might be called an "induced contractual" approach. By adopting an open system architecture, as Apple had done, and by making the operating system information publicly available, a spectacular output of third part software was induced. IBM estimated that by mid-1983, at least 3000 hardware and software products were available for the PC. [8] Put differently, IBM pulled together the complementary assets, particularly software, which success required, without even using contracts, let alone integration. This was despite the fact that the software developers were creating assets that were in part cospecialized with the IBM PC, at least in the first instance.

A number of special factors made this seem a reasonable risk to the software writers. A critical one was IBM's name and commitment to the project. The reputation behind the letters I.B.M. is perhaps the greatest cospecialized asset the company possesses. The name implied that the product would be marketed and serviced in the IBM tradition. It guaranteed that PC-DOS would become an industry standard, so that the software business would not be solely dependent on IBM, because emulators were sure to enter. It guaranteed access to retail distribution outlets on competitive terms. The consequences was that IBM was able to take a product which represented at best a modest technological accomplishment, and turn into a fabulous commercial success. The case demonstrates the role that complementary assets play in determining outcomes.

The spectacular success and profitability of G.D. Searle's NutraSweet is an uncommon story which is also consistent with the above framework. In 1982, Searle reported combined sales of $74 million for NutraSweet and its table top version, Equal. In 1983, this surged to $336 million. In 1985, NutraSweet sales exceeded $700 million [9] and Equal had captured 50 percent of the U.S. sugar substitute market and was number one in five other countries.

NutraSweet, which is Searle's tradename for aspartame, has achieved rapid acceptance in each of its FDA approved categories because of its good taste and ability to substitute directly for sugar in many applications. However, Searle's earnings from NutraSweet and the absence of a strategic challenge can be traced in part to Searle's clever strategy.

It appears that Searle has managed to establish an exceptionally tight appropriability regime around NutraSweet – one that may well continue for some time after the patent has expired. No competitor appears to have successfully "invented around" the Searle patent and commercialized an alternative, no doubt in part because the FDA

[8] F. Gens and C. Christiansen, Could 1,000,000 IBM PC Users Be Wrong, *Byte*, November 1983, 88.

[9] See *Monsanto Annual Report, 1985*.

approval process would have to begin anew for an imitator who was not violating Searle's patents. A competitor who tried to replicate the aspartame molecule with minor modification to circumvent the patent would probably be forced to replicate the hundreds of tests and experiments which proved aspartame's safety. Without patent protection, FDA approval would provide no shield against imitators coming to market with an identical chemical and who could establish to the FDA that it is the same compound that had already been approved. Without FDA approval on the other hand, the patent protection would be worthless for the product would not be sold for human consumption.

Searle has aggressively pushed to strengthen its patent protection. The company was granted U.S. patent protection in 1970. It has also obtained patent protection in Japan, Canada, Australia, U.K., France, Germany, and a number of other countries. However, most of these patents carry a 17-year life. Since the product was only approved for human consumption in 1982, the 17-year patent life was effectively reduced to five. Recognizing the obvious importance of its patent, Searle pressed for and obtained special legislation in November 1984 extending the patent protection on aspartame for another 5 years. The U.K. provided a similar extension. In almost every other nation, however, 1987 will mark the expiration of the patent.

When the patent expires, however, Searle will still have several valuable assets to help keep imitators at bay. Searle has gone to great lengths to create and promulgate the use of its NutraSweet name and a distinctive "Swirl" logo on all goods licensed to use the ingredient. The company has also developed the "Equal" tradename for a table top version of the sweetener. Trademark law in the U.S. provides protection against "unfair" competition in branded products for as long as the owner of the mark continues to use it. Both the NutraSweet and Equal trademarks will become essential assets when the patents on aspartame expire. Searle may well have convinced consumers that the only real form of sweetener is NutraSweet/Equal. Consumers know most other artificial sweeteners by their generic names – saccharin and cyclamates.

Clearly, Searle is trying to build a position in complementary assets to prepare for the competition which will surely arise. Searle's joint venture with Ajinomoto ensures them access to that company's many years of experience in the production of biochemical agents. Much of this knowledge is associated with techniques for distillation and synthesis of the delicate hydrocarbon compounds that are the ingredients of NutraSweet, and is therefore more tacit than codified. Searle has begun to put these techniques to use in its own $160 million Georgia production facility. It can be expected that Searle will use trade secrets to the maximum to keep this know-how proprietary.

By the time its patent expires, Searle's extensive research into production techniques for L-phenylalanine, and its 8 years of experience in the Georgia plant, should give it a significant cost advantage over potential aspartame competitors. Trade secret protection, unlike patents, has no fixed lifetime and may well sustain Searle's position for years to come.

Moreover, Searle has wisely avoided renewing contracts with suppliers when they have expired. [10] Had Searle subcontracted manufacturing for NutraSweet, it would have created a manufacturer who would then be in a position to enter the aspartame market itself, or to team up with a marketer of artificial sweeteners. But keeping manufacturing inhouse, and by developing a valuable tradename, Searle has a good chance of protecting its market position from dramatic inroads once patents expire. Clearly, Searle seems to be astutely aware of the importance of maintaining a "tight appropriability regime" and using cospecialized assets strategically.

6. Implications for R&D strategy, industry structure, and trade policy

6.1. Allocating R&D resources

The analysis so far assumes that the firm has developed an innovation for which a market exists. It indicates the strategies which the firm must

[10] Purification Engineering, which had spent $5 million to build a phenylalanine production facility, was told in January 1985 that their contract would not be renewed. In May, Genex, which claimed to have invested $25 million, was given the same message, A Bad Aftertaste, *Business Week*, July 15, 1985, issue 2903.

follow to maximize its share of industry profits relative to imitators and other competitors. There is no guarantee of success even if optimal strategies are followed.

The innovator can improve its total return to R&D, however, by adjusting its R&D investment portfolio to maximize the probability that technological discoveries will emerge that are either easy to protect with existing intellectual property law, or which require for commercialization cospecialized assets already within the firm's repertoire of capabilities. Put differently, if an innovating firm does not target its R&D resources towards new products and processes which it can commercialize advantageously relative to potential imitators and/or followers, then it is unlikely to profit from its investment in R&D. In this sense, a firm's history – and the assets it already has in place – ought to condition its R&D investment decisions. Clearly, an innovating firm with considerable assets already in place is free to strike out in new directions, so long as in doing so it is cognizant of the kinds of capabilities required to successfully commercialize the innovation. It is therefore rather clear that the R&D investment decision cannot be divorced from the strategic analysis of markets and industries, and the firm's position within them.

6.2. *Small firm versus large firm comparisons*

Business commentators often remark that many small entrepreneurial firms which generate new, commercially valuable technology fail while large multinational firms, often with a less meritorious record with respect to innovation, survive and prosper. One set of reasons for this phenomenon is now clear. Large firms are more likely to possess the relevant specialized and cospecialized assets within their boundaries at the time of new product introduction. They can therefore do a better job of milking their technology, however meager, to maximum advantage. Small domestic firms are less likely to have the relevant specialized and cospecialized assets within their boundaries and so will either have to incur the expense of trying to build them, or of trying to develop coalitions with competitors/owners of the specialized assets.

6.3. *Regimes of appropriability and industry structure*

In industries where legal methods of protection are effective, or where new products are just hard to copy, the strategic necessity for innovating firms to integrate into cospecialized assets would appear to be less compelling than in industries where legal protection is weak. In cases where legal protection is weak or nonexistent, the control of cospecialized assets will be needed for long-run survival.

In this regard, it is instructive to examine the U.S. drug industry (Temin [10]). Beginning in the 1940s, the U.S. Patent Office began, for the first time, to grant patents on certain natural substances that involved difficult extraction procedures. Thus, in 1948 Merck received a patent on streptomycin, which was a natural substance. However, it was not the extraction process but the drug itself which received the patent. Hence, patents were important to the drug industry in terms of what could be patented (drugs), but they did not prevent imitation [10, p.436]. Sometimes just changing one molecule will enable a company to come up with a different substance which does not violate the patent. Had patents been more all-inclusive – and I am not suggesting they should – licensing would have been an effective mechanism for Merck to extract profits from its innovation. As it turns out, the emergence of close substitutes, coupled with FDA regulation which had the de facto effect of reducing the elasticity of demand for drugs, placed high rewards on a product differentiation strategy. This required extensive marketing, including a sales force that could directly contact doctors, who were the purchasers of drugs through their ability to create prescriptions. [11] The result was exclusive production (i.e., the earlier industry practice of licensing was dropped) and forward integration into marketing (the relevant cospecialized asset).

Generally, if legal protection of the innovator's profits is secure, innovating firms can select their

[11] In the period before FDA regulation, all drugs other than narcotics were available over-the-counter. Since the end user could purchase drugs directly, sales were price sensitive. Once prescriptions were required, this price sensitivity collapsed; the doctors not only did not have to pay for the drugs, but in most cases they were unaware of the prices of the drugs they were prescribing.

boundaries based simply on their ability to identify user needs and respond to those through research and development. The weaker the legal methods of protection, the greater the incentive to integrate into the relevant cospecialized assets. Hence, as industries in which legal protection is weak begin to mature, integration into innovation-specific cospecialized assets will occur. Often this will take the form of backward, forward and lateral integration. (Conglomerate integration is not part of this phenomenon.) For example, IBM's purchase of Rolm can be seen as a response to the impact of technological change on the identity of the cospecialized assets relevant to IBM's future growth.

6.4. Industry maturity, new entry, and history

As technologically progressive industries mature, and a greater proportion of the relevant cospecialized assets are brought in under the corporate umbrellas of incumbents, new entry becomes more difficult. Moreover, when it does occur it is more likely to involve coalition formation very early on. Incumbents will for sure own the cospecialized assets, and new entrants will find it necessary to forge links with them. Here lies the explanation for the sudden surge in "strategic partnering" now occurring internationally, and particularly in the computer and telecommunications industry. Note that it should not be interpreted in anti-competitive terms. Given existing industry structure, coalitions ought to be seen not as attempts to stifle competition, but as mechanisms for lowering entry requirements for innovators.

In industries in which technological change of a particular kind has occurred, which required deployment of specialized and/or cospecialized assets at the time, a configuration of firm boundaries may well have arisen which no longer has compelling efficiencies. Considerations which once dictated integration may no longer hold, yet there may not be strong forces leading to divestiture. Hence existing firm boundaries may in some industries – especially those where the technological trajectory and attendent specialized asset requirements has changed – be rather fragile. In short, history matters in terms of understanding the structure of the modern business enterprise. Existing firm boundaries cannot always be assumed to have obvious rationales in terms of today's requirements.

6.5. The importance of manufacturing to international competitiveness

Practically all forms of technological know-how must be embedded in goods and services to yield value to the consumer. An important policy for the innovating nation is whether the identity of the firms and nations performing this function matter.

In a world of tight appropriability and zero transactions cost – the world of neoclassical trade theory – it is a matter of indifference whether an innovating firm has an in-house manufacturing capability, domestic or foreign. It can simply engage in arms-length contracting (patent licensing, know-how licensing, co-production, etc.) for the sale of the output of the activity in which it has a comparative advantage (in this case R&D) and will maximize returns by specializing in what it does best.

However, in a regime of weak appropriability, and especially where the requisite manufacturing assets are specialized to the innovation, which is often the case, participation in manufacturing may be necessary if an innovator is to appropriate the rents from its innovation. Hence, if an innovator's manufacturing costs are higher than those of its imitators, the innovator may well end up ceding the lion's share of profits to the imitator.

In a weak appropriability regime, low cost imitator-manufacturers may end up capturing all of the profits from innovation. In a weak appropriability regime where specialized manufacturing capabilities are required to produce new products, an innovator with a manufacturing disadvantage may find that its advantage at early stage research and development will have no commercial value. This will eventually cripple the innovator, unless it is assisted by governmental processes. For example, it appears that one of the reasons why U.S. color TV manufacturers did not capture the lion's share of the profits from the innovation, for which RCA was primarily responsible, was that RCA and its American licenses were not competitive at manufacturing. In this context, concerns that the decline of manufacturing threatens the entire economy appear to be well founded.

A related implication is that as the technology gap closes, the basis of competition in an industry will shift to the cospecialized assets. This appears to be what is happening in microprocessors. Intel is no longer out ahead technologically. As Gordon Moore, CEO of Intel points out, "Take the top 10 [semiconductor] companies in the world... and it is hard to tell at any time who is ahead of whom.... It is clear that we have to be pretty damn close to the Japanese from a manufacturing standpoint to compete." [12] It is not just that strength in one area is necessary to compensate for weakness in another. As technology becomes more public and less proprietary through easier imitation, then strength in manufacturing and other capabilities is necessary to derive advantage from whatever technological advantages an innovator may possess.

Put differently, the notion that the United States can adopt a "designer role" in international commerce, while letting independent firms in other countries such as Japan, Korea, Taiwan, or Mexico do the manufacturing, is unlikely to be viable as a long-run strategy. This is because profits will accrue primarily to the low cost manufacturers (by providing a larger sales base over which they can exploit their special skills). Where imitation is easy, and even where it is not, there are obvious problems in transacting in the market for know-how, problems which are described in more detail elsewhere [9]. In particular, there are difficulties in pricing an intangible asset whose true performance features are difficult to ascertain ex ante.

The trend in international business towards what Miles and Snow [7] call "dynamic networks" – characterized by vertical disintegration and contracting – ought thus be viewed with concern. (*Business Week*, March 3, 1986, has referred to the same phenomenon as the Hollow Corporation.) "Dynamic networks" may not so much reflect innovative organizational forms, but the disassembly of the modern corporation because of deterioration in national capacities, manufacturing in particular, which are complementary to technological innovation. Dynamic networks may therefore signal not so much the rejuvenation of American enterprise, but its piecemeal demise.

6.6. How trade and investment barriers can impact innovators' profits

In regimes of weak appropriability, governments can move to shift the distribution of the gains from innovation away from foreign innovators and towards domestic firms by denying innovators ownership of specialized assets. The foreign firm, which by assumption is an innovator, will be left with the option of selling its intangible assets in the market for know how if both trade and investment are foreclosed by government policy. This option may appear better than the alternative (no renumeration at all from the market in question). Licensing may then appear profitable, but only because access to the complementary assets is blocked by government.

Thus when an innovating firm generating profits needs to access complementary assets abroad, host governments, by limiting access, can sometimes milk the innovators for a share of the profits, particularly that portion which originates from sales in the host country. However, the ability of host governments to do so depends importantly on the criticality of the host country's assets to the innovator. If the cost and infrastructure characteristics of the host country are such that it is the world's lowest cost manufacturing site, and if domestic industry is competitive, then by acting as a de facto monopsonist the host country government ought to be able to adjust the terms of access to the complementary assets so as to appropriate a greater share of the profits generated by the innovation. [13]

If, on the other hand, the host country offers no unique complementary assets, except access to its own market, restrictive practices by the government will only redistribute profits with respect to domestic rather than worldwide sales.

6.7. Implications for the international distribution of the benefits from innovation

The above analysis makes transparent that innovators who do not have access to the relevant specialized and cospecialized assets may end up ceding profits to imitators and other competitors, or simply to the owners of the specialized or cospecialized assets.

Even when the specialized assets are possessed by the innovating firm, they may be located abroad. Foreign factors of production are thus

[12] Institutionalizing the Revolution, *Forbes*, June 16, 1986, 35.

[13] If the host country market structure is monopolistic in the first instance, private actors might be able to achieve the same benefit. What government can do is to force collusion of domestic enterprises to their mutual benefit.

likely to benefit from research and development activities occurring across borders. There is little doubt, for instance, that the inability of many American multinationals to sustain competitive manufacturing in the U.S. is resulting in declining returns to U.S. labor. Stockholders and top management probably do as well if not better when a multinational accesses cospecialized assets in the firm's foreign subsidiaries; however, if there is unemployment in the factors of production supporting the specialized and cospecialized assets in question, then the foreign factors of production will benefit from innovation originating beyond national borders. This speaks to the importance to innovating nations of maintaining competence and competitiveness in the assets which complement technological innovation, manufacturing being a case in point. It also speaks to the importance to innovating nations of enhancing the protection afforded worldwide to intellectual property.

However, it must be recognized that there are inherent limits to the legal protection of intellectual property, and that business and national strategy are therefore likely to the critical factors in determining how the gains from innovation are shared worldwide. By making the correct strategic decision, innovating firms can move to protect the interests of stockholders; however, to ensure that domestic rather than foreign cospecialized assets capture the lion's share of the externalities spilling over to complementary assets, the supporting infrastructure for those complementary assets must not be allowed to decay. In short, if a nation has prowess at innovation, then in the absence of iron clad protection for intellectual property, it must maintain well-developed complementary assets if it is to capture the spillover benefits from innovation

7. Conclusion

The above analysis has attempted to synthesize from recent research in industrial organization and strategic management a framework within which to analyze the distribution of the profits from innovation. The framework indicates that the boundaries of the firm are an important strategic variable for innovating firms. The ownership of complementary assets, particularly when they are specialized and/or cospecialized, help establish who wins and who loses from innovation. Imitators can often outperform innovators if they are better positioned with respect to critical complementary assets. Hence, public policy aimed at promoting innovation must focus not only on R&D, but also on complementary assets, as well as the underlying infrastructure. If government decides to stimulate innovation, it would seem important to clear away barriers which impede the development of complementary assets which tend to be specialized or cospecialized to innovation. To fail to do so will cause an unnecessary large portion of the profits from innovation to flow to imitators and other competitors. If these firms lie beyond one's national borders, there are obvious implications for the internal distribution of income.

When applied to world markets, results similar to those obtained from the "new trade theory" are suggested by the framework. In particular, tariffs and other restrictions on trade can in some cases injure innovating firms while simultaneously benefiting protected firms when they are imitators. However, the propositions suggested by the framework are particularized to appropriability regimes, suggesting that economy-wide conclusions will be illusive. The policy conclusions derivable for commodity petrochemicals, for instance, are likely to be different than those that would be arrived at for semiconductors.

The approach also suggests that the product life cycle model of international trade will play itself out very differently in different industries and markets, in part according to appropriability regimes and the nature of the assets which need to be employed to convert a technological success into a commercial one. Whatever its limitations, the approach establishes that it is not so much the structure of markets but the structure of firms, particularly the scope of their boundaries, coupled with national policies with respect to the development of complementary assets, which determines the distribution of the profits amongst innovators and imitator/followers.

References

[1] W.J. Abernathy and J.M. Utterback, Patterns of Industrial Innovation, *Technology Review* 80(7) (January/July 1978) 40–47.

[2] Kim B. Clarke, The Interaction of Design Hierarchies and Market Concepts in Technological Evolution, *Research Policy* 14 (1985) 235–251.

[3] G. Dosi, Technological Paradigms and Technological Trajectories, *Research Policy* 11 (1982) 147–162.

[4] Thomas Kuhn, *The Structure of Scientifc Revolutions*, 2nd ed (University of Chicago Press, Chicago, 1970).

[5] R. Levin, A. Klevorick, N. Nelson, and S. Winter, Survey Research on R&D Appropriability and Technological Opportunity, unpublished manuscript, Yale University, 1984.

[6] Regis McKenna, Market Positioning in High Technology, *California Management Review*, XXVII (3) (spring 1985).

[7] R.E. Miles and C.C. Snow, Network Organizations: New Concepts for New Forms, *California Management Review* (spring 1986) 62–73.

[8] David A. Norman, Impact of Entrepreneurship and Innovations on the Distribution of Personal Computers, in: R. Landau and N. Rosenberg (eds.), *The Positive Sum Strategy* (National Academy Press, Washington, DC, 1986).

[9] D.J. Teece, The Market for Know how and the Efficient International Transfer of Technology, *Annals of the American Academy of Political and Social Science*, November 1981.

[10] P. Temin, Technology, Regulation, and Market Structure in the Modern Pharmaceutical Industry, *The Bell Journal of Economics* (autumn 1979) 429–446.

[5]

RICHARD C. LEVIN
Yale University

ALVIN K. KLEVORICK
Yale University

RICHARD R. NELSON
Columbia University

SIDNEY G. WINTER
Yale University

Appropriating the Returns from Industrial Research and Development

To HAVE the incentive to undertake research and development, a firm must be able to appropriate returns sufficient to make the investment worthwhile. The benefits consumers derive from an innovation, however, are increased if competitors can imitate and improve on the innovation to ensure its availability on favorable terms. Patent law seeks to resolve this tension between incentives for innovation and widespread diffusion of benefits. A patent confers, in theory, perfect appropriability (monopoly of the invention) for a limited time in return for a public

We are grateful for the support of the National Science Foundation and especially to Rolf Piekarz of the NSF's Division of Policy Research and Analysis.

We also wish to thank the 650 respondents to our survey and the R&D executives who helped us pretest it—especially Ralph Gomory, Bruce Hannay, and Lowell Steele. Donald DeLuca, Wendy Horowitz, and other members of the Roper Center for Survey Research helped manage the survey. Robert W. Wilson, and Margaret Blair, Marc Chupka, Emily Lawrance, Constance Helfat, Andrew Joskow, Kathleen Rodenrys, Somi Seong, Andrea Shepard, and Hal Van Gieson also provided valuable assistance.

disclosure that ensures, again in theory, widespread diffusion of benefits when the patent expires.

Previous investigations of the system suggest that patents do not always work in practice as they do in theory.[1] On the one hand, appropriability is not perfect. Many patents can be circumvented; others provide little protection because of stringent legal requirements for proof that they are valid or that they are being infringed. On the other hand, public disclosure does not always ensure ultimate diffusion of an invention on competitive terms. For example, investments to establish the brand name of a patented product may outlive the patent itself.[2] And patents may not always be necessary. Studies of the aircraft and semiconductor industries have shown that gaining lead time and exploiting learning curve advantages are the primary methods of appropriating returns. Other studies have emphasized the importance of complementary investments in marketing and customer service.[3]

Evidence on the nature and strength of conditions for appropriability and on the working of the patent system is, however, scattered and unsystematic. Because imperfect appropriability may lead to underinvestment in new technology, and because technological progress is a primary source of economic growth, it would be useful to have a more comprehensive empirical understanding of appropriability, in particular, to identify those industries and technologies in which patents are effective in preventing competitive imitation of a new process or product. It would also be desirable to know where patents can be profitably licensed. Where patents are not effective, it would be useful to understand why they are not and whether other mechanisms are.

1. F. M. Scherer and others, *Patents and the Corporation: A Report on Industrial Technology under Changing Public Policy*, 2d ed. (privately published, 1959); and C. T. Taylor and Z. A. Silberston, *The Economic Impact of the Patent System: A Study of the British Experience* (Cambridge University Press, 1973).

2. See, for example, Meir Statman, "The Effect of Patent Expiration on the Market Position of Drugs," in Robert B. Helms, ed., *Drugs and Health: Economic Issues and Policy Objectives* (Washington, D.C.: American Enterprise Institute, 1981), pp. 140–51.

3. The importance of lead time and learning curve advantages is documented in Almarin Phillips, *Technology and Market Structure: A Study of the Aircraft Industry* (Lexington Books, 1971); and John E. Tilton, *International Diffusion of Technology: The Case of Semiconductors* (Brookings, 1971). For the importance of marketing and customer service, see Marie-Therese Flaherty, "Field Research on the Link between Technological Innovation and Growth: Evidence from the International Semiconductor Industry," working paper 84-83 (Harvard University, Graduate School of Business Administration, no date).

This paper describes the results of an inquiry into appropriability conditions in more than one hundred manufacturing industries. We discuss how this information has been and might be used to cast light on important issues in the economics of innovation and public policy. Our data, derived from a survey of high-level R&D executives, are informed opinions about an industry's technological and economic environment rather than quantitative measures of inputs and outputs.

Although our use of semantic scales to assess, for example, the effectiveness of alternative means of appropriation introduces considerable measurement error, more readily quantifiable proxies would probably not serve as well. Remarkable progress has been made toward developing a methodology to estimate the economic value of patents.[4] But suitable data are as yet unavailable in the United States, and European data lack sufficiently reliable detail to support inferences about interindustry differences in the value of patents. Our judgment was that asking knowledgeable respondents about the effectiveness of patents and alternative means of appropriation was at least as likely to produce useful answers as asking for quantitative estimates of the economic value of a typical patent.

We have taken considerable care to establish the robustness of our findings in the presence of possibly substantial measurement error, but ultimately the value of the data will depend on their contribution to better empirical understanding of technological change and more discriminating discussion of public policy. To view the empirical contribution of the data from the simplest perspective, consider their potential for improving the quality of research that uses patent counts to measure innovative activity.[5] This line of inquiry has shown, among other results, that industries vary significantly in the average number of patents generated by each dollar of R&D investment.[6] Our findings on industry

4. See, especially, Ariel Pakes, "Patents as Options: Some Estimates of the Value of Holding European Patent Stocks," *Econometrica*, vol. 54 (July 1986), pp. 755–84.

5. For a summary of the best of this work, see Zvi Griliches, Ariel Pakes, and Bronwyn H. Hall, "The Value of Patents as Indicators of Inventive Activity," working paper 2083 (Cambridge, Mass.: National Bureau of Economic Research, November 1986). For other perspectives on the usefulness of patent data, see the special issue of *Research Policy*, vol. 16 (August 1987).

6. F. M. Scherer, "The Propensity to Patent," *International Journal of Industrial Organization*, vol. 1 (March 1983), pp. 107–28; and John Bound and others, "Who Does R&D and Who Patents?" in Zvi Griliches, ed., *R&D, Patents and Productivity* (University of Chicago Press for National Bureau of Economic Research, 1984), pp. 21–54.

differences in patent effectiveness may help explain this variation in the apparent productivity of R&D.

More fundamentally, large and persistent interindustry differences in R&D investment and innovative performance have resisted satisfactory explanation, in part for lack of data that adequately represent the theoretically important concepts of appropriability and technological opportunity. Promising but ultimately unsatisfactory results have been obtained in exploratory work that used crude proxy variables and econometric ingenuity to capture the influence of appropriability and opportunity conditions.[7] Our desire to provide a stronger basis for this line of inquiry was a prominent motive for our survey research and helped to shape its design.

Finally, gathering better information on the nature and strength of appropriability is particularly timely in view of the prominence of current debates on the adequacy of laws and institutions to protect intellectual property. One impetus for change has been the need to clarify and perhaps strengthen the system of property rights at various new frontiers of technology. Thus, for example, recent legislation has adapted copyright law to protect the rights of the creator of new computer software, a new legal framework has been constructed to protect intellectual property embodied in semiconductor chip designs, and important court decisions and administrative actions have shaped the development of a property rights system in biotechnology.[8]

Another spur to change has been the need to resolve conflicts between the aims of social regulation and the exercise of intellectual property rights. For example, the Drug Price Competition and Patent Term Restoration Act of 1984 extended patent lives of pharmaceuticals to compensate for regulatory requirements that delay the introduction of new drugs.

7. Richard C. Levin, "Toward an Empirical Model of Schumpeterian Competition," working paper 43 (Yale University, School of Organization and Management, 1981); Richard C. Levin and Peter C. Reiss, "Tests of a Schumpeterian Model of R&D and Market Structure," in Griliches, ed., *R&D, Patents and Productivity*, pp. 175–204; and Ariel Pakes and Mark Schankerman, "An Exploration into the Determinants of Research Intensity," in Griliches, ed., *R&D, Patents and Productivity*, pp. 209–32.

8. See Computer Software Act of 1980; Semiconductor Chip Protection Act of 1984; *Diamond* v. *Chakrabarty*, 447 U.S. 305 (1980), holding that plant and animal life is patentable under U.S. patent law; and D. J. Quigg, memorandum of April 7, 1987, explaining the policies of the U.S. Patent and Trademark Office concerning applications to patent life forms.

Intellectual property rights also figure prominently among policy issues milling under the banner of competitiveness. Recent annual reports of the U.S. trade representative have focused on the difficulties U.S. manufacturers encounter in protecting intellectual property rights in foreign markets. The trade bill passed in 1987 by the House of Representatives contains several provisions that increase the scope of protection and the opportunities for relief available to U.S. manufacturers confronted with imports that infringe these rights.[9] Proposed antitrust legislation, motivated by a concern that courts have kept inventors from reaping rewards that patent laws are intended to provide, stipulates that patent license agreements and similar contracts relating to use of intellectual property "shall not be deemed illegal per se under any of the antitrust laws."[10]

To the extent that all this activity attempts to rectify obvious inadequacies in existing institutions, the case for reform appears strong and straightforward. It is easy to deplore the blatant copying of innovative integrated circuit designs, the importation of "knock off" copies of trademarked or patented U.S. products, and the piracy of copyrighted written matter and audio and video cassettes. But reforms may yield unintended consequences. In its simplest form, this concern translates into wariness about Trojan horses: provisions brought into the law by the rhetorical tug of "competitiveness" and "intellectual property" may harbor instruments of protectionism and price fixing. Other potential consequences are subtler but no less important. For example, seemingly uniform adjustments of intellectual property, antitrust, or trade law may affect some industries quite differently than others.

And it should not be taken for granted that more appropriability is better, that better protection necessarily leads to more innovation, which yields better economic performance—higher standards of living, better competitiveness, and so on. Better protection may yield more innovation at the cost of incrementally increasing resources devoted to producing the innovation: the larger prize may merely encourage duplicative private effort to capture it.[11] Alternatively, better protection may induce inno-

9. See H.R. 3, the Omnibus Trade and Competitiveness Reform Act of 1987, which is currently under consideration by a House-Senate conference committee.

10. H.R. 557 and S. 438, 100 Cong., 1 sess.

11. This is the "free access" externality, first emphasized in the context of innovation in Yoram Barzel, "Optimal Timing of Innovations," *Review of Economics and Statistics*, vol. 50 (1968), pp. 348–55. For a survey of the literature on patent races, see Jennifer

vation of the wrong kind, or it may buy the innovation by further delaying access to it on competitive terms.[12]

The premise that stronger protection will always improve the incentives to innovate is also open to challenge. Unimpeded diffusion of existing technology is immediately beneficial not only for consumers but also for those who would improve that technology. Because technological advance is often an interactive, cumulative process, strong protection of individual achievements may slow the general advance. This would not occur in a hypothetical world without transaction costs, in which efficient contracts to share information would be made. In reality, however, markets for rights to information are subject to major transactional hazards, and strong protection of a key innovation may preclude competitors from making socially beneficial innovations. The semiconductor industry of the 1950s and 1960s provides an excellent example of rapid progress in a cumulative technology that might have been impossible under a regime that strongly protected intellectual property.[13]

The remainder of this paper discusses our survey instrument, the construction of the sample, and the interpretation of the data, then turns to our findings concerning the effectiveness of patents and other means of appropriating the returns from R&D. The results of related work that employs the survey data to reexamine central questions in the empirical literature on R&D are summarized, and we discuss how our findings might contribute to a more discriminating discussion of patent law, antitrust law, and trade policy.

Questionnaire Design and Survey Methods

The content of our questionnaire was shaped with guidance from the conceptual literature on technological change, empirical literature on

Reinganum, "The Timing of Innovation: Research, Development and Diffusion," in Richard Schmalensee and Robert Willig, ed., *Handbook of Industrial Organization* (North-Holland, 1988).

12. Richard R. Nelson, "Assessing Private Enterprise: An Exegesis of Tangled Doctrine," *Bell Journal of Economics*, vol. 12 (Spring 1981), pp. 93–111; and William D. Nordhaus, *Invention, Growth, and Welfare: A Theoretical Treatment of Technological Change* (MIT Press, 1969).

13. Richard C. Levin, "The Semiconductor Industry," in Richard R. Nelson, ed., *Government and Technical Progress: A Cross-Industry Analysis* (Pergamon Press, 1982), pp. 9–100.

the economic impact of the patent system, the work of Mansfield and his associates on imitation costs, and numerous case studies.[14] The questionnaire was aimed at high-level R&D managers with knowledge of both the relevant technology and market conditions. To check the interpretability of the questions and the likely validity and reliability of the responses, we pretested the questionnaire with twelve managers representing diverse businesses.[15]

To understand how appropriability differs across industries, we asked each respondent to report typical experiences or central tendencies within a particular industry. Respondents were thus treated as informed observers of a line of business rather than as representatives of a single firm, an approach that encouraged cooperation (they were not placed in the position of possibly divulging practices or policies of their own firms), but led inevitably to heterogeneity in the responses within a given industry.

The questionnaire contained four parts. Parts 1 and 2 concerned appropriability; parts 3 and 4 concerned technological opportunity and technological advance. Questions in part 1 asked about the effectiveness of alternative means of protecting the competitive advantages of R&D, limits on the effectiveness of patents, and ways of acquiring knowledge of a competitors' technology. Part 2 asked about the cost and time required to imitate innovations of rivals; we distinguished process from

14. Among the sources of ideas for the questions are Paul Allan David, *Technical Choice, Innovation and Economic Growth: Essays on American and British Experience in the Nineteenth Century* (Cambridge University Press, 1975); Richard R. Nelson and Sidney G. Winter, "In Search of Useful Theory of Innovation," *Research Policy*, vol. 6 (Winter 1977), pp. 36–76; Nathan Rosenberg, "Science, Invention and Economic Growth," *Economic Journal*, vol. 84 (March 1974), pp. 90–108; and Devandra Sahal, *Patterns of Technological Innovation* (Addison-Wesley, 1981). For empirical literature on the economic effects of the patent system, see Scherer and others, *Patents and the Corporation;* and Taylor and Silberston, *The Economic Impact of the Patent System*. For imitation costs, see Edwin Mansfield, Mark Schwartz, and Samuel Wagner, "Imitation Costs and Patents: An Empirical Study," *Economic Journal*, vol. 91 (December 1981), pp. 907–18.

15. These managers had experience in communications equipment, industrial inorganic chemicals, metal cutting machine tools, shoe machinery, household electrical appliances, processed foods, computing equipment, semiconductors, copper smelting and refining, radio and TV sets, and industrial organic chemicals. They were asked to complete the questionnaire for a specific line of business, but to keep in mind the suitability of the questions for other lines of business with which they were familiar. After completing the questionnaire, they were interviewed face-to-face or by telephone. Interviews typically lasted one-half hour or more, and each question was discussed to eliminate sources of ambiguity.

product innovations, major from typical, and patented from unpatented.[16] Part 3 explored the links between an industry's technology and other sources of technological contribution. We asked about the importance of scientific research in general and university-based research in particular. We also asked about the extent to which interindustry spillovers are an important source of technological opportunity. Part 4 asked some broad questions about the pace and character of technological advance.[17] This paper analyzes responses to the questions in parts 1 and 2.[18]

SAMPLE CONSTRUCTION

As a sampling frame, we used the lines of business defined by the Federal Trade Commission. In the manufacturing sector, these chiefly correspond to four-digit SIC industries, although some are defined as groups of four-digit or even three-digit industries. The FTC lines provide the most disaggregated level at which data on R&D expenditures are available. An additional consideration was that F. M. Scherer's technology flow matrix, which classifies patents by industry of origin and industry of use, was also constructed at this level of aggregation.[19]

Ultimately, we received responses from 650 individuals representing 130 lines of business, with ten or more responses from eighteen industries and five to nine from twenty-seven industries. The sample was reason-

16. The questions were similar to those in Mansfield, Schwartz, and Wagner, "Imitation Costs and Patents," but covered typical rather than specific innovations. Our industry sample was also broader.

17. One objective was to examine "natural trajectories" of the sort described in Nelson and Winter, "In Search of Useful Theory," p. 56.

18. Data from responses to questions in parts 3 and 4 have been used in Richard C. Levin, Wesley M. Cohen, and David C. Mowery, "R&D Appropriability, Opportunity, and Market Structure: New Evidence on Some Schumpeterian Hypotheses," *American Economic Review*, vol. 75 (May 1985, *Papers and Proceedings, 1984*), pp. 20–24; Cohen, Levin, and Mowery, "Firm Size and R&D Intensity: A Re-examination," *Journal of Industrial Economics*, vol. 35 (June 1987), pp. 543–65; and Richard C. Levin and Peter C. Reiss, "Cost-Reducing and Demand-Creating R&D with Spillovers" (Stanford University, Graduate School of Business, 1986). See Richard R. Nelson, "Institutions Supporting Technical Advance in Industry," *American Economic Review*, vol. 76 (May 1986, *Papers and Proceedings, 1985*), pp. 186–89, for findings on the importance of external sources of technological knowledge.

19. F. M. Scherer, "Inter-industry Technology Flows in the United States," *Research Policy*, vol. 11 (August 1982), pp. 227–45.

ably representative of firms performing R&D, though the exclusion of those without publicly traded securities undoubtedly means that small start-up ventures, important sources of innovation, were underrepresented. The number of respondents in a line of business was positively correlated with the line's R&D spending, sales volume, and R&D intensity. The number of respondents did not increase in strict proportion to the level of industry R&D or sales, but the rate of response within a line of business was not significantly correlated with industry R&D spending, sales, or R&D intensity. The Appendix presents further details of sample construction.

METHODOLOGICAL ISSUES

Given our interest in identifying differences in the appropriability of R&D, it is reassuring that analysis of variance confirmed the presence of significant interindustry variation in the responses to most questionnaire items.[20] There was, however, also substantial intraindustry variation in the responses.

There are several potential sources of intraindustry heterogeneity in the responses to any given question. First, the lines of business as defined by the FTC may be objectively heterogeneous in their products and technologies. For example, if two firms classified as manufacturers of industrial inorganic chemicals produce different products using different technologies, they might differ markedly in their perception of the effectiveness of patents or the time required for imitation in their "industry." To eliminate this source of heterogeneity, we asked respondents to identify two major innovations—a process and a product—within their industries during the past ten to fifteen years. For most industries with ten or more respondents, more than half the respondents agreed on at least one such innovation. We thus believe it unlikely that overly aggregated industry definition was a major source of intraindustry heterogeneity.[21]

20. Interindustry differences are significant at the 0.05 level for approximately 60 percent of the questions in parts 1 and 2 of the questionnaire. If a higher level of aggregation is used to measure industry effects, such as the level at which the National Science Foundation reports R&D spending (a hybrid of two- and three-digit level industries), interindustry differences are significant at the 0.05 level for 70 percent of the questions.

21. Heterogeneity, as anthropologists have long insisted, is, however, in the eye of the

A respondent's perception of the central tendencies within an industry may also be affected by his firm's policies or strategies. Respondents in the same line of business may thus have different perceptions of the common technological environment that they were asked to characterize. A two-way analysis of variance of the responses on the effectiveness of patents, for example, revealed that both firm and industry effects are statistically significant. A representative multi-industry firm, however, tends to be involved in technologically related industries, and thus what appear to be effects attributable to the firm in the data may simply reflect the correlation in responses from related industries.

The third, and probably most important, source of intraindustry heterogeneity is the inherently subjective nature of the semantic scales used in the survey. Most answers were reported on a seven-point Likert scale. The effectiveness of patents in preventing duplication was, for instance, evaluated on a scale ranging from "not at all effective" to "very effective." There is no natural or objective anchor for such evaluative ratings. Individuals may perceive the same environment but simply use the scale differently. Some might systematically favor high scores; others might concentrate responses in the center of the scale; still others might frequently use extreme values.

The numerous techniques available to control for differences among respondents in means and variances generally require abandoning one or more dimensions along which the data might be informative. For example, we were interested in interindustry comparisons of answers to a single question; controlling for fixed effects among respondents would vitiate such comparisons, since we expected a respondent's mean score over all questions to depend on his industry. Standardizing the variance of each respondent's answers raised similar problems: the distribution of "correct" responses was unknown and it almost certainly differed systematically among industries. Rather than impose an arbitrary standardization, therefore, we examined the results for each group of questions using a variety of techniques and perspectives to assess the

beholder. One R&D manager, asked to inform us about the air and gas compressor industry, inquired whether we were interested in large, medium, or small compressors. In his view the technologies were fundamentally different. We asked him to note on his questionnaire where the answers to our questions differed across these size categories. The booklet he returned contained no such notation.

robustness of our principal conclusions. There was undeniably much noise in the data, but several important signals were robust to alternative weightings of the observations, alternative partitions of the sample, and the use of alternative summary statistics.[22]

We sidestepped one methodological difficulty by treating ratings along a seven-point semantic continuum as if they were interval data. The data were, of course, more properly to be regarded as ordinal. It would have been straightforward to treat them as ordinal if we had been interested only in interindustry comparisons of responses to a single question. We also sought, however, to make comparisons among questions (for example, are patents more or less effective than secrecy in protecting process innovations from duplication?), and we therefore treated the data as if they were interval.[23]

One additional methodological concern was whether our level of industry aggregation was appropriate for the problems being studied. The FTC line-of-business level was chosen to facilitate merging the data with disaggregated R&D data and Scherer's classification of patents by industries of origin and use. Our analysis indicated, however, that most of the interesting interindustry distinctions among the 130 lines defined at the FTC level were robust to an aggregation of the data into the 25 industry groups used by the National Science Foundation in its annual survey of R&D spending and employment patterns.

Patents and Other Means of Appropriation

Table 1 shows the pattern of responses, based on a seven-point scale, to questions on the effectiveness of alternative means of capturing and protecting the competitive advantages of new or improved processes and products. The first two columns report the mean response for the

22. One notable consequence arising from the measurement error in the data was that industry mean responses from lines of business with only one or two respondents tended to be disproportionately located near the extremes of the distribution of mean responses to any given question. Most conclusions based on the full sample of 130 lines, and virtually all those emphasized in this paper, were replicated in the smaller sample of 75 lines that had more than two respondents.

23. We designed the questionnaire to ensure that cross-question comparisons would arise naturally in the minds of the respondents. The items were arranged in blocks, with each item in a block rated on the same semantic scale.

Table 1. Effectiveness of Alternative Means of Protecting the Competitive Advantages of New or Improved Processes and Products[a]

	Overall sample means		Distribution of industry means[b]	
Method of appropriation	*Processes*	*Products*	*Processes*	*Products*
Patents to prevent duplication	3.52 (0.06)	4.33 (0.07)	2.6–4.0[c]	3.0–5.0[c]
Patents to secure royalty income	3.31 (0.06)	3.75 (0.07)	2.3–4.0[c]	2.7–4.8[c]
Secrecy	4.31 (0.07)	3.57 (0.06)	3.3–5.0	2.7–4.1
Lead time	5.11 (0.05)	5.41 (0.05)	4.3–5.9[c]	4.8–6.0[c]
Moving quickly down the learning curve	5.02 (0.05)	5.09 (0.05)	4.5–5.7	4.4–5.8
Sales or service efforts	4.55 (0.07)	5.59 (0.05)	3.7–5.5	5.0–6.1

Source: Authors' calculations.
a. Range: 1 = not at all effective; 7 = very effective. Standard errors in parentheses.
b. From the upper bound of the lowest quintile of industries to the lower bound of the highest quintile.
c. Differences in means significant at the .01 level.

entire sample of 650 respondents to each question, as well as the standard error of each estimated mean. These statistics, of course, give equal weight to each respondent and consequently weight each industry in proportion to its number of respondents. The overall pattern across questions, however, is robust to the use of alternative summary statistics, such as the mean of industry means or the median of industry means. This is apparent in columns 3 and 4, which summarize the distribution of industry mean responses to each question. Each pair of numbers represents the range of industry means from the upper bound of the lowest quintile to the lower bound of the highest quintile of industries: 20 percent of the 130 industries had mean responses at or below the bottom of the range indicated for each question, and 20 percent had mean responses at or above the top of the range. Mean responses for the remaining 60 percent (or 78 industries) fell within the reported range.

The picture is striking. For new processes (columns 1 and 3), patents were generally rated the least effective of the mechanisms of appropriation: only 20 percent of the lines of business surveyed rated process patent effectiveness in excess of 4.0. Eighty percent scored the effectiveness of lead time and learning curve advantages on new processes in excess of 4.3. Secrecy, though not considered as effective as lead time and learning advantages, was still considered more effective than patents in protecting processes.

Patents for products were typically considered more effective than those for processes, and secrecy was considered less effective in protecting products than processes. Generally, lead time, learning curves, and sales or service efforts were regarded as substantially more effective than patents in protecting products. Eighty percent of the sample businesses rated the effectiveness of sales and service efforts above 5.0, but only 20 percent considered product patents this effective.[24]

The tendency to regard secrecy as more effective than process patents but less effective than product patents probably reflects the greater ease and desirability of maintaining secrecy about process technology. Firms may sometimes refrain from patenting processes to avoid disclosing either the fact or the details of an innovation.[25] But firms have every incentive to advertise the advantages of new or improved products and to get them into the hands of customers, thereby facilitating direct observation of the product and the technology it embodies. Maintaining secrecy about product innovations is thus likely to be both difficult and undesirable.

Respondents also tended to regard patents to prevent duplication as more effective than patents to secure royalty income. This finding was consistent with the view that licensing arrangements are beset with transactional difficulties.

Only 3 of 130 lines of business rated process patents higher than five on a seven-point scale of effectiveness in preventing duplication. Two of these were concrete and primary copper; the other had only a single respondent.[26] Only 5 of 130 industries rated product patents to prevent

24. This view of the efficacy of sales and service efforts is consistent with the emphasis given to investment in "cospecialized assets" as a means of appropriation in David J. Teece, "Profiting from Technological Innovation: Implications for Integration, Collaboration, Licensing and Public Policy," *Research Policy,* vol. 15 (December 1986), pp. 285–305.

25. See Ignatius Horstmann, Glenn M. MacDonald, and Alan Slivinski, "Patents as Information Transfer Mechanisms: To Patent or (Maybe) Not to Patent," *Journal of Political Economy,* vol. 93 (October 1985), pp. 837–58, for a theoretical treatment of the issue.

26. To preserve confidentiality, we do not identify any industry in which there was only one response. Hereafter, we refer to such cases as singletons.

duplication higher than six points. Two of these were singletons; the other three were drugs, pesticides, and industrial organic chemicals. Twenty other lines rated product patents between five and six. Of those with more than two responses, almost all fell neatly into chemical products (including inorganic chemicals, plastic materials, synthetic fibers, synthetic rubber, and glass) or relatively uncomplicated mechanical equipment (air and gas compressors, power-driven hand tools, and oilfield machinery). The only anomalies were roasted coffee and products of steel rolling and finishing mills.

Table 2 shows additional industry-level detail—the mean rating given for the effectiveness of patents in preventing duplication in eighteen industries with ten or more respondents. These industries tend to be much more research-intensive than the sample average, yet the pattern of interindustry variation was similar to that in the full sample. Except for petroleum refining, product patents were considered more effective than process patents. Only four chemical industries (drugs, plastic materials, inorganic chemicals, and organic chemicals) and petroleum refining rated process patent effectiveness higher than four on a seven-point scale, and only these four chemical industries and steel mills rated product patents higher than five.[27]

The data on these eighteen most heavily sampled industries help to establish the robustness of our conclusion about the limited effectiveness of patents as a means of appropriation. In none did a majority of respondents rate patents—either to prevent duplication or to secure royalty income—as more effective than the most highly rated of the other four means of appropriating returns from new processes, although in drugs and petroleum refining a majority regarded process patents as at least the equal of the most effective alternative mechanism of appropriation. In only one industry, drugs, were product patents regarded by a majority of respondents as strictly more effective than other means of appropriation.[28] In three others—organic chemicals, plastic materials,

27. The same pattern appears when the survey data are aggregated up to the level (roughly two and one-half digit) at which the National Science Foundation reports detailed data on the extent and composition of research and development expenditures. Of the twenty-five industries into which the manufacturing sector is divided, only industrial chemicals, drugs, and petroleum refining rated process patents higher than four points, and only industrial chemicals and drugs rated product patents higher than five.

28. Our results were reinforced by Edwin Mansfield's finding that among the twelve broadly defined industries he studied only in the drug industry were patents considered

Table 2. Effectiveness of Process and Product Patents in Industries with Ten or More Survey Responses

	Process patents		Product patents	
Industry	*Mean*	*Standard error*	*Mean*	*Standard error*
Pulp, paper, and paperboard	2.6	0.3	3.3	0.4
Cosmetics	2.9	0.3	4.1	0.4
Inorganic chemicals	4.6	0.4	5.2	0.3
Organic chemicals	4.1	0.3	6.1	0.2
Drugs	4.9	0.3	6.5	0.1
Plastic materials	4.6	0.3	5.4	0.3
Plastic products	3.2	0.3	4.9	0.3
Petroleum refining	4.9	0.4	4.3	0.4
Steel mill products	3.5	0.7	5.1	0.6
Pumps and pumping equipment	3.2	0.4	4.4	0.5
Motors, generators, and controls	2.7	0.3	3.5	0.5
Computers	3.3	0.4	3.4	0.4
Communications equipment	3.1	0.3	3.6	0.3
Semiconductors	3.2	0.4	4.5	0.4
Motor vehicle parts	3.7	0.4	4.5	0.4
Aircraft and parts	3.1	0.5	3.8	0.4
Measuring devices	3.6	0.3	3.9	0.3
Medical instruments	3.2	0.4	4.7	0.4
Full sample	3.5	0.06	4.3	0.07

Source: Authors' calculations. Mean score on a scale of 1 to 7.

and steel mill products—most respondents rated patents as no less effective than the best alternative.

The exclusion from our sample of firms that offered no publicly traded securities may have biased our findings. For small, start-up ventures, patents may be a relatively effective means of appropriating R&D returns, in part because some other means, such as investment in complementary sales and service efforts, may not be feasible. The patents held by a small, technologically oriented firm may be its most marketable asset. Although our respondents were asked to describe the typical experience of firms in their industries, they may well have overlooked aspects of appropriability that are particularly relevant for new firms.

essential to developing and marketing most inventions. Chemicals was the only other industry that considered patents essential for as many as 30 percent of inventions. See "Patents and Innovation: An Empirical Study," *Management Science*, vol. 32 (February 1986), pp. 173–81.

The most probable explanation for the robust finding that patents are particularly effective in chemical industries is that comparatively clear standards can be applied to assess a chemical patent's validity and to defend against infringement. The uniqueness of a specific molecule is more easily demonstrated than the novelty of, for example, a new component of a complex electrical or mechanical system. Similarly, it is easy to determine whether an allegedly infringing molecule is physically identical to a patented molecule; it is more difficult to determine whether comparable components of two complex systems "do the same work in substantially the same way." To the extent that very simple mechanical inventions approximate molecules in their discreteness and easy differentiability, it is understandable that industries producing such machinery rank just after chemical industries in the perceived effectiveness of patent protection.

The perceived ineffectiveness of patents in most industries raises the question of why firms use them. Further work is needed here, but we offer some speculations informed by the comments of our pretest subjects and by several survey respondents at a conference we held to report preliminary findings. These executives identified two motives for patenting that have little connection with appropriating returns from investment. One is to measure the performance of R&D employees, which is a significant problem because these workers are typically engaged in team production. Legal standards for identifying inventors on a patent application are, however, reasonably rigorous. The second motive is to gain access to certain foreign markets. Some developing countries require, as a condition of entry, that U.S. firms license technology to a host-country firm, and some patents are filed primarily to permit such licensing.[29]

29. Yet another motive discussed in the literature is to gain strategic advantage in negotiation. In the semiconductor industry, for instance, the cumulative nature of the technology makes it difficult to participate legally without access to the patents of numerous firms. In consequence, there is widespread cross-licensing. Established firms, however, rarely license a new entrant until it has established a significant position in the market. As a defense against infringement suits, a prudent new entrant will establish a patent portfolio of its own, thus compelling established firms to negotiate cross-license agreements. See Eric von Hippel, "Appropriability of Innovation Benefit as a Predictor of the Source of Innovation," *Research Policy*, vol. 11 (January 1982), pp. 95–115; and Levin, "Semiconductor Industry," pp. 80–81.

Conditions Affecting Appropriability

Thus far we have focused on the overall strength of various mechanisms of appropriation and on interindustry variations in the effectiveness of patents. The patterns of covariation in the responses, however, suggested that interindustry differences in conditions affecting appropriability might be summarized by a limited number of factors. Moreover, the clear indications that patents are effective in only a few industries suggested that it might be fruitful to classify industries into clusters distinguishable by a primary means of appropriation and perhaps by the overall ease of appropriating returns. Such clusters could prove useful in examining links between appropriability conditions and measures of R&D, innovation, and productivity growth.

Correlations among responses to questions on the effectiveness of alternative means of appropriation revealed some interesting patterns.[30] When patents effectively prevent competitors from duplicating processes and products, they tend also to be effective in securing royalty income. But neither form of effectiveness was strongly correlated with the effectiveness of other means of appropriation. For processes, there was a strong connection among three other mechanisms: lead time, learning curve advantages, and secrecy. For products, superior sales and service efforts were strongly linked to lead time and learning advantages, though not to secrecy.

The correlations suggested that the mechanisms of appropriation may be reduced to two dimensions: one associated with the use of patents, the other related to secrecy, lead time, and learning curve advantages. For product innovations, sales and service efforts may be involved in the second of these dimensions. We investigated this possibility by reducing the data to principal components and employing a variety of factor-analytic techniques. Principal factor analysis and several methods

30. Simple correlation coefficients were calculated using the individual respondent and industry mean responses as the units of observation. Correlations among industry means for the entire sample of 130 lines of business were qualitatively similar to those obtained when the sample was restricted to those with more than two responses. These and other correlation matrices discussed in this paper are available from the authors upon request.

Table 3. Principal Components Analysis of Methods of Appropriation

	Processes and products separately		*Processes and products together*	
Method of appropriation	*Coefficients of 1st principal component*	*Coefficients of 2d principal component*	*Coefficients of 1st principal component*	*Coefficients of 2d principal component*
New Processes				
Patents to prevent duplication	.04	.86	.01	.73
Patents to secure royalties	.12	.86	.08	.78
Secrecy	.59	−.12	.54	.04
Lead time	.84	−.09	.79	−.04
Moving down the learning curve	.84	−.05	.80	−.04
Sales and service efforts	.51	.11	.45	−.06
Cumulative variance explained	.34	.59	n.a.	n.a.
New Products				
Patents to prevent duplication	.06	.87	.06	.73
Patents to secure royalties	.06	.87	.07	.80
Secrecy	.51	.01	.51	.06
Lead time	.84	.00	.79	−.03
Moving down the learning curve	.84	−.07	.82	−.04
Sales and service efforts	.69	−.09	.62	−.11
Cumulative variance explained	.36	.61	.31	.50

Source: Authors' calculations.

of rotation did little to alter the picture presented by the principal components, which are shown in table 3.[31]

The first two columns of the table show the weights associated with the first two principal components when the six questions relating to process appropriability are analyzed separately from the six questions relating to product appropriability. The next two columns report the results of a principal components analysis on the entire set of twelve questions. With both approaches, the first principal component gives near-zero weight to the two patent-related methods of appropriation and heavy weight to the other mechanisms. The weighting is reversed for the second principal component. Thus the first two principal components (and, in the factor analysis, the first two factors) are readily interpreted, respectively, as nonpatent- and patent-related dimensions of appropriability. Despite this clear interpretation, the data do not reduce very

31. The results reported in table 3 are based on a principal components analysis undertaken at the level of individual responses. An analysis at the level of industry mean responses produced similar results.

Table 4. Cluster Analysis of Mechanisms of Appropriation

	Cluster		
Method of appropriation	*1*	*2*	*3*
New Processes			
Number of industries	38	67	25
Mean score			
Patents to prevent duplication	3.1	3.0	4.7
Patents to secure royalties	2.9	2.9	4.8
Secrecy	2.8	4.6	4.7
Lead time	4.2	5.4	5.6
Learning curves	4.3	5.3	5.1
Superior sales or service	4.7	4.5	4.9
New Products			
Number of industries	20	68	42
Mean score			
Patents to prevent duplication	3.1	3.8	5.3
Patents to secure royalties	3.2	3.1	5.0
Secrecy	2.6	3.5	4.0
Lead time	4.0	5.6	5.7
Learning curves	4.2	5.3	5.2
Superior sales or service	5.2	5.7	5.6

Source: Authors' calculations.

satisfactorily to just two dimensions. As table 3 indicates, when the process and product questions are analyzed separately, the first two components explain only 60 percent of the variance in the responses to six questions, and when the two sets of questions are combined, two components explain only 50 percent of the variance.

Our interpretation that the means of appropriation can be grouped into patent and nonpatent mechanisms was nonetheless reinforced by a cluster analysis that classified industries according to mean responses to the relevant questions. The best clustering results were achieved by dividing the industries into three groups, as shown in table 4. Industries assigned to cluster 1 tended to have relatively low scores for all mechanisms of appropriation. Sales and service effort was the most highly rated mechanism and was, in fact, regarded as reasonably effective in capturing returns from new products. Industries in cluster 2 rated lead time and learning curves as relatively effective, but not patents. Secrecy was important in appropriating process returns, and sales and service efforts complemented lead time and learning advantages for products. Only for cluster 3 were product and process patents deemed effective, but still the effectiveness of lead time and learning was no lower than for the industries in cluster 2. Those few industries in which patents were rated as more effective than other mechanisms were all in the third cluster.

The cluster analysis suggested that there was a group of industries in which no appropriation mechanism was particularly effective. As an alternative approach to identifying settings with low appropriability, we considered the maximum score an industry assigned to any of the six mechanisms on the questionnaire. Only 11 of the 130 failed to rate at least one means of appropriating returns from product innovation higher than five on the seven-point scale. The industries in this group with more than two responses were all drawn from the food products and metal-working sectors: milk, meat products, iron and steel foundries, boiler shops, and screw machine products (nuts, bolts, and screws). Many more industries (34 of 130) rated no means of appropriating process returns higher than five. This group contained all the industries (except milk) that ranked low on product appropriability but was otherwise a diverse lot. The heaviest concentration was in fabricated metals and machinery. But several chemical industries were also represented, including the three industries in which product patents were viewed as most effective—organic chemicals, pesticides, and drugs.

The urge to find patterns in the data should not be carried too far. The associations among mechanisms of appropriation revealed by the correlation, principal components, and cluster analyses are suggestive, but there is substantial heterogeneity in the underlying data. As noted, the first two principal components, though readily interpretable, explained an unsatisfactory fraction of the overall variance. A similar lack of good fit characterized the cluster analyses of process and product appropriability. Despite the fairly clear interpretation that could be given to each cluster, the variance within the clusters was almost twice that between clusters.

Limitations on Effectiveness of Patents

To understand why patent protection might be weak in some industries, we asked respondents to rate the importance of possible limitations on patent effectiveness. Table 5 summarizes the responses. The ability

Table 5. Limitations on Effectiveness of Patents for New or Improved Processes and Products[a]

Limitation	*Overall sample means* *Processes*	*Products*	*Distribution of industry means[b]* *Processes*	*Products*
New processes or products patentable	4.32 (0.07)	3.75 (0.07)	3.6–5.4[c]	2.8–4.8
Patents unlikely to be valid if challenged	4.18 (0.06)	3.92 (0.07)	3.5–5.0[c]	3.0–5.0[c]
Firms do not enforce patents	4.29 (0.06)	3.84 (0.07)	3.5–5.0[c]	3.0–4.8[c]
Competitors legally "invent around" patents	5.49 (0.05)	5.09 (0.06)	4.9–6.0	4.4–5.9[d]
Technology moving so fast that patents are irrelevant	3.40 (0.07)	3.34 (0.07)	2.0–4.3[d]	2.0–4.0[d]
Patent documents disclose too much information	4.19 (0.07)	3.65 (0.07)	3.2–5.0	2.8–4.5[c]
Licensing required by court decisions	2.96 (0.06)	2.79 (0.06)	2.0–3.8	2.0–3.3
Firms participate in cross-licensing agreements with competitors	3.08 (0.06)	2.93 (0.06)	2.2–3.9[d]	2.1–3.9[d]

Source: Authors' calculations.
a. Range: 1 = not an important limitation; 7 = very important limitation. Standard errors in parentheses.
b. From the upper bound of the lowest quintile of industries to the lower bound of the highest quintile.
c. Interindustry differences significant at the .10 level.
d. Interindustry differences significant at the .01 level.

of competitors to "invent around" both process and product patents was rated higher than five on a seven-point scale of importance by 60 percent of the responding industries. Only one other constraint—the lack of ready patentability for new processes—was rated this important by more than 20 percent. Limitations on patents were generally considered more severe for processes than for products, which was consistent with our finding that product patents tend to be more effective than process patents. In particular, the lack of patentability was more serious for processes than for products, and so was the disclosure of information through patent documents.[32]

32. Additional evidence of the internal consistency of the survey results was provided by the pattern of negative correlation between responses concerning limitations on patent effectiveness and responses concerning the effectiveness of patents. Using either individual respondents or industry means as the unit of observation, all such correlation coefficients were negative except in the case of compulsory licensing. Most correlations were significant at the .01 level.

The responses concerning limits on patent effectiveness may illuminate and focus policy discussion. In recent years there has been considerable interest in making patent protection more effective. One initiative has been to make the legal requirements for a valid patent claim less stringent.[33] Another has been to vacate court decrees that compel licensing. Our data identified industries in which stringent requirements for patent validity or compulsory licensing were perceived as important limitations on the usefulness of patents in appropriating returns.

Respondents from twenty-two lines of business, mostly in the food processing and fabricated metals sectors, considered the likely inability to withstand challenges to validity as significantly limiting the effectiveness of process patents (scoring the importance higher than five on a seven-point scale); for fourteen of these industries the mean response was six or higher on the scale. This group and the nineteen industries citing invalidity as a constraint on the effectiveness of product patents (again assigning a score higher than five) overlapped considerably. Further investigation would be required to determine just why these two sectors appear to have difficulty establishing valid claims. Perhaps because they are mature industries, opportunities may be limited or novelty may be difficult to achieve or simply difficult to prove.

Compulsory licensing was rarely judged a significant limit on the effectiveness of patents. Only one industry with one respondent rated this constraint higher than five on the scale for products, and only six cited compulsory licensing of process patents as of comparable importance. Two of these industries were not singletons—metal containers and electron tubes. Compulsory licensing decrees were thus perceived as important in only a small subset of the industries that F. M. Scherer indicated were subject to such decrees.[34] The overall lack of impact from compulsory licensing requirements was consistent with Scherer's finding that they did not discourage R&D spending.

33. For example, P.L. 98–622, passed in 1984, modified the previous requirement that each coinventor listed in a patent application also had to be a coinventor on every claim of the patent. The new law allows inventors to apply jointly, even though they may not have physically worked together, made the same level of contribution, or contributed individually to the subject matter of each claim. For a thorough discussion, see Patrick Kelley, "Recent Changes in the Patent Law Which Affect Inventorship and the Ownership of Patents," unpublished manuscript (1985).

34. F. M. Scherer, *The Economic Effects of Compulsory Patent Licensing* (New York University, Graduate School of Business Administration, 1977).

The choice between obtaining a patent and maintaining secrecy may be influenced by the extent to which the disclosures made in the patent document facilitate inventing around the patent. Our data provided some support for this theory. The effectiveness of secrecy was positively correlated with the extent to which disclosures limited the effectiveness of patents. The link was stronger for product patents than for process patents. But patent disclosures represented a substantial limitation on the effectiveness of product patents for only 4 of the 130 industries (scoring as high as six on the scale), and only 16 regarded process disclosures as comparably important. In only one line of business of those with five or more respondents—metal cutting machine tools—did disclosures constrain so substantially the effectiveness of both process and product patents.

Channels of Information Spillover

To the extent that a rival can learn easily about an innovator's technology, the incentive to invest in R&D is attenuated. But to the extent that learning is easy, wasteful duplication or near duplication of R&D effort by rival firms may be avoided. Also, knowledge of an innovator's new technology may complement rival R&D effort by enhancing its productivity. Richard Nelson and Sidney Winter, Michael Spence, and Richard Levin and Peter C. Reiss have developed models that begin to disentangle these offsetting effects, called by Spence the incentive and efficiency effects of interfirm spillovers.[35] A sharper characterization of interindustry differences in the nature and strength of the mechanisms by which firms learn about their competitors' technology should advance these modeling efforts.

Table 6 summarizes the responses to questions about the effectiveness of alternative ways of learning. There is little difference between the pattern of responses for processes and for products, except that, as one would expect, reverse engineering is markedly more effective in yielding

35. Richard R. Nelson and Sidney G. Winter, "The Schumpeterian Tradeoff Revisited," *American Economic Review*, vol. 72 (March 1982), pp. 114–32; Michael Spence, "Cost Reduction, Competition, and Industry Performance," *Econometrica*, vol. 52 (January 1984), pp. 101–21; and Levin and Reiss, "Tests of a Schumpeterian Model," and "Demand-Creating and Cost-Reducing R&D."

Table 6. Effectiveness of Alternative Methods of Learning about New Processes and Products[a]

	Overall sample means		Distribution of industry means[b]	
Method of learning	*Processes*	*Products*	*Processes*	*Products*
Licensing technology	4.58 (0.07)	4.62 (0.07)	3.4–5.6[c]	3.5–5.5[c]
Patent disclosures	3.88 (0.05)	4.01 (0.06)	3.0–4.6[c]	3.0–4.8[c]
Publications or technical meetings	4.07 (0.05)	4.07 (0.05)	3.4–4.7	3.3–4.6[d]
Conversations with employees of innovating firm	3.64 (0.06)	3.64 (0.06)	2.9–4.7[d]	2.9–4.5[d]
Hiring R&D employees from innovating firm	4.02 (0.07)	4.08 (0.07)	2.7–5.0[c]	2.8–5.0[c]
Reverse engineering of product	4.07 (0.07)	4.83 (0.06)	3.0–5.0[c]	4.0–5.7[d]
Independent R&D	4.76 (0.06)	5.00 (0.05)	4.0–5.5	4.4–5.6[c]

Source: Authors' calculations.
a. Range: 1 = not at all effective; 7 = very effective. Standard errors in parentheses.
b. From the upper bound of the lowest quintile of industries to the lower bound of the highest quintile.
c. Interindustry differences in means significant at the .01 level.
d. Interindustry differences in means significant at the .05 level.

information about product technology. On average, independent R&D was rated as the most effective means of learning about rival technology.[36] This may appear to be wasteful duplication, but it need not be. One pretest subject said that R&D effort devoted to determining what a competitor has done may have strong complementarities with a firm's own research program in areas not directly imitative of the innovating competitor. Licensing was also rated, on average, an important way of gaining access to a rival's new technology.

The correlations among individual and industry mean responses show that mechanisms relying on interpersonal communication (publications and technical meetings, informal conversations, and hiring away employees) are strongly intercorrelated. Learning through licensing technology is uncorrelated with nearly all other learning mechanisms except disclosure through patent documents. There are two possible interpre-

36. Wesley Cohen and Daniel Levinthal have studied the incentives to engage in R&D that is directed toward developing absorptive capacity, the ability to make use of technology developed by others. See "Innovation and Learning: The Two Faces of R&D" (Carnegie-Mellon University, Department of Social and Decision Sciences, March 1987).

tations of this last connection. Potential licensees may learn about the opportunity to license through patent documents, or the documents may prove useful in employing new technology once it is licensed. We cannot tell whether the "announcement" effect or the "complementary information" effect of disclosures predominates.[37]

The pattern of correlation suggested that there might be three or four clusters of industries, distinguished in turn by an emphasis on learning through licensing, interpersonal channels, and reverse engineering or independent R&D, or both. The results obtained from cluster analysis were not entirely satisfactory.[38] Nonetheless, table 7 presents the results of grouping the lines of business into three clusters on the basis of responses to the questions on channels of spillover.

For both new processes and products, the largest group of industries typically relied on licensing and independent R&D to learn about competitive technology. Interpersonal channels were relatively unimportant, and reverse engineering was important for products. For both processes and products, there was a second cluster of industries in which interpersonal channels of spillover were most important. In the case of learning about new products, only ten industries were classified in this cluster, and in the case of learning about new processes, other channels—independent R&D and reverse engineering—were nearly as valuable. For processes, a third cluster appeared to find all mechanisms of learning relatively unproductive. For products this group found all mechanisms moderately effective.

Cost and Time Required for Imitation

As part of our investigation we asked respondents to estimate typical costs and time required to duplicate several categories of innovations if

37. The correlations between the effectiveness of particular learning mechanisms and the effectiveness of alternative methods of appropriation are interesting and internally consistent. In particular, when patent protection is effective, learning tends to take place primarily through licensing and patent disclosures. The effectiveness of patents is essentially uncorrelated with the effectiveness of interpersonal channels of learning and of independent R&D, and it is negatively correlated with the effectiveness of reverse engineering.

38. With three clusters the ratio of variance among clusters to variance within clusters was low, but attempts to find more than three clusters were thwarted by the persistent appearance of clusters containing only one or two lines of business.

Table 7. Cluster Analysis of Channels of Learning

Learning mechanism	Cluster 1	Cluster 2	Cluster 3
New Processes			
Number of industries	68	43	19
Mean score			
Licensing technology	5.0	4.3	2.5
Patent disclosures	4.0	4.0	3.2
Publications or technical meetings	3.8	4.6	3.9
Conversations with employees of innovating firm	3.2	4.8	3.0
Hiring R&D employees from innovating firm	3.7	5.1	2.4
Reverse engineering of product	3.8	4.6	4.0
Independent R&D	5.0	4.6	4.3
New products			
Number of industries	68	10	52
Mean score			
Licensing technology	4.7	2.5	4.5
Patent disclosures	3.9	2.9	4.3
Publications or technical meetings	3.7	5.1	4.3
Conversations with employees of innovating firm	3.0	4.6	4.5
Hiring R&D employees from innovating firm	3.2	4.4	4.9
Reverse engineering of product	4.7	3.0	5.2
Independent R&D	5.1	3.7	5.0

Source: Authors' calculations.

they were developed by a competitor. For each category, respondents were asked to identify (within a range) the cost of duplication as a percentage of the innovator's R&D cost. Intervals measured in months or years were used to classify the time required. In light of evidence that there is a time-cost trade-off in certain industries, we asked respondents to estimate the cost and time required "to have a significant impact on the market."[39]

Tables 8 and 9 show frequency distributions of industry median responses.[40] The dispersion of industry medians suggests substantial variations among industries in both the cost and time required to duplicate

39. See Edwin Mansfield, *Industrial Research and Technological Innovation: An Econometric Analysis* (Norton, 1968), for evidence on the time-cost trade-off.

40. Qualitatively identical results and interpretations are obtained from frequency distributions of individual responses and from the distribution of industry means.

Table 8. Cost of Duplicating an Innovation as a Percentage of Innovator's R&D Cost, Frequency Distribution of Median Responses

Type of innovation	*Less than 25 percent*	*26 to 50 percent*	*51 to 75 percent*	*76 to 100 percent*	*More than 100 percent*	*Timely duplication not possible*
New process						
Major patented new process	1	5	19	66	26	10
Major unpatented new process	5	10	55	49	6	2
Typical patented new process	2	15	61	41	6	2
Typical unpatented new process	8	43	58	14	4	0
New product						
Major patented new product	1	4	17	63	30	12
Major unpatented new product	5	13	58	40	7	4
Typical patented new product	2	18	64	32	9	2
Typical unpatented new product	9	58	40	15	5	0

Source: Survey of 127 lines of business.

all categories of innovation. If, however, individual responses to the questions on cost are coded on a six-point interval scale, there is sufficient intraindustry variation to render interindustry differences insignificant at the 0.01 level. Interindustry differences in the time required for duplication are, by contrast, significant at the 0.01 level in every instance except the time required to duplicate a typical patented new process.

Several conclusions are apparent. First, duplicating major innovations tends to cost more and take longer than duplicating typical innovations. (In a sense, this confirms that respondents correctly interpreted the distinction between typical and major innovations.) Second, for a given category of innovation, the cost and time required to duplicate are distributed very similarly for products and processes. Products tend to be slightly cheaper and quicker to duplicate than processes, though this generalization does not hold for major patented innovations. Finally, patents tend to raise imitation costs and time for each category of innovation. These increases can be regarded as alternative indicators of the relative effectiveness of patents in different industries.

To explore this point further, we coded the individual responses to the imitation costs and time questions on a six-point interval scale,

Table 9. Time Required to Duplicate an Innovation, Frequency Distribution of Median Responses

Type of innovation	*Less than 6 months*	*6 months to 1 year*	*1 to 3 years*	*3 to 5 years*	*More than 5 years*	*Timely duplication not possible*
New process						
Major patented new process	0	4	72	37	9	7
Major unpatented new process	2	20	84	17	2	4
Typical patented new process	0	40	73	13	0	3
Typical unpatented new process	8	66	47	6	1	1
New product						
Major patented new product	2	6	64	40	8	9
Major unpatented new product	3	22	89	12	1	2
Typical patented new product	5	39	72	6	4	3
Typical unpatented new product	18	67	39	4	1	0

Source: Survey of 129 lines of business.

calculated the individual and industry mean increases in costs and time associated with the presence of patents, and correlated these, respectively, with individual and industry mean responses to our questions on the effectiveness of patents in preventing duplication. For each category of innovation, the reported effectiveness of patents was positively correlated with the increase in duplication costs and time associated with patents, although the correlations tended to be stronger for products than for processes. We also found some evidence, at the level of the individual respondent, that patent effectiveness was associated with the absolute level of duplication costs for patented processes and products. We found a much stronger association, however, between reported patent effectiveness and the amount of time required to duplicate both patented process and product innovations.

These broad-brush patterns of association conceal some striking anomalies. For particular categories of innovation, at least two and as many as fourteen industries reported that patents actually reduced the costs or time required for duplication. A partial explanation is that a disproportionate number of these industries also reported that disclosure of information through patent documents was a significant limitation on patent effectiveness.

A second anomaly is that, despite the positive correlation between patent effectiveness and the costs of imitating patented products, in several industries patents were relatively ineffective and duplication costs were nonetheless very high, whether or not the innovation was patented. Among these were guided missiles and several types of industrial machinery (food products machinery, electric welding apparatus, and speed changers, drives, and gears). In these instances the relative complexity of the products presumably makes reverse engineering inherently costly despite relatively weak patent protection.

It is interesting to compare our findings with those of Edwin Mansfield, Mark Schwartz, and Samuel Wagner, who studied the effects of patents on imitation costs in three industries.[41] They concluded that patents generally raised imitation costs by 30 percentage points in drugs, 20 points in chemicals, and 7 points in electronics. To render our data comparable, we evaluated each respondent's answer at the mean of the relevant range and computed crude industry average imitation costs for each type of innovation.[42] Our results were consistent with those of Mansfield, Schwartz, and Wagner. We found that patents raise imitation costs by 40 percentage points for both major and typical new drugs, by 30 points for major new chemical products, and by 25 points for typical chemical products. In electronics, our results differed somewhat for semiconductors, computers, and communications equipment, but the range was 7 to 15 percentage points for major products and 7 to 10 for typical products.[43]

41. Mansfield, Schwartz, and Wagner, "Imitation Costs and Patents."

42. The ranges are shown in the headings of table 8. The fifth and sixth column headings are not readily quantified. To permit the comparison discussed in the text, we assigned these categories the values of 112.5 percent and 137.5 percent, respectively, thereby maintaining a constant spacing of 25 percentage points between each pair of categories.

43. Our results on the time required to duplicate a rival's new products or processes were also roughly consistent with recent findings of Edwin Mansfield. In all but one of the ten industries he surveyed, the median respondent indicated that six to twelve months usually elapsed before the nature and operation of a new product were known to a firm's rivals. Effective duplication, as we have defined it, should take as long or longer, and table 9 shows that it typically does. The median and modal industries require one to three years to duplicate a major innovation or a typical patented innovation. A typical unpatented innovation, however, is more often duplicated within six to twelve months. See "How Rapidly Does New Industrial Technology Leak Out?" *Journal of Industrial Economics*, vol. 34 (December 1985), pp. 217–24.

Although the costs and time required for duplication are related to the effectiveness of patents, they do not seem to be linked strongly to any other mechanism of appropriability. In particular, most imitation time and cost measures are uncorrelated with lead time and learning curve advantages, and where such correlations are statistically significant (at the level of the individual respondent), the correlation coefficient is invariably below .15. These results make sense. Lead time and learning advantages may permit appropriation of returns even when duplication is relatively quick and inexpensive. Effective patents, however, presumably require considerable time and expense to be invented around.

Finally, most of our respondents believed only a few firms were capable of duplicating new processes and products. As table 10 shows, the median and modal number of firms judged capable of duplicating a major process or product innovation was three to five. The median and modal number of firms regarded as capable of duplicating a typical process or product innovation was six to ten. The data revealed only the slightest tendency toward a smaller number of capable duplicators for processes than for products.

R&D and Innovation

In this section, we summarize how data derived from our survey have been employed to understand better the sources of interindustry differences in R&D spending and the rate of technological advance. In the first such effort Richard Levin, Wesley Cohen, and David Mowery used several survey-based measures to explain variations in the published Federal Trade Commission data on industry-level R&D spending as a percentage of sales.[44] They also sought to explain interindustry differences in the rate at which new processes and new products were introduced during the 1970s, as reported by our survey respondents.[45]

44. The ratio of company-financed R&D to sales (R&D intensity) varies considerably among industries defined at the FTC line-of-business level of aggregation. In the 1976 data used by Levin, Cohen, and Mowery, R&D intensity ranged from 0.08 percent to 8.5 percent; both the mean and standard deviation were 1.7 percent. See "R&D Appropriability, Opportunity, and Market Structure."

45. Respondents were asked to identify, on a seven-point Likert scale ranging from "very slowly" to "very rapidly," the rate at which new processes and products had been

Table 10. Number of Firms Capable of Duplicating an Innovation, Frequency Distribution of Median Responses

Type of innovation	*None*	*1 or 2*	*3 to 5*	*6 to 10*	*More than 10*
Major new or improved process	2	32	75	18	2
Typical new or improved process	1	7	41	58	22
Major new or improved product	2	25	73	25	4
Typical new or improved product	1	5	33	63	26

Source: Survey of 129 lines of business.

In a subsequent paper, Cohen, Levin, and Mowery studied the extent to which the same survey-based measures explained the powerful industry effects in the confidential FTC data on R&D intensity at the level of the business unit.[46]

The first paper focused on the Schumpeterian hypothesis that R&D intensity and innovation rates are significantly influenced by the level of industry concentration. One common rationale for this hypothesis is that industry concentration enhances the potential for appropriation of R&D returns. A different view is that, in the long run, concentration tends to be a consequence of industry evolution in a regime of abundant technological opportunity and a high degree of uncertainty associated with investment in R&D. Both perspectives suggest that there is no simple, causal relationship between concentration per se and R&D. Concentration may be statistically significant in simple regression specifications because it reflects the influence of the unobserved appropriability and opportunity conditions that directly affect R&D spending and the rate of innovation.

In ordinary least squares and two-stage least squares specifications that included only the four-firm concentration ratio and its square as

introduced in their industries since 1970. Industry mean responses were highly correlated with total factor productivity growth, and the plausibility of the responses was reinforced by the identity of the highest and lowest industries in the sample. Excluding singletons, the lines of business reporting the slowest rates of product introduction were concrete, cement, boiler shops, milk, gypsum, primary copper, grain mill products, and sawmills. Those reporting the most rapid rates of product introduction were electrical equipment for internal combustion engines, radio and TV sets, computers, semiconductors, communications equipment, photographic equipment and supplies, engineering and scientific instruments, and guided missiles. Levin, Cohen, and Mowery used as a dependent variable the average of each industry's reported rates of process and product introduction.

46. Cohen, Levin, and Mowery, "Firm Size and R&D Intensity."

regressors, Levin, Cohen, and Mowery replicated with the industry-level FTC data the familiar inverted-U relationship between concentration and R&D intensity, and they found a strong relationship of the same form between concentration and the rate of innovation.[47] Adding two-digit industry fixed effects weakened slightly the effect of concentration on R&D, but the innovation-rate equation was unaffected.

The results changed dramatically with the addition of measures of appropriability and technological opportunity derived from the survey.[48] Whether or not two-digit industry fixed effects were included, the coefficients on concentration and its square fell by an order of magnitude in the R&D equation, and the effect of concentration was no longer statistically significant at the .05 level in either the R&D intensity or the innovation-rate equation. The vector of survey-based opportunity variables was significant at the .05 level in all specifications, and the opportunity and appropriability variables were jointly significant. The appropriability variables, however, were not individually significant in the R&D equation, although the rate of innovation was positively related to the effectiveness of an industry's most effective means of appropriation.[49]

The paper by Cohen, Levin, and Mowery used the disaggregated FTC data at the level of the business unit to investigate the Schumpeterian hypothesis linking size and R&D intensity. The authors found that when either fixed industry effects (at the level of the line of business) or survey-based industry characteristics were taken into account, firm size had a very small and statistically insignificant effect on R&D intensity. The size of the business unit did have a significant effect on the probability

47. All coefficients in the R&D and innovation-rate equations were statistically significant at the .01 level.

48. To represent appropriability conditions, Levin, Cohen, and Mowery used two survey-based measures: the maximum of the mean scores an industry's respondents assigned to the effectiveness of the six methods of appropriation and the time required to duplicate effectively a patented major product innovation. To represent opportunity conditions, they used a measure of an industry's closeness to science as well as variables summarizing the importance of four other external sources of knowledge for an industry's technological advance: material suppliers, equipment suppliers, users of the industry's products, and government agencies and research labs.

49. It may seem anomalous that the effectiveness of appropriation was positively related to innovation but not to R&D, but the relationship was observed at the level of the industry. Better appropriability may discourage R&D directed toward imitation to an extent that more than compensates for its stimulus to innovative R&D. Such a reallocation of effort would be entirely consistent with the observed positive relationship between appropriability and the rate of innovation.

of engaging in R&D, but there was no perceptible tendency for R&D intensity to increase with size within the group of R&D performers. Size effects, however, explained only two-tenths of 1 percent of the variance in R&D intensity, while industry effects at the line-of-business level explained half this variance.

Cohen, Levin, and Mowery found that industry-level measures of appropriability, opportunity, and demand conditions were consistently significant in ordinary least squares, GLS, and Tobit regressions explaining business unit R&D intensity. Moreover, these industry characteristics explained approximately half the variance in R&D intensity explained by fixed industry effects. When attention was focused on those lines for which there were at least three survey respondents, measured industry characteristics explained 56 percent of the variation in R&D intensity among industries. Within particular two-digit industries (chemicals, machinery, and electrical equipment), measured characteristics explained 78 to 86 percent of the variance explained by fixed effects.

The results obtained in the two papers indicated that survey-based measures can contribute substantially to an explanation of interindustry differences in R&D intensity and innovative performance. Measures derived from the survey, despite their imperfections, have also been found useful for various other purposes.[50]

Remarks on Policy

Our findings suggested some general principles relevant to policies that affect the incentives to engage in innovative activity.

A first principle is that the patent system and related institutions to

50. Levin and Reiss have used survey-based measures of appropriability and opportunity in a simultaneous equation model of R&D spending and market structure that builds on their work in "Tests of a Schumpeterian Model." Cohen and Levinthal use survey-based variables in their work on R&D as investment in absorptive capacity; see "Innovation and Learning." Iain Cockburn and Zvi Griliches are studying the usefulness of our survey measures in estimating the value of patents from stock market data; see "Industry Effects and Appropriability Measures in the Stock Market's Valuation of R&D and Patents," *American Economic Review* (forthcoming, May 1988). Meryl Finkel, "Overseas Research and Development by U.S. Multinationals: Ownership Structure Decisions" (Ph.D. dissertation, Harvard University, 1986), explored the effect of our appropriability variables on the investment decisions of multinational corporations. Franco Malerba is using the survey data to explain interindustry differences in the extent and effectiveness of learning mechanisms.

protect intellectual property should be understood as social structures that *improve* the appropriability of returns from innovation. They are not the only nor necessarily the primary barriers that prevent general access to what would otherwise be pure public goods. Lead time accrues naturally to the innovator, even in the absence of any deliberate effort to enhance its protective effect. Secrecy, learning advantages, and sales and service efforts can provide additional protection, though they require the innovator's deliberate effort. The survey confirmed that these other means of appropriation are typically more important than the patent system. Hence in examining a proposed adjustment of the patent system or related institutions, it is important to recognize that the incremental effect of the policy change depends on the protection other mechanisms provide.

The survey results also confirmed substantial interindustry variation in the level of appropriability and in the mechanisms that provide it. From this follows our second major principle, which is that the incremental effects of policy changes should be assessed at the industry level. For example, in the aircraft industry, where other mechanisms provide considerable appropriability, lengthening the life of patents would tend to have little effect on incentives for innovation. In the drug industry the effect of a longer lifetime would matter more.[51]

Finally, improving the protection of intellectual property is not necessarily socially beneficial. Empirical work has so far indicated a positive cross-sectional relationship between strong appropriability, as measured by variables constructed from our survey, and innovative performance. But the social cost-benefit calculation is not straightforward. Stronger appropriability will not yield more innovation in all contexts and, where it does, innovation may come at excessive cost.

To illustrate how our survey results and general perspective might inform policy discussion, consider the 1987 proposal (S. 438, H.R. 557) that patent license agreements and other contracts relating to the use of intellectual property "shall not be deemed illegal per se under any of the antitrust laws." One consequence would be to eliminate the per se illegality of tie-in arrangements (those in which purchase of one product,

51. For a calculation of the impact of the Drug Price Competition and Patent Term Restoration Act of 1984, see Henry Grabowski and John Vernon, "Longer Patents for Lower Imitation Barriers: The 1984 Drug Act," *American Economic Review*, vol. 76 (May 1986, *Papers and Proceedings, 1985*), pp. 195–98.

the "tying product," is dependent on purchase of other products) where the tying product is covered by a patent or otherwise protected as intellectual property.[52] Our findings have suggested some issues a court should consider in evaluating such a tying arrangement under the rule of reason.

When the rule of reason is applied to tying cases, a relevant consideration is the firm's power in the market for the tying good. Courts have often presumed that intellectual property protection is itself evidence for such power. To the other good reasons for rejecting such a presumption,[53] we add that the mere existence of a patent or other legal protection says nothing about its efficacy in a competitive context. As the survey results showed, the effectiveness of protection varies widely among industries. Thus in deciding a case, a court should inquire into the actual competitive significance of intellectual property protection in the particular market.

Suppose, for example, that a pharmaceutical company were to tie hospital sales of supplies or equipment to its sale of a patented drug. Since patent protection of drugs is generally strong and effective, and a drug is often uniquely suited for particular purposes, skepticism about the reasonableness of the tie-in would be in order. The arrangement could not plausibly be regarded as a straightforward means of appropriating returns to which the firm was entitled as owner of the patent. Given the typical effectiveness of drug patents, the price of the drug should suffice for that purpose. There might, of course, be benign explanations for the tie; for example, if the supplies or equipment were complementary to the use of the drug, the arrangement might be explicable as an attempt to control the quality of treatment. But if no such explanation were

52. We focus on this particular consequence of the proposed legislation and set aside two major considerations regarding its merits in its present form. First, without amendment the legislation is likely to undercut severely the per se treatment of price fixing. Second, it might be more appropriate to consider eliminating per se treatment of all tying arrangements rather than just those involving intellectual property. On this point, see the concurring opinion in *Jefferson Parish* v. *Hyde*, 466 U.S. 2 (1984).

53. See E. W. Kitch, "Patents: Monopolies or Property Rights," in John Palmer, ed., *Research in Law and Economics: The Economics of Patents and Copyrights*, vol. 8 (JAI Press, 1986), pp. 31–47; and the associated commentary of F. M. Scherer, p. 51. *Digidyne Corp.* v. *Data General Corp.*, 743 F. 2d 1336 (9th Cir. 1984) is an example of an application of per se doctrine in a context where the intellectual property (software) does not convincingly convey market power.

supported by the evidence, the tie would seem an unreasonable restraint of trade.

By contrast, consider a producer of a patented product in an industry where no mechanism of appropriability functions particularly well—plywood, for example, where patents, secrecy, lead time, and learning advantages are all rated no higher than four on a seven-point scale of effectiveness. In this instance the low level of appropriability in general and the ineffectiveness of patents in particular should weigh against any presumption that a patent confers market power. The patentee in such an industry should be entitled to some scope for ingenuity in constructing arrangements that maximize the return to the patent, provided that these arrangements are not open to antitrust objections on grounds independent of the role played by the patent.

The intellectual property provisions of the Omnibus Trade and Competitiveness Reform Act also serve to illustrate the relevance of the survey results. One provision requires the U.S. trade representative to identify countries that have been particularly insensitive, as a matter of law or de facto policy, to the need for protection of intellectual property and to initiate unfair trade practice (section 301) investigations against them.[54] This provision of the trade bill would complement the administration's diplomatic efforts to strengthen intellectual property rights throughout the world and particularly in countries that permit firms to copy patented or copyrighted products from the United States.

Since the impact of legal protection of intellectual property depends on the strength of other appropriability mechanisms and varies widely among industries, focused efforts to solve problems in specific markets would be more prudent than a broad attempt to upgrade protection. There is little point in expending diplomatic capital to compel foreign countries to pass or enforce laws that, in most industries, would have minimal impact on the competitive process. By contrast, in those specific industries such as pharmaceuticals—in which patent protection is effective, other means of appropriation are poor substitutes, and foreign governments often restrict, officially or tacitly, the ability of U.S. firms to exploit patents—a more persuasive case could be made for the United States to pressure its trading partners to change their behavior.

54. The trade representative may at his discretion escape this requirement by finding that such an investigation would not be in the national interest.

Appendix: Details of Sample Construction

Our review of the FTC data indicated that several lines of business did not report any R&D activity, and several others were aggregated to prevent violating confidentiality rules. Anticipating difficulty in finding knowledgeable respondents in industries without formal R&D activity, and wishing to avoid industry categories that included technologically disparate products, we eliminated those lines of business from our sampling frame.

The industries eliminated on grounds of heterogeneity were either the FTC's aggregations of technologically disparate industries or those corresponding to SIC industries with four-digit codes ending with 9. Such industries are residual categories within the relevant three-digit groups; their titles usually contain the words "miscellaneous, not elsewhere classified."

Confidentiality requirements prohibited us from using the FTC data as a means of identifying the firms that conduct R&D in each line of business. Instead, we used the *Business Week* annual R&D survey to identify all publicly traded firms that reported R&D expenses in excess of either 1 percent of sales or $35 million. This constitutes a nearly comprehensive list of private firms performing significant R&D. There were 746 such firms in 1981, when our survey design efforts commenced.

We used the information in Dun and Bradstreet's *Million Dollar Directory* to assign each of the *Business Week* firms to its major lines of business. Dun and Bradstreet's does not provide a complete list of each firm's lines of business, but it indicates as many as six four-digit SICs for each firm, in rough order of sales. Since some firms operate in nonmanufacturing industries, in manufacturing industries absent from our sample, or in two or more industries that fall into only one FTC line of business, we had substantially fewer than 746 × 6 observations. Within our sample lines of business, we found a total of 1,928 units operated by 688 firms.

A major design problem was how to obtain responses for business units within the same firm. Of our 688 firms, 470 participated in more than one of our sample lines of business. We initially attempted to identify relevant respondents using *Industrial Research Laboratories of the United States*. But our pretest subjects told us that more than half the people in such a sample were inappropriate. Some had been assigned to the wrong line of business; others had been promoted or had left the relevant division or the firm.

We therefore adopted a two-stage approach in which each firm's senior R&D vice president or chief executive officer was asked to furnish the names of employees with the knowledge to complete the questionnaire for specific lines of business. We sent first-round requests to 470 firms representing 1,710 business units. There was attrition of 332 business units from this sample for three reasons: the firm did not do R&D in the specified line of business, the industry definition did not fit any of its activities, or a respondent could not be located. From this adjusted sample frame of 1,378 business units in firms with multiple units, we received names of respondents for 716. We sent questionnaires to each of these potential respondents as well as to representatives of the 218 firms operating in only one line of business. At this stage, there was some further attrition in the sample. Ultimately, we received 650 completed questionnaires from an overall adjusted sample frame of 1,562—an overall response rate of 41.6 percent.

Comments and Discussion

Richard Gilbert: The authors' research program will have lasting value for people interested in R&D markets and markets for intellectual property. They are correct in focusing on appropriability as a key factor in the incentive to undertake R&D. And their findings are generally consistent with those of other studies, for example, those by Christopher T. Taylor and Z. A. Silberston and Edwin Mansfield, Mark Schwartz, and Samuel Wagner.[1] While this consistency may take a bit of the drama out of what Levin and his colleagues have done, the convergence of knowledge on this subject gives us some reason to believe we might be getting to the truth.

One of the authors' main conclusions is that there are very large differences, both among industries and within them, in the effectiveness of various means of appropriating intellectual property and also in the cost of imitation. It is an important result, but one that may cause some consternation. A main function of microeconomic theory is to form testable generalizations about the way the world works. Some of the work implies—at least, in the market for R&D and intellectual property—that such generalizations are extremely risky. We might be inventing a new field of microeconomic analysis, or "picoeconomics." Picoeconomics would keep us busy for a long, long time. But if we go down that path, our models will soon become as complicated as the world we are trying to explain.

These particular authors do not seem to want to lead us toward

1. Christopher T. Taylor and Z. A. Silberston, *The Economic Impact of the Patent System: A Study of the British Experience* (Cambridge University Press, 1973); and Edwin Mansfield, Mark Schwartz, and Samuel Wagner, "Imitation Costs and Patents: An Empirical Study," *Economic Journal*, vol. 91 (December 1981), pp. 907–18.

picoeconomics. It is apparent from their follow-up work that they intend to draw some general conclusions about how appropriability varies across industries. Clearly, we would like to know how market structure and capital intensity in different industries influence the degree of appropriation and affect incentives to innovate. But merely adding appropriation as another explanatory variable in these regressions does little good. We really want to know whether there are systematic relations between the degree of appropriation and other observable economic variables.

With regard to methodology, I suggest using a weighting scheme based on the amount of R&D a firm does and, perhaps, the number of patents it has produced. The purpose would be to weight responses according to the quality of the information. Some industries seem not to have performed any R&D for twenty years. Although it is important to know why these firms have not been active, their responses should be adjusted to reflect the information they possess. Also, I suggest that in their survey work the authors include a definition of R&D. That is not a trivial task because there is much variation in what is called research and development.

I would have liked the survey to address more directly some of the theoretical issues in the economics of R&D. Various models in the R&D literature have different implications for the simultaneous determination of R&D intensity and market structure. For example, models such as that of Partha Dasgupta and Joseph Stiglitz imply that the current rate of R&D spending should be independent of cumulative R&D expenditures by a firm or the rivals of the firm.[2] This is a consequence of the constant-hazard rate model. Other researchers such as Drew Fudenberg and his colleagues imply that past R&D is crucial to current and future R&D expenditures.[3] The dynamic implications of these models are very different. The preemption-type models also suggest that a history of successful R&D gives a firm a technological advantage that provides some protection from future R&D competition and tends to increase concentration in a market.

2. Partha Dasgupta and Joseph E. Stiglitz, "Uncertainty, Industrial Structure, and the Speed of R&D," *Bell Journal of Economics*, vol. 11 (Spring 1980), pp. 1–28.

3. Drew Fudenberg and others, "Preemption, Leapfrogging and Competition in Patent Races," *European Economic Review*, vol. 22 (June 1983), pp. 3–31.

I would be interested in knowing if the survey could have elicited some kind of response about the way R&D success alters the competitive environment of the firms and, conversely, how the competitive environment influences R&D spending. The questions in which the investigators ask how many firms are viable competitors with a given firm and how many could replicate its R&D bear on this question. It is interesting that the number of serious rivals for each firm was small, somewhere between three and six.

There is an empirical problem with surveys of the relationship between competition and R&D. If R&D really does have an effect on entry and competition, then the sample is necessarily biased. There are potential competitors who were not represented in the sample because the firms failed. How one accounts for the failures and puts them back in the sample is a difficult empirical problem.

In terms of patent policy, the diversity and effectiveness of patent protection across industries raises an obvious and interesting question. How should patent policy deal with the large differences in the values of patents among various industries? The authors point to the example of the semiconductor industry and suggest that the industry need not be any worse off as a consequence of the limited patent protection it has received. But one might also question whether the performance of the pharmaceuticals industry might have been improved if patent protection had been circumscribed to some extent.

I have experimented with a very simple model of optimal patent life with limited appropriability. It is basically a Nordhaus-type model with entry and spillovers.[4] It shows that the optimal patent life is not a function of the size of the innovation, so one does not have to worry that there are big innovations in some industries and small innovations in others. The optimal patent life does, however, depend on elasticities of R&D and its benefits and costs. It also depends on the degree of appropriability, and there is the intuitive answer that the optimal life is inversely related to the extent of private appropriation of the social value of the invention. This result suggests that we need either more protection in semiconductors or less protection in the patent drug industry.

The authors' survey provides a basis for contrasting patents with

4. See William D. Nordhaus, *Invention, Growth, and Welfare: A Theoretical Treatment of Technological Change* (MIT Press, 1969).

other approaches to protect intellectual property. A patent is a peculiar policy instrument. It represents an unnatural barrier to market entry that is erected to facilitate private appropriation. The survey suggests that other factors may be more important as a means of appropriation, and that other unnatural barriers might be more effective in stimulating R&D. As an extreme example (which I am not proposing), a tax on capital could arguably make entry more difficult and therefore stimulate R&D. It would be useful to do a survey of the effectiveness of different laws governing rights for intellectual property in different countries in an industry such as pharmaceuticals, which is one of the few in which patents really do seem to matter.

The survey results suggest that patents are important as a barrier to entry in the semiconductor industry not because they protect an individual innovation or invention but because they provide a hurdle for potential entrants, who have to acquire a package of marketable processes and products that they can cross-license to other firms. While this seems an inappropriate or at least unintended outcome of the patent grant, it might well be that by increasing ordinary barriers to entry in the semiconductor industry, the returns to research and development would be enhanced. This is another illustration of the Schumpeterian hypothesis and the tensions between strong enforcement of the antitrust laws and the desire to provide a stable platform for encouraging investment in R&D. If these observations with regard to the semiconductor industry generalize to other markets, they provide a starting point for further reexamination of the antitrust laws in the context of industrial R&D policy.

Thus this survey has raised some very interesting questions. Now we have to get on with their resolution. Thanks to this project, we have some of the data we need for the job.

Zvi Griliches: We should be grateful to Richard Levin and his associates for providing us with a new and detailed glimpse into a subject that is both very important and also lacking in good data. Far too little fresh economics data is collected, and we all have much to learn from the effort of this endeavor. That I am going to quarrel with some of the authors' assessments does not diminish in my eyes the basic value of this enterprise.

The authors have collected a large set of responses from many individuals located in different industries. This multipurpose survey will have many uses as we learn more about the responses and how to interpret them. I will focus on how these answers can help us learn which industries find patents an effective mechanism for appropriating returns from innovative effort, which ones do not, and whether mechanisms are available instead of or in addition to patents.

"Conditions of appropriability" determine the returns from a given innovative effort and hence the incentive to engage in it. One would expect that in industries in which appropriability is easy, there would be more innovative effort, higher returns, and a faster rate of technological progress. Such conditions may not be a fixed, unchanging characteristic of an industry, however. As more inventive effort is pursued, projects may become less easily appropriable, information may be leaked, and conditions may actually equalize among industries. The problem here is the same as in most empirical research programs: What is exogeneous and what is endogenous?

There are two generic problems with using the responses from such a survey: Are the responses comparable among individuals and do they reflect real differences among industries? Given the use of a scale of one to seven, I remain unsure about whether one person's response of five is equivalent to another's of four or six. Most questions do not have an objective anchor and could, therefore, differ greatly in the meanings attached to them by different respondents. This may account for the large dispersion in responses to most questions even within the same industry. It also leads to the difficulty of deciding whether the responses reflect real differences across industries or just random fluctuations among individuals.

There is a surprising amount of variability within industries in responses to the same question. Some questions, such as those on the effectiveness of patents, are reasonably objective and seem to have a variance among industries. Other questions, such as whether secrecy is effective, do not seem to be particularly industry-specific and do not discriminate well among industries. Questions about lead time, secrecy, sales effort, and service quality are really questions about different ways of succeeding, not about properties of an industry. It is well to have a long lead time or to achieve secrecy, but how is that to be accomplished? These are not characteristics over which either the firm or the policy-makers have clear control. Patents are at least a somewhat better defined

Table 1. Analysis of Variance of Differences among Industries in Responses to Questions on the Effectiveness of Different Appropriability Mechanisms

Question	*Individual response*[a]		*Aggregated industry response*[b]	
	Processes	*Products*	*Processes*	*Products*
Patents to prevent duplication	1.9	2.8	3.4	5.5
Patents to secure royalties	1.6	1.4	2.4	2.3
Secrecy	1.2	1.2	1.3	1.3
Lead time	1.7	1.6	2.2	2.5
Quickly down the learning curve	1.0	1.0	.9	1.2
Superior sales or service	1.2	.9	.7	.9
Average of questions 1 and 2	2.1		4.8	
Average of questions 3 through 6	1.1		1.6	

Source: Authors' calculations.
a. Covers 541 responses for 130 industries; approximate .05 significance level is 1.3.
b. Covers 620 responses for 24 industries; approximate .05 significance level is 1.6.

instrument, and we have some ideas about how the patent system could be tinkered with.

Table 1 presents my analysis of variance results for some of the responses derived from the Yale survey. (I am grateful to Levin for providing me with the original survey data.) It shows that there is more variance among industries in the responses to questions on the effectiveness of patents, especially for products, and very little variance in the other questions on conditions for appropriability, especially for process innovations. Process innovations are clearly less industry-specific and so is the importance of superior sales and service efforts.

Another way of seeing this problem is to look at the authors' table 6, which shows that for many of the nonpatent mechanisms the cross-correlation among methods of appropriation is lower at the aggregated industry level than at the level of the individual response. If industrial classification mattered, one would expect higher correlations for the aggregated variables.[1] This point is illustrated by a very simple model. Assume that two questions effectively measure the same thing. Then a variance components model for responses to these questions would be

$$y_{qij} = m_i + a_j + e_{qij},$$

where m_i is the "true" industry effect perceived by all individuals, a_j is

1. Y. Grunfeld and Zvi Griliches, "Is Aggregation Necessarily Bad?" *Review of Economics and Statistics*, vol. 42 (February 1960), pp. 1–13.

the individual deviation from the average respondent independent of the industry he is in, and e_{qij} is the random response error associated with the particular question q and individual ij. Taking these effects as independent from each other, and assuming that the first component does not average out as one aggregates within industries but that the other ones do in proportion to the average number of respondents per industry, gives us a little model that can be fit to the observed variances and covariances at the micro and macro industry levels. The following material shows the relevant numbers for two pairs of questions: whether product patents are effective against duplication and in securing royalty income, and whether moving quickly down the learning curve and superior sales and service efforts are effective in protecting the competitive advantage of new products. The two-question expected variance-covariance matrix is

Individual level *Industry level*

$$\begin{bmatrix} \sigma_m^2 + \sigma_a^2 + \sigma_1^2 & \sigma_m^2 + \sigma_a^2 \\ \ldots & \sigma_m^2 + \sigma_a^2 + \sigma_2^2 \end{bmatrix} \qquad \begin{bmatrix} \sigma_m^2 + \dfrac{(\sigma_A^2 + \sigma_1^2)}{N_i} & \sigma_m^2 + \sigma_A^2/N_i \\ \ldots & \sigma_m^2 + \dfrac{(\sigma_A^2 + \sigma_2^2)}{N_i} \end{bmatrix}$$

where m_i is the "true" industry effect perceived by all individuals, a_j is the individual deviation from the average respondent independent of the

	Product patents IB1	*Effective IB2*	*Learning curve IB5*	*Secrecy IB6*
Individual	2.860	1.435	1.810	.803
(N = 643)	. . .	2.748	. . .	1.662
Industry	.618	.384	.148	.013
(N = 24)	. . .	.376	. . .	.075
Implied estimates				
($\bar{N}$ = 27)				
σ_A^2	1.09		.820	
σ_m^2	.34		<0	

The numbers imply that the common variance between industries accounts for about one-eighth of the variance at the level of the individual response and more than half at the aggregated industry level. For the two other questions the implied "true" variance between industries is negative. (For the patents question the correlation rises from .51 at the level of the individual respondent to .80 for averages at the NSF industry level, while for the two appropriability questions the numbers go from .46 to .12, implying that such averaging attenuates rather than strengthens the relationship between the responses to such questions.) In short, while these questions might be interesting, they do not seem to be able to pick out significant differences among industries. Coming quickly down the learning curve and providing superior services are about equally effective.

Do the results of this survey help us explain other phenomena besides the relationship between answers to different sets of related questions within the survey? Looking at the responses to the questions evaluating patents and other appropriability mechanisms, the evidence appears to be mixed. Levin, Cohen, and Mowery did not find the appropriability variables significant in explaining differences among industries in R&D intensity, even in the absence of industry dummies.[2] In a forthcoming paper Iain Cockburn and Zvi Griliches use the Yale survey responses, aggregated to fifty-five industries at approximately a 3-digit SIC level, to see whether the stock market values the accumulated patents and the current R&D policy of a firm more or less in industries where the appropriability conditions are better in some sense.[3] Table 2 reproduces typical results from this study. Patent effectiveness measures help in some sense. The equations seem to imply that both accumulated past patents and current R&D moves are valued more by the market when patent protection is effective. Other appropriability measures do not help. But neither set of measures does better than just an interaction with ten higher-level (2-digit) industries dummies. The greater detail available in the Yale survey appears to be counterbalanced by the (inevitably?) greater imprecision of these measures at the detailed industry level. So there is something there but not as much as might be wished. But we should be thankful for there is hope that a more detailed study of these and other responses in this survey will help us understand our world better. In particular, the information on the differential

2. Richard C. Levin, Wesley M. Cohen, and David C. Mowery, "R&D Appropriability, Opportunity, and Market Structure: New Evidence on Some Schumpeterian Hypotheses," *American Economic Review*, vol. 75 (May 1985, *Papers and Proceedings, 1984*), p. 23.

3. Iain Cockburn and Zvi Griliches, "Industry Effects and Appropriability Measures in the Stock Market's Valuation of R&D and Patents," working paper 2465 (National Bureau of Economic Research, December 1987).

Table 2. Stock Market's Valuation of R&D and Patents, 722 U.S. Manufacturing Corporations, 1980[a]

Variables	*Coefficients*						
	(1)	*(2)*	*(3)*	*(4)*	*(5)*	*(6)*	*(7)*
SP/A[b]	.165	.380	.107	.249	.360	.077	.199
	(.100)	(.171)	(.167)	(.155)	(.170)	(.183)	(.161)
K/A[c]	...	...	.932	.335	...	.898	.342
			(.201)	(.178)		(.224)	(.175)
NR[d]	...	...	...	11.96	...	...	12.24
				(1.37)			(1.38)
PPP[e]	...	.034	.019	.019	.035	.023	.019
		(.024)	(.024)	(.023)	(.024)	(.025)	(.024)
PPP · (*SP/A*)	...	.236	.075	.098	.267	.115	.164
		(.116)	(.110)	(.101)	(.133)	(.142)	(.128)
PPP · (*K/A*)	...	...	.365	...	...	.432	...
			(.130)			(.172)	
PPP · *NR*	...	...	...	2.788	...	...	2.60
				(1.231)			(1.39)
NPP[f]	...	...	...	...	.039	.100	.054
					(.079)	(.089)	(.075)
NPP · (*SP/A*)	...	...	...	...	.127	.174	.290
					(.388)	(.432)	(.293)
NPP · (*K/A*)	...	...	...	...	...	.263	...
						(.636)	
NPP • *NR*	...	...	...	...	...	...	−1.89
							(4.77)
$\bar{R}^2$	.166	.172	.200	.310	.170	.198	.309

Source: Adapted from Iain Cockburn and Zvi Griliches, "Industry Effects and Appropriability Measures in the Stock Market's Valuation of R&D and Patents," working paper 2465 (National Bureau of Economic Research, December 1987), tables 3b and 5.

a. Dependent variable is log Q (market value divided by replacement value). All equations contain also ten 2-digit SIC industry dummy variables and a logarithm of total assets variable whose coefficient is small but consistently significant, on the order of −.03 (.01).

b. Stock of patents (30 percent depreciation rate) divided by total net assets.

c. Cumulated R&D "capital" stock (15 percent depreciation rate) divided by total net assets.

d. Net R&D investment divided by net assets $(R - .15K)/A$.

e. Sum of responses to "patents provide protection against duplication" questions for both process and product innovations. Averages at a fifty-five industries (approximately 3.5 digit level) aggregation.

f. Average of responses to all other "effectiveness of nonpatent appropriability mechanisms" questions.

connectedness of science in different industries is very intriguing and may be of help in future analyses of the contribution of science to technological advance.

General Discussion

Richard Levin agreed with Zvi Griliches that the appropriability variables could not discriminate effectively among more than about ten industry groupings, but he suggested that this may be a good thing, especially in light of Richard Gilbert's concern that studies such as this amount to "picoeconomics," from which no generalizations can be drawn. Sidney Winter noted that the results suggest there may be a relatively short list of variables to consider in an analysis of appropriability and incentives for R&D, and that ten industry groupings may provide all the information needed. In other dimensions, a finer division may be important. For example, Levin pointed out that, relative to questions on lead time, learning curves, and duplication costs, the questions on patent effectiveness discriminate better, as do the questions on learning and information spillover and those on duplication time.

Moreover, he added, a principal conclusion, that patents do not matter very much except in the chemical industries and in semiconductors, comes through regardless of problems with questions about other mechanisms of appropriation. In these two industries, the meaning and role of patent protection is different. Chemical products are easy to patent because the structure of the molecule of each product is unique, but patents are easy to invent around because it is often possible to create a discrete but structurally similar product with similar properties. With semiconductors, however, the innovation process is cumulative, with each invention built very distinctly on the previous one. The innovation provided by one firm makes the product invented by another firm more valuable. So the role patents play is to define the property rights (usually through the licensing process) so that the proceeds of this cumulative process can be shared and innovation can be encouraged.

Griliches also raised questions about whether the variables measured in this study are appropriately regarded as exogenous. Sales and service effort, for example, is a choice variable for the firms, and hence effectiveness should be endogenous. Likewise, lead time should be thought of as an outcome of the technology race, rather than an exogenous condition of it. Winter pointed out that even the distinction of product and process may be somewhat endogenous in the sense that firms take into consideration the importance of secrecy and the possibility of reverse engineering in designing their products. Firms often work to make their high-technology products inaccessible to reverse engineering, he noted, which tends to make these products more like processes from the standpoint of appropriability.

Several participants seemed concerned about problems of measurement and scaling biases in the data. Levin responded that various techniques, such as weighting the responses by the inverse of the variances or standard deviations of the individual responses, were tried to correct for these biases. The principal findings were robust to efforts to stretch or squeeze the distribution of responses, he noted, but it was unclear what such "corrections" mean since no one knows what the true distribution should be.

Joseph Farrell took issue with Gilbert's argument that it is not particularly interesting or helpful to worry about the determinants of R&D in industries that do not perform R&D. In fact, he suggested, it would be very useful to know why some industries seem to do so little research while others do so much. Gilbert agreed, but argued that it was still important to assign some sort of weights to individual responses to particular questions, based on the respondent's experience with those issues. Martin Baily argued that weighting the responses by how much R&D the responding firms do would be inappropriate, however, since the amount of R&D is what the authors are trying to explain.

Richard Schmalensee suggested that some of the within-industry variance in the responses about sales and service and other mechanisms of appropriation may be due to the fact that the R&D executives who responded to the questionnaire are less knowledgeable about what happens to the product after it leaves their jurisdiction in the organization.

Responding to a question from Paul Joskow, Levin said if he were doing the survey over again, he would want to do more pretesting. For example, the authors might have picked up on the issue of intraindustry variance earlier if they had pretested multiple respondents from the same industry. Or they might have learned ways to restructure questions to discriminate more carefully between exogenous and endogenous factors.

Robert Litan noted that one implication of the study for trade policy is that issues of intellectual property rights should be dealt with industry by industry. This is, in effect, how section 301 of the Trade Act already works, he added. The section provides a procedure for pursuing complaints about unfair trade practices abroad, but these complaints must be brought product by product. Litan also suggested that mechanisms of appropriating returns from R&D might vary between large and small firms. Levin agreed, but noted that the sampling procedure surveyed only publicly held firms, so that start-up ventures were completely excluded. Patents may be much more important for a start-up company because they provide something tangible to sell if the firm tries to sell out later.

[6]

Appropriating the Gains from Innovation

SIDNEY G. WINTER
The Wharton School

No matter how great the technology or how big the market for it, there is no guarantee that the value from a new technology will go to the innovator. The strategy for appropriating the gains from innovation is critical in taking a successful technological innovation to the bank. Will patents do the job? This chapter points out that managers often place too much emphasis on intellectual property law in protecting the gains from innovations. The author identifies some of the inherent limitations of patent protection and examines other mechanisms for appropriating the gains, including secrecy, control of complementary assets, and lead time.

The path to extraordinary profitability runs through innovation, but to reach this goal, innovative efforts must surmount three hurdles: First, the technology must be successful—that is, the "wonder widget" must work. Second, it must create value. It cannot be something that only a widget engineer could love, but something for which buyers are willing to lay out a purchase price that exceeds the cost of production. If there are enough such buyers, the cost of the innovative effort will be covered, and the prospect arises that someone will benefit substantially from this innovation.

This brings us to the third hurdle: appropriating those gains. The "someone" who benefits may not be the innovator. Most of the gains may go to rivals who have either imitated the innovation or reached similar results on their own. The value could also be captured by buyers or suppliers of

---

The author would like to acknowledge the very helpful comments on an earlier draft provided by the editors, Wes Cohen, and Robert Merges. The errors and limitations that remain are my own.

other resources involved in the processes of production and use. The innovator hopes to find a way to appropriate enough of the total gains to make the whole thing worthwhile—or, preferably, *very much* worthwhile.

This chapter is about the management challenges at the third hurdle, value appropriation. Many discussions of this subject overemphasize the importance of intellectual property law, and patent law in particular, in securing the returns from an innovation. A number of considerations contribute to this overemphasis, the most important being that there is an intellectual property "industry." Like most industries, it promotes its wares. Also, there is a public policy dimension to intellectual property law, and the presence of such a dimension generally creates a higher level of visibility. This overemphasis on intellectual property obscures other mechanisms for appropriating gains from innovations and leads to a focus on the *duration* of the period for which gains are secured. This is probably because the patent law promises more in terms of the duration of protection than rival mechanisms can deliver (also more than patents typically deliver).

Four major classes of appropriability mechanisms are examined here: (1) patents and related legal protections, (2) secrecy, (3) control of complementary assets, and (4) lead time. These four mechanisms are not mutually exclusive (nor are they fully exhaustive). They often are complementary approaches. There are significant interactions among them and opportunities for using one to leverage another. The chapter also provides a framework for examining appropriability issues in emerging technologies and explores the role of dynamic capabilities that yield multiple innovations—creating a "golden goose" rather than a single golden egg.

PUTTING INTELLECTUAL PROPERTY IN PERSPECTIVE[1]

The overemphasis on intellectual property distorts discussion and analysis in three significant aspects. First, it contributes to an impression that patents and other legal protections are the dominant means that innovators employ to capture returns. This is very far from the case. While there are industries in which patents do play a major role, and scattered examples of important patents elsewhere, the general effectiveness of patents varies substantially across industries. Other mechanisms generally play a more important role. The most recent systematic evidence on this point comes from a 1994 survey of R&D managers conducted by researchers at Carnegie Mellon University.[2] For product innovations, the CMU study concludes that "patents are unambiguously the least central of the major appropriability mechanisms overall, reflecting their subordinate role in most industries."[3] For the most part, other mechanisms provide shorter periods of effective protection than patents *promise*—and thus seem inferior to patents *prima facie*. In many cases, the period of protection provided is not, however, inferior to what patents actually deliver.

Second, the overemphasis seems to contribute to a misallocation of attention and resources. Resources and attention are focused on making the most of available legal protection, while energetic and sophisticated use of the alternatives is neglected. Patent litigation can be costly in technical and financial resources, and the effort to obtain legal redress for the injuries of the last competitive round may amount to a self-inflicted injury in the next round. Survey evidence shows that lead time appears to be much more effective across industries than patents. This suggests it may be more effective to get the researchers off the witness stand and back into the lab.

Third, this perceptual distortion leads to mistakes in selecting projects to pursue. It lends superficial credibility to the idea that "if we can't count on getting intellectual property protection for the results of this project, the project is not worth doing." This assumption is quite wrong. Whether the returns are protected by intellectual property law or some other mechanism does not matter. And, in an important sense, it does not really matter how well the returns from the innovation are protected. That is, it does not matter if undeserving imitators leap in and grab 90 percent of the hypothetically available returns, if in fact the costs are only 5 percent of the returns: ten over five is still doubling your money. It may be frustrating to speculate that if those imitators could be fended off to the point where they grabbed "only" 85 percent of the returns, you would be tripling rather than doubling your profits. The unavailability of the tripling is not, however, rational grounds for passing up the doubling.

This last point is certainly appreciated by managers who innovate in ways that the intellectual property system does not significantly protect. Or rather, since they have low expectations about what the legal system will do for them in the first place, they are not subject to the pain of contemplating what a perfect system could do for them. They rightly consider themselves innovative if they succeed in making money by innovating; there is no occasion to focus on the fact that they might be making more money if they had more support from the law. Consider, for example, the comment of Howard Schultz, legendary CEO of the Starbucks coffee chain:

> We had no lock on the world's supply of fine coffee, no patent on the dark roast, no claim to the words *caffé latté* apart from the fact that we popularized the drink in America. You could start up a neighborhood espresso bar and compete against us tomorrow, if you haven't done so already. . . .
>
> What we proposed to do . . . was to reinvent a commodity. We would rediscover the mystique and charm that had swirled around coffee throughout the centuries. We would enchant customers with an atmosphere of sophistication and style and knowledge. . . .
>
> The best ideas are those that create a new mind-set or sense a need before others do, and it takes an astute investor to recognize an idea that not only is ahead of its time but also has long-term prospects.[4]

Starbucks relied not upon patents but on lead time leveraged by complementary assets to appropriate gains from its innovation.

Part of the problem of flawed perspectives is that strong patents, in the relatively rare cases where they can be secured, define a high standard of protection that is (wrongly) sought in other contexts. The absence of such protection leads to the speculation that, if the higher standard of protection were achievable, the prospects for profitable innovation would be much better. This argument is misleading.[5] The fact that a company would profit more from innovation X if it (alone) had stronger protection does not establish that it would profit more from X if protection were *generally* stronger in its industry. That might mean, on the contrary, that a rival firm (responding to the strengthened incentives) comes up with an innovation Y that makes X obsolete. Improving appropriability in a particular activity or industry tends to raise the standard of innovative competition—which may or may not be socially desirable, but cannot be presumed to be more favorable for profits for a particular company than any other rise in competitive intensity.

FOUR MECHANISMS OF APPROPRIABILITY

To create a broader strategy for appropriating gains, managers need to see intellectual property protection as just one of the mechanisms for protecting innovation gains, which are described here under four headings:[6]

- Patents and related legal protection.
- Secrecy.
- Control of complementary assets.
- Lead time.

The discussion here concerns the capture of gains from a potential innovation that is already "in hand." It may be an invention or a cluster of related inventions; it may be a new way of organizing internal functions or relations with customers or suppliers. Whatever else it is, it is not yet an "innovation" because the "putting into practice" that traditionally distinguishes innovation from invention has not yet happened, and in fact the details of putting into practice will be determined in part by the capture strategy. For simplicity, the term "invention" is used here for the generic potential innovation, and alternative possibilities are explicitly referenced when needed. It should be emphasized, however, that innovation very often takes the form of procedural and organizational changes that are not "inventions" as commonly understood, and certainly not patentable inventions.

Patents and Related Legal Protection

History demonstrates the power of patent protection, particularly for inventions related to the early development of new technology-based industries. For example, Alexander Graham Bell's basic patents on telephony permitted the Bell Telephone Company to achieve profitable dominance of the new U.S. telephone industry until those patents expired in 1894—after which the number of independent rivals grew over one hundred fold in less than 10 years.[7] More recently, the Cohen-Boyer patent covering gene-splicing techniques, a basic element of biotechnology, is estimated to have earned over $220 million for its owner, Stanford University. There are also examples of strong and very valuable patents on discrete inventions in established industries.

Patent protection, in the typical form of "utility" patents, is available for four types of subject matter—(1) machines, (2) products of manufacture, (3) compositions of matter, and (4) processes—plus improvements thereon. To be patentable, an invention must meet standards of usefulness (utility), novelty and non-obviousness. In practice, little bureaucratic effort is expended on critical evaluation of the inventor's contention that the invention is useful,[8] but a lot of systematic effort is expended on the "novelty" test, and especially on the narrow question of whether the applicant's invention is familiar or has already been patented.

Since 1995, the duration of a utility patent has been 20 years from the date of application.[9] During that period, the inventor has the legal right to prevent anyone else from using the invention—regardless of how it was obtained. That is, others are precluded from using the invention even if they created it themselves with no reference whatsoever to the patented version, and even if they were entirely unaware of the patent until notified that they should desist from infringing it. It is important also to emphasize the negative character of the right. If two or more legally distinct parties own patents on complementary inventions required in a single innovation, no one of them has the right to carry out the innovation unilaterally. A licensing deal must ordinarily be made if the innovation is to be brought about.

Limitations of Patent Protection. The effectiveness of patents is an issue that is considerably murkier and more ambiguous than the high-profile success stories of Bell Telephone or Cohen-Boyer suggest. Among the limitations are the high legal costs of defending patents and opportunities for rivals to "invent around" the patent. Patents vary greatly in both strength and value; some are valuable only when arrayed in mass formation with other related patents, some are relevant for a period much shorter than that of the patent grant, and some are little more than scarecrows to frighten off the most timid of the poaching imitators. Among the significant limitations of patents are:

- *Legal costs:* Indeed, the Bell patents themselves illustrate one dimension of this complexity. The patent grant per se did not confer dominance on Bell Telephone; that result was accomplished through the defense of that right by very protracted litigation involving more than 600 patent suits.[10] When the second component of a patents-and-litigation strategy turns out to be costly, a significant portion of the rents may be diverted to lawyers' fees and other legal expenses. Another example is Eli Whitney, inventor of the cotton gin, who spent much time and money defending his patent in court while his invention stimulated a boom in the value of cotton land, from which others made great fortunes.[11]
- *Limited effectiveness in some industries:* Economists have long been aware that the effectiveness of patents tends to vary systematically

across industries, as shown in Table 11.1. The table shows the results of two studies that support this picture of inter-industry variation—the "Yale survey" in the early 1980s and the "CMU survey," referenced earlier, conducted in 1994.[12] Table 11.1 illustrates the consistent (and unsurprising) finding that patents are quite effective in the pharmaceutical industry—and also that they are only of middling effectiveness in some industries usually considered highly progressive, such as computers and semiconductors.

- *Inventing around:* When R&D managers in these surveys were asked why they thought patents were of limited effectiveness, managers in both surveys cited the fact that competitors can legally "invent around" a patent as one of their greatest concerns.[13] That is, competitors can come up with their own inventions that do not infringe the patent but do succeed in appropriating some of the potential returns of the focal invention.[14]

The notion of "inventing around" offers insights into the permeability of patent protection and the broader challenges of establishing effective property rights in knowledge. To be invulnerable to inventing around, a patent must protect the key idea of the invention. Ideas, however, are considered too

Table 11.1
Patent Effectiveness Scores for Product Innovations, Selected Industries

Industry	*Yale*[a]	*CMU*[b]
Drugs	6.53	50.2
Plastic products	4.93	32.7
Medical instruments	4.73	54.7
Semiconductors	4.50	26.7
Petroleum refining	4.33	33.3
Aircraft and parts	3.79	32.9
Computers	3.43	41.7

a. Effectiveness of patents to prevent duplication, mean response on a 1 to 7 scale.

b. Percentage of product innovations for which patents considered effective in protecting competitive advantage, mean of responses.

vague to be patentable per se. An idea, unrelated to a specific physical means for carrying it out, cannot be patented. On the other hand, a patent may readily be obtained if such means are detailed very precisely, assuming other criteria of patentability are met. But such precision often leaves abundant room for an imitator to achieve a similar result by somewhat different means, thus, inventing around the patent. In effect, an overly-narrow patent of this sort achieves a result analogous to what is achieved by copyright, which does not protect the substantive content of a work, but only the "form of expression." Borrowing ideas from a copyrighted work without violating the copyright is typically (and intentionally, from a policy viewpoint) easy because there are many ways to express the same idea.

Patents can come closer than copyright to protecting ideas because an imitator's trivial "cosmetic" changes in the specifics of an invention are not sufficient to avoid infringement. This is the import of the "doctrine of equivalents" in patent law. As the Supreme Court said in 1950:

> [I]f two devices do the same work in substantially the same way, and accomplish the same result, they are the same, even though they differ in name, form, or shape.

The ability to recognize that two devices "do the same work in substantially the same way" closely approximates the ability to recognize "the same idea."

In practice, courts are more likely to recognize an "equivalent" when there is greater similarity in the physical expression of an idea. For this reason, patents tend to be relatively invulnerable to inventing around in areas of technology where the physical expression of ideas is relatively inflexible.

Drug discoveries provide important illustrations of this logic. Until the 1970s, pharmaceutical companies pursued new drugs by a "random screening" approach. They explored the medical efficacy of molecules by the thousands, looking for ones that would have therapeutic effect against some condition or other, and with little or no guiding idea for the search. The efficacy of a drug discovered by such a process is virtually an isolated fact; an imitator can attempt to make "cosmetic" changes in the molecule while retaining efficacy, but such attempts tend to be successful only when they are so trivial as to provide no defense against a charge of patent infringement. By contrast, the process of "guided" drug discovery that has become prevalent since the 1970s is guided by scientific ideas about mechanisms of action, such as looking for a molecule that blocks the action of a particular enzyme.[16] If the validity of the idea of treating a particular condition by blocking a particular receptor is demonstrated by a molecule identified by inventor A, inventor B may borrow the idea and develop another molecule that does "the same work" but in a quite different way—thus inventing around A's patent. Thus, "guided" drug search, or "rational drug design" is likely to produce a regime where pharmaceutical patents are more susceptible to inventing around than they were under the "random screening" regime.[17]

The knowledge context and the physical constraints of the subject matter—along with the evolution of the law—determine the ease of "inventing around," with different results in different areas and great variability at the level of the individual invention. While competent patent attorneys can shape the scope of the claims in an individual patent application, they cannot reshape the underlying knowledge conditions and physical constraints. Still less can they stop the leakage of other key information about the innovation that strongly affects the level and quality of rival efforts to encroach upon the gains, such as the demonstration of the existence of a previously unmet need and the identification of where that need resides. Some of this information may leak out in the patent application itself.

Some patents manage to resist this process of inventing around, continuing to be highly valuable for their full legal duration. The end of legal protection then brings a precipitous decline in the rent stream. For example, Glaxo Wellcome's anti-ulcer medication, Zantac, became the world's largest selling prescription medicine, with sales peaking at $3.8 billion in 1994.[18] When the U.S. patents on the drug expired at the end of July 1997, generic competitors captured over half the prescriptions within one quarter. From 1996 to 1998, Glaxo Wellcome's sales of the drug declined from $3.01 billion to $1.26 billion.

In many cases, however, science and technology catch up with the patent in a period substantially shorter than the patent life. Zantac itself was the villian of such a tale for Tagamet, the anti-ulcer medication that Zantac surpassed as the leading prescription drug in 1986, eight years before Tagamet went off patent. Outside the pharmaceutical industry, examples of rent streams shorter than the patent life seem to represent the typical case. Both the Yale and CMU surveys inquired into the lengths of "imitation lags" for both patented and unpatented innovations, and both surveys yielded evidence that these time periods are remarkably short—with "imitation lags" of less than five years, and often much shorter.[19] (It should be noted that the term "imitation lag" is used rather loosely here: The survey did not inquire directly into imitation, but asked how long it was before another firm introduced a "competing alternative" for a recent major innovation of the respondent's firm.) What patents accomplish, most typically, is a marginal improvement in lead time.

Frustration with the imperfections of the patent system, stimulated perhaps by persistent visions of the rents that got away, has led to calls for improvements. A certain utopianism often appears in this commentary. Such radical reforms appear unlikely, given that patent laws are implemented by government bureaucracies like the U.S. Patent and Trademark Office (PTO), and this is a time of budgetary stringency, diminished respect and non-competitive pay for government employees, rapid technological change and burgeoning growth in patent applications. Rather than railing against the weaknesses of the patent system, managers may more profitably put their energies into other mechanisms for appropriating innovation gains.

Secrecy

Secrecy is the analogue in the knowledge realm of a fence around real property—a straightforward but limited type of protection that the owner can provide without aid from a government.[20] As such, it plays a gap-filling role in the strategic repertoire of a private entity: It is the available private means to provide protection that public institutions have failed to afford. Its importance tends to be greatest when the performance of public institutions is the weakest (e.g., in less-developed countries).

In the knowledge strategy of an innovative company, there is virtually always some role for secrecy. Whether it can or should play the central role in appropriability strategy for a particular innovation is generally a harder call. Sometimes it cannot, and sometimes it should not because the sacrifice of other objectives is too great. The potential protection offered by secrecy varies dramatically, depending on attributes of the discovery and the circumstances of its use.

Protecting innovative products through secrecy is difficult. Aspiring imitators often can acquire the product—either by posing as legitimate customers or by inducing legitimate customers to pass along the product or information about it—and then reverse-engineer it. This vulnerability typically emerges only after the product is available in the market, which gives the innovator the advantage of lead time, particularly if secrecy is maintained during product development. Manufacturers also may make their products resistant to reverse-engineering. For example, semiconductor manufacturers have sought to protect their devices from reverse-engineering of circuit design by enclosing them in epoxy resins, so removing the coating destroys the circuit.[21] Other companies retain physical control of the innovative product and sell its services, as may be possible for the equipment in manufacturer-operated repair facilities or for equipment supporting business services such as data processing.

In general, it is easier to keep secrets about processes than about products. Production processes can usually be conducted behind the company's walls, out of sight of potential imitators. Companies also can discourage loose talk or outright defection by the personnel involved. Some business processes are inherently hard to protect, however. Some are "observable in use," for example, the manner in which service personnel relate to customers. Others rely on secrets that can be conveyed very simply, however difficult they may have been to discover in the first place. For example, it has been reported that one of the "secrets" of McDonald's famous french fries is to cook them "until the oil temperature has climbed three degrees above the low temperature reached when the potatoes are put in the oil."[22] Such a simple rule is obviously a hard secret to keep. The situation is only slightly different when the secret information cannot easily be expressed in ordinary language, but can be rendered precisely in an appropriate "code" that numerous experts understand—for example, the reaction path of a chemical process.

At the other extreme, the risks of leakage may be minimal because the secret is too large, too complex or too resistant to articulation to transfer easily. The detailed operating techniques that account for high yields in a semiconductor fab are one example, the combined wisdom and experience of a team of expert consultants are another.

Efforts to facilitate internal knowledge transfer by codifying the key information may have the unintended consequence of facilitating the escape of the secrets to imitators.[23] The severity of this effect depends on the extent to which the knowledge and capabilities required to interpret and exploit the codified account are widely available among potential imitators.[24] Rarely, however, can a discovery be effectively exploited while its principal secrets remain locked up in a big safe or a secure R&D lab. Large scale exploitation is likely to require that many people know something about those secrets, and/or that a large number of documents record them at least in part. It would be self-defeating to make major sacrifices in the breadth of the rent stream in the attempt to extend its length through secrecy. But managing such strategic trade-offs can be painful, especially for a small company aware that large rivals can bring to bear many strengths in exploitation if they once get hold of the discovery itself.

It is not just the current innovation but also the future productivity of the company's R&D activity that can be hampered by too much secrecy. Within the company, attempts to enhance secrecy through compartmentalization of information may reduce R&D efficiency in a variety of ways. Heavy-handed enforcement of secrecy rules may produce disaffection in key employees.[25] Either success or failure in one project may provide useful input to another, but strict isolation of projects will postpone the sharing of success and perhaps sacrifice the learning from failure. Numerous small "wheels" may have to be reinvented time and again. The byproducts that are useless in their project of origin may provide the keys to success in another project—as in the famous case of the failed adhesive that was the technical key to 3M's Post-It Notes.[26] Such benefits will not be exploited if a penchant for secrecy consigns the information about them to the burn bag.

Similar issues arise with respect to communication across company boundaries. Secrecy aims to prevent valuable information from leaking *out,* but the sealants used may be equally effective in preventing valuable information from leaking *in*. There is good reason to believe that informal professional networks contribute importantly to the productivity of scientists and engineers. (See, for example, Lori Rosenkopf's discussion of knowledge networks in Chapter 15.) Informal norms of trust and reciprocity guide interactions in such networks—A's willingness to share information with B depends on A's perception that B won't make him regret it in the short term and might well reciprocate in the long term.[27] The more significant the information, the more prominent these norms are likely to be. As a result, company policies that impede intellectual trade may be beneficial in the short term—because leakage is stopped while inflows temporarily continue—but counter-productive in the long as new ideas cease to flow into the organization.

There appears to be increasing awareness among managers of the importance of secrecy in appropriating innovation gains, as seen in a striking contrast between the results of the Yale survey and those of the CMU survey eleven years later. Secrecy ranked last overall in the Yale survey among the major means for capturing returns from product innovations, but in the CMU survey secrecy ranked at the top, in a virtual tie with lead time. Adjusting for difference in industry definitions, the CMU survey shows secrecy ranked first or second in 24 out of 33 comparison industries, as against none in the Yale survey.[28] This change may reflect a general intensification of competitive pace and pressure, including a heightened awareness of the hazards of industrial espionage. (It is also possible that this seemingly strong result is partly an artifact of differences in the details of the two surveys—for example, differences in the specific question asked—but the survey results do indicate the importance of secrecy in protecting innovations, particularly in the early stages.)

A company's efforts to protect its secrets are "leveraged" by the law of trade secrets. The two key conditions for information to be entitled to protection as a trade secret are that it affords some competitive advantage (if not, who cares?) and that it be treated as a secret by its owner. The latter condition makes improper acquisition of trade secrets an offense akin to the crime of breaking and entering: there is, by definition, no such offense if the door was left unlocked. Thus, by such measures as maintaining physical security and limiting access to sensitive premises, avoiding inadvertent disclosure, insisting on confidentiality, nondisclosure and noncompete agreements with employees and others exposed to the information, a company not only raises a direct barrier to the appropriation of its secrets but also satisfies a crucial precondition for legal action against anyone who might surmount the barrier.

Trade secret law affords no protection, however, if a rival independently conceives of the same information or discovers it by parallel research—legitimate activities that do not require the transgressing of any barrier set by another "owner." This represents an important distinction between secrecy and patent protection, since in principle patent protection does bar the use of the invention by an independent inventor.

Complementary Assets

Controlling complementary assets can help to capture the gains from an innovation. As discussed by Mary Tripsas in Chapter 8, complementary assets include resources such as access to distribution, service capability, customer relationships, supplier relationships, and complementary products. The firm that controls these assets may be more likely to benefit commercially from the innovation.

Perhaps the simplest example of the effect of complementary assets arises when a new product lacking patentable features requires specialized facilities and capabilities for its manufacture. If the innovator is the sole owner of such complementary assets, the lack of patent protection may matter very little. (As this example suggests, the term "assets" is to be understood expansively in this discussion, and includes asset services and organizational capabilities as well as assets in the narrow sense.) Unlike patent protection, which the law offers only to the "original and true inventor" or his assignee, the opportunity to capture returns through complementary assets is open to anyone who holds such assets—whether the innovator or someone else.

An innovator who lacks some of the critical complementary assets may have great difficulty capturing the gains.[29] Suppose, for example, that the innovation is "co-specialized" to assets held by one or a few other companies—not useful at all except in conjunction with those assets. In that event, even if the legal protection for the innovation is secure, those other actors are in a position to insist on a slice of the pie and are not likely to pass it up. On the other hand, when intellectual property protection is strong, complementary assets are of a generic character, or services are available in a reasonably competitive marketplace, innovators who lack complementary assets may not pay too high a price to obtain them.

It would be hard to imagine a more emphatic demonstration of the relevance of complementary assets than is provided by the biotechnology industry, and its pharmaceutical branch in particular.[30] In the earliest days of the industry, new technology emerged from university laboratories by way of a large number of entrepreneurial start-up companies, usually with university links. The first, and one of the most successful, of these new firms was Genentech, co-founded by Herbert Boyer, the Stanford University scientist who was also co-inventor of the recombinant DNA technology basic to the biotechnology industry. These small companies had R&D capabilities and programs, or dreams, and very little else. Some developed new drugs for which they obtained patent protection, but the complementary assets and capabilities required for actually getting new drugs to market lay in the hands of the established pharmaceutical companies. Those companies had the ability to arrange clinical trials, obtain FDA approvals, establish manufacturing facilities and get regulatory approval for processes, and ultimately market and distribute the drugs.

The ability of the established pharmaceutical companies to come into the biotech game relatively late, and profit from it, is a strong confirmation of the role of their complementary assets in appropriating the gains of biotech innovations. At the same time, however, patent protection of the start-ups (and their control over research capabilities that could be valuable in the future) helped them retain the gains that otherwise might easily have been seized by their giant partners.

The power of complementary assets also can be seen in the strength of post-war IBM.[31] IBM developed a strong sales force and service/support organization, and innovation and manufacturing capabilities. By leveraging these pre-existing capabilities and building new ones, IBM arrived as of the mid-1950s at a domestic market share of about 85 percent—which it then maintained for three decades, despite waves of technological change. As product and process innovations flowed through the industry, IBM's position gave it the opportunity to make about five times as much use of any given innovation as the rest of the domestic industry combined. The fact that innovation overall was so rapid meant that an occasional innovative success outside the company could not (for a long time) imperil IBM's position. Other companies and independent inventors found they needed IBM for manufacturing and market access. As a result, IBM was in a position to pick and choose from the streams of innovations, and it captured a substantial portion of the gains from those innovations. Technical and financial success naturally fed back on itself, leading to the accumulation of even more complementary assets.

In this multidecade saga of industrial leadership, patents apparently played no significant subsidiary role. Most of the scholarly accounts of IBM's dominance make no mention at all of patents or other intellectual property considerations, or mention them only for the purpose of discounting their importance.[32]

As the example of IBM illustrates, the most powerful role of complementary assets is not in appropriating gains from a single innovation but across a wide range of innovations. IBM created a strong *dynamic capability*—the ability to make continuing improvements in a related set of products and processes by re-deploying related assets and capabilities in R&D, manufacturing, marketing, and other functional areas.[33] In the context of such strength, successful appropriation of gains from individual innovations is largely pre-ordained—and when it is not, it doesn't matter much because strong dynamic capabilities are likely to prevail again in the next round. In the large firm, dynamic capability is the proverbial goose that lays the golden eggs of innovation. As long as the goose is secure, it doesn't matter much if somebody makes off with a few eggs occasionally. Of course, that's not to say that we *like* having somebody steal our eggs.

Lead Time

The CMU survey ranked "lead time" as the most effective appropriability mechanism for product innovations, slightly higher than secrecy (based on the percentage of innovations for which the mechanism is considered effective).[34] As in the case of other mechanisms, there is substantial variation across industries in the effectiveness of lead time. For example, respondents rated it as effective for almost two-thirds of product innovations in communications equipment, auto parts and cars and trucks, but for only about one-third of innovations in electrical equipment.[35]

The time scale of lead time advantages also varies across contexts. Achievable leads tend to be quite short when the opportunity for a particular type of innovation is, so to speak, on a public agenda that is visible to numerous competent players. When novel resources offer innovative opportunity to all comers on reasonably similar terms—such as a latest-generation microprocessor for PCs—the company that moves fastest in developing the innovative application will be able to gather rents until followers appear to chisel the advantage away. The length of the lead time is determined by the good luck, flexibility and skills of the leader as well as the bad luck, inertia and incompetence of the others.

An innovative leader can obtain longer lead times if the innovator is not merely the inventor of the winning entry in a widely recognized contest, but the creator of the contest itself. Few if any rival efforts get under way until the innovator's experience provides convincing demonstration of the opportunity.[36] Lead times then tend to be longer, to appear in combination with other innovative advantages of the leader, and to provide opportunity for strengthening appropriability further through complementary assets or other means. The extent and durability of such competitive advantages has been the subject of extensive discussion about the strength of "first mover advantages" and the impact of "hypercompetition."[37] Similarly, the question of what being first is worth per se has been difficult to answer.

It is clear, however, that the dramatic successes achieved by many "first movers" do not constitute a convincing case that all aspiring innovators should try harder to be first. On the one hand, leaders often have other advantages that contribute to their success. On the other, extreme efforts to rush a product to market may lead to an excessive sacrifice of quality, reliability, and readiness to provide service support.

Lead time can be coupled with learning capabilities to appropriate gains from innovations for longer periods. For example, when Nucor introduced compact strip production (CSP) process in steel making, its German equipment supplier, SMS Schloemann-Siemag (SMS), controlled the intellectual property for the technology.[38] This created the significant risk that Nucor's learning from the initiative might well be captured by SMS and then transferred along with new equipment installations to Nucor's steel industry rivals. While the new technology would enable Nucor to produce higher quality flat steel and move into higher markup markets, this also had the potential to incite the wrath of large integrated rivals such as USX who dominated those markets. Nucor went ahead nevertheless. Learning proved difficult and expensive, but a few years later Nucor had mastered the technology while the rest of the industry stood aside and watched. CSP and other innovations helped Nucor generate an annual total return to investors in excess of 27 percent from 1984 to 1994, a far better performance than its principal rivals.[39]

Although it initially seemed competitors could easily follow Nucor's highly visible lead, this example shows that the protections of lead time are stronger than an abstract view might suggest. Nucor's ability to profit from CSP rested on its substantial dynamic capabilities as a process innovator in steel production, which endowed it with a potential lead time advantage more fundamental than a mere decision to go first. And in braving the hazards of retaliation from large integrated firms, or of finding the fruits of its costly learning dispensed free to other rivals by SMS, Nucor drew on well-founded confidence in its learning abilities and problem-solving culture.

The impact of lead time also is affected by product characteristics, particularly durability. The more durable the product, the more valuable a given lead tends to be. This is because the appearance of an innovative version of a product stimulates a burst of purchasing, after which demand is largely driven by replacement. Airlines, for example, typically only buy new aircraft to replace planes in their existing fleets, or to support an expanded schedule of flights. When innovation improves the price/performance significantly, the result is a wave of obsolescence of the existing stock, and increase in purchases. Once this bulge in demand works through, the manufacturers confront only replacement demand again. Thus, the company that can be first in bringing to market a major improvement in the product may be able, if its lead time and manufacturing capability allow, to be the supplier for a major fraction of the stock adjustment. Given some willingness of buyers to accept delivery lags from the innovator, the lead time required to capture most of the gain may be only on the order of a year or two, or even less. A series of relatively minor wins of this sort can lead over time to the dominance of an industry, as the history of the commercial aircraft industry illustrates.[40]

Lead time advantages also are increased when the innovator can establish a strong reputation or customer switching costs are high. Particularly in consumer markets, there is enough inertia in demand to confer surprisingly large benefits on the producer who meets the demand first. Aggressive competition from otherwise qualified followers is largely forestalled simply because the price or equivalent concessions that they must make to attract the customers are too large to make it worthwhile.[41] The mutual fund industry provides a striking illustration of this point. No innovative product is more readily "reverse engineered" than a novel type of mutual fund. Yet it appears that the "first mover" advantage in the industry is quite significant,[42] though it seems to rest on nothing more substantial than the inertia and friction that retard the flow of customers toward the best offer.

Finally, lead time may offer not only the opportunity to do some profitable business before rivals appear on the scene, but also to lock up complementary assets—with the result that the rivals find life quite frustrating, even when they do appear on the scene. As mentioned above, Starbucks' ability to make high quality coffee beverages is easily imitated but the company developed a capability to open new stores at a remarkable pace, now on the order of a store a day, while maintaining its high quality standards in all aspects of its operations. That capability, not the ability to make good coffee drinks, was the "inimitable" aspect of its innovative success. By identifying the best store locations, acquiring them and putting attractive and well-functioning stores in them, Starbucks altered innumerable local environments in ways unfavorable to late-coming competition.[43]

Appropriability Issues in Emerging Technologies

How can companies use these four mechanisms to appropriate the gains from emerging technologies? There is no magic formula, no one-size-fits-all answer to the question of how to capture gains, either in general or—even more emphatically—for the specific context of new industries arising from emerging technologies. Much of the art of successful strategy, in this area as elsewhere, lies in the accurate matching of the specifics of the solutions to the specifics of the problems. Because the diversity of the problems arising from the variety of company situations and contexts is enormous, a useful handbook can hardly be provided in a few pages. The exhortations and observations in this section are put forward in the hope of providing useful guidance for this matching process—and in the spirit of this chapter's general warning against locating capture strategy in the cramped quarters of the intellectual property rubric.

Identify the Uncertainties

Uncertainty tends to be pervasive in industries based on emerging technologies and appropriability conditions share that tendency. As one example, intellectual property law evolves with court decisions and legislative actions, which can increase or erode the protection given to technology development already in progress. For example, the Supreme Court decision in *Diamond v. Chakrabarty* was an important legal milestone for the biotechnology industry because it opened the way to patenting of things that, though not naturally occurring, were derived from living things. This decision came in 1980, eight years after the patent application was filed. It may have helped to trigger the boom in biotechnology start-ups that occurred in that year, but a substantial number of such firms were already pursuing research programs whose value was likely to be affected by the ruling.[44]

There is also considerable uncertainty about the effectiveness of secrecy in protecting competitive advantages, particularly in the early stages of emerging technology industries when markets for personnel holding key (and arguably proprietary) know-how are unusually fluid. For example, one leader of a very successful information services start-up bemoaned the loss of a talented young programmer. The programmer was lured away with an annual compensation offer that leaped from $75,000 to $350,000 in a single bound—a jump that may have reflected his possession of key knowledge of the practices of his former employer. Trade secret law may provide some protection against losses by this mechanism, but in the tumultuous context of a young industry, the protection is often weak.

Similarly, substantial uncertainty surrounds the claims that holders of complementary assets may have against innovation rent streams. It is generally fairly clear what sorts of assets are relevant to a particular line of innovative activity—but it is not at all clear how supply and demand for those assets will develop in a highly dynamic context. That is what determines the rents that such assets can command. For example, it is clear that innovative online retailers like Amazon.com need the support of the more conventional assets and capabilities required in the physical warehousing and distribution of the goods. If competition for these assets intensifies, it is quite conceivable that much of the gains from online retailing will be transferred to firms like Ingram's and UPS that are well positioned in the complementary assets—or the value could ultimately go to consumers.

Companies pursuing lead-time advantages in an emerging technology-based industry have even less certainty. A dense fog of uncertainty obscures the positions of the contestants on the various technological racecourses; indeed, some of the racecourses themselves may be invisible. A company trying to establish its position in assets complementary to its own technological approach, in the hope of assuring long-term dominance, may find its approach rendered obsolete and its investments devalued by some entirely different approach. Even the brief lead-time advantage that its approach might have enjoyed may be sacrificed in the quest for long-term advantages. Such scenarios seem to play out recurrently in the telecommunications industry, where the quest continues for more and cheaper bandwidth to support the needs of the Information Age and both technical and regulatory uncertainties abound. Worst of all, the firm that picks what turns out to be the right technological goal and succeeds in getting there first may find that it has arrived too early for the party—when the technology is immature, Wall Street is not yet impressed, and the customers have not caught the fever yet. Its followers will learn from its mistakes.

Although uncertainty is largely inescapable where emerging technologies are concerned, the effort to identify the key uncertainties affecting the gains from a specific innovation is still worthwhile. It is often possible to develop strategies to address at least some of those uncertainties—by buying real options or making hedging investments where they are not too expensive, by devising probes for answers to the most critical questions, or by refining monitoring mechanisms to trace developments more effectively. Beyond that, stay alert and wait for the dust to settle.

Assess and Reassess the Knowledge Environment

The effectiveness of the various appropriability mechanisms depends on the character of the knowledge environment. Managers can ask the following questions to better understand their environment and its impact on appropriating innovation gains:

- *Who are your competitors?* Companies may be the lone runner in a particular new field or face a close race among many rivals seeking to reach the market with a new generation of an established product.

The company's position on this continuum affects its appropriation strategy. Crowded fields tend to generate close races, with obvious implications for lead time and patent decisions. If there are no products out on the market, how can these rivals be identified? Although these competitors are often invisible in the early stages, they can be imagined by careful consideration of the background features that led to the pursuit of the innovation. Specifically, managers can answer the question: "How did it happen that we got involved in this particular quest?" If enough humility can be mustered so that the answer "brilliant creative insight" can be set aside, some more specific explanations may surface. How many other people might have read both of those journal articles that we insightfully juxtaposed? How many others might have suffered a particular frustration or problem in research or in everyday life? How many saw the other company's machine at the trade show and wondered about their design choices? Who also trained with the Nobel Prize winner who dominates the relevant branch of science? By this path, it may be possible to gain a sense of the likely number of invisible rivals in the innovative race—rivals who are out there but perhaps taking care not to make themselves visible.

To take a less defensive approach, managers might ask: "Where should we look for the innovative ideas that are *least* likely to be under active exploration by someone else, and thus most likely to bring us lead time advantages or a successful patent application?" A possible answer is to identify the innovative ideas already known to us but not known to the world. This points to the importance of stepping quickly on the stepping stones of our own achievements, searching carefully for the new possibilities that such achievements open up—before the rest of the world gets a look at them.

- *Where are the headwaters of the new knowledge that flows in this field?* This question is an extension of the sort of analysis just mentioned. Biotechnology is the leading contemporary example of an emerging technology with deep and recent roots in government-funded academic research. Companies hire recently trained PhDs, who then pursue commercially valuable inventions and innovations in areas closely related to—or even identical with—their dissertation research. The aspiring innovator in such a field, where the headwaters are well defined and knowledge at the headwaters is highly relevant to innovation, should be in touch with what is happening at the headwaters. Otherwise, there is the risk that the apparent opportunity spotted floating by a downstream location has already been seen upstream by other more qualified participants, who are either actively working on it, hence ahead in the race, or have convinced themselves that it won't work. Access to the knowledge flows from the headwaters often provides general guidance to a company's innovative efforts that goes beyond the specific information acquired. If an assessment made with good upstream knowledge suggests that the concept under consideration is highly original, a major break with standard thinking, it is unlikely that the field is very crowded.

In contrast, the headwaters of information technology are harder to identify because there are probably more college drop-outs than PhDs active in innovation. The headwaters are everywhere and nowhere (although Silicon Valley might be a good place to start looking). There is little that can be done to generate reasonable confidence that an apparently innovative concept has not occurred to someone else—and such confidence is rarely warranted. The result is that speed and quality of execution count for virtually everything, and rarely are there more than short leads available for the winner to exploit.

- *How fast are the rapids?* When knowledge is advancing rapidly, the rent streams that can be captured from individual innovations tend to be short regardless of the means adopted to try to protect them. Even a position well fortified by strong patents can simply be bypassed by the technological wave front. For example, Kodak and Polaroid were slugging it out in chemical photography while the action moved to digital. Under such conditions, emphasis should be given to effective exploitation and leveraging of whatever lead time can be gained, and to complementary assets and dynamic capability advantages that are more durable—as the IBM example vividly illustrates.
- *What are the prospects for strong patents?* As discussed earlier, patents are not the sine qua non of successful rent capture that they are often made out to be, but strong patents are certainly useful when they are attainable. Strong patents are critically important for a small, highly innovative firm that comes up with a new product so advanced that it may even be ahead of its time. It needs long-lasting protection if it is to capture a reasonable share of the rents, lead-time advantages are likely to be slim because the product is too early relative to the demand, and in any case the potential rent stream is a long one. Secrecy is of little avail; imitators will have time to reverse-engineer the product and still serve the first wave of demand. Being small, the company is likely to have to make deals to access the assets and capabilities complementary to its own innovations. It also is unlikely to have resources

to risk on extended patent litigation or to have the time and resources to accumulate a large portfolio of supporting patents. In this situation, strong patent protection is about the only resource that may be powerful enough to keep the larger, stronger late-coming rivals at bay. The unambiguous "need" for such strong patents, however, by no means guarantees that they are attainable. When the outlook is bleak, the realistic prescription is to prepare to be content with less than a big win.

There are two considerations that may be indicative of the outlook for obtaining strong patents. They both bear, in different ways, on whether the patent is relatively invulnerable to "inventing around." First, if the courts are likely to consider the patent to relate to a "pioneer invention"—defined by the Supreme Court as "a patent covering a function never before performed, a wholly novel device, or one of such novelty and important as to mark a distinct step in the progress of the art"[45]—they will apply the doctrine of equivalents in a manner generous to the patent holder and "stretch to find infringement even by a product whose characteristics lie considerably outside the boundaries of the literal claims."[46] The second consideration is whether, apart from its possibly pioneering character, the invention embodies ideas that are relatively inflexible in their technological expression. The less flexible they are, the harder it is for the imitator or independent inventor to accomplish the result in a manner that does not bear a strong resemblance to the protected manner. This may well be the underlying consideration that accounts for the fact that patents tend to be strong in the chemical industries; in particular, molecules are relatively inflexible ways of expressing the "ideas" associated with their functionality.

Guard the Golden Goose

There is no better hedge against failures of rent capture for a particular innovation than the ability to come up with the next innovation at modest incremental cost. When the fund of technological opportunities is rich and the pace is rapid, it is foolish to focus overmuch on the individual battles and neglect the prosecution of the war. It is wiser to turn attention to building and sustaining the dynamic capabilities that permit the company to track and sometimes lead the portions of the technological frontier that are most relevant to its business. Long-term survival depends on figuring out what to do for an encore, and then for an encore to the encore, and ultimately on building the dynamic capability to support sustained innovative accomplishment. Companies that succeed in doing this create a moving target for the competition that is much harder to hit than the stationary targets represented by the individual innovative achievements. Also, the subtle internal dynamics of a successful R&D organization are much harder to reverse-engineer and imitate than the specific products the lab turns out.

A key strategic question here is the scope of the dynamic capability, the section of the technological frontier in which the company seeks to maintain a significant presence. If the scope is too narrow, there is the risk that a relatively minor shift in the conditions of technological opportunity will make the company's capabilities obsolete. Also, opportunities for fruitful cross-fertilization and recombination among the company's different capabilities are limited. If the scope is too broad, there is the danger of losing focus and flexibility; the managers who make top-level resource allocation decisions become too remote from the action to make informed judgments—or are close to only part of the action and hence find it difficult to make impartial judgments. Large companies tend to fall into this trap, while smaller companies tend to suffer from too narrow a focus on a single innovative success that cannot be sustained forever.

Emerging technologies face an even more basic question: Which goose will produce the golden eggs? Even a very broad and strong dynamic capability may not be worth further investment if the very foundations of competitive advantage are being undercut by "competence-destroying" change.[47] And since it is likely to be impossible to master a new field comprehensively and simultaneously, the beginnings of dynamic capability in a new technology are almost necessarily of narrow scope.

Managing the dynamic capabilities needed for success in emerging technologies raises increasingly complex human resource challenges, as discussed by John Kimberly and Hamid Bouchikhi in Chapter 18. The mobility of workers and growing need for organizational flexibility not only are redefining working relationships, but also mean there is a general tendency for the rents to shift to the technical talent as the whole compensation profile floats upward with the salaries of the stars. Successful response to these contemporary threats to the foundations of sustained innovative achievement may do more for long-term profitability than the effort to surround the quicksilver of intellectual property with legal fences originally designed for real estate.

NOTES

1. The term *intellectual property* is sometimes used quite broadly. It is here used narrowly, to refer to the sorts of things that are (arguably) eligible for recognition and protection under the laws of patent, copyright, trademark and trade secrets. A first bit of perspective is implicit in this usage: for example, ideas are products of the intellect but there is no legally recognized property right in ideas as such; hence it does not seem helpful to think of ideas as intellectual property.
2. W.M. Cohen, R.R. Nelson, and J. Walsh, "Protecting Their Intellectual Assets: "Appropriability Conditions and Why U.S. Manufacturing Firms Patent (or Not)" Working Paper (Pittsburgh: Carnegie Mellon University, 1999).
3. See note 2, p. 5.
4. H. Schultz and D.J. Yang, *Pour Your Heart Into It: How Starbucks Built a Company One Cup at a Time* (New York: Hyperion, 1997), pp. 76–77.
5. From a public policy standpoint, the case for strengthening protection where it is currently weak deserves more serious consideration. The different levels of innovativeness that prevail across sectors and types of innovation undoubtedly reflect to some extent the differential effectiveness of the intellectual property system—though not in the simple way that is sometimes suggested The complexity arises from the fact that enhanced protection of a given innovation generally means partial suppression of knowledge spillovers, which in turn means that the costs of further innovation are increased. See, for example, R.P. Merges and R.R. Nelson, "On the Complex Economics of Patent Scope," *Columbia Law Review*, vol. 90 (1990), pp. 839–916.
6. Other systems of categories have been used; for example, the CMU study cited above treats complexity as a separate category, while it is here treated as contributing to secrecy. Trade secret law is also usefully treated under that heading, although it is obviously a form of legal protection and often viewed as a cousin to patent law. The CMU study also explores the effects of combination of mechanisms, finding unsurprisingly that complementary assets and lead time advantages go together. See note 2, p. 7.
7. W. Barnett, "Telephone Companies," *Organizations in Industry: Strategy, Structure, and Selection*, eds. G.R. Carroll and M.T. Hannan (New York: Oxford University Press, 1995).
8. This lack of attention to utility means there is abundant work for anthologists of "wacky and wonderful" patents. For some of these, consult www.colitz.com/site/wacky/wackyold
9. Previously, protection normally extended for seventeen years from the issue date. This change was enacted pursuant to international agreements on harmonization of intellectual property law made at the Uruguay Round of GATT (1994).
10. See note 7.
11. J. Hirshleifer, "The Private and Social Value of Information and the Reward to Inventive Activity," *American Economic Review*, vol. 61 (1971), pp. 561 574.

12. R.C. Levin, R. R. Nelson, and S. G. Winter, "Appropriating the Returns from Industrial Research and Development," *Brookings Papers on Economic Activity* No. 3783–820 (1987). The surveys differed in the framing of specific questions, and they were conducted eleven years apart, so not too much should be made of the difference in results between the two.
13. See note 12, pp. 802–803.
14. See note 2, p. 13.
15. Graver Tank & Manufacturing. Co. v. Linde Air Products Co., 339 U.S. 605, 608, as quoted by Merges and Nelson, see note 5; see the latter for discussion of the doctrine.
16. R. Henderson, R.L. Orsenigo, and G.P. Pisano, "The Pharmaceutical Industry and the Revolution in Molecular Biology: Exploring the Interactions among Scientific, Institutional and Organizational Change," *Sources of Industrial Leadership: Studies of Seven Industries,* eds. D.C. Mowery and R.R. Nelson (New York: Cambridge University Press, 1999).
17. G.P. Pisano, *The Development Factory: Unlocking the Potential of Process Innovation* (Boston: Harvard Business School Press, 1997), p. 65.
18. Converted from 2.4 billion pounds at 1.6 dollars per pound.
19. See note 2, p. 24.
20. As discussed below, government institutions supplement the "natural" effects of secrecy to some extent by enforcing trade secret laws and contractual obligations relating, for example, to confidentiality or noncompete agreements.
21. D. Shapley, "Electronics Industry Takes to 'Potling' Its Products for Market," *Science,* vol. 202 (1978), pp. 848–849.
22. R.H. Hayes, G.P. Pisano, and D.M. Upton, *Strategic Operations: Competing through Capabilities* (New York: Free Press, 1996), p. 54.
23. S.G. Winter, "Knowledge and Competence as Strategic Assets," *The Competitive Challenge: Strategies for Industrial Innovation and Renewal,* ed. D.J. Teece (Cambridge, MA: Ballinger, 1987), pp. 159–184.
24. U. Zander and B. Kogut, "Knowledge and the Speed of Transfer and Imitation of Organizational Capabilities: An Empirical Test," *Organization Science,* vol. 6, no. 1 (1995), pp. 76–92.
25. J.P. Liebeskind, "Keeping Organizational Secrets: Protective Institutional Mechanisms and Their Costs," *Industrial and Corporate Change,* vol. 6 (September 1997), pp. 623–663.
26. M.J.C. Martin, *Managing Innovation and Entrepreneurship in Technology-Based Firms* (New York: Wiley, 1994), p. 45.
27. E. von Hippel, "Cooperation between Rivals: Informal Know-How Trading," *Research Policy,* vol. 16 (1987), pp. 291–302; see note 25.
28. See note 2, p. 11.

29. D.J. Teece, "Profiting from Technological Innovation," *Research Policy,* vol. 15, no. 6 (1986), pp. 285–305.
30. On this subject, see note 16, on which the present account draws.
31. This discussion of IBM draws heavily on S.W. Usselman, "IBM and Its Imitators: Organizational Capabilities and the Emergence of the International Computer Industry," *Business and Economic History,* vol. 22 (winter 1993), pp. 1–35.
32. For an example of a dismissive comment about the relevance of patents to IBM's success, see G.W. Brock, *The U.S. Computer Industry: A Study of Market Power* (Cambridge, MA: Ballinger, 1975), p. 64. Studies that make essentially no mention of patents include F.M. Fisher et al., *Folded, Spindled, and Mutilated: Economic Analysis and U.S. v. IBM* (Cambridge, MA: MIT Press, 1983); and R.T. DeLamarter, *Big Blue: IBM's Use and Abuse of Power* (New York: Dodd, Mead & Co., 1986). These two represent the *pro-* and *contra*-IBM views of the most recent antitrust case. Also see, B.G. Katz and A. Phillips, "The Computer Industry," *Government and Technical Progress: A Cross-Industry Analysis,* ed. R.R. Nelson (New York: Pergamon Press, 1982); R. Langlois, "Cognition and Capabilities: Opportunities Seized and Missed in the History of the Computer Industry," *Technological Innovation: Oversights and Foresights,* eds. R. Garud, P.R. Nayar, and Z.B. Shapira (Cambridge, MA: Cambridge University Press,1995); T.F. Bresnahan and F. Malerba, "Industrial Dynamics and the Evolution of Firms' and Nations' Competitive Capabilities in the World Computer Industry," *Sources of Industrial Leadership: Studies of Seven Industries,* eds. D.C. Mowery and R.R. Nelson (New York: Cambridge University Press, 1999); Usselman, "IBM. . . ." Although "absence of evidence is not evidence of absence," the uniform neglect of the topic by all these authors must surely tell us something about the actual importance of patents to IBM.
33. D. Teece, G. Pisano, and A. Shuen, "Dynamic Capabilities and Strategic Management," *Strategic Management Journal,* vol. 18, no. 7 (1997), pp. 509–533. Cohen et al. (see note 2, p. 23), point out that patents also may confer greater advantages when massed as patent "fences" than they do for individual innovations.
34. Lead time is referred to in the survey as the advantage of "being first to market."
35. See note 2, Table 1.
36. The innovator in such cases is playing the historical role of the entrepreneur, as that role was described in Joseph Schumpeter's classic book, *The Theory of Economic Development* (Cambridge, MA: Harvard University Press, 1934). First published in German in 1911.
37. For critical reviews of the "first mover" literature, see M.B. Lieberman and D.B. Montgomery, "First-Mover Advantages," *Strategic Management Journal,* vol. 9 (1988), pp. 41–58; and M.B. Lieberman and D.B. Montgomery "First-Mover (Dis)Advantages: Retrospective and Link with the Resource-Based View," *Strategic Management Journal,* vol. 19 (1998), pp. 1111–1125. On "hypercompetition," see R. D'Aveni, *Hypercompetition: The Dynamics of Strategic Maneuvering* (New York: Free Press, 1994).

38. This account draws primarily on R.S. Rosenbloom, *Nucor: Expanding Thin-Slab Capacity* (Boston: Harvard Business School Press, 1991) Case 9-792-023; and P. Ghemawat, *Nucor at a Crossroads* (Boston: Harvard Business School Press, 1992) Case 9-793-039.
39. *Fortune,* vol. 131, "The Fortune 500" (May 15, 1995), p. F-57.
40. A. Phillips, *Technology and Market Structure: A Case Study of the Aircraft Industry* (Lexington, MA: D.C. Health, 1971).
41. The level of the innovator's price affects how much room the followers have to attract customers by cutting price without making loses. This situation illustrates the point that an innovator may be able to protect the length and breadth of the rent stream by sacrificing some of its depth.
42. R. Makadok, "Can First-Move and Early-Mover Advantages Be Sustained in an Industry with Low Barriers to Entry/Imitation?" *Strategic Management Journal,* vol. 19 (1998), pp. 683–696.
43. S.G. Winter and G. Szulanski, "Replication as Strategy," *Organization Science* (2000).
44. A. Ryan et al., "Biotechnology Firms," *Organizations in Industry: Strategy, Structure and Selection,* eds. G.R. Carroll and M.T. Hannan (New York: Oxford University Press, 1995).
45. Quoted in Merges and Nelson, note 5. Whether the last clause adds something beyond what the word "pioneer" itself conveys seems open to doubt.
46. See note 5.
47. M. Tushman and P. Anderson, "Technological Discontinuities and Organization Environments," *Administrative Science Quarterly,* vol. 31 (1986), pp. 439–465.

[7]

ELSEVIER

Research Policy 32 (2003) 333–350

research policy

www.elsevier.com/locate/econbase

The product market and the market for “ideas”: commercialization strategies for technology entrepreneurs

Joshua S. Gans [a], Scott Stern [b,*,1]

[a] Melbourne Business School and Intellectual Property Research Institute of Australia, University of Melbourne, Melbourne, Australia

[b] Kellogg School of Management, Northwestern University, Evanston, IL 60208, USA

Abstract

This paper presents a synthetic framework identifying the central drivers of start-up commercialization strategy and the implications of these drivers for industrial dynamics. We link strategy to the *commercialization environment*—the microeconomic and strategic conditions facing a firm that is translating an “idea” into a value proposition for customers. The framework addresses why technology entrepreneurs in some environments undermine established firms, while others cooperate with incumbents and reinforce existing market power. Our analysis suggests that competitive interaction between start-up innovators and established firms depends on the presence or absence of a “market for ideas”. By focusing on the operating requirements, efficiency, and institutions associated with markets for ideas, this framework holds several implications for the management of high-technology entrepreneurial firms.

Keywords: Product market; Commercialization strategies; Technology entrepreneurs; Ideas market; Appropriability

1. Introduction

The past two decades have witnessed a dramatic increase in investment in technology entrepreneurship—the founding of small, start-up firms developing inventions and technology with significant potential commercial application. Because of their youth and small size, start-up innovators usually have little experience in the markets for which their innovations are most appropriate, and they have at most two or three technologies at the stage of potential market introduction. For these firms, a key management challenge is how to translate promising technologies into a stream of economic returns for their founders, investors and employees. In other words, the main problem is not so much invention but *commercialization*.

Effective commercialization strategies seem to differ across industrial sectors. For example, in the early 1980s computer industry, Sun Microsystems’ commercialization strategy involved direct entry into the workstation market. Sun’s entry was mostly discounted by established firms such as Digital, IBM, and Apollo Systems, giving Sun the time to translate its overall technological vision (“the network is the computer”) into a concrete series of technological, organizational, and market-positioning choices. As a disruptive entrant, Sun emerged as a leading computer hardware firm by building a novel value chain for high-end computer purchasers (Baldwin and Clark, 1997).

On the other hand, many technology entrepreneurs have secured extraordinary returns by integrating their innovations into an existing value chain, often

* Corresponding author.
E-mail address: s-stern2@northwestern.edu (S. Stern).
URL: http://www.mbs.edu/home/jgans/research.htm

[1] Stern is also associated with NBER, and Brookings Institution.

0048-7333/02/$ – see front matter © 2002 Elsevier Science B.V. All rights reserved.
PII: S0048-7333(02)00103-8

involving intimate cooperation with established industry players. For example, in the interconnection technology segment, the profits earned by companies such as American Internet Corporation and Growth Networks are the result of an alliance strategy (and ultimately an acquisition) by the industry market leader, Cisco Systems. In the case of the 20-month-old closely-held Growth Networks, the US$ 355 million acquisition in 1999 secured a return for Growth Networks' stakeholders valued at over *US$ 5 million per employee* (New York Times, 2000). From the perspective of the established firm, Cisco has retained market leadership over several generations of novel technology and despite a turbulent industry slowdown in 2000 and 2001.

This objective of this paper is to offer a synthetic framework identifying the drivers of start-up commercialization strategy and the implications of these drivers on industrial dynamics. This framework links strategy to the *commercialization environment*—the microeconomic and strategic conditions facing a firm translating an "idea" into a valuable proposition for customers. By focusing on the commercialization environment, we can assess why companies like Sun exploit technological leadership to construct a novel value proposition and compete against incumbents, while companies such as Growth Networks work with established firms and leverage the *existing* value proposition. Our analysis suggests that the crucial factor determining patterns of competitive interaction between start-up innovators and established firms is the presence or absence of a "market for ideas". By focusing on the operating requirements, efficiency, and institutions associated with markets for ideas, we offer a framework isolating how start-up commercialization strategy depends on the economic environment.[2]

To understand the role of markets for ideas, consider the experience of Robert Kearns, the independent inventor of the intermittent windshield wiper in the early 1960s. Unable to commercialize on his own, Kearns approached senior engineers at the Ford Motor Company, disclosing both the operating principles and functionality of his invention. After some negotiation, Ford rejected a licensing agreement with Kearns, but introduced a similar technology to the market shortly thereafter. For over 20 years, Ford and other automakers declined to pay Kearns royalties on this invention; it was not until the 1990s that Kearns successfully upheld his patent and extracted a portion of the economic returns (Seabrook, 1994). In this case, the absence of a market for ideas reduced Kearns' ability to earn returns on his invention and, by setting a precedent, eliminated the incentives for start-up innovation in the automotive technology sector.

Markets for ideas play a crucial role in shaping commercialization strategy and industrial dynamics. In order to understand that role, we build upon and advance the agenda laid out in Teece (1986), emphasizing two central elements of the commercialization environment: the nature of the appropriability environment, and the distribution of ownership and control over specialized complementary assets, such as distribution and manufacturing capabilities, or a brand-name reputation. As Teece highlights, the innovator's share of the value created by her innovation will be smaller when appropriability is weak (due to imitation by competitors) or when specialized complementary assets are controlled by other players (due to bargaining among actors in the value chain). Using this framework, Teece illustrates how these factors shape the innovator's strategy choice, highlighting the role that "hold-up" can play in fostering integration between innovators and complementary asset owners.

In this paper, we focus on the specific challenges faced by technology entrepreneurs. Our framework is premised on the insight that, for many start-up innovators, those firms that control key complementary assets are precisely those that are the most likely and/or most effective potential product market imitators. This perspective refines Teece's analysis, where the challenges associated with contracting for complementary asset access were treated distinctly from the potential for imitation. In our reformulation, a principal hazard in pursuing cooperation with complementary asset owners is the possibility that these owners are current product market players with incentives to expropriate the innovator's technology and commercialize it themselves.

[2] This framework builds upon and complement our theoretical and empirical research examining incentives and equilibrium commercialization strategy for start-up firms (Gans and Stern, 2000, in press; Gans et al., in press). In contrast to this earlier research, here we attempt to combine a rich phenomenological understanding of technology entrepreneurship with a nuanced and explicitly strategic framework in order to assess how commercialization choice varies with the economic environment.

As a result, commercialization strategy for start-up innovators often presents a tradeoff between establishing a novel value chain and competing against established firms versus leveraging an existing value chain and earning returns through the market for ideas. For example, when considering how to commercialize the Navigator browser in 1995, Netscape considered a technology licensing agreement with Microsoft before committing itself to competing head-to-head with Microsoft by offering an independent cross-platform browser product. If Netscape had instead collaborated with Microsoft (allowing both firms to avoid the costly standards battle which ensued throughout the late 1990s), this cooperative strategy (and the details of that licensing agreement) would have taken place in the *shadow* of potential product market competition.

By focusing on a commercialization environment where established firms can both control complementary assets and serve as potential imitators, our framework offers several new insights into effective commercialization strategy, including the role that appropriability, reputation and economic institutions plays in shaping strategic choice, and the impact of commercialization strategy on competitive dynamics. For example, whereas most previous strategic analysis highlights the level of appropriability as a driver of strategic choice, our framework suggests that the key driver of effective commercialization strategy should be the "type" of appropriability (e.g. whether appropriability is based on formal intellectual property rights such as patents versus informal mechanisms such as secrecy). Even when tight secrecy offers a strong appropriability environment, transacting in the market for ideas will often undermine that secrecy and increase the potential for expropriation. In contrast, when the innovator controls formal IPR such as a patent, the potential for expropriation will be reduced, and the innovator is likely to find a cooperation strategy more attractive.

As well, this framework highlights the role played by reputation and institutions for ideas trading in "mixed" environments—when the appropriability environment and complementary asset environment place competing pressures on the start-up in terms of commercialization strategy choice. For example, when the established firm controls necessary commercialization assets but negotiation risks expropriation, effective strategy depends on a careful evaluation of the reputation of alternative established players. Moreover, since knowledge of reputation may be difficult for a start-up innovator to observe, the market for ideas may be facilitated by the use of intermediaries, such as venture capitalists, who undertake repeated transactions with incumbent players.

Finally, our framework offers insight into competitive interaction between start-up and established firms. Most prior research on the impact of start-up innovation considers a case where the principal commercialization option for entrepreneurs is to undermine the advantage of established firms in the product market (Foster, 1986; Henderson and Clark, 1990; Christensen, 1997). However, our framework explicitly considers how competing in the product market compares with cooperating established firms through the market for ideas. For example, to the extent that a weak intellectual property environment increases the *relative* returns to competition over cooperation, the potential for disruptive technologies to overturn established sources of market power is higher in environments with weaker intellectual property protection.

The next section describes the commercialization strategy choice facing technology entrepreneurs, highlighting industries and competitive environments where we tend to see one strategy or the other. Sections 3 and 4 then present a simple commercialization strategy framework, which considers how the nature of appropriability and the ownership of complementary assets interact to determine the elements of an effective commercialization strategy and competitive dynamics. We then turn to the strategic implications of this framework for technology entrepreneurs. Section 6 considers implications for public policy and future research in this area.

2. The product market versus the market for "ideas"

For many technology entrepreneurs, the commercialization stage is the first opportunity to truly define a firm's strategy and positioning.[3] Because of their

[3] While we recognize the crucial role played by the motivation and experience of founders in technology entrepreneurship and the challenges associated with accessing initial external financing for entrepreneurial firms, we abstract away from these issues to focus on commercialization strategy drivers. Essentially, our analysis is

limited financial and human resources, start-up innovators usually can only pursue a small number of strategic options at any one time without losing effectiveness in delivering consumer value (Bhide, 2000; Veugelers and Cassiman, 1999). While a start-up can occasionally make the transition from a competition to cooperation strategy and vice versa, there are costs and constraints on switching. For example, entry into the product market requires sunk investments that mitigate the gains from cooperation with established firms, and antitrust laws and regulatory issues may make collaboration more difficult after entry has occurred. Conversely, negotiations in the market for ideas entails substantial risk, requiring costly search and disclosures that confer power towards established firms. Even by testing the waters in ideas markets, start-up innovators may foreclose on the most profitable commercialization route. The choice between cooperation versus competition, therefore, requires a fine-grained ex ante analysis of the costs and benefits of each option.

2.1. Profiting from innovation through the product market

Consider a start-up innovator intending to launch its product independently. Beyond the intrinsic value of the technology, profitability will depend on several factors. First, the start-up must develop key capabilities and acquire complementary assets to ensure that the innovation offers a novel customer value proposition. As well, profitability will be sensitive to the competitive strategies of incumbents, including the potential for aggressive price competition and the ability of established firms to quickly imitate the functionality of the start-ups technology.

Entering the product market sometimes offers an opportunity. Technology entrepreneurs may be able to develop competencies precisely because more established firms may be ineffective at organizing for and marketing new technological opportunities (Foster, 1986; Henderson and Clark, 1990; Christensen, 1997). A key implication is that established firms are at risk of losing their competitive advantage to the "gale of creative destruction". For example, as an early exploiter of the Internet, Amazon used emerging technology to shift the basis of competitive advantage in the bookseller market, posing a threat to dominant market players such as Barnes and Noble. When the complementary assets necessary for effective commercialization are themselves novel, prior market leadership may hinder effective exploitation of new technology.

To be sure, several difficulties confront technology entrepreneurs implementing a product market-focused commercialization strategy. The start-up innovator must undertake aggressive investments (such as in marketing or manufacturing), manage multiple dimensions of uncertainty, and focus scarce organizational resources on establishing a market presence. The start-up must simultaneously persuade customers of their novel value proposition while avoiding "detection" and an aggressive response by established players. For example, in the typesetting industry, several entrants have commercialized new technologies for creating pre-publication formatted written images. However, established market leaders—such as Linotype—have been the ultimate beneficiaries of these innovations in nearly all cases. Linotype's aggressive response to new competitive threats and their utilization of specialized complementary assets, such as font libraries, allow them to maintain continued market leadership in the face of changing technological leadership (Tripsas, 1997). In other words, a product market strategy requires that the technology entrepreneur offers an integrated value proposition and avoids detection and a competitive reaction from established market players.

2.2. Profiting from innovation through the market for ideas

For a start-up innovator, the main alternative to competing directly in the product market is through a "cooperation" strategy. This strategy is composed of identifying and executing agreements with other firms—usually incumbents—who serve as conduits for commercializing technology to the product market. Essentially, the start-up chooses to earn returns on innovation through the market for ideas rather than

premised on the idea that financing has been available to develop a technologically successful invention, and that the entrepreneur is interested in maximizing the economic returns on this innovation. For an introduction to the broader issues associated with technology entrepreneurship see, for example, Roberts (1991), Bhide (2000), Shane (2001) and Lerner and Gompers (2001).

directly through the product market. The value of a cooperation strategy is determined by the "price" the start-up receives through negotiations in the market for ideas. While technology entrepreneurs face several delicate contracting hazards in the market for ideas, negotiations take place in the shadow of *potential* product market competition. That is, the value derived from cooperation increases with the threat posed by the start-up innovator to the product market position of the established firm.

Cooperation strategies take several distinct forms.[4] On the one hand, the start-up can formally license intellectual property to one or more ideas buyers. Under licensing, each ideas buyer has the right to exploit the start-ups innovation, receives technical assistance according to the terms of the agreement, and pays according to a fixed fee, royalty or more complex payment agreement. While the optimal structure of a license depends on features of the technology and contracting environment, the key element of licensing is that both the start-up and licensees cooperate in commercialization while maintaining organizational independence.[5] At another extreme, the markets for ideas may operate through acquisitions of start-up innovator by established firms (Blonigen and Taylor, 2000). Over the past decade, acquisition has come to play an increasingly important role and now accounts for the *majority* of financial returns realized by venture capitalists (Gompers, 1995; Black and Gilson, 1998). Under acquisition, a technology entrepreneur not only foregoes independent commercialization for current technology but cedes control over their organization to incumbent players. As well, "intermediate" contracting relationships are possible, from joint ventures to strategic and "educational" alliances to milestone financing (Roberts and Berry, 1985; Oxley, 1997). Each of these modes for cooperation involve subtle nuances. For example, whereas the returns from licensing are shaped by the value of the technology itself, the returns from acquisition depend also on the quality and coherence of the technical team. The key point is that any form of a cooperation strategy has the impact of limiting investment by the start-up in downstream commercialization, muting potential product market competition between start-up innovators and incumbent firms.[6]

Commercializing through the market for ideas confers several benefits, allowing buyers and sellers of technology to soften downstream product market competition, avoid duplicative investment, and engage in complementary technology development.[7] First, cooperation reinforces established market power and softens market competition. Since the total

[4] An extensive literature in economics and the management of technology addresses the mode of cooperation between a research-oriented innovator and a downstream market player. Much of this literature builds on the more general economic analysis of arms-length contracting versus integration (Williamson, 1985; Hart, 1995; Aghion and Tirole, 1994). A number of researchers in strategic management attempt to gauge the impact of transaction costs and other factors on the management of innovation relationships, particular after a technology has already been developed (Pisano, 1991; Chesbrough and Teece, 1996; Pisano et al., 1988; Arora et al., 2001). Overall, the choice between licensing, acquisition joint venture or alliance depends on an analysis of the incentives to maintain control over the technology for future development versus the benefits of ownership for those with direct control over commercializing the innovation. While the choice of cooperation mode is crucial for earning maximal returns on innovation, this paper highlights the broader choice between cooperation and competition, so detailed analysis of cooperation mode is beyond the scope of the paper.

[5] An extensive literature evaluates the structure of optimal licensing contracts in different environments. While this literature has been mostly theoretical (Katz and Shapiro, 1985; Kamien and Tauman, 1986; Kamien, 1992; Wang, 1998; Saracho, 2002), a systematic empirical literature has emerged in recent years (Lerner and Merges, 1998; Anand and Khanna, 2000).

[6] While firms may attempt to execute on a "dual-track" strategy, technology entrepreneurs usually lack the financial, management, or organizational resources to pursue two tracks simultaneously (Bhide, 2000; Veugelers and Cassiman, 1999). As well, elements of an effective product market strategy (such as avoiding detection by the established players) conflict with key elements of an effective ideas market strategy (such as broadcasting the value of the innovation).

[7] Our analysis of the benefits and costs of profiting from innovation through the market for ideas builds on a large body of important prior work, most of which (a) evaluates the role of technology markets without an explicit comparison to a product market competition strategy and (b) does not focus on the specific challenges facing technology entrepreneurs. Building on the seminal work of Teece (1986), this literature examines a range of issues, including the impact of the contracting environment and the nature of knowledge on the distribution of innovative activity (Pisano, 1991; Pisano and Mang, 1993; Arora and Gambardella, 1994; Arora, 1995). As well, recent theoretical work in economics emphasizes the role of technology markets in shaping innovation incentives and the distribution of rents from innovation (Anton and Yao, 1994; Aghion and Tirole, 1994; Gans and Stern, 2000). For a useful introduction to this literature (see Arora et al., 2001).

profits associated with competition are lower than the profits associated with monopolization, choosing a cooperative path preserves industry rents precisely because it subverts potential competition. When Bell and Western Union engaged in their epic battle for the emerging local telephone service market in the 1870s, neither firm was able to implement a profitable business model while the other was competing. This head-to-head competition continued until 1879, at which time a cooperative agreement was reached in which Western Union ceded control over commercialization of the telephone to Bell in exchange for a 40% equity stake and promises that Bell would stay out of the telegraph business (Smith, 1985).

As well, transactions in the market for ideas allow start-up innovators to avoid sunk investments in complementary assets necessary for commercialization (Teece, 1986). At the same time, established firms avoid investments in imitative research programs necessary for "catching-up" to the new market entrant (Gans and Stern, 2000). Finally, the availability of a market for ideas provides incentives to develop innovations reinforcing the value of current technology. For example, companies such as Intel spend considerable resources explicitly encouraging the *external* development of complementary technology (Gawer and Cusumano, 2002).[8]

However, several forces *counter* the benefits from contracting, discouraging collaboration between technology entrepreneurs and more established firms, and increasing the relative returns to a competition strategy. Perhaps the most fundamental friction arises from the *paradox of disclosure* (Arrow, 1962; Anton and Yao, 1994). Simply put, when trading in ideas, the willingness-to-pay of potential buyers depends on their knowledge of the idea, yet knowledge of the idea implies that potential buyers need not pay in order to exploit it. Disclosure increases the buyer's intrinsic valuation but reduces the inventor's bargaining power In the absence of formal intellectual property, potential buyers can claim that an idea was known, expropriating innovators once they have disclosed their technology. After Robert Kearns demonstrated the operation and functionality of the intermittent windshield wiper, Ford had incentives to exploit this idea without compensating Kearns for his technology. Because the disclosure of ideas shifts bargaining power from the sellers to the buyers of knowledge, the severity of the disclosure problem reduces the returns of technology entrepreneurs in the market for ideas relative to a product market competition strategy (Gans et al., in press).

The disclosure problem can be ameliorated if precise intellectual property rights are available or if the innovator can maintain effective bargaining threats. When intellectual property protection is available (e.g. through strong patent protection), disclosure does not detract from the owner's ability to profit from it. Unfortunately, for most technologies and industries, intellectual property protection is highly imperfect, leaving potential ideas sellers vulnerable to expropriation. However, when many partners are potentially available, start-up innovators can credibly threaten to pursue their idea with a third party when conducting bilateral negotiations with a particular established firm. The start-ups ability to threaten pervasive disclosure increases their bargaining power and so reducing the degree of expropriation (Anton and Yao, 1994, 1995). More generally, the operation and effectiveness of a market for ideas depends crucially on whether start-up innovators can credibly threaten to compete with potential partners.

At the same time, technology entrepreneurs must overcome the costs of identifying and accessing appropriate partners. Established firms are often reluctant to even begin negotiations with start-up innovators, discounting the potential commercial value of external technology. When the costs of gaining an "audience" with established market players is high, the relative returns to independent commercialization increases. Over the past decade, venture capitalists seem to have played an increasingly important role in developing markets for ideas in various market segments, connecting portfolio companies to a network of established firms, and so lowering the costs for start-up innovators of pursuing cooperative commercialization activity (Aldrich and Zimmer, 1986; Robinson and Stuart, 2000; Aoki, 2000; Hsu, 2002).

[8] Our perspective differs from those who suggest that the relationship between start-up innovators and established firms depends simply on the "nature" of technology (Chesbrough and Teece, 1996). Our analysis suggests that technology choice is, at least in part, under the firm's control and so is an endogenous outcome of the commercialization environment. In other words, when entrepreneurs expect to cooperate, their innovations will complement rather than cannibalize the incumbent value proposition.

3. The drivers of start-up commercialization strategy

Overall, both product market competition and cooperation strategies involve substantial risks and confer distinct benefits, raising crucial questions: How do start-up innovators choose between alternative commercialization strategies? What considerations guide the execution of the chosen strategy? How do these strategies impact upon overall competitive dynamics? The remainder of this paper addresses these questions by considering how a technologically successful start-up innovator chooses commercialization strategy when there are strong incumbent firms in the product market. We highlight this case for expositional simplicity in order to derive how the commercialization environment drives commercialization strategy choice and patterns of interaction between start-up innovators and established firms. Our analysis so far suggests focusing on two subtle yet crucial elements of the commercialization environment:

- *Excludability environment*: To what extent can successful technological innovation by the start-up preclude effective development by an incumbent with knowledge of the innovation?
- *Complementary asset environment*: To what extent does the incumbent's complementary assets contribute to the value proposition of the new technology?

3.1. Excludability (can successful technological innovation by the start-up preclude effective development by an incumbent with knowledge of the innovation?)

Perhaps the key potential hazard facing a start-up innovator is the potential for expropriation by potential ideas buyers. To mitigate this hazard, not all appropriability mechanisms are equal. Appropriability mechanisms such as speed-to-market or traditional trade secrecy will usually be ineffective as mechanisms for *excluding* more established firms in the context of a cooperation strategy. Even when trade secrecy is effective for avoiding imitation by competitors, it may not be effectively used to preclude development by a potential partner, since disclosure is necessary for trade in the market for ideas. In contrast, two alternative appropriability mechanisms—intellectual property protection and technology design—may allow a firm to disclose technology while preserving bargaining power. For example, even when patent or copyright protection is modest, the ability to use litigation to temporarily halt the activities of an expropriator provides incentives for potential users to reach an agreement with the entrepreneur (Shapiro, 2001).

Similarly, certain technology designs have the benefit of displaying functionality while masking details that would allow imitation. Consider object-oriented software. For older programming languages, such as Fortran and Cobol, evaluating the logic of a given program necessitates examining the source code of the program. Perhaps not surprisingly, these languages flourished when most software development was either conducted in-house or in the context of ongoing relationships between vendors and users. However, over the past 20 years, object-oriented programming languages such as C++ and Java have achieved dominance. In addition to their substantial engineering benefits, object-oriented programming languages allow the precise logic underlying an application to be demonstrated (by revealing the object code) while withholding the source code necessary for imitation. Though reverse engineering is possible, imitation remains costly. This specific *form* of trade secrecy (incorporating non-imitability into the design) allows a start-up innovator to exploit the cost and market power advantages of cooperation without unduly placing itself in a weak bargaining position. Of course, even with strong intellectual property protection or with a strong design, imitation is often possible after some period of time. However, these mechanisms confer a *relatively* high cost on potential imitators who become aware of new technology, and so increase the attractiveness of strategies which require disclosure of functionality.

3.2. Specialized complementary assets (does the incumbent's specialized complementary assets contribute to the value proposition from the new technology?)

As Teece (1986) has emphasized, the control over costly-to-build complementary assets is a key wedge between the capabilities of the start-up and more established firms in an industry, and the inability to acquire these resources cost-effectively has an important impact on the returns earned by a start-up innovator. Specifically, when specialized complementary assets are required, the sunk costs of product market entry become substantial. This both reduces the returns to competition and weakens the relative bargaining position of the start-up when contracting with established players.

Under a product market competition strategy, the costs associated with duplicating specialized complementary assets held by established firms are entirely borne by the technology entrepreneur. However, under a cooperation strategy, the gains from trade will include the avoidance of costly duplication of investments, and these gains will be shared between the partners in the collaboration. Consequently, when considering commercialization strategy choice, an increase in the importance or concentration of control of complementary assets raises the *relative* returns to cooperation over competition (Gans and Stern, 2000). Thus, even though an increase in the importance of complementary assets reduces the absolute share of total value earned by the innovator (Teece, 1986), this factor will tend to encourage collaboration with more established firms over direct and independent entry into the product market.

4. The impact of the commercialization environment on strategy and competitive dynamics

Effective commercialization strategy results from the interaction between the excludability and complementary asset environment. These two factors define four distinct commercialization environments, each of which is examined in this section. Our analysis is encapsulated in Tables 1 and 2. In Table 1, we highlight the overall pattern associated with each commercialization environment. Table 2 summarizes the strategic choices facing start-up innovators and established firms in each environment, and the implications of these strategies for competitive dynamics.

Overall, our approach to identifying the drivers of commercialization strategy for technology entrepreneurs refines Teece's classical analysis (Teece, 1986), where the hazards associated with contracting for complementary asset access are treated distinctly from the potential for imitation. In our formulation, a principal challenge in pursuing cooperation with complementary asset owners arises from the fact that ownership is concentrated among incumbents with incentives to expropriate the innovator's technology. By focusing on this specific (but empirically common) case, our framework provides insight into the differences across technologies and industries in the strategic challenges facing technology entrepreneurs.

4.1. The attacker's advantage

Consider an environment with poor intellectual property protection and where incumbents do not control the complementary assets necessary for effective commercialization. In this environment, start-ups and established firms face off on a "level" playing field. Start-up investments in the product market need not be duplicative and are often modest in size. However, technological leadership will likely be fleeting: established firms have the opportunity to imitate once they recognize the nascent threat. Under these conditions, competition is likely to be intense, with continual entry challenges by start-ups aimed at undermining the value of existing market leadership positions. While

Table 1
Commercialization strategy environments

		Do incumbent's complementary assets contribute to the value proposition from the new technology? No	Yes
Can innovation by the start-up preclude effective development by the incumbent?	No	The Attacker's Advantage	Reputation-Based Ideas Trading
	Yes	Greenfield Competition	Ideas Factories

Table 2
The impact of the commercialization environment of strategy and competitive

	Overturns Incumbent Asset Value		Reinforces Incumbent Complementary Assets	
Non-Excludable Technology	**ATTACKER'S ADVANTAGE**		**REPUTATION-BASED IDEAS TRADING**	
	Start-Up Strategies • Few opportunities for effective contracting • Opportunity to exploit technical leadership to capture market leadership • Performance depends on 'stealth' product market entry	*Incumbent Strategies* • Competitive advantage in products not competencies • Sustained market position requires continual reinvention and preemption • Constant monitoring and tight integration of value chain	*Start-Up Strategies* • May be few opportunities for contracting • Product market entry risky due to high costs and imitation risk • Performance depends on existence of incumbent commitment to ideas trading	*Incumbent Strategies* • Competitive advantages in both competencies and products • Opportunity for sustainable positioning by developing reputation for ideas trading • Often results in internal R&D focus
	Expected Competitive Dynamics • Market leadership determined by technological leadership • Established firms face competition from entrants in 'niche' markets • Start-ups will make new investments in complementary assets as part of establishing a novel value proposition		*Expected Competitive Dynamics* • Relative market and technological stability • Established firms face few competitive threats from start-up firms • Start-ups may play a greater role if incumbent chooses reputation strategy	
Excludable Technology	**GREENFIELD COMPETITION**		**IDEAS FACTORY**	
	Start-Up Strategies • Ideal opportunity to choose between contracting and product market entry • Opportunity to use temporary monopoly power to build future positioning • Performance depends on strength of technological competition	*Incumbent Strategies* • Competitive advantage is based on products not competencies • Sustained market position requires continual innovation and ceding profits to upstream providers • Develop reputation from strong innovative performance	*Start-Up Strategies* • Contracting with established firms • Product market entry is very costly and perhaps impossible • Performance depends on securing bargaining power	*Incumbent Strategies* • Competitive advantage is in competencies not products • Sustained market position requires securing start-up partners • Find balance between internal development and use of external start-up innovation
	Expected Competitive Dynamics • Technological leadership drives rent distribution along the value chain • Start-ups and incumbents compete for technological priority • Substantial investments in new platforms and complementary assets		*Expected Competitive Dynamics* • Frequent changes in technological but not market leadership • Start-ups compete with one another for priority in negotiations with incumbents • Start-up innovation will reinforce existing platforms	

entrepreneurs have an opportunity to overturn established positions, easy imitability gives most start-ups a very small share of the value over the long-term.

In this environment, start-ups have an opportunity to capture market leadership by effectively developing and diffusing competence-destroying technology. That is, smart attackers have an advantage in the commercialization process (Foster, 1986). At the same time, there are few opportunities for contracting with current market leaders. Not only does the current market leader have little to offer the start-up, the very act of bringing the value of the technology to the attention of the current market leader weakens the position of the initial innovator, reducing its advantage from either cooperation or competition. In this environment, "stealth" is a crucial element of an effective competition-oriented commercialization strategy by a technology entrepreneur.

The key to a stealth strategy is to position the technology in the market in order to exploit the "blind spot" of current market leaders (Foster, 1986; Henderson and Clark, 1990). For example, incumbents often seem to overestimate the potential for further improvement from existing technologies at the expense of recognizing the growth path associated with emerging technologies emphasizing new performance characteristics. In his careful study of the hard disk drive industry, Christensen (1997) finds that while market leaders innovated incrementally in each generation of disk drives, new firms were responsible for key product introductions for each *new* disk drive generation. Though established firms had the means and technical freedom to commercialize new products (e.g. no firm could exclude others from switching to a "smaller" disk drive form factor), start-up innovators were able to exploit low barriers to entry and the availability of multiple market niches. Specifically, Christensen finds that successful commercialization of these disruptive innovations was accomplished by careful positioning of the new technology towards underserved customer groups. Established firms focus on responding to the needs and requirement of current customers made them particularly vulnerable to entry emphasizing new customer segments. In other words, given the commercialization environment in the disk drive industry, effective start-up commercialization was achieved by exploiting the inertia of established firms.

Ultimately, an environment with high imitability and low dependence on existing complementary assets implies tight integration between research and commercialization. Intense competition forces firms to invest in risky R&D and take advantage of "competency traps" in order to establish a novel value proposition for the industry. Technological leadership results in temporary market leadership, which is itself vulnerable to additional waves of entrepreneurial innovation via creative destruction.

4.2. Ideas factories

Standing in complete contrast is an environment where successful invention precludes effective development by more established firms but those firms control the complementary assets required for effective commercialization. In this environment, we expect the emergence of "ideas factories"—technological leaders focusing on research and commercializing through reinforcing partnerships with more downstream players (Pisano and Mang, 1993; Arora and Gambardella, 1994; Gans and Stern, 2000). Not only would the start-up innovator need to undertake duplicative investments under a competition strategy, but negotiations with established firms do not unduly threaten the start-ups control over the technology.

The key issue is no longer whether to pursue a cooperation strategy but when and how. In this environment, the return on innovation will depend on the bargaining power of the start-up innovator, which can be enhanced in several ways. First, the value offered by the technology must be clearly signaled and demonstrated (in contrast, note that disclosure undermines bargaining power when appropriability is weak). Second, the start-up innovator might be able to play established firms against each other in a bidding war. In an ideal case, an ideas factory "auctions" off technology to the highest bidder, with high participation in the auction and low uncertainty over the value of the technology.

Rather than disrupting their advantage, ideas factories reinforce the basis of advantage for established firms by offering a fertile source of new innovation. A supply relationship with these specialized technology producers enhances competitive advantage, particularly when the ideas factory develops technology complementary to the existing value proposition.

Table 3
Pharmaceutical firm market leadership

Sales rank		Company	Date established
1997	1973		
1	2	Merck	17th century
2	9	Bristol-Myers Squibb	1887, 1856
3	6	American Home Products	1926
4	7	Pfizer	1848
5	21	Abbott Labs	1900
6	11	Eli Lilly	1876
7	3	Warner Lambert	1852
8	79	Baxter	1931
9	15	Schering-Plough	1851
10	31	SmithKline Beecham	1830

Sources: various corporate web sites; BioWorld Publishing (1998).

Indeed, established firms face new challenges in this environment. Balancing the commercialization of technologies developed both within and outside the firm requires a capacity for monitoring internal and external innovation (Cohen and Levinthal, 1989; Rosenberg, 1990).

When markets for ideas exist, a high rate of innovation by start-up research-oriented firms is associated not with creative destruction but with the *reinforcement* of incumbent market power, a pattern exemplified by the biotechnology industry. While the radical technical and scientific breakthroughs promised by biotechnology were heralded originally as a force for creative destruction (United States Office of Technology Assessment, 1984), market leadership has remained relatively constant in the pharmaceutical industry over the past 25 years. Not one of the ten largest independent pharmaceutical firms in 1997 have their origins in biotechnology; indeed, seven of the top ten firms in 1997 were among the top 15 leaders by sales in 1973, and the remaining 3—Abbott, Baxter, and SmithKline Beecham—all have their origins in the traditional pharmaceutical business (Table 3). This is not to suggest that biotechnology products have not been commercialized. Indeed, by 1997, over 55% of all new products approved by the FDA are based, at least in part, on discoveries developed with the tools of biotechnology (BioWorld Publishing, 1998). In most cases, these are innovative outputs of research-oriented biotechnology firms in collaboration with at an incumbent pharmaceutical firm in the commercialization process.

Consider the "race" to develop synthetic insulin in the early years of biotechnology (Hall, 1988; Stern, 1995). Eli Lilly, the dominant incumbent supplier of beef and pig insulin, encouraged three separate teams to undertake research to express the insulin gene, a precondition for commercial development of human insulin. Two of the teams were university-based (at Harvard and UCSF), while the third effort was pursued by Genentech, a venture-backed start-up biotechnology firm. In August, 1978, Genentech researchers successfully synthesized the human insulin gene after an intense competition, ending the technology race. Within a day after the research was validated, Genentech filed a patent application and signed an exclusive license agreement with Eli Lilly to cooperate in commercialization.[9]

Despite a competitive race to develop the technology, commercialization itself occurred through cooperation. Even though the role that biotechnology firms would play in the pharmaceutical industry was unclear at that point, Lilly encouraged this external research, willing to risk the potential for competition in order to earn gains in the market for ideas. Indeed, rather than undermining the competitive advantage of pharmaceutical firms, biotechnology firms came to reinforce and sustain those advantages.

4.3. Reputation-based ideas trading

The above environments have well-defined patterns of competitive interaction because both the disclosure and complementary asset environment reinforce the same strategy—either competition (when incumbent complementary assets are not valuable and the disclosure problem is severe) or cooperation (when incumbent complementary assets are valuable and the disclosure problem is less severe). However, when considering the off-diagonal elements of our framework, more subtle forces come into play, reflecting the tradeoffs determining optimal commercialization strategy.

[9] While Lilly had initially encouraged this race between external R&D teams, Lilly was a strong negotiator, de-emphasizing their need for the technology and questioning its commercial viability (Hall, 1988, p. 230). As well, though the collaboration was initially a success, the relationship ultimately soured; Genentech and Lilly engaged in a long and costly litigation battle over the distribution of rents from this innovation.

Consider an environment where the disclosure problem is severe but incumbents possess the complementary assets necessary for effective commercialization. Though a market for ideas would confer a potential mutual gain (since the start-up innovator avoids investing in duplicative assets and the established firm reinforces their advantage by controlling the technology), a cooperative solution is difficult to achieve. In capital-intensive industries such as automobiles or aircraft, established firms are tempted to expropriate technology revealed to them, such as Bob Kearns' intermittent windshield wiper. This expropriation discourages start-ups from pursuing collaboration as a strategy and additionally discourages research in the first place (since competition is also likely to be unprofitable). Though the automobile industry invests heavily in internal R&D, very little innovation results from technology entrepreneurship. Since entrepreneurs and investors (rationally) expect start-up innovators to face great difficulties in appropriating the returns from their innovation, the auto industry has been bypassed in the venture investment boom of the past decade (Gans and Stern, in press).

As a result of these difficulties, established firms in this environment have an incentive to invest in mechanisms to enable a market for ideas. Rather than exploiting all opportunities for gain in each transaction, an established firm can develop a reputation for "fairness", thus, encouraging future start-ups to approach them with promising new technologies. Incumbent firms that succeed in establishing relational research contracts will be able to profit from the commercialization of new technologies at a higher rate than their competitors (Greif, 1994; Baker et al., 2002). In other words, while a "low" equilibrium exists in which established firms expropriate and, thus, discourage start-up innovation (as in the auto industry), there also exists a "high" equilibrium where incumbents foster a reputation for ensuring mutual advantage from the acquisition of external innovation.

Cisco Systems has perhaps undertaken the most ambitious strategy in this regard (Bunnell and Brate, 2000; Charles River Associates, 1999). Since the early 1990s, Cisco has committed to a strategy in which internal R&D activities are modest and the acquisition of new technology is systematized. For example, Cisco first undertakes a strategic alliance or partnership with firms that are ultimately acquired. Moreover, when alliances do not result in an acquisition, Cisco maintains positive relations with the firm and has mostly avoided litigation over the ownership or control of jointly developed technologies. As a result of these investments, Cisco receives unfettered access to the financial information, personnel, and customers of acquirees during the course of evaluating a potential deal. In contrast to other firms undertaking acquisition in the industry, Cisco is able to conduct interviews with key employees and customers without senior management present. While the downturn in the networking industry during 2001 has raised concerns about Cisco's ability to sustain this reputation-based strategy (since they cannot "promise" an ever-rising share price or a no-layoff policy), Cisco's reputation during the late 1990s seems to have allowed them to value start-up technologies more effectively and increase their ability to integrate external technology (Bunnell and Brate, 2000).

In their careful study of the management of external innovation at Intel, Gawer and Cusumano (2002) document a range of practices reinforcing Intel's reputation in the market for ideas. Senior managers at Intel responsible for relationships with external innovators have explicit incentives to encourage the *growth of the worldwide semiconductor industry* (rather than Intel's current profits). By providing industry-based incentives for the managers negotiating contracts in the market for ideas, Intel signals start-up innovators its commitment to attempt to avoid short-term expropriation in the interest of longer-term relational contracting. However, despite substantial investment and attention to this area, Intel has been accused of expropriation by some firms, particularly those with technologies relevant for Intel's core microprocessor business. In other words, even when a firm has substantial incentives to invest in reputation and senior management commits to this strategy, execution may be difficult to achieve across all of the expropriation opportunities available to a dominant market player.

Several complementary mechanisms may also underpin the functioning of markets for ideas in environments where excludability is weak but the gains from collaboration between entrepreneurs and established firms is high. For example, since individual venture capitalists have repeated interactions with established firms in particular sectors, such individuals have come to play a crucial role as brokers, preserving start-up bargaining power during negotiations and allowing for more efficient pursuit of opportunities on the part of established firms (Robinson and Stuart, 2000; Hsu, 2001). Indeed, prior to the rise of in-house industrial laboratories in the early 20th century, patent brokerage was a key route through which new technologies were commercialized (Lamoreaux and Sokoloff, 1999).

Finally, in some sectors, more formal institutions have arisen to overcome the expropriation hazard. For example, in the musical recording industry, individual agreements between songwriters and performers may be difficult to enforce. However, composers and music publishers may participate in licenses negotiated by ASCAP, a non-profit organization which maintains "master" agreements with nearly all organizations that profit from music performance (e.g. broadcast media, concert halls, etc.). Similarly, the rise of university licensing offices following the *Bayh–Dole Act* seems to have increased the commercialization rate of technologies developed within universities (Mowery, 2001). Though universities have always produced ideas with potential commercial application, the establishment of a formal process for technology transfer increased the ability of universities to protect their inventions while allowing firms to evaluate and experiment with potential commercial applications.

Overall, while individual established firms may engage in reputation-based ideas trading, the extent of the market may depend on third-parties and institutions outside the their direct control. Moreover, reputation serves as an asset reinforcing their competitive advantage by increasing their access to external technologies. However, the maintenance of this reputation requires continued commitment. In 1980s, Johnson and Johnson was drawn into a costly litigation with Amgen over their commercialization partnership for Amgen's blockbuster drug EPO. While this litigation battle may have made sense in the context of the Johnson and Johnson/Amgen relationship, it diminished Johnson and Johnson's reputation as a biotechnology collaboration partner, at least for a few years. From the perspective of an established firm, ineffective management of reputation cedes external innovation opportunities to other downstream competitors.

4.4. Greenfield competition

The patterns of commercialization are similarly subtle in the final environment, where incumbent complementary assets are unimportant but start-up innovators can preclude effective imitation. While established firms set the terms for ideas trading when excludability is weak, the power to determine the most effective commercialization strategy lies with the start-up innovator under Greenfield competition. While the potential for returns in the product market are high (since imitation is difficult), this market power will be reflected in increased bargaining power with potential partners. As a result, the *relative* returns to competition over cooperation will depend on factors distinct from the intrinsic value of the technology. In this environment, technology entrepreneurs enjoy freedom to evaluate competition and cooperation options in the absence of the risk of expropriation or the inability to overcome established firm market power.

In this environment, both competition and cooperation strategies may be effective. For example, after Chester Carlson and the Haloid Corporation agreed to commercialize early xerography technology (changing their name to Xerox in the process), the company developed a complete vertical chain to support the technology, from manufacturing to distribution to marketing and servicing. This strategy allowed Xerox to maintain tight control over the development and deployment of subsequent technologies (resulting in close antitrust scrutiny in the 1970s). In contrast, Nintendo's game platform business is based on the widespread licensing of its software development tools to independent game developers. In those platform generations in which Nintendo has established a strong market position, Nintendo has been able to extract tremendous rewards from this system, by encouraging high-quality video game development and encouraging medium-term standardization on the Nintendo platform.

When choosing between cooperation and competition under Greenfield competition, the ability to control the development and evolution of platforms and standards may be decisive. For example, tight control over the technology may allow a technology entrepreneur to ensure compatibility with future generations of the technology; as a result, investments to control the key elements of the value chain may be worthwhile. On the other hand, when there are few opportunities to leverage the current technology as a platform for future generations, an "open" system of independent incremental innovations may be more effective. More generally, technology entrepreneurs must analyze the opportunities and potential for future hazards associated with cooperation or competition at each stage of the value chain, with the result that the optimal strategy may involve an intermediate level of integration.[10]

Overall, this environment offers a tremendous opportunity for start-up innovators. However, this potential raises the possibility of a first-stage "race" to secure a first-mover position. In other words, both technology entrepreneurs and established firms may engage in Schumpeterian competition "for the market" rather than through traditional tactics "within the market" (Evans and Schmalensee, 2002). The potential to dissipate the value of the rents arising from market leadership offers a final cautionary note about the strategic attractiveness of Greenfield competition.

5. Implications for technology entrepreneurs

The commercialization strategy framework suggests that variations in how innovations are introduced across sectors result from differences in firms' commercialization environment. To maximize the returns from a given technological innovation, start-up strategy involves exploiting the specific type of opportunities arising from that environment. This insight holds several implications for the management of high-technology entrepreneurial firms:

5.1. Commercialization strategy should reflect the firm's commercialization environment

For many technology entrepreneurs, the challenge of earning any return on innovation often results in ad hoc strategy development and execution (Bhide, 2000). Firms opportunistically take advantage of potential revenue opportunities as they present themselves, rather than choosing a strategy that focuses resources and attention towards activities most likely to yield the highest long-term return. To best take advantage of an innovation, a start-up innovator should undertake a systematic analysis of the level of excludability and the degree to which key complementary assets are controlled by established firms who could serve as competitive threats. Importantly, given that no intellectual property regime is perfectly secure, the start-up must assess whether the enforcement of intellectual property rights is sufficiently cost-effective to serve as a deterrent to expropriation during negotiations. Further, as Teece (1986) emphasized, strategy choice should be include a complete accounting of the complementary assets required for effective commercialization and the degree to which they are controlled by existing players. While it may be possible to enter a market without investing in all areas, the ability to extract value from innovation ultimately depends on the customer value proposition, rather than the simple offering of the technology by itself. During the Internet boom of the late 1990s, many technology entrepreneurs seem to have underestimated the costs and challenges associated with developing de novo complementary assets (e.g. Webvan's failure was a result in part of their underestimate of how costly it would be to offer a vertically integrated substitute for the supermarket industry value chain). While such investments are a necessary component of a competition strategy, strategy choice depends on evaluating the relative returns to innovation along alternative routes, accounting for the sunk cost of investing in assets along each path.

In the absence of explicit evaluation and strategy choice, technology entrepreneurs often "iterate" towards a position which fits their overall environment. In Gans et al. (in press), we report on the results of a detailed survey of over 100 start-up innovators over five industry segments, examining the relationship between commercialization strategy and the commercialization environment. We relate the probability of cooperation (licensing, strategic alliance, or acquisition) to whether (a) the start-up innovator received a patent on the innovation and (b) the start-up innovator considered *control* over complementary assets to be a cost-effective mechanism for earning profits from their innovation. Our results suggest that cooperation is much more likely to be chosen by firms able to acquire intellectual property protection or for whom control over complementary assets was

[10] For example, Segway, the start-up innovator behind a new transportation technology, is outsourcing the manufacturing of key components but maintaining control over final assembly, branding and distribution.

not cost-effective. While these results suggest that the strategy ultimately reflect the environment, our qualitative assessment, based on our data gathering experience, is that many entrepreneurial firms find choosing and executing on strategy choice among the most difficult of their organizational challenges. Specifically, many entrepreneurs recounted the evolution and experimentation that preceded their current strategy, remarking that an ex ante analysis would have allowed them to earn additional returns from their innovations.

5.2. The value earned on the market for ideas extends beyond the value of access to specialized complementary assets

In most cases, the imperfect excludability of technology implies that start-up innovators should be sensitive to subtle facets of the market for ideas. On the one hand, the ability to trust potential collaborators is at the heart of an effective cooperation strategy. The value of a partnership depends on whether a reputation has been established in the past and whether incentives exist to maintain that reputation into the future. When both of these conditions hold for a potential partner, it is likely that expropriation can be avoided, raising the overall returns to innovation. Beyond reputation, brokers, such as venture capitalists, and more formal institutions may increase the operational effectiveness of market for ideas, allowing technology entrepreneurs to be exposed to a wider and more sophisticated population of potential ideas buyers.

When excludability is particularly strong, relationships forged through the market for ideas will impact the longer-term advantage that may be derived from the technology. Often including follow-on product development and the management of uncertain contingencies, relational contracting allows both technology entrepreneurs and more established firms to overcome the hazards of partnership by committing to a longer-term relationship with each other.

5.3. The returns to cooperation depend on the timing of collaboration

For many technologies, investments in complementary assets can proceed in multiple steps. From the earliest development and refinement of new technology, innovators pass through key hurdles and milestones and finally introduce a product to market by developing a manufacturing capability and distribution chain. Cooperative commercialization strategies potentially involves collaboration with downstream partners at any of these stages. Early on, the uncertain value of the technology, potential inability to secure the strongest intellectual property protection, and the availability of alternatives for established firms weakens the bargaining position of the start-up innovator. So long as assets required at these early stages are not controlled by others, the start-up has an option to invest in these to both improve its bargaining position and avoid hazards associated with disagreement over a technology's value in subsequent ideas trading.

However, independent commercialization can go too far. If cooperation waits until the latest stages, innovators will have incurred substantial sunk investment costs, reducing the gains from trade with established firms. The key to an effective cooperation strategy is to initiate cooperation at a point where technological uncertainty is sufficiently low but sunk investment costs have not yet become substantial. Achieving this delicate balance depends on start-up innovators assessing commercialization strategy at each stage, weighing the bargaining advantages arising from continued independence against the cost advantages associated with collaboration. Within the biotechnology industry, this tension about when to cooperate is pervasive, with many firms attempting to increase the range of their independent activities as their experience with multiple drugs grows over time.

5.4. The returns to product market entry depend on the pacing of competition

At the heart of a competition strategy is the ability to delay the timing of when established firms recognize the threat posed by the novel value proposition offered by the technology entrepreneur. If detection is sufficiently early, established firms can respond and adapt the entrepreneur's technology in order to take advantage of their competencies and specific positioning. Strategies such as targeting niche customer segments allow a start-up innovator to delay detection until they are ready to compete head-to-head with incumbents. As pointed by Cusumano and Yoffie (1998), Netscape's strategy of openly challenging

Microsoft's product market leadership was a principal driver of Microsoft's ability to reverse its course of ignoring browser technology and move to compete head-to-head with Netscape. Cusumano and Yoffie recommend "do not moon the giant"; a stealth commercialization strategy achieves this objective.

5.5. Investment in innovation should be sensitive to the most attractive commercialization environment

Our analysis suggests that each commercialization strategy environment holds its own hazards and potential opportunities. For example, even an environment that may be a poor choice for technology investment in most circumstances—such as where innovation is non-excludable and key complementary assets are held by established firms—may be quite lucrative under the right conditions. To the extent that established players are investing in their reputations as ideas buyers, the returns to technology entrepreneurs may be substantial, precisely because that reputation involves ceding rents to start-up innovators. Conversely, when the rents from commercialization tends to be the most favorable—Greenfield competition, the market may be subject to rent dissipation as multiple firms race for a dominant position. Overall, entrepreneurs can choose to direct scarce resource into those technology projects which have the highest chance of yielding innovations in a commercialization environment favorable for the firm's long-term advantage.

Our analysis suggests that firms pay attention to the *interaction of intellectual property and complementary assets*, and the impact of that interaction on more subtle aspects of the business environment. This focus on the commercialization environment informs strategy choice both after technologies have been developed and during project selection when innovative investments are initially being contemplated.

6. Conclusion

This paper provides a framework for evaluating start-up commercialization strategy and patterns of competitive interaction between start-up and established firms. Our primary argument has been that key aspects of the commercialization environment drive technology entrepreneurs to choose between cooperative or competitive strategies, and these choices impact the evolution of market structure. When intellectual property protection is strong and important specialized complementary assets are held by incumbent firms, start-up firms generate more innovative rents if they pursue cooperative options with incumbent firms rather than competing directly in product markets. As a result, changes in technological leadership need not result in changes in market leadership. In contrast, when weak intellectual property for innovation exists alongside low barriers to entry, competitive commercialization strategies are more likely. A clear understanding of this environment leads start-up innovators to exploit the "blind spots" of incumbent players. As a result, the failure to recognize threats to market leadership may often be the result of an active "stealth" strategy on the part of entrepreneurs.

This framework sheds light on several aspects of technology strategy. Consider the role of intellectual property rights. While intellectual property protection provides a valuable asset, it also serves to enhance the creation of markets for ideas. Consequently, it allows for cooperation between start-ups and incumbents who might otherwise view innovation purely as a competitive threat. This serves as an opportunity for incumbents to tap the high-powered incentives, creativity and flexibility traditionally associated with small firms. Relative to a market with a high rate of creative destruction (and so few stable rents), this pattern may allow for an overall increase in resources to be devoted to innovation, with gains for established firms and entrepreneurs.

Further, our analysis suggests that the strategic management of innovation should focus less on whether a particular technology is 'radical' in an engineering sense (i.e. displacing existing technologies) and more on whether that technology is 'radical' in an organizational or market sense. If entrepreneurial innovation undermines existing incumbent assets, the returns to cooperation are reduced in favor of competition. In contrast, the fact that a technology is disruptive should not be decisive if key complementary assets remain with established players. In this situation, an incumbent's competitive position may often be enhanced rather than threatened by start-up innovation.

This framework also provides insight for public policy. For example, while debates over intellectual property protection concentrate on the costs associated with temporary monopoly power, our analysis reinforces an emerging perspective that one of the critical roles played by the patent system is to enhance the efficiency of the market for ideas. Stronger intellectual property protection allows innovators to earn greater rents by improving their contracting options and not simply by granting them market power. That is, such policies alleviates problems of disclosure allowing start-up firms to consider contracting options without fear of expropriation.

This effect impacts subsequent competition since start-up innovators are likely to reinforce rather than undermine established market power. While this raises important antitrust concerns about the effects of licensing on competitive entry, an analysis of the appropriate tradeoff for public policy is subtle. A start-up innovator with weak intellectual property protection is likely a weak competitor, dampening the innovation incentives of entrepreneurs. At the same time, the precise structure and functioning of the market for ideas will depend on subtle factors, including the incentives for reputation-building on the part of established firms, the existence of institutions and brokers that facilitate trade, and the ability to secure exclusive agreements while simultaneously threatening to cooperate with third-parties (as in Anton and Yao, 1994). An integrated analysis of the interplay among these factors is important for effective policymaking in this area, but this is something we leave for future work.

Acknowledgements

We would like to thank the firms who participate in the MIT Commercialization Strategies survey for their time and effort. Rebecca Henderson, David Hsu, Pierre Azoulay, Scott Shane, two anonymous referees and conference participants at the University of Maryland Technology Entrepreneurship Conference provided valuable comments and suggestions. We also acknowledge funding from the MIT Center for Innovation in Product Development under NSF Cooperative Agreement EEC-9529140 and from an Australian Research Council Large Grant. Stern completed this paper while a Visiting Fellow at the Brookings Institution, whose hospitality is gratefully acknowledged. Finally, we thank Mercedes Delgado for outstanding research assistance.

References

Aghion, P., Tirole, J., 1994. The management of innovation. Quarterly Journal of Economics 109 (4), 1185–1210.

Aldrich, H., Zimmer, C., 1986. Entrepreneurship through social networks. In: Sexton, D.L., Smilor, R.W. (Eds.), The Art and Science of Entrepreneurship. Ballinger, Cambridge, MA.

Anand, B.N., Khanna, T., 2000. The structure of licensing contracts. Journal of Industrial Economics 48 (1), 103–135.

Anton, J.J., Yao, D.A., 1994. Expropriation and inventions: appropriable rents in the absence of property rights. American Economic Review 84 (1), 190–209.

Anton, J.J., Yao, D., 1995. Start-ups, spin-offs, and internal projects. Journal of Law Economics & Organization 11, 362–378.

Aoki, M., 2000. Information and governance in the Silicon Valley model. Finance, Governance, and Competitiveness in Japan. Oxford University Press, Oxford.

Arora, A., 1995. Licensing tacit knowledge: intellectual property rights and the market for know-how. Economics of Innovation & New Technology 4, 41–59.

Arora, A., Gambardella, A., 1994. The changing technology of technological change, research policy: general and abstract knowledge and the division of innovative labour. Research Policy 32, 523–532.

Arora, A., Fosfuri, A., Gambardella, A., 2001. Markets for Technology: The Economics of Innovation and Corporate Strategy. MIT Press, Cambridge, MA.

Arrow, K.J., 1962. Economic welfare and the allocation of resources for invention. In: The Rate and Direction of Inventive Activity. Princeton University Press, Princeton, pp. 609–625.

Baker, G., Gibbons, R., Murphy, K.J., 2002. Relational contracts and the theory of the firm. Quarterly Journal of Economics 117 (1), 39–84.

Baldwin, C., Clark, K., 1997. Sun wars: competition within a modular cluster. In: Yoffie, D.B. (Ed.), Competing in the Age of Digital Convergence. Harvard Business School Press, Boston. MA.

Bhide, A., 2000. The Origin and Evolution of New Businesses. Oxford University Press, New York.

BioWorld Publishing, 1998. Biotechnology State of the Industry Report 1998. BioWorld Publishing, Atlanta, GA.

Black, B.S., Gilson, R.J., 1998. Venture capital and the structure of capital markets: banks versus stock markets. Journal of Financial Economics 47, 243–277.

Blonigen, B.A., Taylor, C.T., 2000. R&D activity and acquisitions in high technology industries: evidence from the US electronics industry. Journal of Industrial Economics 48, 47–70.

Bunnell, D., Brate, A., 2000. Making the Cisco Connection. Wiley, New York (Chapters 6 and 7).

Charles River Associates, 1999. A Stand-Up Double: The Story of American Internet Corporation. VIDEO, Boston, MA.

Chesbrough, H., Teece, D., 1996. When is virtual virtuous: organizing for innovation. Harvard Business Review 74, 65–74.

Christensen, C.M., 1997. The Innovator's Dilemma: When New Technologies Cause Great Firms to Fail. Harvard Business School Press, Boston, MA.

Cohen, W.M., Levinthal, D.A., 1989. Innovation and learning: the two faces of R&D. Economic Journal 99 (397), 569–596.

Cusumano, M.A., Yoffie, B., 1998. Competing on Internet Time: Lessons from Nestcape and its Battle with Microsoft. Free Press, New York.

Evans, D., Schmalensee, R., 2002. Some economic aspects of antitrust analysis in dynamically competitive industries. In: Jaffe, A.B., Lerner, J., Stern, S. (Eds.), Innovation Policy and the Economy, vol. 2. pp. 1–50.

Foster, R.N., 1986. Innovation: The Attacker's Advantage. Summit Books, New York.

Gans, J.S., Stern, S., 2000. Incumbency and R&D incentives: licensing the gale of creative destruction. Journal of Economics and Management Strategy 9 (4), 485–511.

Gans, J.S., Stern, S., in press. When does funding small firms bear fruit? Evidence from the SBIR program. Economics of Innovation and New Technology.

Gans, J.S., Hsu, D., Stern, S., in press. When does start-up innovation spur the gale of creative destruction? RAND Journal of Economics.

Gawer, A., Cusumano, M., 2002. Platform Leadership: How Intel, Palm, Cisco and Others Drive Industry Innovation. Harvard Business School Press, Cambridge, MA.

Gompers, P., 1995. Optimal investments, monitoring, and the staging of venture capital. Journal of Finance 50, 1451–1490.

Greif, A., 1994. Cultural beliefs and the organization of society: a historical and theoretical reflection on collectivist and individualist societies. Journal of Political Economy 102 (5), 912–950.

Hall, S.S., 1988. Invisible Frontiers: The Race to Synthesize a Human Gene. Sidgwick & Jackson, London.

Hart, O., 1995. Firms, Contracts and Financial Structure. Oxford University Press, Oxford.

Henderson, R., Clark, K., 1990. Architectural innovation: the reconfiguration of existing product technologies and the failure of established firms. Administrative Science Quarterly 35, 9–30.

Hsu, D., 2001. Do Entrepreneurs Pay to Access More Experienced Venture Capitalists? MIT Sloan School of Management, Mimeo.

Hsu, D., 2002. Do Venture Capitalists Affect Commercialization Strategies at Start-Ups? MIT Sloan School of Management, Mimeo.

Kamien, M., 1992. Patent licensing. In: Aumann, R.J., Hart, S. (Eds.), Handbook of Game Theory with Economic Applications, vol. 1. North-Holland, Amsterdam, pp. 331–354.

Kamien, M., Tauman, Y., 1986. Fees versus royalties and the private value of a patent. Quarterly Journal of Economics 101, 471–491.

Katz, M.L., Shapiro, C., 1985. On the licensing of innovation. RAND Journal of Economics 16 (4), 504–520.

Lamoreaux, N., Sokoloff, K., 1999. Inventive Activity and the Market for Technology in the United States, 1840–1920. Working Paper No. 7107, NBER.

Lerner, J., Gompers, P., 2001. The Money of Invention: How Venture Capital Creates New Wealth. Harvard Business School Press, Boston, MA.

Lerner, J., Merges, R.P., 1998. The control of technology alliances: an empirical analysis of the biotechnology industry. Journal of Industrial Economics 46 (2), 125–150.

Mowery, D.C., et al., 2001. The growth of patenting and licensing by US universities: an assessment of the effects of the Bayh–Dole Act of 1980. Research Policy 30 (1), 99–119.

Cisco Buys Producer of Chips for Networks. New York Times. 17 February 2000.

Oxley, J., 1997. Appropriability hazards and governance in strategic alliances: a transactions cost approach. Journal of Law Economics and Organization 13 (2), 387–409.

Pisano, G., 1991. The governance of innovation: vertical integration and collaborative arrangements in the biotechnology industry. Research Policy 20, 37–249.

Pisano. G., Mang, P., 1993. Collaborative product development and the market for know-how: strategies and structures in the biotechnology industry. In: Rosenbloom, R.S., Burgelman, R.A. (Eds.), Research on Technological Innovation, Management, and Policy, vol. 5. JAI Press, Greenwich, CT.

Pisano, G., Shan, W., Teece, D., 1988. Joint ventures and collaboration in the biotechnology industry. In: Mowery, D. (Ed.), International Collaborative Ventures in US Manufacturing. Ballinger Press, Cambridge, MA.

Roberts, E.B., 1991. Entrepreneurs in High Technology. Oxford University Press, New York.

Roberts, E.B., Berry, C.A., 1985. Entering new businesses: selecting strategies for success. Sloan Management Review 26 (3), 3–17.

Robinson, D., Stuart, T., 2000. Network Effects in the Governance of Strategic Alliances in Biotechnology. University of Chicago GSB, Mimeo.

Rosenberg, N., 1990. Why do firms do basic research (with their own money)? Research Policy 19 (2), 165–174.

Saracho, A.I., 2002. Patent licensing under strategic delegation. Journal of Economics and Management Strategy 11 (2), 225–251.

Seabrook, J., 1994. The Flash of Genius. The New Yorker. 11 January, pp. 38–52.

Shane, S., 2001. Technological opportunities and new firm creation. Management Science 47 (2), 205–220.

Shapiro, C., 2001. Navigating the patent thicket: cross licenses, patent pools, and standard setting. In: Jaffe, A.B., Lerner, J., Stern, S. (Eds.), Innovation Policy and the Economy, vol. 1.

Smith, G.D., 1985. The Anatomy of a Business Strategy: Bell, Western Electric, and the Origins of the American Telephone Industry. Johns Hopkins Press, Baltimore, MD.

Stern, S., 1995. Incentives and focus in university and industrial research: the case of synthetic insulin. In: Rosenberg, N. (Ed.), Sources of Medical Technology, Universities and Industry. National Academy Press, Washington, DC, pp. 157–187.

Teece, D.J., 1986. Profiting from technological innovation: implications for integration, collaboration, licensing, and public policy. Research Policy 15, 285–305.

Tripsas, M., 1997. Unraveling the process of creative destruction: complementary assets and incumbent survival in the typesetter industry. Strategic Management Journal 18, 119–142.

United States Office of Technology Assessment, 1984. Commercial Biotechnology: An International Analysis. US Congress, OTA-BA-218, Washington, DC.

Veugelers, R., Cassiman, B., 1999. Make and buy in innovation strategies: evidence from Belgian manufacturing firms. Research Policy 28, 63–80.

Wang, X.H., 1998. Fee versus royalty licensing in a Cournot Duopoly model. Economics Letters 60 (1), 55–62.

Williamson, O.E., 1985. The Economic Institutions of Capitalism. Free Press, New York.

[8]

MANAGEMENT SCIENCE
Vol. 52, No. 2, February 2006, pp. 293–308
ISSN 0025-1909 | EISSN 1526-5501 | 06 | 5202 | 0293

informs®

DOI 10.1287/mnsc.1050.0437

Patent Protection, Complementary Assets, and Firms' Incentives for Technology Licensing

Ashish Arora
H. John Heinz III School of Public Policy, Carnegie Mellon University, 5000 Forbes Avenue, Pittsburgh, Pennsylvania 15213, ashish@andrew.cmu.edu

Marco Ceccagnoli
College of Management, Georgia Institute of Technology, 800 West Peachtree Street, NW, Atlanta, Georgia 30308, marcoc@gatech.edu

This paper analyzes the relationship between technology licensing and the effectiveness of patent protection. Using the 1994 Carnegie Mellon survey on industrial research and development (R&D) in the United States, we develop and test a simple structural model in which the patenting and licensing decisions are jointly determined. We find that increases in the effectiveness of patent protection increases licensing propensity, but only when the firm lacks specialized complementary assets required to commercialize new technologies. In contrast, for firms with specialized complementary assets, increases in patent effectiveness increase patenting propensity but reduce the propensity to license. We present systematic cross-industry empirical support for the proposition that intellectual property protection is a key determinant of the vertical boundaries of the firm and the market for technology but that its impact is mediated by a firm's ownership of specialized complementary assets.

Key words: patent; licensing; complementary assets
History: Accepted by Scott Shane, technological innovation, product development, and entrepreneurship; received May 11, 2004. This paper was with the authors $7\frac{1}{2}$ months for 2 revisions.

1. Introduction

Over the last two decades, technology-related alliances, such as contract research and development (R&D), R&D joint ventures, and especially technology licensing, have grown rapidly. Estimates for the 1980s suggest that such alliances account for as much as 10%–15% of total civilian R&D in the Organization for Economic Cooperation and Development (OECD) countries (Arora et al. 2001, p. 43). Both start-up and established firms in industries such as biotechnology and pharmaceuticals, semiconductors, instruments, and chemicals have relied extensively on licensing to appropriate rents from their innovations. In pharmaceuticals, 38% of the new chemical entities approved by the U.S. Food and Drug Administration from 1963 to 1999 were licensed (DiMasi 2000). Market-based licensing of chemical processes by established firms such as Union Carbide, BP, Shell, and even Dow and DuPont is also well documented (Arora and Fosfuri 2003, Fosfuri 2004a).[1]

Similarly, over the last two decades, patenting in the United States has grown rapidly, due in part to the increasing importance of patent-intensive sectors such as pharmaceuticals. However, part of the increase in growth is due to increases in patent propensity in sectors such as semiconductors and electronics (Kim and Marschke 2004), in which patents have not traditionally been seen as very important, partly reflecting the rise of firms specializing in chip design (Hall and Ziedonis 2001). Changes in the legal environment have been important as well. In 1982, the Court of Appeals for the Federal Circuit was established to make patent law more uniform; indirectly, it also strengthened it. Patents have also become a growing preoccupation of management (Grindley and Teece 1997, Rivette and Kline 1999).

It is plausible that these trends are related. In particular, the link between patenting and licensing can be analyzed in light of the challenges faced by sellers and buyers of technology, highlighted by Arrow (1962) and fleshed out in greater detail by Mowery (1983), Williamson (1991), and others. Arrow (1962) pointed out that a potential licensee would naturally wish to verify the quality of the invention before paying for it. However, once the inventor discloses the invention, the potential licensee would have little incentive to pay for it. Patents are a possible solution because the invention can be disclosed without fear of imitation. Licensing is possible, in principle, even without patents. Anton and Yao (1994) provide a theoretical

[1] Patent-based technology transactions are, of course, not new. Lamoreaux and Sokoloff (1998) document the extensive trade in patent rights in the United States in the late 19th century. Also, data availability limits the focus of this paper to technology licensing, but the market for technology has other dimensions as well, including strategic alliances and mergers and acquisition.

model of how an invention protected only through secrecy can be licensed by threatening a licensee who reneges with enhanced competition through further disclosure. Arora (1996) provides a model and supporting empirical evidence wherein tacit know-how can be bundled with other inputs (which could include patents) in a self-enforcing contract. However, although plausible, the empirical importance of these alternative mechanisms remains unknown. Thus, Teece (1986) has argued that licensing is mainly recommended if the innovator enjoys strong patent protection *and* lacks complementary assets such as manufacturing and marketing. Implicit in Teece's prescription is the notion that access to complementary assets is difficult—typically they cannot be "rented," and acquiring them is costly and time consuming.

The available evidence (reviewed in §2) of the impact of patents, or their effectiveness, on licensing is limited, typically confined to studies of individual industries, and the empirical findings are inconclusive. This paper shows that one key reason for the mixed evidence, which has been overlooked by previous studies, is that the impact of patent protection on licensing is mediated by specialized complementary assets required for commercializing innovations: Effective patents are more likely to encourage licensing among firms that lack such assets compared with firms that possess them. Patent protection can affect licensing through two routes: the patenting decision (because nonpatented inventions are difficult to license) and the licensing decision conditional on patenting. By estimating a system of simultaneous equations, where the patenting and licensing decisions are jointly determined, we are able to identify the differential impact of patent effectiveness on both the probability of licensing and licensing conditional on patenting and to disentangle the mediating role of specialized complementary assets. We use data from the 1994 Carnegie Mellon survey on U.S. industrial R&D (Cohen et al. 2000).

This paper also contributes to the literature on the role of patents in appropriating rents from innovation. Surveys by Mansfield (1986), Levin et al. (1987), and Cohen et al. (2000) suggest that patents are featured as an effective appropriability strategy in only a few industries. Arora et al. (2003) estimate that only for a small fraction of innovations do patents provide greater net expected returns. We contribute to this literature by analyzing how the payoffs to patenting and licensing are conditioned by specialized complementary assets. Our results imply that although the strength of patent protection increases the returns to patenting, the source of the increase is more likely to be licensing for firms lacking specialized complementary assets. Our results also imply that the presence of specialized complementary assets enhances the value of patent protection.

However, our paper does not address the long-run configuration of firms and industries. In particular, it cannot address fully when industries can feature a division of innovative labor (Arora et al. 2001). For this, we would have to examine how and when innovating firms strategically choose to acquire (or not) various complementary assets. The resource-based theory of the firm (e.g., Wernerfelt 1984, Dierickx and Cool 1989, Barney 1991, Amit and Schoemaker 1993) suggests that firms are distinguished by capabilities that are valuable, rare, and difficult to imitate. In the same spirit, we assume that firms are endowed with different levels of specialized complementary assets, at least in the short term.

In §2, we present a simple model of patenting and licensing decisions, the hypotheses to be tested, and the empirical specification. Section 3 describes the data and measures used for estimation, and §4 discusses the results. Section 5 examines the robustness of the results to alternative assumptions about specifications and endogeneity. The conclusion follows in §6.

2. Theory Development and Empirical Specification

In this section, we first present a simple empirical model to better structure the development of our hypothesis. In particular, we posit that a firm that has developed an innovation faces four mutually exclusive and exhaustive options: (1) patent and license; (2) patent and not license; (3) not patent and license; and (4) not patent and not license. The probabilities that any given innovation is patented or licensed are as follows:

$$\Pr(\text{Patent}) = \Pr(\text{Patent and License}) + \Pr(\text{Patent and Not License}), \quad (1\text{-}1)$$

$$\Pr(\text{License}) = \Pr(\text{Patent and License}) + \Pr(\text{Not Patent and License}). \quad (1\text{-}2)$$

In our data, firms that do not patent rarely license. In particular, cross-tabulating our data by licensing and patenting propensities shows that less than 10% of licensors do not patent, but nearly 33% of nonlicensors do not patent; and that only 12% of nonpatentees license, whereas about 40% of the patentees license. Therefore, our data strongly suggest that the presence of a patent is almost essential for licensing, and it will be difficult to empirically estimate the determinants of license and not patent. Accordingly, we set Pr(License and Not Patent) equal to 0 in (1-2) and drop the 25 anomalous observations from the analysis.[2]

[2] Section 5.4 discusses the robustness of this assumption, including a more detailed examination of those cases where licensing is apparently not based on a patent.

Arora and Ceccagnoli: *Patent Protection, Complementary Assets, and Firms' Incentives for Technology Licensing*
Management Science 52(2), pp. 293–308, ©2006 INFORMS 295

Let V_L denote the payoff to a firm if it patents and licenses a given innovation, and V_P if it patents but does not license it. Similarly, let V_S denote the payoff if the firm keeps the innovation secret (not patent) and also does not license it. Assuming that firms choose the action with the highest payoff, we obtain

$$\Pr(\text{Patent}) = \Pr(V_L = \text{Max}\{V_L, V_P, V_S\}) + \Pr(V_P = \text{Max}\{V_L, V_P, V_S\}), \quad (2\text{-}1)$$

$$\Pr(\text{License}) = \Pr(V_L = \text{Max}\{V_L, V_P, V_S\}). \quad (2\text{-}2)$$

We use a linear random utility specification, obtained by incorporating an additive stochastic component to each payoff, and assume that the stochastic terms are independent and identically distributed (i.i.d.) across innovations with Type 1 extreme value (Gumbel) distribution and are observed by the firm but not the econometrician. The licensing and patenting probabilities[3] are (cf. McFadden 1973)

$$\Pr(\text{Patent}) = [\exp(V_L) + \exp(V_P)] / [\exp(V_L) + \exp(V_P) + \exp(V_S)], \quad (3\text{-}1)$$

$$\Pr(\text{License}) = [\exp(V_L)] / [\exp(V_L) + \exp(V_P) + \exp(V_S)], \quad (3\text{-}2)$$

$$\Pr(\text{License} \mid \text{Patent}) = [\exp(V_L)] / [\exp(V_L) + \exp(V_P)]. \quad (3\text{-}3)$$

Our data are at the firm level, and we do not observe whether a specific innovation is licensed or patented. Rather, we observe the proportions of innovations a firm patents or licenses. Thus, although our setup is similar to a multinomial logit, our dependent variables are not binary variables but proportions. We treat the probability of an event as the true mean of the observed proportion. It is also obvious that we cannot identify the individual payoffs, only the differences. By dividing both the numerator and denominator of the probabilities by $\exp(V_S)$, we get the estimating equations

$$Y_1 = [\exp(V_L - V_S) + \exp(V_P - V_S)] / [\exp(V_L - V_S) + \exp(V_P - V_S) + 1] + \nu_1, \quad (4\text{-}1)$$

$$Y_2 = [\exp(V_L - V_S)] / [\exp(V_L - V_S) + \exp(V_P - V_S) + 1] + \nu_2. \quad (4\text{-}2)$$

Here, Y_1 and Y_2 are the patenting and licensing propensities, $V_L - V_S$ and $V_P - V_S$ are made functions of observed firm and industry characteristics discussed below, and the econometric error terms, ν_1 and ν_2, represent sampling errors that account for the difference between theoretical and empirical probabilities.[4]

2.1. Determinants of $V_L - V_S$, $V_P - V_S$ and Implications for Patenting and Licensing Propensity

Our empirical specification makes it clear that observed behavior—the decision to patent or license—does not depend on the underlying payoffs in a straightforward manner. For instance, a variable that increases the licensing payoff, V_L, could nonetheless decrease licensing propensity if it increases V_P sufficiently. Accordingly, we develop our hypotheses in two parallel tracks. We first discuss the impacts of patent effectiveness on relative payoffs. We then develop the implications for patent and licensing propensities. The theoretical predictions are summarized in Tables 1a and 1b. The appendix formalizes the hypothesis developed and provides proofs.

2.1.1. Patent Effectiveness. We use the term "patent effectiveness" to mean strength of patent protection. A variety of factors may drive the effectiveness of patents, including increases in length or breadth of protection, greater codifiability of knowledge, decreases in costs of application, and costs of disclosure (Horstmann et al. 1985). Because we cannot empirically distinguish among these, we use effectiveness as a summary measure. An increase in patent effectiveness should increase $V_L - V_S$ as well as $V_P - V_S$, thus unambiguously increasing patent propensity. However, although patents are virtually always required for licensing, an increase in patent effectiveness has conflicting effects on the decision to license: While more effective patents increase V_L, they also increase V_P, so that the impact on licensing propensity is ambiguous. More effective patent protection increases the net benefits from licensing (e.g., by decreasing transaction costs or by increasing the licensor's bargaining power) but may also increase the opportunity cost of licensing by enhancing the payoff from the exclusive commercialization of the innovation.

Indeed, the available empirical evidence is mixed. Using a sample of Massachusetts Institute of Technology inventions, Gans et al. (2002) find that the presence of patents increases the likelihood that an inventor will license to an incumbent rather than

[3] Note that (3-3) can be derived from (3-1) and (3-2). Estimation of (3-3) by itself is feasible but inefficient, because it would require that we drop about 25% of the sample with zero patent propensity (see §5.3).

[4] Our framework assumes that the licensing and the patenting decisions are made at the same time. Firms may patent first and later decide whether to license. Estimates of a sequential model yield similar results (see §5.3), but, with cross-sectional data, identification in the sequential model relies more heavily upon functional form assumptions.

Table 1a The Predicted Effect of Patent Protection and Complementary Assets on Licensing and Patenting Payoffs

	$V_L - V_S$ Licensing payoffs relative to not patenting	$V_L - V_P$ Licensing payoffs relative to patenting and not licensing	$V_P - V_S$ Patenting and not licensing payoffs relative to not patenting
Patent effectiveness (Hypothesis 1a)	+	?	+
Complementary assets × patent effectiveness (Hypothesis 2a)	Null	−	+

Table 1b The Predicted Effect of Patent Protection and Complementary Assets on Licensing and Patenting Propensities

	Probability of patenting	Probability of licensing conditional on patenting	Probability of licensing
Patent effectiveness (Hypothesis 1b)	+	?	?
Complementary assets × patent effectiveness (Hypothesis 2b)	?	−	−

enter the product market by commercializing the invention. Anand and Khanna (2000) find that in the chemicals sector, where patents are believed to be effective, there are more technology deals—a large fraction of these are arm's length and involve exclusive licenses, a large fraction concerns future technologies rather than existing technologies, and a small fraction is among related firms than for all sectors as a whole. In contrast, Cassiman and Veugelers (2002) do not find that more effective patents encourage Belgian firms to enter into collaborative R&D arrangements.

Evidence from cross-national data is also mixed. Some studies find a positive association between patents and licensing. Yang and Maskus (2005) report a strong positive relationship between improved intellectual property rights (IPR) regimes and licensing by U.S. multinational corporations. Analyzing data on international technology licensing contracts of Japanese firms, Nagaoka (2002) finds that weak patent regimes are associated with an increase in the fraction of transfers to an affiliate (such as a subsidiary), rather than to an unaffiliated firm. Smith (2001) finds that U.S. firms are more likely to export or directly manufacture than to license technology in countries with weak patent regimes. A study using French data on the export of technology services finds that such exports are greater for countries with more effective patent protection, but mainly for higher-income countries (Bascavusoglu and Zuniga 2002).

Other studies cast doubts on the link between patent protection and the extent or form of international technology licensing. Fink (2005) finds a very weak relationship using German data. Puttitanun (2003) reports a higher response of direct investment than licensing to changes in the level of IPR protection. Similarly, Fosfuri (2004b) does not find that patent protection significantly affects the extent or composition of technology flow (as joint venture, direct investment, or licensing) in the chemical sector. The mixed nature of the findings is further reflected in a recent study by Branstetter et al. (2006). Using detailed data on the technology royalty payments received by U.S. firms and controlling for country, industry, and firm fixed effects, they find that stronger patent protection does not increase the transfer of technology by U.S. multinationals to unaffiliated parties. However, it *does* increase the flow of technology to affiliates. The above discussion leads us to the following hypotheses:

HYPOTHESIS 1A. *Increases in patent effectiveness increase $V_L - V_S$ and $V_P - V_S$. The impact on $V_L - V_P$ is ambiguous.*

HYPOTHESIS 1B. *Increases in patent effectiveness increase patent propensity. Increases in patent effectiveness may increase or decrease the share of licensing in patented innovations.*

2.1.2. Complementary Assets and Their Conditioning Role. Successful innovation requires manufacturing, marketing, and a variety of other complementary assets (cf. Teece 1986). If these assets were "generic" and hence could be readily accessed through the market, they could be valuable, but their strategic importance would be limited (e.g., Barney 1991, Rothaermel and Hill 2005). However, manufacturing and marketing assets required for commercializing an innovation are often not generic. Instead, they are specialized to the innovation and thus have limited alternative uses. Their market availability is also limited because firms tend to gain control over them to avoid potential bargaining problems. They are difficult to imitate because they result from the interaction of people from different parts of a firm's organization and are built over long periods of time (Teece 1992). Indeed, Shane (2001) finds that specialized complementary assets increase the likelihood that technology will be exploited inside a firm. Conversely, Gans et al. (2002) find that technology startups (which lack complementary assets) are more likely to ally with incumbents in sectors where complementary assets are costly to acquire. In our setup (nongeneric) complementary assets should increase V_P and V_S and, as we argue below, should have little impact on V_L.

Our focus in this paper is not on this direct impact of complementary assets. Rather, we want to highlight an important interaction between a firm's complementary assets and patent effectiveness. In particular, if higher patent effectiveness implies that rivals are unable to produce close substitutes, then more effective patent protection should translate into a "larger market" for the patented innovation. An innovator with specialized complementary assets should be able to profit more from a larger market than an innovator lacking such assets. For instance, if complementary assets imply that the innovative product can be produced at a lower cost or that quality can be increased at a lower cost, then this is more profitable, the greater the volume of demand for the product. Thus, V_P should increase faster with patent effectiveness for firms with specialized complementary assets.

For the other payoffs, this interaction effect is absent. The payoff from not patenting, V_S, cannot, by definition, vary with patent effectiveness. Similarly, the payoff from licensing, V_L, should depend on the complementary assets of the *licensee* rather than those of the licensor. It follows that $V_L - V_P$ should fall faster with patent effectiveness for firms with specialized complementary assets. Similarly, the positive impact of patent effectiveness on $V_P - V_S$ is larger for firms with the specialized complementary assets, but the impact on $V_L - V_S$ is unaffected.

Insofar as the ownership of specialized complementary assets also increases licensee bargaining power in licensing negotiations (e.g., Gans et al. 2002), the impact of complementary assets on V_L is positive (rather than zero) so that the impact of the interaction between patent effectiveness and complementary assets on $V_L - V_P$ is ambiguous. The empirical model does allow for a specialized complementary asset to affect V_L, but the empirical estimates show that the estimate of such impact, if any, is small.[5]

The consideration of the interaction effect on relative payoffs is also critical to evaluate the impact of patent effectiveness on licensing probabilities. If the positive interaction effect on V_P is sufficiently strong, both the propensity to license and the share of licensing in patented innovations will increase faster (or decrease more slowly) with patent effectiveness for firms lacking the relevant specialized complementary assets. To summarize, we will test the following hypotheses:

Hypothesis 2a. *The impact of increasing patent effectiveness on $V_L - V_S$ is the same for firms with different levels of specialized complementary assets. However, the impact of increasing patent effectiveness on $V_P - V_S$ is higher for firms owning specialized complementary assets.*

Hypothesis 2b. *The impact of patent effectiveness on both licensing and the share of licensing in patented innovations is weaker for firms with specialized complementary assets.*

3. Data and Variables

The data used come from the Carnegie Mellon survey (CMS) on industrial R&D (Cohen et al. 2000). The population sampled is that of all R&D labs located in the United States conducting R&D in manufacturing industries as a part of a manufacturing firm. The sample was randomly drawn from the eligible labs listed in the *Directory of American Research and Technology* (Bowker 1995) or belonging to firms listed in Standard and Poor's Compustat, stratified by three-digit SIC industry. R&D lab managers were asked to answer questions with reference to the "focus industry"—defined as the principal industry for which the unit was conducting its R&D. Valid responses were received from 1,478 R&D units, with a response rate of 54%. The data refer to the period 1991–1993. After trimming for outliers and dropping observations with missing data for the variables of interest, we obtain a final sample of 757 observations.[6] Table 2 provides descriptive statistics for the variables used in the analysis. Also, although we use the term "firm" for the unit of analysis, the unit of analysis is the business unit within the parent firm, operating in the "focus industry" of the responding R&D lab.

3.1. Endogenous Variables

Licensing Propensity. The CMS asks respondents to state the percentage of their R&D projects over the last three years that were undertaken with the objective of earning licensing revenues. (Projects could have multiple objectives.) There were five response categories: <10%, 10%–40%, 41%–60%, 61%–90%, and >90%. We used the midpoints of each response category. For the first category, it is likely that most respondents actually meant zero. Thus, we assigned all respondents in the first category zero licensing

[5] Additional support for this assumption comes from the empirical estimates of the direct impact of complementary assets on licensing propensity. If the bargaining power effect were negligible, the direct impact ought to be negative. Our results indicate that the impact of complementary assets on both licensing and the share of patented innovations that are licensed is indeed negative. The structural estimates indicate that complementary assets decrease $V_L - V_P$ as well as $V_L - V_S$. In sum, the results are consistent with a small or negligible impact of complementary assets on V_L.

[6] This also reflects the exclusion of business units with more than 10 employees, R&D units reporting more than 50 patent applications per million dollars of R&D, and those reporting an R&D budget of less than or equal to $100,000. Including very small units does not affect our results but is consistent with Arora et al. (2003) and Cohen et al. (2000).

Table 2 Descriptive Statistics

	Mean	Median	St. dev.	Min	Max
Endogenous variables					
Licensing propensity	0.06	0	0.15	0	0.95
Patent propensity	0.29	0.25	0.27	0	1
Explanatory variables					
Business unit employees (Log)	6.60	6.45	2.08	2.30	13.01
Patent effectiveness	0.35	0.25	0.29	0.05	0.95
Complementary assets	0.50	0	0.50	0	1
% basic research	0.03	0	0.06	0	0.50
Importance of basic science	1.93	2	0.96	0	3
Importance of medical science	0.66	0	1.00	0	3
Number of technological competitors	3.95	4	4.75	0	32
Parent firm is global	0.78	1	0.42	0	1
Parent firm is foreign	0.10	0	0.30	0	1
Parent firm is public	0.66	1	0.47	0	1

$N = 757$

propensity, unless they indicated that the possibility of earning licensing revenue was one of the reasons for patenting.[7] The available measure of licensing is not based on actual licensing deals but on firms' willingness to invest in R&D with the intention to license. Our ex ante measure thus excludes cross-licensing deals and transactions that take place in the shadow of infringement suits (Hall and Ziedonis 2001). Because we lack information on technology buyers, our measure of licensing propensity may be better than the actual number of licensing deals. Note that licensing propensity is not separately available for products and processes.

Patent Propensity. The CMS asks respondents for the percentage of product and process innovations for which they applied for patents in the period 1991–1993 in the United States. We computed a weighted average of product and process patent propensities using the percentage of R&D effort devoted to product and process innovations, respectively, as weights, as reported by the CMS respondents.

3.2. Explanatory Variables

Patent Effectiveness. The CMS asks respondents to indicate the percentage of their product and process innovations for which patent protection had been effective in protecting their firm's competitive advantage from those innovations during the prior three years. There were five mutually exclusive response categories for product and process innovations separately: <10%, 10%–40%, 41%–60%, 61%–90%, and >90%. We computed a weighted average of the product and process scores (using midpoints), with the percentage of R&D effort devoted to product and process innovations as weights, to construct *patent effectiveness*.

To inform our interpretation of this measure, we analyzed the relationship between reasons to patent and not to patent and the respondents' patent effectiveness scores, using an ordered probit model (Arora et al. 2003). The results indicate that patent effectiveness is indeed a broad summary measure of the various costs (such as information disclosure) and benefits of patenting (including preventing imitation, facilitating technology negotiations, or building patent fences). In §5, we report results where we instrument for patent effectiveness to address the possibility that firms that patent a lot also perceive patents to be effective, or that there are unmeasured differences in knowledge that drive both patenting and patent effectiveness.[8]

Complementary Assets. We focus on specialized manufacturing capability. The CMS provides a measure for the frequency of face-to-face interaction between personnel from R&D and production, measured in a four-point Likert scale.[9] We constructed a binary variable, *complementary asset*, which takes value 1 if R&D and manufacturing personnel interact daily (the median value is weekly interaction).

In general, measuring the degree of specialization is difficult but, as Teece (1992) suggests, complementary assets often arise from the interaction and learning over time of people from different parts of a firm's organization. This is especially relevant for the interaction with R&D, which typically requires organizationally embedded interpersonal and interfunctional activities (Zhao et al. 2005). Previous studies have used measures of manufacturing or sales force (e.g., Tripsas 1997, Nerkar and Roberts 2004) or financial assets as measures of complementary capabilities (e.g., Helfat 1997). Our measure does not simply reflect a firm's ownership of complementary manufacturing capabilities, because all firms in our sample have manufacturing capability. Further, we control

[7] This procedure increases the sample probability of licensing conditional on patenting, because only patentees can indicate whether licensing was a motive for patenting. However, using the midpoint of the first category gives similar results.

[8] Our patent effectiveness measure does not take into account the skewed nature of the net returns from patenting (Schankerman and Pakes 1986, Scherer and Harhoff 2000). This is relevant only insofar as one suspects that even as the percentage of innovations for which patents are effective increases, the average return to patenting falls.

[9] Respondents were asked, "How frequently do your R&D personnel talk face to face with personnel from the Production, Marketing or Sales, and Other R&D units functions?"

for size, which also proxies for the manufacturing and marketing assets. Indeed, the extent of interaction between the R&D and manufacturing functions measures how *important* it is for the innovating firm to also have manufacturing capability—the quintessence of the notion of specialized complementary assets.[10]

There are other possible—and broader—measures of specialized complementary assets, such as the importance of complementary manufacturing and marketing assets in appropriating the profits from innovations (cf. Cohen et al. 2000, Shane 2001, Gans et al. 2002). In the online appendix (http://mansci.pubs.informs.org/ecompanion.html) we show that their use in the empirical model yields qualitatively similar—albeit less precise—results than those presented here. Finally, it is plausible that firms that intend to license may choose more modular organizations, with fewer interactions between R&D and manufacturing. We probe the robustness of our results to such endogeneity in §5.

Business Unit Size. We controlled for size using business unit size, measured by the log of the number of business unit employees. We also experimented with firm size, measured by the log of the total employees of the unit's parent firm, as well as both measures of size, and obtained similar results.

Nature of Knowledge. Tacit technology is costly to transfer, implying lower licensing payoffs (Teece 1977). Typically, technologies that are strongly science based, especially in sectors such as chemicals, molecular biology and genetics, and in some cases, semiconductors, are less likely to be tacit, and patents are more likely to be effective (Arora and Gambardella 1994, Winter 1987, Anand and Khanna 2000). The CMS provides various measures of the degree to which the firm's knowledge is science based and thus likely to be codifiable or nontacit.[11] We use three such variables. One measure, labeled *importance of basic science*, represents the maximum score computed across all the fields, except the engineering fields and the medical and health science field. A second measure, labeled *importance of medical science*, represents the importance of the medical and health science field alone. A third measure is the *percentage of R&D effort devoted to basic research*, defined as scientific research with no specific commercial objective.

Basic research is arguably more codifiable than applied research. For example, basic biotech research has highly codifiable output, and there is ample evidence of contract R&D and technology licensing. On the other hand, advances in internal combustion engines are still based on a great deal of trial-and-error type research, the results are difficult to codify effectively, and there is relatively little contract R&D or technology licensing. Even in biopharmaceuticals, fermentation and other production technologies tend to be tacit. The share of basic research (and perhaps the importance of basic science) has an alternative, although not mutually exclusive, interpretation as well. As Nelson (1959) points out, basic scientific research is more likely to result in useful knowledge that a firm is itself unable or unwilling to commercialize or the use of which may be much broader than the firm's domain of operation. In such cases, licensing is a natural means of appropriating the benefits of such research.[12]

Technological Competition. Technological rivalry is likely to raise the payoff to patenting and may also raise the opportunity cost of licensing, particularly if there is a prospect of valuable knowledge leaking out (e.g., Hill 1992, Ceccagnoli 2005). We used the midpoint of the *number of technological rivals*, which in the CMS is reported as a categorical variable in the following ranges: 0, 1–2, 3–5, 6–10, 11–20, or >20.[13]

Industry Fixed Effects. We included seven industry dummy variables to control for industry effects, such as licensing and patenting norms.[14]

Other Controls. We used binary variables indicating whether the firm owning the R&D lab was *global* (also sells products in Japan or Europe), *foreign* (the parent is located abroad), or *public* (publicly traded).

[10] Of course, some industry-specific studies have the advantage of being able to use finer measures of specialized complementary assets: Tripsas (1997) uses font libraries in her study of the typesetting industry; Thomke and Kuemmerle (2002) use therapeutical area-specific chemical libraries in pharmaceuticals, and Penner-Hahn and Shaver (2005) use the stock of fermentation patents as a measure of complementary manufacturing capabilities for Japanese pharmaceutical firms.

[11] Respondents were asked to rate (on a four-point Likert scale) the importance of different fields of science and engineering in terms of their contribution to a firm's R&D activity during the most recent three years.

[12] Licensing agreements in biotechnology indicate that being basic is not an intrinsic barrier to licensing and closeness to science allows commercially relevant milestones to be specified using scientific terms, such as the characterization of a protein or the structure of a receptor.

[13] Technological rivals are defined as the number of U.S. competitors in the focus industry capable of introducing competing innovations that can effectively diminish the respondent's profits from an innovation. This variable represents each respondent's assessment of his or her focus industry conditions, and thus varies across respondents within an industry.

[14] The industry dummies are based on the following groupings: chemicals, petroleum and plastics (SIC 28, 29, 30, excluding 283), biotechnology and pharmaceuticals (SIC 283), computer and electronics (SIC 357 and 36), machinery (SIC 35, excluding 357), transportation (SIC 37), instruments (SIC 38, excluding 384), and medical instruments (SIC 384). Note that the results presented below are robust to the use of more disaggregated industry definitions.

4. Results

4.1. OLS Results

We first present single equation OLS results, reported in Table 3 (with and without interactions). The results are consistent with the theory developed in the previous section, suggesting that patent effectiveness has a positive and significant effect on patent propensity (Hypothesis 1b). We also find a positive and significant effect on licensing propensity, consistent with Hypothesis 1b, as well. The average elasticities of patenting and licensing, with respect to patent effectiveness, are about 0.5 and 0.6, respectively. The effect of the interaction between patent effectiveness and the complementary assets variable on licensing propensity is negative and significant, consistent with Hypothesis 2b.

Table 3 Determinants of Patent and Licensing Propensities: OLS Estimates

	Patent propensity		Licensing propensity	
	I	II	I	II
Intercept	−0.06 (0.038)	−0.049 (0.039)	0.029 (0.026)	0.01 (0.026)
Patent effectiveness	0.495** (0.027)	0.466** (0.037)	0.091** (0.019)	0.138** (0.025)
Complementary assets	−0.037* (0.015)	−0.058* (0.024)	−0.027** (0.01)	0.007 (0.016)
Business unit employees (Log)	0.015** (0.004)	0.015** (0.004)	−0.004 (0.003)	−0.004 (0.003)
No. of technological rivals	−0.003* (0.002)	−0.003[a] (0.002)	0.001 (0.001)	0.0002 (0.001)
% basic R&D	−0.028 (0.124)	−0.036 (0.124)	0.304** (0.084)	0.318** (0.084)
Importance of medical/health science	−0.002 (0.010)	−0.003 (0.01)	0.017* (0.007)	0.018** (0.007)
Importance of basic science	0.001 (0.008)	0.002 (0.008)	0.011[a] (0.006)	0.01[a] (0.006)
Parent firm is global	0.038* (0.018)	0.037* (0.018)	−0.013 (0.012)	−0.011 (0.012)
Parent firm is foreign	0.029 (0.029)	0.03 (0.029)	−0.015 (0.02)	−0.016 (0.019)
Parent firm is public	0.061** (0.019)	0.06** (0.019)	0.002 (0.013)	0.002 (0.013)
Patent effectiveness × complementary assets		0.061 (0.052)		−0.1** (0.035)
Industry fixed effects (7)	Yes	Yes	Yes	Yes
N	757	757	757	757
Adj. R^2	0.44	0.44	0.14	0.15

Note. Standard errors are in parentheses.
**, *, [a]: Significantly different than zero at the 0.01, 0.05, and 0.10 confidence levels, respectively.

4.2. Structural Estimates, Marginal Effects, and Elasticities

The reduced-form estimates have a number of shortcomings. They are inefficient because the residuals are heteroscedastic. The predicted outcome cannot be interpreted as a predicted probability because predictions could lie outside the (0–1) limits. Most of all, the estimated coefficients only provide qualitative information on the payoffs.[15] Thus, these estimates cannot directly test Hypotheses 1a and 2a. In contrast, in addition to being more precise, structural estimates allow us to disentangle the impact of the explanatory variables on $V_L - V_S$ and $V_P - V_S$, separately. This is an advance over the literature because it allows us to understand the channel through which intellectual property protection affects licensing.

Table 4 presents the structural estimates of (4-1) and (4-2), where we impose cross-equation restrictions, namely that $V_L - V_S$ and $V_P - V_S$ are present in the denominators in both equations and that the numerator in both equations shares the coefficients from $V_L - V_S$. We use the general method of moments (GMM), which allows for potential correlation across errors of the two equations and corrects for possible heteroscedasticity (cf. Gallant 1987, pp. 442–451). The estimates for $V_L - V_P$, reported in the second column of Table 4, are obtained as the difference between the estimates of $V_L - V_S$ and $V_P - V_S$.

For the control variables, Table 4 shows that measures of the codifiability of knowledge have a positive and significant effect on licensing payoffs, as expected. We also find that foreign, global, and public firms have substantially higher payoffs from patenting for commercialization, $V_P - V_S$, but lower licensing payoffs, $V_L - V_S$. Technological rivalry has a negative impact on $V_P - V_S$ but a small and insignificant impact on $V_L - V_S$ and $V_P - V_L$. The industry effects (not reported) are jointly not significant in $V_P - V_S$. This is most likely because our science-specific dummies, especially the importance of medical/health science, which vary across respondents, pick up industry effects.[16] Industry effects are, instead, jointly significant in $V_P - V_S$, where the science dummies have no impact. As expected, chemical, pharmaceuticals, biotechnology, and medical instruments sectors are associated with higher payoffs for licensing patenting innovations, $V_L - V_P$.

The results, reported in Table 5a, provide strong support for Hypothesis 1a; that is, patent effectiveness

[15] For instance, the right-hand side of the patenting equation can be thought of as reflecting the weighted average of $V_L - V_S$ and $V_P - V_S$, with firms with higher licensing propensities having higher-than-average weights on $V_L - V_S$, relative to firms with lower licensing propensities.

[16] Excluding biotechnology and pharmaceutical firms from the estimation does not change the results appreciably.

Table 4 GMM Structural Estimates of the System of Patent and Licensing Propensity Equations

	$V_L - V_S$ Licensing payoffs relative to not patenting	$V_L - V_P$ Licensing payoffs relative to patenting and not licensing	$V_P - V_S$ Patenting and not licensing payoffs relative to not patenting
Intercept	−4.164** (0.563)	−0.914 (0.670)	−3.250** (0.307)
Patent effectiveness	2.564** (0.481)	0.464 (0.534)	2.1** (0.256)
Complementary assets	−0.201 (0.311)	0.414 (0.372)	−0.615** (0.183)
Business unit employees (Log)	−0.021 (0.05)	−0.135* (0.059)	0.115** (0.029)
Patent effectiveness × complementary assets	−0.590 (0.568)	−1.657* (0.648)	1.067** (0.368)
No. of technological rivals	−0.007 (0.02)	0.023 (0.026)	−0.03* (0.013)
% basic R&D	2.822* (1.094)	4.355** (1.317)	−1.533 (0.997)
Importance of medical/health science	0.268* (0.105)	0.388** (0.142)	−0.119 (0.077)
Importance of basic science	0.246* (0.124)	0.295* (0.147)	−0.049 (0.061)
Parent firm is global	−0.072 (0.196)	−0.426 (0.246)	0.355* (0.141)
Parent firm is foreign	−0.242 (0.321)	−0.696[a] (0.395)	0.454* (0.208)
Parent firm is public	0.103 (0.256)	−0.457 (0.343)	0.559** (0.164)
Industry fixed effects (7)	Yes	Yes	Yes
$N = 757$			

Notes. Standard errors are in parentheses. For estimation we used the SAS Proc Model GMM default option, which provides standard errors that are robust to potential heteroscedasticity and autocorrelation of the residuals. Adjusted R^2 is equal to 0.19 for the licensing equation and 0.43 for the patent propensity equation.

**, *, [a]: Significantly different than zero at the 0.01, 0.05, and 0.10 confidence levels, respectively.

Table 5a Marginal Effects on Licensing and Patenting Payoffs

	$V_L - V_S$	$V_L - V_P$	$V_P - V_S$
Hypothesis 1a			
Patent effectiveness	2.27** (0.33)	−0.36 (0.36)	2.63** (0.20)
Hypothesis 2a			
Patent effectiveness × complementary assets	−0.59 (0.56)	−1.66* (0.65)	1.07** (0.37)

Table 5b Marginal Effects on the Probabilities of Patenting and Licensing

	Probability of licensing	Probability of licensing conditional on patenting	Probability of patenting
Hypothesis 1b			
Patent effectiveness	0.08** (0.01)	−0.06 (0.06)	0.47** (0.03)
Hypothesis 2b			
Patent effectiveness × complementary assets	−0.06* (0.03)	−0.25* (0.11)	

Notes. Standard errors are in parentheses. Marginal effects and elasticities are obtained using the GMM estimates, evaluated at the mean of the respective sample.

**, *, [a]: Significantly different than zero at the 0.01, 0.05, and 0.10 confidence levels, respectively.

increases the patenting and licensing payoffs. The partial derivative of the net licensing payoff, $V_L - V_S$, with respect to patent effectiveness, is equal to 2.27, significant at the 1% level. The corresponding derivative of $V_P - V_S$ is 2.63, significant at the 1% level.

To examine Hypothesis 1b, we evaluate the marginal impact of patent effectiveness on licensing and patenting propensities, evaluated at the mean of the sample. The impact on patenting propensity is positive and significant at the 1% level, consistent with Hypothesis 1b, implying that a unit increase in patent effectiveness increases the probability of patenting by 0.47. Patent effectiveness, however, decreases the probability of licensing conditional on patenting by 0.06 and increases the unconditional probability of licensing by 0.08 (significant at the 5% level). In other words, more effective patents *do* lead to more licensing, but this is mostly driven by the indirect effect of patent effectiveness on the probability of patenting. This is more easily seen, noting that Elasticity of Pr(Lic) = Elasticity of Pr(Lic | Pat) + Elasticity of Pr(Pat). The elasticity of the conditional probability is −0.08, and the elasticity of patenting is 0.51, so that the elasticity of the probability of licensing with respect to (w.r.t.) patent effectiveness is 0.43.

The central result of this paper relates to the impact of the interaction between patent effectiveness and complementary assets on licensing payoffs and probabilities, related to Hypotheses 2a and 2b. The structural estimates of the parameter related to the interaction term, shown in Table 4, clearly indicate a negative and significant impact on $V_L - V_P$ and a positive and significant impact on $V_P - V_S$. These results confirm Hypothesis 2a, namely that firms with specialized complementary assets derive greater value from more effective patent protection, but not through licensing.

The cross-partial effect of patent effectiveness and complementary assets on the licensing probabilities is shown at the bottom of Table 5b. To compute this effect, we first divide the sample into firms with high and low specialized complementary assets (according to our measure). For each group, we then compute the marginal effects of patent effectiveness on the respective licensing probabilities evaluated at the group mean. Consistent with Hypothesis 2b, we find that licensing is less responsive to patent effectiveness, both for patented innovations and for all innovations, for firms with specialized complementary assets relative to firms that do not possess them (the difference is significant at the 5% level). Because we control for business unit size, these results are not merely reflecting scale effects and suggest that the ownership of specialized complementary assets critically conditions the impact of patent protection on licensing.

Recall that Hypothesis 2b does not require that an increase in patent effectiveness actually decreases licensing for firms with high complementary assets, but merely that any increase be smaller than that for firms with low complementary assets, although such a result is obviously consistent with it. Table 6, which reports the marginal effects and elasticities of patent effectiveness on the licensing probabilities, shows that firms with high specialized complementary assets and higher effectiveness of patent protection are more likely to patent but less likely to license their patented innovations than firms with high specialized complementary assets but low patent effectiveness. Firms with low specialized complementary assets and high patent effectiveness are more likely to license, but they are also more likely to license their patented innovations than firms with low specialized complementary assets and low patent effectiveness.

In sum, the results suggest that firms with specialized complementary assets react to stronger patent protection by patenting more, but using the patents to enhance the payoff from commercialization. Firms lacking specialized complementary assets also respond by increasing patenting but are more likely to use patents as a basis for licensing.

Table 6 The Impact of Patent Protection on Licensing Probabilities by Complementary Assets (Hypothesis 2b)

	Marginal effects and elasticities w.r.t. patent effectiveness of			
	Probability of licensing		Probability of licensing conditional on patenting	
	Marginal effect	Elasticity	Marginal effect	Elasticity
Complementary assets				
Low	0.11** (0.03)	0.62	0.07 (0.08)	0.14
High	0.05** (0.01)	0.24	−0.18* (0.08)	−0.30
Total	0.08** (0.01)	0.43	−0.06 (0.06)	−0.08

Notes. Standard errors are in parentheses. Marginal effects and elasticities are obtained using the GMM estimates, evaluated at the mean of the respective sample.

**, *, a: Significantly different than zero at the 0.01, 0.05, and 0.10 confidence levels, respectively.

5. Sensitivity Analysis[17]

We now evaluate the sensitivity of our results to possible endogeneity and alternative specifications.

5.1. Potential Endogeneity of Complementary Assets

We have assumed, in the spirit of the resource-based theory, that complementary assets are exogenous. However, it is possible that unobserved factors that reduce licensing costs or increase licensing payoff may also reduce complementarity between manufacturing and R&D.[18] To address this concern, we instrument for the presence of complementary manufacturing assets using a dummy variable indicating whether the R&D lab is located near a production facility. Thus, if one accepts that the location of the lab is a long-term decision, colocation between R&D and manufacturing is a source of exogenous variation in the frequency of interactions between R&D and manufacturing.[19]

As Tables 7a and 7b show, all our main results are confirmed, although the standard errors are typically larger. The interaction between complementary assets and patent effectiveness continues to have a negative and significant effect on $V_L - V_P$ and a positive and significant effect on $V_P - V_S$ (Columns VI and X, Table 7a). The marginal effect of patent effectiveness on licensing propensity is positive and decreases from 0.09 to 0.06 between firms with low and high specialized complementary assets (Column II, Table 7b). Similarly, the marginal effect of patent effectiveness on the share of licensing in patented innovations is lower for firms with low specialized complementary assets (Column VI, Table 7b).

5.2. Endogeneity of Patent Effectiveness and Other Possible Biases

It is possible that R&D managers from labs with high patent propensity also tend to report higher patent effectiveness to "justify" their patenting behavior and related costs. However, even after controlling for industry-fixed effects, the average patent effectiveness score when the respondent is from a non-R&D function (about 13% of the sample) is only 0.002 lower than the average score of respondents from the R&D lab, a very small and insignificant difference.

[17] See the online appendix for further details of results presented in this section.

[18] In a linear specification, this would imply that the coefficient on complementary assets would have a negative bias in the licensing equation. In a nonlinear specification such as ours, the bias cannot be judged as readily.

[19] The correlation coefficient between complementary assets and colocation is 0.2, significant at the 1% level.

Arora and Ceccagnoli: *Patent Protection, Complementary Assets, and Firms' Incentives for Technology Licensing*
Management Science 52(2), pp. 293–308, © 2006 INFORMS 303

Table 7a **Marginal Effects on Licensing and Patenting Payoffs Using Instrumental Variables for Patent Effectiveness and Complementary Assets (Hypothesis 2a)**

	$V_L - V_S$				$V_L - V_P$				$V_P - V_S$			
	I Exog. θ and ρ	II Endog. ρ	III Endog. θ	IV Endog. θ and ρ	V Exog. θ and ρ	VI Endog. ρ	VII Endog. θ	VIII Endog. θ and ρ	IX Exog. θ and ρ	X Endog. ρ	XI Endog. θ	XII Endog. θ and ρ
Patent effectiveness (θ) × complementary assets (ρ)	−0.59 (0.56)	−0.25 (1.71)	−0.69 (1.03)	−2.78 (2.29)	−1.66* (0.65)	−3.08 (2.17)	−0.86 (1.5)	−4.97[a] (2.82)	1.07 (0.37)	2.83[a] (1.51)	0.17 (1.15)	2.19 (2.1)

Table 7b **IV Estimates of the Impact of Patent Effectiveness (θ) on Licensing Probabilities by Complementary Assets (ρ) (Hypothesis 2b)**

	Marginal effects w.r.t. patent effectiveness (θ) of							
	Probability of licensing				Probability of licensing patented innovations			
	I Exog. θ and ρ	II Endog. ρ	III Endog. θ	IV Endog. θ and ρ	V Exog. θ and ρ	VI Endog. ρ	VII Endog. θ	VIII Endog. θ and ρ
Complementary assets (ρ)								
Low	0.11** (0.0)	0.09** (0.03)	0.14** (0.03)	0.13** (0.03)	0.07 (0.08)	0.09 (0.1)	−0.03 (0.22)	0.08 (0.17)
High	0.05** (0.01)	0.06* (0.03)	0.07** (0.02)	0.07 (0.05)	−0.18* (0.08)	−0.47 (0.36)	−0.14 (0.15)	−0.95[a] (0.58)
Total	0.08** (0.01)	0.07** (0.02)	0.10** (0.02)	0.11** (0.03)	−0.06 (0.06)	−0.16 (0.08)	−0.09 (0.16)	−0.29 (0.25)

Note. Standard errors are in parentheses.
**, *, [a]: Significantly different than zero at the 0.01, 0.05, and 0.10 confidence levels, respectively.

A different concern is that the reported patent effectiveness simply reflects reporting biases (e.g., managers who, for some reason, favor patenting also report patents to be more effective) or differences in knowledge base or technical opportunity. To address this, we also estimated a specification instrumenting for patent effectiveness, using two sets of instruments. One is the industry average of patent effectiveness at the level of the *primary industry of the respondent's parent firm*.[20] Patent effectiveness, a self-reported measure of the net benefits from patenting, reflects both "endogenous" variation in the strength of patents, such as the patentability of technology, as well as "exogenous" variation related to the business unit's parent firm's patenting culture or to the shared resources available to enforce patents at the firm level. It is the latter that we hope to capture through the instrument. The instrument is valid in the sense that the effectiveness of patents in the industry of the parent firm should not have any *direct* impact on patenting or licensing payoffs of the business unit and should only have an *indirect effect* through patent effectiveness.

[20] Roughly half of the business units belong to an industry different from the primary industry of the parent firm.

Similarly, insofar as location is a long-term choice, differences in the litigation environment across federal districts should affect behavior only through how effective managers perceive patents to be. We use the average time to resolution of patent cases during 1990–1993 and its standard deviation, the average success rate of patent holders, and the number of terminated patent cases in the district court where the respondent lab is located as additional instruments for patent effectiveness.[21]

Tables 7a and 7b present the estimates for payoffs and propensities, respectively, where we instrument for patent effectiveness and complementary assets. The results are qualitatively unchanged, although the standard errors are larger.

5.3. Additional Specifications

Controlling for R&D. We measure licensing propensity as the percentage of R&D invested with the objective of licensing. For a given share of innovations

[21] Source: "Federal Court Cases: Integrated Data Base, 1970–1994," Federal Judicial Center, Interuniversity Consortium for Political and Social Research. The time to resolution is computed as the number of months between the date on which a case was filed in a district court and its termination by any means (e.g., settlement, dismissal, judgment).

that are licensed, the percentage of R&D invested for licensing will decrease with the level of R&D. Similarly, a firm that has more innovations than it can develop in house will be more likely to license. However, controlling for this possibility by including R&D or R&D intensity (R&D/sales) as controls does not change the results. Indeed, R&D or R&D intensity has only a small and statistically insignificant impact on patenting and licensing payoffs.

Sequential Decision Making. Our model assumes that the licensing and patenting decisions are taken simultaneously. However, some patenting decisions may be made before the licensing opportunities have materialized. We also estimated a specification where the licensing decision is made after the patenting decision and obtained similar results to those presented here (see the online appendix). The estimated variance of the difference between random components of the payoffs is small, which suggests that our firm-level cross-sectional data are poorly suited to the task of estimating a model of sequential decision making. The analysis of the sequential stages of patenting and licensing decisions requires analysis with different data and is left for future research.

Correlation in Errors. Our model assumes that the unobserved, stochastic components in payoffs are mutually uncorrelated. We also estimated two different specifications that relax this assumption—the nested logit and the multinomial probit specifications (estimated using proportions and the GMM method). The estimated correlation coefficient between unobserved errors (e.g., allowing correlation between V_L and V_P, or V_P and V_S) is not significantly different from zero, suggesting that the main results of the paper are robust to relaxation of such assumption.

Alternatively, one can estimate the equation for the probability of licensing conditional on patenting (3-3) as a single equation, using the observed licensing propensity divided by the patent propensity as a dependent variable. If V_L and V_P were correlated, the conditional probability would be a function of $(V_L - V_P)/(1-\sigma)$, with σ reflecting the strength of the correlation and varying between 0 and 1 (Maddala 1983, pp. 70–72). The estimated coefficient of the interaction term in the single structural equation is still negative, as predicted by Hypothesis 2a, but is statistically insignificant. The fall in precision in the estimates is likely to be due to the exclusion of the nonpatentees, or about 25% of the sample.

Within-Group Estimation. We also estimated the model within the groups of firms with low and high complementary assets, allowing all coefficients to vary across groups, and found that the central result of the paper, summarized in Table 6, is even stronger. The estimated elasticity of licensing w.r.t. patent effectiveness is 0.7 (0.02) for low (high) complementary asset firms. Similarly, the elasticity of the share of licensing in patented innovations is 0.2 (−0.5) for low (high) complementary asset firms.

5.4. Licensing Without Patents

We further analyzed the 25 cases that involved positive licensing propensity but zero patent propensity. Two-thirds of these involved public firms, for which 10Ks and annual reports were available, and a closer examination revealed that most of these firms typically did not engage in technology licensing, but may be licensing software or trademarks. In a small number of cases, it appeared that firms had applied for patents, but not during 1991–1993, the survey time frame, which implies measurement error as the cause of the apparent anomaly. Including those 25 observations in the sample and assigning them a null propensity to license does not change our results. Although, in principle, firms that patent may still license some of their unpatented innovations, data limitations do not allow us to investigate this possibility further.

6. Conclusion

The ability of firms to appropriate the returns from their innovations is a key driver of the willingness of firms to invest in innovative activity. In recent years, some firms appear to have resorted to technology licensing as a way of appropriating the returns. But even when licensing is feasible, it is well known that the market for technology suffers from various imperfections. Effective patents can ameliorate some of these imperfections. However, more effective patents also increase the payoff from commercialization, making it important to quantify the impact of patent effectiveness on licensing.

We analyze how patenting and licensing strategies are related and how patent effectiveness and complementary assets condition the use of patenting and licensing by firms to appropriate rents from innovation. We find that increases in patent effectiveness indirectly affect licensing by increasing the propensity of firms to seek patent protection. However, this also decreases the proportion of patented innovations that are licensed, implying a smaller net increase in licensing propensity.

Our results highlight the importance of the interaction between patent effectiveness and a firm's ownership of specialized complementary assets in conditioning licensing decisions. We argue that insofar as complementary assets are difficult to acquire or imitate, stronger patent protection will increase the payoff to commercialization relative to licensing. Indeed, we find that higher patent effectiveness elicits much larger increases in licensing from firms lacking specialized complementary assets.

Thus, our study also helps reconcile conflicting empirical evidence of the impact of patent protection on licensing. For instance, studies undertaken using samples of small firms or start-ups, which are less likely to have specialized complementary assets (e.g., Gans et al. 2002), are also more likely to find a stronger impact of patent effectiveness on licensing. Conversely, studies of cross-national technology licensing by large multinationals are less likely to find an impact of patent effectiveness on licensing (e.g., Fosfuri 2004b). Similarly, studies focusing on patented innovations are more likely to find only a small impact of the strength of patents on licensing, because they neglect the effect on the propensity to patent.

These results point to the need to better understand the interplay between different strategic instruments available to firms in their quest to appropriate rents from innovation. In particular, it seems important to move beyond the short-term adjustment model inherent in the present study by relaxing the assumption of exogenous investments in complementary assets. For instance, although our findings are consistent with the view that technology start-ups lacking manufacturing or marketing assets should license their innovations when patent protection is effective, other options may be available. Limited access to such assets could be obtained through alliances with established partners, as in pharmaceuticals, or through independent foundries, as in semiconductors. Alternatively, a start-up may decide to invest in in-house manufacturing and marketing capability, as firms such as Genentech and Amgen appear to have done. When a strategy of acquiring complementary assets is superior to one of licensing or one of alliances is thus an important area for future research. Our findings point to the importance of understanding better when the long-term industry configuration can feature vertical "technology specialists" (Arora and Gambardella 1994) and when, instead, innovators must battle with incumbents.

An online appendix to this paper is available on the *Management Science* website at http://mansci.pubs.informs.org/ecompanion.html.

Acknowledgments

The authors are grateful to Wes Cohen for permission to use data from the Carnegie Mellon University survey and for many helpful comments and suggestions. The authors also thank David Hsu, Michelle Gittelman, Shane Greenstein, Will Mitchell, and seminar participants at the Fuqua School of Business, INSEAD, Copenhagen Business School, Bocconi University, Universitat Pompeu Fabra, Wharton School, Rotman School of Management, London Business School, Academy of Management, the Strategic Management Society, the 4th Annual Roundtable for Engineering Entrepreneurship Research at Georgia Institute of Technology, and two anonymous referees for their suggestions. The usual disclaimers apply.

Appendix. A Model of the Determinants of Licensing

Proof of Hypotheses 1a and 1b. Let θ represent an index of patent effectiveness. As argued in the text, it is intuitive that increases in θ increase both $V_L - V_S$ and $V_P - V_S$. The impact on $V_L - V_P$ is, however, ambiguous (Hypothesis 1a).

$$\Pr(\text{Pat}) = \frac{1}{1+B}, \quad B = \frac{1}{\exp(V_P - V_S) + \exp(V_L - V_S)}.$$

Because both $(V_L - V_S)$ and $(V_P - V_S)$ are nondecreasing in θ, it follows that B is decreasing in θ, and so Pr(Pat) is increasing in θ.

$$\frac{d\Pr(\text{Lic})}{d\theta} = \frac{\Pr(\text{Lic})}{A}\left(\exp(V_P - V_S)\frac{d(V_L - V_P)}{d\theta} + \frac{d(V_L - V_S)}{d\theta} - 1\right),$$

$$A = \exp(V_P - V_S) + \exp(V_L - V_S) + 1.$$

Because the sign of $d(V_L - V_P)/d\theta$ is ambiguous, so is the overall expression. Similarly, it also follows that the share of licenses in patented innovations could either increase or decrease with respect to patent effectiveness. This demonstrates Hypothesis 1b. □

Proof of Hypothesis 2a. By definition, $dV_S/d\theta = 0$, because patent effectiveness can have no effect if the innovation is not patented, and hence $d^2V_S/d\rho d\theta = 0$. Similarly, $dV_L/d\rho = 0$ because if the innovation is licensed, what matters is the complementary assets of the licensee, not those of the innovator. In the text, we argued that $d^2V_P/d\rho d\theta > 0$ because stronger patents are more valuable to firms owning the specialized complementary assets. In what follows, the intuition is formalized using standard models of both product and process innovations with asymmetric firms and imitation.

Process Innovation. In a standard two-stage model of R&D competition (cf. d'Aspremont and Jacquemin 1988, De Bondt et al. 1992, and Ceccagnoli 2005), in stage 1 two firms (i, j) choose the amount of cost reducing R&D, the cost of which is quadratic in R&D, and in stage 2, firms compete à la Cournot in a homogeneous product market with linear demand ($p = 1 - Q$). Marginal costs are equal to $c_i = c - \rho_i - R_i - (1 - \theta_j)R_j$ for firm i, where ρ_i is an index of firm i complementary assets, and $(1 - \theta_j)$ represents the fraction of rival j's R&D, which is imitated by firm i, thus reducing firm i's costs, with $0 \le \theta_j \le 1$. A greater θ_j reflects stronger patent protection for firm j. Profits from commercialization and patenting for firm i are equal to

$$V_i^P = (q_i^c)^2 - \tfrac{1}{2}R_i^2, \quad \text{with } q_i^c = \tfrac{1}{3}[1 - 2c_i + c_j].$$

It is easily verified that, for any given level of R&D,

$$\frac{\partial^2 V_i^P}{\partial \rho_i \partial \theta_i} = \tfrac{4}{9}R_i > 0,$$

which formalizes Hypothesis 2a for the case of process innovations.

Product Innovation. An innovation may also be quality improving (rather than merely cost reducing). This can be dealt with using a standard model of vertical differentiation (e.g., Shaked and Sutton 1982). The buyer's utility function is $vx-p$, where x is the quality of the product, p is the price, and v is the buyer's willingness to pay for quality. Typically, v is assumed to be uniform, and for simplicity, we let it be uniform between $[0, 1]$. Let the innovator's quality be x and the rival's quality be x', $x > x'$. The (constant) marginal costs of production are, respectively, c and c', and the prices chosen are p and p', respectively. Let $\Delta x = x - x'$, and define Δp and Δc similarly. Assuming that the entire market is covered, the demand for the innovator is given by $1-v^*$, where $v^* = \Delta p/\Delta x$. We assume that the innovator can choose the quality level by incurring a product R&D cost $G(x)$, $G_x > 0$, $G_{xx} > 0$, so that higher quality requires increasingly higher product R&D. Complementary commercialization assets, ρ, reduce the "cost of achieving high quality," i.e., $G_\rho < 0$, $G_{\rho x} < 0$. We also assume that through spillovers, rivals can partially imitate the innovator, so that x' is an increasing function of x. Unless spillovers are perfect, the increase in the rival's quality should be less than the increase in the innovator's quality. Further, an increase in patent effectiveness, by reducing spillovers, reduces x'. Specifically, we assume that $\partial x'/\partial x = 1 > \gamma > 0$, and $\partial x'/\partial\theta < 0$. The profit function for the innovator is given by $(p-c)(1-v^*) - G(x)$, and the innovator chooses price and quality to maximize profits. Let $\pi(\theta, \rho)$ represent the maximized profit function. Then,

$$\frac{\partial\pi}{\partial\rho} = -G_\rho, \qquad \frac{\partial^2\pi}{\partial\rho\partial\theta} = -G_{x\rho}\frac{\partial x}{\partial\theta}.$$

Because $G_{\rho x} < 0$, a necessary and sufficient condition for $\partial^2\pi/\partial\rho\partial\theta > 0$ is $\partial x/\partial\theta > 0$, i.e., the innovator's quality increases with patent effectiveness. We show below that this is indeed true.

Case 1: *Rivals Are a Competitive Fringe.* We assume that imitators are a competitive fringe so that they price at marginal cost, i.e., $p' = c'$. Assuming an interior optimum for x (so that $\pi_{xx} < 0$), it follows that $(\partial x/\partial\theta) = (\partial^2\pi/\partial x\partial\theta)/(-\pi_{xx})$. Thus, the sign of $(\partial x/\partial\theta)$ is the same as the sign of $\partial^2\pi/\partial x\partial\theta$. Substituting in the optimum price, one can write the innovator's profits, $\pi = \Delta x/4 + (\Delta c)^2/4\Delta x - \Delta c/2 - G(x)$. Thus, $\partial\pi/\partial x = (1-\gamma)/4\{1-(\Delta c)^2/(\Delta x)^2\} - G_x$, so $\partial^2\pi/\partial x\partial\theta = [(1-\gamma)(\Delta c)^2/2(\Delta x)^3](-\partial x'/\partial\theta) > 0$.

Case 2: *Strategic Price Competition.* We continue to assume that the innovator selects quality, and this also determines the rival's quality through spillovers. Unlike Case 1, the innovator and its rival set prices strategically. Thus, we have a two-stage game, wherein the innovator chooses the quality level (i.e., product R&D level), determining its quality and that of its rival, and in stage 2, both compete by setting prices. One can show that in the second stage, the innovator's profits, π, can be written as $\pi = \Delta x/9 + (\Delta c)^2/9\Delta x - 8\Delta c/9 - G(x)$. Note that this has the same form as the profit function in Case 1, and thus it is easy to show that $\partial^2\pi/\partial x\partial\theta > 0$. This formalizes Hypothesis 2a for the case of product innovation. □

Proof of Hypothesis 2b. As noted in the text, the relationship between payoffs and behavior is nonlinear so that second-order impacts of variables on payoffs do not translate directly into impacts on behavior such as licensing propensity. Note that

$$\frac{d^2\text{Pr(Lic)}}{d\rho d\theta} = \frac{1}{\text{Pr(Lic)}}\frac{d\text{Pr(Lic)}}{d\theta}\frac{d\text{Pr(Lic)}}{d\rho} + \text{Pr(Lic)}\left[\frac{d\text{Pr(Lic)}}{d\rho}\left(\frac{dV_P}{d\theta}-\frac{dV_L}{d\theta}\right) - \frac{dV_P}{d\theta}\frac{d\text{Pr(Pat)}}{d\rho} + [\text{Pr(Lic)}-\text{Pr(Pat)}]\frac{d^2V_P}{d\rho d\theta}\right].$$

The first term on the right-hand side is indeterminate in theory but negative in our empirical results. The second term depends on three subterms inside the square brackets, of which only the third is predicted to be negative by our theory (by Hypothesis 2a); the other two are indeterminate. Given that Pr(Lic) is always lower than Pr(Pat), a sufficient condition for $d^2\text{Pr(Lic)}/d\rho d\theta$ to be negative is that the positive interaction effect, $d^2V_P/d\rho d\theta$, is sufficiently strong.

The impact on the share of licensing in patented innovations is also indeterminate in theory and becomes negative if the interaction effect between patent effectiveness and complementary assets presented in Hypothesis 2a is sufficiently strong:

$$\frac{d^2\text{Pr(Lic | Pat)}}{d\theta d\rho} = \frac{d\text{Pr(Lic | Pat)}}{d\rho}\left(\frac{dV_L}{d\theta}-\frac{dV_P}{d\theta}\right) \cdot [\text{Pr(NotLic | Pat)} - \text{Pr(Lic | Pat)}] - \text{Pr(Lic | Pat)}\,\text{Pr(NotLic | Pat)}\left(\frac{d^2V_P}{d\theta d\rho}\right).$$

The first term of this difference is indeterminate in theory, although our empirical results imply that it is positive but small (insignificantly different from zero). The second term is positive, because the cross-partial is positive. Therefore, a sufficient condition for $d^2\text{Pr(Lic | Pat)}/d\rho d\theta$ to be negative is that the positive interaction effect, represented by $d^2V_P/d\rho d\theta$, is sufficiently strong, so that the second term outweighs the first. □

References

Amit, R., P. Schoemaker. 1993. Strategic assets and organizational rent. *Strategic Management J.* **14** 33–46.

Anand, B. N., T. Khanna. 2000. The structure of licensing contracts. *J. Indust. Econom.* **48** 103–135.

Anton, J., D. A. Yao. 1994. Expropriation and inventions: Appropriable rents in the absence of property rights. *Amer. Econom. Rev.* **84** 190–209.

Arora, A. 1996. Contracting for tacit knowledge: The provision of technical services in technology licensing contracts. *J. Development Econom.* **50** 233–256.

Arora, A., A. Fosfuri. 2003. Licensing the market for technology. *J. Econom. Behavior Organ.* **52** 277–295.

Arora, A., A. Gambardella. 1994. The changing technology of technical change: General and abstract knowledge and the division of innovative labor. *Res. Policy* **23** 523–532.

Arora, A., M. Ceccagnoli, W. M. Cohen. 2003. R&D and the patent premium. Working paper 9431, National Bureau of Economic Research, Cambridge, MA.

Arora, A., A. Fosfuri, A. Gambardella. 2001. *Markets for Technology: Economics of Innovation and Corporate Strategy*. MIT Press, Cambridge, MA.

Arrow, K. 1962. Economic welfare and the allocation of resources for invention. R. R. Nelson, ed. *The Rate and Direction of Inventive Activity*. Princeton University Press, Princeton, NJ.

Barney, J. 1991. Firm resources and sustainable competitive advantage. *J. Management* **17** 99–120.

Bascavusoglu, E., M. P. Zuniga. 2002. Foreign patent rights, technology and disembodied knowledge transfer cross borders: An empirical application. Unpublished manuscript, University of Paris, Paris, France.

Bowker Co., R. R. 1995. *Directory of American Research and Technology*, 30th ed. R. R. Bowker, New Providence.

Branstetter, L. G., R. Fisman, C. F. Foley. 2006. Do stronger intellectual property rights increase international technology transfer? Empirical evidence from U.S. firm-level panel data. *Quart. J. Econom.* Forthcoming.

Cassiman, B., R. Veugelers. 2002. R&D cooperation and spillovers: Some empirical evidence from Belgium. *Amer. Econom. Rev.* **92** 1169–1184.

Ceccagnoli, M. 2005. Firm heterogeneity, imitation, and the incentives for cost reducing R&D effort. *J. Indust. Econom.* **53** 83–100.

Cohen, W. M. 2004. Patents and appropriation: Concerns, evidence and a Schumpeterian perspective. *J. Tech. Transfer* **30** 57–71.

Cohen, W. M., R. R. Nelson, J. P. Walsh. 2000. Protecting their intellectual assets: Appropriability conditions and why U.S. manufacturing firms patent or not. Working paper 7552, National Bureau of Economic Research, Cambridge, MA.

d'Aspremont, C., A. Jacquemin. 1988. Cooperative and noncooperative R&D in duopoly with spillovers. *Amer. Econom. Rev.* **78** 1133–1137.

De Bondt, R., P. Slaets, B. Cassiman. 1992. The degree of spillovers and the number of rivals for maximum effective R&D. *Internat. J. Indust. Organ.* **10** 35–54.

Dierickx, I., K. Cool. 1989. Asset stock accumulation and sustainability of competitive advantage. *Management Sci.* **35** 1504–1513.

DiMasi, J. A. 2000. New drug innovation and pharmaceutical industry structure: Trends in the output of pharmaceutical firms. *Drug Inform. J.* **34** 1169–1194.

Fink, C. 2005. Intellectual property rights and U.S. and German international transactions in manufacturing industries. C. Fink, K. E. Maskus, eds. *Intellectual Property and Development. Lessons from Recent Economic Research*. World Bank and Oxford University Press, New York, 75–110.

Fosfuri, A. 2004a. The licensing dilemma: Understanding the determinants of the rate of licensing. Working paper 04-15, Universidad Carlos III de Madrid, Madrid, Spain.

Fosfuri, A. 2004b. Determinants of international activity: Evidence from the chemical processing industry. *Res. Policy* **33** 1599–1614.

Gallant, R. A. 1987. *Nonlinear Statistical Models*. John Wiley & Sons, New York.

Gans, J., D. Hsu, S. Stern. 2002. When does start-up innovation spur the gale of creative destruction? *RAND J. Econom.* **33** 571–586.

Grindley, P., D. Teece. 1997. Managing intellectual capital: Licensing and cross-licensing in semiconductors and electronics. *California Management Rev.* **39** 8–41.

Hall, B. H. 2004. Exploring the patent explosion. *J. Tech. Transfer* **30** 35–48.

Hall, B. H., R. H. Ziedonis. 2001. The patent paradox revisited: An empirical study of patenting in the U.S. semiconductor industry, 1979–1995. *RAND J. Econom.* **32** 101–128.

Helfat, C. E. 1997. Know-how and asset complementarity and dynamic capability accumulation: The case of R&D. *Strategic Management J.* **18** 339–360.

Hill, C. W. L. 1992. Strategies for exploiting technological innovations: When and when not to license. *Organ. Sci.* **3** 428–441.

Horstmann, I., J. M. MacDonald, A. Slivinski. 1985. Patents as information transfer mechanisms: To patent or (maybe) not to patent. *J. Political Econom.* **93** 837–858.

Kim, J., G. Marschke. 2004. Accounting for the recent surge in U.S. patenting: Changes in R&D expenditures, patent yields, and the high-tech sector. *Econom. Innovation New Tech.* **13** 543–558.

Lamoreaux, N. R., K. L. Sokoloff. 1998. Inventors, firms, and the market for technology in the late nineteenth and early twentieth centuries. N. R. Lamoreaux, D. M. Raff, P. Temin, eds. *Learning by Doing in Firms, Markets, and Nations*. The University of Chicago Press, Chicago, IL.

Levin, R. C., A. K. Klevorick, R. R. Nelson, S. G. Winter. 1987. Appropriating the returns from industrial R&D. *Brookings Papers Econom. Activity* **1987** 783–820.

Maddala, G. S. 1983. *Limited-Dependent and Qualitative Variables in Econometrics*. Cambridge University Press, Cambridge, UK.

Mansfield, E. 1986. Patents and innovation: An empirical study. *Management Sci.* **32** 173–181.

McFadden, D. 1973. Conditional logit analysis of qualitative choice behavior. P. Zarembka, ed. *Frontier in Econometrics*. Academic Press, New York.

Mowery, D. 1983. The relationship between contractual and intrafirm forms of industrial research in American manufacturing, 1900–1940. *Explorations Econom. Hist.* (October) 351–374.

Nagaoka, S. 2002. Impact of intellectual property rights on international licensing: Evidence from licensing contracts of Japanese industry. Unpublished manuscript, Hitotsubashi University, Tokyo, Japan.

Nelson, R. R. 1959. The simple economics of basic scientific research. *J. Political Econom.* **67** 297–306.

Nerkar, A., P. W. Roberts. 2004. Technological and product-market experience and the success of new product introductions in the pharmaceutical industry. *Strategic Management J.* **25** 779–799.

Penner-Hahn, J., M. Shaver. 2005. Does international research and development increase patent output? An analysis of Japanese pharmaceutical firms. *Strategic Management J.* **26** 121–140.

Puttitanun, T. 2003. Intellectual property rights and multinational firms' modes of entry. Unpublished manuscript, San Diego State University, San Diego, CA.

Rivette, K. G., D. Kline. 1999. *Rembrandts in the Attic: Unlocking the Hidden Value of Patents*. Harvard Business School Press, Boston, MA.

Rothaermel, F. T., C. W. L. Hill. 2005. Technological discontinuities and complementary assets: A longitudinal study of industry and firm performance. *Organ. Sci.* **16**(1) 52–70.

Schankerman, M., A. Pakes. 1986. Estimates of the value of patent rights in European countries during the post-1950 period. *Econom. J.* **96** 1052–1076.

Scherer, F. M., D. Harhoff. 2000. Technology policy for a world of skew distributed outcomes. *Res. Policy* **29** 559–566.

Shaked, A., J. Sutton. 1982. Relaxing price competition through product differentiation. *Rev. Econom. Stud.* **49** 3–13.

Shane, S. 2001. Technology regimes and new firm formation. *Management Sci.* **47** 1173–1190.

Smith, P. J. 2001. How do foreign patent rights affect U.S. exports, affiliate sales, and licenses? *J. Internat. Econom.* **55** 411–439.

Teece, D. J. 1977. Technology transfer by multinational firms: The resource cost of transferring technological know-how. *Econom. J.* **87** 242–261.

Teece, D. J. 1986. Profiting from technological innovation: Implications for integration, collaboration, licensing and public policy. *Res. Policy* **15** 285–305.

Teece, D. J. 1992. Competition, cooperation and innovation. *J. Econom. Behavior Organ.* **18** 1–25.

Thomke, S., W. Kuemmerle. 2002. Asset accumulation, interdependence and technological change: Evidence from pharmaceutical drug discovery. *Strategic Management J.* **23** 619–635.

Tripsas, M. 1997. Unraveling the process of creative destruction: Complementary assets and incumbent survival in the typesetter industry. *Strategic Management J.* **18** 119–142.

Wernerfelt, B. 1984. A resource-based view of the firm. *Strategic Management J.* **5** 171–180.

Williamson, O. 1991. Comparative economic organization: The analysis of discrete structural alternatives. *Admin. Sci. Quart.* **36** 269–296.

Winter, S. G. 1987. Knowledge and competence as strategic assets. D. J. Teece, ed. *The Competitive Challenge*. Balling Publishing Company, Cambridge, MA, 159–184.

Yang, G., K. Maskus. 2005. Intellectual property rights and licensing: An econometric investigation. C. Fink, K. E. Maskus, eds. *Intellectual Property and Development. Lessons from Recent Economic Research*. World Bank and Oxford University Press, New York, 111–134.

Zhao, Z., J. Anand, W. Mitchell. 2005. A dual networks perspective on inter-organizational transfer of R&D capabilities: International joint ventures in the Chinese automotive industry. *J. Management Stud.* **42** 127–160.

[9]

Strategic Management Journal, Vol. 18 (Summer Special Issue), 119–142 (1997)

UNRAVELING THE PROCESS OF CREATIVE DESTRUCTION: COMPLEMENTARY ASSETS AND INCUMBENT SURVIVAL IN THE TYPESETTER INDUSTRY

MARY TRIPSAS*
The Wharton School, University of Pennsylvania, Philadelphia, Pennsylvania, U.S.A.

When radical technological change transforms an industry established firms sometimes fail drastically and are displaced by new entrants, yet other times survive and prosper. Drawing upon an unusually rich data set that covers the technological and competitive history of the typesetter industry from 1886 to 1990, this paper uses a combination of quantitative and qualitative analysis to unravel this process of creative destruction. It argues that the ultimate commercial performance of incumbents vs. new entrants is driven by the balance and interaction of three factors: investment, technical capabilities, and appropriability through specialized complementary assets. In this industry, specialized complementary assets played a crucial role in buffering incumbents from the effects of competence destruction, and an analysis that examined investment or technical capabilities in isolation would have led to misleading results. This work thus highlights the importance of considering multiple perspectives when examining the competitive implications of technological change.

INTRODUCTION

Why do incumbent firms sometimes fail drastically in the face of radical technological change, yet other times survive and prosper? This paper explores this question by analyzing the technological and competitive history of the global typesetter industry for a period of over 100 years. Drawing upon an unusually rich data set that includes detailed firm- and product-level data for every firm in the history of the industry, it traces the nature of technological change, firm responses, and product market performance. From its inception in 1886 through 1990, the industry has undergone three waves of 'creative destruction' (Schumpeter, 1950) where competence-destroying, architectural technological change transformed the industry. Incumbents were displaced by new entrants, however, in only one of these three shifts. By exploring the dynamics of each of these shifts in depth, this paper helps to unravel this process of creative destruction.

Two contrasting perspectives on the process of creative destruction are present in the literature. The first, following in the tradition of Schumpeter's early work (Schumpeter, 1934), paints a picture of relatively fluid industries where new entrants innovate with technologically superior products and displace incumbent firms, only to have the cycle repeated. This continual failure of established firms in the face of radical innovation has been documented in a number of empirical studies (Cooper and Schendel, 1976; Majumdar, 1982; Tushman and Anderson, 1986; Henderson and Clark, 1990; Christensen, 1993). In contrast, other research has built on Schumpeter's later work (Schumpeter, 1950), focusing on the advantages that established firms have over new entrants. For example, Teece (1986) argues that when incumbents possess critical specialized

Key words: creative destruction; technological change; typesetter industry

*Correspondence to: Mary Tripsas, Department of Management, The Wharton School, University of Pennsylvania, 2015 Steinberg Hall–Dietrich Hall, Philadelphia, PA 19104-6207, U.S.A.

CCC 0143–2095/97/S10119–24 $17.50

complementary assets, new entrants unable to contract for those assets may be at a disadvantage, despite their potential technological superiority, and Chandler (1990) in an exhaustive examination of the development of industrial economies identifies substantial advantages that accompany the scale and scope of large firms.

This paper sheds light on these two perspectives by breaking out three crucial factors that, together, influence the ultimate commercial performance of incumbents and new entrants: (1) investment in developing the new technology; (2) technical capabilities; and (3) the ability to appropriate the benefits of technological innovation through specialized complementary assets. The balance and interaction among these three factors determine whether incumbents or new entrants are more successful in the face of competence-destroying technological change.

In dissecting the differential success of incumbents and new entrants across multiple waves of technological change in the typesetter industry, this paper examines each of these three factors and the interaction among them. It finds, first, that lack of investment was not responsible for incumbent failure. Consistent with both economic theories of investment behavior (Arrow, 1962; Gilbert and Newberry, 1982; Reinganum, 1983) and theories of resource allocation (Christensen and Bower, 1996), incumbents invested significant amounts in the development of each new generation of technology. However, while incumbents invested in developing new, competence-destroying technology, the technical performance of the products they developed in each new generation of technology proved to be significantly inferior to the performance of new entrant products. In line with previous research, organizational architectures, routines, and procedures fine-tuned to fit with the prior generation of technology appear to have handicapped incumbents (Majumdar, 1982; Tushman and Anderson, 1986; Henderson and Clark, 1990).

Although incumbent products were technologically inferior in all three competence-destroying generations of technology, incumbents were displaced by new entrants in only one of these three generations. Examining the effect of each technological shift on the value of the incumbents' specialized complementary assets helps to explain this disparity. When specialized complementary assets unavailable to new entrants retained their value despite a technological shift, incumbents maintained their market position in the new generation of technology. These assets appear to have buffered incumbents from the effects of competence destruction, enabling them to sustain a high level of commercial performance despite their technological disadvantage. This result highlights the importance of considering multiple perspectives when analyzing the competitive implications of technological change.

The paper begins with a brief literature review followed by a description of the data and of the research setting, the typesetter industry. Each of the major technological shifts in typesetters is then categorized in terms of its effect on investment incentives, technological competence, and specialized complementary assets. The next section uses descriptive data to contrast incumbents vs. new entrants in terms of investment behavior, technical capabilities, and the ability to appropriate the benefits of innovation through complementary assets. The final section uses quantitative analysis of product market share to further understand the intersection between competence destruction and complementary assets and their joint effect on commercial performance.

LITERATURE REVIEW

A large body of literature has analyzed the relationship between technological change and the competitive position of incumbents and new entrants. This section provides a brief synopsis of this work in order to frame our expectations regarding three related questions: (1) What factors drive the investment behavior of incumbents and new entrants? (2) Given competence-destroying technological change, how does the technical performance of incumbents compare to that of new entrants? and (3) What factors drive the ability of incumbents and new entrants to appropriate the benefits of innovation in the product market? While separate research addresses each of these questions, relatively little work has examined the intersection between them. This section first examines each question and then summarizes how, in combination, the balance and interaction between investment, technical capabilities, and appropriability drive the commercial performance of incumbents and new entrants.

Investment behavior

Much theoretical work in economics has examined the differing incentives of incumbents and new entrants to invest in innovative activity. Building on early work by Arrow (1962) this stream of research suggests that when innovation is radical, in the sense that it replaces rather than competes with the old technology (i.e., the monopolist's postinnovation price is less than the preinnovation cost), then incumbent monopolists have less incentive to invest in the new technology than new entrants. In contrast, when innovation is incremental (i.e., it competes with the existing technology) then incumbents have greater incentives than new entrants to invest (e.g., Gilbert and Newberry, 1982; Reinganum, 1983).

An alternative explanation for incumbent failure to invest in new technology is put forth by Christensen and Bower (1996), based on empirical data in the disk drive industry. They argue that established firms fail to invest in developing radically new technology as a result of firms' resource allocation mechanisms. Since resource allocation in established firms is guided by the needs of existing customers, when radically new technologies are 'disrupting' in that they target emerging markets instead of addressing the needs of existing customers, then established firms quite rationally focus their research efforts away from the new technology. As resource dependence (Pfeffer and Salancik, 1978) would predict, the purchasing power of existing customers influences the investment patterns of the established firms. Similarly, if new technology is 'sustaining' in that it meets the needs of the existing customer base, then incumbent firms should rationally invest in the technology.

Technical capabilities

Technological progress in an industry is generally characterized as passing through long periods of incremental innovation punctuated by periods of radical change (Abernathy and Utterback, 1978; Dosi, 1982; Sahal, 1985; Tushman and Anderson, 1986). Dosi likens this pattern to Kuhnian theories of the development of new science. A technology develops incrementally along a given 'technological trajectory' within a given 'technological paradigm' until it is replaced with a new paradigm — a radical innovation.

Different stages of this technology life cycle have major implications for the technical capabilities of incumbents and new entrants. During an incremental period, when technological innovation builds upon the capabilities of established firms, they have an advantage over new entrants. Established firms develop organizational structures, routines, and procedures that enable them to efficiently process information within the context of the existing technological regime (Burns and Stalker, 1961; Galbraith, 1973; Arrow, 1974; Nelson and Winter, 1982). When faced with a radical, competence-destroying technological shift, however, established firms are often at a disadvantage (Tushman and Anderson, 1986). Core competencies during a period of incremental innovation can become 'core rigidities', making it difficult for a firm to adapt (Leonard-Barton, 1992). Even if component technologies remain constant, architectural innovation — changes in the way components interface — can destroy the value of existing beliefs and patterns (Henderson and Clark, 1990). Finally, years of incremental innovation may result in selection-induced inertia, as only firms with stable structures and activities survive (Hannan and Freeman, 1977, 1984); such firms will find change difficult. As a result competence-destroying technological discontinuities often result in inferior technical performance on the part of established firms (Cooper and Schendel, 1976; Majumdar, 1982; Tushman and Anderson, 1986; Henderson and Clark, 1990; Afuah, 1994).

In contrast, Christensen and Bower (1996) found that in the disk drive industry established firms did not have difficulty developing new technology, even when innovation was architectural in nature. Despite the fact that innovation was radical in an organizational sense, incumbents had the resources and ability to develop new capabilities. This result is consistent with other empirical work that has demonstrated the strength of large research labs in coming up with new ideas (e.g., Freeman, 1982; Chandler, 1990). It is also possible that these incumbents possessed what Teece, Pisano, and Shuen (1997) call 'dynamic capability ... the capacity of a firm to renew, augment, and adapt its core competencies over time'. Given these two conflicting perspectives, there is no clear prediction as to whether established firms will have inferior technological performance in competence-destroying technological generations.

Appropriability through complementary assets

When incumbents experience a technological disadvantage in the face of competence-destroying technological change, the extent to which that disadvantage translates into a commercial disadvantage may depend upon the other assets possessed by established firms. While in his early work Schumpeter (1934) argued that entrepreneurs should be responsible for most innovation, later work (Schumpeter, 1950) suggests that large established firms with capital and market power are in a stronger position to exploit innovation. Teece (1986) lays out a framework for identifying when the assets of large, established firms confer them with an advantage. He uses the label complementary assets to describe factors such as specialized manufacturing capability, access to distribution channels, service networks and complementary technologies. Teece distinguishes between generic, specialized, and cospecialized[1] complementary assets. Whereas generic assets have multiple applications and can be easily contracted for, specialized and cospecialized assets are useful only in the context of a given innovation. If a firm has proprietary access to the specialized complementary assets necessary for the commercial exploitation of an innovation, then that firm has a distinct advantage. Under a regime of weak intellectual property protection when an innovation can easily spill over to competing firms, complementary assets become particularly important if a firm is to appropriate the benefits of its innovation.

By explicitly considering the importance of complementary assets, one gains insight into the performance of incumbents and new entrants. As Teece (1986: 301) notes:

> Business commentators often remark that many small entrepreneurial firms which generate new, commercially valuable technology fail while large multinational firms, often with a less meritorious record with respect to innovation, survive and prosper. One set of reasons for this phenomenon is now clear. Large firms are more likely to possess the relevant specialized and cospecialized assets within their boundaries at the time of a new product introduction.

Empirical support for the value of specialized complementary assets is found in the medical diagnostic imaging industry. Mitchell (1989, 1992) finds that when competence-destroying innovations have low transilience (Abernathy and Clark, 1985) in that they do not substantially change the market/customer linkages, then incumbents perform well in the market. The sales/service relationships of the incumbents serve as a specialized complementary asset that new entrants find hard to contract for or imitate. The continued value of these assets can serve as a buffer when firms are faced with competence-destroying technological change, protecting the firm from innovative new entrants. In a similar vein, Rosenbloom and Christensen (1994: 657) introduce the term 'value network . . . the system of producers and markets serving the ultimate user of the products or services to which a given innovation contributes'. They argue that when technological innovation causes a shift in the value network, then established firms are at a disadvantage. In contrast, even when new technology is competence-destroying, if the value network does not change, then established firms are less likely to suffer at the hands of new entrants.

While the ongoing value of specialized complementary assets can provide incumbents a buffer, technological innovation can destroy the value of these assets. The shift from electromechanical to electronic calculators provides an example (Majumdar, 1982). While this shift was competence-destroying in a technological sense, it also destroyed the value of the specialized complementary assets accumulated by the incumbent firms. The sales force and service networks of electromechanical calculator firms—about 1500 individuals per firm—were vital to successful competition in the old regime. 'You don't have a chance in this business without this capability' stated a senior executive of one firm (Majumdar, 1982: 34). Since electronic calculators were more reliable, however, service was less essential, and office equipment dealers became a viable alternative form of distribution. The electronic calculators supplied by new entrants were distributed through these dealers, thereby circumventing the need for a sales and service network. Given this

[1] Teece distinguishes between cospecialized assets, where innovation and assets are mutually specialized, as opposed to specialized assets, where either the innovation or the asset is more dependent upon the other. Since this distinction is not crucial to this analysis, the term 'specialized' is used to refer to both specialized and cospecialized complementary assets.

change, new entrants were successful in displacing established industry players.

It is also possible that new entrants to an industry can possess relevant complementary assets. Previous studies have shown that diversifying entrants are likely to perform better when they enter related markets (Rumelt, 1982; Montgomery, 1985; Montgomery and Harihan, 1991). In fact, Chandler suggests that the majority of creative destruction is the result of this type of diversification:

> The major challengers in the capital-intensive industries of the twentieth century were not smaller firms that took advantage of changes in technologies and markets . . . far more often the successful challengers were long-established companies, usually first movers, from other countries or from other industries in the same country . . . Here the established enterprise became Schumpeter's entrepreneur. (1990: 601–602)

This leads one to expect that new entrants with relevant specialized complementary assets are more likely to be successful than those without them.

Synthesis

In summary, the previous discussion identifies three important elements driving incumbent vs. new entrant performance in the face of radical technological change: investment, technical capabilities, and appropriability through specialized complementary assets. This paper argues that the expected outcome in terms of ultimate commercial performance depends upon the balance and interaction among them. If incumbents choose not to invest in the new technology, then new entrants that make the investment will dominate the market for the new technology. If incumbents do invest, but their technological performance is inferior to that of new entrants, then, assuming a regime of weak intellectual property protection, their commercial performance will depend upon whether the technological shift also devalued the relevant specialized complementary assets necessary to appropriate the benefits of innovation. If incumbents possess these assets, and due to their specialized nature they cannot be acquired by new entrants, then incumbents are likely to dominate the market even if their products are technologically inferior. If, however, the technological shift also decreased the value of these complementary assets, then the incumbents have no buffer from competition and new entrants should dominate.

Finally, if incumbents invest in the competence-destroying technology and their technological performance is on par or superior to that of new entrants, the commercial result is still dependent upon who possesses the necessary specialized complementary assets. If the technological change does not devalue the incumbents' complementary assets, then they will clearly dominate in the market. If, however, incumbents' complementary assets are devalued and diversifying new entrants possess relevant specialized complementary assets, the new entrants can be expected to dominate, even if their technology is initially inferior. If neither incumbents nor new entrants already possess specialized complementary assets, then it is unclear which firms will dominate.

DATA AND RESEARCH SETTING

The data

These issues are examined through a study of the technological and competitive history of the typesetter industry for a period of over 100 years. The industry provides a particularly good setting in which to understand the effect of technological change on competition. Between 1886 and 1990, the industry experienced three waves of radical, competence-destroying technological change, accompanied by a great deal of variation in the competitive landscape. In total 42 different firms participated in the industry, with a maximum of 25 competitors and a minimum of three in the industry at any given point in time. Industry participants have also invested substantial amounts in R&D, with industry R&D/sales averaging 10 per cent from 1985 to 1990.

The core of the data consists of a comprehensive longitudinal data set covering the entire history of the worldwide typesetter industry from the inception of the industry in 1886 through 1990. It was collected during a 14-month field-based study conducted from the fall of 1993 through the winter of 1995. The data set includes the entry date of every firm in the industry and, for those firms that exited, the exit date. Detailed data for 95 per cent of the products introduced by these firms covers product performance characteristics, price, and unit sales over time. Unfortunately quantitative R&D investment data were unavailable for the majority of firms. Firm-level data are supplemented by aggregate industry-level data, including industry size and growth.

Quantitative data were supplemented with qualitative data about how organizations responded to new technology, including in-depth case studies of multiple firms. Detailed schematics of typesetter machines from each generation of technology were reviewed with development engineers in order to understand changes in machine components and architecture. This work enabled a careful determination of the nature of technological change and its effects on organizational competence.

These data come from a combination of primary and secondary sources including company and trade association archives, field interviews, personal records of retired employees, industry consultants, industry historians, government records, as well as industry trade and scientific journals. In total, over 50 interviews were conducted, with interviews lasting from 2 hours to all day. Wherever possible, data have been cross-checked with multiple sources.

The typesetter industry

Typesetting is the process of arranging text as input to the printing process. While this process was accomplished manually for many years based on the invention of moveable type by Johann Gutenberg in about 1440, the invention of the Linotype machine by Ottmar Mergenthaler in 1886 sparked the beginning of the typesetter industry. A typesetter machine generally performs three functions: text input, text formatting, and text output. Text is input by an operator from a prepared manuscript, usually via a typewriter-like keyboard. The text is then formatted either by the operator or automatically. The output of the typesetter, either paper or film, is then used to create a printing plate. This plate is used on a printing press to produce high-volume output. Customers for typesetter machines include newspapers, commercial printers, high-end typographers and corporate 'in-house' publishers.

TECHNOLOGICAL CHANGE IN THE TYPESETTER INDUSTRY

From the discussion above, three effects of technological change are key in understanding how it will drive competition: the effect on (1) investment incentives, (2) technological competence, and (3) specialized complementary assets. In this section each new generation of typesetter technology is categorized in terms of its effect on each of these factors. There have been three generations of radical technological change since the initial invention of mechanical, 'hot metal' typesetter technology in 1886: analog phototypesetting (1949), digital CRT phototypesetting (1965), and laser imagesetting (1976). Each generation of technology is described in the Appendix.

The effect of each generation on investment incentives

Each of these generations was incremental in the economic sense in that the old generation of technology continued to compete with the new generation. This competition is evident in Figure 1, which charts industry sales by generation of technology over time. Even though the first analog phototypesetter was introduced in 1949, it was not until 1968 that half of annual industry sales were comprised of phototypesetters. Similarly, there was a 19-year lag between the introduction of CRT machines in 1965 and their dominance in 1984, and a 12-year lag between the introduction of laser imagesetters in 1976 and their dominance in 1988.

Evidence of the competition among generations of technology is also evident in the trade press and in government publications, with numerous articles comparing the old and new technologies in order to identify the conditions under which a potential customer should purchase one or the other (e.g., U.S. Congress Joint Committee on Printing, 1970; National Composition Association, 1973). Entrants with each new technology were therefore not able to monopoly price, ignoring competition from the old technology. The price/performance of each new technology did improve over time, however, resulting in eventual substitution for the old technology.

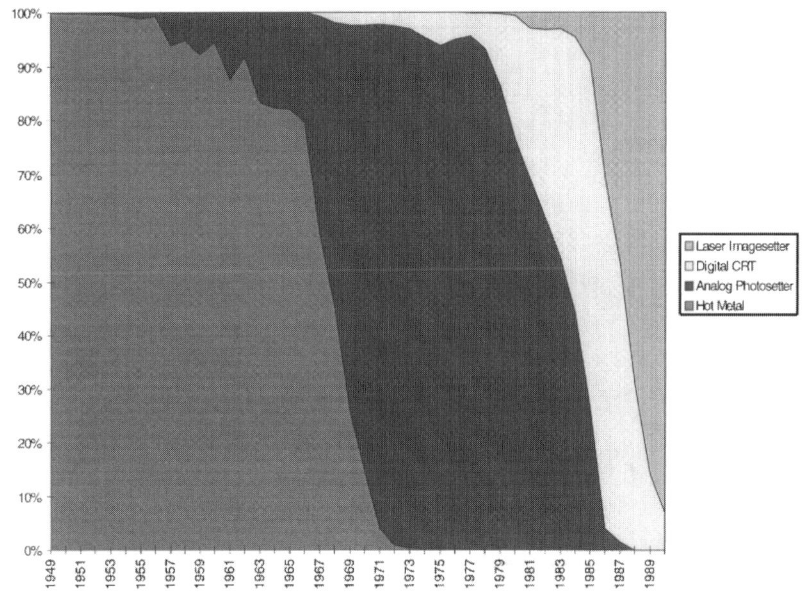

Figure 1. Percentage of typesetter sales by technology, 1949–90

The effect of each generation on technological competence

Two measures were used to gauge how each generation of technology affected the value of the technological competence of established firms. First, changes in the required skills for a product development team were compared for each generation. This measure, however, fails to capture any sense of architectural change in the product. Product components, interfaces, and overall machine logic were therefore also compared in order to gain a better understanding of architectural change.

Changes in development team skill base

The analysis of the skills required to develop a product in each generation of technology was based on an examination of the staffing of a subset of development projects. The goal was to compare the skills of individuals on development teams for each generation of technology in order to gauge how radical a new generation was in terms of new skills required. A number called 'new skills' was computed for each generation of technology (see Table 1) as follows.

For each project team in the sample (5 hot metal; 4 analog phototypesetters; 4 digital CRT phototypesetters; and 3 laser imagesetters), relevant skills of team members were identified. Then, the percentage of the development team with each skill was calculated.[2] These percentages were averaged across projects within a given generation. The average for each generation was then compared to the previous generation to determine the percentage of skills that were new. For instance, the entire development team for a hot metal machine was comprised of mechanical

[2] In the majority of cases archival project staffing records were used to determine development team skills. Where staffing records were unavailable, development engineers were asked for their best recollection of the composition of development teams. Engineers' recollections of research projects going back to the 1950s suffer two potential problems: a lack of memory and potential bias, given the knowledge of the subsequent turbulence resulting from new technology. Interviewees were surprisingly good, however, at recalling the names and backgrounds of team members. To address the potential bias that might accompany a recollection, at least two members of each development team were interviewed.

Table 1. The effect of typesetter technological innovation on development team skills

Generation	Hot metal	Analog phototypesetter	Digital CRT phototypesetter	Laser imagesetter
Development team skills	Mechanical engineer	Lens designer	Lens designer	Lens designer
		Mechanical engineer	Optomechanical engineer CRTs	Optomechanical engineer Lasers
		Optical engineer		
		Electrical engineer Electromechanical	Electrical engineer Solid state CRTs Minicomputers	Electrical engineer Solid state Lasers Microprocessors Raster image processing
			Software Applications	Software Applications Raster image processing Font encoding
Skill loss	n/a	90%	70%	50%
Competence-destroying?	n/a	Yes	Yes	Yes

engineers. In the analog phototypesetter generation, however, on average only 10 per cent of the team were mechanical engineers—the other 90 per cent of the skills of team members were new. The value of 'new skills' for the analog phototypesetter generation was therefore 90 per cent.

While detailed information about development team staffing was available for only a subset of project teams, I believe these projects are representative in that they include multiple firms and span multiple years. There was also a high degree of consistency across projects within a generation in terms of the mix and balance of skills. The mix of skills was the same for all projects within each generation, and no given skill category varied by more than ± 10 per cent across the projects within a generation.

If a large percentage of the required skills in a generation were new, then the relative value of an incumbent's existing skill base decreased. A generation was classified as competence-destroying from the standpoint of skill base if 50 per cent or more of the skills in a generation were new. With new skill requirements of 90 per cent, 70 per cent, and 50 per cent, the second, third, and fourth generations of typesetter technology were all competence-destroying.

Architectural changes

A simple comparison of skills fails to capture any changes in the architecture of the machines from different generations, so this analysis was supplemented by an examination of the characteristics of the overall product architecture relative to the preceding generation. All three generations were also competence-destroying from the standpoint of their effect on architectural knowledge.

Table 2 compares the controlling machine logic, the method of character escapement (spacing of characters), the method of font storage, and the method of character output for each generation. For each of these technological shifts, both product components and interfaces changed. For instance the font output component changed from pouring molten lead, to exposing a photograph of a character with xenon flash, to writing a character digitally with a CRT, to writing a character digitally with a laser. In addition, the final laser generation enabled the output of both text and images, thus requiring a shift to raster

Table 2. The effect of typesetter technological innovation on architectural knowledge

Generation	Controlling machine logic	Escapement (character spacing)	Font format	Character output	Effect on architectural knowledge
Hot metal	Mechanical	Mechanical	Metal matrix	Hot metal	n/a
Analog phototypesetter	Electromechanical	Mechanical	Film	Xenon flash	Destroy
Digital CRT phototypesetter	Electronic/software; programmable minicomputer	Mechanical/ electronic	Digital start/stop pattern	CRT strokes	Destroy
Laser imagesetter	Electronic/software; microprocessor	Electronic	Proprietary digital outline	Laser raster strokes	Destroy

image processing where horizontal strokes were written onto a page. The manner in which component interfaces were managed—i.e., the machine logic—moved from mechanical to electromechanical to electronic to primarily software.

The effect of each generation on specialized complementary assets

Extensive discussions with industry participants identified three salient complementary assets in the typesetter industry: specialized manufacturing capability, a sales and service network, and a font library. The overall effect of each new generation of typesetter technology on the value of incumbents' specialized complementary assets was therefore based on an examination of these three factors. In different industries other specialized complementary assets such as access to distribution channels or exclusive supplier relationships have value. In this industry, however, a direct sales force remained the preferred method of distribution throughout the period under consideration, so access to distribution channels was therefore not relevant. Similarly, exclusive supplier relationships did not play a role in the industry. Table 3 summarizes how each technological generation affected the value of specialized complementary assets relative to their value in the prior generation.

Specialized manufacturing capability

Three firms—Mergenthaler Linotype, Intertype, and Monotype—dominated the original hot metal typesetter era, controlling 99 per cent of the market by 1916. Each of these firms developed very strong specialized manufacturing capabilities and made ongoing investments in continuous improvement. For instance, a 1940 company monograph entitled 'The Autobiography of Capital B' describes the manufacturing of hot metal matrices at the industry leader, Mergenthaler Linotype:

> In the matrix department are 154 machine tools which are necessary for the many operations in the production of Linotype matrices . . . During the last 30 years . . . 90% of the machines and attachments have been redesigned and specially built in our own tool room. (Linotype & Machinery, Ltd., 1940)

The shift to analog phototypesetting destroyed the value of this manufacturing capability. While production of hot metal machines and matrices required highly specialized machine tooling and sophisticated metal shops, phototypesetter production required much less specialized equipment. Electronic components used were similar to those used in other electronic products, and some manufacturing was outsourced. In Teece's (1986) terms, manufacturing capability became a 'generic' complementary asset that could be easily contracted for. It remained a generic complementary asset in the CRT and laser generations.

Sales and service network

A strong sales and service network was a crucial specialized complementary asset developed by hot metal manufacturers given the critical nature of typesetting to buyers. The primary typesetter

Table 3. The effect of typesetter technological innovation on the value of specialized complementary assets

Generation	Specialized manufacturing capacity	Sales and service network	Extensive proprietary font library	Devalue specialized complementary assets?
Hot metal	High value	High value	High value	n/a
Analog phototypesetter	Much lower value than prior generation	Much lower value than prior generation	Same value as prior generation	Yes
Digital CRT phototypesetter	Same value as prior generation	Same value as prior generation	Same value as prior generation	No
Laser imagesetter	Same value as prior generation	Same value as prior generation	Same value as prior generation	No

buyer segments during this era comprised newspapers/magazines, commercial printers, and typographers/advertisers. The typesetter was necessary for each of these segments to do business. Newspapers, for instance, had deadlines to meet, and typesetting was on the critical path for getting the paper to press. Typographers, whose primary task was to translate manuscripts into typeset output, could not carry on their business without a typesetter machine. Commercial printers needed typeset material to print. So in addition to technical purchase criteria (e.g., speed, flexibility, and output quality), reliability and service were also important purchase criteria for each of these segments (see Table 4). In response to these customer needs, all three hot metal incumbents established strong international sales and service networks. Mergenthaler Linotype and Monotype in particular had a strong presence in Third World countries.

The value of the established firms' sales and service networks decreased with the advent of analog phototypesetters due to the emergence of a new buyer segment, the in-house publisher. Corporations with a critical mass of in-house publications now found it worthwhile to bring the typesetting of those publications in-house as opposed to contracting out for the service. Previously the noise levels and safety hazards of molten lead had precluded the operation of typesetter machines in an office environment. Analog phototypesetters eliminated these problems. Since established players had no sales presence in this new segment, they were on an equal footing with new entrants, and perhaps even at a disadvantage relative to diversifying entrants with ties to these customers.

Although this new buyer segment diminished the value of the existing sales/service network, it did not result in a completely new set of buyer purchase criteria in the manner that Christensen and Bower (1996) found with new generations of disk drives. While these new buyers cared more about noise and safety, these were typically minimum criteria. Purchase decisions were still based on more traditional criteria such as technical performance (speed, flexibility, output quality) and reliability/service (see Table 4). In addition, the new segment was still served by a sales force, not alternate distribution channels.

While the second generation of technology resulted in a new market segment, the third and fourth generations—CRT and laser machines—did not. The same set of buyers (now including in-house publishers) continued to be served through the same set of salespeople. As a result, the sales and service networks built up during the second generation remained valuable specialized complementary assets during the third and fourth generations.

Proprietary font library

The size of a manufacturer's font library affected the value of the firm's typesetter machine in much the same way the variety of software available for a computer affects its value to a buyer—a larger variety of available fonts made the typesetter more valuable. For three customer segments—typographers/advertisers, commercial printers, and in-house publishers—customer interviews indicated that the size of the available font library was one of the most critical purchase criteria (see Table 4). In order to compete in these segments,

Table 4. Overview of serving different typesetter market segments

	Newspapers	Typographers/ advertisers	Commercial printers	In-house publishers
Mode of distribution	Direct sales force	Direct sales force	Direct sales force	Direct sales force
Rank-ordered purchase criteria	• Speed • Reliability • Service • Compatibility • Flexibility • Font library • Output quality	• Output quality • Font library • Reliability • Service • Flexibility • Speed • Compatibility	• Speed • Font library • Flexibility • Output quality • Reliability • Service • Compatibility	• Noise level[a] • Safety[a] • Speed • Font library • Flexibility • Output quality • Reliability • Service • Compatibility

[a] Minimum purchase criteria.

customers indicated that a firm needed at least 500 typefaces, and even then was still at a disadvantage relative to firms that had larger selections. While newspapers required a smaller variety of fonts, they were often tied to a manufacturer that had proprietary ownership of a specific font that was important to the 'look and feel' of the paper.

Recognizing the importance of fonts, each of the hot metal incumbents invested heavily in developing proprietary font libraries. As early as 1895 the leading firm's annual report (Mergenthaler Linotype) stressed the need to invest in typeface development, with expenditures for the design of new fonts exceeding the amount spent on R&D. In 1902 Mergenthaler had a library of over 100 typefaces. By 1913 the library had grown to 1000 typefaces, and by 1923 it had reached 2000. This library included not only various roman alphabet fonts but also a number of foreign language fonts such as Greek and Cyrillic. Despite large investments, it took Mergenthaler about a year to develop 100 typefaces. At that rate, it would take an entrant 20+ years to duplicate Mergenthaler's library as it stood in 1923, and 5 years to reach the minimum competitive level of 500 fonts identified by customers.

A proprietary font library retained its value throughout the three subsequent generations of technology. While established players had to transfer typeface designs between formats, the fact that they already had a large library of designs to work from decreased the cost significantly compared to that of potential new entrants. New entrants had to either attempt to license typefaces, slowly design typefaces on their own, or attempt to copy the faces of an established firm. Established firms were not willing to license their typefaces, and designing new typefaces was extremely time-consuming. While some entrants did simply attempt to copy typefaces, this process was much more difficult than originally anticipated. For instance, despite investing significant amounts in font development, it took Compugraphic, the most successful second-generation new entrant, 10 years and approximately $23.8 million to acquire 1000 typefaces, and most of these faces were simply copies of Mergenthaler designs. The quality of these copied typefaces was generally considered inferior to the original faces. In addition, the original faces had brand names which were protected by trademark.[3] Customers preferred the true 'Helvetica' to some 'cheap imitation'.

In addition, if an established customer needed a particular typeface that was only available from its historical vendor then there would be switching costs associated with changing vendors. In order to meet the needs of *its* customers, for instance, a typesetter buyer might feel it essential to provide certain typefaces. As a commercial typographer wrote to a new phototypesetter entrant:

> Naturally, I understand that you have many typefaces available ... My problem, however, is to

[3] Many of today's well-known typefaces, such as Helvetica, were trademarked by Mergenthaler during the early hot metal era. Once it became clear in a 1970 court decision that fonts could not be patented or copyrighted, these trademarks became critical for intellectual property protection of fonts.

> perfectly match our present typestyles — all of which are Mergenthaler faces. We are in the middle of converting from hot to cold type and have many cautious customers who want the same quality that hot type produces. I'm sure you can see our predicament. (Compugraphic correspondence, 1983)

In summary, all three new generations of technology in the typesetter industry should have created an incentive for incumbents to invest. They were all incremental from an economic standpoint, and 'sustaining' from a resource allocation perspective in that they appealed to existing customers. All three generations were competence-destroying, making the technological skills and routines of incumbents obsolete. Of these three competence-destroying technological generations, however, only the second — analog phototypesetters — also decreased the value of specialized complementary assets. Two of the three specialized complementary assets examined lost their value in the shift to the second generation. So, while one asset — a font library — did retain value, the overall value of specialized complementary assets decreased significantly in the second generation.

How then, did the process of creative destruction unfold as each generation of new technology substituted for the old? The next three sections examine each of the three links in the process described earlier. Did new entrants and incumbents invest in the new technology? Given that they invested, how did the technical performance of incumbent machines compare to that of new entrants? Did the ongoing value of specialized complementary assets buffer incumbents from the effects of inferior technical performance, and did possession of relevant specialized complementary assets facilitate the entry of diversifying firms?

CREATIVE DESTRUCTION IN THE TYPESETTER INDUSTRY

Investment behavior of new entrants and incumbents

As both the economic and innovation literature would predict, there was very little investment by new entrants in hot metal typesetter technology during the many years of incremental hot metal innovation. Introduced in 1886, Ottmar Mergenthaler's Linotype architecture became dominant by 1903, with over 50 per cent of annual industry sales. Other than four new entrants (only one of which survived) around 1911 when key patents expired, there was no entry into hot metal typesetting between 1903 and 1970, when production of hot metal machines in the United States ceased.

Entrants did, however, clearly perceive and pursue opportunities created by the three subsequent generations of competence-destroying technology. There were 17, 14, and 8 new entrants in the second, third, and fourth generations of technology. As Chandler (1990) would predict, these entrants were overwhelmingly diversifying firms with related experience. In particular, for the analog phototypesetter generation of technology, 88 per cent of new entrants had prior related market knowledge, having sold other graphic arts products to typesetter buyers. This number was 50 per cent for the CRT generation and 62 per cent for the laser generation.

Since each new typesetter generation was incremental from an economic standpoint and 'sustaining' from a resource allocation standpoint, one would expect to find incumbent investment greater than new entrant investment. Unfortunately, a lack of data precludes a comparison of the levels of incumbent and new entrant investment. All that can be tested is whether and when the incumbents invested.

Almost every firm that established even a moderate presence in a given typesetter generation (at least a 2% market share) invested in developing a machine for the following generation. All three first-generation hot metal firms invested in developing the second generation of technology. Of 11 incumbents in the second generation, 10 invested in the third, and of 11 incumbents in the third generation, 9 invested in the fourth. In addition, qualitative data from interviews with both management and development engineers indicate that the level of investment by incumbents was at least equivalent to that of new entrants.

Moreover, incumbents invested in second-generation machines much earlier than new entrants. The average year that hot metal incumbents announced their initial second-generation machine was 1955 as opposed to 1967 for new entrants. Incumbents and new entrants did not differ significantly in their investment timing for the third and fourth generations. For the third generation, the mean introduction date for incumbents was 1975 as opposed to 1973 for new entrants, and in the fourth generation incumbents' mean introduction date was 1983 vs. 1984 for new entrants.

Technical performance of incumbents vs. new entrants

This section begins by looking in depth at two examples of incumbent development efforts in order to understand the degree to which incumbents were handicapped by their prior experience. It focuses on the development of second-generation analog phototypesetters by hot metal incumbents Intertype and Monotype. In each of these cases, existing capabilities shaped the manner in which incumbents approached the new technology. Given that these organizations had strongly established, efficient routines embedded in the architecture of the prior generation, their natural inclination was to utilize those same routines in the development of the following generation's machines. The first phototypesetter developed by each of the hot metal firms was thus based on its hot metal architecture and, as a result, had significantly inferior performance.

Intertype

Intertype began work on a phototypesetter in 1936 after the firm's president was approached about the idea during a European business trip. He assigned the development task to the hot metal chief engineer. This individual worked on the project with a team of mechanical engineers from the hot metal organization. Since the group had little expertise in optics and lenses, they involved Eastman Kodak in the design of the lens component of the machine. Kodak also developed special film to work with it.

After 10 years of development effort, a prototype was ready to be field tested and was installed at the U.S. Government Printing Office in Washington in 1946. The machine was essentially a modified hot metal typesetter and did not operate any faster than its hot metal counterpart. The basic architecture was identical except that the mechanically circulating matrices were modified to have a film image of a character embedded in them as opposed to a mold of a character. The component of the machine where hot metal used to be injected into the matrix was replaced by a camera that photographed one character at a time. The film carriage advanced mechanically between letters, based on the width of the matrix.

This awkward machine, called the Fotosetter, shipped in 1949 and was the first commercial phototypesetter on the market. Sales were initially slow, with a handful of machines shipping each year as design and production problems were worked out. By 1954, Intertype reported revenues of over half a million dollars from the sale of Fotosetters—about 15 machines—and by 1956 that number had risen to almost a million dollars. The 'success' of the Fotosetter led to additional incremental phototypesetter innovation within the old hot metal architecture. The Intertype 1956 Annual Report (p. 4) states: 'Your Corporation continues its research and development work in the field of photocomposition with special emphasis on enlarging the range of type and adding to the Fotosetter's utility'. In the meantime, new entrants announced truly innovative electromechanical machines that incorporated significantly improved technology, and by 1961 Intertype's share of the phototypesetter market had fallen from 100 per cent to only 12 per cent.

Monotype

Monotype also became involved in photocomposition at an early stage. The manager of the firm's London office commenced research on a phototypesetter in the early 1930s although without the official sanctioning of the corporation. His design utilized many of the concepts embodied in the hot metal Monotype machine. Patents on the machine, called the Rotofoto, date from 1936.

Shortly after World War II Monotype formally initiated a project to develop a commercial phototypesetter. This project was staffed by the same engineering staff responsible for hot metal development, and the resulting Monophoto bore a great deal of resemblance to both the Rotofoto and the Monotype hot metal machine. Rather than cast metal into a mold, a selected character was positioned over a flash and exposed onto film. It ran at the same speed as a hot metal Monotype, and only one size could be set at a time. The machine was not installed commercially until 1957, and by 1959 only 5–10 machines had been placed.

In the mid-1960s while new entrants were developing electromechanical machines, Monotype was continuing to make incremental improvements to the mechanical Monophoto. One Monotype employee at the time made an attempt to increase the firm's interest in electronics and in a 1994 interview commented, 'I became a one-eyed man in a blind kingdom.' Frustrated with his attempts to enlighten the 'blind', he went to work for another firm.

When Monotype did finally develop an electro-mechanical machine—the Monophoto 400—in 1969, it was described by one industry expert as 'an odd amalgam of mechanics, fluidics and electronics ... with anachronistic technical concepts which ran counter to prevailing development trends' (Wallis, 1993: 4).

Overall technical performance of incumbents vs. new entrants

Intertype and Monotype were clearly handicapped by their many years of experience in hot metal technology when they attempted to develop second-generation machines. This section examines whether their experience was representative of incumbent firms in general by comparing the average technical performance of incumbent and new entrant machines for each generation of technology. As a measure of technical performance product speed is used. While many criteria are important to typesetter buyers, speed is the one criterion valued highly by all segments (see Table 4), and in interviews with product development engineers it was consistently identified as a primary goal. In addition, speed has also been used as the primary indicator of technical progress in technology forecasting studies of the typesetter industry (Mohn, 1971).

Table 5 compares the average speed of machines made by incumbents as opposed to new entrants for each competence-destroying generation. A comparison of the average speed of the first machine introduced by each incumbent with the average speed of the first machine of each new entrant indicates that in all three generations incumbents' machines were significantly slower. However, as discussed earlier, incumbents invested in the second generation of technology much earlier than new entrants did. An additional comparison was therefore performed for the analog phototypesetter generation in which incumbents' second machines (average introduction date of 1963) were compared to the first machines of new entrants (average introduction date of 1967). The average speed of the incumbent second machines, 19 newspaper lines per minute, was still much slower than the new entrants' average speed of 41, but no longer significantly so in a statistical sense. Since there were only three incumbents for this generation, however, there is little statistical power, and this result is not surprising.

Finally, a comparison of all machines introduced after the first machine was made in order to determine whether incumbents eventually caught up technologically with new entrants. Data for this analysis were only available for the analog phototypesetter and digital CRT generations. For analog phototypesetters, although the absolute difference between average speeds—26 newspaper lines per minute vs. 55—seems substantial, it is not statistically significant. Once more, the small sample size may be limiting the power of the test. In the case of digital CRT machines, a statistically significant difference between the average speed of incumbent and new entrant machines remains. This analysis provides limited evidence that, even after several years of experience in the new technology, incumbents did not catch up technologically with new entrants. The effect of prior routines and procedures appears to have been quite persistent.

Appropriability and specialized complementary assets

As discussed earlier, when an incumbent's technological competence is destroyed but the incumbent still controls valuable specialized complementary assets, it should be able to protect its competitive position despite developing new technological capability more slowly than new entrants. Similarly, new entrants that possess relevant specialized complementary assets should have an advantage over those that do not. This section examines these expectations in the context of the three competence-destroying generations of technology.

Descriptive data

The descriptive data clearly support the buffering of incumbents by specialized complementary assets that retain their value. Table 6 summarizes the effect of each generation on investment, technical capabilities, and complementary assets and

Table 5. Technical performance of incumbent and new entrant machines: Average speed (newspaper lines per minute) of machines, all incumbents vs. all new entrants (pairwise two-tailed t-test)

Technological generation	Incumbent machines	New entrant machines	Significance
Analog phototypesetters			
First machine	14	41	$p < 0.10$
Incumbent second machine vs. new entrant first machine	19	41	n.s.
All subsequent machines	26	55	n.s.
Digital CRT phototypesetters			
First machine	399	974	$p < 0.05$
All subsequent machines	547	1583	$p < 0.05$
Laser imagesetters			
First machine	381	648	$p < 0.10$

Table 6. Summary of descriptive data on incumbent vs. new entrant performance

Generation	Incumbent incentive to invest?/Did incumbents invest?	Technological competence destroyed?/Incumbent technical performance inferior?	Specialized complementary assets devalued?	Number of new entrants	Market share of new entrants
Analog phototypesetter	Yes/Yes	Yes/Yes	Yes	17	89%
Digital CRT phototypesetter	Yes/Yes	Yes/Yes	No	14	16%
Laser imagesetter	Yes/Yes	Yes/Yes	No	8	12%

then lists the number of new entrants in each generation and how successful they were. In the case of the two generations where specialized complementary assets retained their value, there were a large number of new entrants, 14 in the CRT and eight in the laser generation, but new entrants captured only 16 per cent and 12 per cent respectively of cumulative CRT and laser unit sales. New entrants perceived an opportunity due to the changing technology, but the ownership of specialized complementary assets by incumbents, a sales/service network and font libraries in particular, appear to have protected the incumbents from competition. In contrast, new entrants to the analog phototypesetter generation, where both technological competence and specialized complementary assets were devalued, captured 89 per cent of the cumulative market. Although font libraries did retain value in this generation, other important complementary assets — specialized manufacturing capability and sales/service networks — lost value, and the combined effect of these factors was stronger than the protection provided by a font library.

Given that incumbents to the analog phototypesetter generation lacked some of the relevant specialized complementary assets, one would expect that diversifying new entrants that possessed them would perform well. The next section examines one such entrant, AM Varityper.

The case of new entrant Varityper

As discussed earlier, the second generation of technology — analog phototypesetters — created a new market segment: in-house publishers. AM Varityper, a subsidiary of diversified conglomerate AM (Addressograph-Multigraph), was ideally positioned to take advantage of this emerging segment given its prior relationships with corporate buyers interested in graphic arts products. The firm captured 28 per cent of the cumulative unit share of the analog phototypesetter generation.

As a corporation, AM had a great deal of related market experience in the graphic arts. The Varityper division sold justifying typewriters; the Multigraph division sold offset duplicators; the Bruning division sold copiers and diazo printers, and the Buckeye division sold printing supplies.[4] The firm had a strong presence in corporate offices where Varityper justifying typewriters were used in conjunction with Multigraph offset duplicators, allowing for an inexpensive in-house publishing operation.

The commercialization of phototypesetters provided an attractive option for these customers to upgrade and produce higher-quality internal publications. Phototypesetters were generally less expensive than hot metal typesetters and, lacking a pot of molten lead, were much better suited to an office environment. AM recognized the threat of substitution for its existing business, and the firm therefore evaluated the feasibility of entry into the typesetter market.

Despite its own lack of relevant technological expertise, the firm perceived an opportunity created by the changing technology. Hot metal incumbents did not have any established reputation with 'in-house' customers and also lacked a sales presence in that segment of the market. Varityper had both a good reputation and a strong sales and service presence. Varityper also had a moderate font library since it offered a variety of fonts for use in its justifying typewriters. In addition, the firm clearly understood the need to invest in additional fonts and to signal to buyers that fonts were a high priority.

The only thing the firm lacked was technology, so management contracted with the technological leader in the industry: a new entrant, Photon. AM Varityper's first machine, shipped in 1969, was designed and manufactured by Photon. Given the strength of Varityper's sales organization the machine did moderately well, and three more models were announced in 1970 and 1971 with similar results. At the same time, the firm was gradually developing its own technological expertise. Announced in 1972, the first internally developed product was a low-cost version of the Photon machine. While only moderately successful, the effort gave the development organization crucial experience in phototypesetter development. Work on a totally home-grown machine had begun in late 1971, and this time the result was a huge success. The Comp/Set 500, announced in 1974, was the first machine with a full-size video screen for text input and editing.[5] It was also relatively low cost and targeted at the in-house segment where AM Varityper was strong. Over 11,000 Comp/Sets were sold over the life of the product, compared to only 3500 units sold for the most successful incumbent machine in this generation.

THE ROLE OF SPECIALIZED COMPLEMENTARY ASSETS

The prior section provided descriptive data in support of the importance of specialized complementary assets. When new technology was competence-destroying, but the value of incumbent specialized complementary assets was preserved (the CRT and laser generations), then incumbents maintained a strong market presence. In contrast, when complementary assets did not retain their value (analog phototypesetters), new entrants dominated the market. New entrants with relevant specialized complementary assets would be expected to perform well, and qualitative data about one such successful entrant were discussed. This section explores the degree to which quantitative analysis is consistent with these results.

Given a lack of available data on profits, product market share in the relevant generation of technology is used as a measure of commercial performance. The model controls for the amount of competition in the technological segment,

[4] Other divisions included Addressograph, which sold business-tabulating/data-processing systems, and Emeloid, which sold custom plastics.

[5] Interestingly, the lead engineer on the Comp/Set 500 was a mechanical engineer who had previously worked on the design of justifying typewriters. When the focus of the organization shifted to electronics and optomechanical engineering, this individual spent some time at different universities (e.g., MIT) retraining. How a firm should handle human resources, and the trade-off between retraining existing employees as opposed to hiring all new employees, is an interesting agenda item for future research.

growth of the segment, and whether a product is a firm's first in the generation, and then uses dummy variables to measure the effect of incumbency and of complementary assets on performance. Of course, this model necessarily omits a variety of factors that might have also shaped market share, such as advertising levels or the size of the sales force. Unfortunately, data limitations preclude their inclusion.

Dollar market share is used as the dependent variable. This measure incorporates the two sources of advantage associated with new product introduction: an increase in realized demand and the ability to charge high prices. Since I am not interested in understanding the relation between these two factors *per se*, but in their combined effect, dollar market share is an appropriate measure. To carefully distinguish the relation between these two factors, one would need to estimate a differentiated product demand model (e.g., Trajtenberg, 1990; Bresnehan, Stern and Trajtenberg, 1996), and limitations on the data set preclude this.[6]

The incumbent dummy variable in the model is designed to capture the effect of differences between the technical performance of incumbent and new entrant machines since product performance is not explicitly included in the model. Given that all products in the sample come from competence-destroying generations of technology, one would initially expect incumbency to have a negative effect on commercial performance. Controlling for the effect of specialized complementary assets, however, a negative coefficient is expected only when both competence is destroyed and specialized complementary assets lose value. When complementary assets retain their value, they should buffer incumbents from the effects of competence destruction. One would also expect that new entrants with relevant complementary assets would perform better than new entrants without them. The expected coefficient of the dummy for new entrants with complementary assets is therefore positive.

The basic model is as follows (note the omitted dummy is for New Entrants that do not possess complementary assets):

$$\begin{aligned} Ln\ (Market\ Share_{it}) = {} & \alpha + \beta_1\ Competition_t \\ & + \beta_2\ Segment\ Growth_t \\ & + \beta_3\ First\ Product_{it} \\ & + \beta_4\ Incumbent_{it}{*}Complementary\ Assets\ Devalued_{it} \\ & + \beta_5\ Incumbent_{it}{*}Complementary\ Assets\ Not\ Devalued_{it} \\ & + \beta_6 New\ Entrant_{it}{*}Possess\ Complementary\ Assets_{it} \\ & + \epsilon_{it} \end{aligned}$$

An additional model that substitutes years of experience in the prior generation for the incumbent dummy variable is also tested in order to determine the effect of the length of incumbency. Variable definitions follow. Definitions and descriptive statistics are summarized in Tables 7 and 8.

Measures

The dependent variable: Dollar Market Share

In order to capture the initial effect of a product in the marketplace, dollar market share is measured for the first 3 full years of a product's life. Each product which lasts 3 years therefore has three observations. Products withdrawn after 1 or 2 years have only one or two observations. Market share is calculated within a given generation of technology. This allows for a distinction between incumbent performance in different generations and also controls for the stage of diffusion of the technology.

Competition

The effect of competition is captured by the number of products competing in a generation of technology in a given year. Clearly the more competitors there are, the lower one would expect market share to be.

[6] A more limited model that uses unit share as the dependent variable and relative price as an additional explanatory variable was examined. Since relative price is endogenous in that it is correlated with unobserved quality in the error term, two-stage least-squares techniques were used. Unfortunately, good instruments for relative price, such as cost data, were not available, so lagged relative price was used. The results of this model were qualitatively identical to the results obtained using dollar market share.

Table 7. Product market share model: Variables and measures

Variable	Measure
Market Share	The dollar market share of a product in a given year and generation of technology
Measures of technical experience	
Incumbent	Dummy variable set equal to one if a firm was present in the prior generation
Stock of Prior Experience	The number of years of experience a firm has in the prior generation of technology
Measures of specialized complementary assets	
Complementary Assets Devalued	Dummy variable set equal to one if this generation resulted in the devaluation of specialized complementary assets that had value in the prior generation (see Table 3)
Possess Complementary Assets	Dummy variable set equal to one if a typesetter new entrant had prior experience selling other graphic arts products to typesetter buyers
Controls	
Number of Competing Products	The total number of products competing in the generation at the beginning of the period
Segment Growth	Average unit growth for the prior 3 years for the given generation
First Product	Dummy variable set equal to one if this is the first product in the current generation developed by this manufacturer

Table 8. Market share model descriptive statistics

	Mean	S.D.	Minimum	Maximum
Market Share	15.0	18.0	0.06	78.8
Number of Competing Products	10.1	3.3	3	16
Segment Growth	0.53	0.46	−0.07	1.97
First Product	0.47	0.50	0	1
Incumbent	0.51	0.50	0	1
Stock of Prior Experience	18.2	28.0	0	85
Complementary Assets Devalued	0.50	0.50	0	1

Growth

Growth each year is measured as the average unit growth in a technological generation for the prior 3 years. It is expected that a product announced during a high growth period has a higher likelihood of gaining market share. It is generally easier to gain new sales as opposed to stealing existing customers, especially given the high switching costs in the industry.

First product

This dummy variable is set equal to one if a product is the first a firm has shipped in a given generation of technology. Since a firm has little experience in the new technology, one might expect its first product to be less successful than subsequent efforts. The expected sign of First Product is therefore negative.

Incumbent dummy

The incumbent dummy is set equal to one if the manufacturer of a product also shipped products in the prior generation.

Prior experience

An incumbent's prior experience is measured as the number of years of experience in the preceding generation at the time the current product is shipped. This measure should capture the routines and procedures embedded in the product development organization of incumbent firms. The longer the firm has participated in the prior generation, the more difficult these routines should be to change. This measure is simply a richer way of capturing the potential liability of incumbent experience, and the expected effects on performance are the same as for the incumbent dummy variable.

Complementary assets devalued

If a product is in a technological generation that destroyed the value of specialized complementary assets relative to the prior generation (i.e., the analog phototypesetter generation), then this dummy is coded as one. The construction is based upon an in-depth examination of the industry and the technologies as discussed above.

Possess complementary assets

Dummy variable for diversifying new entrants set equal to one if they had related market experience, specifically prior experience selling graphic arts products to potential typesetter buyers. In an ideal case, one would have a scalar estimate of market relatedness, based on a number of factors including, for instance, whether the purchase decision maker was the same for the new market. Unfortunately, this level of detailed data was not available.

Results of modeling market share

Controls

The results of the analysis of market share are displayed in Table 9. The coefficient of the control variable *Number of Competing Products* is in the expected direction and the variable is highly significant. For each additional competing product in the market, there is about a 0.33 per cent decrease in market share. The *Growth* control variable, however, is not significant in any specification. Apparently faster growth did not enable new products to gain substantially more market share than products entering slower growth environments. The final control variable, *First Product*, was significant with a negative coefficient in all but one specification. A firm's first product in a given generation did not perform as well as subsequent products. Using the coefficients from specifications 2 and 3, being the first product in a generation resulted in a 0.56 per cent decrease in market share. One might expect that the negative effect of a first product would be even stronger for incumbents than for new entrants since all technological generations included in the sample are competence-destroying. The interaction of the *First Product* and *Incumbent* dummies, however, was consistently insignificant when tested in alternative specifications.

The effect of specialized complementary assets

This section examines whether specialized complementary assets are in fact buffering incumbents from the effect of inferior technological performance. Specification 1 tests for the effect of competence destruction on incumbent market share through the inclusion of an incumbent dummy. This is the test that might be performed in a traditional analysis of competence destruction and does not control for other effects. The coefficient of the incumbent dummy is not significant. If one went no further, one might assume incumbents were not generally disadvantaged by competence destruction. In specification 2, however, when the interaction between *Incumbent* and the dummy variable *Complementary Assets Devalued* is included, the results change dramatically. When specialized complementary assets are devalued in addition to competence being destroyed, the effect of being an incumbent is significant and negative. Being an incumbent (vs. a new entrant) decreases market share by 1.01 per cent. When specialized complementary assets retain their value, however, the effect of being an incumbent no longer hurts performance significantly. In fact, it has a signifi-

Table 9. Understanding the effect of specialized complementary assets: determinants of market share, dependent variable = ln (Market share). Sample includes only competence-destroying technological generations, $n = 154$

	Model					
	1	2	3	4	5	6
CONSTANT	4.82***	4.99***	5.07***	5.33***	5.66***	5.9***
	(0.44)	(0.38)	(0.52)	(0.42)	(0.38)	(0.42)
Number of Competing Products	–0.33***	–0.32***	–0.32***	–0.33***	–0.26***	–0.26***
	(0.03)	(0.03)	(0.03)	(0.03)	(0.03)	(0.03)
Segment Growth	0.004	0.001	0.001	0.005**	–0.003	–0.003
	(0.003)	(0.002)	(0.002)	(0.003)	(0.002)	(0.002)
First Product	–0.34	–0.56***	–0.56***	–0.65**	–0.42***	–0.44**
	(0.24)	(0.21)	(0.21)	(0.24)	(0.20)	(0.20)
Incumbent	0.20					
	(0.22)					
Incumbent Complementary Assets Devalued*		–1.01***	–1.11***			
		(0.25)	(0.36)			
Incumbent Complementary Assets Not Devalued*		1.09***	0.99***			
		(0.22)	(0.33)			
New Entrant Possess Complementary Assets*			–0.12			–0.34
			(0.32)			(0.26)
Stock of Prior Experience				–0.014***		
				(0.004)		
Stock of Prior Experience Complementary Assets Devalued*					–0.008***	–0.012***
					(0.003)	(0.004)
Stock of Prior Experience Complementary Assets Not Devalued*					0.007	0.006
					(0.014)	(0.017)
Complementary Assets Devalued					–1.55***	–1.56***
					(0.28)	(0.28)
Adjusted R^2	0.40	0.55	0.55	0.46	0.61	0.61

Standard errors in parentheses; *** $p < 0.01$; ** $p < 0.05$; * $p < 0.10$

cant and positive effect on market share, now resulting in a 0.99 per cent increase in share. So despite the technological disadvantage imposed by competence destruction, it appears that specialized complementary assets sheltered firms in the industry—being an incumbent did not handicap them as long as specialized complementary assets retained their value.

Specification 3 examines whether new entrants with related specialized complementary assets, in this case related market experience, had an advantage over those without them. If related specialized complementary assets are valuable, one would expect the dummy for new entrants with specialized complementary assets to have a significant, positive coefficient. It is, however, insignificant. Given that actual relatedness is difficult to measure, this result is not surprising. As Mitchell (1992) points out, the sales method for related products serving the same customer can vary substantially. In this case, it is not clear that prior experience selling printing ink to typesetter buyers provides an advantage in entering the typesetter market. The individual making the purchase decision for the products and the entire purchase process may be quite different. Unfortunately, more refined measures of relatedness were not available.

Specifications 4 through 6 substitute for the incumbent dummy variable with a more precise measure of the incumbent's prior experience—years in the prior technological generation. The results of using this measure are qualitatively the same as those obtained when using the incumbent dummy. Specification 4 examines the effect of competence destruction without controlling for complementary assets. In this case, an incumbent's prior experience in terms of years has a negative and significant effect on market share. On the surface, this result provides evidence of incumbents being at a disadvantage. In specification 5, however, one sees that when the interaction with whether specialized complementary assets lose value is included, the significant negative effect only holds if both competence is destroyed and specialized complementary assets lose value. Each additional year of experience in the prior generation leads to a 0.01 per cent decrease in market share. Given that some firms spent over 80 years in the prior generation this small annual effect is consistent with the effect of the incumbent dummy, which led to about a 1 per cent decrease in market share. When specialized complementary assets remain valuable, the estimated coefficient is positive and no longer significant. Once more specialized complementary assets appear to be buffering incumbents from the effects of competence destruction. Consistent with the prior result (specification 3), when a related market experience dummy variable is added in specification 6, to evaluate the effect of related specialized complementary assets for new entrants, it is not significant.

Interestingly, the control dummy variable for whether complementary assets are devalued or not is significant and negative in specifications 5 and 6. This implies that, on average, market shares were lower in the generation where incumbent complementary assets were devalued. Since there was a great deal of successful new entry, the market apparently became less concentrated and more competitive, resulting in relatively lower shares.

CONCLUSIONS

This paper has improved our understanding of the process of creative destruction through an examination of over 100 years of technological and competitive history in the typesetter industry. The industry has undergone three waves of creative destruction, where competence-destroying technological change has shaken the industry. In only one of these three cases, however, were incumbents displaced by new entrants. Using a data base that includes sales, price, technical characteristics, and organizational effects for almost every product introduced by every firm in the industry between 1886 and 1990, this paper explores how the balance and integration of three factors — investment, technical capabilities, and specialized complementary assets — drove the commercial performance of incumbents vs. new entrants.

This study contributes to our understanding of creative destruction in two ways. First, it adds to the limited number of detailed longitudinal industry studies that have attempted to understand the role of multiple waves of technological change in shaping the competitive landscape. Studies of photolithographic alignment equipment (Henderson and Clark, 1990; Henderson, 1993), disk drives (Christensen and Bower, 1996; Christensen, 1993), and medical diagnostic imaging equipment (Mitchell, 1989, 1992) each suggest factors that influence the relative position of incumbents and new entrants. This study provides additional empirical support for some of their findings.

While a lack of investment is sometimes responsible for incumbent failure (e.g., Christensen and Bower, 1996), other times incumbents invest substantial amounts in new technology (Henderson, 1993). Incumbents in the typesetter industry invested overwhelmingly in each new generation of competence-destroying technology. Since each new generation was incremental in the economic sense, and 'sustaining' in that it met the needs of existing customers, this result is consistent with theoretical expectations.

Despite timely investments, both qualitative and quantitative analysis confirmed that established firms were handicapped by their prior experience in that their approach to new product development was shaped by that experience. The initial products developed by established firms were consistently inferior to those of new entrants. The need for both new technical skills and new architectural knowledge proved difficult for incumbents to manage. This result is consistent with findings in many industries (Cooper and Schendel, 1976; Majumdar, 1982; Tushman and Anderson, 1986; Henderson and Clark, 1990; Afuah, 1994).

Incumbents did not, however, necessarily suffer commercial consequences as a result of their inferior technological positions. When incumbent firms possessed specialized complementary assets (Teece, 1986) that retained their value despite the technological shift, these assets were found to buffer incumbents from the effects of competence destruction. Incumbents only suffered in the market when both competence was destroyed and the value of specialized complementary assets was diminished. This result is consistent with Mitchell's (1989, 1992) findings in the medical diagnostic imaging industry.

In addition to providing support for prior findings, this study takes prior work one step further by explicitly examining how the balance and interaction among investment, technical performance, and complementary assets drives commercial performance. With the exception of Henderson (1993), most prior work has focused on one dimension or another, without attempting to disentangle their differential effects. By explicitly distinguishing among the three, this work helps to pinpoint which factors drive ultimate commercial performance. In the typesetter industry, the importance of specialized complementary assets was paramount, and an analysis of the effect of competence-destroying technological change that focused only on investment or technical performance and ignored the role of complementary assets would have led to misleading results. This work therefore highlights the importance of integrating multiple perspectives when analyzing competition.

While this work takes an important first step towards understanding creative destruction, much work remains. Some of the measures in this paper move beyond the usual one/zero classifications of technological change, but additional work on improved ways to measure the effect of shifts in technology would be welcome. In particular, improved measures of competence and competence destruction are needed.

Perhaps the most serious limitation of this paper is its treatment of incumbents as a class of firms without distinguishing between individual firms within that class. There is great variation in the performance of incumbent firms, and understanding that variation is crucial. There is evidence, for instance, that intraorganizational ecological processes can help firms to survive shifts in their external environment (Burgelman, 1991, 1994). In addition, one might expect that differences in 'absorptive capacity', a firm's ability to exploit outside knowledge (Cohen and Levinthal, 1990), might explain the ability of some incumbents to traverse shifts. The degree of organizational separation of new technology development can also play a crucial role in an established firm's technological performance (Cooper and Smith, 1992). Finally, differences in country institutional settings might also play an important role in firm performance (Porter, 1990). Levinthal (1992) suggests that differential selection environments may be responsible for the survival of some incumbents and not others, and different country institutional settings could be responsible for differential selection. Government subsidies may, for instance, make firms that might otherwise succumb to the competition of new entrants viable players. Further work exploring these issues should inform our understanding of creative destruction.

ACKNOWLEDGEMENTS

My thanks to Rebecca Henderson, Spencer Palocz, Eric Rebentisch, Nancy Rose, Richard Rosenbloom, Anjali Sastry, Stephan Schrader, Nicholas Souleles, Scott Stern, and three anonymous referees for comments on an earlier version of this paper. Errors and omissions remain entirely my own responsibility. Financial support from the MIT International Center for Research on the Management of Technology and the Consortium on Cooperation and Competitiveness is gratefully acknowledged.

REFERENCES

Abernathy, W. J. and K. B. Clark (1985). 'Innovation: Mapping the winds of creative destruction', *Research Policy*, **14**, pp. 3–22.

Abernathy, W. J. and J. Utterback (1978). 'Patterns of industrial innovation', *Technology Review*, pp. 40–47.

Afuah, A. (1994). 'Strategic adoption of innovation: The case of reduced instruction set computer (RISC) technology', unpublished Ph.D. dissertation, MIT, Cambridge, MA.

Arrow, K. (1962). 'Economic welfare and the allocation of resources for inventions'. In T. Nelson (ed.), *The Rate and Direction of Inventive Activity: Economic and Social Factors*. Princeton University Press, Princeton, NJ, pp. 609–625.

Arrow, K. (1974). *The Limits of Organization*. Norton, New York.

Bresnehan, T., S. Stern and M. Trajtenberg (1996). 'Market segmentation and the sources of rents from innovation: Personal computers in the late 1980s', NBER Working Paper no. 57–26.

Burgelman, R. (1991). 'Intraorganizational ecology of strategy making and organizational adaptation theory and field research', *Organization Science*, 2, pp. 239–262.

Burgelman, R. (1994). 'Fading memories: A process theory of strategic business exit in dynamic environments', *Administrative Science Quarterly*, **39**, pp. 24–56.

Burns, T. and G. M. Stalker (1961). *The Management of Innovation*. Free Press, New York.
Chandler, A. (1990). *Scale and Scope: The Dynamics of Industrial Capitalism*. Harvard University Press, Cambridge, MA.
Christensen, C. (1993). 'The rigid disk drive industry: A history of commercial and technological turbulence', *Business History Review*, **67**, pp. 531–588.
Christensen, C. and J. Bower (1996). 'Customer power, strategic investment, and the failure of leading firms', *Strategic Management Journal*, **17**(3), pp. 197–218.
Cohen, W. M. and D. A. Levinthal (1990). 'Absorptive capacity: A new perspective on learning and innovation', *Administrative Science Quarterly*, **35**, pp. 128–152.
Cooper, A. C. and D. Schendel (1976). 'Strategic responses to technological threats', *Business Horizons*, pp. 61–69.
Cooper, A. C. and C. Smith (1992). 'How established firms respond to threatening technologies', *Academy of Management Executive*, **6** (2), pp. 55–70.
Dosi, G. (1982). 'Technological paradigms and technological trajectories', *Research Policy*, **11**(3), pp. 147–162.
Freeman, C. (1982). *The Economics of Innovation*. MIT Press, Cambridge, MA.
Galbraith, J. (1973). *Designing Complex Organizations*. Addison-Wesley, Reading, MA.
Gilbert, R. J. and H. M. Newberry (1982). 'Preemptive patenting and the persistence of monopoly', *American Economic Review*, **72**, pp. 514–526.
Hannan, M. and J. Freeman (1977). 'The population ecology of organizations', *American Journal of Sociology*, **82**, pp. 929–964.
Hannan, M. and J. Freeman (1984). 'Structural inertia and organizational change', *American Sociological Review*, **49**, pp. 149–164.
Henderson, R. M. (1993). 'Underinvestment and incompetence as responses to radical innovation: Evidence from the photolithographic alignment equipment industry', *Rand Journal of Economics*, **24**(2), pp. 248–270.
Henderson, R. M. and K. B. Clark (1990). 'Architectural innovation: The reconfiguration of existing product technologies and the failure of established firms', *Administrative Science Quarterly*, **35**, pp. 9–30.
Leonard-Barton, D. (1992). 'Core capabilities and core rigidities: A paradox in managing new product development', *Strategic Management Journal*, Summer Special Issue, **13**, pp. 111–125.
Levinthal, D. (1992). 'Surviving Schumpeterian environments: An evolutionary perspective', *Industrial and Corporate Change*, **1**(3), pp. 427–443.
Linotype & Machinery, Ltd. (1940). *The Autobiography of Capital B*. Linotype & Machinery, Ltd., London.
Majumdar, B. A. (1982). *Innovations, Product Developments and Technology Transfers: An Empirical Study of Dynamic Competitive Advantage, the Case of Electronic Calculators*. University Press of America, Washington, DC.
Mitchell, W. (1989). 'Whether and when? Probability and timing of incumbent's entry into emerging industrial subfields', *Administrative Science Quarterly*, **34**, pp. 208–234.
Mitchell, W. (1992). 'Are more good things better: Or will technical and market capabilities conflict when a firm expands?', *Industrial and Corporate Change*, **1**(2), pp. 327–346.
Mohn, N.(1971). 'Statistical analyses with progress trend concepts in forecasting technology', unpublished Ph.D. dissertation, University of Texas at Austin.
Montgomery, C. (1985). 'Product market diversification and market power', *Academy of Management Journal*, **28**, pp. 789–798.
Montgomery, C. and S. Hariharan (1991). 'Diversified expansion by large established firms', *Journal of Economic Behavior and Organization*, **15**, pp. 71–89.
National Composition Association (1973). *Guidelines for Making the Phototypesetting Equipment Decision*. Printing Industries of America, Washington, DC.
Nelson, R. and S. Winter (1982). *An Evolutionary Theory of Economic Change*. Harvard University Press, Cambridge, MA.
Pfeffer, J. and G. R. Salancik (1978). *The External Control of Organizations: A Resource Dependence Perspective*. Harper & Row, New York.
Porter, M. E. (1990). *The Competitive Advantage of Nations*. Free Press, New York.
Reinganum, J. F. (1983). 'Uncertain innovation and the persistence of monopoly', *American Economic Review*, **73**, pp. 741–748.
Rosenbloom, R. and C. Christensen (1994). 'Technological discontinuities, organizational capabilities and strategic commitments', *Industrial and Corporate Change*, **3**(3), pp. 655–686.
Rumelt, R. (1982). 'Diversification strategy and profitability', *Strategic Management Journal*, **3**(4), pp. 359–369.
Sahal, D. (1985). 'Technological guideposts and innovation avenues', *Research Policy*, **14**(2), p. 61.
Schumpeter, J. A. (1934). *The Theory of Economic Development*. Harvard University Press, Cambridge, MA.
Schumpeter, J. A. (1950). *Capitalism, Socialism and Democracy*. Harper Brothers, New York.
Teece, D. (1986). 'Profiting from technological innovation: Implications for integration, collaboration, licensing and public policy', *Research Policy*, **15**, pp. 285–305.
Teece, D., G. Pisano and A. Shuen (1997). 'Dynamic capabilities and strategic management', *Strategic Management Journal*, forthcoming.
Trajtenberg, M. (1990). *Economic Analysis of Product Innovation*. Harvard University Press, Cambridge, MA.
Tushman, L. M. and P. Anderson (1986). 'Technological discontinuities and organizational environments', *Administrative Science Quarterly*, **31**, pp. 439–465.
U.S. Congress Joint Committee on Printing (1970). *Electronic Composition Cost Comparison*. U.S. Government Printing Office, Washington, DC.

Wallis, L. (1993). *Typomania*. Severnside Printers, Upton-upon-Severn, U.K.

APPENDIX: DESCRIPTION OF TECHNOLOGICAL GENERATIONS

Hot metal typesetters

The original typesetter technology, 'hot metal typesetters', was based on Ottmar Mergenthaler's invention in 1886 and dominated the industry for over 70 years. These machines were entirely mechanical and worked as follows. An operator entered text on a typewriter-like keyboard and molds of letters (called matrices) fell from a case into a row in the appropriate sequence. When an entire row was finished, molten lead was injected into the molds to form a 'line of type'. After the metal cooled, the lines of type with raised letters, called slugs, were ejected from the machine, and the matrices were recirculated. Slugs were then arranged in a frame for letterpress printing.

Analog phototypesetters

In analog phototypesetters, the metal mold of a character was replaced with a photographic image of the character. These images were often stored on a spinning film disk that carried multiple fonts. Using varying electromechanical mechanisms, these images were exposed onto film, generally with a xenon flash. The film was then used to create a printing plate for high-volume printing.

Digital CRT typesetters

As the name implies, digital CRT machines took the analog images of characters and digitized them. Characters could then be stored in electronic format on a magnetic disk. Instead of a xenon flash, a CRT was used to write a character onto film by alternating 'on and off' positions using vertical strokes.

Laser imagesetters

Laser imagesetters differed significantly from the CRT machines in that, rather than exposing just characters onto film, these machines could integrate text and graphics on the same page. Dots were written onto the page using a raster scan. This meant that rather than write out a character at a time using vertical strokes, characters and images were created by painting a set of horizontal strokes across a page. In addition, rather than use a CRT to expose the film a laser was used.

Part III
Delivering Innovative Value Through Resource Allocation and Organization Activity

[10]

Absorptive Capacity: A New Perspective on Learning and Innovation

Wesley M. Cohen
Carnegie Mellon University
Daniel A. Levinthal
University of Pennsylvania

In this paper, we argue that the ability of a firm to recognize the value of new, external information, assimilate it, and apply it to commercial ends is critical to its innovative capabilities. We label this capability a firm's absorptive capacity and suggest that it is largely a function of the firm's level of prior related knowledge. The discussion focuses first on the cognitive basis for an individual's absorptive capacity including, in particular, prior related knowledge and diversity of background. We then characterize the factors that influence absorptive capacity at the organizational level, how an organization's absorptive capacity differs from that of its individual members, and the role of diversity of expertise within an organization. We argue that the development of absorptive capacity, and, in turn, innovative performance are history- or path-dependent and argue how lack of investment in an area of expertise early on may foreclose the future development of a technical capability in that area. We formulate a model of firm investment in research and development (R&D), in which R&D contributes to a firm's absorptive capacity, and test predictions relating a firm's investment in R&D to the knowledge underlying technical change within an industry. Discussion focuses on the implications of absorptive capacity for the analysis of other related innovative activities, including basic research, the adoption and diffusion of innovations, and decisions to participate in cooperative R&D ventures.•

INTRODUCTION

Outside sources of knowledge are often critical to the innovation process, whatever the organizational level at which the innovating unit is defined. While the example of Japan illustrates the point saliently at the national level (e.g., Westney and Sakakibara, 1986; Mansfield, 1988; Rosenberg and Steinmueller, 1988), it is also true of entire industries, as pointed out by Brock (1975) in the case of computers and by Peck (1962) in the case of aluminum. At the organizational level, March and Simon (1958: 188) suggested most innovations result from borrowing rather than invention. This observation is supported by extensive research on the sources of innovation (e.g., Mueller, 1962; Hamberg, 1963; Myers and Marquis, 1969; Johnston and Gibbons, 1975; von Hippel, 1988). Finally, the importance to innovative performance of information originating from other internal units in the firm, outside the formal innovating unit (i.e., the R&D lab), such as marketing and manufacturing, is well understood (e.g., Mansfield, 1968).

The ability to exploit external knowledge is thus a critical component of innovative capabilities. We argue that the ability to evaluate and utilize outside knowledge is largely a function of the level of prior related knowledge. At the most elemental level, this prior knowledge includes basic skills or even a shared language but may also include knowledge of the most recent scientific or technological developments in a given field. Thus, prior related knowledge confers an ability to recognize the value of new information, assimilate it, and apply it to commercial ends. These abilities collectively constitute what we call a firm's "absorptive capacity."

0001-8392/90/3501-0128/$1.00.

•

We appreciate the comments of Kathleen Carley, Robyn Dawes, Mark Fichman, Tom Finholt, Sara Kiesler, Richard Nelson, Linda Pike, and three anonymous referees. The representations and conclusions presented herein are those of the authors. They have not been adopted in whole or in part by the Federal Trade Commission, its Bureau of Economics, or any other entity within the commission. The FTC's Disclosure Avoidance Officer has certified that the data included in this paper do not identify individual company line-of-business data.

Absorptive Capacity

At the level of the firm—the innovating unit that is the focus here—absorptive capacity is generated in a variety of ways. Research shows that firms that conduct their own R&D are better able to use externally available information (e.g., Tilton, 1971; Allen, 1977; Mowery, 1983). This implies that absorptive capacity may be created as a byproduct of a firm's R&D investment. Other work suggests that absorptive capacity may also be developed as a byproduct of a firm's manufacturing operations. Abernathy (1978) and Rosenberg (1982) have noted that through direct involvement in manufacturing, a firm is better able to recognize and exploit new information relevant to a particular product market. Production experience provides the firm with the background necessary both to recognize the value of and implement methods to reorganize or automate particular manufacturing processes. Firms also invest in absorptive capacity directly, as when they send personnel for advanced technical training. The concept of absorptive capacity can best be developed through an examination of the cognitive structures that underlie learning.

Cognitive Structures

The premise of the notion of absorptive capacity is that the organization needs prior related knowledge to assimilate and use new knowledge. Studies in the area of cognitive and behavioral sciences at the individual level both justify and enrich this observation. Research on memory development suggests that accumulated prior knowledge increases both the ability to put new knowledge into memory, what we would refer to as the acquisition of knowledge, and the ability to recall and use it. With respect to the acquisition of knowledge, Bower and Hilgard (1981: 424) suggested that memory development is self-reinforcing in that the more objects, patterns and concepts that are stored in memory, the more readily is new information about these constructs acquired and the more facile is the individual in using them in new settings.

Some psychologists suggest that prior knowledge enhances learning because memory—or the storage of knowledge—is developed by associative learning in which events are recorded into memory by establishing linkages with pre-existing concepts. Thus, Bower and Hilgard (1981) suggested that the breadth of categories into which prior knowledge is organized, the differentiation of those categories, and the linkages across them permit individuals to make sense of and, in turn, acquire new knowledge. In the context of learning a language, Lindsay and Norman (1977: 517) suggested the problem in learning words is not a result of lack of exposure to them but that "to understand complex phrases, much more is needed than exposure to the words: a large body of knowledge must first be accumulated. After all, a word is simply a label for a set of structures within the memory system, so the structures must exist before the word can be considered learned." Lindsay and Norman further suggested that knowledge may be nominally acquired but not well utilized subsequently because the individual did not already possess the appropriate contextual knowledge necessary to make the new knowledge fully intelligible.

The notion that prior knowledge facilitates the learning of new related knowledge can be extended to include the case in which the knowledge in question may itself be a set of learning skills. There may be a transfer of learning skills across bodies of knowledge that are organized and expressed in similar ways. As a consequence, experience or performance on one learning task may influence and improve performance on some subsequent learning task (Ellis, 1965). This progressive improvement in the performance of learning tasks is a form of knowledge transfer that has been referred to as "learning to learn" (Ellis, 1965; Estes, 1970). Estes (1970: 16), however, suggested that the term "learning to learn" is a misnomer in that prior experience with a learning task does not necessarily improve performance because an individual knows how to learn (i.e., form new associations) better, but that an individual may simply have accumulated more prior knowledge so that he or she needs to learn less to attain a given level of performance. Notwithstanding what it is about prior learning experience that may affect subsequent performance, both explanations of the relationship between early learning and subsequent performance emphasize the importance of prior knowledge for learning.

The effect of prior learning experience on subsequent learning tasks can be observed in a variety of tasks. For instance, Ellis (1965: 4) suggested that "students who have thoroughly mastered the principles of algebra find it easier to grasp advanced work in mathematics such as calculus." Further illustration is provided by Anderson, Farrell, and Sauers (1984), who compared students learning LISP as a first programming language with students learning LISP after having learned Pascal. The Pascal students learned LISP much more effectively, in part because they better appreciated the semantics of various programming concepts.

The literature also suggests that problem-solving skills develop similarly. In this case, problem-solving methods and heuristics typically constitute the prior knowledge that permits individuals to acquire related problem-solving capabilities. In their work on the development of computer programming skills, Pirolli and Anderson (1985) found that almost all students developed new programs by analogy-to-example programs and that their success was determined by how well they understood why these examples worked.

We argue that problem solving and learning capabilities are so similar that there is little reason to differentiate their modes of development, although exactly what is learned may differ: learning capabilities involve the development of the capacity to assimilate existing knowledge, while problem-solving skills represent a capacity to create new knowledge. Supporting the point that there is little difference between the two, Bradshaw, Langley, and Simon (1983) and Simon (1985) suggested that the sort of necessary preconditions for successful learning that we have identified do not differ from the preconditions required for problem solving and, in turn, for the creative process. Moreover, they argued that the processes themselves do not differ much. The prior possession of relevant knowledge and skill is what gives rise to creativity, permitting the sorts of associations and linkages that may have never been considered before. Likewise, Ellis (1965: 35) suggested that Harlow's (1959) findings on the development of learning sets provide a possible explanation for the behavioral phenomenon of "insight" that typically refers to the rapid solution of a problem. Thus, the psychology literature suggests that creative capacity and what we call absorptive capacity are quite similar.

To develop an effective absorptive capacity, whether it be for general knowledge or problem-solving or learning skills, it is insufficient merely to expose an individual briefly to the relevant prior knowledge. Intensity of effort is critical. With regard to storing knowledge in memory, Lindsay and Norman (1977: 355) noted that the more deeply the material is processed—the more effort used, the more processing makes use of associations between the items to be learned and knowledge already in the memory—the better will be the later retrieval of the item. Similarly, learning-set theory (Harlow, 1949, 1959) implies that important aspects of learning how to solve problems are built up over many practice trials on related problems. Indeed, Harlow (1959) suggested that if practice with a particular type of problem is discontinued before it is reliably learned, then little transfer will occur to the next series of problems. Therefore, he concluded that considerable time and effort should be spent on early problems before moving on to more complex problems.

Two related ideas are implicit in the notion that the ability to assimilate information is a function of the richness of the preexisting knowledge structure: learning is cumulative, and learning performance is greatest when the object of learning is related to what is already known. As a result, learning is more difficult in novel domains, and, more generally, an individual's expertise—what he or she knows well—will change only incrementally. The above discussion also suggests that diversity of knowledge plays an important role. In a setting in which there is uncertainty about the knowledge domains from which potentially useful information may emerge, a diverse background provides a more robust basis for learning because it increases the prospect that incoming information will relate to what is already known. In addition to strengthening assimilative powers, knowledge diversity also facilitates the innovative process by enabling the individual to make novel associations and linkages.

From Individual to Organizational Absorptive Capacity

An organization's absorptive capacity will depend on the absorptive capacities of its individual members. To this extent, the development of an organization's absorptive capacity will build on prior investment in the development of its constituent, individual absorptive capacities, and, like individuals' absorptive capacities, organizational absorptive capacity will tend to develop cumulatively. A firm's absorptive capacity is not, however, simply the sum of the absorptive capacities of its employees, and it is therefore useful to consider what aspects of absorptive capacity are distinctly organizational. Absorptive capacity refers not only to the acquisition or assimilation of information by an organization but also to the organization's ability to exploit it. Therefore, an organization's absorptive capacity does not simply depend on the organization's direct interface with the external environment. It also depends on transfers of knowledge across and within subunits that may be quite removed from the original point of entry. Thus, to understand the sources of a firm's absorptive capacity, we focus on the structure of communication between the external environment and the organization, as well as among the subunits of the organization, and also on the character and distribution of expertise within the organization.

Communication systems may rely on specialized actors to transfer information from the environment or may involve less structured patterns. The problem of designing communication structures cannot be disentangled from the distribution of expertise in the organization. The firm's absorptive capacity depends on the individuals who stand at the interface of either the firm and the external environment or at the interface between subunits within the firm. That interface function may be diffused across individuals or be quite centralized. When the expertise of most individuals within the organization differs considerably from that of external actors who can provide useful information, some members of the group are likely to assume relatively centralized "gatekeeping" or "boundary-spanning" roles (Allen, 1977; Tushman, 1977). For technical information that is difficult for internal staff to assimilate, a gatekeeper both monitors the environment and translates the technical information into a form understandable to the research group. In contrast, if external information is closely related to ongoing activity, then external information is readily assimilated and gatekeepers or boundary-spanners are not so necessary for translating information. Even in this setting, however, gatekeepers may emerge to the extent that such role specialization relieves others from having to monitor the environment.

A difficulty may emerge under conditions of rapid and uncertain technical change, however, when this interface function is centralized. When information flows are somewhat random and it is not clear where in the firm or subunit a piece of outside knowledge is best applied, a centralized gatekeeper may not provide an effective link to the environment. Under such circumstances, it is best for the organization to expose a fairly broad range of prospective "receptors" to the environment. Such an organization would exhibit the organic structure of Burns and Stalker (1961: 6), which is more adaptable "when problems and requirements for action arise which cannot be broken down and distributed among specialist roles within a clearly defined hierarchy."

Even when a gatekeeper is important, his or her individual absorptive capacity does not constitute the absorptive capacity of his or her unit within the firm. The ease or difficulty of the internal communication process and, in turn, the level of organizational absorptive capacity are not only a function of the gatekeeper's capabilities but also of the expertise of those individuals to whom the gatekeeper is transmitting the information. Therefore, relying on a small set of technological gatekeepers may not be sufficient; the group as a whole must have some level of relevant background knowledge, and when knowledge structures are highly differentiated, the requisite level of background may be rather high.

The background knowledge required by the group as a whole for effective communication with the gatekeeper highlights the more general point that shared knowledge and expertise is essential for communication. At the most basic level, the relevant knowledge that permits effective communication both within and across subunits consists of shared language and symbols (Dearborn and Simon, 1958; Katz and Kahn, 1966; Allen and Cohen, 1969; Tushman, 1978; Zenger and Lawrence, 1989). With regard to the absorptive capacity of the firm as a whole, there may, however, be a trade-off in the efficiency of internal communication against the ability of the subunit to assimilate and exploit information originating from other subunits or the environment. This can be seen as a trade-off between inward-looking versus outward-looking absorptive capacities. While both of these components are necessary for effective organizational learning, excessive dominance by one or the other will be dysfunctional. If all actors in the organization share the same specialized language, they will be effective in communicating with one another, but they may not be able to tap into diverse external knowledge sources. In the limit, an internal language, coding scheme, or, more generally, any particular body of expertise could become sufficiently overlapping and specialized that it impedes the incorporation of outside knowledge and results in the pathology of the not-invented-here (NIH) syndrome. This may explain Katz and Allen's (1982) findings that the level of external communication and communication with other project groups declines with project-group tenure.

This trade-off between outward- and inward-looking components of absorptive capacity focuses our attention on how the relationship between knowledge sharing and knowledge diversity across individuals affects the development of organizational absorptive capacity. While some overlap of knowledge across individuals is necessary for internal communication, there are benefits to diversity of knowledge structures across individuals that parallel the benefits to diversity of knowledge within individuals. As Simon (1985) pointed out, diverse knowledge structures coexisting in the same mind elicit the sort of learning and problem solving that yields innovation. Assuming a sufficient level of knowledge overlap to ensure effective communication, interactions across individuals who each possess diverse and different knowledge structures will augment the organization's capacity for making novel linkages and associations—innovating—beyond what any one individual can achieve. Utterback (1971), summarizing research on task performance and innovation, noted that diversity in the work setting "stimulates the generation of new ideas." Thus, as with Nelson and Winter's (1982) view of organizational capabilities, an organization's absorptive capacity is not resident in any single individual but depends on the links across a mosaic of individual capabilities.

Beyond diverse knowledge structures, the sort of knowledge that individuals should possess to enhance organizational absorptive capacity is also important. Critical knowledge does not simply include substantive, technical knowledge; it also includes awareness of where useful complementary expertise resides within and outside the organization. This sort of knowledge can be knowledge of who knows what, who can help with what problem, or who can exploit new information. With regard to external relationships, von Hippel (1988) has shown the importance for innovation of close relationships with both buyers and suppliers. To the extent that an organization develops a broad and active network of internal and external relationships, individuals' awareness of others' capabilities and knowledge will be strengthened. As a result, individual absorptive capacities are leveraged all the more, and the organization's absorptive capacity is strengthened.

The observation that the ideal knowledge structure for an organizational subunit should reflect only partially overlapping knowledge complemented by nonoverlapping diverse knowledge suggests an organizational trade-off between diversity and commonality of knowledge across individuals. While common knowledge improves communication, commonality should not be carried so far that diversity across individuals is substantially diminished. Likewise, division of labor promoting gains from specialization should not be pushed so far that communication is undermined. The difficulties posed by excessive specialization suggest some liabilities of pursuing production efficiencies via learning by doing under conditions of rapid technical change in which absorptive capacity is important. In learning by doing, the firm becomes more practiced and hence more capable at activities in which it is already engaged. Learning by doing does not contribute to the diversity that is critical to learning about or creating something that is relatively new. Moreover, the notion of "remembering by doing" (Nelson and Winter, 1982) suggests that the focus on one class of activity entailed by learning by doing may effectively diminish the diversity of background that an individual or organization may have at one time possessed and, consequently, undercut organizational absorptive capacity and innovative performance.

It has become generally accepted that complementary functions within the organization ought to be tightly intermeshed, recognizing that some amount of redundancy in expertise may be desirable to create what can be called cross-function absorptive capacities. Cross-function interfaces that affect organizational absorptive capacity and innovative performance include, for example, the relationships between corporate and divisional R&D labs or, more generally, the relationships among the R&D, design, manufacturing, and marketing functions (e.g., Mansfield, 1968: 86–88). Close linkages between design and manufacturing are often credited for the relative success of Japanese firms in moving products rapidly from the design stage through development and manufacturing (Westney and Sakakibara, 1986). Clark and Fujimoto (1987) argued that overlapping product development cycles facilitate communication and coordination across organizational subunits. They found that the speed of product development is strongly influenced by the links between problem-solving cycles and that successful linking requires "direct personal contacts across functions, liaison roles at each unit, cross-functional task forces, cross-functional project teams, and a system of 'product manager as integrator' " (Clark and Fujimoto, 1987: 24). In contrast, a process in which one unit simply hands off the design to another unit is likely to suffer greater difficulties.

Some management practices also appear to reflect the belief that an excessive degree of overlap in functions may reduce the firm's absorptive capacity and that diversity of backgrounds is useful. The Japanese practice of rotating their R&D personnel through marketing and manufacturing operations, for example, while creating knowledge overlap, also enhances the diversity of background of their personnel. Often involving the assignment of technical personnel to other functions for several years, this practice also suggests that some intensity of experience in each of the complementary knowledge domains is necessary to put an effective absorptive capacity in place; breadth of knowledge cannot be superficial to be effective.

The discussion thus far has focused on internal mechanisms that influence the organization's absorptive capacity. A question remains as to whether absorptive capacity needs to be internally developed or to what extent a firm may simply buy it via, for example, hiring new personnel, contracting for consulting services, or even through corporate acquisitions. We suggest that the effectiveness of such options is somewhat limited when the absorptive capacity in question is to be integrated with the firm's other activities. A critical component of the requisite absorptive capacity for certain types of information, such as those associated with product and process innovation, is often firm-specific and therefore cannot be bought and quickly integrated into the firm. This is reflected in Lee and Allen's (1982) findings that considerable time lags are associated with the integration of new technical staff, particularly those concerned with process and product development. To integrate certain classes of complex and sophisticated technological knowledge successfully into the firm's activities, the firm requires an existing internal staff of technologists and scientists who are both competent in their fields and are familiar with the firm's idiosyncratic needs, organizational procedures, routines, complementary capabilities, and extramural relationships. As implied by the discussion above, such diversity of knowledge structures must coexist to some degree in the same minds. Moreover, as Nelson and Winter's (1982) analysis suggests, much of the detailed knowledge of organizational routines and objectives that permit a firm and its R&D labs to function is tacit. As a consequence, such critical complementary knowledge is acquired only through experience within the firm. Illustrating our general argument, Vyssotsky (1977), justifying the placement of Bell Labs within AT&T, argued: "For research and development to yield effective results for Bell System, it has to be done by . . . creative people who understand as much as they possibly can about the technical state of the art, and about Bell System and what System's problems are. The R&D people must be free to think up new approaches, and they must also be closely coupled to the problems and challenges where innovation is needed. This combination, if one is lucky, will result in insights which help the Bell System. That's why we have Bell Labs in Bell System, instead of having all our R&D done by outside organizations."

Path Dependence and Absorptive Capacity

Our discussion of the character of absorptive capacity and its role in assimilating and exploiting knowledge suggests a simple generalization that applies at both the individual and organizational levels: prior knowledge permits the assimilation and exploitation of new knowledge. Some portion of that prior knowledge should be very closely related to the new knowledge to facilitate assimilation, and some fraction of that knowledge must be fairly diverse, although still related, to permit effective, creative utilization of the new knowledge. This simple notion that prior knowledge underlies absorptive capacity has important implications for the development of absorptive capacity over time and, in turn, the innovative performance of organizations. The basic role of prior knowledge suggests two features of absorptive capacity that will affect innovative performance in an evolving, uncertain environment (Cohen and Levinthal, 1989b). Accumulating absorptive capacity in one period will permit its more efficient accumulation in the next. By having already developed some absorptive capacity in a particular area, a firm may more readily accumulate what additional knowledge it needs in the subsequent periods in order to exploit any critical external knowledge that may become available. Second, the possession of related expertise will permit the firm to better understand and therefore evaluate the import of intermediate technological advances that provide signals as to the eventual merit of a new technological development. Thus, in an uncertain environment, absorptive capacity affects expectation formation, permitting the firm to predict more accurately the nature and commercial potential of technological advances. These revised expectations, in turn, condition the incentive to invest in absorptive capacity subsequently. These two features of absorptive capacity—cumulativeness and its effect on expectation formation—imply that its development is domain-specific and is path- or history-dependent.

The cumulativeness of absorptive capacity and its effect on expectation formation suggest an extreme case of path dependence in which once a firm ceases investing in its absorptive capacity in a quickly moving field, it may never assimilate and exploit new information in that field, regardless of the value of that information. There are two reasons for the emergence of this condition, which we term "lockout" (Cohen and Levinthal, 1989b). First, if the firm does not develop its absorptive capacity in some initial period, then its beliefs about the technological opportunities present in a given field will tend not to change over time because the firm may not be aware of the significance of signals that would otherwise revise its expectations. As a result, the firm does not invest in absorptive capacity and, when new opportunities subsequently emerge, the firm may not appreciate them. Compounding this effect, to the extent that prior knowledge facilitates the subsequent development of absorptive capacity, the lack of early investment in absorptive capacity makes it more costly to develop a given level of it in a subsequent period. Consequently, a low initial investment in absorptive capacity diminishes the attractiveness of investing in subsequent periods even if the firm becomes aware of technological opportunities.[1] This possibility of firms being "locked-out" of subsequent technological developments has recently become a matter of concern with respect to industrial policy. For instance, Reich (1987: 64) declaims Monsanto's exit from "float-zone" silicon manufacturing because he believes that the decision may be an irreversible exit from a technology, in that ". . . each new generation of technology

1
A similar result emerges from models of adaptive learning. Levitt and March (1988: 322) noted that "a competency trap can occur when favorable performance with an inferior procedure leads an organization to accumulate more experience with it, thus keeping experience with a superior procedure inadequate to make it rewarding to use."

builds on that which came before, once off the technological escalator it's difficult to get back on."

Thus, the cumulative quality of absorptive capacity and its role in conditioning the updating of expectations are forces that tend to confine firms to operating in a particular technological domain. If firms do not invest in developing absorptive capacity in a particular area of expertise early on, it may not be in their interest to develop that capacity subsequently, even after major advances in the field. Thus, the pattern of inertia that Nelson and Winter (1982) highlighted as a central feature of firm behavior may emerge as an implication of rational behavior in a model in which absorptive capacity is cumulative and contributes to expectation formation. The not-invented-here syndrome, in which firms resist accepting innovative ideas from the environment, may also at times reflect what we call lockout. Such ideas may be too distant from the firm's existing knowledge base—its absorptive capacity—to be either appreciated or accessed. In this particular setting, NIH may be pathological behavior only in retrospect. The firm need not have acted irrationally in the development of the capabilities that yields the NIH syndrome as its apparent outcome.

A form of self-reinforcing behavior similar to lockout may also result from the influence of absorptive capacity on organizations' goals or aspiration levels. This argument builds on the behavioral view of organizational innovation that has been molded in large part by the work of March and Simon (1958). In March and Simon's framework, innovative activity is instigated due to a failure to reach some aspiration level. Departing from their model, we suggest that a firm's aspiration level in a technologically progressive environment is not simply determined by past performance or the performance of reference organizations. It also depends on the firm's absorptive capacity. The greater the organization's expertise and associated absorptive capacity, the more sensitive it is likely to be to emerging technological opportunities and the more likely its aspiration level will be defined in terms of the opportunities present in the technical environment rather than strictly in terms of performance measures. Thus, organizations with higher levels of absorptive capacity will tend to be more proactive, exploiting opportunities present in the environment, independent of current performance. Alternatively, organizations that have a modest absorptive capacity will tend to be reactive, searching for new alternatives in response to failure on some performance criterion that is not defined in terms of technical change per se (e.g., profitability, market share, etc.).

A systematic and enduring neglect of technical opportunities may result from the effect of absorptive capacity on the organization's aspiration level when innovative activity (e.g., R&D) contributes to absorptive capacity, which is often the case in technologically progressive environments. The reason is that the firm's aspiration level then depends on the very innovative activity that is triggered by a failure to meet the aspiration level itself. If the firm engages in little innovative activity, and is therefore relatively insensitive to the opportunities in the external environment, it will have a low aspiration level with regard to the exploitation of new technology, which in turn implies that it will continue to devote little effort to innovation. This creates a self-reinforcing cycle. Likewise, if an organization has a high aspiration level, influenced by externally generated technical opportunities, it will conduct more innovative activity and thereby increase its awareness of outside opportunities. Consequently, its aspiration level will remain high. This argument implies that reactive and proactive modes of firm behavior should remain rather stable over time. Thus, some organizations (like Hewlett-Packard and Sony) have the requisite technical knowledge to respond proactively to the opportunities present in the environment. These firms do not wait for failure on some performance dimension but aggressively seek out new opportunities to exploit and develop their technological capabilities.[2]

The concept of dynamically self-reinforcing behavior that may lead to the neglect of new technological developments provides some insight into the difficulties firms face when the technological basis of an industry changes—what Schumpeter (1942) called "the process of creative destruction." For instance, the change from electromechanical devices to electronic ones in the calculator industry resulted in the exit of a number of firms and a radical change in the market structure (Majumdar, 1982). This is an example of what Tushman and Anderson (1986) termed competence-destroying technical change. A firm without a prior technological base in a particular field may not be able to acquire one readily if absorptive capacity is cumulative. In addition, a firm may be blind to new developments in fields in which it is not investing if its updating capability is low. Accordingly, our argument implies that firms may not realize that they should be developing their absorptive capacity due to an irony associated with its valuation: the firm needs to have some absorptive capacity already to value it appropriately.

Absorptive Capacity and R&D Investment

The prior discussion does not address the question of whether we can empirically evaluate the importance of absorptive capacity for innovation. There is a key insight that permits empirical tests of the implications of absorptive capacity for innovative activity. Since technical change within an industry—typically incremental in character (Rosenberg and Steinmueller, 1988)—is often closely related to a firm's ongoing R&D activity, a firm's ability to exploit external knowledge is often generated as a byproduct of its R&D. We may therefore consider a firm's R&D as satisfying two functions: we assume that R&D not only generates new knowledge but also contributes to the firm's absorptive capacity.[3] If absorptive capacity is important, and R&D contributes to it, then whatever conditions the firm's incentives to learn (i.e., to build absorptive capacity) should also influence R&D spending. We may therefore consider the responsiveness of R&D activity to learning incentives as an indication of the empirical importance of absorptive capacity. The empirical challenge then is to understand the impact of the characteristics of the learning environment on R&D spending.

We construct a simple static model of firm R&D intensity, which is defined as R&D divided by sales. Normalization of R&D by firm sales controls for the effect of firm size, which

2

This argument that such reactive and proactive behavior may coexist in an industry over the long run assumes that there is slack in the selection environment and that technologically progressive behavior is not essential to survival. One can, alternatively, identify a number of industries, such as semiconductors, in which it appears that only firms that aggressively exploit technical opportunities survive.

3

We refer readers interested in the details of the theoretical and subsequent empirical analysis and results to Cohen and Levinthal (1989a), from which the following discussion is drawn.

affects the return per unit of R&D effort. This model is developed in the broader context of what applied economists have come to believe to be the three classes of industry-level determinants of R&D intensity: demand, appropriability, and technological opportunity conditions (Cohen and Levin, 1989). Demand is often characterized by the level of sales and the price elasticity of demand. The latter indicates the degree to which a firm's revenue will increase due to a reduction in price. For example, in the case of a process innovation that reduces the cost of production and, in turn, the product price, the price elasticity of demand reflects the associated change in total revenue that influences the economic return to innovative effort. Appropriability conditions refer to the degree to which firms capture the profits associated with their innovative activity and are often considered to reflect the degree to which valuable knowledge spills out into the public domain. The emphasis here is on valuable knowledge, because if a competitor's knowledge spills out but the competitor has already exploited a first-mover advantage in the marketplace, this knowledge is no longer valuable to the firm and does not constitute a spillover by our definition. The level of spillovers, in turn, depends on the strength of patents within an industry, the efficacy of secrecy, and/or first-mover advantages. Technological opportunity represents how costly it is for the firm to achieve some normalized unit of technical advance in a given industry. As typically conceived, there are two dimensions of technological opportunity (Cohen and Levin, 1989). The first, incorporated in our model, refers simply to the quantity of extraindustry technological knowledge, such as that originating from government or university labs, that effectively complements and therefore leverages the firm's own knowledge output. The second dimension of technological opportunity is the degree to which a unit of new knowledge improves the technological performance of the firm's manufacturing processes or products and, in turn, the firm's profits. For example, given the vitality of the underlying science and technology, an advance in knowledge promises to yield much larger product-performance payoffs in the semiconductor industry than in steel.[4]

The basic model of how absorptive capacity affects the determination of R&D expenditures is represented diagramatically in Figure 1. We postulate that learning incentives will have a direct effect on R&D spending. We also suggest that where the effect of other determinants, such as technological opportunity and appropriability, depend on the firm's or rivals' assimilation of knowledge, absorptive capacity—and therefore learning incentives—will mediate those effects. Finally, we suggest that the effect of appropriability conditions (i.e., spillovers) will be conditioned by competitor interdependence. In this context, we define interdependence as the extent to which a rival's technical advances diminish the firm's profits.

There are two factors that will affect a firm's incentives to learn, and, therefore, its incentives to invest in absorptive capacity via its R&D expenditures. First, there is the quantity of knowledge to be assimilated and exploited: the more there is, the greater the incentive. Second, there is the difficulty (or, conversely, the ease) of learning. Some types of information are more difficult to assimilate and use than others. We inter-

4 This second dimension is incorporated in the model developed in Cohen and Levinthal (1989a). We do not incorporate this second dimension in the present model because all the qualitative theoretical and empirical results associated with this second dimension of technological opportunity are the same as those associated with the first considered here.

Figure 1. Model of absorptive capacity and R&D incentives.

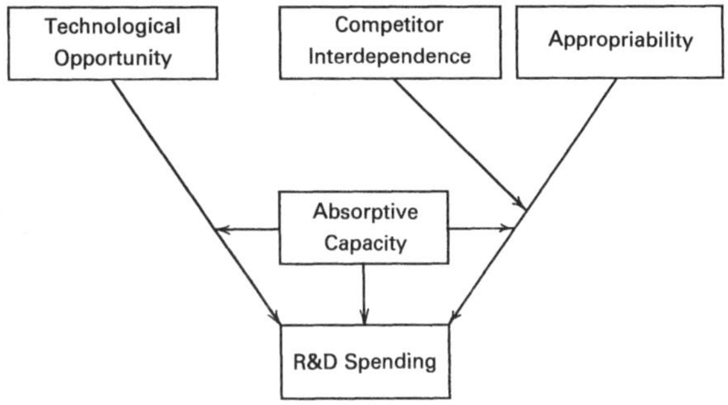

pret this to mean that per unit of knowledge, the cost of its absorption may vary depending on the characteristics of that knowledge. As learning is more difficult, more prior knowledge has to have been accumulated via R&D for effective learning to occur. As a result, this is a more costly learning environment. In such a setting, R&D is more important to building absorptive capacity and the more R&D effort the firm will need to have expended to achieve some level of absorptive capacity. Thus, for a given level of a firm's own R&D, the level of absorptive capacity is diminished in environments in which it is more difficult to learn. In addition, we are suggesting that a more difficult learning environment increases the marginal effect of R&D on absorptive capacity. In contrast, in environments in which learning is less demanding, a firm's own R&D has little impact on its absorptive capacity. In the extreme case in which external knowledge can be assimilated without any specialized expertise, a firm's own R&D would have no effect on its absorptive capacity.

We have argued that the ease of learning is in turn determined by the characteristics of the underlying scientific and technological knowledge. Although it is difficult to specify a priori all the relevant characteristics of knowledge affecting the ease of learning, they would include the complexity of the knowledge to be assimilated and the degree to which the outside knowledge is targeted to the needs and concerns of the firm. When outside knowledge is less targeted to the firm's particular needs and concerns, a firm's own R&D becomes more important in permitting it to recognize the value of the knowledge, assimilate, and exploit it. Sources that produce less targeted knowledge would include university labs involved in basic research, while more targeted knowledge may be generated by contract research labs, or input suppliers. In addition, the degree to which a field is cumulative, or the field's pace of advance, should also affect how critical R&D is to the development of absorptive capacity. The more that findings in a field build on prior findings, the more necessary is an understanding of prior research to the assimilation of subsequent findings. The pace of advance of a field affects the importance of R&D to developing absorptive capacity because the faster the pace of knowledge generation, the larger the staff required to keep abreast of new developments. Finally, following Nelson and Winter (1982), the less explicit and codified the relevant knowledge, the more difficult it is to assimilate.

To structure the analysis, we assumed that firms purposefully invest in R&D to generate profit and take into account R&D's dual role in both directly generating new knowledge and contributing to absorptive capacity. Knowledge is assumed to be useful to the firm in that increments to a firm's own knowledge increase the firm's profits while increments to rivals' knowledge diminish them. We posit a simple model of the generation of a firm's technological knowledge that takes into account the major sources of technological knowledge utilized by a firm: the firm's own R&D knowledge that originates with its competitors' R&D, spillovers, and that which originates outside the industry. Figure 2 provides a stylized representation of this model in which, first, the firm generates new knowledge directly through its own R&D, and second, extramural knowledge, drawn from competitors as well as extraindustry sources such as government and university labs, also contribute to the firm's knowledge. A central feature of the model is that the firm's absorptive capacity determines the extent to which this extramural knowledge is utilized, and this absorptive capacity itself depends on the firm's own R&D. Because of this mediating function, absorptive capacity influences the effects of appropriability and technological opportunity conditions on R&D spending. Thus, the effects of appropriability and technological opportunity are not independent of R&D itself.

Figure 2. Model of sources of a firm's technical knowledge.

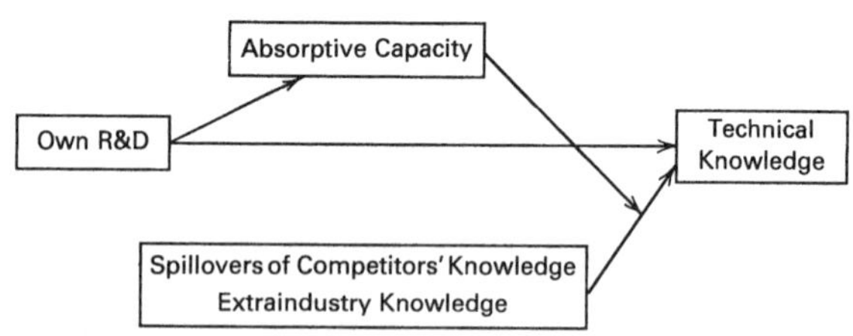

A key assumption in the model is that exploitation of competitors' research findings is realized through the interaction of the firm's absorptive capacity with competitors' spillovers. This interaction signifies that a firm is unable to assimilate externally available knowledge passively. Rather, to utilize the accessible R&D output of its competitors, the firm invests in its absorptive capacity by conducting R&D. Figure 2 also illustrates that, like its assimilation of competitors' R&D output, a firm's assimilation of extraindustry knowledge—the dimension of technological opportunity considered here—is constrained by its absorptive capacity. According to our model, therefore, the factors that affect learning incentives (i.e., the ease of learning and the quantity of available knowledge) influence the effects of appropriability and technological opportunity conditions on R&D.

Direct effect of ease of learning. As shown formally in Cohen and Levinthal (1989a), this model implies that as the ease of learning diminishes, learning becomes more dependent on a firm's own R&D, and R&D spending increases because of two effects. First, the marginal impact of R&D on absorptive capacity is greater in more difficult learning environments. As the learning environment becomes more difficult, however, there is a second, more subtle effect. Since, ceteris paribus, a more difficult learning environment lowers firms' absorptive capacities, R&D activity becomes more of a private good in the sense that competitors are now less able to tap into the firm's R&D findings that spill out.

Technological opportunity. We predict that an increase in technological opportunity—the amount of available relevant external technical knowledge—will elicit more R&D in more difficult learning environments. Greater technological opportunity signifies greater amounts of external information, which increase the firm's incentive to build absorptive capacity, and a more challenging learning environment increases the level of R&D necessary to build absorptive capacity.

Appropriability. We predict that spillovers will provide, in part, a positive incentive to conduct R&D due to the interaction of spillovers with an endogenous absorptive capacity. Traditionally, spillovers have been considered only a deterrent to R&D activity (e.g., Nelson, 1959; Arrow, 1962; Spence, 1984). In the standard view, a firm's incentive to invest in R&D is diminished to the extent that any findings from such activities are exploited by competitors and thereby diminish the innovator's own profits. In our framework, however, this negative appropriability incentive associated with spillovers is counterbalanced by a positive absorptive-capacity-building incentive. The more of its competitors' spillovers there are out there, the more incentive the firm has to invest in its own R&D, which permits it to exploit those spillovers.

We have shown elsewhere (Cohen and Levinthal, 1989a) that when this absorption incentive is large, as when learning is difficult, spillovers may actually encourage R&D. The relative magnitude of the absorption incentive is greater when firms within an industry are less interdependent in the sense that rivals' technical advances have less of an effect on the firm's own profits. With less interdependence, the degree to which rivals gain from the firm's R&D spillovers at the firm's expense diminishes relative to the benefit of being able to exploit the rivals' spillovers. Either a more competitive market structure or a higher price elasticity of demand for the firm's product can diminish interdependence in an industry.

METHODS

Data and Measures

To test the predictions of our framework for R&D activity, we used cross-sectional survey data on technological opportunity and appropriability conditions in the American manufacturing sector collected from R&D lab managers by Levin et al. (1983, 1987), and the Federal Trade Commission's Line of Business Program data on business unit sales, transfers, and R&D expenditures. The dependent variable, R&D intensity, was defined as company-financed business-unit research and development expenditures, expressed as a percentage of business unit sales and transfers over the period 1975 through 1977. The data on interindustry differences in technological opportunity and appropriability are industry (line of business) mean scores computed as an average over all respondents within a given industry. The sample consists of 1,719 business units representing 318 firms in 151 lines of business.

The data pose two estimation issues. First, some 24 percent of the firms performed no R&D in at least one year. If the independent variables reflect both the probability of conducting R&D, as well as the amount of R&D spending, then a Tobit analysis would be appropriate. Alternatively, a firm may require some initial level of absorptive capacity before it is influenced by the characteristics of the learning environment. In this case, the variables reflecting the ease of learning only affect the amount of R&D conducted by firms engaging in R&D activity and not the probability of engaging in R&D activity. In light of the uncertainty over the appropriate estimation technique, we explored the robustness of the results by analyzing a Tobit and an OLS (or GLS) specification. The second estimation issue is the presence of heteroscedasticity. We found the assumption of homoscedasticity to be violated, with the logarithm of the error variance being a linear function of the exogenous variables and the number of respondents to Levin et al.'s (1983, 1987) survey. Unless otherwise noted, the results we report in this section reflect robust effects that hold across three different estimation methods, including ordinary least squares (OLS), generalized least squares (GLS) in which we adjust for heteroscedasticity, and Tobit, which was used when we included the observations for which R&D expenditures were zero.

We tested our predictions in the context of an empirical model of business unit R&D intensity in which technological opportunity, appropriability, and demand conditions are considered as the principal industry-level determinants of firms' R&D spending. While data constraints do not permit observation of the direct effect of the ease of learning or its determinants on firms' R&D spending, we were able to examine how these variables condition the influence on R&D of technological opportunity and appropriability conditions.

Technological opportunity was assessed with variables measuring the "relevance" or "importance" for technological progress in each line of business of what are considered to be two critical sources of technological opportunity—the science base of the industry and extraindustry sources of knowledge (Cohen and Levin, 1989). These measures are drawn from Levin et al.'s survey, in which R&D managers indicated on a 7-point Likert scale the relevance of eleven basic and applied fields of science and the importance of external sources of knowledge to technological progress in a line of business. The basic fields of science include biology, chemistry, mathematics, and physics, and the applied fields of science include agricultural science, applied math/operations research, computer science, geology, materials science, medical science, and metallurgy.[5] The five extraindustry

5

Although geology was classed as a basic science by Levin et al., we classed it as an applied science because of its inductive methodology and intensive use by firms in the extractive sector.

sources of knowledge considered here included equipment suppliers (EQUIPTECH), materials suppliers (MATERIALTECH), downstream users of the industry's products (USERTECH), government laboratories and agencies (GOVTECH), and universities (UNIVTECH). We interpreted the measures of the relevance or importance of each field or knowledge source to index the relative quantity of knowledge generated by that field or source that is potentially useful. We then distinguished across the eleven scientific fields and the five extraindustry knowledge source variables on the basis of the ease of learning associated with each. We suggested above that one important determinant of the ease of learning is the degree to which outside knowledge is targeted to a firm's needs and concerns. One can readily distinguish among both the eleven fields and the five extraindustry knowledge sources on that basis. The knowledge associated with the basic sciences is typically less targeted than that associated with the applied sciences. We also distinguished among the extraindustry knowledge sources on the same basis. A priori, we ranked university labs, government labs, materials suppliers, and equipment suppliers as providing increasingly more targeted knowledge to firms. We did not rank the relative effect of knowledge originating from users because, as suggested by von Hippel (1978), users will often provide a product idea to potential suppliers, but the informativeness of the "solution concept" is quite variable. Therefore, the targeted quality of the information is variable as well.

To represent intraindustry spillovers of R&D, we employed measures from Levin et al.'s survey of the effectiveness of six mechanisms used by firms to capture and protect the competitive advantages of new processes and new products: patents to prevent duplication, patents to secure royalty income, secrecy, lead time, moving quickly down the learning curve, and complementary sales and service efforts. We employed the maximum value of the effectiveness scores attained by these mechanisms as our measure of appropriability or spillovers, and label this variable APPROPRIABILITY; a high level of APPROPRIABILITY reflects a low level of spillovers.

In our theory, we predicted an interaction effect by which, as the ease of learning diminishes, or firms become less interdependent, the effect of spillovers on R&D spending should become more positive (or less negative). In the absence of any direct measure of the ease of learning, we distinguished categorically between those industries in which basic science was more relevant to technical progress than the relatively more targeted applied sciences and assumed that learning was generally less difficult in industries that fell into the latter category. Thus, we created a dummy variable, DUMBAS, that equals one when the average value of the relevance scores associated with the basic fields exceeds that associated with the applied fields and that equals zero otherwise. We specified the dummy variable, DUMAPP, analogously. To capture the interdependence of firms, we employed measures of industries' competitiveness as represented by each industry's four-firm concentration ratio (C4) and industry-level estimates of the price elasticity of demand (PELAS).

To further control for industry demand conditions, we used industry estimates developed by Levin (1981) of price elasticity (PELAS) and income elasticity (INCELAS) and a demand time-shift parameter (DGROWTH). Finally, we included another control variable that may also reflect technological opportunity, industry maturity. We used a somewhat crude measure of industry maturity, NEWPLANT, that measures the percentage of an industry's property, plant, and equipment installed within the preceding five years.

RESULTS

Technological opportunity. Our theory suggests that when the targeted quality of knowledge is less (i.e., learning is more difficult), an increase in the relevance (i.e., quantity) of knowledge should have a more positive effect on R&D intensity. Therefore, the coefficient estimates of the variables measuring the relevance of the four basic scientific fields should exceed those of the variables measuring the relevance of the seven applied scientific fields. Confirming the prediction, Table 1 indicates that the estimated coefficients for the applied sciences are, with the exception of computer science, lower than that for the basic sciences. The similarity of the estimate of the effect of the relevance of computer science, an applied science, to those of some of the basic sciences suggests that the assumption may not be correct that only one determinant of the ease of learning, the targeted quality of the field, varies systematically across the fields of applied and basic science. Another determinant of the ease of learning postulated above is a field's pace of advance, where faster pace should require more R&D to permit assimilation, and the pace of advance in computer science has been relatively rapid over the past two decades.

To further test the prediction that the coefficient values of the less targeted, basic science field variables would exceed those of the applied fields, we estimated a specification, otherwise identical to the first, in which we constrained the coefficients of the basic sciences to be the same and the coefficients of the applied sciences to be the same. This shows the effect on R&D spending as the overall technological opportunity associated with basic science and applied science, respectively, change. The constrained coefficient estimates of the effect of the technological opportunity associated with the basic and applied sciences are significantly different (at the $p < .01$ level) across all estimation methods, with the former equal to .189 and the latter equal to $-.080$ in the GLS estimation. Therefore, relative to the effect of an increase in the technological opportunity associated with applied science, an increase in that associated with basic science elicits more R&D.

Our predicted ranking of the coefficient magnitudes associated with the extraindustry sources of knowledge, reflecting increasingly targeted knowledge from these sources, is largely confirmed. The coefficient estimate for the importance of knowledge originating from universities exceeds that for government labs, which, in turn, is greater than that for materials suppliers, which exceeds that for equipment suppliers. The difference between coefficient values is statistically significant in the case of government sources versus materials suppliers for both the OLS and Tobit results ($p < .01$) and in the case of materials suppliers versus equipment suppliers in

Table 1

Analysis of R&D Intensity*

Variable	OLS (N = 1302)	Regression Coefficient GLS (N = 1302)	Tobit (N = 1719)
Intercept	−5.184••	−2.355•	−4.086••
	(1.522)	(1.037)	(1.461)
APPROPRIABILITY × C4	.213	.342••	.368••
	(.128)	(.103)	(.130)
APPROPRIABILITY × PELAS	−.192	−.200•	−.176
	(.106)	(.091)	(.103)
APPROPRIABILITY × DUMAPP	.448•	.248	.211
	(.202)	(.143)	(.194)
APPROPRIABILITY × DUMBAS	.302	.174	.094
	(.208)	(.144)	(.206)
USERTECH	.470••	.397••	.612••
	(.104)	(.069)	(.107)
UNIVTECH	.374••	.318••	.395••
	(.131)	(.091)	(.147)
GOVTECH	.221•	.069	.137
	(.106)	(.079)	(.107)
MATERIALTECH	−.258••	−.074	−.303••
	(.098)	(.070)	(.100)
EQUIPTECH	−.401••	−.484••	−.574••
	(.111)	(.077)	(.117)
Biology	.314••	.185••	.276•
	(.102)	(.071)	(.114)
Chemistry	.289••	.081	.191•
	(.084)	(.062)	(.088)
Math	.184	.151	.123
	(.131)	(.097)	(.143)
Physics	.373••	.323••	.310•
	(.117)	(.091)	(.128)
Agricultural Science	−.441••	−.273••	−.308••
	(.088)	(.064)	(.099)
Applied Math/Operations Research	−.237	−.117	−.366•
	(.148)	(.102)	(.152)
Computer Science	.294•	.116	.433••
	(.124)	(.090)	(.122)
Geology	−.363••	−.240••	−.365••
	(.084)	(.061)	(.097)
Materials Science	−.110	−.150	.116
	(.125)	(.095)	(.118)
Medical Science	−.179	−.133	−.133
	(.093)	(.070)	(.103)
Metallurgy	−.315••	−.195••	−.393••
	(.077)	(.053)	(.089)
NEWPLANT	.057••	.049••	.045••
	(.008)	(.006)	(.007)
PELAS	.936	1.082•	.892
	(.611)	(.527)	(.573)
INCELAS	1.077••	.587••	1.112••
	(.170)	(.131)	(.188)
DGROWTH	.068	−.074	.004
	(.090)	(.053)	(.105)
R^2	.287		

•$p < .05$; ••$p < .01$.
* Reproduced from Cohen and Levinthal (1989a: 590–591, 569–596). Standard errors are in parentheses.

the GLS results ($p < .01$). While we had no prediction regarding the coefficient value for USERTECH, the consistently high value of the coefficient estimate may reflect some element of demand conditions. Consistent with this, we have observed the variable USERTECH to be significantly correlated with measures of the importance of product differentiation (cf. Cohen and Levinthal, 1989a).

Appropriability. The results largely support the prediction that the ease of learning conditions the effect of knowledge spillovers. The effect on R&D intensity of increasing appropriability (i.e., diminishing spillovers) was significantly greater ($p < .05$) in those industries in which the applied sciences are more relevant to innovation than the basic sciences. This result suggests that the positive absorption incentive associated with spillovers is greater in industries in which the difficulty of learning is greater. Second, there is a significant positive effect ($p < .01$) of the interaction between market concentration and the appropriability level. As market concentration increases (indexing a diminution in competitiveness), the positive effect of a given appropriability level on R&D intensity increases, as predicted. Likewise, the effect of the interaction of the price elasticity of demand and the level of appropriability is negative (but only significant at $p < .05$ in the GLS estimate), providing additional support for the proposition that the positive effect of spillovers will increase in industries in which firms are less interdependent. The results suggest that the learning environment affects the impact of spillovers on R&D spending and that the importance of the positive absorptive-capacity-building incentive relative to that of the negative appropriability incentive is conditioned by the degree of competitor interdependence.

While we have shown that the learning environment modifies the effect of appropriability conditions, the question remains whether spillovers may, on balance, actually encourage R&D in some industries. To explore this possibility, we examined the effect of spillovers in the four two-digit SIC code level industries for which our sample contains enough lines of business to permit separate industry regressions. These include SICs 20 (food processing), 28 (chemicals), 35 (machinery), and 36 (electrical equipment). Due to the reduction in the degrees of freedom for industry-level variables, we simplified the estimating equation to consider only the direct effect of APPROPRIABILITY, and the science field variables were summarized as the maximum relevance scores attained by the basic and applied fields, respectively. In SICs 28 and 36, the effect of the APPROPRIABILITY variable was negative and significant at conventional levels, implying that R&D intensity rises with spillovers. In the Tobit results, the sign was also positive for SICs 28 and 36, but the coefficient estimates were not quite significant at the .05 confidence level. Thus, in SICs 28 (chemicals) and 36 (electrical equipment), R&D intensity rose with spillovers when we controlled for other industry-level variables conventionally thought to drive R&D spending, including technological opportunity and demand conditions. Although the analyses showing a positive effect of spillovers in these two industry groups do not represent a direct test of our model, the results suggest, particularly when considered with the interaction results, that the positive absorption incentive associated with spillovers may be sufficiently strong in some cases to more than offset the negative appropriability incentive.

IMPLICATIONS FOR INNOVATIVE ACTIVITY

Drawing on our prior work (Cohen and Levinthal, 1987, 1989a), we offer some implications of absorptive capacity for the analysis of other innovative activities, including basic research, the adoption and diffusion of innovations, and decisions to participate in cooperative R&D ventures, that follow from the preceding analyses.

The observation that R&D creates a capacity to assimilate and exploit new knowledge provides a ready explanation of why some firms may invest in basic research even when the preponderance of findings spill out into the public domain. Specifically, firms may conduct basic research less for particular results than to be able to provide themselves with the general background knowledge that would permit them to exploit rapidly useful scientific and technological knowledge through their own innovations or to be able to respond quickly—become a fast second—when competitors come up with a major advance (see also Rosenberg, 1990). In terms of our discussion of the cognitive and organizational aspects of absorptive capacity, we may think of basic research as broadening the firm's knowledge base to create critical overlap with new knowledge and providing it with the deeper understanding that is useful for exploiting new technical developments that build on rapidly advancing science and technology.

This perspective on the role of basic research offers a rather different view of the determinants of basic research than that which has dominated thinking in this area for the thirty years since Nelson's (1959) seminal article. Nelson hypothesized that more diversified firms will invest more heavily in basic research because, assuming imperfect markets for information, they will be better able to exploit its wide-ranging and unpredictable results. Nelson thus saw product-market diversification as one of the key determinants of basic research.[6] Emphasizing the role of basic research in firm learning, our perspective redirects attention from what happens to the knowledge outputs from the innovation process to the nature of the knowledge inputs themselves. Considering that absorptive capacity tends to be specific to a field or knowledge domain means that the type of knowledge that the firm believes it may have to exploit will affect the sort of research the firm conducts. From this vantage point, we would conjecture that as a firm's technological progress becomes more closely tied to advances in basic science (as has been the case in pharmaceuticals), a firm will increase its basic research, whatever its degree of product-market diversification. We also suggest, with reference to all firm research, not just basic research, that as the fields underlying technical advance within an industry become more diverse, we may expect firms to increase their R&D as they develop absorptive capacities in each of the relevant fields. For example, as automobile manufacturing comes to draw more heavily on newer fields such as microelectronics and ceramics, we expect that manufacturers will expand their basic and applied research efforts to better evaluate and exploit new findings in these areas.

The findings on the role of absorptive capacity and the ways in which it may be developed also have implications for the analysis of the adoption and diffusion of innovations. Our perspective implies that the ease of learning, and thus technology adoption, is affected by the degree to which an innovation is related to the pre-existing knowledge base of

6
Markets for information often fail because they inherently represent a situation of information asymmetry in which the less informed party cannot properly value the information he or she wishes to purchase, and the more informed party, acting self-interestedly, attempts to exploit that inability (Williamson, 1975).

prospective users. For example, personal computers diffused more rapidly at the outset among consumers and firms who had prior experience on mainframes or minicomputers. Likewise, software engineering practices seem to be adopted more readily by programmers with previous Pascal rather than Fortran experience because the structure of Pascal more closely reflects some of the underlying principles of software engineering (Smith et al., 1989). Our argument also suggests that an innovation that is fully incorporated in capital equipment will diffuse more rapidly than more disembodied innovations that require some complementary expertise on the part of potential users. This is one of the anticipated benefits of making computers more "user friendly."

The importance of absorptive capacity also helps explain some recent findings regarding firms' cooperative research ventures. First, Link (1987) has observed that cooperative research ventures are actually found more typically in industries that employ more mature technologies rather than in industries in which technology is moving ahead quickly—as seems to be suggested by the popular press. Second, it has been observed that cooperative ventures that have been initiated to pursue basic research, as well as more applied research objectives, have been subject over the years to increasing pressure to focus on more short-term research objectives (Mowery and Rosenberg, 1989). The simple notion that it is important to consider the costs of assimilating and exploiting knowledge from such ventures provides at least a partial explanation for these phenomena. Many cooperative ventures are initiated in areas in which the cost to access the output of the venture is low, or they often gravitate toward such areas over time. Conversely, those who are attempting to encourage cooperative research ventures in quickly advancing fields should recognize that the direct participation in the venture should represent only a portion of the resources that it will take to benefit from the venture. Participating firms also must be prepared to invest internally in the absorptive capacity that will permit effective exploitation of the venture's knowledge output.

CONCLUSION

Our empirical analysis of R&D investment suggested that firms are in fact sensitive to the characteristics of the learning environment in which they operate. Thus, absorptive capacity appears to be part of a firm's decision calculus in allocating resources for innovative activity. Despite these findings, because absorptive capacity is intangible and its benefits are indirect, one can have little confidence that the appropriate level, to say nothing of the optimal level, of investment in absorptive capacity is reached. Thus, while we have proposed a model to explain R&D investment, in which R&D both generates innovation and facilitates learning, the development of this model may ultimately be as valuable for the prescriptive analysis of organizational policies as its application may be as a positive model of firm behavior.

An important question from a prescriptive perspective is When is a firm most likely to underinvest in absorptive capacity to its own long-run detriment? Absorptive capacity is more likely to be developed and maintained as a byproduct of routine activity when the knowledge domain that the firm wishes to exploit is closely related to its current knowledge base. When, however, a firm wishes to acquire and use knowledge that is unrelated to its ongoing activity, then the firm must dedicate effort exclusively to creating absorptive capacity (i.e., absorptive capacity is not a byproduct). In this case, absorptive capacity may not even occur to the firm as an investment alternative. Even if it does, due to the intangible nature of absorptive capacity, a firm may be reluctant to sacrifice current output as well as gains from specialization to permit its technical personnel to acquire the requisite breadth of knowledge that would permit absorption of knowledge from new domains. Thus, while the current discussion addresses key features of organizational structure that determine a firm's absorptive capacity and provides evidence that investment is responsive to the need to develop this capability, more research is necessary to understand the decision processes that determine organizations' investments in absorptive capacity.

REFERENCES

Abernathy, William J.
1978 The Productivity Dilemma. Baltimore: Johns Hopkins University Press.

Allen, Thomas J.
1977 Managing the Flow of Technology. Cambridge, MA: MIT Press.

Allen, Thomas J., and Stephen D. Cohen
1969 "Information flows in R&D labs." Administrative Science Quarterly, 20: 12–19.

Anderson, John R., Robert Farrell, and Ron Sauers
1984 "Learning to program in LISP." Cognitive Science, 8: 87–129.

Arrow, Kenneth J.
1962 "Economic welfare and the allocation of resources for invention." In R. R. Nelson (ed.), The Rate and Direction of Inventive Activity: 609–625. Princeton, NJ: Princeton University Press.

Bower, Gordon H., and Ernest R. Hilgard
1981 Theories of Learning. Englewood Cliffs, NJ: Prentice-Hall.

Bradshaw, Gary F., Patrick W. Langley, and Herbert A. Simon
1983 "Studying scientific discovery by computer simulation." Science, 222: 971–975.

Brock, Gerald W.
1975 The U.S. Computer Industry. Cambridge, MA: Ballinger.

Burns, Tom, and George M. Stalker
1961 The Management of Innovation. London: Tavistock.

Clark, Kim B., and Takahiro Fujimoto
1987 "Overlapping problem solving in product development." Technical Report, Harvard Business School.

Cohen, Wesley M., and Richard C. Levin
1989 "Empirical studies of innovation and market structure." In R. C. Schmalensee and R. Willig (eds.), Handbook of Industrial Organization: 1059–1107. Amsterdam: Elsevier.

Cohen, Wesley M., and Daniel A. Levinthal
1987 "Participation in cooperative research ventures and the cost of learning." Technical Report, Dept. of Social and Decision Sciences, Carnegie Mellon University.
1989a "Innovation and learning: The two faces of R&D." Economic Journal, 99: 569–596.
1989b "Fortune favors the prepared firm." Technical Report, Dept. of Social and Decision Sciences, Carnegie Mellon University

Dearborn, R., and Herbert A. Simon
1958 "Selective perception in executives." Sociometry, 21: 140–144.

Ellis, Henry Carlton
1965 The Transfer of Learning. New York: MacMillan.

Estes, William Kaye
1970 Learning Theory and Mental Development. New York: Academic Press.

Hamberg, Daniel
1963 "Invention in the industrial research laboratory." Journal of Political Economy, 71: 95–115.

Harlow, H. F.
1949 "The formation of learning sets." Psychological Review, 56: 51–65.
1959 "Learning set and error factor theory." In S. Koch (ed.), Psychology: A Study of Science, 2: 492–537. New York: McGraw-Hill.

Johnston, R., and M. Gibbons
1975 "Characteristics of information usage in technological innovation." IEEE Transactions on Engineering Management, 27–34, EM-22.

Katz, Daniel, and Robert L. Kahn
1966 The Social Psychology of Organizations. New York: Wiley.

Katz, Ralph
1982 "The effects of group longevity on project communication and performance." Administrative Science Quarterly, 27: 81–104.

Katz, Ralph, and Thomas J. Allen
1982 "Investigating the not invented here (NIH) syndrome: A look at the performance, tenure, and communication patterns of 50 R&D project groups." R&D Management, 12: 7–12.

Lee, Denis M. S., and Thomas J. Allen
1982 "Integrating new technical staff: Implications for acquiring new technology." Management Science, 28: 1405–1420.

Absorptive Capacity

Levin, Richard C.
1981 "Toward an empirical model of Schumpeterian competition." Technical Report, Dept. of Economics, Yale University.

Levin, Richard C., Alvin K. Klevorick, Richard R. Nelson, and Sidney G. Winter
1983 "Questionnaire on industrial research and development." Dept. of Economics, Yale University.
1987 "Appropriating the returns from industrial R&D." Brookings Papers on Economic Activity, 783–820.

Levitt, Barbara, and James G. March
1988 "Organizational learning." Annual Review of Sociology, 14: 319–340.

Lindsay, Peter H., and Donald A. Norman
1977 Human Information Processing. Orlando, FL: Academic Press.

Link, Albert N.
1987 "Cooperative research activity in U.S. Manufacturing." Technical Report, University of North Carolina, Greensboro, Final report submitted to the National Science Foundation under grant PRA 85-212664.

Majumdar, Bodiul Alam
1982 Innovations, Product Developments and Technology Transfers: An Empirical Study of Dynamic Competitive Advantage, The Case of Electronic Calculators. Lanham, MD: University Press of America.

Mansfield, Edwin
1968 Economics of Technological Change. New York: Norton.
1988 "The speed and cost of industrial innovation in Japan and the United States: External vs. internal technology." Management Science, 34(10): 1157–1168.

March, James G., and Herbert A. Simon
1958 Organizations. New York: Wiley.

Mowery, David C.
1983 "The relationship between intrafirm and contractual forms of industrial research in American manufacturing, 1900–1940." Explorations in Economic History, 20: 351–374.

Mowery, David C., and Nathan Rosenberg
1989 Technology and the Pursuit of Economic Growth. New York: Cambridge University Press.

Mueller, Willard F.
1962 "The origins of the basic inventions underlying DuPont's major product and process innovations, 1920 to 1950." In R. R. Nelson (ed.), The Rate and Direction of Inventive Activity: 323–358. Princeton: Princeton University Press.

Myers, Sumner, and Donald C. Marquis
1969 "Successful industrial innovations." Washington, DC: National Science Foundation, NSF 69-17.

Nelson, Richard R.
1959 "The simple economics of basic research." Journal of Political Economy, 67: 297–306.

Nelson, Richard R., and Sidney Winter
1982 An Evolutionary Theory of Economic Change. Cambridge, MA: Harvard University Press.

Peck, Merton J.
1962 "Inventions in the postwar American aluminum industry." In R. R. Nelson (ed.), The Rate and Direction of Inventive Activity: 279–298. Princeton: Princeton University Press.

Pirolli, Peter L., and John R. Anderson
1985 "The role of learning from example in the acquisition of recursive programming skill." Canadian Journal of Psychology, 39: 240–272.

Reich, Robert B.
1987 "The rise of techno-nationalism." Atlantic, May: 63–69.

Rosenberg, Nathan
1982 Inside the Black Box: Technology and Economics. New York: Cambridge University Press.
1990 "Why do firms do basic research (with their own money)?" Research Policy (in press).

Rosenberg, Nathan, and W. Edward Steinmueller
1988 "Why are Americans such poor imitators?" American Economic Review, 78(2): 229–234.

Schumpeter, Joseph A.
1942 Capitalism, Socialism and Democracy. New York: Harper and Row.

Simon, Herbert A.
1985 "What we know about the creative process." In R. L. Kuhn (ed.), Frontiers in Creative and Innovative Management: 3–20. Cambridge, MA: Ballinger.

Smith, Gordon, Wesley M. Cohen, William Hefley, and Daniel A. Levinthal
1989 "Understanding the adoption of Ada: A field study report." Technical Report, Software Engineering Institute, Carnegie Mellon University.

Spence, A. Michael
1984 "Cost reduction, competition, and industry performance." Econometrica, 52: 101–122.

Tilton, John E.
1971 International Diffusion of Technology: The Case of Semiconductors. Washington, DC: Brookings Institution.

Tushman, Michael L.
1977 "Special boundary roles in the innovation process." Administrative Science Quarterly, 22: 587–605.
1978 "Technical communication in R&D laboratories: The impact of project work characteristics." Administrative Science Quarterly, 21: 624–644.

Tushman, Michael L., and Philip Anderson
1986 "Technological discontinuities and organizational environments." Administrative Science Quarterly, 31: 439–465.

Utterback, James M.
1971 "The process of technological innovation within the firm." Academy of Management Journal, 12: 75–88.

von Hippel, Eric
1978 "Successful industrial products from customer ideas." Journal of Marketing, 42: 39–49.
1988 The Sources of Innovation. New York: Oxford University Press.

Vyssotsky, V. A.
1977 "The innovation process at Bell Labs." Technical Report, Bell Laboratories.

Westney, D. Eleanor, and Kiyonori Sakakibara
1986 "The role of Japan-based R&D in global technology strategy." In M. Hurowitch (ed.), Technology in the Modern Corporation: 217–232, London: Pergamon.

Williamson, Oliver E.
1975 Markets and Hierarchies: Analysis and Antitrust Implications. New York: Free Press.

Zenger, Todd R., and Barbara S. Lawrence
1989 "Organizational demography: The differential effects of age and tenure distributions on technical communication." Academy of Management Journal, 32: 353–376.

[11]

Strategic Management Journal
Strat. Mgmt. J., **25**: 909–928 (2004)
Published online in Wiley InterScience (www.interscience.wiley.com). DOI: 10.1002/smj.384

SCIENCE AS A MAP IN TECHNOLOGICAL SEARCH

LEE FLEMING[1]* and OLAV SORENSON[2]
[1] *Harvard Business School, Boston, Massachusetts, U.S.A.*
[2] *Anderson Graduate School of Management, University of California at Los Angeles, Los Angeles, California, U.S.A.*

A large body of work argues that scientific research increases the rate of technological advance, and with it economic growth. The precise mechanism through which science accelerates the rate of invention, however, remains an open question. Conceptualizing invention as a combinatorial search process, this paper argues that science alters inventors' search processes, by leading them more directly to useful combinations, eliminating fruitless paths of research, and motivating them to continue even in the face of negative feedback. These mechanisms prove most useful when inventors attempt to combine highly coupled components; therefore, the value of scientific research to invention varies systematically across applications. Empirical analyses of patent data support this thesis. Copyright © 2004 John Wiley & Sons, Ltd.

A long line of work supports the idea that scientific research stimulates technological innovation, thereby accelerating economic growth. Though one can trace this notion as far back as Adam Smith (Stephan, 1996), researchers did not begin empirically testing it until the second half of the twentieth century. One set of studies has focused on establishing the relationship between the investments in, or outcomes of, scientific research and their effects on economic growth (e.g., Mansfield, 1972; Rosenberg, 1974; Sveikauskas, 1981). For instance, a paper by James Adams (1990) demonstrates that cumulative research output, in the form of published papers, appears to accelerate growth. Another set—typically drawing on either rich case histories or patent data—has investigated the more proximate linkages between scientific research and technological innovation. Adam Jaffe (1989), for example, finds a positive relationship between university research expenditures and local patenting rates (cf. Jaffe and Trajtenberg, 1996). Together, these complementary lines of research provide strong empirical support for a link between scientific research, technological innovation, and economic growth.

Despite this evidence for the value of scientific research to innovation, inventors across fields vary tremendously in the degree to which they make use of science. Universities, for example, disproportionately patent inventions in two broad sectors: drugs and medicine, and chemicals (Henderson, Jaffe, and Trajtenberg, 1998). Similarly, most studies detailing the beneficial effects of spillovers from academic research to the private sector focus on biotechnology or pharmaceutical firms (e.g., Henderson and Cockburn, 1994; Gambardella, 1995; Zucker and Darby, 1996; Zucker, Darby, and Brewer, 1998).

Why do inventors draw more heavily on scientific research in these areas? Can institutional factors account for the links between academics and

Keywords: science; invention; recombinant search; coupling
* Correspondence to: Lee Fleming, Harvard Business School, Morgan T95, Boston, MA 02163, USA.
E-mail: lfleming@hbs.edu

firms in particular sectors, or do industries differ in the potential returns available to the application of scientific research, thereby affecting the incentives faced by firms? Though institutional factors almost certainly play a role, Mansfield (1995) contends that the benefits of applying science to invention differ across sectors; for instance, pharmaceutical firms in his sample asserted that 27 percent of their inventions required the application of science to avoid costly delays, while electronics firms made the same claim only 6 percent of the time. How do these fields differ such that the application of science appears highly beneficial in one, while of limited use in the next? Answering this question requires us to examine technology at a finer grain of detail; progress in our understanding of the value of science, in the words of Richard Nelson, 'would appear to require a better understanding of what knowledge actually does for inventing' (Nelson, 1982: 455).

To address this issue, this paper investigates the importance of science at the level of the individual invention, explicating one factor—the difficulty of the inventive problem—that might influence the returns to the application of science at that level. We suggest that science proves most useful when inventors seek to combine tightly coupled components. Following a long tradition in the study of innovation (e.g., Gilfillan, 1935; Schumpeter, 1939), we conceptualize invention as a process of searching for better combinations of existing components. When inventors seek to combine relatively independent components (i.e., those with a low degree of coupling), finding useful new configurations proves relatively easy—any search algorithm can locate the most useful combinations. As the space searched becomes increasingly complex, however, local search routines break down, failing to identify the best combinations (Fleming and Sorenson, 2001). In these cases, we posit that science may transform invention from a relatively haphazard search process to a more directed identification of useful new combinations, thus mitigating the complications typically encountered when combining coupled components.

Our empirical analyses estimate the returns to the application of science using patent data. Following a common practice in studying patents, the future citation count provides our measure of the value of the resultant invention. We measure coupling by observing the ease with which subclasses have previously been recombined, and identify the use of science by noting when patents reference published articles in refereed scientific journals. Our results broadly support our expectations: science has no apparent effect when inventors work with relatively independent pieces; it only appears beneficial when inventors seek to combine highly coupled components—a particularly difficult task.

INVENTION AS RECOMBINANT SEARCH

Before considering how science alters the process of invention, one must first ask what inventors actually do. One popular view in the history of technology conceptualizes invention as a process of recombination.[1] Research out of this tradition holds that invention comes either from combining technological components in a novel manner (Gilfillan, 1935; Schumpeter, 1939; Usher, 1954; Nelson and Winter, 1982; Basalla, 1988; Weitzman, 1996), or through reconfiguring existing combinations (Henderson and Clark, 1990). By 'technological components,' we mean any fundamental bits of knowledge or matter that inventors might use to build inventions. For example, Gilfillan (1935) describes the steamship as a combination of the boat with a steam engine, or one might think of the computer workstation as a novel aggregation of several components and subsystems: a CPU, a motherboard, virtual storage and memory, a display, graphics processors, as well as systems and applications software.

Though complex physical systems quite clearly combine multiple elements, many inventions—such as nylon, a polymer—intuitively seem like one component, unbreakable without resorting to atomic-level decomposition. A closer look, however, typically reveals that these inventions also arise from the recombination of discrete components and processes. For example, polymers involve the linkage of several molecules into a substance with new properties (Smith and Hounshell, 1985). Moreover, innovation in these substances frequently occurs in the processes for producing them—often themselves recombinations of existing manufacturing steps. For instance, after the

[1] Other theorists view invention as exogenous shocks (e.g., Klepper, 1996). Though this point of view undoubtedly has validity in many cases, it also relegates the invention process to a black box without first considering whether recombination might usefully characterize the process.

initial synthesis of polymers, Wallace Carothers greatly increased polymer length by applying gentle heat in a molecular still, thereby eliminating interference from water molecules (Smith and Hounshell, 1985). Alternatively, an invention might involve the application of an existing technology to a new purpose. Nylon once again provides a salient example. Though immediately recognized as a replacement for silk, the advent of World War II saw nylon's application to parachutes, flight suits, and glider towropes. Thus, conceiving of inventions as recombinations of existing components does not greatly limit our scope. The question remains, though, how do inventors search for these new inventions?

Local search

Researchers most commonly point to local search —also referred to as exploitation—as the predominant algorithm used in innovation (March and Simon, 1958; Cyert and March, 1963; Hansen and Lovas, 2004; Nelson and Winter, 1982; Stuart and Podolny, 1996). Local search implies that inventors typically alter one component at a time, either reconfiguring it relative to the other components or replacing it with a different component; in other words, inventors search incrementally. Traditional drug discovery, for example, follows this pattern; after identifying a promising candidate through high-throughput screening or luck, researchers attempt to find commercially viable drugs by slightly altering this initial candidate (Drews, 2000). The label 'local' refers to the fact that research activities relate quite closely to prior research activities, by definition implying some experience with the technologies being developed.

Explanations for the prevalence of local search frequently focus on the value of experience. These accounts assume that inventors have an extremely limited understanding of the elements being recombined and the ways in which those elements interact—due to either cognitive limits or fundamental uncertainty (March and Simon, 1958; Nelson, 1982; Vincenti, 1990). As inventors gain experience, they develop some (possibly mistaken) understanding of the components that allows them to invent with greater reliability by avoiding elements that did not work in the past (Vincenti, 1990). Nevertheless, local search also has a downside: focusing on familiar combinations can preclude the inventor from investigating more distant—and potentially more useful—possibilities; as inventors continue to work with a particular set of components, they may exhaust the set of useful combinations (March, 1991; Fleming, 2001).

Empirical research supports the prevalence and effects of local search across a variety of levels. At the firm level, Stuart and Podolny (1996), in an analysis of semiconductor patents, show that firms' new patents concentrate in areas technologically proximate to their existing patent portfolios—a tendency that increases as firms mature, potentially leading to obsolescence (Sørensen and Stuart, 2000; see also Ahuja and Lampert, 2001). Similar results appear in research on new products: Audia and Sorenson (2001) find that computer workstation manufacturers tend to introduce new products with features similar to their existing offerings (cf. Martin and Mitchell, 1998), a practice that retards future sales growth. In research at the community of practice level, Fleming (2001) demonstrates that experience at this level does indeed appear to enhance search, both by increasing the average value of inventions and by decreasing the variability of outcomes, but this beneficial effect eventually succumbs to exhaustion.

Science and search

Scientific knowledge, by contrast, may lead to a very different type of search; in effect, it provides inventors with the equivalent of a map—a stylized representation of the area being searched. Though local search portrays inventors as rummaging around, blindly bumping along in the search for new technologies, they can also proceed with greater foresight (Vincenti, 1990). For example, studies of engineers have found that they sometimes consult the scientific literature to resolve technological difficulties (Gibbons and Johnston, 1974; Allen, 1977). Scientific knowledge differs from that derived through local search, in particular, because the scientific endeavor attempts to generate and test *theories*. Science attempts to explain *why* phenomena occur, providing a means of predicting the results of untried experiments and the usefulness of previously uncombined configurations of technological components. Having an understanding of the fundamental problem—a map—likely modifies the search process in multiple, complementary ways. Notably, it might lead inventors quite directly to the proper combinations of components to solve a particular technical problem. Even absent such a direct linkage, science could usefully increase the effectiveness of search by identifying useless directions of search beforehand, and by providing a glimpse of the possible.

Theoretical understanding of the underlying properties of technological components and their interactions may facilitate effective search. Lippman and McCall (1976) argue that this advantage stems from the ability to assess alternatives 'offline'—in essence, running the experiment in one's mind or on paper without incurring the costs of actually performing the test. For example, laying the groundwork for a future generation of chip technology, researchers predicted that single-wall carbon nano-tubes would either conduct or semiconduct, depending on the diameter of the tube and the angles of the carbon bonds (Dresselhaus, 2001). This prediction pointed to two specific configurations (out of an essentially infinite number of possibilities) that resulted in the successful fabrication of conducting and semiconducting nano-tubes (Ouyang *et al.*, 2001). The theory did not predict everything and researchers must still engage in trial-and-error search to produce working transistors. Theoretical guidance nonetheless accelerated the researchers' arrival at the proper diameter and angles. Maps typically do not detail every rock and crevice dotting the landscape; similarly, scientific theories do not predict every property and interaction of a new combination of technological components. Theory may still get researchers 'in the ballpark,' thereby leading them to a promising new invention more efficiently.

By eliminating fruitless avenues, science may also play the more modest role of reducing the size of the combinatorial search pace (Nelson, 1982). In other words, science can tell inventors how to avoid wasted effort. Consider the classic goal of alchemy: transforming lead into gold. Once people understood the structure and properties of atoms, it became quite clear that anything short of a nuclear reaction would not yield the desired results. More recently, researchers at Lord Manufacturing Company switched from working with electro-rheological materials to magneto-rheological materials in their effort to develop controllable fluids. Even after 10 years of fruitless effort with electrical approaches, the switch occurred only after the lead researcher, Dave Carlson, investigated the physics of the materials and calculated that the potential power available from the use of magnetic fields exceeded the potential of electrical fields by two orders of magnitude (Carlson, 2001; Stix, 2001). Without realizing the futility of the search, researchers at Lord may have pursued an unobtainable goal indefinitely; thus, science improves the efficacy of search by preventing inventors from wasting valuable effort on the impossible.

Even when science has an inaccurate or incomplete understanding of the problem, it might still alter the search process in a useful manner. When inventors experiment with combinations and configurations of components, most of the iterations that they try will fail to yield useful outcomes. In the absence of a predictive model of the problem, inventors constantly face the question of how long they should persist with a failing approach. Should they quit after 10, 100, or 1000 failures? By suggesting that an avenue of solution might succeed theoretically, science can encourage inventors to continue working with a seemingly unfruitful set of components. Consider the case of Prozac. Bryan Molloy began Eli Lilly's research with local search, deriving hundreds of benadryl compounds (Kramer, 1997: 62). They all failed. His colleague David Wong continued the search, however, because he believed a theory that blocking serotonin uptake, a property of benadryl derivatives, held the key to treating depression. Using a new experimental technique that enabled more accurate evaluation, Wong tested the same compounds as Molloy and demonstrated the efficacy of one of them, fluoxetine (the chemical name for Prozac). In addition to motivating inventors in the face of failure, the belief that something better might exist could also lead them to continue searching even after having found acceptable solutions (Gavetti and Levinthal, 2000, provide a similar argument for the usefulness of managerial frameworks in strategic search).

Understanding the strengths and weaknesses of these search processes also requires an understanding of the spaces being searched; thus, we turn to a discussion of technological landscapes.

TECHNOLOGY LANDSCAPES

Landscapes offer a useful heuristic for thinking about the space that inventors must search when attempting to discover useful new inventions. Wright (1932) first introduced the idea of a fitness landscape to describe the evolution of traits in species. Since then, the idea has been applied usefully to topics ranging from spin-glasses in physics (Weinberger, 1991) to organizational routines (Levinthal, 1997; Rivkin, 2000). This same metaphor can help us understand technological search. Think of each possible set of components as corresponding to a particular technological landscape that inventors search; positions on these landscapes represent different configurations of components. Inventors search new landscapes when they combine new sets of components and move across these landscapes when they reconfigure a particular set of components. The peaks and higher elevations represent more useful configurations. Technological landscapes can take a variety of shapes. At one extreme, the landscape might smoothly slope up to a single peak; at the other, it might jump around with treacherous valleys and soaring zeniths. Thus, one might ask, what determines the topography of these technological landscapes?

One important factor influencing the shape of these landscapes is coupling. By 'coupling,' we mean the degree to which components interact in determining the functionality of the overall invention.[2] A rugged landscape implies that adjacent points of the terrain differ dramatically in their usefulness. Coupled technologies exhibit precisely such a characteristic: the functionality of coupled inventions overall becomes highly sensitive to minor changes in the individual components (Ulrich, 1995). For example, a change of one particle in 10 million of semiconductor dopant can change its conductivity by a factor of 10 thousand (Millman, 1979).[3] In contrast, independent components—those with no coupling—gradually change the functionality of the overall system, with each component contributing individually to overall usefulness. Kauffman (1993) presents this relationship in stark simplicity. His simulations include two parameters: the number of components and the degree of interaction between those components. These simulations characterize the landscapes produced by different types of components; interacting components create more jumbled, uncorrelated landscapes, while independent elements generate a landscape with a single peak. In short, the ruggedness of the technological landscape increases as the degree of coupling among the components intensifies.

Landscapes and search

Combining these landscapes with a search algorithm allows us to make predictions regarding the likely outcomes of inventive search. Consider local search. When inventors search locally, they move across adjacent positions on the fitness landscape. Although gradually climbing uphill will eventually bring the inventor to an optimum, this peak might not represent the 'best,' or even a particularly 'good,' configuration. Rather, the topography of the landscape importantly affects the efficacy of the search algorithm. Local search processes work best on smooth, correlated landscapes because incremental improvements more frequently find the global maximum. In contrast, local search processes operate less effectively on jumbled, uneven, and uncorrelated landscapes. On these landscapes, local search processes yield unpredictable outcomes and often trap the searcher on a local peak because any incremental change yields less desirable outcomes (March, 1991; Kauffman, 1993). Inventors who search incrementally with strongly coupled components make slow progress and have no assurance of ultimate success.

Although coupling between components makes the search process more difficult, it also creates the potential for very useful inventions. On flat landscapes, inventors searching locally can find the optima, but few exist and those that do rise only modestly above alternative configurations (i.e., they offer minimal gains in usefulness). Competitors may also find these innovations easy to imitate (Rivkin, 2000). Rugged landscapes contain a large number of potentially useful configurations, if inventors can only find them. Together these processes generate a non-monotonic relationship between the usefulness of

[2] This idea goes by a variety of monikers in different literatures. Milgrom and Roberts (1990) refer to it as complementarity. Kauffman (1993) and Sorenson (2002) call it interdependence, and many use the term modularity to refer to the inverse of interdependence (e.g., Baldwin and Clark, 2000). Given our technological context, we prefer Ulrich's (1995) definition of coupling: 'Two components are coupled if a change made to one component requires a change to the other component for the overall product to work correctly.'

[3] Other examples abound. The fluoxetine molecule resembles other benadryl derivatives quite closely, yet it provides much more effective relief against depression (Kramer, 1997). Superconductor research continues to find unexpectedly high temperature breakthroughs with novel combinations of common materials, such as boron (Campbell, 2001).

inventions found through local search and the coupling of the components combined (Fleming and Sorenson, 2001). On flat landscapes—those with little coupling—local search can locate the highest peak, but that apex offers limited functionality. At low levels of coupling, somewhat rugged landscapes provide a larger number of taller peaks that inventors can still find through local search. As the coupling increases, however, the difficulty of search rises, and the average usefulness of inventions declines. Coupling also increases the uncertainty of invention. As the number of unpredictable interactions rises, the uncertainty, or variability, of the expected outcomes also increases. Fleming and Sorenson (2001) find strong support for this model, using patent citations as a measure of invention usefulness.

By contrast, inventors can proceed quite differently when science provides them with some understanding of the underlying landscape. The exact effect of this process on inventive outcomes depends on how science influences the search process. Consider the first mechanism: cheap offline experimentation. If science allows a rough prediction of the expected interactions between coupled components, it should allow inventors to move quite directly toward the highest peaks—the most useful configurations—on the landscape, thereby reducing the variability of inventive outcomes. As the potential for useful inventions increases with coupling, it should also allow inventors to exploit synergies more effectively; thus, the average usefulness of inventions should rise with coupling when inventors use science. It might also reduce the number of times that inventors use the same combinations of components since they can efficiently focus on the most useful configurations while avoiding the rest.

Even if science plays the more modest role of allowing inventors to rule out unfruitful directions, our second mechanism, this narrowing of options should have a similar—though less pronounced—beneficial effect. One would still expect an improvement in the average usefulness of inventions, as researchers avoid the worst regions of the solution space. The variability of outcomes would likely remain high, however, as even the best regions in fitness landscapes of highly coupled components include peaks and valleys (Kauffman, 1993).

Under the third mechanism, motivation, the expected effects of science on search may differ. Having some sense of the height and location of the tallest peak may lead inventors to better outcomes by preventing them from becoming trapped in local optima (Gavetti and Levinthal, 2000). However, in this case, inventors might actually search *more* when directed by science because they will continue to look for better alternatives even after finding 'good' ones. In the words of Gavetti and Levinthal (2000), they may begin 'chasing cognitive rainbows.'

EMPIRICAL ANALYSIS

Studying how science affects the process of invention can prove quite difficult and, as a result, much of the work on this topic relies on careful case studies. Though useful, these studies lack the statistical power necessary to distinguish potential patterns linking science and invention from random noise. Thus, we analyzed patent data to examine the role of science.

Patent data admittedly offer imperfect measures of invention. Companies sometimes avoid patenting, and industries vary in their propensities to patent (Levin *et al.*, 1987). Patents also do not allow us to observe all points on the fitness landscape. Because inventors may limit their patent applications to their most successful inventions, patents likely represent only the higher positions on these landscapes. This implies that we must infer the topography of the underlying landscape from truncated data.[4] Despite these imperfections, patents do offer a window into the process of invention, allowing us to develop a quantitative measure of fitness across a broad range of technologies, to identify those inventions in which science likely played a role, and to develop a measure of component coupling—the variables needed to investigate our proposition empirically. We collected information on all U.S. patents granted in May and June of 1990 ($n = 17{,}264$)[5] listed in the Micro-Patent product; from these, we excluded the

[4] Simulations demonstrate, however, that increases in the mean of a normal distribution generate corresponding increased means in right truncated observations; thus, these truncated observations should still allow us to observe the rough structure of the technology landscapes (Fleming and Sorenson, 2001).

[5] We chose 1990 to allow sufficient time to observe future citation patterns, our dependent variable; we used a random number generator to select the starting month, May, and included only 2 months of data due to resource limitations (calculation of the coupling measure required months of CPU time and the

442 patents that do not cite any prior art (previous work in the area) because these patents may lie beyond our theory, involving the discovery of fundamental new components rather than the recombination of previous technologies (their inclusion does not change the results). Thus, our analyses include 16,822 patents granted in May and June of 1990.

Key variables

Science

We began by identifying when science likely influenced the process of invention. Much of the work linking science to invention focuses only on research that occurs at a university or public research institution (e.g., Henderson *et al.*, 1998). Since science also gets applied outside the university context (Narin, Hamilton, and Olivastro, 1997; Cockburn, Henderson, and Stern 2000), we took a different approach, identifying the usage of science through non-patent references (Narin *et al.*, 1997; Perko and Narin, 1997).[6]

We took a reference to the scientific literature as an indication that the inventor made use of—or at least had some awareness of—scientific knowledge in the process of invention. In addition to citing the prior art (previous patents), U.S. patents also cite a variety of non-patent literature. Our sample made 16,698 references to the non-patent literature (Table 1). The majority of these non-patent references (67%) cite journals listed in the *Scientific Index*: a publication that covers peer-reviewed scientific journals.[7] Unlike references to patent prior art, often added by the patent examiner, non-patent references more frequently come from the inventors themselves; Tijssen (2001) reports that 81 percent of patents with non-patent references include an inventor's citation to their own scientific work. Thus, non-patent references provide a reasonable indicator of the influence of science; as Tijssen notes, 'these citations at the very least indicate an awareness of scientific results with some indirect bearing on elements of the invention. In the best case, they reflect strong evidence of substantial direct contributions of scientific inputs to breakthrough technological innovations' (Tijssen, 2001: 52).[8] A dummy variable indicates if a patent referenced one of these publications.[9]

coding of the non-patent references involved hundreds of hours of research assistance).

[6] Though Schmoch (1993) and Meyer (2000) question the use of the non-patent references as indicators of the direct application of science, their contention that these references simply denote a relationship between the technology and science fits well with our assumption of their meaning.

[7] References to science most commonly appear in the following classes: 505 Superconductor technology (85%), 530 Natural resins and derivatives (81%), 435 Molecular biology (78%), 512 Perfumes (71%), 548 Organic compounds (69%).

[8] In our own survey of inventors (detailed below), 62 percent of inventors indicated an awareness of the specific scientific papers cited on their patents and 71 percent reported a awareness of the broader scientific literature on the subject. Though our results suggest a weaker link than Tijssen (possibly because we draw a representative sample of all patents while Tijssen selects on patents citing Dutch science), non-patent references to the scientific literature still appear a good proxy for the availability of scientific knowledge to inventors.

[9] Using a count of the number of scientific articles referenced in place of this dummy variable produced qualitatively equivalent results; including both a dummy and the count reveals that the count adds no significant information to the models. We

Table 1. Descriptive statistics of references to non-patent literature

	Number of patents making references	Conditional statistics			
		Mean	S.D.	Min.	Max.
Scientific Index journal (reference to science)	2,919	3.56	4.90	1	75
Non-Index journal (non-science reference)	290	1.56	1.19	1	8
Corporate non-technical (non-science reference)	700	2.11	2.42	1	34
Corporate technical	331	1.91	2.26	1	27
Book	870	1.59	2.24	1	55
Technical report	320	1.54	1.30	1	12
Conference proceedings	465	1.66	1.30	1	11

Coupling

Next we characterized the difficulty of the inventive problem according to the degree of coupling among the components. Our measure essentially observes the degree to which an invention's components have been previously recombined.[10] To calculate this measure, we used the subclass references on each patent. In effect, we treated the subclass assignments as proxies for the underlying components that combine to create that invention. The U.S. Patent Office uses subclass references to indicate which technologies relate to the patent, developing and updating these subclasses such that they consistently track technology back to 1790.[11] The approximately 100,000 subclasses allow for fine-grained classification of inventions (Trajtenberg, Henderson, and Jaffe, 1997). Though in many cases subclasses correspond quite closely to physical components (such as the example in Figure 1), they do not always match so well. Our measure, however, only requires that these subclasses define pieces of knowledge rather than identifiable physical components. Combining some pieces which interact sensitively to each other proves more difficult than connecting relatively independent chunks of knowledge.

We calculated our measure of coupling in two stages. Equation 1 details our measurement of the observed ease of recombination, or inverse of coupling, of an individual subclass i used in patent j. The score increases as a particular subclass combines with a wider variety of subclasses, controlling for the total number of applications; thus, it captures the observed ease with which a component has been recombined. Since most patents (92%) belong to more that one subclass of technology, we averaged the inverse of these subclass scores to create a patent-level measure

also investigated classifying technical references together with *Scientific Index* journal cites as indicators of science; using this more inclusive definition yielded weaker, but substantively identical, results.

[10] Economists seeking to identify complementarity—a closely related concept—have also proposed using the realized combinations of activities to identify the pattern and strength of these interdependencies (Athey and Stern, 1998).

[11] The Patent Office often redraws the boundaries of patent subclasses and retroactively reassigns patents to subclasses based on these changes. This system has the advantage of making subclasses usable over time; however, this reclassification likely introduces noise into our measure of coupling, making our tests of its effects more conservative. We used the concordance of 1996 for our calculations.

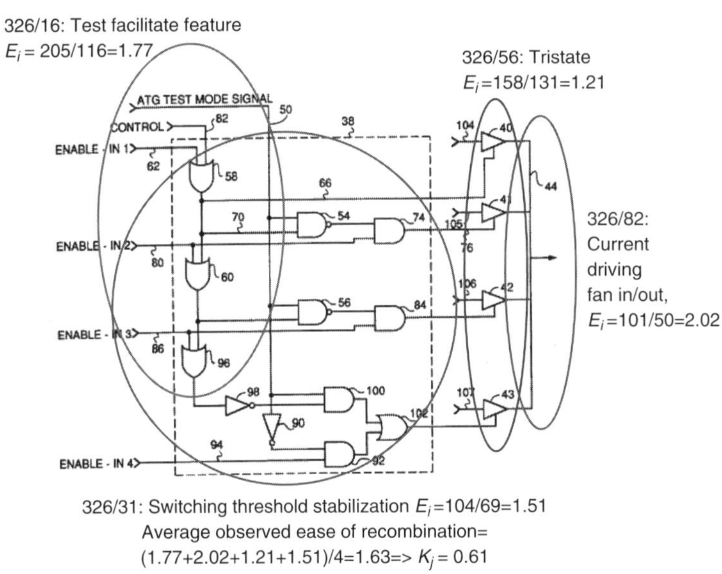

Figure 1. Calculation of K for patent 5,136,185

(Equation 2). We computed this measure using the entire 200 years of pre-sample history and tested its robustness to temporal bias with a 10-year window.

$$\text{Observed ease of recombination of subclass } i \equiv E_i = \frac{\text{Count of subclasses previously combined with subclass } i}{\text{Count of previous patents in subclass } i} \quad (1)$$

$$\text{Coupling of patent } j \equiv K_j = \frac{\text{Count of subclasses on patent } j}{\sum_{j \in i} E_i} \quad (2)$$

As an example of the coupling calculation, consider one of the first author's patents, #5,136,185. Figure 1 illustrates calculation of the measure and the correspondence between the USPTO classification scheme and the components used, all standard elements described in digital design textbooks at the time of invention (see, for example, McCluskey, 1986). 326/16 identifies the 'Test facilitate feature' subclass, which implements a testing mode within a semiconductor chip. Prior to the author's use of this component (i.e., subclass), it had been recombined 116 times with 205 other components, implying an observed ease of recombination score of 205/116 = 1.77. 326/56 indicates the 'Tristate subclass', and 326/82 points to 'Current driving fan in/out.' 326/31 identifies the 'Switching threshold stabilization' subclass (essentially a priority encoder). Figure 1 illustrates the location of these components on the circuit, and the calculation of their ease of recombination scores, as well as the calculation of the patent's coupling score of 0.61. This score, slightly below the mean value of 0.63, seems reasonable as digital hardware recombines with relative ease.

To validate the measure across a wider range of technologies, we surveyed inventors about the coupling of the components of their inventions. Randomly choosing 0.2 percent of U.S. patents granted to U.S. inventors in the year 2000 gave us a sample of 199 patents. Of these, we located current contact information for the patent holders for 189, from which we received responses in 64 cases. To assess the coupling between an invention's components, we asked (from Ulrich, 1995), 'Modules are said to be *coupled* when a change made to one module requires a change to the other module(s) in order for the overall invention to work correctly. How coupled were the modules of your invention?' The Pearson correlation coefficient between the responses and our measure of coupling is 0.30 ($p \sim 0.03$); though not perfect, the relationship is highly significant.[12] Most importantly, our automated procedure allows us to analyze a large set of patents, and appears to provide at least a rough estimate of coupling.

Invention usefulness

We defined usefulness as the number of citations that a patent receives in the 5 years following its grant date.[13] Each patent by law must cite previous patents that relate closely to its own technology. Research demonstrates that the number of citations a patent receives correlates highly with its technological importance, as measured by expert opinions, social value, and industry awards (Trajtenberg, 1990; Albert *et al.*, 1991). It also corresponds closely to the economic value of an invention (Harhoff *et al.*, 1999, Hall, Jaffe, and Trajtenberg, 2000). Thus, citation counts offer a means of measuring inventive usefulness across a broad range of technologies.[14] Descriptive statistics for these variables, as well as the others used in the models, appear in Table 2.

Science, coupling, and citations

One can see the relationship between science, coupling, and the usefulness of the inventions even without running a complex statistical analysis. Figure 2 depicts this relationship graphically: the points plot the mean number of citations received against the average level of coupling for each quintile of the coupling distribution. It plots this relationship separately for patents that cite a scientific

[12] While we provided inventors with specific definitions of how they should think about their subsystems and components, they may still have placed different boundaries in the organization of their inventions. This would increase the error of our measure. Increased error, however, should only dampen our estimates of the effects of coupling, making our empirical tests more conservative.

[13] We chose this period to capture the bulk of the citations to a patent, as citations typically peak about 3 years after the grant date (Jaffe, Trajtenberg, and Henderson, 1993). Although citation rates to patents appear quite consistent over time, we tested the sensitivity of this assumption by also using a 10-year window of citation counts; the results remain robust across both.

[14] The propensity to cite does vary across technologies as a function of the level of activity in that technology. We introduce multiple methods for controlling for this heterogeneity below.

Table 2. Descriptive statistics of the variables used in the models

	Full sample $N = 16{,}822$		Cite science $N = 2{,}919$		Do not cite science $N = 13{,}903$	
	Mean	S.D.	Mean	S.D.	Mean	S.D.
Cites	2.75	3.54	3.43	4.25	2.60	3.34
Mean technology control	1.19	0.41	1.27	0.45	1.17	0.39
Number of prior art cites	7.63	6.99	7.90	8.66	7.27	6.04
Single subclass dummy	0.08	0.27	0.08	0.28	0.08	0.27
Number of subclasses	4.31	3.31	4.97	4.86	4.02	2.79
Number of classes	1.78	0.95	1.90	1.06	1.74	0.91
Number of trials	2.94	15.23	2.55	18.01	2.98	14.07
Recent technology	3.96	0.65	4.17	0.49	3.92	0.67
Science stock	0.21	0.22	0.45	0.24	0.16	0.17
Reference to science	0.18	0.38	1.00	0.00	0.00	0.00
Coupling	0.63	0.35	0.50	0.28	0.66	0.36

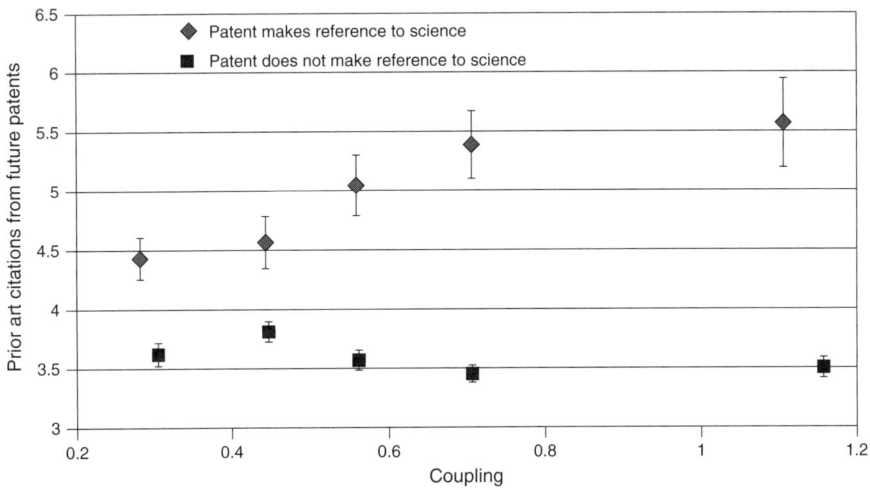

Figure 2. Mean citations (±1 S.E.M.) across quintiles of the coupling variable

article and for those that do not. Patents that reference science receive more cites on average. More importantly, the figure illustrates the predicted positive interaction effect. One can clearly see that while the relationship between coupling and citations first rises and then falls for patents that do not cite science, those that do reference an article in the *Scientific Index* show a positive relationship between coupling and citations. Even at this cursory level, the data support the claim that science helps most when inventors attempt a difficult recombination task.

Multivariate analysis

To investigate the process in more detail and to control for a variety of factors, we estimated fixed effects negative binomial models of future citations. The dependent variable of citation counts takes on only whole number values (i.e. 0, 1, 2, 3...). Using linear regression on such data can yield inefficient, inconsistent, and biased coefficient estimates. Count models can avoid these problems (Cameron and Trivedi, 1998). We employed one particular variant suggested by Hausman, Hall, and Griliches (1984): the fixed effects negative binomial. This procedure usefully allows for overdispersion—where the variance exceeds the mean—which these data exhibit. It also allows us to control broadly for differences across technological domains by defining fixed effects at the level of the USPTO class.[15] In this case, using the fixed effects negative binomial amounts to taking the number of citations received by all patents in a particular class as a given and estimating which factors predict the distribution of those citations among the patents within the class.

Several measures captured other potential sources of heterogeneity. First, we developed a *technology control*, which essentially refines the fixed effects. In the first stage, we calculated the average number of citations that each patent in a particular USPTO class received from patents granted between January of 1985 and June of 1990 (Equation 3).[16] We weighted these parameters according to the patent's class assignments (Equation 4), where p indicates the proportion of patent k's subclass memberships that fall in class i.

$$\text{Average citations in patent class } i \equiv \mu_i = \frac{\sum_{j \in i} \text{Citations}_j \text{ (before 7/90)}}{\text{Count of patents } j \text{ in subclass } i} \quad (3)$$

$$\text{Technology mean control patent } k \equiv M_k = p_{ik}\mu_i \quad (4)$$

To augment our technology control, we included several additional variables. We incorporated a count of *the number of major classes*. Patents that cover a broad range of technologies may carry a higher risk of being cited simply because future inventions from each field might potentially cite them—similar to what happens when an academic article spans several literatures. This measure also controls for the interdisciplinary breadth of the invention.[17] We also included the *number of prior art citations*—references to prior patents assigned by the U.S. Patent Office—as a control for the degree of local search (Podolny and Stuart, 1995); it may also capture idiosyncratic differences in citation propensity that our technological class controls miss. We additionally measured previous local search as the *number of repeated trials* on a particular landscape (Fleming and Sorenson, 2001). Specifically, we counted the number of previous patents that combined exactly the same set of subclasses.

We control for the number of components or problem domains in an invention by including the *number of subclasses*. Eight percent of the patents in our data belong to only one subclass. Although we suspect that these inventions also arise from a process of recombination—as the prior art references in these patents reveals, they do build on previous inventions—this combination occurs at a finer grain than our measures can capture. Therefore, we added a *single subclass dummy* variable to capture any systematic differences between these inventions and those assigned to multiple subclasses. A *recent technology* control accounts for any differences in citation patterns for technologies closer to the 'cutting edge' (Katila, 2002); we calculated this variable by averaging the patent numbers of the prior art cited.[18] Finally, we controlled for the cumulative *science stock* in a field of technology by calculating the percentage of patents up to 1990 that cite a non-patent reference in the same primary subclass as the focal patent. Table 3 presents the correlations between these variables, as well as those of theoretical interest.

The results of the fixed effects negative binomial models appear in Table 4. Model 1 provides a baseline. After controlling for a variety of factors, strong coupling generates an even stronger negative effect on citations than suggested by Figure 2. Science clearly has a positive effect on citation counts. The fact that this effect persists while controlling for the science stock reveals that these benefits accrue specifically to the patents drawing

[15] In unreported analyses, we also defined fixed effects at the level of the inventing firm, with qualitatively equivalent results. Thus, firm-level differences cannot account for our results.

[16] We allow all patents issued between January 1985 and June 30, 1990 to enter the estimation of the technology control, meaning that the patents used to calculate it vary in the time during which they can receive citations. Alternatively, we could select a small set of patents from 1985 and base the measures on the subsequent 5 years of citations; however, this approach would ignore the patent activity just prior to our sample.

[17] Unreported models indicated that coupling actually decreases the likelihood that a patent receives citations from other classes. Hence, our measure of coupling does not appear to pick up the breadth of application.

[18] Because patenting occurs at a relatively constant rate, this measure correlates at roughly 0.99 with the average age of the patent's prior art measured in terms of days.

Table 3. Correlation matrix

Correlation matrix	Cites	Mean	Priors	Single	Subclass	Classes	Trials	Recent	Science
Mean technology control	0.298								
Number of prior art cites control	0.117	0.020							
Single subclass dummy control	−0.056	0.002	−0.065						
Number of subclasses control	0.110	0.013	0.079	−0.288					
Number of classes control	0.074	−0.021	0.063	−0.244	0.513				
Number of trials control	−0.028	−0.028	−0.007	0.448	−0.168	−0.143			
Recent technology	0.208	0.377	−0.160	−0.005	0.066	0.044	−0.012		
Reference to science	0.095	0.089	0.018	0.006	0.107	0.061	−0.012	0.146	
Coupling	−0.029	−0.057	0.017	0.194	−0.309	−0.275	0.364	−0.135	−0.178

Table 4. Negative binomial estimates of citation counts (5-year window, standard errors in parentheses)[a]

	Model 1 baseline	Model 2	Model 3 10-year variable	Model 4
Mean technology control	**0.246**	**0.245**	**0.212**	**0.246**
	(0.036)	**(0.036)**	**(0.036)**	**(0.036)**
Number of prior art cites	**0.015**	**0.015**	**0.015**	**0.015**
	(0.001)	**(0.001)**	**(0.001)**	**(0.001)**
Single subclass dummy	**−0.186**	**−0.187**	**−0.201**	**−0.185**
	(0.036)	**(0.036)**	**(0.036)**	**(0.038)**
Number of subclasses	**0.027**	**0.027**	**0.029**	**0.027**
	(0.003)	**(0.003)**	**(0.002)**	**(0.003)**
Number of classes	**0.059**	**0.059**	**0.062**	**0.059**
	(0.009)	**(0.009)**	**(0.009)**	**(0.009)**
Number of trials	−0.000	−0.000	−0.001	−0.000
	(0.001)	(0.001)	(0.001)	(0.003)
Recent technology	**0.316**	**0.316**	**0.290**	**0.316**
	(0.016)	**(0.016)**	**(0.016)**	**(0.016)**
Science stock in subclass	0.083	0.085	0.073	0.083
	(0.052)	(0.051)	(0.051)	(0.052)
Cite to scientific publication	**0.098**	0.105	0.079	**0.098**
	(0.022)	(0.058)	(0.052)	**(0.022)**
Coupling	**0.462**	**0.485**	**0.632**	**0.471**
	(0.071)	**(0.076)**	**(0.079)**	**(0.078)**
Coupling^2	**−0.116**	**−0.136**	**−0.155**	**−0.121**
	(0.028)	**(0.030)**	**(0.036)**	**(0.032)**
Coupling × cite to scientific publication		−0.097	−0.023	
		(0.127)	(0.125)	
Coupling^2 × cite to scientific publication		**0.118**	**0.099**	
		(0.049)	**(0.051)**	
Number of trials × coupling				−0.000
				(0.003)
Number of trials × coupling^2				0.000
				(0.001)
Constant	**−1.891**	**−1.893**	**−1.783**	**−1.894**
	(0.084)	**(0.085)**	**(0.078)**	**(0.085)**
Fixed effects	363 Classes	363 Classes	363 Classes	363 Classes
Log-likelihood	−33,619.4	−33,614.5	−33,571.3	−33,619.4
N	16,822	16,822	16,822	16,822

[a] Bold type indicates that the coefficient estimate differs signification from zero with 95% confidence.

on science, rather than to all patents in a particular technological area. The primary test of the role of science appears in Model 2; this model interacts coupling with whether or not the patent cites a journal in the *Scientific Index*. The interaction significantly improves the model ($\chi^2 = 9.8$, d.f. = 2)[19] and shows that while coupling has a non-monotonic relationship with citations for patents that do not cite science, the relationship remains positive for patents citing science. Because of this fact, the benefits to science increase with coupling: at the mean level of coupling, patents that reference science receive 9.5 percent more citations than those that do not; at one standard deviation above the mean on coupling, the advantage increases to 13.1 percent; and at two standard deviations it reaches 20.3 percent. Model 3 tests the robustness of our measure of coupling by substituting a measure based only on the 10-year window preceding our sample rather than on the entire history of the U.S. Patent Office. The relationship between science and the effects of coupling remain the same.

Still, one might also ask whether scientific knowledge influences the process of search in a different manner from empirical knowledge (experience). Model 4 investigates this proposition by interacting the number of trials—a measure of the technological community's experience with a set of components—with coupling. The results reveal that, unlike science, experience does not mitigate the detrimental effects of high coupling, strongly corroborating the notion that search with the aid of science differs fundamentally from local search. Unreported models also attempted interactions, following Podolny and Stuart (1995), with the number of prior art citations as an indicator of local search; these models similarly failed to find an interaction between local search and interdependence.

If science helps inventors predict the results of their recombinant search, it should also reduce the variance of search outcomes—the uncertainty of invention. Thus, we analyzed the residuals from the negative binomial models to determine whether science reduced the variation in the usefulness of inventive outcomes. Table 5 presents ordinary least squares estimates of the absolute value of the residuals from Model 2.[20] We included the number of future citations in the model to account for the relationship between the mean and the variance of a count variable. As one would expect if science helps direct inventors, science reduces the dispersion of outcomes as the degree of coupling rises. At the mean level of coupling, the application of science reduced variability by 6.5 percent; at one standard deviation above the mean, this reduction in risk increases to 9.0 percent.

Table 5. OLS analysis of absolute residuals (standard errors in parentheses)[a]

	Model 5
Citation count	**0.967**
	(0.001)
Coupling	**0.068**
	(0.007)
Cite to scientific publication	−0.008
	(0.013)
Coupling × cite to scientific publication	**−0.257**
	(0.022)
Constant	**−0.076**
	(0.006)
R^2	0.99

[a] Bold type indicates that the coefficient estimate differs signification from zero with 95% confidence.

Robustness checks

Some additional robustness checks appear in Table 6. Models 6 and 7 examine the importance of the definition of the dependent variable. In particular, Model 6 excludes self-citations (when the future citing patent belongs to the same assignee) from the dependent variable. Self-citations may differ in their meaning from citations by competitors—notably Sørensen and Stuart (2000) interpret self-citations as an indicator that the organization's patents may not fit the current state of the market; thus, self-citations have an ambiguous role as a measure of patent importance. The results hold, however, even when excluding self-citations from the dependent variable. Model 7 increases to 10 years the length of the future

[19] Haberman's (1977) chi-squared, two times the difference in log-likelihoods, provides a test of the statistical significance of the difference between two nested models.

[20] In unreported analyses, we estimated the mean and variance effects simultaneously, using the variance decomposition negative binomial (King, 1989). Simultaneous estimation produced equivalent results. The variance decomposition negative binomial, however, has the disadvantage of not being able to incorporate fixed effects; thus, we report the two-stage estimation.

Table 6. Negative binomial estimates of citation counts (standard errors in parentheses)[a]

	Model 6 no self-citations	Model 7 10-year citations	Model 8	Model 9 Selection
Mean technology control	**0.256**	**0.135**	**0.248**	**0.242**
	(0.038)	**(0.028)**	**(0.036)**	**(0.036)**
Number of prior art cites	**0.014**	**0.015**	**0.015**	**0.014**
	(0.001)	**(0.001)**	**(0.001)**	**(0.001)**
Single subclass dummy	**−0.181**	**−0.176**	**−0.188**	**−0.200**
	(0.038)	**(0.031)**	**(0.036)**	**(0.037)**
Number of subclasses	**0.029**	**0.024**	**0.026**	**0.027**
	(0.003)	**(0.002)**	**(0.003)**	**(0.003)**
Number of classes	**0.061**	**0.050**	**0.059**	**0.059**
	(0.010)	**(0.008)**	**(0.009)**	**(0.009)**
Number of trials	−0.000	−0.000	−0.000	0.000
	(0.001)	(0.001)	(0.001)	(0.001)
Recent technology	**0.275**	**0.218**	**0.318**	**0.297**
	(0.016)	**(0.013)**	**(0.016)**	**(0.019)**
Science stock in subclass	**0.145**	−0.013	0.068	0.087
	(0.054)	(0.045)	(0.052)	(0.052)
Cite to scientific publication	0.085	0.062	**0.097**	0.118
	(0.062)	(0.053)	**(0.022)**	(0.060)
Cite to non-scientific publication			0.106	
			(0.112)	
Coupling	**0.423**	**0.418**	**0.459**	**0.441**
	(0.080)	**(0.064)**	**(0.072)**	**(0.100)**
Coupling^2	**−0.123**	**−0.121**	**−0.112**	**−0.109**
	(0.032)	**(0.025)**	**(0.028)**	**(0.035)**
Coupling × cite to scientific publication	−0.072	−0.016		−0.133
	(0.136)	(0.119)		(0.127)
Coupling^2 × cite to scientific publication	**0.108**	**0.094**		**0.121**
	(0.053)	**(0.048)**		**(0.047)**
Coupling × cite to non-scientific publication			0.044	
			(0.254)	
Coupling^2 × cite to non-scientific publication			−0.078	
			(0.112)	
Selection (likelihood of citing scientific publication)				**−0.469**
				(0.239)
Selection × coupling				**2.045**
				(0.609)
Selection × coupling^2				**−1.104**
				(0.403)
Constant	**−1.785**	**−1.304**	**−1.898**	**−1.837**
	(0.089)	**(0.069)**	**(0.084)**	**(0.091)**
Fixed effects	363 classes	363 classes	363 classes	363 classes
Log-likelihood	−31,411.3	−46,458.2	−33,613.9	−33,607.8
N	16,822	16,822	16,822	16,822

[a] Bold type indicates that the coefficient estimate differs signification from zero with 95% confidence.

citation window observed. If knowledge diffuses at different rates across technologies, and these rates correlate to the usage of science, this process might influence our results. Regardless, our models continue to show a non-monotonic relationship between coupling and citations for patents that do not cite a *Scientific Index* journal, and a strictly positive relationship between coupling and future citations for those referencing science.

One potential alternate explanation for these results involves the diffusion of knowledge. References to non-patent sources might speed the dissemination of information, allowing communities of practice to build knowledge more rapidly.

This account implies that the mere act of publishing increases the flow of information, thereby increasing citation rates. Model 8 tests this possibility directly by interacting the effects of coupling with whether or not the patent references a *non*-scientific publication.[21] A reference to these publications does nothing to mitigate the effect of coupling, failing to support this account. The diffusion story further suggests that references to science should uniformly increase citations, rather than primarily increasing citation rates for inventions combining coupled technologies. Since the citation to science main effect has no impact on future citations, the models also fail to support this account on that basis. Variation in knowledge diffusion does not appear to account for our findings.

Another potential explanation for our findings concerns the effort invested in solving the inventive problem. Science may denote a greater level of effort that proves particularly useful in the face of difficult technological challenges. Although we cannot measure effort directly, the number of inventors may proxy for this factor.[22] Nevertheless, in both cases using and not using science, coupling correlates *negatively* and significantly with the number of inventors listed on the patent. The evidence, therefore, suggests that we can dismiss this alternative.

A more serious critique might contend that patents that cite science differ in some other systematic way from those that do not. To investigate this possibility, we first estimated a logistic regression of whether or not a patent cited a *Scientific Index* journal (see Table 7). Even after controlling for differences across technologies using class fixed effects (Model 11), these estimates reveal some differences between those patents that reference science and those that do not. First, patents that draw on science tend to come from highly active research domains, as the positive and significant coefficients for both the number of prior art citations and the number of trials reveal. This finding suggests that science does not differ from non-science by identifying fundamental new components. Second, patents drawing on science more likely come out of academia, though university patents account for only a small proportion (4.8%) of the total number of patents citing science in our sample. Notably, the models fail to show that science produces knowledge of any greater generality; neither the number of classes nor subclasses has a significant relationship with the use of science in the invention process. In Models 12 and 13, we compared the factors that predict a citation to science to those that correlate to university ownership (Model 12) and citations to non-scientific references (Model 13).

Though this analysis increases our confidence that non-patent references to the scientific literature capture a connection to science, it also reveals one difference that requires further investigation: patents that cite science exhibit lower coupling. Our theory actually accommodates this fact. Because we measure coupling as a function of the history of recombination, the prior application of scientific knowledge in a technological domain should facilitate prior recombination, thereby lowering the observed coupling score. One might worry, however, that this correlation represents some other type of selection rather than an unfortunate (but necessary) weakness in our measure of coupling. To test the possible influence of this selection on our models, we created a selection term from the logistic regression results (Model 10) representing the probability that each patent would cite science given its characteristics.[23] We then interacted this selection parameter with the coupling score and its quadratic term to determine whether selection effects might account for our previous findings. The results—see Model 9—demonstrate that the effects of science remain robust even after explicitly accounting for potential selection issues.

DISCUSSION

The results provide strong evidence that the returns to science depend on the difficulty of the inventive

[21] We coded any patent that referenced a journal not in the *Scientific Index* (e.g., *Time*) or a non-technical corporate publication—usually a product catalog or advertisement—as having a non-scientific reference because these publication types most clearly do *not* involve the application of scientific theory.

[22] Thanks go out to Riitta Katila and Steven Klepper for suggesting this test.

[23] The logic for this specification follows that of a Heckman selection model. The *application period*—the number of years between the application and granting of the patent—allows us to identify the procedure. Although this factor positively relates to both university assignees and the likelihood of citing science, we have no reason to expect that it would influence the number of future patents that would cite the focal patent following the grant date and the lack of correlation ($r = 0.02$) between application period and future cites supports this notion.

Table 7. Logit models of the correlates of citing scientific and non-scientific publications and of university ownership (standard errors in parentheses)[a]

	Model 10 cite science	Model 11 cite science (fixed effects)	Model 12 university	Model 13 cite non-science
Number of prior art cites	**0.035**	**0.041**	0.009	**0.057**
	(0.003)	**(0.003)**	(0.008)	**(0.003)**
Single subclass dummy	−0.041	−0.097	0.322	0.055
	(0.100)	(0.106)	(0.246)	(0.144)
Number of subclasses	0.013	0.010	−0.031	0.015
	(0.008)	(0.009)	(0.022)	(0.012)
Number of classes	−0.028	0.015	**0.175**	−0.020
	(0.029)	(0.031)	**(0.078)**	(0.042)
Number of trials	**0.004**	**0.004**	0.005	0.001
	(0.002)	**(0.002)**	(0.003)	(0.002)
Recent technology	**0.296**	0.028	0.164	**−0.345**
	(0.047)	(0.053)	(0.141)	**(0.052)**
Science stock in subclass	**5.594**	**4.806**	**3.010**	**1.830**
	(0.122)	**(0.159)**	**(0.300)**	**(0.176)**
Application period	**0.113**	**0.071**	**0.109**	0.008
	(0.022)	**(0.023)**	**(0.053)**	(0.032)
Coupling	**−0.613**	**−0.297**	−0.401	0.171
	(0.100)	**(0.127)**	(0.283)	(0.126)
University dummy	**0.921**	**0.900**		−0.324
	(0.137)	**(0.169)**		(0.287)
Cite to non-scientific publication	**0.274**	**0.538**	−0.115	
	(0.092)	**(0.097)**	(0.289)	
Cite to scientific publication			**1.004**	0.116
			(0.166)	(0.095)
Constant	**−4.253**		**−6.415**	**−2.518**
	(0.122)		**(0.632)**	**(0.233)**
Fixed effects	None	249 Classes	None	None
Log-likelihood	−5743.1	−4976.3	−1021.0	−3392.7
N	16,822	14,778[b]	16,822	16,822

[a] Bold type indicates that the coefficient estimate differs signification from zero with 95% confidence.
[b] 2,044 cases drop out of Model 11 because they belong to classes in which none of the patents cite science.

problem being addressed. When inventors work with relatively independent components, science offers little or no advantage. On the other hand, science can allow inventors recombining highly coupled components to avoid the difficulties inherent in local search on a rugged landscape. Though the size of this effect may appear modest, a 10 percent increase in citations at the mean level of coupling, it likely corresponds to a much larger economic benefit. If the exponential relationship between citations and economic value found by Harhoff *et al.* (1999) for German patents holds in the U.S. market, this 10 percent increase in citations would correspond to a 214 percent increase in economic value.[24] At one standard deviation above the mean of coupling, the expected value of applying science would increase the average value of a patent by 281 percent, at two standard deviations, 473 percent. Hence, the economic value of science may increase considerably with the difficulty of the inventive problem.

A more nuanced consideration of the results allows us to investigate which proposed mechanisms likely play the most important roles in generating this effect. The results do not, however, allow us to rule out any of the mechanisms through which science might influence the search process of invention. As one would expect if science either points inventors more directly to the most useful configurations of components or simply allows them to avoid swaths of less productive solution space, science increases the average usefulness of more tightly coupled inventions (see

[24] Harhoff *et al.* (1999) find the following relationship for value: Dollars $= 1\,000\,000\, e^{\left(\frac{\ln \text{cites} - 2.076}{0.119}\right)}$.

Tables 4 and 6). Moreover, the strong effects that science has on reducing the variance in inventive outcomes using coupled components suggests that science operates more by leading inventors directly to useful outcomes (see Table 5); reducing the solution space would have a relatively weak effect on the unpredictability in outcomes because even the 'best' regions of rugged landscapes still have highly varied terrain. Inconsistent with this direct link, however, inventors calling on science appear to work in regions with more search activity. Both Models 10 and 11 indicate that inventions that call on science work with components that have been recombined more frequently in the past, whereas a direct path to the best combinations should reduce the number of iterations. This finding suggests that science also aids invention by providing researchers with motivation to continue searching in a particular direction despite negative feedback.

These two mechanisms, one directing researchers to the most useful regions of space, the other encouraging them in the face of failure, likely complement each other in the process of inventing with coupled components. Even in areas with well-developed science, inventors often do not have sufficient knowledge to predict all of the interactions that might occur among a highly coupled set of components. Hence, even after embarking on the right path, they must typically engage in trial-and-error learning (probably through local search) to fill in these missing pieces. Many qualitative accounts, such as Fleming's (2002) history of ink-jet printing, point to just such an interaction.[25] Nucleation physics greatly enhanced the inventors' understanding of the fundamental processes underlying the formation of ink bubbles, but they still went through hundreds of iterations between science and empirical search to find the right combination of materials to produce consistently high-quality printing. As in this case, science likely operates through multiple mechanisms in improving inventive outcomes.

Science may even influence the shape of technological innovation along other dimensions beyond the process that inventors use in recombining components. For example, science might aid in the discovery of new components (Smith and Hounshell, 1985). Although the recombination of elements implies that the potential number of inventions expands very rapidly with the number of fundamental components (Weitzman, 1996), completely new components must obviously arise from somewhere. Some exogenous process might govern these arrivals, but science may also stimulate technological progress by identifying new fundamental components for recombination. Though outside the scope of our analysis, it seems interesting to note that of the 442 patents excluded from our analysis because they had no prior art, 46 percent reference a scientific journal article (vs. 17% in the sample as a whole).

Regardless of the particular mechanisms, the differential returns to the application of science likely influence when inventors call on science and when they do not. Although the analysis of when non-patent science references appear on a patent suggests that these patents involve the recombination of less coupled elements (see Models 10 and 11), this result likely arises as an artifact of the calculation of our coupling measure. We estimated coupling by observing the historical ease of recombining patent subclasses. If researchers have used science in the past to facilitate the use of this component, this practice would bias our coupling measure downward (indeed, the stock of science in a particular subclass correlates negatively, $r = -0.28$, with its observed difficulty of recombination). The effects of the selection terms in Model 9 corroborate this notion; the types of patents on which science references appear exhibit larger marginal effects to coupling. Hence, inventors appear to use science most frequently in those domains that, because they work with highly coupled components, offer the greatest returns to its application. To some degree, then, variation in the application of science across industries likely stems from the types of inventive challenges participants in these sectors face.

These results extend prior research on the link between science and innovation. Much of the existing work of the role of science in stimulating technological advance investigates the research and patenting activities of universities, identifying such features as changes in federal law,[26] the

[25] Rosenberg (1974) offers several additional examples.

[26] Prior to 1980, the property rights to the results of research funded by government grants belonged to the federal government. Thus, university researchers funded by federal grants had little incentive to apply for patents. The Bayh-Dole Act, passed in 1980, eliminated this provision and allowed researchers to retain the property rights to federally funded research.

establishment of university offices for technology transfer, and increased industry funding of university research as factors influencing the propensity of universities to patent (Henderson *et al.*, 1998; Mowery and Shane, 2002). Much of the application of science, however, occurs outside the university; in our sample, only 4.5 percent of the patents referencing an article in the *Scientific Index* had university assignees. Though private firms also face incentives to use science (Cockburn *et al.*, 2000), most prior research has failed to address the question of how these incentives might vary systematically according to the firm's technology. By focusing on precisely how science might aid inventors, this research identifies one technological factor—the difficulty of recombining coupled components—that appears to influence the potential returns to the application of science.

Our findings also inform research on complex systems—particularly research applying Kauffman's NK model to the social sciences. Kauffman's model predicts a non-monotonic relationship between coupling and the outcomes of local search. When actors do not use science—and presumably therefore engage in local search—the results conform closely to Kauffman's expectations. Nevertheless, our results clearly show the *inapplicability* of this model to situations in which the searchers have some understanding or cognitive model of the landscape they search; inventors following science do not behave like local searchers. Thus, researchers wishing to apply the NK model to cognitively aware agents should revisit the model and modify it to account for the search heuristics that people actually use (cf. Gavetti and Levinthal, 2000; Rivkin, 2000).

For the individual inventor, the policy implications seem straightforward. Science offers limited benefits to inventors working with modular components—given the cost of searching and digesting the scientific literature, the price tag for using science likely exceeds its benefits for those working with uncoupled components. In contrast, science offers large potential rewards to inventors operating with highly coupled components. Without science, inventors that search these landscapes must rely on exhaustive search techniques. Although inventors could focus resources on developing methods for testing large numbers of combinations cheaply, efforts along these lines have met with limited success; for example, combinatorial synthesis strategies remain problematic, as demonstrated by the recent controversy in drug discovery over the effectiveness of combinatorial chemistry and high-throughput screening (Drews, 2000). An understanding of the forces creating these interactions through the development of basic science offers an attractive means for accelerating invention in these highly coupled technologies.

In conclusion, we wish to reiterate that this paper proposed and tested one explanation for why the returns to the application of scientific knowledge to invention might vary across technological fields. In particular, by envisioning invention as a process of searching through combinations of technological components for new and useful configurations, we argue that science acts like a map—providing inventors with a sense of the underlying technological landscape they search—thereby allowing them to avoid the difficulties inherent in trying to combine highly coupled components. Our results demonstrate that scientific knowledge mitigates the negative effect that coupling typically has on the outcomes of invention. In doing so, this research elaborates the more proximate role that science plays in technological advance and how it influences the rate and quality of invention. Thus, it brings a somewhat more nuanced view of the benefits of science in stimulating technological advance, and with it economic growth, to the already extensive literature on the subject.

ACKNOWLEDGEMENTS

We acknowledge the Harvard Business School Department of Research for their financial support, Corey Billington and Ellen King of Hewlett Packard for donation of their patent database, and Evelina Fedorenko for her historical research. We especially wish to thank Sue Borges for her painstaking analysis of over 16,000 non-patent citations. Comments from Pierre Azoulay, Michael Darby, Riitta Katila, Steven Klepper, Dan Levinthal, Sue McEvily, Jan Rivkin, and the participants of seminars at UCLA and NBER improved this paper substantially. The usual disclaimer applies.

REFERENCES

Adams J. 1990. Fundamental stocks of knowledge and productivity growth. *Journal of Political Economy* **98**: 673–702.

Ahuja G, Lampert J. 2001. Entrepreneurship in the large corporation: a longitudinal study of how established firms create breakthrough inventions. *Strategic Management Journal*, Special Issue **22**(6–7): 521–543.
Albert M, Avery D, Narin F, McAllister P. 1991. Direct validation of citation counts as indicators of industrially important patents. *Research Policy* **20**: 251–259.
Allen T. 1977. *Managing the Flow of Technology*. MIT Press: Cambridge, MA.
Athey S, Stern S. 1998. An empirical framework for testing theories about complementarity in organizational design. Working paper #6600, National Bureau of Economic Research.
Audia P, Sorenson O. 2001. Aspirations, performance and organizational change: a multi-level analysis. Working paper, London Business School.
Baldwin C, Clark K. 2000. *Design Rules: The Power of Modularity*. MIT Press: Cambridge, MA.
Basalla G. 1988. *The Evolution of Technology*. Cambridge University Press: Cambridge, U.K.
Cameron A, Trivedi P. 1998. *Regression Analysis of Count Data*. Cambridge University Press: Cambridge, U.K.
Campbell A. 2001. Superconductivity: how could we miss it? *Science* **292**: 65–66.
Carlson D. 2001. Personal interview, June 5: Cary, NC.
Cockburn I, Henderson R, Stern S. 2000. Untangling the origins of competitive advantage. *Strategic Management Journal*, Special Issue **21**(10–11): 1123–1145.
Cyert R, March J. 1963. *A Behavioral Theory of the Firm*. Prentice-Hall: Englewood Cliffs, NJ.
Dresselhaus M. 2001. Burn and interrogate. *Science* **292**: 650–651.
Drews J. 2000. Drug discovery: a historical perspective. *Science* **287**: 1960–1964.
Fleming L. 2001. Recombinant uncertainty in technological search. *Management Science* **47**: 117–132.
Fleming L. 2002. Finding the organizational sources of technological breakthroughs: the story of Hewlett Packard's thermal inkjet. *Industrial and Corporate Change* **11**: 1059–1084.
Fleming L, Sorenson O. 2001. Technology as a complex adaptive system: evidence from patent data. *Research Policy* **30**: 1019–1039.
Gambardella A. 1995. *Science and Innovation: The U.S. Pharmaceutical Industry During the 1980s*. Cambridge University Press: Cambridge, U.K.
Gavetti G, Levinthal D. 2000. Looking forward and looking backward: cognitive and experiential search. *Administrative Science Quarterly* **45**: 113–137.
Gibbons M, Johnston R. 1974. The roles of science in technological innovation. *Research Policy* **3**: 220–242.
Gilfillan S. 1935. *Inventing the Ship*. Follett: Chicago, IL.
Haberman S. 1977. Log-linear model and frequency tables with small expected cell counts. *Annals of Statistics* **5**: 1148–1165.
Hall B, Jaffe A, Trajtenberg M. 2000. Market value and patent citations: a first look. NBER working paper 7741.
Hansen MT, Lovas B. 2004. How do multinational companies leverage technological competancies? Moving from single to interdependent explanations. *Strategic Management Journal*, Special Issue **25**(8–9): 801–822.
Harhoff D, Narin F, Scherer FM, Vopel K. 1999. Citation frequency and the value of patented inventions. *Review of Economics and Statistics* **81**: 511–515.
Hausman J, Hall B, Griliches Z. 1984. Econometric models for count data with an application to the patents–R&D relationship. *Econometrica* **52**: 909–938.
Henderson R, Clark K. 1990. Architectural innovations: the reconfiguration of existing product technologies and failure of established firms. *Administrative Science Quarterly* **35**: 9–30.
Henderson R, Cockburn I. 1994. Measuring competence? Exploring firm effects in drug discovery. *Strategic Management Journal*, Winter Special Issue **15**: 63–84.
Henderson R, Jaffe A, Trajtenberg A. 1998. Universities as a source of commercial technology: a detailed analysis of university patenting, 1965–1988. *Review of Economics and Statistics* **80**: 119–127.
Jaffe A. 1989. Real effects of academic research. *American Economic Review* **79**: 957–970.
Jaffe A, Trajtenberg M. 1996. Flows of knowledge from universities and federal labs: modeling the flow of patent citations over time and across institutional and geographic boundaries. *Proceedings of the National Academy of Sciences* **93**: 12 671–12 677.
Jaffe A, Trajtenberg M, Henderson R. 1993. Geographic localization of knowledge spillovers as evidenced by patent citations. *Quarterly Journal of Economics* **108**: 577–598.
Katila R. 2002. New product search over time: past ideas in their prime. *Academy of Management Journal* **45**: 995–1010.
Kauffman S. 1993. *The Origins of Order*. Oxford University Press: New York.
King G. 1989. Event count models for international relations: generalizations and applications. *International Studies Quarterly* **33**: 123–147.
Klepper S. 1996. Entry, exit, and innovation over the product life-cycle. *American Economic Review* **86**: 562–583.
Kramer P. 1997. *Listening to Prozac*. Penguin: New York.
Levin R, Klevorick A, Nelson R, Winter S. 1987. Appropriating the returns from industrial research and development: comments and discussion. *Brookings Papers on Economic Activity* **3**: 783–831.
Levinthal D. 1997. Adaptation on rugged landscapes. *Management Science* **43**: 934–950.
Lippman S, McCall J. 1976. The economics of job search: a survey. *Economic Inquiry* **14**: 155–189.
Mansfield E. 1972. Contribution of R&D to economic growth in the United States. *Science* **175**: 477–486.
Mansfield E. 1995. Academic research underlying industrial innovations: sources, characteristics and financing. *Review of Economics and Statistics* **77**: 55–65.

March J. 1991. Exploration and exploitation in organizational learning. *Organization Science* **2**: 71–87.
March J, Simon H. 1958. *Organizations*. Blackwell: Cambridge, MA.
Martin X, Mitchell W. 1998. The influence of local search and performance heuristics on new design introduction in a new product market. *Research Policy* **26**: 753–771.
McCluskey E. 1986. *Logic Design Principles with Emphasis on Testable Semi-custom Circuits*. Prentice-Hall: Englewood Cliffs, NJ.
Meyer M. 2000. Does science push technology? Patents citing scientific literature. *Research Policy* **29**: 409–434.
Milgrom P, Roberts J. 1990. The economics of modern manufacturing: technology, strategy and organization. *American Economic Review* **80**: 511–528.
Millman J. 1979. *Microelectronics*. McGraw-Hill: New York.
Mowery D, Shane S. 2002. Introduction to the special issue on university entrepreneurship and technology transfer. *Management Science* **48**: v–ix.
Narin F, Hamilton K, Olivastro D. 1997. The increasing linkage between U.S. technology and public science. *Research Policy* **26**: 317–330.
Nelson R. 1982. The role of knowledge in R&D efficiency. *Quarterly Journal of Economics* **97**: 453–470.
Nelson R, Winter S. 1982. *An Evolutionary Theory of Economic Change*. Belknap: Cambridge, MA.
Ouyang M, Huang J, Cheung C, Lieber C. 2001. Energy gaps in 'metallic' single-walled carbon nanotubes. *Science* **292**: 702–705.
Perko J, Narin F. 1997. The transfer of public science to patented technology: a case study in agricultural science. *Journal of Technology Transfer* **22**: 65–72.
Podolny JM, Stuart TE. 1995. A role-based ecology of technological change. *American Journal of Sociology* **100**: 1224–1260.
Rivkin J. 2000. Imitation of complex strategies. *Management Science* **46**: 824–844.
Rosenberg N. 1974. Science, invention, and economic growth. *Economic Journal* **84**: 90–108.
Schmoch U. 1993. Tracing the knowledge transfer from science to technology as reflected in patent indicators. *Scientometrics* **26**: 193–211.
Schumpeter J. 1939. *Business Cycles*. McGraw-Hill: New York.
Smith J, Hounshell D. 1985. Wallace H. Carothers and fundamental research at DuPont. *Science* **229**: 436–442.
Sørensen J, Stuart T. 2000. Aging, obsolescence and organizational innovation. *Administrative Science Quarterly* **45**: 81–112.
Sorenson O. 2002. Interorganizational complexity and computation. In *Companion to Organizations*, Baum JAC (ed). Blackwell: Oxford; 664–685.
Stephan P. 1996. The economics of science. *Journal of Economic Literature* **34**: 1199–1235.
Stix G. 2001. Project Skyhook: a 'smart' material that transforms from a liquid to a solid state on cue is beginning to show up in prosethetics, automobiles, and other applications. *Scientific American* May: 28–29.
Stuart T, Podolny J. 1996. Local search and the evolution of technological capabilities. *Strategic Management Journal*, Summer Special Issue **17**: 21–38.
Sveikauskas L. 1981. Technological inputs and multifactor productivity growth. *Review of Economics and Statistics* **63**: 275–282.
Tijssen R. 2001. Global and domestic utilization of industrial relevant science: patent citation analysis of science–technology interactions and knowledge flows. *Research Policy* **30**: 35–54.
Trajtenberg M. 1990. A penny for your quotes: patent citations and the value of innovations. *Rand Journal of Economics* **21**: 172–187.
Tratjenberg M, Henderson R, Jaffe A. 1997. University versus corporate patents: a window on the basicness of invention. *Economic Innovation and New Technology* **5**: 19–50.
Ulrich K. 1995. The role of product architecture in the manufacturing firm. *Research Policy* **24**: 419–440.
Usher A. 1954. *A History of Mechanical Invention*. Dover: Cambridge, MA.
Vincenti W. 1990. *What Engineers Know, and How They Know It*. Johns Hopkins University Press: Baltimore, MD.
Weinberger E. 1991. Local properties of Kauffman's N-K model: a tunably rugged energy landscape. *Physical Review A* **44**: 6399–6413.
Weitzman M. 1996. Hybridizing growth theory. *American Economics Review* **86**: 207–212.
Wright S. 1932. The roles of mutations, inbreeding, crossbreeding and selection in evolution. *Proceedings of the 11th International Congress of Genetics* **1**: 356–366.
Zucker L, Darby M. 1996. Star scientists and institutional transformation: patterns of invention and innovation in the formation of the biotechnology industry. *Proceedings of the National Academy of Sciences* **93**: 709–716.
Zucker L, Darby M, Brewer M. 1998. Intellectual human capital and the birth of U.S. biotechnology enterprises. *American Economic Review* **88**: 290–306.

[12]

Organization Science
Vol. 15, No. 6, November–December 2004, pp. 617–632
ISSN 1047-7039 | EISSN 1526-5455 | 04 | 1506 | 0617

informs®

DOI 10.1287/orsc.1040.0093

A Knowledge-Based Theory of the Firm—The Problem-Solving Perspective

Jack A. Nickerson, Todd R. Zenger
John M. Olin School of Business, Washington University, Campus Box 1133, One Brookings Drive. St. Louis, Missouri 63130-4899
{nickerson@wustl.edu, zenger@wustl.edu}

In this paper we develop a knowledge-based theory of the firm. While existing knowledge-based theory focuses on the efficiency of hierarchy in economizing on knowledge exchange, we develop a theory of the firm that focuses on the efficiency of alternative organizational forms in *generating* knowledge or capability. Our theory begins with the problem as the basic unit of analysis, arguing that a problem's complexity influences the optimal method of solution search and the optimal means of organizing that search. The distinguishing feature that differentiates among organizational alternatives is the different way each resolves conflict over the selection of solution trials, that is, the way it chooses the path of search. Our theory predicts that efficiency demands that these governance alternatives be matched in a discriminating way to problems based on their associated benefits and costs in governing solution search. Thus, our theory is among the first to simultaneously treat both the boundary choice (i.e., internal versus external) and the choice among alternative internal approaches to organizing.

Key words: knowledge-based view; knowledge management; problem solving; innovation; firm boundaries; organizational boundaries

1. Introduction

In the strategy literature, a key task of the manager is to accumulate and protect valuable knowledge or capability (Rumelt 1984, Barney 1984, Wernerfelt 1984, Teece et al. 1997). Such knowledge or capability defines a firm's capacity to efficiently convert its inputs into valuable outputs (Arrow and Hahn 1971; Debreu 1959; Nelson and Winter 1982, pp. 59–60). Thus, managers enhance the firm's capacity to produce efficiently by updating or advancing knowledge. Therefore, a common assumption and prescription in the strategy literature is that the boundaries of the firm should encompass these valuable competencies and core knowledge (Argyres 1996, Prahalad and Hamel 1990). By internalizing valuable knowledge or keeping this knowledge internal, the firm positions itself to both exploit and protect knowledge. Arguably, however, the key knowledge-based question the manager faces is not how to organize to exploit already developed knowledge or capability, but rather how to organize to efficiently *generate* knowledge and capability.

Recently, scholars within the knowledge- or resource-based perspective have focused their attention on this question. In the process, scholars have sought to develop what some call a knowledge-based view or knowledge-based theory of the firm (Conner 1991; Demsetz 1988; Conner and Prahalad 1996; Kogut and Zander 1992, 1996; Grant 1996; Madhok 1996; Nahapiet and Ghoshal 1998). This work seeks to explain how the choice of organization—particularly the choice of whether to integrate or outsource an activity—influences the efficient production and protection of valuable knowledge and capabilities. In so doing, this literature seeks a theory of the firm that is independent of transaction cost logic and its behavioral assumption of opportunism. A common argument in much of this literature is that firms as organizational forms exist to economize on the exchange of knowledge rather than to attenuate opportunism. As Conner (1991, p. 139) argues, the integration of an activity within the boundaries of the firm thereby becomes a "creator of a positive" rather than an "avoider of a negative."

While this existing literature on the knowledge-based view of the literature makes important contributions to our understanding of the firm, it has significant shortcomings. Others have convincingly contested the claim that a knowledge-based theory of the firm can be independent of opportunism (Foss 1996a, b; Heiman and Nickerson 2002; Mahoney 2001; Williamson 1999), but our critique and contribution addresses other shortcomings. First, this literature has primarily focused on the role of firms in providing efficient knowledge exchange rather than their role in efficiently producing knowledge or capabilities. Second, the two fundamental arguments within the literature that support the efficiency of firms in knowledge exchange relative to markets are fully contradictory. One claims that hierarchies exist to essentially *avoid* knowledge transfer (Demsetz 1988, Conner 1991, Conner and Prahalad 1996), emphasizing the firm's capacity to exercise authority in directing others' actions; the other view claims that hierarchies exist instead to *facilitate* knowledge transfer (Arrow 1974; Kogut and Zander 1992, 1996; Nahapiet and Ghoshal 1998), emphasizing the firm's capacity to support the formation of shared language and identity. Finally, the literature has yet to provide a theory of alignment—a theory that predicts when hierarchies are preferred to markets or when markets prevail, let alone use knowledge-based logic to do so. Indeed, the knowledge-based arguments tend to highlight the virtues of hierarchy, not its limits in forming and transferring knowledge. Hierarchy is argued to be superior in transferring and forming knowledge, leaving other factors to constrain firm boundaries (see Conner and Prahalad 1996). In summary, we do not yet have a knowledge-based theory of the firm.

In this paper, we develop a knowledge-based theory of the firm that addresses these shortcomings. We explain how a firm's prospective objectives for knowledge formation dictate the choice of how to organize. Here the critical question is not whether knowledge should be owned or acquired in the market or how the exchange of knowledge should be facilitated, but rather how a manager should organize individuals to generate knowledge that the firm seeks.

We begin with the assumption that the manager's knowledge-based objective is to create valuable new knowledge. The manager, however, cannot simply choose new knowledge to acquire, because more often than not the desired knowledge does not exist. Instead, the manager must choose valuable problems—those which, if successfully solved, yield desirable knowledge or capability. The value of a particular problem therefore depends on two factors: (1) the values in the array of possible solutions and (2) the costs of discovering a particularly valuable one. Once a problem is selected, the manager then organizes a search for solutions that optimizes the likelihood, speed, and cost with which valuable solutions are discovered. Finally, the manager seeks to appropriate a portion of the solution's value. While we present these steps sequentially for purposes of exposition, in truth the costs of solution search and the capacity to appropriate value must be factored into the initial choice of problems as well. Thus, the unit of analysis in our theory is the problem, and profitable formation of new knowledge (or new capability) is the central goal in choosing a form of governance.

Following previous work, we assume that solutions to complex problems represent unique combinations or syntheses of existing knowledge (Fleming and Sorenson 2000, Henderson and Clark 1990). Recently, scholars have depicted this matrix of unique combinations of knowledge as a landscape. Some solution landscapes are rugged, with many high-value solutions scattered across the terrain. Others are smooth, with a single, high-value solution or with high-value solutions concentrated in a particular region of the landscape. Consequently, problems differ in the optimal form of solution search. For some, directional search or search based on trial and error leads efficiently to a high-value solution (Gavetti and Levinthal 2000). For others, search guided by a cognitively developed heuristic of the landscape's topography is preferred (Gavetti and Levinthal 2000).

After identifying a problem and assessing the ideal form of search, the manager must decide how to organize this search. In particular, the manager must decide how to access relevant knowledge either within or outside the firm. We explicitly explore the capacity of three alternative governance modes to support these differing forms of search: markets, authority-based hierarchy, and consensus-based hierarchy. The differentiating feature among these governance alternatives is the different way in which each resolves conflict over the selection of solution trials—that is, the way it chooses the path of search. Our theory predicts that efficiency demands that these governance alternatives be matched in a discriminating way to problems based on their associated benefits and costs in governing solution search. Thus, our theory is among the first to simultaneously treat both the boundary choice (i.e., internal versus external) and the choice among alternative internal approaches to organizing.

Our theory development proceeds as follows. In §2, we introduce the problem as the unit of analysis, introducing the concept of solution landscapes that differ based on problem complexity. In that section we match problem characteristics to search strategies. In §3, we discuss knowledge-exchange hazards that hinder efforts to conduct one form of search. In §4, we discuss the capacity of alternative governance structures to conduct differing forms of search. In §5 we finally match problems to alternative governance forms. Section 6 discusses our theory and describes future avenues for research.

2. Capability and Knowledge Formation as Problem Solving

Problem solving and knowledge formation are at the center of our theory. The manager's fundamental knowledge-based objective is to sustain above-normal profits by continually discovering new knowledge or new solutions that form from unique combinations of existing knowledge. Nelson and Winter (1982) explicitly define a firm's knowledge (or capability) as the "input-output combinations achievable with all possible mixes and levels of activities known to the firm" (pp. 63–64). In this literature, the state of a firm's knowledge can be advanced by either absorbing existing knowledge external to the firm or by developing new knowledge by first identifying a problem and then discovering a valuable new solution. Therefore, if a firm is to develop unique knowledge or a unique new capability through any manner other than luck, it must identify a valuable problem and conduct an efficient solution search. Valuable solutions deliver value to the firm, either through enhancement or development of a product or service or by reducing the cost of production or delivery.

In our theory, managers choose problems while identifying knowledge sets or existing technology either within or outside the firm that are potentially useful in searching for solutions to that problem (Nelson and Winter 1982). The choice of problems reflects an assessment of the expected value of potential solutions and an assessment of the firm's capacity to profitably reach high-value solutions. A firm's reservoir of knowledge sets and prior experience informs this choice. In choosing problems, managers in essence choose an unknown set of potential solutions, but once a problem is chosen, the task becomes identifying relevant knowledge and then maximizing the probability of discovering a high-value solution. This is achieved by choosing organizational mechanisms that efficiently govern search. To optimally organize search first necessitates an understanding of the solution space under exploration.

2.1. Dimensionalizing Solution Landscapes

Our conceptualization of problems and solution search stems from Simon's (1962) work on complex systems and Kauffman's (1993) broad work on *NK* modeling. Simon (1962) conceptualized a complex system as one "made up of a large number of parts that interact in a nonsimple way" (p. 486). Complex problems represent complex systems in which the value of recombinations of existing technology represents solutions (Fleming 2001, Schumpeter 1939, Henderson and Clark 1990, Weitzman 1996, Fleming and Sorenson 2000). Thus, as Fleming and Sorenson (2000) argue, the invention of the automobile combined technology from "the bicycle, the horse carriage, and the internal combustion engine," while the microprocessor combined knowledge "of a computer's central processing unit with integrated circuit fabrication processes." Continuing with Simon's logic, the complexity of a problem is a function of the degree to which the individual design choices, which define a solution, are either independent or interdependent in their contribution to solution value.

We conceptualize this level of interaction using Kauffman's *NK* modeling. This modeling approach has been used extensively in recent years as a simple mathematical construction of the interaction of elements within systems of varying complexity. In these models N represents the number of parts of the system (in our case, design choices) and K, which ranges from zero to one, represents the degree of interaction among the parts (or design choices). These two parameters then combine to describe landscapes that are either rugged, reflecting a high degree of interaction, or smooth, representing a low level of interaction.

Thus, suppose we identify that a particular problem may benefit from five distinct knowledge sets, with each knowledge set informing a distinct binary design choice. Each unique combination of these design choices yields a distinct solution value. The level of interaction among these binary design choices defines the topography of a solution landscape. The greater this interaction, the more rugged the landscape is, the higher the value of the highest valued solutions is, and the lower the spatial autocorrelation among high-value solutions is. The lower the interaction is, the smoother the landscape (and the lower the value of the highest solution).

For any given problem, the set of all possible combinations of relevant existing knowledge (i.e., solutions) can be represented as a landscape the topography of which defines the value associated with any given combination. Peaks on such solution landscapes represent valuable combinations of knowledge sets or technologies that are highly complementary. Valleys on such landscapes represent low-value combinations of existing knowledge. When knowledge sets are highly interdependent, solution landscapes are more rugged and unpredictable (Fleming and Sorenson 2000). Under these conditions, the value of the global maximum rises, but the average height of peaks declines. On these more rugged landscapes, a series of incremental changes in design is unlikely to lead to the discovery of highly valuable solutions (Fleming and Sorenson 2000).

Problems and their corresponding solution landscapes thus can be arrayed according to the level of interaction among knowledge sets. Borrowing from Simon's typology of complex systems, we provide an analogous typology of problems—decomposable or low-interaction problems, nondecomposable or high-interaction problems, and nearly decomposable problems with moderate levels of knowledge interaction.[1]

Low-interaction/decomposable problems are problems in which the value of solutions depends very little on the interactions among knowledge sets and design choices. With such problems, groups of individuals possessing rather distinct knowledge sets can independently apply their knowledge to unique design choices with a reasonable expectation that the aggregation of their independent efforts, along with the independent efforts of others with distinctly different knowledge sets, will uncover valuable problem solutions. For instance, the problem of discovering a higher-performing desktop personal computer is currently a rather low-interaction problem in knowledge formation. Within a certain range, computer performance can be increased by actors independently improving any number of subsystems (e.g., disk drive, monitor, Ethernet card) or components (e.g., microprocessor, memory). To use Simon's language, low-interaction problems are "decomposable" in the sense that these problems can be subdivided into subproblems, each of which draws from rather specialized knowledge sets. Those possessing knowledge set A can independently address the subproblem of selecting from among choices a_i, while those possessing knowledge set B can independently address the problem of optimizing among b_i, where $A \cap B = \{\phi\}$. In Kauffman's *NK* model, such low-interaction problems correspond to gradual and relatively smooth variations in solution landscapes with few peaks and high spatial autocorrelation, implying that high-value solutions tend to cluster in the same region.

High-interaction problems, by contrast, have solution landscapes in which the value of solutions depends on interactions among design choices. For instance, consider one of the subproblems above, designing a leading-edge microprocessor. To design and build a leading-edge microprocessor circuit demands numerous knowledge sets that extensively interact in determining the value of solutions (Macher 2001). With such high-interaction problems, an actor familiar with a particular technology cannot predictably enhance the value of the product design based solely on the knowledge he or she possesses. In this case, the value of any particular design change will interact with a host of other potential design changes determined by actors possessing distinctly different knowledge sets. In Kauffman's *NK* logic, the more extensive the interactions among knowledge sets are, the more rugged the landscape with abruptly jagged peaks and valleys. With high-interaction problems, knowledge sets cannot be separated into subproblems. Interactions among distinct knowledge sets are so extensive that any attempt to define subproblems and discover corresponding subordinate solutions offers no predictable improvement over randomly selecting trials. In Simon's language such problems are *nondecomposable*.

Moderate-interaction problems are problems that, in Simon's language, are *nearly decomposable*. The level of interaction among design choices is intermediate in the sense that subproblems associated with distinctive knowledge sets can be defined, but the value of a design choice within one subproblem is not fully independent of the design choices made in another subproblem. Thus, interactions among knowledge sets are not trivial. Nonetheless, near decomposability implies that interactions among knowledge sets within subproblems are greater than among subproblems.[2]

Staying with our computer illustration, the problem of enhancing the performance of notebook computers is generally regarded as a problem involving significantly more interdependence among design choices than the problem of refining desktop systems (Hoetker 2003). While the overall problem can clearly be separated into subproblems associated with designing the screen, keyboard, motherboard, etc., there are significant trade-offs and interdependencies in design choices for these components as they relate to performance dimensions of weight, size, and battery life (Hoetker 2003). Thus, the design of a leading-edge notebook computer is currently a nearly decomposable problem in the sense that nontrivial interactions exist among the knowledge sets that support subproblems, but these are significantly less than the interactions among knowledge sets applicable within subproblems. In such cases, Kauffman's *NK* model yields intermediate levels of ruggedness with such an intermediate level of interaction among knowledge sets.

2.2. Efficiently Searching Solution Landscapes

The search for solutions is necessarily uncertain. The likelihood, speed, and cost of arriving at a valuable solution therefore depend both on luck and on the pattern of trials that actors undertake (Simon 1962), with each trial reflecting a unique combination of knowledge or a unique set of design choices. If trials are chosen randomly, the probability of arriving at a valuable solution to a complex, high-interaction problem is very low. More efficient orderings utilize knowledge to direct this search process. There are two fundamental sources of such knowledge that correspond to two distinct approaches to search: directional and heuristic.[3]

Directional or local search[4] is search guided solely by feedback or experience from prior trials. New combinations of knowledge are pursued by altering one design element at a time (Gavetti and Levinthal 2000), observing the resulting change in solution value, and then either continuing with the same path of search or—in the wake of declining performance—restoring the design to its original form and adjusting a different element. The key feature of this form of search is that independent actors, either individuals or groups, independently pursue trials and independently observe performance. This form of experiential search or search through trial and error has been described in a wide range of literature (March and Simon 1958, Cyert and March 1963, Nelson and Winter 1982).

Directional search is well suited to low-interaction problems—problems where there is little interaction among design choices in defining solution performance. On such solution landscapes, the only relevant feedback is whether performance is increasing or declining in response to a specific design change. Thus, searching for solutions to such problems is a fully decomposable activity.

Directional search is, of course, quite inefficient when exploring solutions to high-interaction problems. Design choices here are highly interdependent. The rugged, multipeaked topography of complex solution landscapes ensures that directional search through independent design adjustments is unlikely to discover a valuable peak. Indeed, on this rugged terrain, small changes in design may result in substantial shifts in solution value. Thus, the efficacy of directional search diminishes as

Nickerson and Zenger: *A Knowledge-Based Theory of the Firm*
Organization Science 15(6), pp. 617–632, © 2004 INFORMS 621

problems become less decomposable and knowledge interactions increase.

Heuristic or cognitive search is a form of solution search in which an actor or a group of actors cognitively evaluate the probable consequences of design choices rather than relying solely on feedback after design choices are made (Simon 1991, March and Simon 1958). Trials are thus selected based on a cognitive map or implicit theory of how knowledge sets and specific design choices relevant to the problem interact to determine solution performance. These heuristics can be thought of as simplified representations of the solution landscapes or, as Gavetti and Levinthal describe, "cognitive representations... of lower dimensionality than the actual landscapes" (2000, p. 121). These "theory-driven structures" speed problem, solving by "furnishing a basis for evaluating information" (Walsh 1995). These cognitive representations of the solution landscape are then used to select trials that maximize the probability of quickly discovering a high-value solution. Of course, cognitive maps are not static. As trials are undertaken and knowledge is gleaned through feedback, managers update their heuristics.

Heuristics can be of two essential forms, based either on a single individual's beliefs and cognitions or on a group's collective beliefs. If each actor possessed unlimited cognitive capacity, then each actor could quickly absorb all relevant knowledge, develop theories about the structure of interactions, and choose an optimal pattern of search. However, human minds are limited in the rate at which knowledge can be assimilated, accumulated, and applied (Simon 1945, pp. 40–41). Hence, distinct knowledge sets needed for solving complex problems are likely to be widely dispersed and reside in the minds of many agents who have specialized in unique knowledge, which is why we focus our attention on groups of actors engaged in heuristic search.

Thus, consider three actors with different and distinct knowledge sets, all of which are relevant to a particular problem. Assume that these knowledge sets reflect a set of design choices that define a solution landscape to a particular problem and that the landscape is at least somewhat rugged. Each actor can independently develop an ordering for trials based on his or her respective knowledge. However, differences in knowledge are likely to yield differences in cognitive maps and hence the recommended order of trials. Moreover, absent knowledge sharing, these independently developed cognitive maps are unlikely to be well matched to the topography of the solution landscapes. Only by developing heuristics that encompass the knowledge of all actors can the probability of discovering highly valued solutions be enhanced.

While individual beliefs are the basis of heuristics that guide group decisions, developing group heuristics requires the resolution of inherent conflicts in beliefs (Walsh and Fahey 1986, Fiol 1994). These group heuristics are negotiated beliefs that are shaped not only by the quality of actors' logic, but also by each actor's self-interest and political position (Walsh and Fahey 1986). Thus, while individual beliefs form the basis of group beliefs or heuristics, not all individual beliefs are equally important or influential in this process.

Heuristic search first necessitates knowledge transfer to facilitate the development of heuristics that derive from multiple and dispersed knowledge sets. Knowledge transfer in turn necessitates the development of a shared language to support it. Finally, the development of group heuristics requires the reconciliation of the divergent beliefs about the proper shape of the search heuristic. By contrast, directional search requires little more than choosing an initial set of design choices and allowing each agent to independently choose trials.

These two search processes are easily matched to problem types. Heuristic search is necessary when problems are complex, with high levels of interaction among knowledge sets and design choices. Directional search is warranted when problems are decomposable, involving limited knowledge interaction because heuristics offer no clear benefit in this setting. Nearly decomposable, moderate-interaction problems benefit from both directional and heuristic search: Heuristic search essentially defines the area of the solution landscape where directional search will take place. Thus, as problems become more complex and solution landscapes more rugged, efficient search necessitates extensive knowledge exchange to craft search heuristics. However, because these intermediately complex problems yield landscapes with significant autocorrelation and thus locational clustering of high-value solutions, directional search in regions targeted by cognitive maps may prove particularly fruitful. Cognitive maps in this setting aid in discovering regions of the solution landscape particularly attractive for search, while directional search involving independent actors effectively explores these regions. Table 1 summarizes the relationship between problem type and the relative benefit of directional and heuristic search.

Table 1 Relative Benefit of Trial Ordering and Selection by Problem Type

Trial ordering and selection mechanisms	Attributes of knowledge formation: Decomposable	Nearly decomposable	Nondecomposable
Directional search	++	+	0
Heuristic search	0	+	++

Nickerson and Zenger: *A Knowledge-Based Theory of the Firm*
Organization Science 15(6), pp. 617–632, © 2004 INFORMS

3. Governance Choices and Knowledge Formation Hazards

If differing problems demand differing search processes, then the manager's task is to identify relevant knowledge sets and to craft a mechanism of governance that supports or enhances the method of search appropriate for the chosen problem. For simple, decomposable problems involving limited interaction among design choices, directional search based on feedback is desired. For nondecomposable problems involving extensive interactions among design choices, heuristic search is desired. However, as noted, heuristic search involving several specialized knowledge sets demands that individuals share the knowledge they possess to develop common cognitive maps. Two conditions, however, impede such knowledge sharing: (1) humans are cognitively constrained in the speed with which they can learn (i.e., assimilate, accumulate, and apply knowledge) and (2) they are prone to self-interest, which may include opportunistic forms of self-interest. The former ensures that knowledge relevant to exploring complex problems is widely distributed among individuals. No one individual has all the knowledge relevant to crafting an effective heuristics or cognitive map to guide search on a complex solution landscape. In and of itself, this wide dissemination of knowledge does not constrain the efficient organization of search because individuals could willingly share knowledge when heuristic search is demanded and collectively agree on an optimal path of search. However, the wide distribution of knowledge in conjunction with self-interest, in particular an individual's desire to capture value from the knowledge he or she accumulates, leads to two knowledge-related exchange hazards. These hazards plague efforts to support knowledge sharing necessary for heuristic search.

Hazard 1: Knowledge Appropriation

Incentives alter an individual's willingness to share knowledge. Unfortunately, with independent actors, incentives for knowledge sharing are plagued by a rather simple paradox: The value of knowledge to its potential acquirer is not known until after the knowledge is revealed; however, once that value is revealed, the potential acquirer has no need to pay for it and can resell it at near zero marginal cost (Arrow 1973, p. 171). Thus, the incentives to simply assimilate others' knowledge and thereby extract its value without payment plague efforts to sell knowledge. While property rights and contracts provide some protection, cognitive limits render contractual protections difficult to identify and costly to draft. Thus, opportunism in knowledge exchange discourages actors from sharing knowledge, and knowledge is therefore not placed into the hands of those who will find it of most value. Instead, self-interest encourages actors to hoard knowledge and embed it into saleable products. Embedding knowledge into saleable products provides a mechanism for extracting its value. If, however, the value of an actor's knowledge is revealed only through recombination with other actors' knowledge, then knowledge sharing is necessary, which—absent a means of protection—may simply not happen.

The knowledge-appropriation hazard therefore has clear implications for organizing search. While directional search is unaffected because knowledge sharing is not required for this form of search, heuristic search is clearly damaged by this disincentive to share knowledge. Heuristic search requires extensive knowledge sharing, and hence a remedy for the knowledge-appropriation hazard is required.

Hazard 2: Strategic Knowledge Accumulation

Individuals not only possess incentives to hoard knowledge; they also possess incentives to shape its development. Individuals possess incentives to strategically alter the path of search and the shape of the heuristic that guides search. In particular, individual actors may prefer to shape the search heuristic in ways that optimize the value of the knowledge that they personally accumulate as solutions are explored. The search process itself yields new knowledge that may be useful in a host of different applications. The value of knowledge that actors derive through solution search will depend on the complementary knowledge that they already possess. Actors prefer to see solutions explored that draw heavily from their own knowledge base because such solutions heighten the potential value that they can extract personally. If the organization is more dependent on the knowledge possessed by an actor, then that actor is in a stronger position to bargain for a large portion of the value generated by a solution.

Thus, actors have incentives to strategically influence the pattern of trials in ways that enhance their specialized knowledge or complement knowledge that they already possess, while avoiding efforts that require knowledge sharing. Consequently, absent governance remedies, efforts to explore problems requiring heuristic search are likely to lead to attempts to distort cognitive maps, to conflicts regarding the proper ordering of trials, and more generally to an underinvestment in knowledge-sharing activities that facilitate the development of common heuristics.

We assert that both hazards are essentially irrelevant when problems are decomposable. When problems are decomposable, searching for solutions does not benefit from knowledge transfer and heuristic search. However, nondecomposable, high-interaction problems require knowledge transfer to enable heuristic search. Consequently, as problems become more complex and nondecomposable, efficient search demands mechanisms that mitigate knowledge-exchange hazards.

Nickerson and Zenger: A Knowledge-Based Theory of the Firm
Organization Science 15(6), pp. 617–632, © 2004 INFORMS

4. Governance Alternatives for Mitigating Knowledge-Formation Hazards

To this point we have argued that problems differ in complexity and in optimal forms of solution search. Complex problems require solution search that involves extensive knowledge transfer, which is discouraged by knowledge-exchange hazards. Efficiency thus dictates the selection of governance forms, which support the appropriate level of knowledge transfer. At the same time, efficiency dictates the maintenance of high-powered incentives that encourage actors to actively seek knowledge through solution search. As we will see, there is a fundamental trade-off between governance forms that powerfully motivate search effort and those that more effectively support heuristic development.

In the discussion below, we explore three prototypical alternative governance forms as mechanisms for governing solution search: markets, authority-based hierarchies, and consensus-based hierarchies.[5] Following Williamson (1991), we contend that these three governance forms differ fundamentally in their use of three organizational features: (1) decision rights over the path of solution search, (2) communication channels to support knowledge transfer, and (3) incentives to motivate search. In the discussion below, we argue that these organizational features are configured in unique and complementary ways to achieve three organizational forms or prototypes. While scholars have touted the virtues of each form as a solution to the governance of knowledge exchange, these treatments are generally not comparative and provide no basis for discriminating among them.[6] Thus, we develop here an alignment between governance forms and problem types.

We begin with a discussion of markets, focusing particularly on Hayek's discussion of its virtues as a device for efficiently governing knowledge exchange. Markets, however, have a very low capacity for remedying knowledge-exchange hazards and thus fail as problems become more complex. We argue that hierarchy fundamentally comes in two forms: authority based and consensus based. Each possesses a unique resolution to knowledge exchange hazards. These two governance forms correspond roughly to the distinct and fundamentally competing representations of hierarchy within the knowledge-based theories (see Demsetz 1988, Conner and Prahalad 1996 versus Arrow 1973, Kogut and Zander 1996). Both more efficiently cope with knowledge-formation hazards than markets and in the process support heuristic search more effectively than markets. Nonetheless, these two forms of hierarchy also differ significantly in their approach to supporting heuristic search, with one promoting knowledge transfer and the other essentially economizing on it. In our discussion below, we match these governance solutions and their support for directional or heuristic search to the complexity of problems.

4.1. Markets

Our conceptualization of the knowledge-based governance choice is very consistent with Hayek's (1945) view that the fundamental economic problem is determining the "best way of utilizing knowledge initially dispersed among all people" (p. 520). From this perspective, the manager must decide the best way to organize the creation of knowledge when the existing knowledge is widely dispersed among actors. In addressing this problem of governing knowledge, Hayek espoused the "marvel" of the market, arguing that markets provide "inducements which... make the individuals do the desirable things without anyone having to tell them what to do." Markets, thereby, "dispense with the need of conscious control" (Hayek 1945, p. 527). In our context of problem solving, market governance determines the path of search by decentralizing control to those in possession of valuable, specialized knowledge. In markets, prices provide high-powered incentives that motivate actors to search for solutions that both exploit and enhance their specialized knowledge and that can be encapsulated into saleable products. For a given problem, the path of search is merely an aggregation of the individual search decisions made by a set of actors autonomously choosing paths of search that they anticipate will lead to improved and marketable solutions. Thus, markets achieve unique combinations of knowledge using fully decentralized decision making.

Markets are ideally suited for governing directional search. As previously discussed, effective directional search involves individual actors altering design features based on knowledge they possess and then observing the resulting change in solution performance. Markets also provide high-powered incentives that motivate actors to pursue trials that expand their knowledge or the sale of products and services based on that knowledge. At the same time, however, markets provide weak support for knowledge sharing. They provide little protection against knowledge appropriation and no clear disincentives against strategic manipulation of search heuristics. Further, knowledge sharing requires the formation of a common language by which to communicate knowledge. Markets provide weak incentives to invest in the formation of such language.

However, when problems are decomposable and directional search is desired, knowledge sharing is largely unnecessary. Indeed, it is precisely on knowledge transfer that markets are designed to economize. As Hayek (1945) contends regarding markets: "The most significant fact about this system is the economy of knowledge which it operates, or how little the individual participants need to know [about other actors] in order to be able to take the right action" (p. 527). Thus, the efficiency of markets in governing decomposable problems is that they avoid knowledge-exchange hazards by severely restricting knowledge exchange. Instead, markets offer powerful incentives for these individual actors to make optimal use of the knowledge. Each possesses incentives to engage in local search and thus to focus on trials that enhance their own accumulation of specialized knowledge.

Markets, however, provide a rather cumbersome and costly mechanism (i.e., the courts and classical contracting law[7]) for developing the heuristics necessary for searching complex solution landscapes. Independent actors governed by the market could in theory contractually agree to particular patterns of search, but disputes over performance in pursuing these search patterns would potentially require the intervention of courts and contract law. When solutions to complex problems are sought and a search heuristic is required to enhance the probability of discovering a valuable solution, conflicts among independent actors in developing this heuristic may be intense because of the knowledge hazards discussed above. Thus, the capacity for markets, even when supported by contracts, to manage heuristic search development is quickly exhausted. However, when markets are matched to decomposable problems where actors possessing independent knowledge sets can make independent design choices, commonly shared heuristics are unnecessary and such disagreements are unlikely to arise.

As governance devices, markets quickly fail as problems become more complex and demand heuristic search. Indeed, markets exacerbate the knowledge-exchange hazards that contaminate efforts to perform heuristic search. Markets' high-powered incentives discourage knowledge sharing and instead promote knowledge hoarding. Further, these incentives encourage individual actors to strategically shape heuristics and subsequent patterns of trials in ways that benefit them individually. Moreover, these incentives actively discourage investments in the formation of common language required for knowledge sharing and heuristic search.

Given the set of organizational features that commonly define markets, we posit markets are poorly suited for governing the process of solution search for a complex, nondecomposable problem. The necessary knowledge sharing is undermined by the absence of both appropriate incentives and shared language. Efforts to manage heuristic search through markets or contracts are limited by the high costs of resolving through legal means the disputes that would emerge as individual actors strategically appropriate knowledge and seek to opportunistically alter the trajectory of search in ways that benefit them individually. Thus, markets are quite efficient in guiding directional search for solutions to decomposable problems but quickly fail as problems become more complex and demand heuristic search. They fail because knowledge-exchange hazards are poorly managed through contracts, and the high-powered incentives of markets simply promote these hazards.

4.2. Authority-Based Hierarchy

Many argue that the capacity to use centralized authority provides the distinguishing advantage of hierarchy over markets (e.g., Arrow 1974, pp. 68–70; Milgrom and Roberts 1990, p. 72; Simon 1951, p. 294; Williamson 1985). The efficiency properties of authority have been particularly highlighted in the context of problem solving and knowledge formation. As Arrow (1974) argues, "Authority, the centralization of decision making, serves to economize on the transmission and handling of knowledge" (p. 69).[8] Demsetz (1988) concurs, noting the potential inefficiency of a market in which individual actors must first be convinced of the appropriateness of each action they take. As he contends, the virtue of hierarchy is that "[d]irection substitutes for education (that is, for the transfer of the knowledge itself)." Thus, in solving identified problems, the advantage of authority in hierarchy is the capacity for one actor to identify the precise order of trials, thereby circumventing the need to contractually manage the order of trials or to spontaneously achieve some type of consensus through extensive knowledge sharing.

Because Hayek argued similarly for the efficiency of markets in economizing on knowledge transfer, how do we reconcile the advocated knowledge-based efficiency of both autonomy in markets and authority in hierarchy? The answer, of course, centers on the problem for which efficient governance is sought. Markets are efficient when knowledge transmission is directed at solving decomposable problems where there is little interaction among design choices. When problems are decomposable, knowledge is embedded in products and services and knowledge transmission is largely limited to what can be contained in prices and bundled into products and services. However, bundling knowledge sets within a single firm and exercising authority to direct search becomes efficient when problems become complex and efficient search demands shared heuristics to sequentially guide design choices. In this circumstance, authority economizes on the extensive and costly knowledge sharing and education that would need to occur were the governance of solution search organized through a market interface. As described below, authority-based hierarchy more efficiently resolves the knowledge-based exchange hazards that arise in markets when problems are complex and demand knowledge sharing.

Authority-based hierarchy is composed of a set of complementary features distinct from those that support markets. It is well suited to supporting some degree of heuristic search, though at the expense of dampening incentives for directional search. The key feature of authority-based hierarchy is a central figure who invests in understanding critical knowledge interactions and then composes suitable heuristics to guide search. For instance, the central figure defines and structures subproblems by imposing constraints or design rules on each subproblem, which greatly narrows the area of the solution landscape to be searched.[9] Other features of the governance form also support this form of search. Within the boundaries of hierarchy, the courts exercise forbearance. They refuse to hear disputes internal to the firm, including disputes about the proper path of problem search. It is precisely this forbearance that grants authority within the bounds of hierarchy. Because of this forbearance, individual actors have limited capacity and motivation to strategically manipulate the path of search.

Other features of authority-based hierarchy also facilitate centrally directed heuristic search. The high-powered incentives that motivate and reward directional search and knowledge accumulation in markets are limited within hierarchies. Although workers can exit an employer and take with them some portion of the knowledge they accumulate, the incentive structure within the firm is quite different. Employees grant managers authority in exchange for a wage. This act severs the direct linkage between the knowledge they accumulate and the wage they receive. This departure from market incentives also encourages investments in communication channels and codes that lower the cost of a central authority assimilating, accumulating, and applying knowledge. However, authority-based hierarchy does not promote horizontal communication channels that would support broad knowledge sharing among peers. It is precisely such knowledge transmission and broad education for which the authority and direction of this governance form provide a substitute. Authority-based hierarchy is thus superior to markets in supporting heuristic search but inferior in supporting directional search due to incentive damage. It dampens incentives to strategically manipulate the path of search and provides a central authority that economizes on costly knowledge transmission.

The use of authority within hierarchy to guide search is ideally suited to a range of problems that are moderate in their complexity. Authority has significant limits that become apparent as problems become either too complex and nondecomposable or too simple and decomposable. An authority's heuristic is only valuable when it more efficiently guides search than actors searching independently, each basing their own search on their own specialized knowledge. However, managers have a propensity to meddle in subordinates' decisions (Williamson 1985), in part due to overconfidence in their own judgment (Bazerman 1994). As problems become more complex and nondecomposable, the cognitive limits of managers to develop useful search heuristics, combined with a manager's propensity to meddle, contaminates the efficiency of search. As Weber and other organizational theorists remind us, "Authority does not imply expertise" (see Hammond and Miller 1985, p. 2). When authority is exercised in the absence of knowledge, it contaminates rather than accelerates search. Direction effectively substitutes for knowledge transfer only when managers have valuable knowledge with which to direct subordinates. When search dictates extensive recombining of knowledge, it is quite unlikely that authority is well suited to governing solution search. If many actors possess the required knowledge to define an optimal search heuristic and a central authority is incapable of absorbing that knowledge in a timely manner, then the development of a commonly shared search heuristic is required. Hayek noted precisely this limit to authority when he commented, "We cannot expect that [the problem of coordinating knowledge] will be solved by first communicating all this knowledge to a central board which, after integrating *all* knowledge, issues its orders" (1945, p. 524, emphasis in original).

If firms could constrain managers to intervene only when they possess knowledge that improves the direction of search, then the contributions of authority could extend to searching highly complex problems. However, managers appear to have great difficulty constraining their use of authority within hierarchies. In part, this reflects managers' inability to discern whether they possess information useful in directing search, and a strong bias to assume that they do. Managers are simply prone to meddle, wielding authority where they should delegate. Thus, we conclude that authority-based hierarchy provides an important advantage over markets in directing the search for solutions the more nondecomposable the problem. However, the limits to this governance solution are reached as the level of knowledge interactions escalate and the cognitive capacity of a single individual to assemble the required specialized knowledge reaches its limits.

The effectiveness of authority-based hierarchy relative to market-based control also diminishes rapidly as problems become more decomposable. When problems involve low interaction among design choices, using a manager's authority to dictate search may again damage search performance. An authority is simply unlikely to develop the necessary specialized knowledge required to direct solution search for all relevant subproblems and design choices. Again, managers are prone to meddling, directing the path of search when they lack the knowledge to effectively do so. The low-powered incentives that accompany hierarchy are also detrimental when autonomous searching by independent actors is desired. The weak incentives of hierarchy limit specialized knowledge formation and thereby constrain the efficiency of search. Further, the use of hierarchy is simply more costly than markets due to the costs of paying and supporting the central authority. If no clear value is added, then the added costs discourage its use. Thus, as displayed in Table 2, the combination of greater cost

Table 2 Comparative Analysis of Alternative Governance Mechanisms

		Hierarchy	
	Market	Authority based	Consensus based
Instruments			
Incentive intensity	+	0	0
Communication channels and codes	0	Vertical	Horizontal
Dispute resolution regime	Classical contract law	Forbearance-authority	Forbearance-relational
Performance			
Directional search	++	+	0
Heuristic search	0	+	++

and contaminated search renders authority-based hierarchy inefficient for decomposable problems.

4.3. Consensus-Based Hierarchy

Scholars also have argued for the knowledge-based efficiency of an alternative hierarchical form of governance, one that emphasizes extensive knowledge sharing and commonly shared search heuristics to guide decision making. It is this form of governance that Kogut and Zander (1992, 1996) emphasize when they describe the firm as providing a specialized social community that creates and transfers knowledge more quickly and efficiently than markets. In this conception of the firm, hierarchy is a device that supports knowledge transfer. However, this conception contrasts sharply with authority-based hierarchy, which minimizes knowledge transfer. Rather than hierarchy substituting direction for education as Demsetz (1988) and Conner and Prahalad (1996) suggest, hierarchy in this case substitutes education (i.e., knowledge transfer) for direction. Actors must first educate one another regarding knowledge relevant to defining a collective search heuristic.

This form of hierarchy achieves extensive knowledge transfer by enhancing the efficiency with which knowledge transfer occurs among actors within the firm. Extensive information sharing potentially permits actors within the firm to collectively agree on a path of search that is a consensus reflection of the specialized knowledge sets housed within the firm. However, as Arrow notes, consensus-based decision making, herein referred to as consensus-based hierarchy, is the near polar opposite of authority. Authority in this form of governance is used only to select a project, not the path of search. How then does a manager choose between the efficiency of these alternative internal governance choices of authority- and consensus-based hierarchy? The answer, of course, lies in the nature of the problem to which a manager seeks to find a solution. Consensus-based hierarchy is a potential solution to the failure of authority in governing heuristic search as problems increase in complexity.

Consensus-based hierarchy is supported by a distinct configuration of features that support active knowledge exchange. As Arrow (1974) argues, consensus can substitute for authority so long as knowledge transfer is inexpensive and actors within the group have "a sufficiently overriding commonly valued purpose..." (pp. 69–70). Under these circumstances, each agent within the firm perceives the correct decision based on his or her knowledge and interests. Because interests and knowledge are commonly shared, consensus is spontaneous.

Achieving consensus, however, first demands extensive knowledge sharing, and knowledge sharing requires a commonly shared language. Hierarchy possesses a distinct advantage over markets in promoting the formation of shared language. Arrow, in his treatise *The Limits of Organization*, recognizes the distinct advantages of hierarchy in facilitating knowledge transfer through firm-specific language and identity. He contends that a primary component of firm-specific capital formation is "learning the information channels within a firm and the codes for transmitting information through them" (Arrow 1974, p. 56). He further notes that such investments are not only individually specific, but that they in aggregate represent "irreversible capital accumulation for the organization" that leads to organizations having "distinct identities" (pp. 55–56).[10] Kogut and Zander similarly argue that communication costs are lower within the firm due to the presence of "higher order organizing principles" that "[establish] the context of discourse and coordination among individuals with disparate expertise" (Kogut and Zander 1996, p. 503). In their view, the boundaries of the firm define qualitative differences in the scope of social knowledge available to individuals. The shared identity that exists within a firm lowers the cost of communication and establishes "rules of coordination and influences the direction of search and learning" (Kogut and Zander 1996, p. 503).[11]

Incentives and dispute resolution mechanisms within consensus-based hierarchy are also configured to support knowledge transfer and consensus decision making. Very low-powered incentives are essential to consensus, because such incentives encourage (or more accurately do not discourage) knowledge sharing.[12] High-powered incentives within the firm would encourage knowledge hoarding and strategic manipulation of search.

Nickerson and Zenger: *A Knowledge-Based Theory of the Firm*
Organization Science 15(6), pp. 617–632, © 2004 INFORMS

Dispute resolution in consensus-based hierarchies, of course, also differs from authority-based hierarchy. While forbearance by the courts remains the central characteristic of hierarchy compared to markets, dispute resolution in consensus-based hierarchy involves individual actors collectively deciding this path. Such group decision processes equate to Williamson's (1985, pp. 246–247) relational team and approximate Ouchi's (1980) clan form of organization. In this case, the firm "will engage in considerable social conditioning to help insure that employees understand and are dedicated to the purposes of the firm and employees be provided with considerable job security, which gives them assurance against exploitation" (Williamson 1985, p. 247). Such efforts build relationships among agents and facilitate the formation of a common identity. These relationships and shared identity in turn ease knowledge transfer, facilitate agreement, and discourage the exploitation by other agents of knowledge transferred into the firm (Allen 1967, Tushman 1978).[13]

While consensus-based hierarchies possess distinct advantages over authority-based hierarchies in facilitating heuristic search to highly complex problems, consensus fails in comparison to authority as problems diminish in complexity. The costs associated with supporting extreme levels of knowledge transfer are substantial and become unwarranted as problems diminish in complexity. Further, low-powered incentives constrain the motivation to develop specialized knowledge and dampen incentives for solution search. The scope of investment in shared language and socialization and the efforts involved in the transfer of knowledge can be excessive when problems are only moderately complex. Such investments and effort slow the accumulation of specialized knowledge necessary for effective directional search. Moreover, social attachments and idiosyncratic language that accompany consensus-based hierarchy can increase the cost of search by generating search heuristics that are limited in the knowledge sets that they incorporate. While firm-specific language and social attachments lower the cost of communication among coworkers, they encourage actors to oversearch their channels for knowledge rather than searching out knowledge not contained within the firm. Thus, not only are such social attachments costly to maintain (Hanson et al. 1999), but they also may misguide the process of search. Such oversocialization may reduce the infusion of new ideas and result in "parochialism and inertia" (Adler and Kwon 2002), or as Powell and Smith-Doerr (1994, p. 393) argue, the "ties that bind may become the ties that blind." Individual choices of search trajectory may become increasingly determined by the knowledge that workers and their close friends possess. Thus, social attachments may bias decisions toward continuing patterns of search that extensively utilize existing knowledge sets within the firm and may limit the firm's capacity to search and absorb new forms of knowledge.[14]

Figure 1 Discriminating Alignment Between Governance Alternatives and Problem Complexity

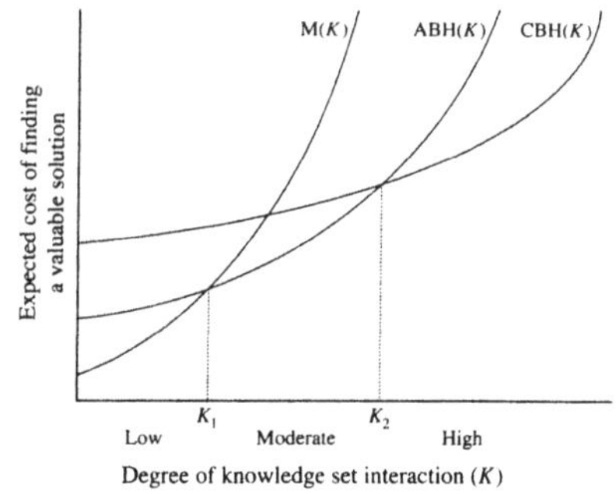

This discussion suggests that the organizational costs of generating consensus are high. Hence, consensus-based hierarchy should only be adopted when the benefits of consensus are high—when problems are highly complex. Thus, for decomposable problems, consensus-based hierarchy is more costly than either market governance or authority-based hierarchy, as it is likely to take more time and trials and hence cost to identify a valuable solution. These costs increase, but slowly, as problems become increasingly nondecomposable, leaving consensus-based hierarchy as a more efficient governance solution for complex, nondecomposable problems.[15]

5. Discriminating Alignment

Our theory suggests an alignment between the search needs of problems and the search costs and performance of governance alternatives. We summarize the proposed alignment in Table 3 and Figure 1. Table 3 summarizes the set of relationships from problem type to governance choice. Figure 1 compares the stylized costs we described above for each governance alternative over a range of problems that vary in the degree to which knowledge sets interact to define the solution landscape. For simplicity of presentation, we represent the horizontal axis as a continuous measure of this degree of knowledge set interaction, holding the actual number of knowledge sets constant (this assumption equates to holding constant the N in an NK model and varying K). The vertical axis represents the expected cost of finding a valuable solution. Also, while the cost curves are all upward sloping, which might suggest that the absolute lowest cost (e.g., the cost of the market for low-interaction problems) is most desirable, it is important to remember that the highest peak of the landscape, and hence the value of the optimal solution, increases with

Table 3 Summary of Theoretical Relationships in the Problem-Solving Approach to Governance Choice

Type of problems	Degree of knowledge set interaction (K) among subproblems	Solution landscape characteristics	Appropriate search mode	Need for knowledge transfer	Knowledge hazards	Predicted governance form
Decomposable (*A problem can be subdivided into subproblems, each of which draws from rather specialized knowledge sets.*)	Low	Few peaks, gradual and smooth variations	Directional	Low	Low threat	Market
Nearly decomposable (*A problem can be subdivided into subproblems that draw on specialized knowledge sets, but interaction among subproblems is sufficient so that value of design choices within one subproblem is not fully independent of the design choices made in another subproblem.*)	Moderate	Intermediate levels of ruggedness	Directional and heuristic	Moderate	Moderate threat	ABH
Nondecomposable (*Interactions among distinct knowledge sets are so extensive that any attempt to define subproblems and discover corresponding subordinate solutions offers no predictable improvement over randomly selecting trials.*)	High	Rugged with abruptly jagged peaks and valleys	Heuristic	High	High threat	CBH

knowledge set interaction. Thus, the relevant issue is finding the lowest governance cost for a given problem.

Our hypothesis is that markets efficiently govern low-interaction problems, authority-based hierarchy efficiently governs moderate-interaction problems, and consensus-based hierarchy efficiently governs high-interaction problems. With low-interaction problems, markets promote specialization and directional search. However, knowledge-exchange hazards ensure that the cost of the market (indicated by $M(K)$) accelerates rapidly as the degree of knowledge set interaction increases. For low-interaction problems, the cost of authority-based hierarchy (indicated by $ABH(K)$) is greater than markets, reflecting the inability of authority to efficiently motivate the intense specialization required to facilitate directional search or for those in authority to constrain their predilection to meddle.[16] As problems become increasingly complex, the costs of markets accelerate rapidly, reflecting their inability to cope with knowledge-exchange hazards. By contrast, the cost of authority-based hierarchy accelerates less quickly, leaving authority as the efficient governance choice for problems with moderate levels of interactions, which are between the points K_1 and K_2. Nonetheless, authority-based hierarchy's costs continue to accelerate, eventually leading to organizational failure as problems display high levels of interaction and the central authority lacks the relevant knowledge to guide search. The cost of consensus-based hierarchy (indicated by $CBH(K)$) is higher than the cost of authority-based hierarchy for problems with low and moderate levels of interaction. Thus, the organization provided by consensus-based hierarchy fails with these types of problems. Nonetheless, the costs of consensus do not accelerate until it is used to organize problems with high levels of interaction, which makes it the economic choice for problems with levels of interaction greater than K_2.

6. Discussion and Conclusion

Our theory addresses several deficiencies in the existing literature on the knowledge-based view of the firm. In particular, while more recent work (Conner and Prahalad 1996, Kogut and Zander 1996) explains from the knowledge-based view why firms exist by articulating the knowledge-based advantages of hierarchy, our theory focuses on *when* firms exist and in what form. Our theory articulates the knowledge-based advantages and disadvantages of both markets and hierarchy and matches these to desired types of knowledge formation. In addition to answering the question of what the knowledge-based reasons for a firm are, our theory informs the boundary choice between and among firms. Equally important, our theory provides some basis for reconciling and choosing between the two distinct and contradictory knowledge-based explanations for the existence of firms. While one theory claims that firms exist to economize on knowledge transfer (Demsetz 1988, Conner 1991, Conner and Prahalad 1996) and the other argues firms facilitate knowledge transfer (Arrow 1974; Kogut and Zander 1992, 1996; Nahapiet and Ghoshal 1998), we contend that a manager has three distinct governance choices in supporting knowledge formation: markets, authority-based hierarchies, and consensus-based hierarchies. These latter two forms correspond roughly to the competing conceptions of hierarchy within the literature with a knowledge-based view of the firm.

It is important to note that our theory is inherently based on a probabilistic assessment of solution discovery. Correctly aligning governance with problem complexity does not guarantee the rapid discovery of a valuable solution. Nonetheless, we expect that over a large number of solution searches, efficient alignment yields superior performance by enhancing the probability of discovering a valuable solution. Our theory, of course, assumes that there are multiple valuable solutions. On reaching a valuable solution, the firm must weigh the probable gain from further search against the added cost of further search.

While our theory aligns governance forms to problems, our theory was developed by implicitly assuming that successively chosen problems are independent. Clearly they are not. Interdependence of problems has important implications for governance choice. Managers likely choose new problems reflecting upon the composition of knowledge sets already in the firm's possession. Therefore, for any given knowledge set, managers are concerned not only with its proper governance, but also with its development and protection. These latter concerns about development and protection place constraints on governance choices not addressed in our theory. Thus, selling off a knowledge set to accommodate its appropriate governance for a highly decomposable problem may leave the firm poorly positioned to govern a subsequent or concurrent nondecomposable problem that requires this knowledge set. Nonetheless, for a knowledge set housed within the boundaries of the firm with wide application to a set of problems pursued by the firm, important governance differences across problems can still occur. Thus, those possessing a particular knowledge set can be tightly integrated with a team in pursuing a nondecomposable problem, receive simple design instructions from an authority in pursuing a nearly decomposable problem, and be granted enormous latitude in contributing to a decomposable problem.

While our theory proposes a rather static alignment between problem choice and governance choice, the impetus for dynamic changes in the composition of the firm are inherent in the model. In our knowledge-based theory of the firm, firms shift their boundaries in response to changes in the problems that they address. Moreover, problems have life cycles. A problem may begin as a nondecomposable task and necessitate heuristic search using consensus-based hierarchy. However, once a valuable solution is achieved and the resulting product widely sold, refined solutions to the original problem may be a decomposable problem that warrants governance through authority-based hierarchy or, with further time, the market. For instance, this may be the case with the emergence of so-called dominant designs. Such evolutionary changes in the nature of the problem to be solved imply dynamic changes in organization structure when viewed through the lens of the problem-solving perspective. A shift from nondecomposable to decomposable problems implies a shift in organizational form from consensus-based hierarchy to markets. Yet, as described above, management may instead take advantage of knowledge and capabilities within the firm and seek solutions to more complex problems. Such strategic and dynamic choices represent a substantial arena for further inquiry.

Obvious questions that arises from our theory are how the theory can account for large organizations with many levels of hierarchy or how it can account for complex linkages among multiple firms. Our theory is developed by discussing problems that may be separable into subproblems if complexity is not too severe. While not fully unpacked here, we anticipate that our theory can be used in a recursive way in the sense that problems with a large N might be decomposed into subproblems, which in turn might be decomposable into more subproblems, etc., until the problems are no longer decomposable. Then our theory can be applied to each level of subproblem to predict not only the organization of the lowest level of subproblems, but also entire branches of subproblems. If this line of thinking is further developed, our theory might provide deeper insights into the structure of firms as well as into the complex set of interorganizational relationships that characterize many industries.

Our theory highlights both the perils and virtues of authority in managing knowledge formation. Managers must recognize that the domain of problems for which their direction is of value is bounded on two sides. Effective knowledge management requires managers to recognize both limits. On the one hand, when problems are quite decomposable, managers must resist the temptation to assume value in their direction by integrating relevant knowledge sets. Instead, they must trust the market's intense motivation to specialize and guide directional search when problems are decomposable. Managers must recognize that cognitive constraints limit their capacity to possess the specialized knowledge to guide directional search and integration will simply dull the incentives of those specialists who do possess the required knowledge. On the other hand, when problems are nondecomposable, managers must also resist the temptation to assume value in their directing the path of search. Again, the manager must recognize that cognitive constraints limit his or her capacity to understand the wide-ranging knowledge interactions required to develop heuristics useful in guiding search. Instead, the manager must trust that a culture of widespread knowledge sharing and consensus decision making is the organizational approach most likely to yield a valuable solution. Therefore, depending on the set of problems pursued by the firm, those gaining from direction (substituting for education) may be quite limited.

Acknowledgments
The authors would like to thank Carliss Baldwin, Michael Leiblein, Glenn MacDonald, Jeff Macher, Laura Poppo, Lee Fleming, Brian Silverman, and seminar participants at Berkeley, Dartmouth, Copenhagen Business School, Harvard Business School, and Ohio State for helpful comments. They also thank Anne Miner, Jim Wade, Claudia Bird Schoonhoven, and anonymous referees at *Organization Science*.

Endnotes

[1]By a complex system, Simon roughly meant a system "made up of a large number of parts that interact in a non-simple way. In such systems, the whole is more than the sum of the parts, not in an ultimate, metaphysical sense, but in the important pragmatic sense that, given the properties of the parts and the laws of their interaction, it is not a trivial matter to infer the properties of the whole" (1962, p. 486). Kauffman's NK model provides a simple mathematical construction that arguably provides one parameterization of Simon's characterization. Like Levinthal (1997), these two parameters in our framework combine to describe a solution landscape. We develop our discussion implicitly assuming that N is nontrivial and constant (although we relax this view toward the end of our theory development). Thus, our focus is on variations in K. Following Simon, we develop our logic using a typology of complexity instead of continuous variables because the typology captures the central issues. Identifying the precise degree of interaction ex ante is more likely to confront cognitive limits than a typology when moving toward operationalization. Nonetheless, also we resort to employing the continuous variable K when describing how the costs of governance alternatives vary with complexity to make clear the relationships we predict.

[2]Ex ante identification that problems are either decomposable or nondecomposable may be relatively straightforward. Yet the degree to which a problem is nearly decomposable raises a question of operationalization. While this operationalization question deserves more attention than can be given here, we will point out that others have grappled with this issue. From a mathematical perspective, there is now a substantial literature on nearly decomposable systems and matrices. This mathematical definition forms the basis of implementable measures of near decomposability. For instance, McCord and Eppinger (1993) develop a task structure matrix for the design process of a laptop computer. This system, which is nearly decomposable, forms the basis for much on Baldwin and Clark's (2000) discussion on degrees of modularity. While surely it is the case that additional work is needed to provide refined measures of near decomposability, this prior work suggests that implementable measures can be constructed.

[3]Simon similarly identified two basic kinds of trials selectivity. One kind of selectivity, feedback, is where "various paths are tried out, the consequences of following them are noted, and this information is used to guide further search" (1962, p. 473). The second kind of trial selectivity is based on previous experience, which greatly reduces search based on trial and error. While this classification resonates with our own, Simon's terminology is less descriptive because feedback ultimately is used for both kinds of selectivity and because it does not relate to the theory of algorithms, which provides mathematical approaches for solving complex problems. In contrast, the terminology we introduce assumes feedback in both kinds of trial selectivity and parallels two broad classifications of solution algorithms. Directional search parallels hill-climbing or gradient methods of maximization, whereas heuristic search parallels combinatorial methods and probabilistic methods such as TABU, simulated annealing, and genetic algorithms for solving complex problems. Of course, we do not anticipate that such heuristics yield tractable algorithmic solutions for problems in which we are interested.

[4]We thank an anonymous referee for pointing out that the terms "directional search" and "heuristic search" are used here in a more narrow sense than is found in the broader organizational literature. The key difference between our usage and that in the broader literature is our unit of analysis. In particular, these terms are used here in reference to searching for a solution to a particular problem.

[5]It is well known that many hybrid organizational structures fall between the polar modes of market and hierarchy and that they may offer a variety of intermediate governance mechanisms. We believe that considering these alternatives here complicates theory development without the promise of adding deep insights to our main hypothesis. Thus, we anticipate that the theory development offered herein will be useful in future research that considers the knowledge-creation costs and competencies of hybrid organizational structures.

[6]Arrow and Demsetz are exceptions here.

[7]Classical contracting law refers to—

> "thick" markets [that] are ones in which individual buyers and sellers bear no dependency relation to each other. Instead, each party can go its own way at negligible cost to another.... Such transactions are monetized in extreme degree; contract law is interpreted in a very legalistic way: more formal terms supersede less formal should disputes arise between formal and less formal features (e.g., written agreements vs. oral amendments), and hard bargaining, to which the rules of contract law are strictly applied, characterizes these transactions (Williamson 1995; 1991, p. 95).

[8]Demsetz (1988) and Conner and Prahalad (1996) make the identical point: For instance, authority provides a "low cost

method of communicating between specialists and the large number of persons who either are nonspecialists or who are specialists in other fields" (Demsetz 1988, p. 172).

[9]Some have referred to such heuristics as "design rules." Baldwin and Clark (2000) provide an overview of this literature and describe the role of design rules set by a central authority to coordinate design of complex products. More generally, their analysis discusses the relationship between modularity and organizational structure.

[10]Hierarchy may not be the only institution able to support investments in communication channels and codes. For instance, channels and codes may emerge for groups of agents such as professional associations, regional cultures, industrial districts, etc. Such communication channels and codes may shift the relative costs and benefits of alternative governance modes. Nonetheless, we maintain that hierarchy remains unique in its ability to choose channels and codes and that the codes have the ability to provide greater richness for knowledge transfer than those that arise endogenously from other institutions.

[11]Monteverde (1995, p. 1,628) similarly contends that organizations are distinguished by their unique communication codes, by the "specific business dialect[s] their members use." He contends that firm-specific language is the "defining essence of what we recognize as organizations." Thus, firm boundaries, according to this logic, "should congeal around transactions rich in such technically necessary, unstructured dialog" (Monteverde 1995, p. 1,629).

[12]While both authority-based and consensus-based hierarchy rely on incentive structures that are much lower powered than those available in the market, we anticipate that they are not identical. Incentives associated with consensus-based hierarchy are more likely than those associated with authority-based hierarchy to be structured to encourage collective action.

[13]Strong social ties that emerge within hierarchies facilitate such knowledge transfer and consensus. Research on technical communication highlights the importance of close social ties in facilitating knowledge flows within organizations (Allen 1967, Tushman 1978). Similarly, the product innovation literature suggests that close, frequent interactions are necessary for timely integration of knowledge (Clark and Fujimoto 1991, Henderson and Cockburn 1994). In a study of knowledge transfer around best practices, Szulanski (1996) finds that knowledge transfer is facilitated by positive personal relationships. Hansen et al. (2001) find that rich networks of social ties are instrumental in pursuing exploratory tasks, which demand extensive knowledge transfer.

[14]Discontinuing patterns of local search often involves shutting down a particular research or development stream. Such disruption requires severing existing social commitments and damaging social attachments. However, loyalty to such social commitments then clouds the process of search. Thus, an inability to sever or ignore social attachments may lead to inefficient rather than efficient search.

[15]This does not mean that the net benefit of consensus-based hierarchy is positive for all nondecomposable problems. Indeed, for any problem where $K > 0$ the number of interactions and the number of relevant knowledge sets may be sufficiently large to increase the cost of search beyond the value of a potential solution (i.e., in terms of an NK model, assuming $K > 0$, then the expected value of finding a solution decreases with N). Moreover, with consensus-based hierarchy, the larger the group of specialists, the more difficult it will become to maintain a commonality of purpose, thereby increasing the cost of arriving at consensual decisions.

[16]By efficiency we mean that the economic benefit from the speed and likelihood of finding a valuable solution, less the cost of doing so, will be the greatest, on average, when problem types are paired with governance modes as hypothesized. Pairings that differ from these predictions will be less efficient, thereby increasing the expected cost and the expected time, or decreasing the likelihood of finding a valuable solution. Because knowledge formation is an uncertain activity, our notion of efficiency is a probabilistic assessment of net benefits including both technical and behavioral costs of knowledge formation.

References

Adler, P. S., S.-W. Kwon. 2002. Social capital: Prospects for a new concept. *Acad. Management Rev.* **27**(1) 17–40.

Allen, T. J. 1967. Communications in the research and development laboratory. *Tech. Rev.* **70**(1) 31–37.

Argyres, N. 1996. Evidence on the role of firm capabilities in vertical integration decisions. *Strategic Management J.* **17** 129–150.

Arrow, K. J. 1973. Information and economic behavior. *Stockholm, Federation Swedish Indust.* **28**. Reprint of Nobel Prize paper.

Arrow, K. J. 1974. *The Limits of Organization.* Norton and Co., New York.

Arrow, K. J., F. H. Hahn. 1971. *General Competitive Analysis.* Holden-Day, San Francisco, CA.

Baldwin, C. Y., K. B. Clark. 2000. *Design Rules: The Power of Modularity.* MIT Press, Boston, MA.

Barney, J. B. 1984. Strategic factor markets: Expectations, luck, and business strategy. *Management Sci.* **32**(10) 1231–1241.

Bazerman, M. H. 1994. *Judgment in Managerial Decision Making.* Wiley, New York.

Clark, K. B., T. Fujimoto. 1991. *Product Development Performance: Strategy, Organization, and Management in the World Auto Industry.* Harvard Business School Press, Boston, MA.

Conner, K. R. 1991. A historical comparison of resource-based theory and five schools of thought within industrial organization economics: Do we have a new theory of the firm? *J. Management* **17**(1) 121–154.

Conner, K. R., C. K. Prahalad. 1996. A resource-based theory of the firm: Knowledge versus opportunism. *Organ. Sci.* **7**(5) 477–501.

Cyert, R. M., J. G. March. 1963. *A Behavioral Theory of the Firm.* Blackwell Business, Cambridge, MA.

Debreu, G. 1959. *Theory of Value.* Wiley, New York.

Demsetz, H. 1988. The theory of the firm revisited. *J. Law Econom. Organ.* **4**(1) 141–162.

Fiol, C. M. 1994. Consensus, diversity, and learning in organizations. *Organ. Sci.* **5**(3) 403–420.

Fleming, L. 2001. Recombinant uncertainty in technological search. *Management Sci.* **47**(6) 117–132.

Fleming, L., O. Sorenson. 2000. Science as a map in technological search. Manuscript, Harvard Business School, Boston, MA.

Foss, Nicolai J. 1996a. Knowledge-based approaches to the theory of the firm: Some critical comments. *Organ. Sci.* **7**(5) 470–476.

Foss, Nicolai J. 1996b. More critical comments on knowledge-based theories of the firm. *Organ. Sci.* **7**(5) 519–523.

Gavetti, G., D. E. Levinthal. 2000. Looking forward and looking backward: Cognitive and experiential search. *Admin. Sci. Quart.* **45** 113–137.

Grant, R. M. 1996. Toward a knowledge-based theory of the firm. *Strategic Management J.* **17** 109–122.

Hammond, T., G. Miller. 1985. A social choice perspective on expertise and authority in bureaucracy. *Amer. J. Political Sci.* **29** 1–28.

Hansen, M. T., J. Podolny, J. Pfeffer. 2001. So many ties, so little time: A task contingency perspective or social capital. *Res. Sociology Organ.*

Hayek, F. 1945. The use of knowledge in society. *Amer. Econom. Rev.* **35**(September) 519–530.

Heiman, B., J. A. Nickerson. 2002. Towards reconciling transaction cost economics and the knowledge-based view of the firm: The context of interfirm collaborations. *Internat. J. Econom. Bus.* **9**(1) 97–116.

Henderson, R. M., K. B. Clark. 1990. Architectural innovation: The reconfiguration of existing product technologies and the failure of established firms. *Admin. Sci. Quart.* **35** 9–30.

Henderson, R. M., I. Cockburn 1994. Measuring competence? Exploring firm effects in pharmaceutical research. *Strategic Management J.* **15**(Special Issue) 63–84.

Hoetker, G. 2003. Do modular products lead to modular organizations? Manuscript, Department of Business Administration, University of Illinois, Urbana-Champaign, IL.

Kauffman, S. 1993. *The Origins of Order.* Oxford University Press, New York.

Kogut, B., U. Zander. 1992. Knowledge of the firm, combinative capabilities and the replication of technology. *Organ. Sci.* **3**(August) 383–397.

Kogut, B., U. Zander. 1996. What firms do: Coordination, identity, and learning. *Organ. Sci.* **7**(5) 502–518.

Levinthal, D. 1997. Adaptation on rugged landscapes. *Management Sci.* **43** 934–950.

Macher, J. 2001. Vertical disintegration and process innovation in semiconductor manufacturing: Founderies v. integrated device manufacturers. Manuscript, McDonough School of Business, Georgetown University, Washington, D.C.

Madhok, A. 1996. The organization of economic activity: Transaction costs, firm capabilities, and the nature of governance. *Organ. Sci.* **7**(5) 577–590.

Mahoney, J. 2001. A resource-based theory of sustainable rents. *J. Management* **27**(6) 651–666.

March, J. G., H. Simon. 1958. *Organizations.* John Wiley, New York.

McCord, K., S. Eppinger. 1993. Managing the integration problem in concurrent engineering. Working Paper 3594-93-MSA, Sloan School of Management, MIT, Cambridge, MA.

Milgrom, P., J. Roberts. 1990. Bargaining costs, influence costs, and the organization of economic activity. J. Alt, K. Shepsle, eds. *Perspectives on Positive Political Economy.* Cambridge University Press, New York, 57–89.

Monteverde, K. 1995. Technical dialog as an incentive for vertical integration in the semiconductor industry. *Management Sci.* **41**(10) 1624–1638.

Nahapiet, J., S. Ghoshal. 1998. Social capital, intellectual capital, and the organizational advantage. *Acad. Management Rev.* **23** 242–266.

Nelson, R. R., S. G. Winter. 1982. *An Evolutionary Theory of Economic Change.* Belknap Press, Cambridge, U.K.

Ouchi, W. 1980. Markets, bureaucracies, and clans. *Admin. Sci. Quart.* **25** 129–141.

Powell, W. W., L. Smith-Doerr. 1994. Networks and economic life. N. J. Smelser, R. Swedberg, eds. *The Handbook of Economic Sociology.* Princeton University Press, Princeton, NJ, 368–402.

Prahalad, C. K., G. Hamel. 1990. The core competence of the corporation. *Harvard Bus. Rev.* (May–June) 79–91.

Rumelt, R. 1984. Toward a strategic theory of the firm. *Competitive Strategic Management* 556–570.

Schumpeter, J. 1939. *Business Cycles.* McGraw-Hill Book Company, Inc., New York.

Simon, H. A. 1945. *Administrative Behavior.* The Free Press, New York.

Simon, H. A. 1951. A formal theory of the employment relationship. *Econometrica* **19**(3) 293–305.

Simon, H. A. 1962. The architecture of complexity. *Proc. Amer. Philos. Soc.* **106**(6) 467–482.

Simon, H. A. 1991. Bounded rationality and organizational learning. *Organ. Sci.* **2**(1) 125–134.

Szulanski, G. 1996. Exploring internal stickness: Impediments to the transfer of best practice within the firm. *Strategic Management J.* **17** 27–43.

Teece, D., G. Pisano, A. Shuen. 1997. Dynamic capabilities and strategic management. *Strategic Management J.* **18**(7) 509–533.

Tushman, M. L. 1978. Technical communication in R&D laboratories: The impact of project work characteristics. *Acad. Management J.* **21** 624–644.

Walsh, J. P. 1995. Managerial and organizational cognition: Notes from a trip down memory lane. *Organ. Sci.* **6**(3) 280–321.

Walsh, J. P., L. Fahey. 1986. The role of negotiated belief structures in strategy making. *J. Management* **12**(3) 325–338.

Weitzman, M. 1996. Hybridizing growth theory. *Proc. Amer. Econom. Association* (May) 207–212.

Wernerfelt, B. 1984. A resource-based view of the firm. *Strategic Management J.* **12** 75–94.

Williamson, O. E. 1985. *The Economic Institutions of Capitalism: Firms, Markets, Relational Contracting.* The Free Press, New York.

Williamson, O. E. 1991. Comparative economic organization: The analysis of discrete structural alternatives. *Admin. Sci. Quart.* **36**(June) 269–296.

Williamson, O. E. 1996. *The Mechanisms of Governance.* Oxford University Press, New York.

Williamson, O. E. 1999. Strategy research: Governance and competence perspectives. *Strategic Management J.* **20**(12) 1087–1103.

[13]

Strategic Management Journal
Strat. Mgmt. J., **21**: 1123–1145 (2000)

UNTANGLING THE ORIGINS OF COMPETITIVE ADVANTAGE

IAIN M. COCKBURN[1], REBECCA M. HENDERSON[2]* and SCOTT STERN[2]
[1]*School of Management, Boston University, Chestnut Hill, Massachusetts, U.S.A. and National Bureau of Economic Research, Cambridge, Massachusetts, U.S.A.*
[2]*Sloan School of Management, Massachusetts Institute of Technology, Cambridge, Massachusetts, U.S.A. and National Bureau of Economic Research, Cambridge, Massachusetts, U.S.A.*

This paper begins to reconcile competing perspectives on the origins of competitive advantage by examining the adoption of 'science-driven' drug discovery, a performance-enhancing organizational practice. Science-driven drug discovery diffused slowly, allowing us to disentangle alternative theories of organizational heterogeneity. Adoption is driven by initial conditions, time-varying internal and external environmental conditions, and convergence (firms positioned least favorably adopt most aggressively). While accounting for initial conditions is critical, managers are sensitive to idiosyncratic environmental cues. The origins of competitive advantage may therefore lie in the ability to identify and respond to environmental cues well in advance of observing performance-oriented pay-offs. Copyright © 2000 John Wiley & Sons, Ltd.

INTRODUCTION

> Strategy studies ... is systematic, like the system of touts at the race track ... (Stinchcombe, 2000)

What are the origins of competitive advantage? Although this question is fundamental to strategy research, it is one to which we lack a clear answer. As strategy researchers we believe that some firms consistently outperform others, and we have some evidence consistent with this belief (Rumelt, 1991; McGahan and Porter, 1997). We also have a number of well-developed theories as to why, at any given moment, it is possible for some firms (and some industries) to earn supranormal returns. As of yet, however, we have no generally accepted theory—and certainly no systematic evidence—as to the origins or the dynamics of such differences in performance. We know, for example, why high barriers to entry coupled with a differentiated product positioning obtained through unique organizational competencies may provide a firm with competitive advantage. But we know much less about how barriers to entry are built: about why this firm and not that one developed the competencies that underlie advantage, and about the dynamic process out of which competitive advantage first arises and then erodes over time.

This conceptual ambiguity has always been problematic for many economists, who have tended to view persistent differences in performance as a function of 'unobserved heterogeneity' (Mundlak, 1961; Griliches, 1986). For example, empirical work in industrial organization routinely controls for 'firm fixed effects.' These are usually statistically significant and often account for a substantial fraction of the total variation in firm productivity or performance. Whereas strategy researchers tend to emphasize the degree to which these kinds of results offer support for the importance of 'capabilities' or 'positioning' (Rumelt, 1991; Henderson and Cockburn, 1994; McGahan and Porter, 1997; Lieberman and Dhawan, 2000),

---

Key words: competitive advantage; pharmaceuticals; R&D; diffusion
*Correspondence to: Rebecca M. Henderson, Sloan School of Management, Massachusetts Institute of Technology, 50 Memorial Drive, Cambridge, MA 02139, U.S.A.

economists tend to emphasize the possibility that fixed effects are simply controlling for a series of much more mundane measurement problems, ranging from the difficulty of computing appropriately depreciated capital stocks and of measuring firm-specific input and output price schedules, to the problem of controlling for difficult-to-observe factors such as worker effort or worker quality. In short, the evidence which strategy researchers view as the motivation for their intellectual agenda are interpreted by many economists in terms of 'nuisance' parameters—things which must be controlled for but which are not of intrinsic interest.

This implicit critique has been reinforced by theoretical and empirical research in the tradition of population ecology (see, for example, Hannan and Freeman, 1989). In summarizing the contributions of this literature and its application to strategy, Stinchcombe (2000) charges that the preponderance of strategy scholars have simply failed to understand (and certainly to systematically account for) the implications of population dynamics for performance heterogeneity. Stinchcombe suggests that if superior performance arises from the degree to which a firm's resources and/or strategy 'match' the competitive environment, and if resources are randomly distributed at 'birth' (or if the environment which firms face at the time strategies are chosen and resource investments are made is sufficiently uncertain), then performance heterogeneity simply reflects the fact that the realized competitive environment favors some strategies and some resource bundles over others. Such a critique implies that the cases which motivate so much of strategy research, and indeed even some of our theoretical frameworks, are roughly equivalent to *ex post* accounts of the way in which a winning gambler chose to put her money on red rather than on black at the roulette table.

In this paper we argue that grappling with this problem should be of central concern to strategy researchers: that while many of us are aware of the issue that Stinchcombe raises, without a more detailed understanding of the origin and dynamics of the development of competitive advantage we run the grave risk of meriting Stinchcombe's taunt that we are indeed 'touts at the racetrack.' We suggest that empirical strategy researchers need to move beyond studies of differential performance to more integrated studies which not only identify those factors that are correlated with superior performance but also attempt to explore the origins and the dynamics of their adoption.

We begin the paper with a brief literature review. By and large, Stinchcombe's critique—and population ecology more generally—suggests that most performance differentials, particularly differences in the probability of survivorship, can be explained by differences in a firm's initial conditions, and, moreover, that differences in initial conditions are largely the result of difficult-to-explain (and even harder-to-measure) differences in each firm's initial allocation of resources and capabilities. In contrast, strategy is centrally concerned with the process of how firms and managers *respond* to and exploit environmental signals. For example, in the last 20 years, there has been an explosion of powerful frameworks for evaluating the determinants of differential performance, from Porter's five forces framework to the resource-based view to transaction-cost economics. While each of these frameworks offers a somewhat different explanation for heterogeneous performance, all share two assumptions: that competitive advantage arises through earlier or more favorable access to resources, markets, or organizational opportunities; and that exploiting such opportunities reflects some degree of active interpretation of internal and external environmental signals by managers. Indeed, we argue that these literatures often share the implicit view that the *origins* of competitive advantage lie in the unusual foresight or ability of the firm's managers.

Of course, this characterization of the two schools of thought is unrealistically stark: population ecologists recognize the possibility that firms may be able to adapt to their environment and their competitive experience (Barnett and Hansen, 1996), and strategy researchers understand that firm strategy and capabilities are subject to powerful inertial forces (Christensen and Bower, 1996). But as we discuss in some detail in the next section, substantial differences in focus do exist: while population ecology is principally focused on exploring the performance implications of strong organizational inertia, one of the core agendas of strategy is understanding which organizational structures allow firms to first identify and then exploit opportunities offered by their environment and so (potentially) overcome organizational constraints.

We then turn to a discussion of an empirical methodology that might enable us to explore the relative importance of initial conditions and strategic choice in shaping competitive advantage. Our goal is to lay out an empirical framework that might allow us to compare and contrast the distinctive implications of the strategic perspective alongside the predictions of the population ecology literature, and thus to both offer a preliminary response to Stinchcombe's critique and to begin disentangling the origins of competitive advantage.

We begin by distinguishing between differences among firm in terms of their 'initial conditions' vs. differences in the rate at which they adopt a particular performance-enhancing practice (or strategy) that has been linked to superior performance. Using Stinchcombe's analysis, we would expect the primary determinant of each firm's degree of adoption at any particular point in time to be the initial condition or founding state of that firm. This is not to imply that Stinchcombe's analysis suggests that firms cannot change, only that we believe that in general he would argue that change will not be systematically tied to the kinds of environmental cues which would be readily amenable to empirical analysis by strategy researchers.

We contrast this hypothesis with two explanations for diffusion which are consistent with a strategic perspective, and are more difficult to reconcile with Stinchcombe's hypothesis. First, a strategic orientation would suggest that, in the absence of organizational failure, those firms whose early history places them in a particularly unfavorable position will tend to be the firms that respond most aggressively in terms of the rate at which they adopt particular performance-enhancing practices or strategies. We label this the 'convergence' hypothesis, under which the key empirical question is not so much whether differences between firms will persist but how long they take to erode. From this perspective, the dynamics of competitive advantage are driven by two distinct processes: the exploitation of particularly favorable combinations of practices and/or market positions by firms whose initial positions 'match' their environment, and the erosion of these rents as competitors 'catch up' by mimicking the successful strategies of market leaders.

Second, we suggest that, beyond initial conditions and the process of convergence, adoption rates may be associated with environmental cues which are more intensively experienced by some firms rather than others. Firm strategy may thus be responsive to factors which provide information about a distinctive opportunity to invest in resources or strategies which will ultimately be associated with competitive advantage. For example, if firms are in different 'niche' markets during the early stage of an industry, then the information these firms will possess about the future evolution of the industry may be different and their strategy should in principle respond to these signals. From an empirical perspective, this hypothesis suggests that strategy will be a function of the firm's *current or recent* environment; of course, a second issue exists as to whether we should focus on environmental cues which tend to be associated with internal (organizational) factors, external (market) factors, or both.

To illustrate these ideas, we apply this empirical framework to an evaluation of the adoption and diffusion of one particular strategy—the use of science-driven drug discovery—in the context of the worldwide pharmaceutical industry. We begin by briefly motivating the study of the adoption of science-driven drug discovery as a useful context in which to think about the origins of competitive advantage, and then turn to a discussion of our data sources and variable construction. We explore the measurement of both initial conditions and convergence, and then use our qualitative knowledge of the industry to identify five distinct 'environmental cues' that might drive 'strategic' adoption: a firm's distance from public science, whether or not the CEO is a scientist, its accumulated stock of scientific knowledge, its market position, and the composition of its sales portfolio.

Our analysis is by no means definitive. We offer it as a preliminary descriptive analysis of a complex phenomenon that raises as many questions as it answers. However, we believe that it illustrates concretely one approach to taking Stinchcombe's hypothesis seriously while also evaluating the origins of competitive advantage in a systematic manner.

Our findings are consistent with a perspective in which both population ecology and strategy have an important role to play in explaining patterns of organizational heterogeneity. On the one hand, 'initial conditions'—or the position of each firm at the beginning of the period covered by our data—play an important role. Firms differ very significantly in the degree to which they had adopted the techniques of science-driven drug discovery in the first years of our period, and these differences persisted for many years. But there is also evidence for a powerful convergence effect, with the firms which were furthest from best practice at the beginning of the period moving most aggressively to adopt it. Finally, after controlling for initial conditions and convergence, additional time-varying characteristics of the firm play a modest role in explaining patterns of diffusion: both the composition of the firm's sales portfolio and its market share are significantly correlated with the rate at which the practice of science-driven drug discovery is adopted by the firm. Our results therefore provide substantial support for both Stinchcombe's view of the sources of competitive advantage and for a more traditional, 'strategic' view, the implications of which are addressed in the conclusion.

THE ORIGINS OF COMPETITIVE ADVANTAGE

Early studies of competitive advantage were rooted firmly in historical analyses and careful qualitative research. This work could be interpreted as suggesting that competitive advantage was a complex phenomenon, that depended crucially on the active presence of superior leadership (Andrews, 1971; Selznick, 1957; Chandler, 1962). For example, Chandler's early work can be read as implying that those firms who adopted the new M-form before their competitors gained a strategic advantage, and, moreover, that the choice to adopt the new organizational form reflected the structure and leadership qualities of a company's top management. Through the 1960s and 1970s, the study of 'strategy' was thus the study of what general managers or 'leaders' should do and it was generally assumed that doing these things would make a difference: firms with better leaders would make better choices and would ultimately do better than their competitors.

Porter turned this paradigm on its head (Porter, 1980). In transforming the study of 'imperfect competition' into the analysis of 'competitive advantage,' Porter shifted the focus of strategy research *outward*, towards the analysis of the firm's microeconomic environment. Porter's approach yielded sharply defined tools for understanding exactly why some firms (and industries) were likely to be more profitable than others. A 'five forces' analysis is essentially a structural map of the underlying economics of an industry: a map of the degree to which competitors, entrants, substitutes, and vertical bargaining power exert pressure on the margins of a firm in a particular industry. A firm operating in an industry in which there are substantial returns to scale coupled with opportunities to differentiate, that buys from and sells to perfectly competitive markets and that produces a product for which substitutes are very unsatisfactory (e.g., the U.S. soft drink in the 1980s), is likely to be much more profitable than one operating in an industry with few barriers to entry, and a large number of similarly sized firms who are reliant on a few large suppliers and who are selling commodity products to a few large buyers (e.g., the global semiconductor memory market).[1]

Structural analysis is a powerful tool for understanding why a particular strategic action (e.g., branding or investment in complementary product areas) may be associated with supranormal returns, but in and of itself says nothing about the role of senior management—or the process of strategic choice—in determining profitability. Consider the case of Crown Cork and Seal.[2] This classic HBS case describes the metal can industry: an industry that has a classically unfavorable structure and in which, in consequence, the vast majority of firms are relatively unprofitable. A structural analysis provides concrete insight into why it is so difficult for most firms to make supranormal returns in steel can production. In addition, the case can be read as suggesting that Crown itself earns supranormal returns because it has developed unique capabilities that have allowed it to differentiate its product in a way that is difficult for competitors to imitate. Does the case therefore imply that Crown was particularly well managed or that it had chosen a 'good' strategy? Analogously, does it also imply that those metal can producers that are not per-

[1] Structural analysis has, of course, moved much beyond Porter's original book, and we do not even attempt to summarize this literature here. For some recent contributions, see Brandenberger and Nalebuff (1998) and Besanko, Dranove, and Shanley (2000).

[2] HBS #: 9-378-024.

forming well are managed by 'bad' managers who have chosen 'bad' strategies? Certainly this is the way that the case is often taught: as if the vision and determination of John Connelley, who developed Crown's strategy, was a critical determinant of its success. To take Stinchcombe's critique seriously, however, is to wonder whether Crown was simply lucky: whether it happened to be managed by Connelley, who happened to be obsessed with customer service. Stinchcombe forces us to ask: would Connelley have been as successful, as 'visionary,' in choosing a strategy in another environment?

Porter's analysis was fundamentally agnostic on this point, but a simplistic interpretation of his work seemed to imply that structural analysis could be used prospectively: that doing strategy was about 'choosing good industries' or 'rebuilding industry structure.' The literature filled up with 'five force analyses', and it was tempting to use them prescriptively: build these kinds of barriers to entry, structure rivalry along these lines, and your firm and perhaps your industry would become more profitable. Notice that at its roots this stream of work returns to the founding assumption of the field: good strategy is about leadership, about foresight. Managers who are smart enough to understand the implications of structural analysis and to make the commitments that it requires are likely to outperform those who do not (Ghemawat, 1991; Shapiro and Varian, 1998). Certainly it is the case that most teaching in strategy (at least implicitly) assumes this to be true. We act as if we believe that if we teach our students to analyze industry structure they will be better positioned to allocate resources into the 'right' industries, or to influence industry structure in favorable directions. But we should recognize that while a variety of field studies certainly suggest that this is the case, there is, at least to our knowledge, no compelling quantitative evidence supporting this view: no broad-based statistical study showing that firms in which senior management actively used analytical tools to understand industry structure outperformed those that did not.

It was against this background that the 'resource-based view' of the firm emerged (Wernerfelt, 1984; Barney, 1991; Peteraf, 1993). At one level, the resource-based view (hereafter 'RBV') is simply a reinterpretation of the environmental perspective. Where the latter describes analytically why a differentiated position within an industry coupled with high entry barriers can lead to profitability, the former redirects attention towards the underlying heterogeneity making such a position sustainable. For example, while early environmental analyses seemed to suggest that competitive advantage arose from purely technological factors (such as economies of scale) or from unique assets (such as a brand name reputation), the RBV emphasized the idea that these technological or market positions reflect internal organizational capabilities, such as the ability to develop new products rapidly, to understand customer needs profoundly, or take advantage of new technologies cheaply. Proponents of the RBV suggested that strategic investments directed towards these *internal* activities might be of equal (or even greater) importance in generating supranormal returns.

More importantly, however, the RBV deepened the discussion of causality by focusing attention on two key insights about the sources of competitive advantage. First, in many cases, an industry's 'structural' features are the result of the organizational capabilities of its constituent firms: a powerful brand name, for example, may reflect years of successful new product introduction and superb (and unique) marketing skills. Second, there are good reasons for thinking that the market for organizational capabilities may be imperfect in exactly the kinds of ways likely to lead to the existence of supranormal returns. In part because of these two insights (which are implicit but not always manifest in environmental analyses), the RBV is often positioned as an 'alternative' to the environmental perspective. In our view, such a positioning reflects a significant misconception, since the RBV and the environmental perspective are complementary in many important respects. Each proposes a model of why firms may sustain superior performance, but the two models are not mutually exclusive, at least in terms of their empirical predictions: while the environmental view focuses attention on external industry structure, the RBV directs us towards the fact that internal capabilities and investments provide the instruments and tools to shape this external environment.[3]

[3] For example, from an environmental perspective, the pharmaceutical industry has historically been profitable because buyers and suppliers are weak, entry is costly and

Moreover, both literatures offer a similar theory about the process of strategic choice. In the case of environmental analysis, while economics is used to explain what kinds of strategic positions are likely to be most profitable, the theory is essentially agnostic as to how firms come to assume them. Firms can be lucky, they can be fast, or they can be far sighted: as long as they deal with the 'five forces' they will be profitable, and the power of the tools lies in explaining exactly what these forces are and what kinds of mechanisms will help a firm deal with them. At least at some level, many (perhaps most) empirical treatments within the RBV tradition follow a similar logic, with the caveat that rather than emphasizing the fact that one firm rather than another 'chose' a particular market position or production technology, the analyses focus on the fact that one firm 'chose' to develop a certain set of internal capabilities or unique organizational assets (see, for example, Henderson and Clark, 1990; Clark and Fujimoto, 1991; Eisenhardt and Tabrizi, 1995).

At a more subtle level, however, the RBV literature begins to address causality and the ultimate origins of competitive advantage more deeply. It has implicit within it a significantly different view of the dynamics of strategic advantage, and, in particular, of exactly what managers can and cannot do. While the canonical reference for the RBV literature is usually taken to be Penrose (1959), the RBV perspective on the strategy *process* seems to be influenced more by Stinchcombe (1965), on the one hand, and Nelson and Winter (1982), on the other. Specifically, RBV scholars often seem to suggest that organizations are fundamentally different from each other for reasons that may have very little to do with any kind of 'strategic logic,' and that they can only change through limited, local search. To the extent that competencies are built on organizational routines that are only tacitly understood—indeed if tacit understanding and complexity are a prerequisite for any competence to be a source of competitive advantage—then there may exist a fundamental tension between the fact that competencies lie at the heart of competitive advantage and the use of this insight to guide strategy choice (Leonard-Barton, 1998). From this perspective, the simple observation that competencies may lead to advantage is only half the battle: the other half is understanding where competencies come from.

The theoretical RBV literature thus makes explicit a view of 'strategy' that has long been latent—the sense that strategy is not all, or not only, about the cognitive ability of the senior management and their ability to make the 'right' decisions, but also about their ability to work creatively with the raw material presented by their firm and their environment (Quinn, 1978; Mintzberg, 1987); to respond appropriately when their firm's organizational structure finds 'good' strategies (Burgelman, 1994); and to create decision structures and procedures that allow a firm to respond to its environment adaptively (Bower, 1974; Levinthal, 1997). In short, in focusing on the dynamics of competence and resource creation, the RBV is centrally concerned with the degree to which successful firms are indeed 'lucky'—since it suggests that many of the competencies underlying advantage are the result of investments made under a heavy cloud of uncertainty and that they are subject to local but not globally adaptive evolution.

This approach promises a bridge between the insights of those who stress 'luck' or 'initial heterogeneity' in shaping firm performance and the central insight of the strategy field: that what managers do matters. But we believe that these implications have not been fully worked out. Most importantly, the RBV has not generated the kinds of empirical studies of adoption that are crucial both to fleshing out a full response to Stinchcombe's critique and to building a richer understanding of the origins of competitive advantage. While there are studies suggesting that the possession of unique organizational competencies is correlated with superior performance (Henderson and Cockburn, 1994; Powell, Koput, and Smith-Doerr, 1996), and others suggesting that competitive advantage may be heavily influenced by conditions at a firm's founding (Holbrook *et al.*, 2000; Eisenhardt, 1988; Eisenhardt and Schoonhoven, 1990), so far as we are

difficult, and substitutes and rivalry are quite muted. At least in part, this structure reflects external forces, such as the fact that the industry's products can be securely protected through patents, and so is not the result of a specific organizational competence on the part of a particular firm. But the industry's attractiveness also reflects the unique competencies developed by the larger pharmaceutical firms, including, among other factors, their years of investment in sophisticated research capabilities, their knowledge of regulatory systems around the world, and their extensive distribution and physician networks.

aware there are few careful studies of the hypothesis that successful strategy is the successful management of the evolution of organizational skills and its changing environment.[4] Indeed, some of the most elegant and widely cited studies of the diffusion of organizational innovation do not control for firm founding conditions (see, for example, Davis, 1991). At its worst, just as overenthusiastic readings of Porter led to the prescription 'choose a good industry,' overenthusiastic readings of the RBV have had the flavor of 'build the right resources/competencies.' Even Barney's original piece has something of this tone, as for example when he suggests that:

> To be a first mover by implementing a strategy before any competing firms, a particular firm must have insights about the opportunities associated with implementing a strategy that are not possessed by other firms in the industry, or by potentially entering firms ... (Barney, 1991: 104)

Such an analysis clearly places the source of competitive advantage in the insight of the firm's managers, and leaves unanswered the critical empirical question: *how does one know?* Is the success of Sony in introducing the Playstation the result of a strategic competence possessed by Sony's senior managers? Or is it the result of a complex history, an evolutionary process by which Sony was 'lucky' enough to be in the right place at the right time? In short, what are the concrete and distinct implications of Stinchcombe's and Nelson and Winter's insights for our understanding of strategy?

EMPIRICAL METHODOLOGY

In the remainder of this paper, we begin to address the implications of these questions for empirical research in strategy, at least in a preliminary way. Distinguishing between alternative theoretical perspectives in evaluating the origins of competitive advantage is a massive and difficult task. The empirical work in the remainder of this paper should thus be interpreted as outlining one type of empirical approach which might yield fruitful analysis of the origins of competitive advantage rather than as a definitive test of one particular theory in the specific context that we examine.

Our empirical analysis is designed to tease apart the issues that we identified above. We identified two core hypotheses: on the one hand, the view that competitive advantage is largely determined by factors put in place at the organization's founding; and, on the other, the view that competitive advantage results from a firm's 'strategic' response to changes in this environment or to new information about profit opportunities. We believe that one way in which one might plausibly begin to separate these two effects is to identify an organizational practice that is closely associated with competitive advantage, and then to explore the determinants of the diffusion of this practice across a population of firms.[5]

To do this, we build on a long tradition of modeling the process of diffusion of new technologies. Our analysis consists of the estimation and interpretation of a 'diffusion equation' $y_{j,t} = f(I_{j,0}, Z_{j,t}, t; \Omega)$, where $y_{j,t}$, our dependent variable, is a measure of the extent to which firm j has adopted the practice as of time t, and the explanatory variables are chosen to illuminate the competing hypotheses. $I_{j,0}$ is a measure of the initial, or founding conditions of firm j, while $Z_{j,t}$ is a vector of variables which reflect the opportunities and environmental conditions affecting firm j at time t, and thus highlight the existence of potential sources of differences across firms. For example, $Z_{j,t}$ might include measures of the nature of firm decision making, of absorptive capacity or other strategies associated with seeking external sources of knowledge, of market position, or of the talent or focus of managers in the firm. Time, t, plays a critical role as an explanatory variable here, since it allows us to capture the *rate* of diffusion of the practice, and so allows us to test whether firms in alternative environments tend to have faster or slower rates of adoption of the practice in question. Ω is the vector of parameters to be estimated.[6]

[4] There is of course a powerful literature demonstrating that changes in top management team compensation are correlated with major strategic reorientations (see, for example, Fletcher and Huff, 1990, and work by Tushman and his collaborators, including Murmann and Tushman, 1997). While this literature is suggestive, in general it leaves the question of causality unaddressed.

[5] Note that our focus on diffusion contrasts with the traditional emphasis on explaining *performance* as a function of initial conditions, environmental factors, or strategic choices.

[6] In the remainder of this discussion we suppress the firm-specific subscript j on all the variables.

Within this general framework, choices about functional form or about which of these explanatory variables to include correspond to different perspectives on the sources of competitive advantage. Three possibilities are of particular interest. First, perhaps the simplest model one can imagine would be to regress y_t on a favored set of Z_t's, from which one might conclude that the level of adoption of the practice was satisfactorily 'explained' by the Z's, providing support for a 'strategic response' perspective.

Alternatively a second approach would be to regress y_t on I_0 by itself, perhaps to test the conclusion that 'it's all initial conditions.' Finally, our focus on diffusion allows us to evaluate a third (somewhat more subtle) implication of the 'environmental' perspective, namely that firms who find themselves to have chosen strategies or organizational practices which are particularly *unfavorable* will be likely to have a higher rate of adoption than firms who are near the 'frontier' of profitability. In other words, even if the level of y_t is positively associated with I_0 (there is 'persistence' in the impact of initial conditions), a strategy-oriented perspective suggests that y_t will be *negatively* associated with $I_0 \times t$. That is, firms which begin at the 'lowest' level will have the highest rate of increase. This strategic response, if it occurs, will manifest itself in the data in the form of *convergence* over time among (surviving) organizations in terms of the organizational practices which they exhibit—the laggards will 'catch up' with firms who were 'lucky' enough to have chosen favorable practices at their founding (or at the beginning of the observed sample period).

To assess the relative salience of 'initial conditions' and environmental heterogeneity in driving patterns of adoption of a performance-enhancing practice, we require an empirical model of the dynamics of diffusion which allows for both effects to work factor simultaneously. We therefore specify a model which includes a firm's initial conditions, I_0, a firm's environment and opportunity set at a given point in time, Z_t, and interactions between each of these elements and time, t.

$$y_t = \alpha + \beta t + \gamma I_0 + \delta I_0 \cdot t + \varphi Z_t + \rho Z_t \cdot t \quad (1)$$

The parameters of this model nest a number of key hypotheses about the determinants of y_t which can be formulated as restrictions on the parameters.[7] For example, a naive reading of population ecology would suggest that, beyond a firm's initial conditions, there should be no systematic pattern to diffusion relating to the firm's changing environmental circumstance: some firms will be able to execute the practice as a function of their founding conditions and others will not. Thus for a given set of firms, current levels of adoption should be a function only of initial conditions, implying that $\varphi = \rho = \delta = 0$.

Conversely, the simple environmental or 'strategic' hypothesis implies that y_t should also be driven by Z_t: firms may change their behavior in response to changes in the environment with levels of adoption not wholly determined by a firm's starting position. This would imply $\varphi \neq 0$ and/or $\rho \neq 0$. For example, we believe that one could interpret a finding that y_t was correlated with the current market position of the firm as suggesting it is indeed reacting to exogenous shocks in its environment, independent of its founding conditions. Similarly a finding that y_t was correlated with the experience of the firm's CEO would suggest that the firm is changing its strategy as it develops new capabilities.

Finally, the 'convergence' hypothesis suggests that, in addition to a firm's initial conditions, and its responsiveness to objective measures of its environment, firms who face unfavorable initial conditions may be those who are most likely to have the highest rate of adoption, that is, $\delta < 0$. This would imply that initial conditions are important in shaping strategy, though hardly in the way that Stinchcombe hypothesized.

Before turning to the data, two further modeling issues must be addressed. The first is that of functional form: if we interpret y_t as 'fraction adopted' then it would be appropriate to follow the diffusion literature in recognizing that the

[7] Note that this is very much a 'reduced form' model of adoption. It is not our goal (at least in this paper) to develop or estimate a fully specified structural model of optimizing adoption behavior. Indeed, given the diverse set of plausible hypotheses about the drivers of adoption behavior, it is not clear that estimating such a model is currently feasible without preliminary empirical work, which substantially narrows down the potential range of theories to be accommodated. Consequently, we confine ourselves here to identifying the principal covariates of adoption, recognizing that adoption is a dynamic process. Distinguishing the relative empirical salience of competing hypotheses may provide an initial guide to specifying a more fully articulated structural model.

diffusion of most practices or technologies follows an S-shaped (or sigmoid) pattern through time. There are a variety of functional forms which allow for this type of nonlinear time path, but the simplest transformation involves taking the log-odds ratio ($\log(y_t/(1 - y_t))$) of the dependent variable measuring the extent of adoption (Griliches, 1957). On the other hand, if y_t is a more general indicator of the use of a practice, then theory offers us little guidance as to functional form.[8]

A second issue relates to the measurement of initial conditions, I_0. Two possibilities present themselves: either to take on the formidable task of identifying and explicitly measuring all the relevant aspects of heterogeneity in firm capabilities, or, more straightforwardly, to capture their empirical content with y_0—the presample value of y_t. This second option is attractive not just because it is easy to implement, but also because it allows us to be agnostic about the precise nature of the initial characteristics of firms which drive their subsequent evolution.[9] Actual estimation of these models therefore requires data only on y and Z. We now turn to our specific application of this framework, the diffusion of science-driven drug discovery in the worldwide pharmaceutical industry.

DATA AND VARIABLE CONSTRUCTION

Competitive advantage from science-driven drug discovery

We focus here on the adoption of 'science-driven drug discovery' as a primary strategy in the management of pharmaceutical research. As we have laid out in a number of papers, prior to the late 1970s pharmaceutical research was conducted largely through a process of so-called 'random' search (see, for example, Henderson, Orsenigo, and Pisano, 1999, and papers referenced therein).[10] From the late 1970s on, firms began to respond to the acceleration in the growth of publicly available biological knowledge by adopting a new mode of so-called 'science-driven' drug discovery. This move appears to have been tightly associated with the creation of competitive advantage: those firms adopting the new techniques appear to have been significantly more productive (Gambardella, 1995; Henderson and Cockburn, 1994) in an industry in which the introduction of blockbuster drugs is a source of tremendous value (Grabowski and Vernon, 1990).[11]

Despite the power of this strategy, its diffusion across the industry was surprisingly slow. As late as 1991, some firms continued to view scientific publication and participation in the broader community of public science—a key requirement for the full adoption of science-driven drug discovery—as a diversion from the more 'important' business of finding drugs. As the research manager at one of these companies put it:

> Why should I let my people publish? It's just a waste of time that could be spent in the search for new drugs. (Quoted in Henderson, 1994)

Here we argue that exploring the factors that drove these differences in the rate of adoption across firms not only concretely illustrates the difficulty of distinguishing between explanations, but also serves as a springboard for thinking through the relationship between theories of adop-

---

[8] Our empirical results are surprisingly robust to variation in the functional form imposed on the adoption-intensity variable. Our robustness checks (not presented here, but available on request) included using the level of adoption intensity as well as the natural logarithm.

[9] Conditioning on the starting value of the dependent variable (or some function of it) is a well-understood procedure in the econometrics literature on dynamic panel data models (see, for example, Arellano and Bond, 1991). This also allows us to understand the implications of our results in the framework the heterogeneity/state dependence distinction used by econometricians (see Heckman, 1991).

[10] Notice that this terminology can be deceptive. The major pharmaceutical firms began investing in fundamental science after the First World War, and by the 1950s many of them employed significant numbers of well-trained scientists (Parascandola, 1985). 'Random' drug discovery was 'random' in that while research was often guided or informed by fundamental science, the precise mechanism of action of many drugs was not understood, and new compounds were screened against animal models rather than against particular disease pathways (Henderson, 1994; Henderson, Orsenigo, and Pisano, 1999).

[11] The question of just *why* 'science-driven drug discovery' should be a major source of competitive advantage is a fascinating one. The interested reader is referred to our related work (Henderson, Orsenigo, and Pisano, 1999; Cockburn and Henderson, 1998; Cockburn, Henderson and Stern, 1999a, 1999b; Stern, 1999). There are, of course, other sources of competitive advantage in the pharmaceutical industry, including the possession of strong regulatory capabilities, an extensive distribution network or well-developed marketing capabilities.

tion and theories of the source of strategic advantage. The remainder of this section describes the sources of the data from which we construct our empirical tests, our alternative measures of the practice itself, y_t, and measures associated each firm's environment at a given point in time, Z_t.

Data sources and sample construction

We draw on two primary data sets. The first contains detailed qualitative data and quantitative data for 10 major pharmaceutical firms for the period 1965–90. These data formed the basis for our previous work and are described in detail there (Henderson and Cockburn, 1994, 1996). The second data set consists of a sample of 16 large research-oriented pharmaceutical firms for the period 1980–97.[12] Overall the average firm in both samples is somewhat larger than the average publicly traded pharmaceutical firm. While they do not constitute a random sample in a statistical sense, they are reasonably representative of the larger pharmaceutical firms who invest substantially in original research and comprise a substantial portion of the industry (the firms in the sample account for approximately 50 percent of U.S. pharmaceutical sales in 1993). While we were able to obtain detailed, internal data for the first sample, the second relies on comprehensive data collected from secondary sources, including information on the composition and background of senior management, firms' geographical location and patenting activity, scientific papers, and product sales compiled at the therapeutic class level (e.g., cardiovascular therapies, anti-infectives, or cancer). Data about senior management were collected from annual reports, while geographical data were derived from the publication record of each firm. Our source for patent data is Derwent Inc.'s *World Patent Index*; scientific publications are drawn from ISI's *Science Citation Index* and *Web of Science*. Sales data are from various publications of IMS America, a market research firm.

Measuring the adoption of 'science-driven drug discovery' (y_t)

A quantitative analysis of the adoption of science-driven drug discovery requires the construction of a reliable measure of the extent to which firms are engaged in the practice.

Our initial measure of this practice, PROPUB, identifies the degree to which researchers were given incentives to establish standing in the scientific public rank hierarchy. In prior work (Cockburn and Henderson, 1998; Cockburn, Henderson, and Stern, 1999b), we have argued that the extent to which a firm has adopted this practice is an indicator of how thoroughly they have embraced the techniques of science-driven drug discovery.[13] This measure was derived from detailed qualitative interviews conducted within 10 major pharmaceutical firms. While it has the great virtue of being a direct measure of the organizational practice in which we are interested, it unfortunately suffers from two important limitations. First, it was derived from qualitative interviews conducted by a single researcher, which may raise questions as to its reliability and replicability. Second, PROPUB is not currently available beyond 1990 or for the 11 firms in the sample that were not included in our earlier data collection efforts.

We therefore also evaluate the diffusion of science-based drug discovery using three quantitative measures derived from public sources. These variables all draw on bibliographic information collected for every paper published in the journals indexed in the Institute for Scientific Information's *Science Citation Index* between 1980 and 1994 for which at least one author's address was at one of our sample of firms. The most straightforward, PAPERS, is a simple publication count: the number of papers published in any given year for which at least one author was affiliated with our sample firms. As prior research has established, pharmaceutical companies publish heavily, with annual counts of papers compa-

[12] The firms are: Abbott, Bristol-Myers Squibb, Burroughs-Wellcome, Ciba-Geigy, Glaxo, Fujisawa, Hoechst, Hoffman La-Roche, Lilly, Merck, Pfizer, Sandoz, Searle/Monsanto, SmithKline Beecham, Takeda, and Upjohn. Nine of these firms are also included in the earlier sample: the remainder were selected to include the industry's leading R&D performers and to obtain worldwide geographical representation.

[13] In the ideal case we would like to be able to use a measure based on the actual nature of the science conducted within each firm over time: in the absence of reliable longitudinal survey data about research methodology this is impossible to construct from secondary sources. Instead we use here a number of measures derived from the publication behavior and incentives provided by research-oriented pharmaceutical firms.

rable to the output of similarly sized universities and research institutes (Koenig, 1982; Hicks, 1995). Publication counts are clearly an important indicator of research activity, and have been previously interpreted as capturing the level of investment in 'basic science' (Gambardella, 1995). Of course, since the volume of publication activity may simply reflect the scale of the firm's research effort, when we use any of these measures as a dependent variable we control for size by including total U.S. pharmaceutical sales in the regression.[14]

While PAPERS has several attractive properties in terms of capturing the degree to which firms have adopted the science-based drug discovery techniques, PAPERS is also fundamentally an output measure, and may therefore also measure factors such as the success of the firm's research program: the discovery of interesting compounds also generates interesting papers. Using the same data, we therefore constructed AUTHORS: a variable which identifies the number of science-oriented researchers at each firm. AUTHORS is a count of the number of distinct individuals publishing at the firm in any given year. We hypothesize that at those firms in which the scientists face high-powered incentives to raise their standing in the public eye the marginal incentives to publish will be higher and the 'marginal' scientist will make more of an effort to get their name on the papers to which they have contributed.[15]

Although counting authors may give us a better measure of the adoption of the new practice than a raw publication count, it remains an output measure. Our preferred measure of the degree to which the firm has adopted propublication incentives is PUBFRAC: the fraction of individuals whose names appear on a patent who also appear as an author on papers published within 2 years of the patent application. By explicitly tying publishing and patenting together, this measure incorporates the degree to which a firm is encouraging those researchers who are directly involved in the firm's drug discovery process to participate in scientific publication. In addition, since this is measured as a share rather than an absolute number of papers or authors, in principle it captures the *propensity* to publish, independent of the scale of the firm, either in terms of sales or number of employed scientists.[16]

Table 1 provides summary statistics and the correlation coefficients for these four variables. In addition to the relatively high levels of objective measures such as PAPERS and AUTHORS, it is worth noting that all four of the measures are

Table 1. Descriptive statistics for alternative dependent variables

(a) Summary statistics

	N	Mean	S.D.
PROPUB	83	3.6	1.3
AUTHORS	153	854	550
PAPERS	153	299	189
PUBFRAC	153	0.65	0.14

(b) Correlation coefficients

	PROPUB	PUBFRAC	AUTHORS
PUBFRAC	0.11		
AUTHORS	0.56	0.42	
PAPERS	0.55	0.38	0.35

[14] Total spending on research would clearly be preferable, but for the complete sample we have only data on total R&D expenditures. This measure is very noisy in that it includes development expenditures in addition to discovery expenditures, and we know from our prior work that the ratio of discovery to development spending is not constant across firms. It is a particularly noisy measure when the firm is diversified beyond the pharmaceutical industry.

[15] Another possibility would be to count 'stars': the number of individuals in a firm who publish prolifically. Zucker, Darby, and Brewer (1998) have suggested that this 'higher standard' is critical for understanding successful performance in, at least, biotechnology firms. In preliminary work we have found that this variable performs much like AUTHORS.

[16] Constructing PUBFRAC was far from straightforward. For each firm in the sample, we first identified all papers (a) which were published between 1980 and 1994 in journals indexed by the *Science Citation Index* and (b) in which the name of the firm, or one of its subsidiaries, appears in at least one of the authors' addresses. We had then to attempt to match the names of many thousands of individuals across two different data sets and consistently apply rules for ambiguous cases. Much of this matching was accomplished straightforwardly in software, but a number of difficulties did induce some measurement error into the process. These included typographical errors in the source data, and differences across papers by the same author in the use of initials, surnames, and given names; extensive hand-coding was necessary to complete the task. A consistent matching procedure was applied to all firms, and so we are reasonably confident that bias in the measurement of PUBFRAC is limited to differences across firms in the severity of these problems.

quite closely correlated. The time path of these variables indicates the gradual diffusion of the practice across the industry over time. Table 2 shows the evolution of both the mean values and the coefficient of variation (standard deviation/mean) for each of our four measures of science-driven drug discovery. Not only does the mean level of each measure rise, but the coefficient of variation steadily falls, suggesting that there was considerable 'convergence' towards a common level of intensity of use of the practice at the same time that its average level increased. Note that this diffusion is not driven primarily by the exit of firms from the sample.[17] In itself, this immediately raises questions about the extreme form of Stinchcombe's hypothesis: the firms in the sample appear to be actively adopting the practice of science-driven drug discovery over the period covered by our data, an observation which suggests that there is more going on than a simple initial random distribution of capabilities followed by an exogenous environmental shock. Indeed, it is consistent with a 'strategic' view in which managers gradually recognize the opportunities presented by the new techniques and move to adopt them. These simple statistics highlight the empirical salience of the convergence hypothesis: that initial conditions engender persistence over the medium term, but that laggard firms aggressively move to adopt these techniques, erasing the importance of the past in the long term.

Measuring initial conditions, I_0

As discussed above, rather than attempt to characterize initial conditions directly, we use y_0, or the firm's initial level of adoption of the practice of science-driven drug discovery, as the summary statistic for its initial condition. This is calculated from average level of y in the 2 years preceding the sample period.

Defining *Z*: measures of firm heterogeneity

We draw on the literature and on our qualitative data to construct six distinct measures of heterogeneity in the environment and opportunity sets of each firm. Some of these focus on general

Table 2. Means and coefficients of variation over time, alternate dependent variables

	PROPUB		PAPERS		AUTHORS		PUBFRAC	
Year	Mean	C.V.	Mean	C.V.	Mean	C.V.	Mean	C.V.
1976	2.40	0.71						
1977	2.40	0.71						
1978	2.50	0.66						
1979	2.70	0.55						
1980	2.70	0.55						
1981	3.10	0.51	182.4	0.63	393.6	0.60	0.54	0.30
1982	3.30	0.43	197.2	0.65	431.7	0.61	0.58	0.24
1983	3.30	0.43	210.1	0.67	478.3	0.64	0.53	0.31
1984	3.40	0.42	199.3	0.69	489.9	0.66	0.56	0.29
1985	3.40	0.42	252.0	0.74	615.8	0.68	0.58	0.25
1986	3.50	0.41	257.9	0.67	656.3	0.62	0.66	0.18
1987	3.50	0.41	283.9	0.67	746.3	0.55	0.68	0.17
1988	3.60	0.37	279.8	0.68	800.1	0.59	0.67	0.21
1989	3.67	0.36	309.1	0.50	913.9	0.46	0.64	0.16
1990	4.00	0.23	370.6	0.46	1128.8	0.38	0.68	0.13
1991			376.8	0.47	1230.1	0.41	0.68	0.16
1992			452.5	0.46	1436.0	0.45	0.73	0.12
1993			460.1	0.49	1553.9	0.45	0.80	0.13

[17] Of our original sample of 10 firms, two exited through merger. Of our later sample of 19, only two have (at the time of writing) exited through merger.

Table 3. Descriptive statistics for independent variables

Variable	Mean	S.D.	Min.	Max.
Cardiovascular patents share of firm patents (%)	12.47	4.40	5.18	27.98
Cardiovascular therapies as share of firm sales (%)	18.54	19.50	0	73.78
Share of firm in U.S. cardiovascular market (%)	5.27	8.71	0	44.92
Total firm sales ($ million)	938.8	701.1	99.9	3653.7
Scientist–CEO	0.065	0.248	0	1
Distance from science	0.363	0.515	0.001	2.290

(150 observations)

characteristics of the firm: others focus on the role of cardiovascular therapies—a leading application for science-driven drug discovery—in the firm's research and sales portfolio and on the firm's share of the market for cardiovascular therapies. This list of variables includes both factors that might 'rationally' drive differences in the diffusion of science-driven drug discovery—including the firm's closeness to public science and its technological experience and market position—and factors that might shape the attention of senior managers, including whether the firm's CEO has a scientific background and the composition of its sales portfolio. Our discussion here focuses largely on the qualitative and theoretical evidence that motivates each variable: full details of the ways in which each was operationalized are included in the Appendix. Table 3 presents descriptive statistics for these variables.

(a) Scientist–CEO

Qualitative interviews with senior industry participants often explained variations across firms in the rate at which they moved to science-driven drug discovery in terms that resonate deeply with a 'strategic' picture of adoption.[18] Our informants suggested that differences in 'leadership' or 'vision' across firms had a very significant effect on the decision to adopt the new mode of research. One senior researcher remembered:

> We spent most of the seventies going to T groups. It was fun, but we didn't get much science done ... The managers of the firm were largely focused on using the cash generated by the pharmaceuticals group to look for diversification opportunities ...

Managers in firms that failed to adopt the new techniques looked back on their failure as an embarrassment:

> By now you will have worked out that we didn't really do any real research in the eighties. We talked about it, but X didn't understand what it required. We kept doing what we had always done ...

The decision to invest in the science-driven approach was often identified with the hiring of key individuals:

> Oh, that's when we hired Y. He built a completely new research facility and started to hire aggressively. He had a fundamental belief that if we did leading edge science we would find breakthrough drugs...

Capturing this insight empirically is a challenging task. One option is to explore the correlation between changes in research strategy and changes in senior management personnel. Unfortunately we have not yet been able to collect this information in rich enough detail for our full sample. Here we use a binary variable (Scientist–CEO) measuring whether or not the firm's CEO has a scientific background as a first attempt at capturing the idea. We hypothesize that firms with

[18] In order to assist us in framing our hypotheses, between September 1998 and April 1999 we conducted interviews at seven pharmaceutical firms. All but one were based in the United States. The interviews were loosely structured discussions that explored, first, whether it was plausible that the adoption of science-driven drug discovery had significant effects on research productivity and, second, why it might have been the case that despite their plausible impact on productivity many firms were slow to adopt them.

scientifically trained CEOs will be more likely, all else equal, to adopt the techniques of science-driven drug discovery. While the classical environmental approach implicitly suggested that top managers who can deeply understand the value associated with novel strategies is important, the RBV literature has placed much greater emphasis on these kinds of factors.

(b) Cardiovasculars as a share of total firm sales

Our qualitative evidence suggests that the speed with which the new techniques were adopted was a function of the balance of power within the firm. Those firms whose sales portfolios were dominated by therapeutic classes in which the new techniques were likely to be particularly important—particularly by sales of cardiovascular drugs—appear to have become convinced that the new techniques were likely to be important much faster than managers in those firms in which sales were much more heavily concentrated in therapeutic areas for which the new techniques were much less useful. Thus we hypothesize that firms whose sales are dominated by cardiovascular therapies will adopt the new techniques faster than their rivals.

Notice that because of the very long lead times characteristic of the pharmaceutical industry, share of the sales portfolio is *not* highly correlated with share of the patent portfolio. Thus we interpret any effect of the share of sales portfolio on the diffusion of science-driven drug discovery as a reflection of the focus of attention of senior management attention, rather than as a response to any difference in the firm's research base.[19] It may also, of course, reflect the presence of marketing and sales assets.

(c) Distance from public science

Our qualitative evidence suggests that differences in geographic location may also have had systematic effects on the decision to adopt the new research techniques since the benefits of adopting the new approach may well have been significantly higher (and the costs significantly lower) for those firms whose research laboratories were located in reasonably close proximity to large communities of publicly funded researchers. Close proximity not only made it easier to attract the best-quality researchers, but probably also made it much easier to maintain close connections to the public sector once these researchers had joined the firm (Jaffe, 1986; Zucker *et al*., 1998). Indeed, both the classical environmental perspective and the RBV have emphasized the role of geographic location in shaping the ability of firms to more aggressively adopt practices associated with superior performance. We have developed a measure, DISTANCE, of the geographical position of a firm's research activities using the publications data from ISI.

(d) Knowledge capital: cardiovascular patents as a percentage of total firm patents

One of the key claims of the RBV is that a firm's particular experience with specific technologies or technological trajectories provides it with a firm-specific capital stock which it can both exploit as well as providing a mechanism to draw upon outside knowledge. In the current context, the techniques of science-driven drug discovery were not immediately applicable to all areas of drug research. In the late 1970s and early 1980s, for example, many scientists believed that the new techniques would be much more useful in the search for cardiovascular therapies than they would be in the search for new anti-infective drugs. Scientific understanding of cardiovascular disease was much further advanced, and it was thought that this understanding would provide a particularly rich basis for drug discovery. Firms with more experience in cardiovascular therapies might therefore have expected to benefit disproportionately from adoption of the new techniques. Thus, in the RBV tradition, we hypothesize that adoption will be positively correlated with experience in cardiovascular disease.

(e) Market position: share of the cardiovascular market

Following the classical environmental perspective, we also expect a firm's position in the product

[19] Share of the sales portfolio might also shape the choice of research technique if firms have specialized marketing or distribution assets. These assets certainly exist—Eli Lilly, for example, has historically had a very strong position in the distribution and marketing of treatments for diabetes. However, we believe that in the majority of cases they are relatively fungible.

market to have an effect on its decision to adopt the new techniques. A large literature in economics links the degree to which a firm possesses monopoly power in the product market to its decision to adopt an innovation (Gilbert and Newbery, 1982; Reinganum, 1981; Fudenberg and Tirole, 1985; Henderson, 1993; Karshenas and Stoneman, 1993). Unfortunately this literature yields no very clear predictions—under some circumstances monopolists appear to have incentives to adopt innovations *before* entrants; under others they have an incentive to wait. Moreover, it has not directly addressed the question of the drivers of adoption of a research technique, as opposed to the adoption of a new product. Nonetheless, we believe that in this context the most plausible interpretation of this literature is that firms with monopoly power in the relevant product market will have higher incentives to adopt the new techniques, since they are more likely to benefit from them. Thus, consistent with an externally oriented environmental perspective, we evaluate whether firms that have a high share of the cardiovascular market will adopt the new techniques faster than others.

(f) Firm sales

Finally, total U.S. sales is included to control for the fact that there may be economies of scale in the adoption of the new techniques. Zucker and Darby (1997) found evidence for economies of scale in the adoption of the techniques of biotechnology, and to the degree that 'science-driven' drug discovery also takes advantage of scientific resources that can be maintained centrally and exploited throughout the organization one would expect economies of scale to be important in this case as well. Indeed, one of the key tenets of most strategy perspectives (either the classical environmental perspective or RBV) is that *scale matters*, if only in terms of having sufficient liquidity or slack resources to aggressively adopt performance-enhancing organizational practices.

Other potential sources of heterogeneity

This list of variables could be expanded almost indefinitely. We do not explore, for example, the degree to which either a firm's position in its network or wider institutional context determines adoption. Determining the network position of a major pharmaceutical firm turns out to be an exceptionally difficult undertaking[20] and, while we are sure that the wider institutional context plays a role in shaping the adoption of science-driven drug discovery, distinguishing between the effects of institutional isomorphism or management fashion and the simple effects of time and convergence turns out to be very difficult. We also do not address the extent to which governance structures and incentive systems at the most senior levels of the firm affect adoption decisions. While this is an intriguing hypothesis, preliminary analysis highlighted both the difficulty of constructing consistent measures of governance problems across national boundaries and the fact that they appeared to be only very weakly correlated, if at all, with the firm's choice of research technique.

RESULTS

In this section we discuss the results from two sets of regression models. The first set focuses on the 'initial conditions' and convergence model. The second set brings in the measures of environmental heterogeneity.

Our analysis begins by considering a simple model of the adoption of science-driven drug discovery which focuses exclusively on the dynamic impact of 'initial conditions':

$$y_t = \alpha + \beta t + \gamma y_0 + \delta y_0 \cdot t + \epsilon_t$$

This equation is estimated using each of our alternate measures of the extent of diffusion of science-driven drug discovery (with the appropriate functional forms for each measure). Table 4 presents the results, which we find quite striking. In the first place, they show that at any point in time the degree to which a firm has adopted the techniques of science-driven drug discovery is to

[20] To begin with, there is the question of which network to focus on. In the context of a single industry study, it does not make sense to focus on the degree to which there is interlock across the board of directors, for example, since within a single industry directors are forbidden from serving on multiple boards to avoid potential conflicts of interest. We explored the degree to which we could use patterns of joint publication to construct a plausible network, but discovered that constructing such a network requires the collection (and coding) of data about hundreds of thousands of papers and would result in an intractably large network containing thousands of nodes. We are currently exploring whether we can usefully construct such a network by tracing the movement of senior executives across the industry.

Table 4. Initial conditions and convergence at determinants of the adoption of science driven drug discovery

Dependent variable: N	Log-odds (PROPUB) 135	Ln (PAPERS) 150	Ln (AUTHORS) 150	Log-odds (PUBFRAC) 150
$PROPUB_0$	1.00** (0.08)			
$PROPUB_0$ * time	−0.04** (0.01)			
Ln($PAPERS_0$)		1.02** (0.07)		
Ln($PAPERS_0$) * time		−0.05** (0.01)		
Ln($AUTHORS_0$)			1.04** (0.05)	
Ln($AUTHORS_0$) * time			−0.04** (0.01)	
$PUBFRAC_0$				3.57** (0.71)
$PUBFRAC_0$ * time				−0.29** (0.11)
Intercept	−3.87** (0.61)	−1.03* (0.04)	−0.85* (0.40)	−2.45** (0.60)
Time	0.18** (0.03)	0.32** (0.06)	0.35** (0.05)	0.26** (0.06)
Ln (Fmsales)	0.21* (0.10)	0.15** (0.05)	0.10** (0.03)	0.09 (0.08)
Adj. R^2	0.75	0.82	0.91	0.38

*$p<0.05$; **$p<0.01$

a very great extent a function of the firm's position at the beginning of the period. The estimated coefficient on the initial value of the dependent variable is large and very significant in all of the regressions. Second, the highly significant *negative* coefficient estimated for the interaction of y_0 with time provides evidence in favor of the convergence hypothesis. The firms that were 'furthest away' from best practice (with lower initial levels of the dependent variable) moved *fastest* to adopt. However, while convergence seems to be an important aspect of these data, the coefficient estimates imply a relatively slow diffusion process: 10 years or more for the laggard firms to catch up with the leaders.

These results are important. They illustrate concretely a model of competitive advantage that is implicit in much of our teaching and research but that is rarely demonstrated. Understanding the origins of competitive advantage requires a focus on two distinct processes: *creation*, the means whereby first adopters develop and exploit new techniques or strategies; and *imitation*, whereby laggard firms respond to their unfavorable positions and move to imitate market leaders.

Note that at one level these results are consistent with a Stinchcombian view of the sources of competitive advantage. Early movers, who happen to be endowed with the 'right' set of organizational practices, reap the returns from being in the right place at the right time. Laggards manage to change, but changing does not create advantage, it merely re-levels the playing field. At another level, though, convergence provides evidence for a more 'strategic' perspective on the dynamics of competitive advantage. Competitive advantage is as much about responding to unfavorable positioning as it is about exploiting those opportunities which present themselves. As such, aggressive imitation by those who are initially positioned most unfavorably is consistent with a process whereby thoughtful managers actively interpret their environment and use their discretion to implement a *strategic* response.

Table 5 builds on this analysis and develops evidence relating to the full diffusion model fol-

Table 5. Initial conditions and contemporaneous heterogeneity. Dependent variable: log-odds (PUBFRAC), 150 observations throughout

	(5.1)	(5.2)	(5.3)	(5.4)	(5.5)
$PUBFRAC_0$	3.57** (0.71)				7.11** (0.92)
$PUBFRAC_0$ * time	−0.29** (0.11)				−0.79** (0.15)
Scientist–CEO$_{(t-1)}$		0.84 (0.53)		0.87 (0.54)	0.33 (0.46)
Scientist–CEO$_{(t-1)}$ * time		−0.07 (0.13)		−0.07 (0.13)	−0.01 (0.11)
Share of cardiovasculars in firm sales$_{(t-1)}$		−0.00 (0.01)		0.00 (0.01)	0.05** (0.01)
Share of cardiovasculars in firm sales$_{(t-1)}$ * time		0.00 (0.00)		0.00 (0.00)	−0.006** (0.001)
Distance from science$_{(t-1)}$			0.20 (0.26)	0.29 (0.27)	−0.60* (0.25)
Distance from science$_{(t-1)}$ * time			−0.01 (0.04)	−0.02 (0.04)	0.08 (0.04)
Share of cardiovasculars in firm patents$_{(t-1)}$			0.01 (0.03)	0.00 (0.03)	−0.03 (0.03)
Share of cardiovasculars in firm patents$_{(t-1)}$ * time			0.00 (0.01)	0.00 (0.01)	0.01 (0.01)
Share of U.S. cardiovascular market$_{(t-1)}$			−0.03 (0.02)	−0.03 (0.02)	−0.12** (0.02)
Share of US cardiovascular market$_{(t-1)}$ * time			0.004* (0.002)	0.01* (0.00)	0.02** (0.00)
Intercept	−2.45** (0.60)	−1.22** (0.56)	−1.43* (0.74)	−1.52 (0.83)	−4.99** (0.83)
Time	0.26** (0.06)	0.08** (0.02)	0.04 (0.07)	0.03** (0.07)	0.44** (0.10)
Ln(total firm sales$_t$)	0.09 (0.08)	0.21* (0.09)	0.23* (0.11)	0.24* (0.12)	0.22* (0.10)
Adj. R^2	0.38	0.28	0.28	0.33	0.51

*$p<0.05$; **$p<0.01$

lowing Equation 1. Each column includes either (a) only variables associated with initial conditions (5.1), (b) only variables associated with time-varying environmental heterogeneity ((5.2) through (5.4)), or (c) both types of variables (5.5). For each regression, the dependent variable is PUBFRAC.[21] The environmental variables are lagged by one period to allow time for firms to adjust their level of PUBFRAC, and each regression includes both a time trend and the log of total firm sales as control variables.

After replicating the 'initial conditions plus convergence' result in (5.1) for easy reference,

[21] We have experimented extensively both with PUBFRAC as well as our other measures of the intensity of science-driven drug discovery (PUBFRAC, PAPERS, and AUTHORS). Overall, the key results as they relate to distinguishing between alternative sources of competitive advantage are robust.

we present our initial analysis of the impact of environmental heterogeneity by dividing the explanatory variables into two groups corresponding to 'behavioral' and 'economic' hypotheses about the drivers of adoption. The two 'behavioral' variables are Scientist–CEO and the within-firm share of cardiovascular drugs, which are included by themselves in (5.2). The 'economic' variables are those measuring the firm's distance from science, its knowledge capital, and its competitive positioning, which are included by themselves in (5.3). (5.4) includes both the behavioral and economic variables together but still does not include initial conditions. These results are quite weak. Among the 11 explanatory variables (including firm sales), the only (marginally) significant coefficients are associated with firm sales and the interaction between the firm's market position in cardiovasculars and TIME; the small positive parameter suggests faster 'catch-up' for firms with a larger share of the cardiovascular market.

However, we believe that each of the first four regressions in Table 5 are likely misspecified. For example, by excluding the impact of initial conditions in (5.2) through (5.4), we introduce a great deal of noise into the regression, likely limiting our ability to distinguish the 'true' impact of the environmental drivers. Model (5.5) therefore includes both the initial conditions and environmental heterogeneity variables, with a dramatic impact on the explanatory power of the regression and on the coefficient estimates. For example, the regressions suggest that, after controlling for initial conditions and convergence (the coefficients for which remain large and significant as in earlier regressions), those firms whose sales were concentrated in cardiovascular therapies were associated with higher levels of PUBFRAC (though their *rate* of adoption was slower) and firms more distant from scientific centers are associated with lower levels of PUBFRAC. Those firms that had a larger share of the cardiovascular market were also faster to adopt the new techniques (though the impact on the level is negative (in contrast to our initial hypotheses)). Finally, the measures of Scientist–CEO and knowledge capital (share of firm patents associated with cardiovascular therapies) remain insignificant, perhaps due to the noisy nature of each of these proxy measures and their imperfect correlation with the underlying concepts they are associated with. More generally, (5.5) suggests that the inclusion of variables to incorporate the impact of initial conditions and convergence allows for a more precise and systematic evaluation of the impact of environmental heterogeneity on the diffusion process, as well as providing an important confirmation for our earlier results about the impact of initial conditions and convergence.

The alternative specifications offered in Table 5 also provide some evidence about the relative importance of initial conditions vs. environmental heterogeneity, at least in terms of goodness-of-fit. Of course evaluating the differences in R^2 from each regression is of only limited value since it cannot provide evidence for how a change in one of the explanatory variables changes the expected value of the dependent variable. However, this comparison does suggest two things in the current context. First, the regressions which focus on the firm's initial conditions explain a significantly higher share of the overall variance than the regressions which evaluate environmental heterogeneity by itself. Second, and perhaps more importantly, initial conditions and environmental heterogeneity are complementary in their explanatory power—there is over a 33 percent increase in adjusted R^2 in the combined model (5.5) compared with either (5.1) or (5.4).

CONCLUSIONS AND IMPLICATIONS FOR FURTHER RESEARCH

These results are intriguing. At face value, Table 5 appears to offer support for Stinchcombe's critique, in that initial conditions play a very important role in explaining the diffusion of science-driven drug discovery.[22] However, by developing a dynamic model of adoption which allows for both convergence and change associated with local environmental cues, our results suggest that conscious strategic adjustment is also quite important for understanding the

[22] One possibility, of course, is that y_0 is capturing not the characteristics of the firm's founding, but a complex (unobserved) history of strategic adjustment. As a very preliminary cut at whether this is plausibly the case, we compared the usefulness of our measures of heterogeneity as measured at the beginning of the period, Z_0, as explanatory variables for *y*. The regression performed quite poorly, with an even lower proportion of the variance in the dependent variable explained than in the regressions that included contemporaneous measures of heterogeneity.

dynamics of competitive advantage: convergence is a key factor in explaining the overall pattern of adoption, many of the 'strategic' variables are significant after controlling for initial conditions, and the explanatory power of the diffusion model substantially improves when all of the effects are included simultaneously.

In other words, our results imply that—at least in this context—Stinchcombe's hypothesis is powerful but not all encompassing. One of the most striking features of our data is *convergence*: firms that were initially 'behind' at the beginning of the period move more rapidly to 'catch up' to their more advanced competitors. While associated with the same 'initial conditions' that Stinchcombe emphasizes, the process of convergence and imitation is nonetheless interestingly different from a simple model of competition in which competitive advantage reflects the failure or exit of those firms that are not initially well positioned. Our results also suggest that, after controlling for initial conditions, variables traditionally associated with 'strategic intent' are important determinants of the level and rate of adoption. While Stinchcombe's hypothesis is well worth taking seriously (and we would encourage future researchers to be careful to control for it), it seems to us that there is sufficient evidence here to suggest that firms can and do change in purposeful ways in response to exogenous shocks. For example, convergence suggests that managers who are initially associated with poorly positioned firms may be the most proactive in altering their practices and strategies. In other words, there seems to be a clear role for understanding the processes of strategic adjustment which occur in response to the internal and external environmental cues. In this sense, our framework and results provide evidence consistent *both* with Stinchcombe's hypothesis that competitive advantage is importantly driven by exogenous initial variation and with the more 'rational' or 'strategic' perspective that views firms as responding with foresight to changes in their environment.

Of course, as we have emphasized several times, our empirical results are quite preliminary, and perhaps raise as many questions as they answer. Indeed the statistical analysis is largely descriptive, and we have tried to highlight throughout the paper some of the difficult and interesting statistical questions that are raised by the attempt to distinguish among the central hypotheses. Our paper also highlights the need for richer longitudinal data. One could plausibly argue, for example, that the significance of initial conditions in our results is simply a consequence of 'left truncation:' that if we knew more about the history of the firms in our sample—about the origins of each firm's 'initial conditions'— we might be able to uncover additional evidence for the importance of managerial vision and action in shaping competitive advantage. For example, we know from a number of early histories of the industry that pharmaceutical firms differed enormously in the extent to which they invested in basic or scientific research in the 1940s and 1950s. Some firms, such as Parke-Davis, Lilly, and Merck, made early and extensive commitments to pure research, while other firms delayed making these investments for many years (Parascandola, 1985). Perhaps extending our empirical study of this industry further back in time would allow us to conclude that the adoption of science-driven drug discovery stemmed from far-sighted decisions made decades before the practice became widespread.

This suggestion brings us back to the discussion with which the paper began. During the 1980s, firms that used science-driven drug discovery techniques outperformed those that did not. As such, one might be tempted to conclude that adopting this practice was a 'good' strategy that led, through the creation of unique capabilities, to significant competitive advantage in the sense of superior ability to generate new drug products. We have, ourselves, recommended to pharmaceutical firms that they adopt these techniques and the managers of many firms have by and large agreed with us. But one of the key conclusions to be drawn from our analysis here is that positively responding to such a recommendation is only half the battle. Incorporating best practice is necessary, but imitation based on the experience of rivals can usually only level the playing field, and in this case, at least, imitation took many years. While key aspects of the adoption process—in particular firms' rate of convergence to best practice—reflected purposive 'strategic' responses to a changing environment, these responses were greatly constrained by factors that were in place long before many of these firms even began to think seriously about the issue. The RBV has emphasized the importance of long-lived 'sticky' assets, and the powerful influence of historical circumstances seen here indicates that these can be very long lived and very sticky indeed.

The advent of science-driven drug discovery had a dramatic impact on the innovation process in the pharmaceutical industry. The basis of competitive advantage was changed in important ways, and firms in the industry responded to this shock by making appropriate changes to their internal structure, and by investing in the necessary capabilities and resources. But some firms entered this period of transition much better equipped than others to exploit the potential of the new techniques. Were they lucky or were they smart? The difficulty of answering this question empirically points to a deeper research agenda. We believe that a fuller appreciation of the origins of competitive advantage is to be found in a better analytical and empirical understanding of the types of managerial processes which allow some firms to be ahead, and stay ahead, of the game. *Ex post*, it is clear that some firms actively identify, interpret, and act upon early signals from their internal and external environment, and so position themselves to effectively exploit these opportunities well in advance of others' demonstration of the pay-off from the strategies which emerge later on as 'best practice.' These firms are *creating* new sources of competitive advantage. Understanding how they organize *ex ante* to do this is, in our view, a central question for strategy research.

ACKNOWLEDGEMENTS

This study was funded by POPI, the Program for the Study of the Pharmaceutical Industry at MIT and by the MIT Center for Innovation in Product Development under NSF Cooperative Agreement Number EEC-9529140. Stern completed this paper while a Health and Aging Fellow at the National Bureau of Economic Research. Their support is gratefully acknowledged. Jeff Furman and Pierre Azoulay provided outstanding research assistance, and we would also like to thank Bob Gibbons, Connie Helfat, David Hounshell, Zvi Griliches, and participants at the Dartmouth conference focused on this special issue for their many helpful comments.

REFERENCES

Andrews KR. 1971. *The Concept of Corporate Strategy*. Dow–Jones Irwin: Homewood, IL.

Arellano M, Bond S. 1991. Some tests of specification for panel data: Monte Carlo evidence and an application to employment equations. *Review of Economic Studies* **58**(194): 277–320.

Barnett WP, Hansen MT. 1996. The Red Queen in organizational evolution. *Strategic Management Journal*. **17**: 139–158.

Barney J. 1991. Firm resources and sustained competitive advantage. *Journal of Management* **17**(1): 99–120.

Besanko D, Dranove D, Shanley M. 2000. *Economics of Strategy*. 2nd edn, Wiley: New York.

Bower JL. 1974. *Managing the Resource Allocation Process*. Harvard Business School Press: Boston, MA.

Brandenberger A, Nalebuff B. 1998. *Co-opetition*. Harvard Business School Press: Boston, MA.

Burgelman RA. 1994. Fading memories: a process study of strategic business exit in dynamic environments. *Administrative Science Quarterly* **39**(1): 24–56.

Chandler A. 1962. *Strategy and Structure*. MIT Press: Cambridge, MA (reprinted January 1990).

Christensen CM, Bower JL. 1996. Customer power, strategic investment, and the failure of leading firms. *Strategic Management Journal* **17**(3): 197–218.

Clark KB, Fujimoto T. 1991. *Product Development in the World Automobile Industry*. Harvard Business School Press: Boston, MA.

Cockburn IM, Henderson RM. 1998. Absorptive capacity, coauthoring behavior, and the organization of research in drug discovery. *Journal of Industrial Economics* **46**(2): 157–182.

Cockburn IM, Henderson RM, Stern S. 1999a. Balancing incentives: the tension between basic and applied research. NBER working paper no. 6882, January.

Cockburn IM, Henderson RM, Stern S. 1999b. The diffusion of science driven drug discovery: organizational change in pharmaceutical research. NBER working paper no. 7359, December.

Davis G. 1991. Agents without principles? The spread of the poison pill through intercorporate networks. *Administrative Science Quarterly* **36**: 583–613.

Eisenhardt KM. 1988. Agency- and institutional-theory explanations: the case of retail sales compensation. *American Management Journal* **31**(3): 488–511.

Eisenhardt KM, Schoonhoven CB. 1990. Organizational growth: linking founding team, strategy, environment, and growth among U.S. semiconductor ventures, 1978–1988. *Administrative Science Quarterly* **35**: 504–529.

Eisenhardt KM, Tabrizi BN. 1995. Accelerating adaptive processes. *Administrative Science Quarterly* **40**(1): 84–110.

Fletcher K, Huff AS. 1990. Strategic argument mapping: a study of strategy reformulation at AT&T. In *Mapping Strategic Thought*, Huff AS (ed). Wiley: New York; 165–194.

Fudenberg D, Tirole J. 1985. Pre-emption and rent equalization in the adoption of new technology. *Review of Economic Studies* **52**: 383–401.

Gambardella A. 1995. *Science and Innovation in the*

US Pharmaceutical Industry. Cambridge University Press: Cambridge, UK.
Ghemawat P. 1991. *Commitment: The Dynamics of Strategy*. Free Press: New York.
Gilbert RJ, Newbery DM. 1982. Pre-emptive patenting and the persistence of monopoly. *American Economic Review*, 514–595.
Grabowski HG, Vernon JM. 1990. A new look at the returns and risks to pharmaceutical R&D. *Management Science* **36**(7): 804–821.
Griliches Z. 1957. Hybrid corn: an exploration in the economics of technical change. *Econometrica* **25**(4): 501–522.
Griliches Z. 1986. Data problems in econometrics. In *Handbook of Econometrics*, Intriligator M, Griliches Z (eds). North-Holland: Amsterdam; 1466–1514.
Hannan MT, Freeman JH. 1989. *Organizational Ecology*. Harvard University Press: Cambridge, MA.
Heckman JJ. 1991. Identifying the hand of the past: distinguishing state dependence from heterogeneity. *American Economic Review* **81**(2): 75–79.
Henderson RM. 1993. Underinvestment and incompetence as responses to radical innovation: evidence from the semiconductor photolithographic alignment equipment industry. *Rand Journal of Economics* **24**(2): 248–270.
Henderson RM. 1994. The evolution of integrative competence: innovation in cardiovascular drug discovery. *Industrial and Corporate Change* **3**(3): 607–630.
Henderson RM, Clark KB. 1990. Architectural innovation: the reconfiguration of existing product technologies and the failure of established firms. *Administrative Science Quarterly* **35**: 9–30.
Henderson RM, Cockburn IM. 1994. Measuring competence? Exploring firm effects in pharmaceutical research. *Strategic Management Journal*, Winter Special Issue **15**: 63–84.
Henderson RM, Cockburn IM. 1996. Scale, scope and spillovers: the determinants of research productivity in drug discovery. *Rand Journal of Economics* **27**(1): 32–59.
Henderson RM, Orsenigo L, Pisano G. 1999. The pharmaceutical industry and the revolution in molecular biology: interactions among scientific, institutional and organizational change. In *The Sources of Industrial Leadership*, Mowery D, Nelson R (eds). Cambridge University Press: New York; 267–311.
Hicks D. 1995. Published papers, tacit competencies and corporate management of the public/private character of knowledge. *Industrial and Corporate Change* **4**(2): 401–424.
Holbrook K, Cohen WM, Hounshell DA, Klepper S. 2000. The nature, sources and consequences of the early history of the semiconductor industry. *Strategic Management Journal*. **21**(10–11): 1017–1041.
Jaffe AB. 1986. Technological opportunity and spillovers of R&D: evidence from firms' patents, profits and market value. *American Economic Review* **76**(5): 984–1001.
Karshenas M, Stoneman P. 1993. Rank, stock, order and epidemic effects in the diffusion of new process technologies: an empirical model. *Rand Journal of Economics* **24**(4): 503–528.
Koenig M. 1982. A bibliometric analysis of pharmaceutical research. *Research Policy* **12**: 15–36.
Lieberman MB, Dhawan R. 2000. Assessing the resource base of U.S. and Japanese auto producers: a stochastic production frontier approach. Mimeo, UCLA Anderson School of Management, February.
Leonard-Barton D. 1998. *Wellsprings of Knowledge: Building and Sustaining the Sources of Innovation*. Harvard Business School Press: Boston, MA.
Levinthal DA. 1997. Adaptation on rugged landscapes. *Management Science* **43**: 934–950.
McGahan AM, Porter ME. 1997. How much does industry matter, really? *Strategic Management Journal*, Summer Special Issue **18**: 15–30.
Mintzberg H. 1987. Crafting strategy. *Harvard Business Review* **65**(4): 66–75.
Mundlak Y. 1961. Empirical production function free of managerial bias. *Journal of Farm Economics* **43**: 44–56.
Murmann P, Tushman ML. 1997. Impacts of executive team characteristics and organization context on organization responsiveness to environmental shock. In *Technological Innovation: Oversights and Foresights*, Garud R, Nayyar P, Shapira Z (eds). Cambridge University Press: New York; 260–278.
Nelson RR, Winter SG. 1982. *An Evolutionary Theory of Economic Change*. Harvard University Press: Cambridge, MA.
Parascandola J. 1985. Industry research comes of age: the American pharmaceutical industry, 1920–1940. *Pharmacy in History* **27**(1): 12–21.
Penrose ET. 1959. *The Theory of the Growth of the Firm*. Wiley: New York.
Peteraf MA. 1993. The cornerstones of competitive advantage: a resource-based view. *Strategic Management Journal* **14**(3): 179–191.
Porter ME. 1980. *Competitive Strategy*. Free Press: New York.
Powell WW, Koput KW, Smith-Doerr L. 1996. Interorganizational collaboration and the locus of innovation: networks of learning in biotechnology. *Administrative Science Quarterly* **41**(1): 116–145.
Quinn JB. 1978. Strategic change: logical incrementalism. *Sloan Management Review* **20**(1): 7–23.
Reinganum JF. 1981. Market structure and the diffusion of new technology. *Bell Journal of Economics* **12**: 618–624.
Rumelt RP. 1991. How much does industry matter? *Strategic Management Journal* **12**(3): 167–185.
Selznick P. 1957. *Leadership in Administration: A Sociological Interpretation*. Harper & Row: New York.
Shapiro C, Varian HR. 1998. *Information Rules*. Harvard Business School Press: Boston, MA.
Stern S. 1999. Do scientists pay to be scientists? NBER working paper no. 7410, October.
Stinchcombe AL. 1965. Social structure and organizations. In *Handbook of Organizations*, March J (ed). Rand-McNally: Chicago, IL; 142–193.
Stinchcombe AL. 2000. Social structure and organizations: a comment. In *Economics Meets Sociology in Strategic Management: Advances in Strategic*

Management, Baum J, Dobbin F (eds). JAI Press: Greenwich, CT.
Wernerfelt B. 1984. A resource-based view of the firm. *Strategic Management Journal* **5**(2): 171–180.
Zucker LG, Darby MR. 1997. Present at the revolution: transformation of technical identity for a large incumbent pharmaceutical firm after the biotechnological breakthrough. *Research Policy* **26**(4): 429–447.
Zucker LG, Darby MR, Brewer M. 1998. Intellectual human capital and the birth of U.S. biotechnology enterprises. *American Economic Review* **88**(1): 290–306.

APPENDIX A

Data appendix: construction of our measures of *Z*

Scientist–CEO

Scientist–CEO is a binary variable set to 1 if the firm's CEO has a scientific background. We constructed this variable by systematically checking the Annual Report of each company in our sample for the title or background of the CEO. Those CEOs who were listed as having either a PhD or an MD were coded as scientists.

Distance from science

For any single paper, ISI lists the addresses of each author. From these data we constructed a count of the paper authorships by city for each firm for each year. We designated every city that constituted more than 2 percent of all the firm's authorships over the entire period a 'research node' for the company, and obtained its longitude and latitude. We then weighted each node by the fraction of authorships it hosted relative to all authorships attributed to the company that year. These data gave us a good sense of the changing geographical location of the firm's research over time.

In order to measure the degree to which the firm was 'close' to publicly funded science we needed to construct a weighted measure of the location of publicly funded research. This is difficult to do without full data on public spending by geographical location across the world, and data for Europe and Japan are relatively difficult to obtain. As a first approximation we constructed a data set consisting of the addresses of all those authors who coauthored papers with NIH researchers over our period. We designated each city listed in these data that constituted over 2 percent of total NIH coauthorships an NIH node, and obtained its latitude and longitude. We weighted these nodes using two schemes: as the fraction of NIH coauthorships that they obtained in a given year and as a rank from 1 to 10 on the basis of the scientific reputation of each node.

A firm's 'closeness to science' was then calculated as: $\Sigma_{ij} w_i * w_j * 1/d_{ij}$, where d_{ij} is the distance from each firm node$_i$ to NIH node$_j$, w_i is the weight assigned to the company node and w_j to each node.

Firm sales

Firm sales are measured as total pharmaceutical sales in the United States, as reported to IMS. We excluded sales of OTC or 'over-the-counter' medications that can be obtained without a prescription. Note that the United States is only about 50 percent of the world pharmaceutical market, and that many of the firms in our sample have substantial global sales. U.S. sales are thus a rather noisy measure of total sales.

Knowledge capital: share of patents in cardiovascular therapies

We used Derwent patent data to construct a measure of the firm's experience or 'knowledge capital' in cardiovasculars. For all of the firms in our data set, Derwent generously donated to us complete data about each patent family granted to each firm over our period. Each patent family includes complete information about the patents granted to the firm across the entire world and a complete set of Derwent 'manual codes'.[23] For each patent that had been granted in two out of three major world markets (Japan, the United States, and Europe) we used the 'manual codes' to assign patents to particular therapeutic areas. We then calculated a 'knowledge stock' in the standard manner assuming a 20 percent deprecation rate.

[23] The U.S. patent office classifies pharmaceutical patents largely on the basis of chemical structure: a classification that contains very little information about therapeutic intent. The Derwent manual codes are assigned by specialists in the field who classify each patent on the basis of its therapeutic implications.

Market position: share of the cardiovascular market

We measure share of sales using detailed data at the product level obtained from IMS America. Note that the pharmaceutical market is a global one. The U.S. market represents roughly 50 percent of the world market, and nearly every firm in our sample has extensive global operations. Unfortunately our share measures will be distorted by the fact that the distribution of sales across therapeutic classes is *not* uniform across the world.

Managerial attention: cardiovascular therapies as a share of the sales portfolio

We measure cardiovascular sales as a share of total firm sales using IMS America data. Note that it is subject to the same source of error as our measure of market position, above.

Part IV
Delivering Innovative Value Through Inter-Organizational Drivers

[14]

The R&D Boundaries of the Firm: An Empirical Analysis

Gary P. Pisano
Harvard University

This paper examines how two sources of transaction costs, small-numbers-bargaining hazards and appropriability concerns, may affect established firms' choices between in-house and external sources of R&D when technological change shifts the locus of R&D expertise from established enterprises to new entrants, and established firms face a make-or-buy decision for R&D projects. The relationships of other organizational characteristics to the R&D procurement decision are also considered. Hypotheses are tested with data on 92 biotechnology R&D projects that major pharmaceutical companies have sponsored either in-house or through external contractual arrangements. The results suggest that small-numbers-bargaining problems motivate firms to internalize R&D. Evidence is also found that a firm's R&D experience, its dependence on the pharmaceutical business, and its national origin affect R&D procurement decisions.•

The ability of firms to develop and exploit technological know-how is an important dimension of competition in many industries. Firms' internal research and development (R&D) capabilities are often viewed as the critical determinant of such abilities. However, while in-house R&D has traditionally been an important source of technical know-how for firms (Mowery, 1983), it is not the only possible source. Firms can tap the R&D capabilities of competitors, suppliers, and other organizations through such contractual arrangements as licenses, R&D agreements, and joint ventures. Various empirical studies have documented significant interfirm, interindustry, and intertemporal differences in the degree to which firms obtain R&D services from in-house versus external sources (e.g., von Hippel, 1982; Bozeman and Link, 1983; Pavitt, 1986).

R&D procurement decisions become particularly relevant for established enterprises confronting broad and rapid changes in their core technologies. During such gusts of "creative destruction" (Schumpeter, 1975: 83), established firms' in-house laboratories may lack the relevant technological skills to perform R&D competitively with new entrants. Decisions about which new technical capacities to develop internally and which ones to access through collaborative and contractual links with external sources may affect the firm's long-term viability in the new technological environment. For example, electronics firms that decided in the early 1960s to license-in the new integrated circuit technologies rather than to develop the requisite R&D capabilities in-house had difficulty competing for many years afterward (Malerba, 1985).

Many factors may affect a firm's choice between internal (hierarchical) and contractual (market) modes for organizing a product R&D program. This paper examines the role that transaction costs may play in make-versus-buy decisions for product R&D. While transaction-cost theory has been used in empirical analyses of production and supply activities (Monteverde and Teece, 1982a, 1982b; Stuckey, 1983; Masten, 1984; Walker and Weber, 1984; Balakrishnan and Wernerfelt, 1986; Joskow, 1987) as well as marketing (Anderson and Schmittlein, 1984), its application to the study of the R&D boundaries of the firm has been limited. Williamson (1975),

0001-8392/90/3501-0153/$1.00.

•

The author would like to thank Glenn Carroll, Robert Harris, Michael Katz, Kaz Mishina, Jeffrey Perloff, David Teece, Oliver Williamson, Richard Nelson, Michael Tushman, and the anonymous reviewers for their helpful comments on earlier drafts of this paper. This paper draws from the author's doctoral research, which was supported in part by the National Science Foundation, grant #SRS-841056. Additional support from the Harvard Business School Division of Research is also gratefully acknowledged.

Teece and Armour (1977), and Teece (1988) have argued that R&D is more efficiently governed by hierarchies than by markets. However, the various propositions about the best organizational location for R&D have never been tested with project-level data. This situation stands in contrast to transaction-cost analyses of make-or-buy decisions for production, which have used data at the level of individual components and subassemblies (Monteverde and Teece, 1982a; Masten, 1984; Walker and Weber, 1984).

This paper develops and tests transaction-cost hypotheses for R&D procurement at the project level. While there are many potential sources of transaction costs, including engineering intensity and design specialization (Monteverde and Teece, 1982a, 1982b; Masten, 1984), technological uncertainty (Walker and Weber, 1984; Balakrishnan and Wernerfelt, 1986), and the co-location of specialized assets (Joskow, 1987), this analysis focuses on the following two: (1) small-numbers-bargaining hazards stemming from specialized R&D capabilities and (2) appropriability problems arising from competition in product markets.

Small-numbers-bargaining hazards play a central role in the transaction-cost theory of the firm. However, theoretical and empirical analysis of their effects has been largely confined to intermediate product markets. Whether small-numbers-bargaining hazards also influence R&D markets is a question that remains to be answered. The question is important, because transaction-cost theory must be applicable to R&D as well as production activities if it is to be considered a robust theory of the firm. Appropriability problems have always been an issue of central concern in the economics of innovation (Schumpeter, 1975; Arrow, 1962). More recently, Teece (1986, 1988) argued that appropriability problems are a critical factor impeding market contracting for R&D. Despite its salience, little empirical research has been done on the causes of appropriability problems and their consequences for the organization of R&D.

The empirical analysis in this paper examines the decisions of established pharmaceutical companies to develop new biotechnology-based pharmaceutical products through in-house R&D versus through contractual arrangements with outside firms. The case of biotechnology is particularly interesting because it represents a relatively dramatic change in the technological environment of established pharmaceutical firms. It thus provides an opportunity to examine how transaction costs and other factors may affect the way established firms adjust their R&D boundaries in the wake of such changes.

The biotechnology industry has been the context of other studies on the organization of innovation. Shan (1989) examined the organizational modes through which start-up firms in the biotechnology industry have chosen to commercialize their technologies. While his study and the present one draw data from some of the same sources, they differ in fundamental ways. Shan's analysis focused on how a new firm's competitive position affects its decision whether to commercialize its technologies on its own or through a cooperative arrangement with a partner. The determinants of cooperative commercialization strategies for these start-up firms was the central concern of his study. The present study is concerned with the R&D procurement behavior of large, established firms already capable of commercializing the technology. The transactional difficulties of R&D contracts and how these affect established firms' R&D procurement form the conceptual heart of this analysis.

TECHNOLOGICAL CHANGES AND THE BOUNDARIES OF THE FIRM

Episodes of "competence-destroying" technological change (Tushman and Anderson, 1986), which suddenly make existing skills and capabilities obsolete, are often characterized by a shift in the locus of technical expertise from industry incumbents to newly formed ventures and firms from other industries. Competition from these new entrants and their eventual triumph over established firms is what Schumpeter (1975: 83) had in mind when he referred to the process of "creative destruction." However, a competitive clash between new entrants and incumbents is not an inevitable outcome of such technological upheavals. In some cases, technological change that weakens or destroys established firms' advantages in R&D and production has little or no effect on their relative advantage in distribution and marketing. Such "revolutionary" technological change (Abernathy and Clark, 1985) generates differences in distinctive competences. The new entrants capture a relative advantage in R&D while the established enterprises maintain their advantage in commercializing products. This situation creates opportunities for beneficial trade, rather than competition, between the new sources of technical expertise and the established firms.

The emergence of biotechnology during the late 1970s is an example of a revolutionary technological change (in Abernathy and Clark's sense) for the pharmaceutical industry. Basic research in molecular biology during the 1970s yielded a number of discoveries about how the genetic structure of cells could be manipulated to induce them to produce specific proteins. These discoveries created a methodology for synthesizing potential therapeutic proteins that were too complex to be synthesized through the chemical methods traditionally used in pharmaceutical research. As a means of synthesizing therapeutic compounds, biotechnology represented a competence-destroying technology because it required technical skills that were fundamentally different from those with which established pharmaceutical firms were familiar. For this reason, most of the early commercial biotechnology R&D was conducted by new ventures that formed in the United States between 1976 and 1982 and not by the established pharmaceutical firms (Office of Technology Assessment, 1984).

While biotechnology was competence destroying on the R&D end, it was competence preserving at the commercialization end. The potential new protein drugs made possible by biotechnology must all go through the same clinical tests and regulatory approval process and are sold through the same distribution channels as traditional drugs. With years of commercial experience and existing organizational capabilities in these "downstream" functions, established pharmaceutical companies had an advantage over new entrants in bringing new drugs from the laboratory to the market.

An established pharmaceutical firm that wanted to commercialize a biotechnology-based pharmaceutical product faced the following choice: develop the required R&D capabilities in-house (vertically integrate) or procure the necessary R&D services from a new biotechnology firm. Vertically integrating (on any given project) would mean forgoing the opportunity to tap the distinctive R&D competence of a new biotechnology firm. Because new biotechnology firms could do R&D in this technology more effectively than an inexperienced in-house team, external R&D procurement would seem to be a more economically rational alternative. Because it was difficult for the new firms to commercialize their own products, they might find it equally attractive to collaborate with established enterprises.

In light of these apparent economies of functional or vertical specialization, the incidence with which established pharmaceutical companies undertook biotechnology R&D projects internally is striking. The data presented in this paper suggest that, as of 1986, established pharmaceutical firms were conducting slightly less than half (47 percent) of their biotechnology R&D projects in-house. This paper examines some of the factors that have affected the extent and direction of backward integration into biotechnology R&D by individual pharmaceutical companies and, specifically, how transaction costs may have impeded markets for biotechnology R&D.

Transaction-Cost Theory

As an economic approach to organization, transaction-cost theory attempts to explain why institutional structures other than markets are necessary for the efficient governance of economic activity. It assumes that, due to economies of specialization and the administrative and incentive limits of hierarchies (Williamson, 1985; Grossman and Hart, 1986), markets are a more efficient governance structure, unless a transaction is surrounded by special circumstances. What these special circumstances are, how they arise, and what implications they have for institutional design have been the focus of research in transaction-cost economics over the past fifteen years (e.g., Williamson, 1975, 1985; Klein, Crawford, and Alchian, 1978; Teece, 1982). Transaction-cost theory posits that the costs of market (contractual) governance increase when the terms of exchange are surrounded by uncertainty and require a party to invest in transaction-specific assets. Uncertainty over the terms of trade arises when the contingencies affecting the execution of the agreement are complex and difficult for the trading partners to understand, predict, or articulate. Contracts made under such conditions are necessarily incomplete and may require renegotiation when unexpected contingencies occur. Renegotiation, however, represents a hazardous proposition for a party that has limited exchange alternatives. This situation, known as the small-numbers-bargaining problem (Williamson, 1975), can occur when a firm invests in assets that are costly to transfer to alternative transactions or uses. Because such transaction-specific assets limit the firm's ability to switch partners, they make it vulnerable to opportunistic recontracting.

The hazards of repeatedly entering into contractual agreements that involve uncertainty and transaction-specific assets provide an incentive for vertical integration. Internalization eliminates the ex post bargaining problems that might arise with an outside partner and thus improves incentives to commit specialized capital. Of course, internalization should only take place if the governance economies exceed the incremental economic costs due to additional administrative burdens, incentive distortions, and losses in production efficiency that occur when a firm's in-house capabilities are inferior to those of outside sources (Williamson, 1985; Grossman and Hart, 1986).

Transaction-cost theory can help us to understand R&D boundary choices that occur in the wake of technological changes that make existing R&D capabilities obsolete but preserve capabilities needed to commercialize the new technology. As such, they create opportunities for trade between firms that are masters of the new R&D field and those that continue to be strong in commercial functions. Transaction-cost theory can help us understand why firms may vertically integrate despite the benefits of trading with external parties.

Transaction Costs in Markets for Biotechnology R&D

Uncertainty pervades the process of discovering, synthesizing, and formulating a therapeutic compound through biotechnology R&D. At the outset of a project, accurate assessments of costs, duration, and outcomes are virtually impossible. The uncertainties of R&D in general can only be resolved sequentially as the project progresses (Nelson and Winter, 1977). As a result, parties to an R&D contract usually recognize that it is futile to attempt to lock-in the terms of trade at the outset. Instead, they generally agree to renegotiate the contract as uncertainties are resolved and unexpected contingencies arise. The following passage from an actual R&D agreement between Agricultural Genetic Sciences (AGS) and AC Biotechnics is typical:

> It is anticipated by the parties that the program will require adaptation to actual development. Consequently, the parties intend currently to review experiences gained during the program implementation and to agree on all significant needs for adjustment of the program. . . . (contract disclosed in exhibit to AGS's 1984 10-K Report)

Due to the need to adapt sequentially, parties to an R&D agreement can expect repeated rounds of negotiation during a single project. Because a biotechnology R&D project may run for several years, it is more characteristic of a long-term recurrent transaction than a one-shot exchange. Transaction-cost theory is relevant here because it recognizes that start-up costs of vertical integration make it desirable only for recurring transactions and not for one-time exchanges. Such recurrent transactions within the biotechnology industry are hypothesized to involve two types of contractual hazards that may lead to vertical integration: (1) small-numbers bargaining and (2) appropriability problems.

Small-numbers bargaining. At various stages in the project, the sponsor (a pharmaceutical company) and its R&D partner (usually a new biotechnology firm) must renegotiate such contractual terms as reimbursement of R&D costs, distribution of property rights, and completion dates. The contractor is usually reimbursed for its R&D costs, plus some margin of profit when the project reaches specific "milestones." Biotechnology R&D projects are costly, and the difficulty of reviewing the progress of the project very frequently means that the sponsor has generally invested significant amounts in a project by the time renegotiation occurs. Whether such investments are transaction-specific and allow the R&D supplier to bargain opportunistically depends on the sponsor's options for transferring the project to an alternative supplier.

In R&D agreements, the sponsor is generally entitled to whatever technology had been developed with its funds up to the point of contract termination. The technology is usually "delivered" in the form of molecules, cells, experimental data, models, and other written descriptions. However, such media cannot capture the tacit dimensions of the relevant technological know-how (Teece, 1976). Transferring a project from one R&D partner to another will therefore involve some losses in progress because the new partner will not have access to all of the tacit knowledge gained by the original R&D team during the project. However, the magnitude of these losses will depend on whether the new R&D partner has some experience or capabilities in the same product application area.

Codified descriptions and other technological artifacts provide more valuable insights to an experienced firm than to a novice firm about how the technology was developed and what future paths might be fruitful. The project's progress will be impeded much more if the project is transferred to a supplier who lacks experience and must start the project from the beginning. Thus, if a sponsor has alternative suppliers with relevant R&D experience, it can better preserve the value of its original R&D investment in the event that the agreement with the initial contractor is terminated. If the sponsor cannot find alternative partners with experience, the R&D funding provided to the original contractor is a transaction-specific investment. It has much less value if the project is not completed by the original contractor.

In the procurement of biotechnology R&D, a pharmaceutical company's ability to find alternative partners during a project depends on the degree to which the know-how being procured is specialized to the supplier. Some product areas are more specialized in the sense that the relevant R&D capabilities are concentrated within a relatively small group of suppliers. The accumulation of expertise and proprietary know-how impedes the entry of other suppliers into the same product R&D area. For R&D in a relatively specialized product area, a pharmaceutical company that contracts with an outside supplier has limited options should the supplier bargain opportunistically during one of the renegotiation cycles. Because the sponsor could not credibly threaten to switch partners, it would be stuck in a small-numbers-bargaining situation, which itself creates an incentive for the R&D supplier to bargain opportunistically.

The small-numbers-bargaining problem can also extend into future projects. Once a technological paradigm is established, one generation of product development provides the starting point for subsequent efforts (Nelson and Winter, 1977; Dosi, 1982; Tushman and Anderson, 1986). The firms that have

R&D Boundaries

R&D expertise in one generation of product development are likely to become the experts in the next. If a pharmaceutical company contracts out for R&D capabilities that are relatively specialized to a small group of R&D suppliers, it may have difficulty finding or switching partners for future projects aimed at the same product application area. Under these circumstances, investments in durable assets that are specialized to the product market (such as distribution capabilities or reputation) are a source of "quasi-rents" (Monteverde and Teece, 1982b) and leave the pharmaceutical company vulnerable to rent extraction in negotiating agreements for future projects.[1]

When alternative partners are available, partner switching is a credible action because R&D contracts generally allow the sponsor to retain or share with the contractor the rights to the technology. In addition, the sponsor generally gets to keep (or share with the contractor) technical data, materials, and other descriptive artifacts that were generated in the course of the contract. As noted earlier, these provide a valuable starting point for an experienced supplier. Thus, the key factor limiting the sponsor's ability to switch partners is the availability of alternative partners with relevant experience.

The above discussion suggests that the costs of market governance for a biotechnology R&D project are related to the number of R&D suppliers with R&D programs in the same product application area. In markets where R&D capabilities are specialized to a few suppliers, procurement of R&D through contracts involves small-numbers-bargaining problems. These hazards provide an incentive for internalization. This leads to the following hypothesis:

Hypothesis 1: A pharmaceutical company will be more likely to internalize R&D in those biotechnology product areas in which R&D capabilities are concentrated in fewer R&D suppliers.

Appropriability problems. Firms invest in R&D (both internal and external) to generate and gain proprietary access to specific products and to the more general know-how related to those products. For example, an R&D project that yields a specific therapeutic compound may also create knowledge and build R&D capabilities that are valuable for discovering other drugs for the same disease. The firm not only has an interest in gaining access to both the specific and more general know-how, but also in restricting rivals from using it.

When a project is done internally, the R&D organization can be prohibited (by fiat) from transferring know-how to competitors. When the project is sponsored externally, however, the restriction must be incorporated into the contractual agreement. Such restrictions require a clear delineation of the relevant property rights and a mechanism for enforcing those rights. The sponsor must be able to claim ownership of specific technologies and, by virtue of its ownership, restrict the contractor's right to transfer them to third parties.

An obvious problem arises in specifying all the relevant intellectual property rights. While it may be possible to identify specific elements (such as a particular molecule), much of the broader, applications-level know-how and R&D capabilities that are generated cannot be clearly defined or adequately described in a contract. The know-how generated from one

1
Quasi-rents are the difference between an asset's value in its first-best use and its value in its next-highest value use. Pharmaceutical companies often spend heavily on promoting a new drug and establishing a recognized brand name in the product area. The difference between these assets' value in their use for one product versus their next-best use for another product represents quasi-rents.

project may be so inextricably linked with that created in other projects that it may be impossible to agree on who owns what. As a result, the sponsor is unable to attach claims to much of the valuable intellectual property that is created during the R&D project. Even when specific intellectual property rights claims can be negotiated, enforcement may be problematic, given the uncertain scope and efficacy of biotechnology patents (Cooper, 1987).

An inability either to define or enforce intellectual property rights creates a hazard of expropriation. The R&D contractor may be able to sell the know-how arising from the project to the sponsor's product-market rivals. One of the determinants of the expropriation hazard is the number of rivals with an interest in commercializing applications of the technology. While new biotechnology firms often do not work on the exact same product with more than one commercial sponsor, they commonly work on R&D projects in the same therapeutic application for different clients. Each rival competing in the same therapeutic market represents a possible buyer for such technology in either current or future generations of product competition. Thus, highly competitive therapeutic markets provide an incentive for the pharmaceutical firm to internalize the development of technology. This leads to the second hypothesis:

Hypothesis 2: A pharmaceutical company will be more likely to internalize R&D in those biotechnology product markets in which it faces greater competition from other established pharmaceutical companies.

Other Factors Affecting R&D Procurement Decisions

Going back to Coase (1937), a fundamental axiom of transaction-cost economics is that the boundaries of the firm are determined by the trade-off between the transaction costs of using the market and the organizational costs of using hierarchies. Organizational costs within the economics literature have typically included those factors that make the firm less efficient at either performing or governing certain activities internally. If firms act rationally, they will adjust their boundaries when the trade-off between transaction costs and internal organizational costs change. However, behavioral theories of the firm (e.g., Simon, 1955; Cyert and March, 1963; Nelson and Winter, 1982) posit that bounded rationality prevents firms from making optimal adjustments to environmental changes. These theories suggest that there are internal organizational factors that affect the way firms decide to adjust their organizational boundaries. Incorporating into the model these behavioral factors, as well as those factors affecting the costs of internalizing R&D, will help isolate the effects of transaction costs on R&D boundary choices. The factors discussed below represent the set for which adequate data were available.

Historical factors. A firm's historical pattern of R&D procurement should affect its R&D procurement decisions in the new technological regime in two ways. First, according to a behavioral perspective of the firm, historical patterns of R&D procurement reflect deeply engrained repertoires (Simon, 1947), search rules (Cyert and March, 1963), operating procedures (Allison, 1971), and routines (Nelson and Winter, 1982).

R&D Boundaries

These organizational elements affect how firms perceive changes in the environment, the menu of possible responses they consider, and the choices they ultimately make. According to a behavioral perspective, "firms may be expected to behave in the future according to the routines they have employed in the past" (Nelson and Winter, 1982: 134). Thus, we would expect that routines underlying a historical pattern of internal R&D will lead the firm to internalize R&D even when technology changes.

The second way historical R&D procurement patterns affect current R&D procurement decisions is through their effect on the relative organizational costs of using either markets or hierarchies for R&D. Firms lacking experience with R&D contracting are likely to be less able to search for and select R&D partners and to absorb technologies from external sources (Westney, 1986). Due to their inexperience, these firms may be more vulnerable to the contractual hazards discussed earlier or at least perceive a greater degree of risk. As a result, for any given level of external factors affecting transaction costs, a firm inexperienced with purchasing technology from outside sources is more likely to find internalization a more attractive option. The above discussion suggests the following hypothesis:

Hypothesis 3: A pharmaceutical firm that tended to use in-house R&D for its traditional pharmaceutical products will have a greater propensity to internalize R&D for biotechnology-based products.

R&D experience in the relevant technology. Previous research on technological innovation suggests that when technical advance is cumulative (e.g., Nelson and Winter, 1977) a firm's efficiency in performing a particular type of R&D project depends on its prior experience with similar R&D projects. An R&D "learning curve" results, because the know-how needed to innovate at time $t + 1$ depends on what is learned in the course of R&D at time t. Therefore, a firm's ability to internalize new projects may depend on the number of its previous in-house projects in the relevant technology. A firm that has begun biotechnology R&D relatively early will have accumulated skills and experience that would reduce the costs of undertaking subsequent projects internally. In terms of Coase's (1937) original formulation of the problem, this R&D experience reduces the costs of internalizing new R&D projects; holding transaction costs constant, it should increase the likelihood of internalization:

Hypothesis 4: A pharmaceutical firm will be more likely to undertake a biotechnology R&D project in-house when it has accumulated more in-house R&D experience in the relevant area of biotechnology.

Focus. Decisions to expand the R&D boundaries of the firm into a new technological field are generally made at the highest levels of the corporation. Even when a pharmaceutical company is a division of a diversified corporation, corporate-level management normally plays an important role in allocating the additional resources required for expanding its R&D boundaries. Corporate management will be more likely to approve of internalization when it views the emerging technology as strategically important to the corporation. Many factors are likely to affect whether a technology is labeled as strategically important. One factor is the size of the pharmaceutical business relative to other businesses managed by the corporation. Where pharmaceuticals account for a small share of an enterprise's total activities, the relative impact of biopharmaceuticals on the enterprise may be small. As a result, corporate management is likely to focus its scarce attention (Simon, 1978) and corporate resources on the other lines of business. In contrast, biopharmaceutical technologies are more likely to be viewed as strategically important (to corporate management, which allocates the necessary resources) as pharmaceuticals become a greater share of the corporation's total business activities. This leads to the following hypothesis:

Hypothesis 5: A firm with a higher percentage of its business in pharmaceuticals will be more likely to internalize biotechnology-pharmaceutical R&D projects.

Firm size. Previous research suggests that firm size affects organizational behavior (Scherer, 1980; Miles, 1980) in general and organizational boundary decisions in particular. Anderson and Schmittlein (1984) found that in the electronics industry, the likelihood that a firm will use an internal sales force (rather than independent representatives) increases with firm size. Theoretically, the effect of firm size on R&D boundaries is unclear. On one hand, when expanding R&D boundaries involves fixed costs (such as new equipment and laboratory facilities), larger firms may have scale-economy advantages. They also have more resources to invest in the development of internal capabilities. On the other hand, because increasing size adds complexity to the administrative process, size can have a negative effect on R&D performance. Transaction-cost theorists, beginning with Coase, have generally identified decreasing returns to scale in administration as one of the limiting factors of vertical scope. Empirical work by Clark, Chew, and Fujimoto (1987) suggested that firms can reduce the complexity of managing projects and improve project performance by using suppliers for some development tasks.

National origin. A firm's national-geographic origin may influence its R&D boundary choices by affecting its relative costs of undertaking R&D. When, as in the case of biotechnology, there are national-regional differences in technological expertise, firms from one country may find it more costly to pursue R&D themselves than to purchase R&D from firms in another country better endowed with the relevant personnel and scientific expertise. To control for these effects, a set of national origin dummies was included in the model tested in this study.

METHOD

Sample. To test the hypotheses presented above, data were gathered on R&D project-level procurement decisions of pharmaceutical companies. To provide a starting point for the search, a list of the world's top 50 pharmaceutical companies in 1982 was obtained from *Scrip World Pharmaceutical News* (December 20, 1982: 23). The ranking was based on the total number of new pharmaceutical products a firm had in development in 1982; this number included new products originating from both internal and external R&D sources. Of the 50 firms on this initial list, 43 were also ranked in the world's top 50 by sales.

R&D Boundaries

The largest pharmaceutical firms in the world are an appropriate sampling frame for the analysis because these firms all have established positions in pharmaceutical markets. Because I was interested in modeling backward vertical integration from marketing to R&D, I needed to confine my observations to firms with established commercial positions. I also needed to observe firms that had at least some interest in commercializing biotechnology-based products. As one descends the ranks of pharmaceutical firms, there is a much lower probability that either criterion (commercial position or interest in commercializing biotechnology products) would be met. Preliminary sample selection also indicated that the information needed for the relevant independent and dependent variables would be very difficult to obtain for firms below the top 50. Between them, the 50 firms on the initial list accounted for 53 percent of all of the world's pharmaceutical products in development in 1982 (*Scrip,* December 20, 1982: 23). Thus, while these 50 firms represented a small share of the total number of pharmaceutical firms worldwide, they accounted for a very significant share of the world's total efforts to commercialize new pharmaceutical products.

The next task was to generate a sample of pharmaceutical biotechnology R&D projects in which these 50 firms were involved, either alone or in collaboration with an external source of technology. Two main data sources were used for this purpose: (1) Paine Webber's 1986 *Biotechnology Fact Book* and (2) a data base compiled by a California biotechnology firm. The latter data base was provided to the author under conditions of confidentiality. However, the data contained in it are drawn from public primary sources (annual reports, the industry trade press, industry newsletters, etc.). Thus, data used in this analysis are reproducible. This data base contained information on the R&D and commercial activities of over 700 firms throughout the world that are involved in biotechnology. Both data sources contained information on whether a particular firm was doing a specific R&D project alone or in collaboration with an external partner. For purposes of the analysis, R&D was defined as the activities needed to synthesize, formulate, and test a pharmaceutical product prior to human clinical trials. Thus, the sample did not include projects that had already progressed to the first phase of human clinical trials.

Projects that had progressed to clinical trials and beyond were not included in the sample because it was difficult to obtain consistent and verifiable data on whether a pharmaceutical company obtained a particular product technology through an R&D contract or a technology licensing agreement. R&D contracts and technology licensing agreements are fundamentally different transactions. In an R&D contract, a sponsoring firm is funding an R&D project undertaken by its partner. As discussed earlier, it is a relatively long-term transaction that involves significant contractual uncertainties. Technology licensing agreements are more characteristic of one-time exchanges in which the rights to an already-developed technology are transferred from one firm to another. In the pharmaceutical industry, technology licensing agreements are often struck when the product is ready to begin clinical trials. Therefore, many of the biotechnology products in clinical trials and beyond could have been obtained through technology licensing agreements and not R&D contracts. Because the hypotheses developed above explicitly concerned R&D arrangements, it was necessary to limit the sample to R&D projects.

Additional verification was provided by examining the 10-K reports of newly formed biotechnology firms. These documents commonly discuss the firm's contracts and joint ventures. It was therefore possible to use them to verify particular collaborations between new biotechnology firms and established firms from the sample.

This procedure uncovered a total 94 R&D projects across 30 of the sample firms. However, for two projects, there were no new biotechnology firms working in the relevant area of product R&D. Because external procurement of R&D would be almost impossible in these cases (it would be a "no-numbers" bargaining situation), their inclusion in the sample would bias the results in favor of hypothesis 1. They were therefore excluded, and the final sample consisted of 92 cases. Although precise information on timing was not available, it is presumed that most of these projects began after 1982. Because of the time frame of the data bases, none could have begun after 1986.

The sample of projects could hardly be considered homogenous. Undoubtedly, the projects differed in terms of costs, riskiness, technological uncertainty, degree of difficulty, and other dimensions. Unfortunately, such project characteristics are generally considered by firms to be highly proprietary and thus the relevant data were not available. The degree of unmeasurable heterogeneity, however, is limited by the sampling procedure. All of the projects were in roughly the same technological stage of development; all were based on the same basic technologies and scientific principles; all were for pharmaceutical products; and all had met some minimum threshold for potential technological and commercial viability. Table 1 lists the pharmaceutical firms and their R&D projects in the final sample.

The dependent variable. The column on the far right of Table 1 represents the dependent variable in the analysis, the *i*th pharmaceutical firm's choice of undertaking the *j*th R&D project through in-house or collaborative modes. It was coded as follows: $M_{ij} = 1$ if the *i*th pharmaceutical firm's R&D on the *j*th project was completely internal; or $M_{ij} = 0$ if the *i*th pharmaceutical firm's R&D on the *j*th project involved an external source.

In 47 percent of the cases in the sample, the pharmaceutical firm was doing the R&D project internally (i.e., without an external partner). It should be noted that in most of the cases involving an external source, the partner was a new biotechnology firm. My research and discussions with industry personnel suggest that in these cases, R&D is almost always the sole responsibility of the new biotechnology firm partner. Thus, the dependent variable corresponds very well to the dichotomy between internal and external R&D. For the few cases involving two established firms, the proprietary data base mentioned above was consulted to determine the primary R&D supplier. In the five cases in which there was evi-

R&D Boundaries

Table 1

The Sample of R&D Projects

Case	Firm	Project (Main Application)	In-house R&D (yes/no)
1.	Hoffman La Roche	Interleukin-1 (cancer)	yes
2.	Hoffman La Roche	Immunoagents (cancer)	no
3.	Hoffman La Roche	Herpes II vaccine (herpes II)	yes
4.	Hoffman La Roche	Anti-inflammatory protein (immune modifier)	yes
5.	Bristol-Myers	Alpha-Interferon (immune modifier)	no
6.	Bristol-Myers	Beta-Interferon (immune modifier)	no
7.	Bristol-Myers	Gamma-Interferon (immune modifier)	yes
8.	Bristol-Myers	Immunotoxin (cancer)	yes
9.	Bristol-Myers	Cartilage Inducing Factor (bone growth)	no
10.	Hoechst	Burst Promoting Activity (immune modifier)	no
11.	Hoechst	Colony Stimulating Factor (CSF) (immune modifier)	no
12.	Hoechst	CSF-granulocyte (immune modifier)	no
13.	Hoechst	CSF-1 (immune modifier)	no
14.	Hoechst	Beta-Interferon (immune modifier)	yes
15.	Hoechst	Interleukin-2 (immune modifier)	yes
16.	Hoechst	Hemopoeitin (anemia)	no
17.	Hoechst	Tissue Plasminogen Activase (tPA) (blood clot dissolving)	no
18.	Hoechst	Human Growth Hormone (hGh) (growth regulator)	yes
19.	Merck	Gamma-Interferon (immune modifier)	yes
20.	Merck	Interleukin-1 (Beta) (immune modifier)	yes
21.	Merck	Somatostatin Analog (growth regulator)	yes
22.	Merck	Herpes II vaccine (herpes II)	yes
23.	American Home Products	Renin Inhibitor (hypertension)	no
24.	Upjohn	Interleukin-1 (Beta) (immune modifier)	yes
25.	Upjohn	Albumin (blood plasma)	yes
26.	Upjohn	Protein C (blood coagulation regulator)	yes
27.	Upjohn	Renin Inhibitor (hypertension)	yes
28.	Upjohn	tPA (blood clot dissolving)	yes
29.	Upjohn	Atrial Natriuretic Factor (ANF) (hypertension)	yes
30.	Johnson & Johnson	Interleukin-2 (immune modifier)	no
31.	Johnson & Johnson	Factor VIII-C (blood clotting)	no
32.	Eli Lilly	Immunoagent (cancer)	no
33.	Eli Lilly	Immunocytotoxic agent (cancer)	no
34.	Eli Lilly	Protein C (blood clot regulator)	yes
35.	Eli Lilly	Gram-Negative MAb (gram negative infections)	yes
36.	Ciba Geigy	Alpha-Interferon (immune modifier)	no
37.	Ciba Geigy	Gamma-Interferon (immune modifier)	no
38.	Ciba Geigy	Renin Inhibitor (hypertension)	yes
39.	Ciba Geigy	tPA (blood clot dissolving)	no
40.	Ciba Geigy	Somatomedin C (growth regulator)	no
41.	Sandoz	GM-Colony Stimulating Factor (immune modifier)	no
42.	Sandoz	Alpha-Interferon (immune modifier)	yes
43.	Sandoz	Pro-urokinase (blood clot dissolving)	no
44.	Boehringer Ingelheim	Beta-Interferon (immune modifier)	no
45.	Rhone Poulenc	Immunotoxin agent (organ transplants)	no
46.	Rhone Poulenc	Factor IX (hemophaelia)	no
47.	Bayer	Factor VIII-C (hemophaelia)	no
48.	Bayer	Pseudomonas MAb (septic shock)	no
49.	Schering-Plough	CSF (immune modifier)	yes
50.	Schering-Plough	CSF-other (immune modifier)	yes
51.	Schering-Plough	IgE Peptide (allergies)	yes
52.	Schering-Plough	Beta-Interferon (immune modifier)	no
53.	Schering-Plough	Interleukin-4 (immune modifier)	yes
54.	Schering-Plough	Erythropoietin (EPO) (anemia)	no
55.	Meiji Seika	Beta-Interferon (immune modifier)	no
56.	Takeda	IgE Peptide (allergies)	yes
57.	Takeda	Immunoagent (cancer)	no
58.	Smithkline	IL-1 Antagonist (cancer)	yes
59.	Smithkline	Alpha-1-Antitrypsin (emphysema)	yes
60.	Smithkline	tPA (blood clot dissolving)	no
61.	Pfizer	Renin Inhibitor (hypertension)	yes
62.	Pfizer	ANF (hypertension)	yes
63.	Syntex	Interleukin-1 (Beta) (immune modifier)	no
64.	Syntex	Immunotoxin (cancer)	yes

(continued on next page)

Table 1

The Sample of R&D Projects *(continued)*

Case	Firm	Project (Main Application)	In-house R&D (yes/no)
65.	Schering AG	Somatostatin (growth regulator)	yes
66.	Wellcome	Gamma-Interferon (immune modifier)	yes
67.	Wellcome	tPA (blood clot dissolving)	no
68.	Wellcome	Malaria vaccine (malaria)	yes
69.	Kyowa Hakko	Alpha-Interferon (immune modifier)	yes
70.	Kyowa Hakko	Immunotoxin (cancer)	no
71.	Fujisawa	Tumor Necrosis Factor (Beta) (cancer)	no
72.	Fujisawa	tPA (blood clot dissolving)	no
73.	Fujisawa	Renin Inhibitor (hypertension)	yes
74.	Sankyo	Macrophage Activating Factor (immune modifier)	no
75.	Sankyo	Tumor Necrosis Factor-Alpha (cancer)	no
76.	Sankyo	tPA (blood clot dissolving)	no
77.	Sankyo	ANF (hypertension)	no
78.	Sankyo	Calcitonin (bone diseases)	no
79.	Sanofi (Elf)	tPA (blood clot dissolving)	yes
80.	Sanofi (Elf)	hGH (growth regulator)	yes
81.	Sanofi (Elf)	Somatomedin C (growth regulator)	yes
82.	American Cyanamid	Immunoagent (cancer)	no
83.	American Cyanamid	Immunocytotoxic agent (cancer)	no
84.	American Cyanamid	Herpes II vaccine (herpes II)	no
85.	Green Cross	CSF-1 (immune modifier)	no
86.	Green Cross	Albumin (blood plasma)	no
87.	Green Cross	tPA (blood clot dissolving)	yes
88.	ICI	Alpha-Interferon (immune modifier)	yes
89.	Yamanouchi	Anti-Inflammatory Protein (immune modifier)	no
90.	Yamanouchi	tPA (blood clot dissolving)	no
91.	G. D. Searle	Beta-Interferon (immune modifier)	no
92.	G. D. Searle	Renin Inhibitor (hypertension)	yes

dence that both firms were contributing R&D, the case was treated as external/collaborative ($M_{ij} = 0$).[2]

The independent variables. Hypothesis 1 concerns the effect of small-numbers-bargaining conditions in the supply of R&D services. As noted above, new biotechnology firms are the overwhelmingly dominant source of external biotechnology R&D for established firms. Thus, for each project, the number of new biotechnology firms with R&D programs in the same therapeutic application was used to measure the extent to which small-numbers-bargaining conditions were present. Information on the therapeutic application of each product under development was available in Paine Webber's 1986 *Biotechnology Fact Book.* Data on the number of new biotechnology firms ($SUPPLIERS_j$) working in each of these therapeutic markets were available from the same source. The therapeutic application of each project is shown in Table 1 in parentheses after the product name or type.

A therapeutic area of application is broader than a specific project area. Multiple products can be aimed at a given therapeutic application (e.g., cancer treatment). Constructing the variable at this level allows for the possibility of substituting R&D sources within a class of therapeutically similar products. For example, if a collaborative project for a specific cancer drug terminates, it is not necessary for the sponsor to find another partner working on the exact same product. An R&D contractor with other cancer product projects would presumably be more willing and able to pick up the unfinished

2
An odd number of these cases is possible because the partners are not included in the sample.

project than one outside of the cancer area. The fact that a new biotechnology firm is working on projects within a therapeutic area means that it has "disease-specific" know-how. Thus, taking on the project would not require it to build up biomedical expertise in a new field.

In the model developed here, the variable $SUPPLIERS_j$ is specified in log form because an increase in the number of suppliers in an R&D market does not always have the same effect on small-numbers-bargaining hazards. For example, if there is only one new biotechnology firm supplying R&D services in a particular area of therapeutic application, a significant small-numbers-bargaining hazard exists. This hazard is much lower if there is an alternative supplier. A third supplier will further decrease the potential hazards, although its effect will be less than having only the second supplier in the market. In contrast, small-numbers-bargaining problems are almost nonexistent in an R&D market with 20 suppliers: an additional supplier in this market does little to reduce the already trivial small-numbers-bargaining hazards.

Hypothesis 2 concerns the effects of product-market rivalry among established pharmaceutical companies. For each project, the number of established pharmaceutical companies (i.e., all firms except new biotechnology firms) attempting to develop or commercialize the same product or other products within the same area of therapeutic application was used to measure rivalry. New biotechnology firms were not counted as being rivals in the commercialization stage because they have generally lacked the capabilities to commercialize products on their own. As of 1989, only one new biotechnology firm (Genentech) has commercialized a human therapeutic biotechnology product (tissue plasminogen activase) on its own; however, that firm uses partners to market the product outside of the United States. The measure of rivalry was aggregated over the entire world market. This world aggregation was viewed as appropriate in light of the increasing globalization of pharmaceutical product competition among the world's first-tier pharmaceutical companies (Thomas, 1988). This measure was also constructed from information in the Paine Webber report.

A logic similar to that explained above for the SUPPLIERS variable also justifies the use of a log specification for the RIVALS variable. An additional firm in a relatively concentrated market represents a major change in the number of rivals and, presumably, the intensity of competition. An additional rival in an already crowded market, however, may be barely noticeable.

To measure the historical propensity of each firm in the sample to obtain product R&D from its own laboratories (versus external sources), the following ratio was used:

$$HISTORY_i = \frac{\text{Number of Own R\&D products in development (1982)}}{\text{All R\&D products in development (1982)}}$$

where "Own R&D products" refers to pharmaceutical products in development that originated from the firm's own R&D laboratories and "All R&D products" refers to the total number of pharmaceutical products (from both internal and external sources) the firm had in development. The difference between the numerator and the denominator is the number of products in development contracted from external sources. The year 1982 was chosen because it is far enough in the past not to include a significant number of biotechnology R&D projects. It thus avoids simultaneity with the dependent variable. Because of the long research and development cycles associated with pharmaceuticals, this construct picks up the effects of decisions made over the previous five-to-seven-year period. A firm with a HISTORY variable closer to 1 has tended to be more vertically integrated into traditional (non-biotechnological) pharmaceutical R&D. According to hypothesis 3, we would expect such procurement behavior to carry over into the procurement of biotechnology R&D. Data on the number of own R&D pharmaceutical products and total number of pharmaceutical products under development in 1982 were available from *Scrip* (December 20, 1982).

To test hypothesis 4, it was necessary to construct a variable that would measure a firm's biotechnology R&D experience in the technological area in which each project was conducted. Because the sample contained only ongoing R&D projects, the number of completed projects could be taken as an indicator of each firm's accumulated experience in biotechnology R&D. These prior in-house projects were identified as follows. Using the data sources described earlier, it was possible to generate a list of biotechnology products for each firm that were, as of 1986, either in clinical development, awaiting regulatory approval, or already on the market. These same data sources could also be used to determine whether a specific product originated from in-house R&D or an external source. Through this method, it was possible to generate a list of biotechnology R&D projects that each firm had completed on its own, without an external partner.

Using information in the Paine Webber report, each project was classified into one of the following six technological categories: (1) immune modifiers, (2) anticancer agents, (3) blood proteins and enzymes, (4) hormones, (5) anti-infectives, or (6) vaccines. The effects of prior experience are hypothesized to be relatively specific within categories of technology. To capture these category-specific effects, a variable ($\text{BIO-EXPERIENCE}_{iJ}$) was created that equaled the number of completed R&D projects by firm *i* in biotechnology category *J*. *J* would vary across projects in the sample. For example, if an observation corresponded to Bristol-Myers' choice of developing an anticancer product in-house (versus externally), the independent variable would take on a value that reflected the number of previous in-house anticancer biotechnology projects undertaken by Bristol-Myers.

To test hypothesis 5, data were gathered from annual reports on the percentage of each firm's total sales attributable to its pharmaceutical business. Data for the Japanese firms were drawn from *The Japan Company Handbook* (Tokyo: Toyo Keizai Shinposha) for the years 1982–1985. The variable was termed FOCUS_i.

Firm size was measured in terms of pharmaceutical sales. Pharmaceutical sales was used instead of total corporate sales because economies and diseconomies of internal orga-

R&D Boundaries

nization should occur at the business unit (or divisional) level rather than at the corporate level. Data on sales were available from annual reports and *The Japan Company Handbook*. A quadratic specification for the SALES variable was used to allow for the possibility that the direction of the effect may change with the size of the firm. Because theory does not provide unambiguous prior expectations, however, three other specifications are shown: SALES only, log(SALES), and no size variable at all.

To capture the effects of national origin, firms in the sample were classified geographically as (1) American-based, (2) European-based, or (3) Japanese-based. In the model, dummies on European-based and Japanese-based firms were included.

Descriptive statistics. Table 2 lists the frequency with which the therapeutic applications appear in the sample. Table 3 presents the means, standard deviations, minimums, and maximums of the independent variables. Frequencies are presented for the dependent variable, the national origin dummies, and BIO-EXPERIENCE$_{i,J}$.

Table 2

Frequency of Therapeutic Applications

Therapeutic Application	Cases
1. Immune system modifiers	31
2. Cancer	14
3. Blood clot dissolving	11
4. Hypertension	9
5. Growth regulators	6
6. Vaccine for herpes II	3
7. Anemia	2
8. Blood plasma	2
9. Blood clotting (surgical applications)	2
10. Blood clotting (hemophaelia)	2
11. Allergies	2
12. Bone growth	1
13. Bone diseases	1
14. Blood coagulation regulators	1
15. Gram-negative infections	1
16. Organ transplant rejection	1
17. Septic shock	1
18. Emphysema	1
19. Vaccine for malaria	1

The relative inexperience of established pharmaceutical companies with in-house biotechnology R&D is apparent. In most cases, the pharmaceutical firms had no in-house R&D experience in the relevant technological category. The maximum experience any firm had in any one major class of biotechnologies was two prior R&D projects. Although this appears to be a small number of R&D projects, pharmaceutical R&D projects typically take from one to five years (before clinical trials) and can cost several million dollars (Schwartzman, 1976). Thus, a firm that had already completed two R&D projects in a particular area of biotechnology would have accumulated significant technical experience.

Table 4 presents the correlations among the continuous independent variables. There is a very high correlation between

Table 3

Means, Minimums, Maximums, and Standard Deviations of the Variables

Variable	Mean	S.D.	Min.	Max.
$SUPPLIERS_j$	19.14	13.08	1	35
Log $(SUPPLIERS_j)$	2.58	1.001	0	3.5
$RIVALS_j$	27.64	17.65	2	49
Log $(RIVALS_j)$	2.99	.95	.69	3.9
$HISTORY_i$	.77	.11	.48	.89
$BIO\text{-}EXPERIENCE_{i,j}$	.26	.57	0	2
$FOCUS_i$	49.24	23.16	7.0	95.0
$SALES_i$ ($ in thousands)	1251	596	333	2452

Frequencies

Variable	Number of Cases per Category of Discrete Variables 0	1	2	Total
M_{ij}	49	43	N/A	92
EUROPEAN FIRM	58	34	N/A	92
JAPANESE FIRM	74	18	N/A	92
AMERICAN FIRM	52	40	N/A	92
$BIO\text{-}EXPERIENCE_{i,j}$	74	12	6	92

the independent variables log($SUPPLIERS_j$) and log($RIVALS_j$). This positive correlation should make it more difficult to get significant estimates on the corresponding parameters (which are hypothesized to have opposite signs).

Table 4

Correlations among Continuous Independent Variables (N = 92)

Variable	1	2	3	4	5	6
1. Log $(SUPPLIERS_j)$						
2. Log $(RIVALS_j)$	.92					
3. $HISTORY_i$	−.11	−.16				
4. $FOCUS_i$	−.11	−.03	−.03			
5. $SALES_i$	−.009	−.04	.10	−.50		
6. $SALES_i^2$	.009	−.02	.06	−.47	.98	
7. Log $(SALES_i)$	−.03	−.07	.17	−.49	.96	.89

The low correlation between the share of a firm's total sales from pharmaceuticals and its historical propensity to procure R&D from in-house (versus external) sources is noteworthy in light of hypothesis 5. Hypothesis 5 postulates that firms with a greater share of their sales from pharmaceuticals will be more likely to internalize R&D in new pharmaceutical technologies. The low correlation mentioned above suggests that the link between the firm's share of sales from pharmaceuticals and its tendency to procure R&D internally does not hold historically.

Analysis. Probit was used to estimate the effects of the independent variables on the probability of doing R&D internally. A probit model assumes that the underlying probabilities follow a cumulative normal distribution. Except at its tails, this distribution is similar to the cumulative logistic function used in a logit model (Amemiya, 1979: 1487; Maddala, 1983: 23). However, to ensure the results were not

R&D Boundaries

sensitive to the probability distribution chosen, a logit model was also estimated. No differences in signs or levels of statistical significance resulted. Only the probit results are shown here. The model is specified in the following form:[3]

$$P(M_{ij} = 1) = F[a_0 + B_1\log(\text{SUPPLIERS}_j) + B_2\log(\text{RIVALS}_j) + B_3\text{HISTORY}_i + B_4\text{BIO-EXPERIENCE}_{i,j} + B_5\text{FOCUS}_i + B_6\text{SALES}_i + B_7\text{SALES}_i^2 + B_8(\text{NATIONAL ORIGIN DUMMIES}) + e_{ij}],$$

where $P(M_{ij} = 1)$ is the probability that pharmaceutical firm i will undertake R&D project j in-house, and F [.] is the cumulative normal distribution. As noted above, three alternative specifications of size were also estimated.

RESULTS AND DISCUSSION

The probit estimates are presented in Table 5. They support the transaction-cost hypothesis that a more concentrated supply side of the R&D market increases the likelihood of internalization. The hypothesis that rivalry in the therapeutic market would also lead to internalization (due to potential appropriability problems) was not supported at statistically significant levels. Despite the high correlation between these two variables, there was statistically significant evidence that at least one type of transactional hazard (small-numbers bargaining) affects R&D procurement decisions.[4]

The effect of a firm's historical propensity to procure R&D internally on its R&D procurement decisions was not statistically significant. Thus, the results do not suggest a strong history effect. One possible interpretation is that biotechnology created a significant enough break with past methods to dilute the effects of historical capabilities and traditional behavioral patterns. The costs of internalizing new biotechnology capabilities may have forced firms with traditionally high internal procurement biases to procure a relatively greater share of their biotechnology R&D needs outside the firm.

The effect of biotechnology R&D experience was positive, in the expected direction. These results show that there is an association between R&D procurement decisions at one time and prior R&D experience in the relevant technical area. Such a result suggests that the accumulation of specific technical capabilities may be an important determinant of a firm's incremental R&D boundary decisions. A longitudinal analysis is required to explore these dynamic effects in more detail.

The effect of a firm's dependence on the pharmaceutical business was significant and positive, in the expected direction. Companies more dependent on pharmaceutical sales seemed to be more likely to internalize biotechnology R&D projects. As noted previously, the relationship between a company's dependence on pharmaceutical sales and its tendency to procure traditional (chemical-based) pharmaceutical R&D internally does not seem to hold historically (see Table 4, correlation = −.03). That is, firms with a relatively high percentage of their sales in pharmaceuticals did not, historically, develop a higher proportion of their traditional pharmaceutical products through in-house R&D. The results of the probit analysis indicate, however, that firms with a relatively high

3
The inclusion of technological category dummies was suggested to me in the review process. In a pretest of the model, however, the set of technological category dummies was not statistically significant; there are therefore no statistical grounds for their inclusion.

4
The effect of $\log(\text{SUPPLIERS}_j)$ is also robust. When the model is run without the $\log(\text{RIVALS}_j)$ variable, the coefficient on $\log(\text{SUPPLIERS}_j)$ remains negative and statistically significant. See column 5 of Table 5.

Table 5

Parameter Estimates for the Probit Model*

Variables	Model (1)	(2)	(3)	(4)	(5)
CONSTANT	−.432	.221	2.982	−.563	.568
	(1.653)	(1.405)	(3.168)	(1.275)	(1.331)
Log ($SUPPLIERS_j$)	−.962•	−.953•	−.944•	−.860•	−.459••
	(.427)	(.424)	(.424)	(.411)	(.160)
Log ($RIVALS_j$)	.566	.552	.545	.505	
	(.436)	(.430)	(.430)	(.427)	
$HISTORY_i$	.705	.760	.826	.731	.577
	(1.401)	(1.402)	(1.407)	(1.393)	(1.365)
$BIO\text{-}EXPERIENCE_{iJ}$	.762•	.687•	.632•	.534•	.687•
	(.320)	(.300)	(.286)	(.271)	(.296)
$FOCUS_i$	.017•	.017•	.017•	.019•	.018•
	(.008)	(.008)	(.008)	(.008)	(.008)
$SALES_i$	.64E-03	−.47E-03			−.42E-03
	(.002)	(.0003)			(.0003)
$SALES_i^2$	−.41E-06				
	(.56E-06)				
Log ($SALES_i$)			−.483		
			(.395)		
$EUROPE_{(dummy)}$	−.262	−.281	−.296	−.320	−.258
	(.364)	(.363)	(.361)	(.358)	(.360)
$JAPAN_{(dummy)}$	−1.885••	−2.046••	−2.108••	−1.734••	−1.965••
	(.595)	(.585)	(.595)	(.490)	(.539)
Log-likelihood	−48.812	−49.083	−49.234	−50.003	−49.931
Chi-square	29.52•••	28.98•••	28.68•••	27.14•••	27.29•••
% Cases predicted correctly	74	75	74	73	75

•$p < .05$; ••$p < .01$; •••$p < .001$.
* Standard errors are in parentheses.

percentage of their sales in pharmaceuticals have a proportionally higher likelihood of developing new biotechnology pharmaceutical products through in-house R&D. One plausible hypothesis is that a radical change in the core technology of a particular line of business (pharmaceuticals) alters firms' perceptions about the strategic importance of technology. Of course, confirmation of this hypothesis would require analysis of the product and supplier markets of the nonbiotechnology R&D projects as well as other factors included in the present model.

The signs of the SALES and $SALES^2$ variables in equation 1 were in the expected direction (positive for SALES and negative for $SALES^2$). However, the individual coefficients were not statistically significant. A log-likelihood ratio test indicated that the coefficients on SALES and $SALES^2$ were not jointly significant. The log-likelihood ratio test statistic is $Z = -2(LL_1 - LL_2)$, where LL_1 is the log of likelihood function for the constrained model (equation 4, $LL_1 = -50.003$) and LL_2 is the log of the likelihood function for the unconstrained model (equation 1, $LL_2 = -48.812$). Since Z is a distributed chi-square with degrees of freedom equal to the number of restrictions, the critical value is 5.99 ($p < .05$). Since in this case, $Z = 2.382$, the coefficients are not jointly significant. The size effects were not statistically significant in either the SALES-only specification or the log(SALES) specification. Exclusion of the size effects, however, had little effect on the estimates of the other parameters in the model (equation 4).

R&D Boundaries

The data thus do not provide evidence that firm size affects decisions whether or not to internalize new R&D programs.

A log-likelihood ratio test indicated that the national origin dummies were significant. The log of the likelihood function of the unconstrained model (equation 1) was −48.812. If the two national origin dummies are constrained to be 0 (i.e., eliminated from the model), the log of the likelihood becomes −54.22. Since $-2(-54.22 + 48.812) = 10.82$, and the critical value of a chi-square distribution with 2 degrees of freedom is 5.99, we can reject the null hypothesis: national origin does appear to affect firms' decisions whether to undertake R&D in-house or procure it through contractual arrangements. As discussed earlier, various national-regional differences in "endowments" of indigenous biotechnology know-how may be responsible for these effects. To the extent that size creates different advantages and liabilities in different national and regional contexts, there may also be interactions between national origin and size variables, although such interactions were not significant in any pretests of the model in this study. A study of specific international differences might clarify these results.

Only the sign and level of statistical significance of the parameters can be interpreted directly from probit estimates. Unlike linear regression, probit coefficients do not have a direct interpretation. The effect of a change in independent variable X_k on the probability of internalizing the project can be represented as follows:

$$dP/dX_k = B_k\, f(XB),$$

where B_k = the estimated coefficient on variable k, and $f(XB)$ is the value of the normal density function at the point XB. This formula translates the estimated coefficients into a more intuitive number. It can be evaluated at any point (or several points) in the distribution.

The instantaneous rate of change in the probability of internalization will always be greatest at the mean of a cumulative normal distribution. Evaluating the instantaneous rate of change at the mean therefore indicates just how strong the hypothesized effect may be. This formula shows that a very small change in log(SUPPLIERS) from its mean will change the probability of internalization by a factor of .37. This suggests that in the region (around the mean) where they are strongest, small-numbers-bargaining effects can be substantively significant.

CONCLUSION

Technological change can affect organizations and the environment in which they compete in many different ways. In some cases, technological change shifts the locus of R&D expertise from established enterprises to new entrants. Previous research has suggested that such episodes of technological change result in a competitive struggle between an industry's new and incumbent firms (Schumpeter, 1975; Abernathy and Clark, 1985; Tushman and Anderson, 1986). Established firms are presumed to be able to resist the gusts of "creative destruction" only by adopting the new R&D and production skills required to compete with the new entrants. However, an alternative response for established firms is to procure some R&D projects from external sources and to focus internal resources on those functions such as marketing, in which they have a distinctive advantage.

This paper posited that transaction-cost factors influence whether an established firm would attempt to expand its R&D boundaries into a particular subfield of the new technology or procure the relevant capabilities from an outside source. The hypothesis that small-numbers-bargaining hazards in R&D markets motivate internalization of R&D was supported by the data. The other hypothesis, that rivalry among established firms would lead to internalization, was not supported. The data also suggest that firm-level factors (R&D experience, dependence on the industry affected by the technological change, and location) influence R&D procurement patterns. However, a firm's procurement patterns in the old technological regime did not seem to affect its procurement behavior in the new technology. Size was also not a significant factor.

By providing insights about the factors influencing R&D boundaries of established firms, this study can help us to better understand how the structure of competition between new entrants and established firms may evolve in the wake of technological change. Conditions that make R&D contracting hazardous can be expected to create competition rather than cooperation between new entrants and established firms. In this environment, the success of new firms will depend on their ability to build capabilities in such commercial activities as marketing and distribution; the survival of established firms will depend on their ability to acquire and develop new R&D skills. A different structure, one involving cooperation between vertically or functionally specialized firms, may evolve when R&D can be efficiently governed by contracts. In this environment, survival may depend much more on a firm's ability to select partners and manage cooperative relationships than on its ability to develop new R&D capabilities.

This study has examined R&D boundary decisions of established firms within a context in which it is reasonable to assume that new R&D-intensive entrants are unlikely to integrate into product markets. As a result, it has been able to address only the transaction costs of buying technology. Future research should examine other contexts, such as biotechnology-based diagnostics, in which new entrants have lower barriers to integration into product markets. This would provide insights about the factors contributing to the transaction costs of selling as well as buying technology. Other potential sources of transaction costs, such as technological uncertainty and the efficacy of patent protection (Levin et al., 1984; Teece, 1986) need to be explored before we have a complete picture of how R&D markets work and how they may influence the organizational environment in the wake of radical technological changes.

REFERENCES

Abernathy, William, and Kim Clark
1985 "Innovation: Mapping the winds of creative destruction." Research Policy, 14: 3–22.

Allison, Graham
1971 The Essence of Decision: Explaining the Cuban Missile Crisis. Boston: Little, Brown.

Amemiya, T.
1979 "Qualitative response models: A survey." Journal of Economic Literature, 19 (December): 1483–1536.

R&D Boundaries

Anderson, Erin, and David Schmittlein
1984 "Integration of the sales force: An empirical examination." Rand Journal of Economics, 15 (3): 385–395.

Arrow, Kenneth
1962 "Economic welfare and the allocation of resources for invention." In Richard Nelson (ed.), The Rate and Direction of Inventive Activity: 609–625. Princeton, NJ: Princeton University Press.

Balakrishnan, Srinivasan, and Birger Wernerfelt
1986 "Technical change, competition and vertical integration." Strategic Management Journal, 7: 347–359.

Bozeman, Barry, and Albert Link
1983 Investments in Technology: Corporate Strategies and Public Policy Alternatives. New York: Praeger.

Clark, Kim, W. Bruce Chew, and Takahiro Fujimoto
1987 "Product development in the world auto industry." Brookings Papers on Economic Activity, 3: 729–771.

Coase, Ronald
1937 "The nature of the firm." Econometrica, 4: 386–405.

Cooper, Ivor
1987 Biotechnology and the Law, 2d ed. New York: Clark Boardman Company.

Cyert, Richard, and James G. March
1963 A Behavioral Theory of the Firm. Englewood Cliffs, NJ: Prentice-Hall.

Dosi, Giovanni
1982 "Technological paradigms and technological trajectories: A suggested interpretation of the determinants and directions of technical change." Research Policy, 11: 147–162.

Grossman, Sanford, and Oliver Hart
1986 "The costs and benefits of ownership: A theory of vertical and lateral integration." Journal of Political Economy, 94: 691–719.

Joskow, Paul
1987 "Contract duration and relationship-specific investments: Empirical evidence from coal markets." American Economic Review, 77: 168–185.

Klein, Benjamin, Robert Crawford, and Armen Alchian
1978 "Vertical integration, appropriable rents and the competitive contracting process." Journal of Law and Economics, 21 (October): 297–326.

Levin, Richard, Alan Klevorick, Richard Nelson, and Sidney Winter
1984 "Survey research on R&D appropriability and technological opportunity." Unpublished manuscript, Economics Department, Yale University.

Maddala, Gregory
1983 Limited-Dependent and Qualitative Variables in Econometrics. Cambridge: Cambridge University Press.

Malerba, Franco
1985 The Semiconductor Business. London: Frances Pinter.

Masten, Scott
1984 "The organization of production: Evidence from the aerospace industry." Journal of Law and Economics, 27 (October): 403–417.

Miles, Raymond
1980 Macro Organizational Behavior. Santa Monica, CA: Goodyear.

Monteverde, Kirk, and David Teece
1982a "Supplier switching costs and vertical integration." Bell Journal of Economics, 13: 206–213.
1982b "Appropriable rents and quasi-vertical integration." Journal of Law and Economics, 25 (October): 321–328.

Mowery, David
1983 "The relationship between intrafirm and contractual forms of industrial research in American manufacturing, 1900–1940." Explorations in Economic History, 20: 351–374.

Nelson, Richard, and Sidney Winter
1977 "In search of useful theory of innovation." Research Policy, 6 (Summer): 36–76.
1982 An Evolutionary Theory of Economic Change. Cambridge, MA: Harvard University Press.

Office of Technology Assessment, U.S. Congress
1984 Commercial Biotechnology: An International Analysis. Washington, DC: U.S. Government Printing Office.

Pavitt, Keith
1986 " 'Chips' and 'trajectories': How does the semiconductor influence the sources and directions of technical change?" In Roy MacLeod (ed.), Technology and the Human Prospect: Essays in Honour of Christopher Freeman: 31–54. London: Frances Pinter.

Scherer, Frederick
1980 Industrial Structure and Economic Performance. Chicago: Rand McNally.

Schumpeter, Joseph
1975 Capitalism, Socialism, and Democracy, 3rd ed. New York: Harper & Row (Originally published in 1942 by Harper & Brothers).

Schwartzman, David
1976 Innovation in the Pharmaceutical Industry. Baltimore: Johns Hopkins University Press.

Shan, Weijian
1989 "An empirical analysis of organizational strategies by entrepreneurial high technology firms." Working Paper, The Wharton School, University of Pennsylvania.

Simon, Herbert
1947 Administrative Behavior. London: MacMillan.
1955 "A behavioral model of rational choice." Quarterly Journal of Economics, 69: 99–118.
1978 "Rationality as process and as product of thought." American Economic Review, 68 (2): 1–16.

Stuckey, John
1983 Vertical Integration and Joint Ventures in the Aluminum Industry. Cambridge, MA: Harvard University Press.

Teece, David
1976 The Multinational Corporation and the Resource Costs of Technology Transfer. Cambridge, MA: Ballinger.
1982 "Towards an economic theory of the multiproduct firm." Journal of Economic Behavior and Organization, 3: 39–63.
1986 "Profiting from technological innovation: Implications for integration, collaboration, licensing, and public policy." Research Policy, 15: 285–305.
1988 "Technological change and the nature of the firm." In G. Dosi et al. (eds.), Technical Change and Economic Theory: 256–281. London: Frances Pinter.

Teece, David, and Henry Armour
1977 "Innovation and divestiture in the U.S. oil industry." In D. J. Teece (ed.), R&D in Energy: 7–94. Stanford, CA: Stanford University Institute for Energy Studies.

Thomas, Lacey Glenn
1988 "Multifirm strategies in the U.S. pharmaceutical industry." In D. Mowery (ed.), International Collaborative Ventures in U.S. Manufacturing: 147–182. Cambridge, MA: Ballinger.

Tushman, Michael, and Philip Anderson
1986 "Technological discontinuities and organizational environments." Administrative Science Quarterly, 31: 439–465.

von Hippel, Eric
1982 "Appropriability of innovation benefit as a predictor of the source of innovation." Research Policy, 11 (2): 95–115.

Walker, Gordon, and David Weber
1984 "A transaction cost approach to make-or-buy decisions." Administrative Science Quarterly, 29: 373–379.

Westney, D. Eleanor
1986 "Domestic and foreign learning curves in managing international cooperative strategies." Paper presented at "Cooperative Strategies in International Business," colloquium of the Wharton School and Rutgers Graduate School of Management, October 24–26.

Williamson, Oliver
1975 Markets and Hierarchies. New York: Free Press.
1985 The Economic Institutions of Capitalism. New York: Free Press.

[15]

Interorganizational Collaboration and the Locus of Innovation: Networks of Learning in Biotechnology

Walter W. Powell
Kenneth W. Koput
Laurel Smith-Doerr
University of Arizona

We argue in this paper that when the knowledge base of an industry is both complex and expanding and the sources of expertise are widely dispersed, the locus of innovation will be found in networks of learning, rather than in individual firms. The large-scale reliance on interorganizational collaborations in the biotechnology industry reflects a fundamental and pervasive concern with access to knowledge. We develop a network approach to organizational learning and derive firm-level, longitudinal hypotheses that link research and development alliances, experience with managing interfirm relationships, network position, rates of growth, and portfolios of collaborative activities. We test these hypotheses on a sample of dedicated biotechnology firms in the years 1990–1994. Results from pooled, within-firm, time series analyses support a learning view and have broad implications for future theoretical and empirical research on organizational networks and strategic alliances.•

In recent decades, there has been unprecedented growth in corporate partnering and reliance on various forms of external collaboration (Hergert and Morris, 1988; Mowery, 1988; Hagedoorn, 1990, 1995; Badaracco, 1991; Hagedoorn and Schakenraad, 1992; Gulati, 1995). Historically, firms organized research and development (R&D) internally and relied on outside contract research only for relatively simple functions or products (Mowery, 1983; Nelson, 1990a). Today, companies in a wide range of industries are executing nearly every step in the production process, from discovery to distribution, through some form of external collaboration. These various types of interfirm alliances take on many forms, ranging from R&D partnerships to equity joint ventures to collaborative manufacturing to complex co-marketing arrangements. The most common rationales offered for this upsurge in collaboration involve some combination of risk sharing, obtaining access to new markets and technologies, speeding products to market, and pooling complementary skills (Kogut, 1989; Kleinknecht and Reijnen, 1992; Hagedoorn, 1993; Mowery and Teece, 1993; Eisenhardt and Schoonhoven, 1996).

We do not doubt that the need to combine complementary assets has played a role in the growth of interfirm alliances. Nonetheless, we want to explore a different argument, one that we think has more explanatory power in industries in which knowledge is developing rapidly. A key finding from a diverse set of studies is that the R&D intensity or level of technological sophistication of industries is positively correlated with the intensity and number of alliances in those sectors (C. Freeman, 1991; Hagedoorn, 1995). Viewed broadly, technological change occurs in two forms. When advances build on existing know-how, established firms reap the bulk of the benefits. But when new discoveries create technological discontinuities, or radical breaks from previously dominant methods, incumbents can be robbed of many of their advantages. Moreover, new kinds of organizational practices may emerge to exploit these novel developments (Schumpeter, 1934; Abernathy and Clark,

0001-8392/96/4101-0116/$1.00.

•
This research was supported by grants provided to the first author by the Social and Behavioral Sciences Research Institute, University of Arizona, and the Aspen Institute Nonprofit Sector Research Fund and by grants to the second author by the College of Business and Public Administration, University of Arizona. We have benefited from productive exchanges with numerous audiences to whom portions of this paper have been presented: a session at the 1994 Academy of Management meetings, the Social Organization workshop at the University of Arizona, the Work, Organizations, and Markets workshop at the Harvard Sociology Department, the 1994 SCOR Winter Conference at Stanford University, and colloquia at the business schools at the University of Alberta, UC-Berkeley, Duke, and Emory, and the JFK School at Harvard. For detailed comments on an earlier draft, we are extremely grateful to Victoria Alexander, Ashish Arora, Maryellen Kelley, Peter Marsden, Charles Kadushin, Dick Nelson, Christine Oliver, Lori Rosenkopf, Michael Sobel, Bill Starbuck, Art Stinchcombe, and anonymous reviewers at *ASQ*. We thank Dina Okamoto for research assistance and Linda Pike for editorial guidance.
Address correspondence to Walter W. Powell, Department of Sociology, University of Arizona, Tucson, AZ 85721.

1985; Tushman and Anderson, 1986; Tushman and Rosenkopf, 1992). Such radical new developments have the potential to restructure a mature industry, hence Schumpeter's phrase: "gales of creative destruction." The most apt recent exemplars are the effects of first the transistor and later the integrated circuit on the electronics industry (J. Freeman, 1990) and the effects of biotechnology on the mature pharmaceutical industry. Biotechnology represents a competence-destroying innovation because it builds on a scientific basis (immunology and molecular biology) that differs significantly from the knowledge base (organic chemistry) of the more established pharmaceutical industry. Consequently, biotech provides enhanced research productivity, with less risk and with more speed and potentially higher rewards (Weisbach and Moos, 1995).

The purpose of this paper is to examine the organizational arrangements that have arisen in response to the technological ferment generated by biotechnology. We focus on forms of collaboration undertaken by dedicated biotechnology firms and assess the contribution of cooperative ventures to organizational learning. In short, we seek to map the network structure of this emerging industry and explain the purposes served by the extensive connections that typify the field.

COLLABORATION AND ORGANIZATIONAL LEARNING

When there is a regime of rapid technological development, research breakthroughs are so broadly distributed that no single firm has all the internal capabilities necessary for success. Many groups of competitors are likely to be working on the same targets; the rewards go to the swiftest. Thus, new technologies are both a stimulus to and the focus of a variety of cooperative efforts that seek to reduce the inherent uncertainties associated with novel products or markets. Running throughout the literature on partnering is an argument that collaboration enhances organizational learning (Hamel, 1991; Dodgson, 1993). We discern, however, two rather different strands of thinking about collaboration and learning.

One approach is largely strategic (Teece, 1986; Williamson, 1991). The choice to pool resources with another organization depends on calculations involving risk versus return. Obviously, reliance on external partners involves hazards (Powell, 1990; Sabel, 1993). A lack of trust between the parties, difficulties in relinquishing control, the complexity of a joint project, and differential ability to learn new skills are all barriers to effective collaboration. Moreover, in those industries in which interfirm agreements are relatively frequent, there can be competitive confusion about who is an ally and who is not. The partnering decision thus depends on each partner's size and position in the "value-chain," the level of technological sophistication, resource constraints, and prior experiences with alliances. The form of collaboration is purported to vary according to the specific types of skills and resources to be exchanged (Hennart, 1988; Pisano, 1989; Parkhe, 1993). Posed this way, the decision to collaborate is a variant of the make-or-buy decision, framed largely in terms of transaction cost economics. Firms thus turn to collaboration to acquire resources and skills they cannot produce internally, when the hazards of cooperation can be held to a tolerable level.

According to an alternative argument, learning is a social construction process (Brown and Duguid, 1991). In this view, what is learned is profoundly linked to the conditions under which it is learned. Knowledge creation occurs in the context of a community, one that is fluid and evolving rather than tightly bound or static. The canonical formal organization, with its bureaucratic rigidities, is a poor vehicle for learning. Sources of innovation do not reside exclusively inside firms; instead, they are commonly found in the interstices between firms, universities, research laboratories, suppliers, and customers (Powell, 1990). Consequently, the degree to which firms learn about new opportunities is a function of the extent of their participation in such activities (Levinthal and March, 1994). Brown and Duguid (1991: 48) summarized this view nicely by stating that learning is about becoming a practitioner, not learning about a practice. Von Hippel (1988) has shown that the trading of know-how often requires the establishment of long-term relationships in which exchange occurs within a learned and shared code.

March (1991) captured these divergent views of learning in his discussion of the differences between exploration and exploitation in organizational learning. He argued that the "essence of exploitation is the refinement and extension of existing competencies, technologies and paradigms . . . [the] essence of exploration is experimentation with new alternatives" (March, 1991: 85). Exploitation generates predictable returns, while the returns from exploration are much more uncertain. Exploration is costly, often unfruitful, but "the only way to finish first" (Levinthal and March, 1994: 106).

Yet the messy world of practice often blurs the neat distinctions of theory. Exploitation and exploration, and calculation and community are intertwined. Organizational learning is both a function of access to knowledge and the capabilities for utilizing and building on such knowledge. We follow Nelson (1990b) and Stinchcombe (1990) in arguing that organizational arrangements that provide access to knowledge quickly and reliably produce competitive advantage. But rather than seeing such activity as calculative or strategic, we draw on a long line of research that stresses the centrality of building skills and exercising routines in organizations (Cyert and March, 1963; Nelson and Winter, 1982; Stinchcombe, 1990).

The complex reality of rapidly developing fields, in which knowledge is both sophisticated and widely dispersed, transcends the simple calculation of a make-or-buy decision. Research breakthroughs demand a range of intellectual and scientific skills that far exceed the capabilities of any single organization, as illustrated by two notable recent discoveries in biotechnology. The development of an animal model for Alzheimer's disease appeared in a report (*Nature,* Feb. 9, 1995) coauthored by 34 scientists affiliated with two new biotech companies, one established pharmaceutical firm, a leading research university, a federal research laboratory, and a nonprofit research institute. Similarly, a publication identifying a strong candidate for the gene determining susceptibility to breast and ovarian cancer (*Science,* Oct. 7, 1994) featured 45 coauthors drawn from a biotech firm, a U.S. medical school, a Canadian medical school, an established pharmaceutical company, and a government research laboratory. More important than the number of authors are the diversity of sources of innovation and the wide range of different organizations involved in these breakthrough publications.

Learning through Networks

We argue that when knowledge is broadly distributed and brings a competitive advantage, the locus of innovation is found in a network of interorganizational relationships (Powell and Brantley, 1992). To stay current in a rapidly moving field requires that an organization have a hand in the research process. Passive recipients of new knowledge are less likely to appreciate its value or to be able to respond rapidly. In industries in which know-how is critical, companies must be expert at both in-house research and cooperative research with such external partners as university scientists, research hospitals, and skilled competitors. In examining whether research collaborations increase the subsequent likelihood of other types of cooperation, we build a network analog to Cohen and Levinthal's (1989, 1990) concept of "absorptive capacity." A firm with a greater capacity to learn is adept at both internal and external R&D, thus enabling it to contribute more to a collaboration as well as learn more extensively from such participation. Internal capability and external collaboration are not substitutes for one another, but complementary (Mowery and Rosenberg, 1989; Arora and Gambardella, 1994). Internal capability is indispensable in evaluating research done outside, while external collaboration provides access to news and resources that cannot be generated internally (Nelson, 1990b). A network serves as a locus of innovation because it provides timely access to knowledge and resources that are otherwise unavailable, while also testing internal expertise and learning capabilities.

Our concept of networks of learning highlights two key observations: (1) Interorganizational collaborations are not simply a means to compensate for the lack of internal skills, (2) nor should they be viewed as a series of discrete transactions. A firm's value and ability as a collaborator is related to its internal assets, but at the same time, collaboration further develops and strengthens those internal competencies. Firms deepen their ability to collaborate not just by managing relations dyadically, but by instantiating and refining routines for synergistic partnering. To illustrate, Richard DiMarchi, Vice President for Endocrine Research at Eli Lilly and Company, emphasizes that the biggest mistake his company could make in managing research alliances is to treat them as "one-offs"—independent relationships pursued separately (personal communication with the first author). The development of cooperative routines goes beyond simply learning how to maintain a large number of ties. Firms must learn how to transfer knowledge across alliances and locate themselves in those network positions that enable them to keep pace with the most promising scientific or technological developments.

In our view, collaborations in high-tech industries typically reflect more than just a formal contractual exchange. When the first author presented the chief executive officer (CEO) of Centocor with a list of his firm's formal agreements, he observed that it was "the tip of the iceberg—it excludes dozens of handshake deals and informal collaborations, as well as probably hundreds of collaborations by our company's scientists with colleagues elsewhere." Beneath most formal ties, then, lies a sea of informal relations. Many alliances—no matter what their ostensible function—reflect a relationship that carries benefits beyond the particular exchange designated in a formal agreement. Nonetheless, R&D alliances, which are unambiguously explorative, play a critical role in allowing firms to stay abreast of rapidly changing developments. Knowledge facilitates the use of other knowledge. What can be learned is crucially affected by what is already known. A vice president for corporate research at Xerox put this point nicely: "In order for industrial research organizations to be in close contact with new advances in basic science, it is important . . . to be an active *participant* at the leading edge of world science. Effective technical interchange requires that the industrial organization have its own basic research results . . . to use as a currency of exchange" (Pake, 1986:36, *emphasis in original*).

Accumulated findings at the frontiers of research provide leverage to access, assimilate, and exploit additional ideas and information. R&D collaboration is both an admission ticket to an information network and a vehicle for the rapid communication of news about opportunities and obstacles. The CEO of Chiron remarked that his biotech firm has more than 100 short-term, small-scale (less than two years and $200,000) relationships with university scientists.[1] When such projects lead to the possibility of a new medicine, relationships become formal and contractual. Innovative activities, then, cannot be reduced to a simple process of information acquisition. Because extensive contacts typically cross-knit research communities, involvement in collaborative R&D expands the horizons of a firm's personnel and increases their awareness of additional projects that might be undertaken. Thus, R&D alliances serve as a platform for diverse network activity.

Knowledge also requires other knowledge. When the sources of expertise are disparate, collaborative R&D opens an organization's eyes to the need for accessing ideas and information from a variety of sources, to exploit the research findings in a commercial context. Both skill and experience are needed to accumulate the capability to benefit from the interdependencies across diverse collaborative behaviors. In addition, experience at collaborating is necessary to manage a diverse portfolio of ties. Hence, we argue that firms learn from exploration and experience how to recognize and

1
Comments by Edward Penhoet, CEO of Chiron, on Koput, Powell, and Smith-Doerr (1995).

structure synergies across different types of alliances. Thus we posit:

Hypothesis 1: The greater the (a) number of research and development alliances and (b) experience at managing R&D and other types of collaborations a firm has at a given time, the greater the number of non-R&D collaborations it subsequently pursues; and, in turn, the more diverse its future portfolio of ties will become, controlling for prior levels.

An organization simultaneously learns which collaborations to pursue and how to function within a context of multiple cooperative ventures. The dynamics of cooperation are endogenous in high-tech fields in which scientific advances fuel new discoveries that in turn require novel forms of collaboration to develop. When collaboration stems from membership in a common technological community, as we are arguing, partnering is routinized and occurs more readily, with less effort. Collaboration becomes emergent—stemming from ongoing relationships—informal, and nonpremeditated (Von Hippel, 1988; Hakansson, 1990).

Thus, once a firm begins collaborating, it develops experience at cooperation and a reputation as a partner. Over time, firms develop capabilities for interacting with other firms. Experience with collaborative networks proves a fertile ground for both further formal partnerships and an expanding array of informal relationships. A broader range of collaborative efforts provides greater opportunity to refine organizational routines for cooperating and render them more versatile.

The information that passes through networks is influenced by each participant's position in the industry structure. Firms with access to a more diverse set of activities and those with more experience at collaborating are better able to locate themselves in information-rich positions. We contend that firms with more experience have more ties and that the ties they have provide more central connectedness. Moreover, experience with diverse research-driven networks and the connections that experience brings determine how well situated a firm becomes. Consequently, we predict:

Hypothesis 2: The greater (a) the number of R&D alliances, (b) the diversity of ties, and (c) the experience at managing R&D collaborations or other ties that a firm has at a given time, the more centrally connected the firm subsequently becomes, controlling for the total number of ties and prior connectedness.

Differential location in a network of partnerships results in firms having divergent capabilities for benefiting from collaboration. More formally, we argue that central connectedness shapes a firm's reputation and generates visibility, producing access to resources via benefit-rich networks. Such a reputation can greatly enhance a firm's ability to attract talented new employees. Network location thus shapes the nature of competition. Firms more centrally located should have more timely access to promising new ventures, while those with more collaborative experience should be better positioned to exploit them. Experience at managing ties allows a firm to move quickly in identifying new projects and funneling them inside the organization, enabling growth to occur in a fashion conversely analogous to absorptive capacity. Put colloquially, a firm grows by being a player; it does not become a player by growing. Therefore, we contend:

Hypothesis 3: The greater a firm's (a) centrality in a network of relationships and (b) experience at managing ties at a given time, the more rapid its subsequent growth, controlling for prior growth.

Finally, we expect that these returns from experience with diverse collaborative activity should elicit positive feedback. The information that passes through networks is influenced by each participant's position in the overall network structure. Differential location in a network of partnerships results in firms having divergent capabilities for benefiting from collaboration. Firms that are located in extensive networks of collaboration are enmeshed in complex and shifting patterns of rivalry and cooperation. Moreover, a firm's experience with networks and its position within them alter the nature of competition. No longer can the goal be to vanquish your opponent, lest you eliminate your collaborator on another project. Centrality in a network facilitates common understandings and shared principles of cooperation, thus enhancing further exchange. Hence, we posit:

Hypothesis 4: The greater a firm's centrality in a network of relationships at a given time, the greater its number of subsequent R&D collaborations, controlling for prior collaborative R&D activity.

The Biotechnology Industry

We explore the hypotheses just derived in the context of the field of biotechnology, a young science-based industry. The pioneering work of Watson and Crick in the early 1950s, which described the structure of DNA as a double helix, laid the foundation for the development of the science of molecular biology. The core technologies used in biotechnology—DNA synthesizing and sequencing, cell fusion methodologies for producing hybridomas—are approximately twenty years old. Yet despite its youth, biotechnology is a burgeoning field. Commentators suggest that molecular/cellular biology has displaced physics as the most prominent of the sciences, pushing "biology beyond the descriptive stage into the development of powerful models and experimental techniques that are helping us to understand the most fundamental of life processes" (Keller, 1990: 124).

Similar to the development of physics in the first half of this century, basic research in the biosciences has been spurred by its exceptional technological potential. To a student of organizations, it is the speed and scope of commercial development in biotechnology that is remarkable. While biotech had its origins in the laboratories of universities and research institutes, it was commercially exploited by small, science-based companies, the first of which went public in 1980. In the decade and a half since, hundreds of companies were created in the U.S., and many more abroad. Investors of all kinds poured billions into biotech, by some accounts, more than $60 billion by 1993. These entrepreneurial companies face stiff obstacles; many companies stumble on the long and winding road from drug development through regulatory approval to the medical marketplace. The process of creating new biotech drugs is research-intensive, very protracted, and extraordinarily expensive ($100–300 million per product). Nevertheless, by the close of 1994, twenty-five biotechnology-based drugs were approved by the U.S. Food and Drug Administration (FDA). At present, more than 220 medicines are in various stages of clinical testing, and some two dozen drugs await FDA approval. Drug development timelines average 7 to 11 years from discovery to launch for pharmaceuticals, while biotech firms have brought new medicines to market in a time frame of 4 to 8 years (Powell, 1996a). Biotech industry sales reached $7 billion in 1993, an impressive sum for a young field, but still two billion less than the sales of pharmaceutical giant Merck (Gupta, 1994: C4).

In many respects, biotechnology is not an industry per se, but a set of technologies with the potential to transform various fields—pharmaceuticals, chemicals, agriculture, veterinary science, medicine, even waste disposal. Many researchers (e.g., Barley and Freeman, 1992; Amburgey, Shan, and Singh, 1994) treat the wide array of biotechnology companies as comparable. In contrast, we intentionally restrict our attention to only those for-profit firms engaged principally in human therapeutics and diagnostics, hereafter referred to as dedicated biotech firms, or DBFs. We are persuaded that the therapeutics sector is driven by a different research agenda and operates within a distinctive regulatory regime.[2] We intentionally omit firms engaged only in agricultural, veterinary, or bioremediation activity and exclude peripheral companies that produce only equipment, materials, or test kits for the industry.

As the human biotechnology industry developed in the 1980s, it became clear that the full range of required skills (e.g., basic research, applied research, clinical testing procedures, manufacturing, marketing and distribution, and knowledge of and experience with the regulatory process) could not be easily assembled under one roof. The basic science and applied research skills needed to create new products were based in universities, research institutes, and DBFs (Smith-Doerr, 1994; Zucker, Darby, and Brewer, 1994; Powell, 1996b). In a field in which scientific excellence is paramount, biotech companies must establish their bona fides with research scientists. They do so by investing heavily in R&D, organizing in a fashion that is, comparatively, similar to a university laboratory, allowing their scientists considerable autonomy to work on their own projects and to publish and participate in the scientific community, and by creating postdoctoral research programs. These practices help promote a common technological community between universities and DBFs. Professors take sabbaticals at biotech companies; postdoctoral fellows move back and forth between universities and firms; and top universities compete for industry scientists. The 1993 Nobel Prize in Chemistry was given to Kary Mullis for work done at a biotech firm.

Venture capitalists fueled most of the initial discoveries and guided many firms through their early years. But moving from basic research to product development required not

2
Agricultural and veterinary biotechnology has encountered considerable public opposition in the U.S. to genetically altered food and animals. As a biotechnology attorney put it to the first author, "the decision to take a new cancer drug is a choice a patient makes with his or her doctor, but a new breed of strawberry is not a private decision but rather one with broad externalities."

only lots of money, it also demanded expertise in conducting extensive clinical trials and securing federal regulatory approval. Neither universities nor DBFs have been well equipped for these tasks, but large pharmaceutical firms most certainly have. Large pharmaceuticals were flush with cash and controlled marketing channels worldwide. But established pharmaceutical firms have been unable to create internally the kind of research environment that fosters constant innovation and discovery. So the various participants in biotech have turned to joint ventures, research agreements, minority equity investments, licensing, and various kinds of partnerships to make up for their lack of internal capabilities and resources (Pisano, 1989, 1991; Arora and Gambardella, 1990, 1994; Powell and Brantley, 1992). Because product development can easily last a decade, firms compete for the intellectual and financial resources that are needed to sustain the discovery and development process.

Perhaps the most profound difference between biotech firms and large pharmaceutical companies is the management and organization of the research process, with the former much closer to basic sciences than the latter. Biotechnology, however, has proven to be an unusual case of competence destruction. Scientific discoveries have profoundly reshaped the nature of the drug discovery process, but once a new medicine is developed, the key uncertainties concern the development of the technology into a safe and effective medical product that can be targeted widely, a competency at which established pharmaceutical firms are very good. Hence, the technological breakthroughs that level the playing field on the exploration front also create new opportunities for established firms in exploitation. Consequently, circumstances of mutual need develop. Small firms require large firms' financial support and regulatory savvy, while larger corporations desire access to the research prowess of smaller companies.

METHOD

Data

We sought to explain the pattern of interorganizational agreements that structure learning in the field of biotechnology. These agreements are formal ties, frequently involving very expensive investments by both parties. We built a relational database that contains separate files for (1) DBFs in human therapeutics and diagnostics, (2) the formal contractual, interorganizational agreements involving DBFs, and (3) the partners to these agreements. The information gathered for each DBF includes founding date, employment levels, sources of financing, and collaborative agreements, which we treated as ties. We coded each tie for its purpose and duration, using an implicit logic of production to classify them into categories, as described in Table 1: R&D, clinical trials, manufacturing, marketing and licensing, and so forth.[3] The "partner" datafile for all organizations that appear as partners on any tie with a DBF is large, expanding annually (numbering more than 1000 organizations active in 1994), and exceptionally diverse in both form and nationality, including multinational corporations, government agencies, hospitals, universities, and biopharmaceutical companies.

3

Several readers have questioned whether learning occurs in the context of a licensing agreement, an arrangement that could be handled by the lawyers for the two parties. But we think licensing is almost always a part of a broader collaborative strategy. For instance, in 1990, out of 210 licensing ties in our sample, just three firms pursued only licensing agreements. The preponderance of licensing deals were made by firms with multiple forms of network activity. Moreover, licensing rarely occurs without prior contact between the two parties to explore the viability of the project. Other readers have wondered whether finance ties involve the transfer of knowledge. In biotech, the answer is clearly "yes." Because of their youth and academic origins, companies have been very short on experienced managers; consequently, venture capital and other sources of finance have played a vital role in providing management advice and leadership.

Networks of Learning

Table 1

Description of Biotechnology Agreements

Type of tie	Typical partners
R&D: DBF develops research program with another organization for a specific target	Other DBFs, pharmaceutical corps., research institutes, university labs
Outside investor: Partner invests funds in DBF	Venture capital firms
Clinical trials/evaluation: DBF has partner conduct trials of product on subjects for FDA approval	Research hospitals, firms specializing in clinical trials
Manufacturing: DBF contracts with partner to manufacture its product	Pharmaceutical corporations, chemical companies
Marketing/Licensing: DBF licenses idea to marketer	Pharmaceutical corporations
DBF purchases rights to partner's idea	Universities
Supply/Distribution: Agreement to receive materials or to supply products to distributors	Large chemical or pharmaceutical corporations
Investment/Joint Venture: DBF invests funds (and usually human/scientific capital) in a partner	Other biotechnology firms
Complex: DBF tie that contains more than one of the above-listed activities (i.e., R&D and marketing)	Any partner (except venture capital firms)

We began assembling the database in 1990, using *Bioscan,* an independent industry directory published six times a year that lists a great range of organizations (domestic and foreign, commercial, nonprofit, or government-owned, biotech, and diversified health care corporations). *Bioscan* lists information on a firm's ownership, its current products, and its research in progress. Thus it is reasonably easy to code firms according to our criteria. When we needed help in understanding the science, we consulted colleagues at our university's cancer center. We collected data for the five-year period 1990–1994. When information was missing from *Bioscan,* we consulted numerous other industry directories, such as various editions of *Genetic Engineering and Biotechnology Related Firms Worldwide,* Dun & Bradstreet's *Who Owns Whom?,* and listings in *Moody's* and *Standard & Poor's.* In addition, we consulted annual reports, Securities and Exchange Commission filings, and, when necessary, made phone calls to companies.

Our focus on research-driven DBFs in human therapeutics resulted in a sample of approximately 225 independently owned companies. Various industry estimates suggest there are between 500 and 1,300 companies operating in the general area of biotechnology, but this number includes all segments of the industry, public and private companies, as well as subsidiaries, joint ventures, and branches of firms in other industries. The consulting firm Ernst & Young estimates approximately 1,300 companies are involved in some aspect of biotechnology but contends that less than 20 percent of those are independent, publicly traded, dedicated biotech firms (Burrill and Lee, 1993). We did not exclude firms from the sample due to small size or non-U.S.

location. Our sample is largely U.S. based, but not exclusively so. Naturally, data availability is considerably better for publicly traded companies than privately held ones, and this is reflected in the *Bioscan* listings.

In focusing on contractual agreements, we omitted the myriad informal arrangements out of which formal relationships emerge. Consequently, our analysis is a strict test of the learning argument because the database does not include widely used less formal relationships that promote the transfer of knowledge. Moreover, in the following analyses we do not focus on year-to-year changes in the content of particular agreements; thus the many instances in which relationships are deepened with the passage of time are not analyzed. In this sense, the analyses are a conservative test of our basic argument.

Operationalizations and Measures

Dependent variables. Hypotheses 1–4 predict the subsequent number and diversity of ties, network position in terms of central connectivity, and rates of growth for firms in our sample. We take these in turn.

Number of R&D ties at time t + 1: The number of research and development ties a firm has captures the extent of its involvement in the core activities of the industry. The number of a firm's R&D alliances reflects its network profile for the purposes of exploration. As noted earlier, R&D has a unique status in theories of organizational learning, and so we treat it separately.

Number of ties of each type (other than R&D) at time t + 1: We used the disaggregated number of ties for each stage of the product life-cycle (see Table 1) to capture the non-R&D network activity in which firms are involved. We view these efforts as a way to exploit R&D discoveries and thus as a complement to R&D networks.

Network portfolio diversity at time t + 1: The range of ties that a firm is engaged in at any given time reflects a firm's portfolio of collaborative activities. Portfolio diversity is computed for each firm in each year as follows. For firm i in year t, denote the number of ties of type j as $n_{it,j}$ and the total number of ties aggregated over all types ($j = 1 \ldots J$; $J = 8$) as n_{it}. The proportion of firm i's ties of type j, out of the total number of ties, is denoted $p_{it,j}$ and given by $p_{it,j} = n_{it,j}/n_{it}$. Each $p_{it,j}$ is squared and then the sum is taken over all j and subtracted from 1, resulting in the index of diversity, y_{it}, so that:

$$y_{it} = 1 - \sum_{j=1}^{J} p_{it,j}^2 .$$

This is equivalent to Blau's index of heterogeneity (Blau, 1977). Diversity can be treated as a continuous random variable, though bounded in the interval [0, 7/8].

Central connectivity at time t + 1. Before describing the measures of network centrality that we used, a few remarks are in order. A tie is a link between organizations, i.e., between a DBF and a partner. The partner may be another DBF or, more commonly, one of many organizations in the biopharmaceutical field. In counting ties above, we used only direct agreements between biotech firms and their partners. In computing centrality, however, we need to account for that fact that we do not have a closed network. In this respect, our measure of interfirm networks is somewhat unconventional. We wished to examine the structure of the network linking our sample of DBFs, but we had to define a closed set of firms to compute measures of connectivity. Yet many of the ties that structure the field involve parties outside the scope of our definition of a DBF—the overall universe of partners is open, diverse, and expanding. To capture the information flows that may take place between DBFs through non-DBF partners, we counted a connection between two DBFs when there was a direct tie (degree one) and when the DBFs were linked (at degree distance two) through a common partner. Graphically, Figure 1 diplays both forms of connections—direct ties and indirect linkages.

For example, in Figure 1, Biogen has two ties (to Genentech and Lilly) and two connections (one direct, to Genentech, and one indirect, to Centocor through its tie with Lilly). The direct tie to Lilly is not counted as a connection for the purposes of the DBF network measures, since Lilly is not a DBF. As we argued earlier, each tie reflects underlying, ongoing relationships, and thus a common connection through an outside partner means that the DBFs share a structurally equivalent position with that partner. For similar reasons, in measuring centrality we did not discriminate among connections involving different functions. The various types of collaborative activities each play a comparable role in creating a firm's overall set of relationships. Measures of central connectivity were computed using UCINET IV (Borgatti, Everett, and Freeman, 1992).

Figure 1. Ties and connections between biotechnology firms.

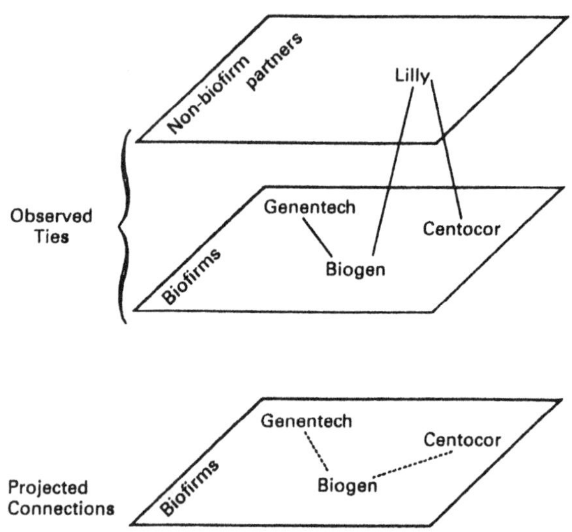

Membership in the main component at time t + 1: A component is generally defined as a maximally connected subgraph, or a set of points that are connected to one another by paths of any distance (Scott, 1991: 104). We defined a component as a group of *n* firms, each of which has a connection to at least one other firm in the group. In our components, connections are either direct or degree distance two, rather than paths of any length (for an even more stringent definition of components, see L. Freeman, 1992). Components define disconnected subgroups. Such a measure captures the degree of fragmentation in an industry's network structure. In the biotech industry for each year of our sample, the field is characterized by a main component, to which the majority of firms with ties are connected. The remainder are fragmented into many very small components, typically composed of isolates. We created a dummy variable, *MainComp,* that takes the value of 1 if a firm is in the main connected component and 0 if not. A firm that is in the main component is considered much more centrally connected than an isolate.

Degree centrality at time t + 1: Centrality is a measure of how well connected, or active, a firm is in the overall network. We gauged centrality of a firm locally rather than globally, in network parlance, such that a firm's centrality is the number of other firms connected to that firm, ignoring how well those partners are connected. Degree centrality is a common measure of the centralization of power in organizational studies of interlocking directorates, where researchers using this operationalization have often noted the centrality of banks in corporate networks (Mizruchi, 1982; Useem, 1984; Mintz and Schwartz, 1985).

Closeness centrality at time t + 1: We also used a measure of centrality based on the concept of closeness (L. Freeman, 1979), which captures independence from the control of others. Closeness centrality was computed for each firm as the reciprocal of the sum of the degree distance to each other firm. In our context, a high closeness score means a firm has access to many other DBFs and, ergo, is not dependent on specific others for access to information.

Growth at time t + 1. We indexed growth in two ways. We used the reported number of *employees at time t + 1* as a measure of size. We also created a dummy variable, *public,* that takes on the value of 1 if the firm is *publicly traded at time t + 1* and 0 otherwise.

Independent variables. We based our predictions on prior measures of collaborative research activity and experience, non-R&D network experience, diversity of alliances, and network centrality. We used the number of R&D ties, network portfolio diversity, and central connectivity as defined above, but at time *t*. We also introduced measures of network experience, defined below.

Collaborative R&D experience at time t was measured as the time since inception of a firm's first R&D alliance. This was computed for each firm in each year as the current date minus the date at which the firm's initial R&D tie was formalized. *Non-R&D network experience at time t,* an additional measure of experience at managing ties, is the time since inception of a firm's first tie for any purpose other than R&D. This was computed for each firm at time *t* as the current date minus the date at which the firm's first nonresearch alliance was formalized.

Control variables. In predicting diversity and centrality, we explicitly incorporated controls for the prior total number of ties. In addition, we controlled for the number of ties of each type (other than R&D), as well as centrality, in all of the predictions in which the number of R&D ties was used as an independent variable. We did so to demonstrate that R&D uniquely drives network behavior. Finally, we controlled for alternative explanations that involve firm age or size as predictors, rather than as outcomes, of network behavior. Age appears as a predictor in ecological and life-cycle theories of organization, while greater size, indicating a more extensive hierarchy, is seen as an alternative to alliances in the transaction-cost literature.

As an alternative to direct experience, we used a firm's calendar *age at time t* to capture vicarious experience or advantages due to the establishment of internal routines as a control in all predictions of the effects of collaborative R&D experience or non-R&D network experience. Age was computed for each firm in each year as the date of founding subtracted from the current date. We used the number of employees to incorporate firm *size at time t* as a control in all predictions of network behavior and position. *Total ties at time t* is the aggregate number of ties of all types and serves as a control in our predictions involving central connectivity.

Statistical Methods

Our data consist of five years of cross-sectional records. In each cross section, the variables were measured at the firm level. To test the predictions of our learning perspective, we used a panel regression model. The selection of this technique involves two primary theoretical considerations and the need to address a number of statistical issues that stem from these concerns. The first theoretical consideration is that learning resides within firms and occurs over time. We argue that while learning occurs through network relationships, firms are both the actors and the recipients of the skills and expertise that learning brings. This presents two related statistical concerns: unobserved heterogeneity and autocorrelation.

Unobserved heterogeneity arises due to differences among firms in omitted variables that are constant over time and may affect both independent and dependent variables (as a common cause). For example, larger firms may have more R&D ties and a more diverse portfolio because of past successes. To eliminate any spurious effects due to unobserved differences among firms, we included fixed firm effects by entering a dummy variable for each firm. Consequently, the estimated coefficients will be interpretable as the amount by which the within-firm deviation on the dependent variable shifts in response to a preceding change in the deviation of the independent variable. This interpretation captures the dynamic and firm-centered nature of learning. Further, we essentially have the population of dedicated biotech firms over our observed time period, and not a random sample.

We were interested in estimating a dynamic model, in which the independent variables are lagged one year. Firms may be "imprinted" or otherwise start on developmental trajectories before the start of our observation period. These trajectories may have naturally evolving patterns that change over time in coherent ways, but ones that we cannot foresee or measure. The omission of an important factor that changes over time within firms will result in autocorrelated errors and may bias estimates of the parameters in which we are most interested. One way of breaking the correlation over time, so as not to overestimate the effects of our hypothesized independent variables, is to include a lagged dependent variable, $y_{i,t-1}$, as a predictor. For each of the models reported, we estimated the effect of a serial correlation term in a first-order autoregressive model, as described in Hsiao (1986: 54–55). After controlling for the lagged dependent variable, we found no significant residual autocorrelation (at the .10 level). When a lagged dependent variable appears as an explanatory variable, however, the fixed effects estimator of the parameter on the lagged dependent variable may not be consistent, because our dataset involves a large number of firms observed over a short period of time. To check the robustness of our reported results to this potential problem, we obtained estimates from three other specifications of the model. First, we omitted the lagged dependent variable and included a first-order serial correlation term. Second, we ran the model with the lagged dependent variable included, but using random effects to control for unobserved firm heterogeneity. Third, we used an instrumental variables estimator (for details, see Hsiao, 1986: ch. 2). The results we report below for our hypothesized predictor variables are systematically lower than the effects from the first two alternative models and about the same magnitude as those from the instrumental variables approach. Hence, the reported estimates are conservative. Most importantly, the pattern of significance is identical across all four specifications.

The second theoretical consideration is that the dynamics of learning involve the co-evolution of firms and networks. This leads to an additional source of statistical nonindependence across our observations: For each firm we measure various properties of its position in the industry's network, but each firm's position can only be assessed in terms of the network positions of others. For instance, a particular firm might have a high centrality score, because a number of outside firms act as partners to multiple firms in the network. Hence, we need to control for effects that vary over time but are constant across firms, such as the overall number of outside partners, the density of the industry's network, government budgets for medical research, or the economic circumstances of pharmaceutical companies. To do so, we included fixed year effects: a dummy variable for each year. This gives each year its own mean on each of the measured variables and should break the nonindependence in network measures.

In using what is termed a fixed-effects specification (Judge, 1985) for both the firm and year controls, with a lagged dependent variable as one of the predictors, the dependent variable, $y_{i,t}$, is modeled as:

$$y_{i,t} = \alpha_i + \gamma_t + \lambda(y_{i,t-1}) + \sum_{j=1}^{J} \beta_j(x_{i,t-1,j}) + \varepsilon_{i,t}.$$

In this equation, α_i is the effect of firm i: $i = 1 \ldots N$; γ_t is the effect of year t: $t = 1 \ldots T$; β_j is the within-firm slope for x_j, pooled over all firms and years; and $\varepsilon_{i,t}$ is a normally distributed error term.

The strict assumptions of the normal regression model are violated, because our primary dependent variables are skewed and only approximately continuous. Concentration and closeness are defined as continuous, though bounded. The number of ties, centrality, and size can only take on non-negative integer values. Their ranges are quite large (approximately 70, 400, 120, and 3000, respectively), however, making the continuity approximation reasonable. While the truncation and skewness are potentially problematic, these features are shared by the independent variables, allowing us to make the a priori working assumption of symmetric disturbances. We confirmed the validity of this assumption with diagnostic plots in a post-hoc residual analysis (not reported here). Nonetheless, we checked the robustness of our results in several ways. First, we applied a square-root transformation to the dependent variables and reestimated the normal regression model. Second, for those dependent variables that are integer numbers, we also estimated panel models for count data. Third, there are two dependent variables that are constrained to the values of 0 and 1: *MainComp* and *public*. For these, we also ran panel models for binary outcomes (i.e., panel logit estimators). All three of the additional models confirmed the findings of the normal regressions. We report only the results of the normal regressions for all variables, to ease interpretation and allow comparison of effects across the models.

The remaining issue in obtaining the fixed-effects estimates is that of colinearity among the predictor variables. We naturally expected some of our independent variables to be correlated within firms over time. We anticipated that as a firm ages, it would grow and gain experience. The result would be a strong correlation among age, size, and experience that could introduce bias or inefficiency into the estimation. In addition, we used a number of measures for some of our concepts, some of which were likely to be colinear. For instance, we used degree centrality, closeness centrality, and membership in the main component to operationalize the notion of central connectivity. The difficulty in operationalizing a concept such as central connectivity is obvious: The a priori restriction to just one measure would be unwise.

Table 2 shows the within-firm correlations among our explanatory variables that might cause colinearity problems

Table 2

Within-firm Correlations among Explanatory and Control Variables

Variable	1	2	3	4	5	6	7	8	9	10
1. R&D ties										
2. Total ties	.5701									
3. Diversity	.2291	.3821								
4. Collaborative R&D experience	.2702	.2390	.1835							
5. Non-R&D network experience	.2394	.2831	.3212	.7824						
6. Degree centrality	.4744	.6817	.2762	.2649	.2574					
7. Closeness	.3060	.4159	.3369	.5434	.5443	.3922				
8. MainComp	.2323	.3450	.2256	.1637	.1337	.3065	.8762			
9. Age	.2510	.3130	.3202	.7678	.8795	.2907	.6168	.2425		
10. Size	.0838	.0380	.0867	.3401	.3407	.0160	.2578	.0165	.3242	
11. Public	.2164	.3105	.3254	.3573	.3769	.2730	.3627	.2134	.4071	.0698

in the statistical estimations. We find three interrelated sets of variables. First, the time-incremental variables—age, collaborative R&D experience, and non-R&D network experience—are, as expected, intercorrelated. The second group of correlated variables are the numbers of ties, total and R&D, along with degree centrality. The third group arises from the various measures we used to locate a firm within industry networks: Closeness centrality and MainComp are highly correlated.

To deal with the problems of colinearity and operationalization, we embedded the panel regression estimator in a variable-selection procedure, programmed in MATLAB (The Mathworks, 1994). For each dependent variable, we began with a model consisting of all the hypothesized predictor and control variables for which there are no colinearity or operationalization problems. We then performed a series of estimations, each adding one of the theoretically justifiable combinations of interrelated hypothesized predictor or control variables contained in our hypotheses or important alternatives for that dependent variable. Within each of the three groups of interrelated variables, we began with each predictor variable on its own, then ran all combinations of two such predictors, and so on as needed until we arrived at a final combination consisting of all the interrelated predictor variables in that group for the dependent variable being explained. For each dependent variable, we analyzed the explanatory power of all such subsets and selected the subset providing the best model fit, adjusted for degrees of freedom. We then assessed the robustness of each selected subset using inclusion criteria of $p < .05$ for both the t-test of the coefficient for each term and the F-test for the improvement in model fit due to each term, compared with all nested models with one fewer predictors, and with exclusion criteria of $p > .10$ for both the t-test of the coefficient for each term and the F-test for the improvement in model fit due to each term, compared with all nested models with one greater predictor. This procedure for assessing fit ensures that the results are not ad-hoc: No variables are either included or excluded in the results due to small changes in attributed variance. We then repeated the steps just described across the selected subsets for all three groups of interrelated variables, to obtain for each dependent variable a final set of predictor and control variables that is meaningful and robust.

RESULTS

We present firm demographics, by year, in Table 3. The results are broken down, within each year, for both firms with ties and those without. The table gives the number of DBFs in each category, along with means and standard deviations for age, size, *public,* and collaborative R&D experience and non-R&D network experience (in years). Firms with ties are slightly older, for most years, though the gap narrows over time and reverses in 1994. More interestingly, firms with ties are substantially larger ($p < .0001$ in all years), and the gap expands (though not significantly) over the years observed. This difference suggests that firms with ties are growing faster (in raw numbers) than those without ties. The means for the indicator variable *public* are the percentages (in decimals) of firms publicly traded. A much larger percentage of firms with ties are publicly traded ($p < .0001$), in all years, and the raw difference remains nearly constant.

Table 3

Firm Demographics by Year*

	1990		1991		1992		1993		1994	
Variable	No ties	Ties	No ties	Ties	No ties	Ties	No ties	Ties	No ties	Ties
Age	5.70	7.99	6.91	8.41	7.07	8.78	8.65	9.38	12.47	9.93
	(5.81)	(5.83)	(7.15)	(5.64)	(6.73)	(6.08)	(7.17)	(6.22)	(6.39)	(6.35)
Size	46.33	141.41	38.63†	128.63	43.36	155.58	41.18	166.70	58.46	174.83
	(52.50)	(278.27)	(49.58)	(233.79)	(44.68)	(331.59)	(44.63)	(356.08)	(81.60)	(392.50)
Public	.08	.48	.12	.49	.14	.59	.19	.62	.25	.68
Collaborative R&D experience	0.15	1.82	0.03	2.34	0.28	2.63	0.29	3.16	1.18	3.70
	(1.07)	(2.31)	(0.10)	(2.60)	(1.73)	(2.75)	(1.13)	(3.06)	(2.29)	(3.40)
Non-R&D network experience	.06	4.18	0.21	4.74	0.37	5.07	0.49	5.66	1.54	6.29
	(.42)	(3.44)	(0.91)	(3.71)	(1.43)	(3.95)	(1.59)	(4.14)	(2.87)	(4.39)
Total firms	63	179	56	184	44	190	37	192	31	195

* Means presented first, standard deviations in parentheses below.
† Mean and standard deviation trimmed of one outlier that had ties in all other years.

The descriptive measures for collaborative R&D experience and non-R&D network experience point to a distinctive pattern of organizational development for firms that rely on network ties. While firms always have the option of eliminating their reliance on external collaborations, it turns out that firms that have ties for any length of time rarely do so. We estimate that roughly 15 percent of ties are terminated each year, but this does not necessarily end a relationship. An R&D alliance on a specific project may conclude, for example, and be replaced by a new research venture, a complex manufacturing and marketing agreement, or some other new arrangement. This finding is seen from the means for those firms with no ties, which are quite small. Firms with ties have significantly more competence and experience than firms without ties ($p < .0001$) in all years. Over the five-year period, the number of firms without ties drops by exactly 50 percent, from 62 to 31. The number of firms with ties grows incrementally, both as a result of new entrants to the field and firms without ties embarking on collaborations.

Table 4 summarizes the network activity of firms, by year. We present means and standard deviations for the number of ties (total and disaggregated by types), partners, portfolio diversity, and network measures of central connectivity. The number of ties and partners, on average, are growing only slightly (p = .33 and .19, respectively). At the same time, the measures of centrality (p = .0069 for degree centrality, p < .0001 for closeness) are expanding substantially. Membership in the main component (MainComp) is also steadily increasing (p = .065). This suggests that firms are not "promiscuous" in their use of ties; rather, they are deepening their connectedness without adding substantial numbers of new ties. On average, firms have ties of each variety, save for manufacturing and clinical ties. These findings have important implications for the strategy that is apparently pursued by biotechs. In contrast to product life-cycle or transaction-cost thinking, as firms grow older and larger they do not appear to reduce the number and type of collaborations in which they are engaged. Diversity is increasing over the period (p = .0027), although R&D (p = .10) and finance (p = .031) ties are the only categories showing any growth in numbers.

Table 4

Network Activity for Biotechnology Firms with Ties by Year*

Variable	1990	1991	1992	1993	1994
Ties	9.56	9.68	9.31	10.04	10.02
	(11.06)	(10.15)	(9.30)	(9.73)	(8.68)
Degree centrality	18.41	20.03	18.85	22.25	24.50
	(23.76)	(23.74)	(21.52)	(24.24)	(24.00)
Closeness	1.02	1.11	1.24	1.41	1.91
	(0.24)	(0.27)	(0.34)	(0.38)	(0.44)
MainComp	.86	.86	.85	.86	.91
Partners	8.11	8.38	8.10	8.76	8.87
	(8.87)	(8.30)	(7.69)	(8.04)	(7.36)
R&D	1.48	1.47	1.56	1.77	1.74
	(1.97)	(1.96)	(2.04)	(2.25)	(2.47)
Finance	2.22	2.40	2.47	2.76	2.89
	(3.30)	(3.26)	(3.58)	(4.03)	(3.67)
Clinical trials	.05	.03	.02	.03	.03
	(0.24)	(0.19)	(0.14)	(0.17)	(0.25)
Manufacturing	.09	.09	.06	.07	.07
	(0.37)	(0.41)	(0.35)	(0.36)	(0.37)
Marketing/licensing	2.19	2.34	2.25	2.39	2.40
	(3.84)	(3.96)	(3.69)	(3.72)	(3.49)
Supply/distribution	.66	.66	.67	.66	.57
	(1.47)	(1.50)	(1.59)	(1.59)	(1.50)
Investment/joint venture	1.14	1.14	.96	.97	.95
	(2.52)	(2.33)	(2.06)	(2.05)	(1.97)
Complex	1.64	1.49	1.27	1.36	1.34
	(3.13)	(2.70)	(2.43)	(2.49)	(2.23)
Diversity	.41	.44	.44	.48	.49
	(0.30)	(0.29)	(0.26)	(0.25)	(0.25)

* Means presented first, standard deviations in parentheses below.

Networks of Learning

The results of our panel regressions are shown in Tables 5a, 5b, and 5c. All reported effects for our hypothesized predictor variables are significant at or beyond the .05 level in both coefficient *t*-tests and model-improvement *F*-tests versus nested subset or superset models, as described in the methods section. Control effects reported are also all significant, except for the effects of number of ties in predictions involving centrality, which are included whether significant or not. Each column reports the results for a dependent variable measured at time $t + 1$. Each row contains the effects of an explanatory variable measured at time *t*. One set of panel regressions was performed for each dependent variable (columns) in which all possible subsets of theoretically justifiable explanatory variables (rows) were entered. Where effects are not reported, the row variable was not included in the best-fitting subset for lack of significance or explanatory power (*p*-values in excess of .10). All models included fixed firm and year effects (dummy variables) that, even when significant, are not reported to conserve space and enhance the readability of the tables. These results and details on the complete variable-selection process are available from the authors. For all of the prediction equations, age was included among the explanatory variables entered but was weaker in its effect than experience when entered alone and showed no improvement in fit when added to experience. Thus, age was eliminated. Similarly, the number of ties of all types other than R&D were also eliminated.

Results in Tables 5a–5c provide support for most of our specific hypotheses and strong overall support for the

Table 5a

Determinants of Network Portfolios: Results of Panel Regressions*

Predictor variables (at time *t*)	Dependent variables (at time $t + 1$)									
	R&D ties	Finance	Clinical trials	Manufacturing	Marketing/ licensing	Supply/ distribution	Investment/ joint venture	Complex	Total ties	Diversity
R&D ties		.0963 (.0443)	.0069 (.0037)		.1625 (.0317)	.0249 (.0102)	.0997 (.0221)	.0589 (.0228)	.4662 (.0025)	.0133 (.0047)
Non-R&D network experience					.1148 (.0309)	.0799 (.0315)				.0186 (.0087)
Closeness centrality	.2225 (.1005)									
Control variables†										
Lagged dependent variable	.4871 (.0339)	.2535 (.0339)	.3632 (.0330)	.2977 (.0301)	.3234 (.0308)	.3755 (.0284)	.3439 (.0301)	.4009 (.0311)	.4194 (.0373)	.1818 (.0351)
Size			.0198 (.0045)							
Public								.1829 (.0933)		.0561 (.0210)
Total ties	.0307 (.1311)									
MainComp							.1429 (.0774)		.5374 (.2516)	
R-squared	.8424	.8471	.8197	.9172	.9505	.9404	.9235	.9387	.9348	.8214
Total firm-years	904	904	650	904	904	904	904	904	904	699

* Standard errors are in parentheses. All reported coefficients are significant at or beyond the .05 level, except the effect of total ties on R&D ties.

† Age and the number of ties of each non-R&D type were included as control variables in all models but were eliminated due to nonsignificance. All models included fixed firm and year effects (dummy variables), in addition to the lagged dependent variable. These estimates are not reported to conserve space.

Table 5b

Determinants of Network Centrality: Results of Panel Regressions*

Predictor variables (at time *t*)	Dependent variables (at time *t* + 1) Degree centrality	Closeness centrality	MainComp
R&D ties	1.5073		
	(.3539)		
Collaborative R&D experience		.0578	
		(.0191)	
Non-R&D network experience	.8371	.0391	
	(.0726)	(.0188)	
Diversity	6.3366	.2202	.2509
	(2.4455)	(.0746)	(.0573)
Control variables†			
Lagged dependent variable	.2416	.2096	.0419
	(.0387)	(.0532)	(.0375)
Total ties	.3810	.0052	.0030
	(.1555)	(.0036)	(.0028)
R-squared	.8958	.7531	.6823
Total firm-years	718	707	718

* Standard errors are in parentheses. All reported coefficients are significant at or beyond the .05 level.

† Age, size and the number of ties of each non-R&D type were included as control variables in all models but were eliminated due to nonsignificance. All models included fixed firm and year effects (dummy variables), in addition to the lagged dependent variable. These estimates are not reported to conserve space.

Table 5c

Determinants of Growth: Results of Panel Regressions*

Predictor variables (at time *t*)	Dependent variables (at time *t* + 1) Size	Public
Collaborative R&D experience		.0353
		(.0098)
Non-R&D network experience	.0618	.0429
	(.0261)	(.0183)
Degree centrality	.0108	.0018
	(.0024)	(.0008)
Control variables†		
Lagged dependent variable	.9507	.3066
	(.0277)	(.0323)
Total ties	.0076	.0031
	(.0099)	(.0033)
R-squared	.9789	.8606
Total firm-years	639	886

* Standard errors are in parentheses. All reported coefficients are significant at or beyond the .05 level.

† Age was included as a control variable in both models but was eliminated due to nonsignificance; similarly for size in the model predicting public. All models included fixed firm and year effects (dummy variables), in addition to the lagged dependent variable. These estimates are not reported to conserve space.

learning perspective. Hypothesis 1 predicted positive effects of the number of R&D ties and network experience at time *t* on the number of ties of each other type at time $t + 1$ and, in turn, on the diversity of ties at time $t + 1$. The row labeled R&D in Table 5a shows the effects of R&D ties on finance, marketing, clinical, complex, investment, and supply ties to be positive and significant. Further, the effects of R&D on the diversity measure are as predicted. Results in the row labeled Non-R&D network experience support the hypothesized experience effect on a firm's portfolio diversity but provide only partial support for the prediction that experience would increase the number of ties in each non-R&D category (.1148, $p < .0001$; .0799, $p = .0057$ for marketing and supply ties, respectively). Interestingly, the amount of experience with non-R&D ties is critical to a firm's ability to manage a more diverse portfolio. Collaborative R&D experience adds no explanatory power once the number of R&D ties is known. Finally, all the other tie types at *t* do not predict R&D ties at $t + 1$, nor do they predict each other or diversity. Hence, as hypothesized, it is collaborative R&D and network experience that drive firms' portfolios.

Hypothesis 2 predicted positive effects of the number of R&D alliances, portfolio diversity, and network experience at time *t* on how centrally connected a firm becomes at time $t + 1$. Central connectedness was measured variously by degree and closeness centrality and membership in the main component (MainComp). Results in Table 5b show significant positive effects of R&D ties on degree centrality, collaborative R&D experience on closeness centrality, non-R&D network experience on both degree and closeness centrality, and portfolio diversity on all three measures of central connectivity. Hence, all aspects of hypothesis 2 receive support. The amount of network R&D activity, along with skill at managing R&D alliances and other forms of collaboration (and, perhaps, the reputational benefits they may bring), help determine how quickly and deeply a firm moves into the core of the industry's network.

Hypothesis 3 predicted positive effects of the measures of central connectedness and network experience at time *t* on size and public at time $t + 1$. This prediction receives support, as shown in Table 5c. Degree centrality positively predicts increases in size and going public, though effects were not found for closeness centrality or MainComp. Size is also positively predicted by non-R&D network experience, while both measures of network experience predict going public. Our last prediction, in hypothesis 4, was that central connectivity at time *t* would enhance a firm's collaborative R&D activity at time $t + 1$. Returning to Table 5a, above, and looking at the first column, we see that this is where closeness centrality has its primary effect.

The overall pattern of results is quite consistent with a learning argument. We stress two general findings. First, age has no effect, while size is an outcome, not a predictor of network behavior. Growth is a process that requires time. Unlike biotic species, however, organizational growth is not programmed from age. Rather, it is the initiation of collaboration that sets the growth clock in motion, with centrality as a further stimulus. Second, network position (central connectedness) has reciprocal influences on R&D alliances, investment ties, and total collaboration. We have argued and shown that R&D ties, experience, and diversity produce central connectedness. But the process does not stop there. Central connectedness cycles back to intensify a firm's commitment to exploring through its network.

Taken together, the effects shown in Tables 5a–5c can be summarized graphically by the learning model displayed in Figure 2, which we label cycles of learning. Firms can enter via R&D ties or by some other type of tie. Initial collaborative relationships trigger the development of experience at managing ties. R&D ties, directly and through increased experience, enable firms to access more diverse sources of collaboration. Both R&D and non-R&D ties provide experience at managing networks. The development of experience enables a firm to become more central, which in turn has two effects. First, regardless of the pathway, centrally located DBFs are connected into the main component of the industry, providing access to critical information and resource flows needed for internal growth. The second effect is a feedback process in which centrality leads to the initiation and continuance of R&D alliances, thus sustaining the dynamics of learning. R&D ties and other types of collaborations are the admission ticket, while diversity, experience, and centrality are the main drivers of a dynamic system in which disparate firms join together in efforts to keep pace in high-speed learning races.

Figure 2. Cycles of learning in the biotechnology network.

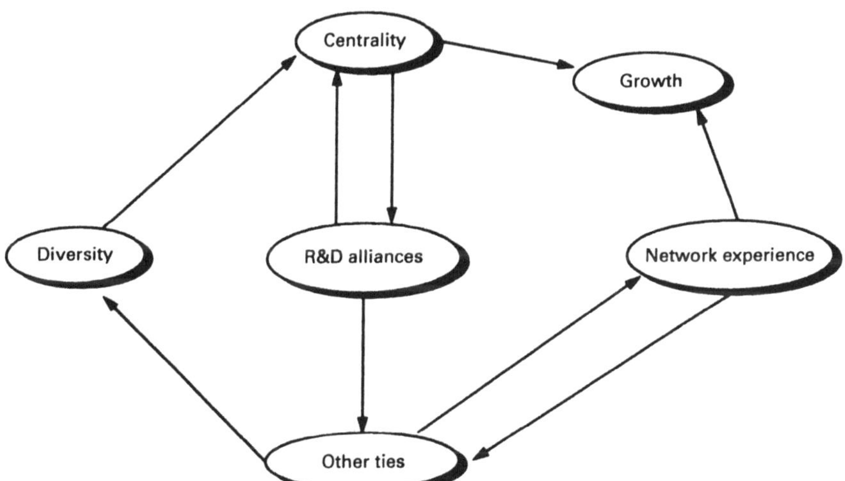

DISCUSSION

We find in our results ample support for the view that networks of collaboration provide entry to a field in which the relevant knowledge is widely distributed and not easily produced inside the boundaries of a firm or obtained through market transactions. We argue that biotech firms grow by being connected to benefit-rich networks. What, then, are the tangible consequences of these network ties? We cannot, of course, answer this question definitively. We have stressed that this is a young industry, with many of the founding firms not quite 20 years old. We lack data on firms that were founded and subsequently failed before 1990. Thus, patterns that have emerged over a five-year period, albeit a crucial one in the industry's emergence from infancy to adolescence, do not provide the full story. Nevertheless, we think our findings are important for several reasons. First, numerous analysts have commented on the industry's very low mortality rate in its early years (Barley and Freeman, 1992; Burrill and Lee, 1993). Second, the rules of the industry have developed and become elaborated during precisely the period we are studying. The majority of the industry's initial 25 products were brought to market and met with considerable success during this period. Finally, we think there are interesting and suggestive points of commonality between our measures of network experience and centrality and various, albeit less precise rankings of success in the industry.

Two illustrations, presented in Table 6, highlight a possible linkage between networks of learning and firm performance. We list sales in 1993 for the top-ten biotech products. Four firms are responsible for developing these medicines, although for six of the cases, a larger company is responsible for sales and marketing. (This division of labor points to the different roles biotechs play in generating new knowledge and pharmaceuticals take in commercializing this knowledge.) Using our network measures as rankings, we find three of these firms on the list of the most central DBFs, and all four are among those firms most steeped in experience.

We also present a list, drawn from Ernst & Young, of biotech firms with a 1994 market value in excess of $500 million (Lee and Burrill, 1994). All of these companies appear on either our experience or centrality rankings, and four of the eight rate highly on both measures. Moreover, three of these four firms—Biogen, Chiron, and Genentech—appear on all four lists: most products to market, greatest market valuation, highest centrality, and most extensive experience.

We next look at how biotech firms compare with firms in other industries, as well as with universities and research laboratories. Again we assess how well our network measures relate to other measures of firm behavior and performance. *Business Week* publishes an annual R&D scorecard, examining the level of R&D investment by U.S. firms. In Table 7, we use this scorecard to list the top R&D spenders as a percentage of sales. The first six are all biotechs, and of this group, five of the companies appear on either our experience or centrality lists, and three companies (once again, Biogen, Chiron, and Genentech) appear on both lists.

Significant investments in R&D can produce varied outcomes, ranging from a new generation of products to novel innovations. The bulk of biotech research is aimed at

Table 6

DBF Centrality, Ten Highest Rated Firms, 1990–94 (means)*

Firm name	Degree	Closeness
Centocor	110.20	1.4791
Biogen	101.40	1.4788
Cambridge Biotech	93.80	1.4796
Genentech	100.20	1.4786
Chiron	82.40	1.4781
Athena Neurosciences	78.00	1.4767
Scios Nova	71.80	1.4767
DNX	71.40	1.4773
ALZA	69.0	1.4765
Genetics Institute	80.5	1.1788

DBF Collaborative Research Experience, Top Eleven Firms, 1990–94 (means, in years)

Firm name	Collaborative R&D experience
Collaborative Research, Inc.	12.33
BioMeasure, Inc.	11.33
Chiron	9.33
Unigene Labs	8.33
PanLabs	8.33
Biogen	8.33
Applied DNA	8.33
Amgen	8.33
Genzyme	7.33
Genentech	7.33
ALZA	7.33

DBFs with Market Value in Excess of $500 Million, June 1994†

Firm name	Value
Amgen	$5,704
Genentech	5,678
ALZA	1,920
Chiron	1,802
Genetics Institute	1,093
Biogen	928
Genzyme	636
Centocor	572

Top Ten Biotechnology Drugs on the Market, 1993 Net Sales (in millions)‡

Product	Firm name	Marketer	Sales
Neupogen	Amgen	Amgen	$719
Epogen	Amgen	Amgen	587
Intron A	Biogen	Schering-Plough	572
Humulin	Genentech	Eli Lilly	560
Procrit	Amgen	Ortho Biotech	500
Engerix-B	Genentech	SmithKline Beecham	480
Recombinax HB	Chiron	Merck	245
Activase	Genentech	Genentech	236
Protropin	Genentech	Genentech	217
Roferon-A	Genentech	Roche	172
Total sales, top ten/industry			$4,288/$7,700

* Listed in order of combined rank (degree and closeness)
† Source: Lee and Burrill (1994: 14).
‡ Source: Lee and Burrill (1994: 16).

developing medicines that are fundamentally new. This focus brings biotech into close contact with basic reseach in molecular biology and genetics. We draw on citation data to emphasize biotechnology's position in the research community. In Table 7, we present a ranking of the most

Table 7

Top Ten R&D Spenders (in relation to sales) in U.S. Industry, 1993*

Firm	R&D as % of sales
	340.9%
Immunex	97.6%
Genetics Institute	58.4%
Chiron	58.1%
Biogen	48.5%
Genentech	36.0%
Genzyme	35.3%
Continuum	28.1%
MacNeal-Schwendler	26.8%
Encore Computer	25.3%
Knowledgeware	

Top Ten Most Visible Scientific Institutions in Molecular Biology and Genetics, 1988–92†

Institution	Number of papers	Cites per publication
Salk Institute	403	41.6
Cold Spring Harbor Labs	359	40.8
Whitehead Institute	392	39.7
Genentech	225	33.1
Chiron	200	32.8
Institute Chemie Biologique	261	31.8
Fred Hutchinson Cancer Center	413	27.1
MIT	1,060	25.8
Princeton	369	24.0
MRC lab Molecular Biology	430	23.7

* Source: COMPUSTAT, reported in *Business Week,* June 27, 1994, p. 79. Biotechnology firms are shown in boldface type.
† Source: Institute for Scientific Information, *Science Watch,* 4:7 (July/August, 1993). Reprinted with permission of ISI. Biotechnology firms are shown in boldface type.

significant scientific institutions in molecular biology and genetics over the period 1988–1992, measured by number of citations per paper. Two of the most central and experienced biotech firms by our measures, Genentech and Chiron, rank fourth and fifth on this prestigious and diverse list—in the company of world-class research institutes, elite private universities, and a leading cancer center. In citation data collected over a longer period of time, from 1981 to 1992, the average citations per paper for biotechs was 27.5, as compared with rates of 31.25 for independent and university labs and 10.8 for pharmaceutical companies.[4] We mention these results for two reasons. First, they illustrate the closeness of the biotech and university communities. Second, the four highest cited biotechs (Genentech, Genetics Institute, Biogen, and Chiron, with scores of 39.55, 37.54, 35.67, and 32.82, respectively) are, once more, highly rated on our experience and centrality measures.

The data in Tables 6 and 7 combine with our statistical results to suggest that being centrally connected is necessary to achieve valued organizational outcomes. Nevertheless, the use of networks is not a guarantor of success. More work is needed before we fully understand the heterogeneous pathways firms take in our cycles-of-learning model and why some lead to visible indicators of success while others do not. In addition, we underscore in the foregoing that the relationship between our measures of

4
Data provided by the Institute for Scientific Information (Philadelphia, PA), as reported in *Biotechnology,* 10 (1992): 1517.

learning and the varied indicators of performance are merely suggestive. Nonetheless, we offer these signposts of organizational performance because they illustrate the critical stages in the process of developing biotechnology medicines. Highly visible publications attract scientific attention and serve as signals to investors and intellectual talent. Obtaining product approval for a new medicine and generating high-volume sales show that science-based companies can take ideas from the laboratory to successful commercialization. A market value in excess of $500 million is evidence of staying power, a robust sign that firms organized around networks of learning are capable of producing enviable results. On all these dimensions of production, our measures of learning are associated with those firms that have, thus far, been industry leaders.

CONCLUSION

We have argued that in a field of rapid technological development, such as biotechnology, the locus of innovation is found within the networks of interorganizational relationships that sustain a fluid and evolving community. Learning occurs within the context of membership in a community and may require different kinds of organizations and organizational practices to access that community. Our empirical analyses allow us to flesh out the picture of a firm under these conditions. Several standard organizational characteristics, such as age and size, appear to be ancillary in accounting for patterns of collaboration. Neither growth nor age reduced the propensity to engage in external relationships. Instead, age, per se, proved unimportant in the context of network experience, and size was an outcome rather than a determinant of partnerships.

We found a path-dependent (Arthur, 1990) cycle of learning in which an early choice of exploration elicited positive feedback. In part, this feedback involved anticipated learning with the project at hand, but many of the gains stem from consequences that are harder for firms to foresee. Knowledge is garnered from collaboration on a specific project, but this participation has unanticipated results not apparent at the outset of the relationship. Science does not follow an orderly path; it has a nasty habit of spiraling off into multiple, uncharted directions. In our view, the development of absorptive capacity (Cohen and Levinthal, 1989, 1990) and skill at managing collaborations, as well as the increased awareness of new projects and reputation as a valuable partner, are all serendipitous benefits of collaboration. Such advantages make it unlikely that firms will retreat from their overall use of alliances, although individual agreements may come and go. When the locus of innovation is found in an interorganizational network, access to that network proves critical. R&D alliances are the admission ticket, the foundation for more diverse types of collaborations, and the pivot around which firms become more centrally connected.

Equally important are changes at the network and industry levels. In our sample, firms without ties are becoming increasingly rare; the modal firm has multiple partnerships. Perhaps our most interesting descriptive result is that the field is becoming more tightly connected not in spite of, but because of a marked increase in the number of partners involved in alliances with DBFs. Network density (based on connections) has increased 50 percent from .06 to .09, while the number of firms has dropped slightly (from 241 to 226). We take this increasing connectivity within an expanding universe as further evidence that two processes of learning are occurring simultaneously and recursively. First, firms are increasingly using ties to enhance the inflow of specific information, resources, and products. Second, firms are becoming much more adept at and reputed for the general practice of collaboration with diverse partners.

As a result of this reciprocal learning, both firm-level and industry-level practices are evolving, with boundaries becoming ever more permeable. In contrast to the much-discussed liability of newness hypothesis (Stinchcombe, 1965; Hannan and Freeman, 1989), there appears to be a liability of unconnectedness (Baum and Oliver, 1992) at work in biotechnology, and other fields in which intellectual developments are expanding rapidly. Rather than using external relations as a temporary mechanism to compensate for capabilities a firm has not yet mastered, firms use collaborations to expand all their competencies. Firms opt for sustaining the ability to learn, via interdependence, over independence by means of vertical integration. This, in turn, promotes a sense of community-level mutualism (Barnett, 1990). Competition is no longer seen as a game with a zero-sum outcome (Thurow, 1980), but as a positive-sum relationship in which new mechanisms for providing resources develop in tandem with advances in knowledge. At the core of this relationship is a vital need to access relevant knowledge: knowledge of a sort that is sophisticated and widely dispersed and not easily produced or captured inside the boundaries of a firm. These conditions are not limited to biotechnology. In fields as diverse as ceramics and software, much of the relevant know-how is neither located inside an organization nor readily available for purchase. When the sources of knowledge are disparate and the pathways of technological development uncharted, we would expect the emergence of networks of learning.

REFERENCES

Abernathy, William, and Kim Clark
1985 "Innovation: Mapping the winds of creative destruction." Research Policy, 14: 3–22.

Amburgey, Terry, Wei-Jan Shan, and Jitendra Singh
1994 "Cooperative strategy and technological competition in biotechnology." Paper presented at the Academy of Management Annual Meeting, Dallas, TX.

Arora, Ashish, and Alfonso Gambardella
1990 "Complementarity and external linkages: The strategies of large firms in biotechnology." Journal of Industrial Economics, 38: 361–379.
1994 "Evaluating technological information and utilizing it: Scientific knowledge, technological capability, and external linkages in biotechnology." Journal of Economic Behavior and Organization, 24: 91–114.

Arthur, W. Brian
1990 "Positive feedbacks in the economy." Scientific American, 262: 92–99.

Badaracco, Joseph L.
1991 The Knowledge Link: How Firms Compete through Strategic Alliances. Boston: Harvard Business School Press.

Barley, Stephen, and John Freeman
1992 "Strategic alliances in commercial biotechnology." In N. Nohria and R. Eccles (eds.), Networks and Organizations: 311–347. Boston: Harvard Business School Press.

Baum, Joel, and Christine Oliver
1992 "Institutional embeddedness and the dynamics of organizational populations." American Sociological Review, 57: 540–559.

Blau, Peter
1977 Inequality and Heterogeneity: A Primitive Theory of Social Structure. New York: Free Press.

Borgatti, S. P., M. G. Everett, and L. C. Freeman
1992 UCINET IV Version 1.00. Columbia, SC: Analytic Technologies.

Brown, John S., and Paul Duguid
1991 "Organizational learning and communities-of-practice: Toward a unified view of working, learning, and innovation." Organization Science, 2: 40–57.

Burrill, Steven G., and Kenneth B. Lee
1993 Biotech 94: Long-Term Value, Short-Term Hurdles. San Francisco: Ernst & Young.

Cohen, Wesley, and Daniel Levinthal
1989 "Innovation and learning: The two faces of R&D." Economic Journal, 99: 569–596.
1990 "Absorptive capacity: A new perspective on learning and innovation." Administration Science Quarterly, 35: 128–152.

Cyert, Richard, and James G. March
1963 A Behavioral Theory of the Firm. Englewood Cliffs, NJ: Prentice-Hall.

Dodgson, Mark
1993 "Learning, trust, and technological collaboration." Human Relations, 46: 77–95.

Eisenhardt, Kathleen, and Claudia Bird Schoonhoven
1996 "Strategic alliance formation in entrepreneurial firms: Strategic needs and social opportunities for cooperation." Organization Science, vol. 7 (in press).

Freeman, Christopher
1991 "Networks of innovators: A synthesis." Research Policy, 20: 499–514.

Freeman, John
1990 "Ecological analysis of semiconductor firm mortality." In Jitendra Singh (ed.), Organizational Evolution: 53–77. Newbury Park, CA: Sage.

Freeman, Linton C.
1979 "Centrality in social networks: Conceptual clarification." Social Networks, 1: 215–239.
1992 "The sociological concept of 'group': An empirical test of two models." American Journal of Sociology, 98: 152–166.

Gulati, Ranjay
1995 "Does familiarity breed trust? The implications of repeated ties for contractual choice in alliances." Academy of Management Journal, 38: 85–112.

Gupta, Udayan
1994 "Now or never: Can the biotech industry deliver?" Wall Street Journal, May 20: C4.

Hagedoorn, John
1990 "Inter-firm partnerships and co-operative strategies in core technologies." In C. Freeman and L. Soete (eds.), New Explorations in the Economics of Technical Change: 3–37. London: Pinter.
1993 "Understanding the rationale of strategic technology partnering: Interorganizational modes of cooperation and sectoral differences." Strategic Management Journal, 14: 371–385.
1995 "Strategic technology partnering during the 1980s: Trends, networks, and corporate patterns in non-core technologies." Research Policy, 24: 207–231.

Hagedoorn, John, and Joseph Schakenraad
1992 "Leading companies and networks of strategic alliances in information technologies." Research Policy, 21: 163–190.

Hakansson, Hans
1990 "Technological collaboration in industrial networks." Engineering Management Journal, 8: 371–379.

Hamel, Gary
1991 "Competition for competence and inter-partner learning within international strategic alliances." Strategic Management Journal, 12: 83–103.

Hannan, Michael T., and John Freeman
1989 Organizational Ecology. Cambridge, MA: Harvard University Press.

Hennart, Jean Francois
1988 "A transaction costs theory of equity joint ventures." Strategic Management Journal, 9: 361–374.

Hergert, Michael, and Deigan Morris
1988 "Trends in international collaborative agreements." In F. Contractor and P. Lorange (eds.), Cooperative Strategies in International Business: 99–109. Lexington, MA: Lexington Books.

Hsiao, Cheng
1986 Analysis of Panel Data. New York: Cambridge University Press.

Judge, George
1985 The Theory and Practice of Econometrics, 2nd ed. New York: Wiley.

Keller, Kenneth H.
1990 "Science and technology." Foreign Affairs, 69 (4): 123–128.

Kleinknecht, Alfred, and J. O. N. Reijnen
1992 "Why do firms cooperate on R&D? An empirical study." Research Policy, 21: 347–360.

Kogut, Bruce
1989 "The stability of joint ventures." Journal of Industrial Economics, 38: 1–16.

Koput, Kenneth W., Walter W. Powell, and Laurel Smith-Doerr
1995 "The role of elite sponsorship in biotechnology's emergence." Paper presented at NIH Economics Roundtable on Biomedical Research, Bethesda, MD, October.

Lee, Kenneth B., and G. Steven Burrill
1994 Biotech 95: Reform, Restructure, and Renewal. Palo Alto, CA: Ernst & Young.

Levinthal, Daniel A., and James G. March
1994 "The myopia of learning." Strategic Management Journal, 14: 95–112.

March, James G.
1991 "Exploration and exploitation in organizational learning." Organization Science, 2: 71–87.

The Mathworks, Inc.
1994 MATLAB version 4.2c. Natick, MA: The Mathworks, Inc.

Networks of Learning

Mintz, Beth, and Michael Schwartz
1985 The Power Structure of American Business. Chicago: University of Chicago Press.

Mizruchi, Mark S.
1982 The American Corporate Network, 1904–1974. Beverly Hills, CA: Sage.

Mowery, David C.
1983 "The relationship between intrafirm and contractual forms of industrial research in American manufacturing, 1900–1940." Explorations in Economic History, 20: 351–374.

Mowery, David C. (ed.)
1988 International Collaborative Ventures in U.S. Manufacturing. Cambridge, MA: Ballinger.

Mowery, David C., and Nathan Rosenberg
1989 Technology and the Pursuit of Economic Growth. New York: Cambridge University Press.

Mowery, David C., and David J. Teece
1993 "Strategic alliances and industrial research." Unpublished manuscript, Haas School of Business, University of California, Berkeley.

Nelson, Richard R.
1990a "U.S. technological leadership: Where did it come from and where did it go?" Research Policy, 19: 119–132.
1990b "Capitalism as an engine of progress." Research Policy, 19: 193–214.

Nelson, Richard R., and Sydney Winter
1982 An Evolutionary Theory of Economic Change. Cambridge, MA: Harvard University Press.

Pake, G. E.
1986 "Business payoff from basic science at Xerox." Research Management, 29: 35–40.

Parkhe, Arvind
1993 "Strategic alliances structuring: A game theoretic and transaction cost examination of interfirm cooperation." Academy of Management Journal, 36: 794–829.

Pisano, Gary
1989 "Using equity participation of support exchange: Evidence from the biotechnology industry." Journal of Law, Economics and Organization, 5: 109–126.
1991 "The governance of innovation: Vertical integration and collaborative arrangements in the biotechnology industry." Research Policy, 20: 237–249.

Powell, Walter W.
1990 "Neither market nor hierarchy: Network forms of organization." In L. L. Cummings and B. M. Staw (eds.), Research in Organizational Behavior, 12: 295–336. Greenwich, CT: JAI Press.
1996a "Inter-organizational collaboration in the biotechnology industry." Journal of Institutional and Theoretical Economics, 151 (1) (in press).
1996b "The social construction of an organizational field: The case of biotechnology." International Journal of Technology Management (in press).

Powell, Walter W., and Peter Brantley
1992 "Competitive cooperation in biotechnology: Learning through networks?" In N. Nohria and R. Eccles (eds.), Networks and Organizations: 366–394. Boston: Harvard Business School Press.

Sabel, Charles
1993 "Learning by monitoring: The institutions of economic development." In N. J. Smelser and R. Swedberg (eds.), The Handbook of Economic Sociology: 137–165. Princeton, NJ: Princeton University Press.

Schumpeter, Joseph A.
1934 The Theory of Economic Development. Cambridge, MA: Harvard University Press.

Scott, John
1991 Social Network Analysis: A Handbook. London: Sage.

Smith-Doerr, Laurel
1994 "Science within organizations: Differences in the organization of science between a biotechnology firm and a university laboratory." Unpublished master's thesis, University of Arizona.

Stinchcombe, Arthur
1965 "Social structure and organizations." In James G. March (ed.), Handbook of Organizations: 142–193. Chicago: Rand McNally.
1990 Information and Organization. Berkeley: University of California Press.

Teece, David J.
1986 "Profiting from technological innovation: Implications for integration, collaboration, licensing and public policy." Research Policy, 15: 785–805.

Thurow, Lester
1980 The Zero-Sum Society. New York: Basic Books.

Tushman, Michael, and Philip Anderson
1986 "Technological discontinuities and organizational environments." Administrative Science Quarterly, 31: 439–465.

Tushman, Michael, and Lori Rosenkopf
1992 "Organizational determinants of technological change: Towards a sociology of technological evolution." In B. M. Staw and L. L. Cummings (eds.), Research in Organizational Behavior, 14: 311–347. Greenwich, CT: JAI Press.

Useem, Michael
1984 The Inner Circle. New York: Oxford University Press.

Von Hippel, Eric
1988 Sources of Innovation. New York: Oxford University Press.

Weisbach, Jerry A., and Walter H. Moos
1995 "Diagnosing the decline of major pharmaceutical research laboratories: A prescription for drug companies." Drug Development Research, 34: 243–259.

Williamson, Oliver E.
1991 "Comparative economic organization: The analysis of discrete structural alternatives." Administrative Science Quarterly, 36: 269–296.

Zucker, Lynne G., M. R. Darby, and M. B. Brewer
1994 "Intellectual capital and the birth of U.S. biotechnology enterprises." National Bureau of Economic Research Working Paper #4653.

[16]

Strategic Management Journal, Vol. 17(Winter Special Issue), 77–91 (1996)

STRATEGIC ALLIANCES AND INTERFIRM KNOWLEDGE TRANSFER

DAVID C. MOWERY
Haas School of Business, University of California, Berkeley, California, U.S.A. and Canadian Institute for Advanced Research, Toronto, Ontario, Canada

JOANNE E. OXLEY
University of Michigan Business School, Ann Arbor, Michigan, U.S.A.

BRIAN S. SILVERMAN
Faculty of Management, University of Toronto, Toronto, Ontario, Canada

This paper examines interfirm knowledge transfers within strategic alliances. Using a new measure of changes in alliance partners' technological capabilities, based on the citation patterns of their patent portfolios, we analyze changes in the extent to which partner firms' technological resources 'overlap' as a result of alliance participation. This measure allows us to test hypotheses from the literature on interfirm knowledge transfer in alliances, with interesting results: we find support for some elements of this 'received wisdom'—equity arrangements promote greater knowledge transfer, and 'absorptive capacity' helps explain the extent of technological capability transfer, at least in some alliances. But the results also suggest limits to the 'capabilities acquisition' view of strategic alliances. Consistent with the argument that alliance activity can promote increased specialization, we find that the capabilities of partner firms become more divergent in a substantial subset of alliances.

INTRODUCTION

Recent literature on the role of firm-specific knowledge in competitive strategy has spawned several theoretical perspectives. The 'resource-based' view of the firm describes the business enterprise as a collection of sticky and difficult-to-imitate resources (Penrose, 1959; Barney, 1986; Wernerfelt, 1984), stressing the capture of rents through the protection and deployment of these resources. A related literature on 'dynamic capabilities' emphasizes the importance of change in the capabilities underpinning these resources (Teece, Pisano, and Shuen, 1997; Teece and Pisano, 1994), focusing in particular on the development, more than the exploitation, of firm-specific resources. Recent 'knowledge-based' views of the firm focus on knowledge as a key competitive asset, and emphasize the capacity of the firm to integrate tacit knowledge (Grant and Baden-Fuller, 1995; Conner and Prahalad, 1996).

A central factor in the 'dynamic capabilities' view of firm strategy is the acquisition of new capabilities through organizational learning. One device cited in this literature as an important tactic for organizational learning is the strategic alliance,[1] now an important element of contemporary firms' competitive strategies (Harrigan,

Key words: alliances; patents; technology transfer

CCC 0143–2095/96/S20077–15

[1] 'The concept of dynamic capabilities as a coordinate management process opens the door to the potential for interorganizational learning. Researchers . . . have pointed out that collaborations and partnerships can be vehicles for new organizational learning, helping firms to recognize dysfunctional routines, and preventing strategic blindspots' (Teece and Pisano, 1994: 545).

1988; Hagedoorn and Schakenraad, 1990; Eisenhardt and Schoonhoven, 1996). A number of scholars have described the use of alliances by firms to acquire technology-based capabilities from alliance partners, and an extensive literature discusses the features of alliances and their participants that facilitate the flow of technology-based capabilities and other knowledge among partners (e.g., Kogut, 1988; Hamel, Doz, and Prahalad, 1989; Çohen and Levinthal, 1990; Hamel, 1991). Alliances may serve other purposes, however, and recent work on alliances and the role of firm-specific knowledge in firm strategy suggests additional motives and effects of alliance formation. Rather than using alliances to acquire capabilities, scholars suggest that firms use interfirm collaboration to gain *access* to other firms' capabilities, supporting more focused, intensive exploitation of existing capabilities within each firm (Grant and Baden-Fuller, 1995; Nakamura, Shaver, and Yeung, 1996).[2]

Empirical assessment of the importance and validity of these contrasting views, as well as broader empirical research on the role of knowledge within the firm and alliances within firm strategy, has been hampered by the widespread reliance on anecdotes and assertion, rather than statistical evidence. The lack of empirical work documenting the effects of participation in alliances on firms' technological capabilities is attributable in part to the difficulty of measuring the technological and other capabilities of firms. This paper analyzes the effects of interfirm knowledge transfers within strategic alliances on partner firms' technological capabilities, using a new measure of change in these capabilities. Using the citation patterns of partner firms' patent portfolios, we measure changes in the extent to which their technological resources 'overlap' with their partners' technological portfolios as a result of participation in an alliance. This methodology represents a significant advance on previous analysis of changing firm capabilities, which have relied on broad firm-level measures such as R&D spending or raw counts of patents issued to partner firms.

Our new measure allows us to test a number of hypotheses from the literature on interfirm transfer of capabilities in alliances. We find support for some elements of this 'received wisdom'—equity arrangements support greater transfer of technological capabilities (Kogut, 1988) and 'absorptive capacity' (Cohen and Levinthal, 1990) helps explain the effectiveness of technology-based capability transfer, at least in some types of alliances. But the empirical analysis also suggests that there are limits to the 'capabilities acquisition' view of alliances. Consistent with the view that alliance activity can lead to increased specialization, as firms access others' capabilities (rather than acquiring them or developing them internally), we find that the capabilities of partner firms become more divergent in a substantial subset of alliances.

Immediately below, we describe recent trends in alliance formation, summarizing current understanding of the motives for collaboration and the prominence of knowledge acquisition in recent discussions of strategic alliances. We then survey the literature on interfirm knowledge transfer in alliances, outlining some hypotheses based on this literature. This section is followed by a discussion of measures of technological capabilities and a description of the patent citation data used in our study. The following section briefly describes our data and empirical methods, and the next two sections of the paper present the empirical results and a concluding discussion of their implications.

TRENDS AND MOTIVES IN STRATEGIC ALLIANCES

Strategic alliances are not new phenomena—they have been important in international business since the turn of the century (Harrigan, 1986) when joint ventures were formed primarily as a vehicle for the exploitation of natural resources. Nonetheless, the rate of formation of alliances has increased significantly over the last two decades and the motives for their establishment have shifted, as alliances have become widespread in technology-intensive industries (e.g., semiconductors, computers, software, commercial aircraft) in which they were of little or no importance prior to 1975. In addition, the activities included in

[2] As Hamel points out, 'the crucial distinction between acquiring such skills in the sense of gaining *access* to them ... and actually *internalizing* a partner's skills has seldom been clearly drawn ... For the partners, an alliance may be not only a means for trading access to each other's skills—what might be termed *quasi-internalization*, but also a mechanism for actually acquiring a partner's skills—*de facto internalization*' (1991: 84; emphasis in original).

many of the alliances of the past 20 years, such as joint R&D and product development, can involve higher levels of knowledge exchange and technology transfer among participants.

Motives for the formation of these more recent alliances include the need to spread the costs and risks of innovation, as capital requirements for development projects in industries such as pharmaceuticals, telecommunications and commercial aircraft have risen (Mowery, 1988). Higher development costs and risks, along with a perceived shrinkage in product life-cycles, also mean that rapid penetration of foreign markets is more important than ever in many technology-intensive industries—something which may be more easily achieved through an alliance. Still other alliances focus on collaboration between users and suppliers of new products as a means of coordinating and formulating technical standards and 'dominant designs' (Grindley, 1995). Strategic motives also play a role in alliances that are formed to facilitate strategic coordination among competitors to increase market power (e.g., Porter and Fuller, 1986; Hagedoorn, 1993).

One of the most widely cited motives for collaboration, linked to many of those just described, is the acquisition of new technical skills or technological capabilities from partner firms (Mariti and Smiley, 1983; Hamel *et al.*, 1989; Shan, 1990; Hamel, 1991; Powell and Brantley, 1992; Mody, 1993; Khanna, 1996). Alliances have advantages over conventional contracts or markets for this task because firm-specific technological capabilities frequently are based on tacit knowledge and are subject to considerable uncertainty concerning their characteristics and performance. These features make it difficult to draft simple contracts governing the sale or licensing of such capabilities (Mowery, 1983; Pisano, 1990).[3] By combining some of the incentive structures of markets with the monitoring capabilities and administrative controls associated with hierarchy (internal organization), alliances can provide a superior means to gain access to technological and other complex capabilities. In fact, alliances run the gamut from fairly simple unilateral (i.e., 'technology for cash') contracts, such as licensing, through more complex contractually based arrangements, such as technology sharing and joint development agreements (which often include joint ownership or other organizational mechanisms for oversight and management), to 'pure' equity joint ventures, where ownership in a separately incorporated entity is shared by the partner firms (see Figure 1).[4]

The received wisdom on interfirm capabilities transfer implicitly assumes that the acquisition of technology-based capabilities is an important goal and effect of interfirm collaboration, and scholars have examined the factors that facilitate knowledge transfer among partner firms. This literature yields a number of testable empirical implications, which form the basis for the empirical hypotheses on the role of alliances in interfirm knowledge transfer that we discuss immediately below.

INTERFIRM KNOWLEDGE TRANSFER IN STRATEGIC ALLIANCES: RECEIVED WISDOM AND HYPOTHESES

Our empirical investigation focuses on transfer of technological capabilities among alliance partners. Specifically, we are interested in how collaboration changes the relationship between a firm's technological portfolio and those of its alliance partner(s). If collaboration results in the interfirm transfer of technological capabilities, then alliances should produce higher levels of technological 'overlap' among partner firms and increased similarity in their technology portfolios (Mowery, Oxley, and Silverman, 1997).

The previous literature provides a number of testable statements about factors influencing the extent of interfirm knowledge transfer in alliances. Kogut (1988) argues that the opportunities for interfirm transfer of capabilities afforded by different alliance structures influences the choice among them, since equity-based joint ventures are more effective vehicles for the transfer of tacit knowledge between the partners: 'Other forms of

[3] Other firm-specific capabilities include knowledge of specific markets or user needs, idiosyncratic, firm-specific 'routines', such as decision-making techniques or management systems, and complex networks for handling the marketing and distribution of products that include procedures for the documentation and analysis of user feedback.

[4] The 'other' category of alliance forms in Figure 1 includes second sourcing and coproduction agreements, joint research pacts and research corporations. 'Mixed modes' are primarily licensing agreements combined with equity exchanges.

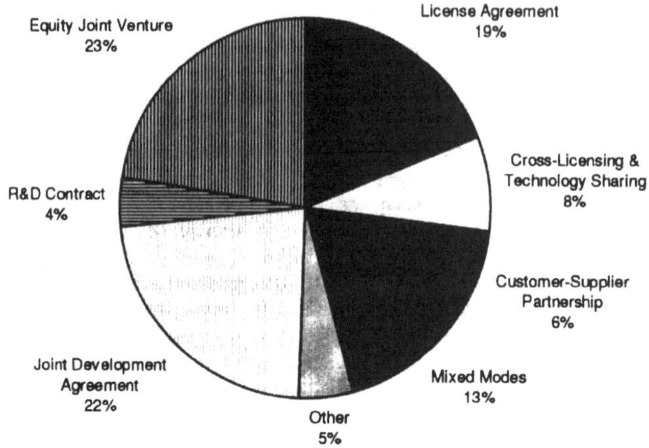

Figure 1. Variety in alliance forms
Source: CATI Database

transfer, such as through licensing, are ruled out . . . because the very knowledge that is being transferred is organizationally embedded' (Kogut, 1988: 323). This observation underpins our first hypothesis:

> *Hypothesis 1a: Interfirm transfer of technological capabilities will be enhanced in equity joint ventures; participants will display higher increases in technological overlap after formation of equity joint ventures than after formation of contract-based alliances.*

An extension of this logic suggests that interfirm knowledge transfers should be more limited in 'unilateral' contract-based alliances, such as licensing agreements. The technology that is exchanged for cash payments in these undertakings generally is more tightly 'packaged' than is the case in bilateral contractual arrangements such as technology sharing or joint development agreements. As a result, 'unilateral' alliances create fewer opportunities for interfirm knowledge transfer.

> *Hypothesis 1b: Participants in unilateral contract-based alliances will display a lower increase in technological overlap after the formation of the alliance than participants in bilateral contract-based alliances.*

However, the transfer of technological capabilities is by no means an assured outcome, even within equity-based joint ventures. Cohen and Levinthal (1990) argue that a necessary condition for a firm's successful exploitation of technological capabilities or knowledge outside its boundaries is development within the firm of the ability to absorb such capabilities. This 'absorptive capacity' requires that a firm have considerable in-house expertise that complements the technology activities of its alliance partner. Absorptive capacity results from a prolonged process of investment and knowledge accumulation within the firm, and its development is path-dependent; a firm's current absorptive capacity is influenced by its historic participation in specific product markets, lines of R&D, and other technical activities.[5]

There are few direct tests of the hypothesized influence of absorptive capacity, but the results of such tests (e.g., Gambardella, 1992) are broadly

[5] '. . . prior knowledge permits the assimilation and exploitation of new knowledge. Some portion of that prior knowledge should be very closely related to the new knowledge to facilitate assimilation . . . Accumulating absorptive capacity in one period will permit its more efficient accumulation in the next. By having already developed some absorptive capacity in a particular area, a firm may more readily accumulate what additional knowledge it needs in the subsequent periods in order to exploit any critical external knowledge that may become available' (Cohen and Levinthal, 1990: 135–136).

supportive of the argument that higher levels of absorptive capacity improve a firm's ability to exploit sources of technical knowledge outside its boundaries. Moreover, a parallel line of research in the broader technology transfer literature suggests that possession of relevant technical skills facilitates inward technology transfer (e.g., Rosenberg and Frischtak, 1991; Agmon and von Glinow, 1991). Still other research indicates that firms tend to establish alliances with firms that have overlapping technological capabilities (Mowery *et al.*, 1997).

A firm's absorptive capacity for learning from its alliance partners should depend on its endowment of relevant technology-based capabilities upon entering an alliance. R&D investment is a necessary (although not necessarily sufficient) condition for the creation of absorptive capacity—indeed, both Cohen and Levinthal's original test and Gambardella's subsequent examination of the issue use R&D intensity as a proxy for absorptive capacity. It is also plausible that larger firms have more diverse technological portfolios and therefore are more likely to possess technology that is 'relevant' to the alliance in question. Although both of these measures merit empirical testing, neither one provides the level of disaggregation and specificity of our patent-based measure of interfirm knowledge transfer (see below). We therefore include in our tests of the importance of absorptive capacity a more narrowly focused measure of absorptive capacity: the prealliance levels of technological overlap among the partners. Since absorptive capacity develops over time in a path-dependent fashion, substantial prealliance overlap between partners' technological portfolios should enhance their capacity to absorb new competencies from one another. Taken together, these arguments lead to the following hypotheses:

Hypothesis 2a: The extent of a firm's absorption of technological capabilities from its alliance partners will be positively related to its prealliance level of technological overlap wth partner firms.

Hypothesis 2b: The extent of a firm's absorption of technological capabilities from its alliance partners will be positively related to its R&D intensity.

Hypothesis 2c: The extent of a firm's absorption of technological capabilities from its alliance partners will be positively related to its size.

Another aspect of interfirm learning that is closely related to absorptive capacity and receives attention in the alliance literature concerns the extent to which firms enter into an alliance with the 'intent to learn' (Hamel, 1991). This argument appears in a number of critical commentaries on the alleged tendencies of U.S. firms to weaken their technological capabilities in alliances with foreign—primarily Japanese—firms (e.g., Reich and Mankin, 1986). Although there is evidence to the contrary (Mowery, 1988), some anecdotal and case-based research suggests the presence of asymmetries or differential rates of capability acquisition by U.S. and Japanese firms in alliances between them. Hamel *et al.* assert that in several U.S.–Japanese alliances, where the Japanese company emerged from an alliance stronger than its partner, 'the Japanese company had made a greater effort to learn' (1989: 134). Discussing a 20-year alliance with a Japanese firm, an executive of a U.S. industrial products company claimed: 'We established them in their core business. They learned the business from us, mastered our process technology ... and today challenge us outside Japan' (Hamel 1991: 86). Other evidence from Mansfield (1988), drawing on a survey of U.S. and Japanese firms, suggests that Japanese firms are more effective in commercializing innovations based on external sources of technology than are U.S. firms, a finding that is consistent with the presence within Japanese firms of a greater 'intent to learn' from external sources.

These examples may not be representative of the broader historical experience of U.S.–Japanese alliances, nor are the mechanisms that underpin the development and maintenance of such 'national traits' in business firms clearly articulated in this literature. The argument nevertheless has been extended to non-Japanese foreign firms: Hamel (1991) suggests that British firms share the alleged arrogance and lack of receptivity of U.S. firms and that French firms' ability to build competencies and learn from alliances approaches that of the Japanese. These arguments are based on extraordinarily broad generalizations from minimal evidence, but they occupy a prominent position in the received wisdom on alliances, and yield the following hypothesis:

Hypothesis 3: Japanese companies will absorb more of the technological capabilities of alliance partners than will firms from other countries.

Before discussing our methods for testing these hypotheses, we must address another issue. Significant interfirm knowledge transfer should be reflected in higher levels of technological overlap following the formation of an alliance. But participation in alliances need not always increase technological overlap. Instead, an alliance may enable one firm to gain access to key knowledge-based capabilities of another without internalizing or acquiring that capability, e.g., in an alliance in which one firm designs and the other manufactures an advanced semiconductor device. Among these alliances, interfirm knowledge transfer may be limited to only the codified information necessary to coordinate otherwise separable activities that draw on different 'knowledge domains' (Grant and Baden-Fuller, 1995). Thus, if an alliance enables its members to specialize in different but complementary areas of technology, partner firms may experience divergence in technological capabilities over the course of the alliance (reflected in lower levels of technological overlap), rather than convergence. Indeed, Nakamura *et al.* (1996) find evidence that in some cases joint venture partners display 'convergent development' (i.e., increasingly similar capabilities), but that other alliances produce divergence in the capabilities of partner firms.

If some alliances increase technological overlap while others promote specialization, then tests that assume convergent development through interfirm knowledge transfer (and consequently an unambiguous increase in postalliance technological overlap) are likely to produce inconclusive results, as was the case in Mowery *et al.* (1997). One test for alliance effects in the presence of both convergent and divergent development examines the *absolute value* of changes in technological overlap. If allying firms exhibit stronger changes in the extent of their technological overlap than nonallying firms—in either a positive (convergent) or negative (divergent) direction—then the changes in technological overlap by convergent and divergent alliances might offset one another, but the absolute value of these changes should be significantly larger than that of nonallying firms:

Hypothesis 4a: The presence of divergence and convergence within the alliance population will prevent the observation of a consistent postalliance increase in the technological overlap of alliance partner firms.

Hypothesis 4b: The absolute value *of the pre- vs. postcollaboration changes in technological overlap will be greater for alliance partners than for a similar sample of nonallying firm pairs over the same time period.*

MEASURING TECHNOLOGICAL CAPABILITIES AND INTERFIRM KNOWLEDGE TRANSFER

There is little empirical research on interfirm transfer of technological capabilities in strategic alliances, and most such work emphasizes case studies or small-scale surveys (for other examples, see Lyles, 1988; Sobrero and Roberts, 1996). These gaps reflect the lack of reliable measures of the technology-based capabilities of partner firms and a corresponding absence of measures of change in capabilities. Recent work on the resource-based view of the firm has faced similar difficulties. Although much of that research focuses on narrowly defined firm-specific capabilities in its conceptual discussion, most empirical tests of propositions in the resource-based view rely on broad measures that are subject to many interpretations. Corporate R&D intensity, for example, has frequently been used as a proxy for technological resources (e.g., Montgomery and Hariharan, 1991; Nakamura *et al.*, 1996), despite the fact that R&D intensity measures inputs to the creation of capabilities and indicates little if anything about resultant changes in capabilities.[6] More recent empirical

[6] For example, Nakamura *et al.* (1996) rely on measured changes in alliance members' R&D intensity and foreign sales (as a share of corporate sales) in assessing the 'divergent' or 'convergent' effects of alliance membership. However, it is difficult to interpret changes in such corporate-wide measures as tracking the effects of alliances that may span only a single line of business or a single product within a line of business.

work in this tradition has measured firms' technical resources with patent data (Silverman, 1996; Mowery *et al.*, 1997), which offers significant advantages over alternative measures such as R&D spending; patents are better measures of the output of R&D activities, the key concern for capabilities development. Moreover, patents provide a more disaggregated measure of changing technological portfolios for examining the effects of alliance activities (see Griliches, 1990; and Silverman, 1996, for further discussion).

Our empirical approach focuses on the citation patterns in a firm's patent portfolio, which allows us to observe changes in the relationship of one firm's technology portfolio to that of a partner firm. The measure shares some of the limitations of other patent-based measures (such as raw patent counts): the commercial importance of patents varies among industries and technologies, and firms therefore may display systematic differences in their propensity to seek patent protection for important technical advances. More importantly, patents are by definition examples of codified knowledge, and citation measures therefore may not capture flows of the tacit knowledge that often forms the basis for firm-specific capabilities. Tacit knowledge flows are virtually impossible to measure, however, and we assume that the codified knowledge represented by patents and tacit knowledge are complements, rather than substitutes, and that codified knowledge flows and the tacit knowledge flows of interest are closely linked. There is considerable support for this assumption (Patel and Pavitt, 1994).

When the U.S. Patent and Trademark Office grants a patent, the granting officer includes a list of all previous patents on which the granted patent is based.[7] Citations of prior patents thus serve as an indicator of the technological lineage of new patents, much as bibliographic citations indicate the intellectual lineage of academic research. As Firm_i acquires technological knowledge from its partner in an alliance, Firm_j, we should see a higher rate of citation of Firm_j's patents in new patents applied for by Firm_i. We refer to this as 'cross-citation rate (Firm_i, Firm_j).' More precisely, the measure is defined as follows:

$$\text{Cross-citation rate } (\text{Firm}_i, \text{Firm}_j) = \frac{\text{Citations to Firm}_j \text{ patents in Firm}_i\text{'s patents}}{\text{Total citations in Firm}_i\text{'s patents}}$$

The cross-citation rate provides a measure of the relative importance of Firm_j in Firm_i's external technology 'pool'. An increase in this measure is an indication of the degree to which Firm_i is acquiring technology-based capabilities from Firm_j, i.e., of the extent of interfirm knowledge transfer in the alliance and of the 'technological overlap' between the two companies.

EMPIRICAL ANALYSIS: SAMPLE AND METHODS

We tested our hypotheses by examining cross-citation rates for partners in bilateral alliances that involved at least one U.S. firm and were established during 1985 and 1986. The sample was taken from the Cooperative Agreements and Technology Indicators (CATI) data base, a comprehensive data set that contains information on over 9000 alliances involving some 5000 firms in many industries and countries (Hagedoorn and Schakenraad, 1990). This data set is based on systematic examination of secondary reports of alliance formation, primarily during the 1980s. Although coverage of the overall population is inevitably incomplete and significant biases remain, it is the most comprehensive and reliable source available for information on alliance activity in the global economy.

Each alliance in our sample involves at least one U.S. partner, since we expect interfirm knowledge transfers in these alliances to be more reliably associated with changing patterns of citation to U.S. patents. The years 1985–86 were chosen as the sample period because these were years in which alliance formation was at its height, and together the 2 years yield a sample size sufficient for the necessary statistical tests. Restricting the period to as few years as possible is important, in order to minimize the impact of overall trends in patenting and citation behavior. In addition, selecting a period in the mid-1980s allows straightforward computation of 'before' and 'after' patent citation rates from the available

[7] The patent officer is aided in this task by the patent applicant, whose application usually provides a list of all patents for relevant 'prior art'. It is in the applicants' interest to be forthcoming in this list because a more complete description of prior art is likely to reduce the prospects of an interference being declared during processing of a patent application.

patent data, which covers patents granted between 1975 and 1994 (see below).

The resulting sample contains a total of 792 alliances. Of these, 132 (16%) are equity joint ventures, 226 (29%) are unilateral contract-based alliances (i.e., technology licenses, R&D contracts and second-source agreements) and 434 (55%) are bilateral contract-based alliances (cross-licensing, joint development and technology sharing agreement, etc.). In 280 (35%) of the alliances, both partners are U.S. firms; 102 (13%) have a Japanese firm partnered with a U.S. firm; the remaining 410 (52%) involve a country from elsewhere in the world (but primarily Europe) partnered with a U.S. company.

In addition to our sample of alliance partners, we constructed a 'control sample' of nonallied firms by generating random pairings of firms in the CATI sample and eliminating any pairs which were listed as alliance partners anywhere in the CATI data base. This control sample of 858 firm pairs thus includes firms known to be active in alliances, and allows us to compare the change in citation patterns of alliance partners with those of a similar sample of nonallying firms, a procedure that is essential to tests of Hypotheses 4a and 4b.[8]

The patent data are drawn from the Micropatent data base, which contains all information recorded on the front page of every patent granted in the United States since 1975. This information includes the patent number, date of application, date of grant, company to whom the patent is assigned (if any) and references to prior patents for each granted patent. All patents assigned to the firms appearing in the sample were extracted from the Micropatent data base and corporate patent portfolios were constructed for each firm.

Many of the firms in the sample of alliances are multinational or multidivisional firms (or are subsidiaries of these firms), which poses a challenge when constructing the relevant patent portfolios. Since we focus on firm-specific capabilities, the relevent patent portfolio for the analysis is the portfolio of the entire firm, rather than for any single subsidiary. Since firms do not always assign a patent to the subsidiary in which the innovation took place and/or where it is used, construction of a firm-level patent portfolio is especially important. The first stage in the construction of our sample was to match firms in the sample with their relevant parent company, subsidiaries and/or 'sister' subsidiaries, using the 1985 edition of *Who Owns Whom* (North American Edition). The patent portfolio for each firm over the period 1975–94 was then created, by collecting the information on each patent issued to the relevant parent and all its subsidiaries. The 838 firms in the resulting data set controlled approximately 14,500 subsidiaries and more than 275,000 patents.

The result of the data collection and tabulation process is a series of firm-specific patent portfolios, detailing (for each patent held) the patent number, application date, issue date, and U.S. references (U.S. patents cited in the application). The number and vintage of the patents in each firm's portfolio vary considerably, and the number of patents cited in any single patent ranges from 0 to approximately 100, with a mean around 10. The total number of patents cited in Firm_i's patents during the sample period thus varies from 0 to over 10,000. The patent cross-citation rates were computed for each partner in each alliance before and after collaboration.[9] Other data on firm characteristics in 1985 that are featured in the hypotheses (i.e., R&D spending and firm size, based on sales) or used as control variables (primary SIC code) were drawn from Compustat for U.S. firms and from Compact Disclosure's 'Worldscope Global' data base for non-U.S. firms.

[8] The limitations of this approach to control sample construction and alternative approaches are discussed at length in our earlier paper (Mowery *et al.*, 1997) where three different control samples, including both random pairing and matched pairs, were used. Here we use the firms from our sample of alliances to construct the control sample. This means that there are no significant differences in the overall characteristics of the two samples with respect to such features as industries or nationalities represented, the size distribution of firms within the sample, overall patenting and citation rates or changes in these rates. Full descriptive statistics are available from the authors upon request.

[9] The 'before' cross-citation rates are computed based on all patents applied for after January 1, 1979 and issued before December 31, 1984. Only citations to patents issued after 1975 are included in the calculation, since this is the earliest year for which patent data are available, and so for patents issued before that data, the assignee is unknown. 'After' citation rates refer to patents with applications dated after January 1, 1987 and issued prior to December 1, 1994. These cutoff dates were chosen to ensure that all applications for patents included in the 'before' calculation were in fact made before the collaboration began (i.e., before 1985) and those in the 'after' were after the latest alliance in the sample was established (i.e., after 1986).

Hypotheses 1–3 relate to the extent of changes in the cross-citation rates of alliance partners that occur as a consequence of interfirm transfer of capabilities. But Hypotheses 4a and 4b suggest that such transfer will not necessarily take place in divergent alliances, for which an entirely different logic of capability development may apply. Our research therefore proceeds in two stages: we first test Hypotheses 4a and 4b to establish whether the sample includes both divergent and convergent alliances, and then test the remaining hypotheses on the convergent sample. Table 1 contains definitions for all the variables used in the OLS and Tobit regression specifications and summarizes the models to be estimated for each hypothesis. The descriptive statistics in Table 2 reveal no systematic differences (except those related to citation rates) between the sample of convergent alliances and the overall alliance sample.

RESULTS

Our empirical results are summarized in Tables 3–5. They provide support for some but not all of the hypotheses, and some results challenge the received wisdom on interfirm transfer of capabilities in international alliances. We first tested for the presence of interfirm knowledge transfer or 'convergent development' in alliances (Hypotheses 4a and 4b) before going on to examine the factors that determine the extent of such transfers. Divergent development should be reflected in negative values of *DPCTCRSS* for the alliance in question, and these observations may cancel out the positive values of *DPCTCRSS* expected in convergent alliances. We investigated this issue by comparing citation rate changes for alliance partners with changes in the citation patterns of our control sample of 'nonallying' firms. Equation 1 in Table 3 shows the results

Table 1. Definition of variables and model specifications

Variable name	Definition
$PCTCRSS_{ij}$	Summed pre-1985 cross-citation rate for a given pair of firms *i* and *j*
$DPCTCRSS_{ij}$	Post-1986 cross-citation rate minus pre-1985 cross-citation rate for firms *i* and *j*
$ABSDCRSS_{ij}$	Absolute value of $DPCTCRSS_{ij}$
$FIRMLRN_{ij}$	Post-1986 cross-citation rate minus pre-1985 citation rate for firm *i* (citing to patents owned by firm *j*)
$ALLIES_{ij}$	dummy variable, = 1 if firms *i* and *j* are alliance partners, 0 otherwise
$EQUITY_{ij}$	dummy variable, = 1 if alliance involves equity, 0 otherwise
$UNILAT_{ij}$	dummy variable, = 1 if alliance is a unilateral contractual agreement, 0 otherwise
$US\text{-}nonUS_{ij}$	dummy variable, = 1 if alliance involves a non-U.S. partner, 0 otherwise
$SAMESIC_{ij}$	dummy variable, = 1 if alliance partners have the same 4-digit primary SIC code, 0 otherwise
$FORGNCO_i$	dummy variable, = 1 if firm i is non-U.S., 0 otherwise
$JAPANCO_i$	dummy variable, = 1 if firm i is Japanese, 0 otherwise
$PRECROSS_i$	Pre-1985 cross-citation rate for firm *i* (citing to patents owned by firm *j*)
$PRECRSS2_i$	Square of $PRECROSS_i$
$RNDINT_i$	R&D intensity (i.e., 1985 R&D expenses/sales) for firm *i*
$LNSALES_i$	Natural log of U.S. dollar value of 1985 sales for firm *i*

Hypothesis	Model specification(s)
H1a	$DPCTCRSS_{ij} = CONSTANT + EQUITY_{ij} + US\text{-}nonUS_{ij} + SAMESIC_{ij}$
H1b	$DPCTCRSS_{ij} = CONSTANT + UNILAT_{ij} + US\text{-}nonUS_{ij} + SAMESIC_{ij}$
H2a and H3	$FIRMLRN_{ij} = CONSTANT + EQUITY_{ij} + US\text{-}nonUS_{ij} + SAMESIC_{ij} + FORGNCO_i + JAPANCO_i + PRECROSS_i$
H2b and 2c	$FIRMLRN_{ij}$ = CONSTANT + EQUITY$_{ij}$ + $US\text{-}nonUS_{ij} + SAMSESIC_{ij} + FORGNCO_i + JAPANCO_i + PRECROSS_i + RNDINT_i + LNSALES_i$
H4a	$DPCTCRSS_{ij} = CONSTANT + ALLIES_{ij} + US\text{-}nonUS_{ij} + SAMESIC_{ij}$
H4b	$ABSDCRSS_{ij} = CONSTANT + ALLIES_{ij} + US\text{-}nonUS_{ij} + SAMESIC_{ij}$

Table 2. Descriptive statistics

(a) Descriptive statistics for alliance sample (*ALLIES* = 1)

Variable	Mean	S.D.	Minimum	Maximum
PCTCRSS	0.568	2.532	0.000	50.000
DPCTCRSS	0.222	3.277	-50.000	40.005
ABSDCRSS	0.771	3.192	0.000	50.000
FIRMLRN	0.111	2.297	-50.000	40.000
EQUITY	0.168	0.374	0.000	1.000
UNILAT	0.283	0.451	0.000	1.000
US-nonUS	0.653	0.489	0.000	1.000
SAMESIC	0.075	0.264	0.000	1.000
FORGNCO	0.326	0.469	0.000	1.000
JAPANCO	0.104	0.306	0.000	1.000
PRECROSS	0.284	1.755	0.000	50.000
RNDINT	6.169	17.391	0.000	28.960
LNSALES	7.706	2.633	-0.646	11.529

(b) Descriptive statistics for 'convergent' alliance sample (i.e., *DPCTCRSS* > 0)

Variable	Mean	S.D.	Minimum	Maximum
PCTCRSS	0.376	1.125	0	10.166
DPCTCRSS	2.059	4.826	0.005	40.005
ABSDCRSS	2.059	4.826	0.005	40.005
FIRMLRN	1.030	3.514	-0.667	40.000
EQUITY	0.204	0.404	0	1
UNILAT	0.204	0.404	0	1
US-nonUS	0.670	0.471	0	1
SAMESIC	0.079	0.272	0	1
FORGNCO	0.335	0.473	0	1
JAPANCO	0.168	0.374	0	1
PRECROSS	0.186	0.762	0	10.000
RNDINT	7.371	14.596	0.007	31.739
LNSALES	8.005	2.489	0.861	11.529

Table 3. Results on divergence/convergence hypotheses (*t*-statistics in parentheses)

	1 (*DPCTCRSS*)	2 (*ABSDCRSS*)
INTERCEPT	0.358***	0.364***
	(3.305)	(3.442)
ALLIES	0.058	0.619***
	(0.433)	(4.697)
US-nonUS	–0.326*	–0.201*
	(–2.634)	(–1.662)
SAMESIC	–0.060	–0.179
	(–0.298)	(–0.916)
n	1650	1650
F-statistic	2.343*	8.131***

*** Significant at 0.01 level; ** significant at 0.05 level; * significant at 0.1 level.

Table 4. Results on alliance structure hypotheses (chi-squares in parentheses)

	3 (*DPCTCRSS*)	4 (*DPCTCRSS*)
INTERCEPT	–2.223***	–2.782***
	(61.493)	(36.422)
EQUITY	1.403***	
	(11.385)	
UNILAT		–1.325**
		(6.311)
US-nonUS	–0.503	–0.091
	(2.304)	(0.034)
SAMESIC	–1.10***	–1.273**
	(8.032)	(4.603)
Scale parameter	5.658	4.946
n	191	147
Log likelihood	–1046.9	–857.1

of OLS estimation on a pooled sample of 792 alliances (*ALLIES* = 1) and 858 nonallying pairs (*ALLIES* = 0).

A control variable, *SAMESIC*, which denotes alliances made up of firms from the same 4-digit SIC, is included in this and other specifications to control for the effects of competition among alliance members in product markets and potential similarity in alliance members' product lines, which otherwise could result in spurious high levels of cross-citation. We also anticipate that 'domestic' alliances, in which all member firms share a common home country, are likely to produce different patterns of interfirm technology transfer and learning than that found in 'international' alliances. The less forbidding barriers of culture, language, educational background, and distance associated with domestic alliances should result in higher levels of knowledge transfer. Since our sample includes a large number of 'U.S.–U.S.' and 'U.S.–non-U.S.' alliances, we inserted another control variable. *Us-nonUS*, to distinguish alliances in which U.S. firms were teamed with foreign enterprises.

Consistent with Hypothesis 4a, we find that the coefficient on *ALLIES* in Equation 1 is very

Table 5. Results on absorptive capacity hypotheses (*t*-statistics in parentheses)

	5 (*FIRMLRN*)	6 (*FIRMLRN*)
INTERCEPT	1.431*** (4.521)	4.506*** (5.481)
EQUITY	1.076** (2.448)	0.301 (0.563)
US-nonUS	−1.280*** (−2.944)	−0.648 (−1.401)
SAMESIC	0.110 (0.252)	0.088 (0.089)
FORGNCO	0.608 (1.133)	0.647 (0.594)
JAPANCO	−0.700 (−1.133)	−0.492 (−0.349)
PRECROSS	0.681*** (2.871)	1.812*** (5.228)
RNDINT		−0.015 (−1.228)
LNSALES		−0.443*** (−4.590)
n	382	155
F-statistic	4.327***	6.857***

small and stastistically insignificant.[10] *US-nonUS* has a negative and significant coefficient, indicating that international alliances produce less interfirm exchange of technological capabilities, and the coefficient for *SAMESIC* is insignificant. There is no consistently positive pattern of interfirm learning in our overall alliance sample. The absence of any consistent pattern of change in citation rates can be explained, however, by the presence of both 'convergent' and 'divergent' alliances in our sample (Equation 2). When the *absolute value* of the difference between post- and prealliance cross-citation rates, *ABSDCRSS*, is regressed on the same variables in an OLS estimation, we obtain a larger and statistically significant coefficient for *ALLIES*, along with a negative and marginally significant coefficient for *US-nonUS* and an insignificant coefficient for *SAMESIC*.

Participation in alliances thus produces significant absolute-valued changes in firms' cross-citation of partner-firm patents, supporting Hypothesis 4b. These results provide some support for the arguments in Nakamura *et al.* (1996) that participation in an alliance may produce either convergence of capabilities through interfirm knowledge transfer or divergence through complementary specialization. In addition, these results suggest that international alliances result in less interfirm knowledge exchange or specialization, reflecting the greater logistical and cultural complexities of managing such undertakings.

The remainder of our empirical analysis focuses on factors that influence the extent of interfirm transfer of technology-based capabilities. Accordingly, we restrict our analysis to alliances in which such transfers take place, i.e., those exhibiting technological 'convergence' (*DPCTCRSS* > 0). The determinants of the presence and degree of 'convergence' or 'divergence' in alliance partners' technology-based capabilities are important issues, but lie beyond the scope of this paper. The results reported in Table 4 exclude negative values of the dependent variable (*DPCTCRSS*) and therefore are estimated with the Tobit technique. The specifications address the impact of alliance structure on patterns of interfirm knowledge and capabilities transfer (Hypotheses 1a and 1b) in the 191 alliances characterized by convergent development.

Equation 3 regersses *DPCTCRSS* on *EQUITY* (a dummy variable indicating alliances that involve shared equity by partner firms) as well as *US-nonUS* and *SAMESIC*. Consistent with Hypothesis 1a, the positive and significant coefficient for *EQUITY* implies that equity-based joint ventures support higher levels of interfirm knowledge and capabilities transfer than contract-based alliances. The Tobit coefficient for *SAMESIC* negative and significant, suggesting that convergent alliances involving firms in the same product lines or markets experience lower levels of interfirm knowledge transfers than those spanning 4-digit SICS. The coefficient for *US-nonUS* remains negative, but is no longer significant at even the 0.1 confidence level.[11]

[10] This result is similar to findings from our earlier paper, which analyzed a small sample of alliances (Mowery *et al.*, 1997). A separate test (not reported here) of another hypothesis from the earlier paper on these data also replicated the finding that prealliance cross-citation rates positively and significantly influence partner choice in alliances.

[11] Estimation of the same model and a similar one using *ABSDCRSS* as the dependent variable for the full sample of 792 alliances (using OLS) produces interesting results (not

Equation 4 in Table 4 tests Hypothesis 1b, which argues that 'unilateral' contracts will result in lower levels of interfirm transfer of technological capabilities than bilateral contract-based (nonequity) alliances. Using the sample of 147 nonequity alliances exhibiting convergent development, we regressed *DPCTCRSS* on *UNILAT*, *SAMESIC* and *US-nonUS*. The results are supportive of our hypothesis; *UNILAT* has a negative, statistically significant coefficient (at the 0.05 level), and the coefficients for *US-nonUS* and *SAMESIC* are similar to the results reported in earlier specifications. 'Unilateral' alliances, which can be placed at the 'market' end of a 'market–hierarchy' continuum of alliance forms, thus seem to support lower levels of interfirm knowledge transfer, consistent with the earlier discussion of the difficulties of acquiring technological capabilities through market-based mechanisms. The results support other empirical and theoretical work that argues that the structure and governance of alliances are influenced by and in turn influence their content and activities (Oxley, 1996).[12]

Results for the remaining hypotheses (Hypotheses 2a–c and 3) are contained in Table 5. The dependent variable in each of the two specifications is *FIRMLRN*, which measures changes in the cross-citation behavior of a single firm in an alliance (see Table 1); negative values are included, and Equations 5 and 6 therefore are estimated with OLS. Equation 5 tests for the effects of 'absorptive capacity' on firms' ability to learn from alliance partners. Since R&D and sales data are available for a small subset of our 838 observations (due to the presence of a large number of foreign or privately held firms), this model uses only *PRECROSS* (firm$_i$'s pre-1985 cross-citations of its partner firm's patents) as a measure of absorptive capacity. The coefficient for *PRECROSS* is positive and statistically significant, supporting Hypothesis 2a concerning the importance of absorptive capacity for firms' ability to extract technological capabilities and knowledge from alliance partners (the size and significance of the coefficients for the other 'control' variables, *SAMESIC* and *US-nonUS*, are largely consistent with previous OLS results).[13]

Equation 6 includes measures of firm$_i$'s R&D intensity (*RNDINT*) and the natural logarithm of its sales (*LNSALES*) as additional measures of absorptive capacity. The coefficient for *PRECROSS* remains positive and significant, but the other results for Equation 6 do not support Hypothesis 2b or 2c—the coefficient for *RNDINT* is negative and nonsignificant, while that for *LNSALES* is negative and significant. Larger firms within this sample thus appear to absorb fewer capabilities from their alliance partners, controlling for other characteristics of absorptive capacity and alliances, while relatively R&D-intensive firms do not exhibit superior capabilities absorption in alliances, *ceteris paribus*. Due in part to the reduction in sample size, the coefficients for both *EQUITY* and *US-nonUS* also are not significant in Equation 6, although their signs remain unchanged.

Our final hypothesis on the determinants of the extent of interfirm knowledge transfer (Hypothesis 3) tested the arguments of Hamel *et al.* (1989) that Japanese firms exhibit consistently superior abilities to learn 'more' through alliances with U.S. firms than is true of firms from other countries. Equations 5 and 6 shed light on this issue: once we control for the lower overall

---

reported here in detail; available from the authors on request) that bear on the convergence/divergence dichotomy. The specification that uses *DPCTCRSS* as the dependent variable again shows a positive and significant effect for *EQUITY*. Estimation of a model using the absolute value of the change in partner firms' cross-citations (*ABSDCRSS*) as the dependent variable yields a coefficient for *EQUITY* that is slightly smaller in magnitude and has a larger standard error. Taken together, these results suggest that firms participating in equity- based joint ventures tend to exhibit 'convergent', rather than 'divergent' development of technological capabilities.

[12] The negative coefficient obtained when this model was run on the complete sample of nonequity alliances (results not shown) suggests that unilateral alliances may in fact support specialization, as defined by Nakamura *et al.* (1996). Firms participating in 'unilateral' alliances exhibit divergence in their patent cross-citation behavior, meaning that participation in such alliances is consistent with a decision to buy (i.e., license), rather than develop the capabilities necessary to 'make' components of their technology portfolios.

---

[13] Other work with patent cross-citations (Mowery *et al.*, 1997) found that firms tend to team with alliance partners that have similar technological portfolios, but there are 'decreasing returns to similarity', i.e., we obtained a negative coefficient for the square of prealliance cross-citations in a model analyzing alliance partner choice. This tendency may apply to absorptive capacity as well: the effects of prealliance technological overlap on knowledge transfer among partner firms may be nonlinear, as opportunities for such transfer diminish at higher levels of similarity in the technological portfolios of alliance partners. We explored this possibility by adding *PRECRSS2* (the square of *PRECROSS*) to our model. Although we obtained the 'correct' (i.e., negative) coefficient on *PRECRSS2*, the coefficient failed all tests of statistical significance.

knowledge transfers achieved within *US-nonUS* alliances (compared with those involving only U.S. firms), we find that foreign firms (*FORGNCO*) do not display significantly lower levels of capabilities acquisition than U.S. firms. Furthermore, the results in Table 5 provide no support for the supposedly superior learning abilities of Japanese companies. Japanese companies (designated by the dummy variable *JAPANCO*) have, if anything, a lower ability to acquire capabilities from alliances with U.S. firms than do other U.S. firms' foreign alliance partners: the coefficient on *JAPANCO* is consistently negative, although it fails to pass conventional tests of statistical significance. Cross-national differences in the ability to acquire capabilities through alliances thus appear to be quite small, and if such a difference exists at all for Japan, it is precisely the opposite of that articulated in the 'received wisdom'.

CONCLUSION

Research on resource- and knowledge-based views of the firm, along with related work on interfirm alliances, has been hampered by the lack of measures of firm-specific capabilities. These difficulties have meant that discussion of the motives and effects of alliance activity has proceeded in a virtual empirical vacuum, and competing views of alliance activity have rarely been brought into sharp focus. This paper uses a novel technique for measuring change in firms' technological capabilities that allows us to track the effects of alliance activity on interfirm knowledge transfers and the transfer of technology-based capabilities from one partner to another.

Our empirical results support two propositions from the 'received wisdom' on interfirm knowledge transfers in alliances. We find that equity joint ventures appear to be more effective conduits for the transfer of complex capabilities than are contract-based alliances such as licensing agreements, consistent with Kogut (1988). Furthermore, lower levels of transfer occur in unilateral contracts than in bilateral nonequity arrangements. This result in turn provides support for the argument that the structure and content of alliances are jointly determined, and that alliances nearer the 'hierarchy' end of the 'market–hierarchy' continuum (Oxley, 1996) outperform alternatives in supporting interfirm learning, *ceteris paribus*.

The analysis also provides some support for the importance of 'absorptive capacity' in the acquisition of capabilities through alliances and bolsters the argument that experience in related technological areas is an important determinant of absorptive capacity (Cohen and Levinthal, 1990). A firm's ability to absorb capabilities from its alliance partner depends on the prealliance relationship between the two firms' patent portfolios, consistent with Cohen and Levinthal's characterization of absorptive capacity as a quality that is both firm-specific and path-dependent. Other measures of absorptive capacity perform poorly, however, and further research on this issue requires better measures of the structure and activities of individual alliances.

Our results also suggest the need for a richer conceptual framework in considering the effects of alliance activity on firm-specific knowledge and capabilities. Significant interfirm transfer of knowledge and technological capabilities occurs in only a subset of alliances, characterized by 'convergent development' (Nakamura *et al.*, 1996). The presence of some alliances in our sample in which the firms display 'divergent development', i.e., declining technological overlap, suggests that some alliances are vehicles for accessing rather than acquiring capabilities. The 'learning' that takes place within alliances thus appears to be more complex than most of the literature on this topic suggests, underlining the need for better definitions of learning in theoretical discussions of alliance activity and highlighting this as an area ripe for further study.

Finally, with respect to an issue that is widely cited in popular commentary on international alliances, we find little evidence that Japanese firms have siphoned off important technological capabilities from their U.S. alliance partners, thereby contributing to the 'hollowing out' of U.S. corporations. Perhaps reflecting their new status of technological leadership (or parity) in many industries, Japanese partners now offer their alliance partners valuable opportunities for learning that our data suggest have been exploited by U.S. firms. In one other respect, however, our results do tend to support the conventional wisdom: U.S. firms' alliances with non-U.S. firms seem to result in lower levels of interfirm knowledge transfer than those involving only U.S. companies. This finding is consistent with arguments made by Gulati (1996) and others about the obstacles to interfirm knowledge transfer created by distance, cultural differences, and other factors.

Despite a substantial literature on these topics, the empirical analysis of strategic alliances, firm-specific capabilities, and interfirm knowledge transfers continues to rely heavily on case studies and imperfect indicators of the underlying phenomena. The empirical analysis reported here is one of the first systematic tests of various assertions in this literature. In addition to shedding light on these specific issues, this study's measures of changing firm-specific technological capabilities have considerable promise for broader application to the analysis of firm strategy and technological innovation, and may provide a stronger empirical underpinning for this provocative conceptual literature.

ACKNOWLEDGEMENTS

We gratefully acknowledge the comments and suggestions of Bernard Yeung and the Guest Editors. Research for this paper was supported by the Alfred P. Sloan Foundation, the U.S. Air Force Office of Scientific Research, and the Canadian Institute for Advanced Research.

REFERENCES

Agmon, T. and M. von Glinow (eds.) (1991). *Technology Transfer in International Business*. Oxford University Press, Oxford, U.K.

Barney, J. B. (1986). 'Strategic factor markets: Expectations, luck, and business strategy', *Management Science*, **32** (10), pp. 1231–1241.

Cohen, W. M. and D. A. Levinthal (1990). 'Absorptive capacity: A new perspective on learning and innovation', *Administrative Science Quarterly*, **35**, pp. 128–152.

Conner, K. and C. K. Prahalad (1996). 'A resource-based theory of the firm: Knowledge versus opportunism', *Organization Science*, **7**(5), pp. 477–501.

Eisenhardt, K. M. and C. B. Schoonhoven (1996). 'Resource-based view of strategic alliance formation: Strategic and social effects in entrepreneurial firms', *Organization Science*, **7** (2), pp. 136–148.

Gambardella, A. (1992). 'Competitive advantages from in-house scientific research: The U.S. pharmaceutical industry in the 1980s', *Research Policy*, **21**, pp. 391–407.

Grant, R. M. and C. Baden-Fuller (1995). 'A knowledge-based theory of inter-firm collaboration', *Best Paper Proceedings 1995*, Academy of Management, pp. 17–21.

Griliches, Z. (1990). 'Patent statistics as economic indicators', *Journal of Economic Literature*, **28**, pp. 1661–1707.

Grindley, P. (1995). *Standards, Strategy and Policy: Cases and Stories*. Oxford University Press, New York.

Gulati, R. (1996). 'Social structure and alliance formation patterns: A longitudinal analysis', *Administrative Science Quarterly*, **40**, pp. 619–652.

Hagedoorn, J. (1993). 'Understanding the rationale of strategic technology partnering: Interorganizational modes of cooperation and sectoral differences', *Strategic Management Journal*, **14** (5), pp. 371–385.

Hagedoorn, J. and J. Schakenraad (1990). 'Inter-firm partnerships and co-operative strategies in core technologies'. In C. Freeman and L. Soete (eds.), *New Explorations in the Economics of Technical Change*. Frances Pinter, London, pp. 3–37.

Hamel, G. (1991). 'Competition for competence and inter-partner learning within international strategic alliances', *Strategic Management Journal*, Summer Special Issue, **12**, pp. 83–103.

Hamel, G., Y. Doz and C. K. Prahalad (1989). 'Collaborate with your competitors—and win', *Harvard Business Review*, **67** (1), pp. 133–139.

Harrigan, K. (1986). *Managing for Joint Venture Success*. Lexington Books, Lexington, MA.

Harrigan, K. (1988). 'Joint ventures and competitive strategy', *Strategic Management Journal*, **9** (2), pp. 141–158.

Khanna, T. (1996). 'Winner-take-all alliances', Harvard Business School Working Paper, No. 96-033.

Kogut, B. (1988). 'Joint ventures: Theoretical and empirical perspectives', *Strategic Management Journal*, **9** (4), pp. 319–332.

Lyles, M. A. (1988). 'Learning among joint venture sophisticated firms', *Management International Review*, **28**, pp. 85–98.

Mansfield, E. (30 September 1988). 'Industrial innovation in Japan and the United States', *Science*, pp. 1769–1774.

Mariti, P. and R. H. Smiley (1983). 'Co-operative agreements and the organization of industry', *Journal of Industrial Economics*, **31** (4), pp. 437–451.

Mody, A. (1993). 'Learning through alliances', *Journal of Economic Behavior and Organization*, **20**, pp. 151–170.

Montgomery, C. A. and S. Hariharan (1991). 'Diversified expansion by large established firms', *Journal of Economic Behavior and Organization*, **15**, pp. 71–89.

Mowery, D. C. (1983). 'The relationship between intrafirm and contractual forms of industrial research in American manufacturing, 1900–1940', *Explorations in Economic History*, **20**, pp. 351–374.

Mowery, D. C. (ed.) (1988). *International Collaborative Ventures in U.S. Manufacturing*. Ballinger, Cambridge, MA.

Mowery, D. C., J. E. Oxley and B. S. Silverman (1997).

'Technological overlap and interfirm cooperation: Implications for the resource-based view of the firm', *Research Policy*, forthcoming.

Nakamura, M., J. M. Shaver and B. Yeung (1996). 'An empirical investigation of joint venture dynamics: Evidence from US–Japan joint ventures', *International Journal of Industrial Organization*, **14**, pp. 521–541.

Oxley, J. E. (1996). 'Appropriability hazards and governance in strategic alliances: A transaction cost approach', University of Michigan working paper, #9602-23.

Patel, P. and K. L. R. Pavitt (1994). 'Technological competencies in the world's largest firms: Characteristics, constraints and scope for managerial choice', unpublished manuscript, Science Policy Research Unit, University of Sussex, Brighton, U.K.

Penrose, E. T. (1959). *The Theory of the Growth of the Firm*. Wiley, New York.

Pisano, G. P. (1990). 'The R&D boundaries of the firm: An empirical analysis', *Administrative Science Quarterly*, **35** (1), pp. 153–176.

Porter, M. E. and M. B. Fuller (1986). 'Coalitions and global strategy'. In M. E. Porter (ed.), *Competition in Global Industries*. Harvard Business School Press, Boston, MA, pp. 315–344.

Powell, W. and P. Brantley (1992). 'Competitive cooperation in biotechnology: Learning through networks?' In N. Nohria and R. Eccles (eds.), *Networks and Organizations: Structure, Form and Action*. Harvard Business School Press, Boston, MA, pp. 365–394.

Reich, R. B. and E. D. Mankin (1986). 'Joint ventures with Japan give away our future', *Harvard Business Review*, **64** (2), pp. 78–86.

Rosenberg, N. and C. Frischtak (1991). In N. Rosenberg and C. Frischtak (eds.), *International Technology Transfer: Concepts, Measures and Comparison*. Praeger, New York.

Shan, W. (1990). 'An empirical analysis of organizational strategies by entrepreneurial high-technology firms', *Strategic Management Journal*, **11** (2), pp. 129–139.

Silverman, B. S. (1996). 'Technical assets and the logic of corporate diversification', unpublished doctoral dissertation, Haas School of Business, University of California, Berkeley, CA.

Sobrero, M. and E. B. Roberts (1996). 'The trade-off between efficiency and learning in inter-organizational relationships', MIT Sloan School of Management working paper, WP#3896.

Teece, D. J., G. Pisano and A. Shuen (1997). 'Dynamic Capabilities and Strategic Management', *Strategic Management Journal*, **18**, forthcoming.

Teece, D. J. and G. Pisano (1994). 'The dynamic capabilities of firms: an introduction', *Industrial and Corporate Change*, **3**, pp. 537–556.

Wernerfelt, B. (1984). 'A resource-based view of the firm', *Strategic Management Journal*, **5** (2), pp. 171–180.

[17]

Strategic Management Journal
Strat. Mgmt. J., **23**: 817–833 (2002)
Published online 2 July 2002 in Wiley InterScience (www.interscience.wiley.com). DOI: 10.1002/smj.259

DO MAKE OR BUY DECISIONS MATTER? THE INFLUENCE OF ORGANIZATIONAL GOVERNANCE ON TECHNOLOGICAL PERFORMANCE

MICHAEL J. LEIBLEIN,[1]* JEFFREY J. REUER[1] and FRÉDÉRIC DALSACE[2]
[1] *Fisher College of Business, Ohio State University, Columbus, Ohio, U.S.A.*
[2] *HEC School of Management, Jouy en Josas, France*

This paper investigates how firms' decisions to outsource or internalize production affect their technological performance. While several popular arguments and some anecdotal evidence suggest a direct association between outsourcing and technological performance, the effects of firms' governance decisions are likely to be contingent upon several specific attributes underlying a given exchange. This paper first demonstrates how standard performance models can improperly suggest a positive relationship between firms' outsourcing decisions and their technological performance. Models that account for firm- and transaction-specific features are then presented, which indicate that neither outsourcing nor internalization per se result in superior performance; rather, a firm's technological performance is contingent upon the alignment between firms' governance decisions and the degree of contractual hazards.

INTRODUCTION

The decision to outsource or vertically integrate a value-chain activity represents one of the more complex choices facing a firm's managers. On the one hand, integration requires management to commit significant resources to a course of action, the effects of which may be costly to reverse, while forgoing numerous advantages associated with the marketplace. On the other hand, integration may be required for a firm to accumulate resources necessary to generate or maintain a competitive advantage (Dierickx and Cool, 1989). The complexity of this decision has been exacerbated in recent years by the increasing rapidity of technological change and the geographic and organizational dispersion of knowledge (Clark and Fujimoto, 1991; Teece, 1992).

A broad variety of outsourcing arrangements has emerged in response to these environmental changes. Whereas outsourcing used to be confined to fairly specialized, repetitive tasks such as facilities management or logistics, it has now spread to more visible and sensitive functions such as customer service, R&D, and manufacturing (Greco, 1997; Liebeskind *et al.*, 1996). Moreover, this trend is expected to continue for the foreseeable future. For instance, the outsourcing market for U.S.-based corporations is expected to grow more than threefold between 1996 and 2001, from \$100 to \$318 billion (Anderson, 1997).

Although there is widespread consensus on the importance of firms' vertical integration decisions, it remains unclear whether or how these boundary decisions affect various dimensions of firm performance. For example, it has been asserted that outsourcing capital-intensive production activities improves a firm's ability to respond flexibly to changes in technology or demand, to accumulate external knowledge, to avoid coordination inefficiencies, and to compress product

Key words: outsourcing; vertical integration; innovation
*Correspondence to: Michael J. Leiblein, Fisher College of Business, The Ohio State University, 700 Fisher Hall, 2100 Neil Ave., Columbus, OH 43210-1144, U.S.A.

Received 8 December 2000
Final revision received 28 March 2002

development cycle times (e.g., Clark and Fujimoto, 1991; D'Aveni and Illinitch, 1992; Harrigan, 1983; Helper and Sako, 1995; Nishiguchi, 1994; Womack, Jones, and Roos, 1990). However, outsourcing may also contribute to the hollowing of corporations, resulting in the depreciation of existing capabilities *vis-à-vis* actual or would-be competitors as well as in the failure to seize opportunities to develop new capabilities (e.g., Bettis, Bradley, and Hamel, 1992; Quinn and Hilmer, 1994; Reich and Mankin, 1986).

An intermediate position is that the effects of individual sourcing decisions are contingent upon features of the individual transaction and the contracting environment. In fact, a substantial body of influential research initiated in the 1970s has identified various factors that influence firms' decisions to select alternative governance structures, be they horizontal or vertical in nature (e.g., Klein, Crawford, and Alchian, 1978; Williamson, 1975). Recent extensions of this research have noted that internalization enhances coordinated adaptation across contiguous value chain activities, just as firms relying on weaker forms of governance can benefit instead from the marketplace's high-powered incentives (e.g., Chesbrough and Teece, 1996; Monteverde, 1995; Teece, 1996). In spite of the wealth of studies considering the factors influencing firms' governance choices, however, very little empirical research has gone on to examine the performance and competitive implications of firms' governance decisions in general, or their production sourcing strategies in particular (e.g., D'Aveni and Ravenscraft, 1994; Masten, 1993).

In response to the tension between the potential benefits and risks associated with outsourcing relative to vertical integration, as well as to redress the limited evidence on the actual consequences of firms' vertical boundary decisions, this study focuses on the performance effects of firms' production sourcing strategies. More specifically, we investigate the technological performance implications of production internalization versus outsourcing by firms in the global semiconductor industry. Our focus on a precise, disaggregated measure of firm performance as well as our modeling approach mitigate the potential confounding effects that impede the identification of organizational performance's antecedents (March and Sutton, 1997).

Beyond considering popular arguments concerning the absolute benefits or shortcomings of outsourcing and vertical integration, we also assess comparative perspectives on the performance implications of firms' vertical integration decisions. Specifically, we examine the fundamental 'discriminating alignment' tenet of transaction cost theory (Williamson, 1985), which focuses on the performance implications of the fit between firms' governance choices and a set of specific attributes of the transaction at hand (e.g., Anderson, 1988; Sampson, 2001; Silverman, Nickerson, and Freeman, 1997). The paper also recognizes that even after one accounts for the influence of the specific transactional features highlighted by transaction cost theory, other unobserved transactional or even firm-level attributes may shape firms' vertical boundary choices and their consequences. We explicitly account for the potential influence of unobserved heterogeneity by employing a series of two-stage, self-selection models. Relative to prior work examining the relationship between vertical integration and its antecedents or the relationship between vertical integration and its consequences, this two-stage design permits an integrative model that simultaneously captures firms' vertical integration and outsourcing decisions as well as both the observed and unobserved antecedents of these decisions and their performance implications. As a result, we are able to address a criticism of existing empirical research in transaction cost economics that focused on the effects of transactional characteristics on the choice of governance, rather than on the performance of alternative modes of organization (e.g., Masten, 1993).

The paper proceeds as follows: We first offer a brief review of the literature and present background theoretical material to motivate two research questions. The next section presents the research design, provides methodological details on self-selection, and describes the resulting two-stage models we employ in the analyses. This is followed by a presentation of results from a sample of over 700 decisions involving the production of semiconductor devices. These results imply that outsourcing positively influences firms' technological performance when one relies on standard multivariate analyses that do not address the endogeneity of firms' outsourcing and vertical integration decisions. However, models that account for self-selection indicate that such effects are driven by unobserved firm and transactional characteristics that lead firms to choose outsourcing over vertical integration. Moreover, the findings demonstrate that the alignment between firms' governance choices and the features of exchange identified by transaction cost theory affects firms' technological performance. For example, firms that outsource production activities in the face of various contractual hazards identified by transaction cost theory tend to experience technological performance penalties. A concluding section discusses the implications of these findings and identifies a number of promising avenues for future research that integrates firms' governance choices with the antecedents and consequences of those decisions. This discussion also serves to illustrate the broader applicability of our modeling approach in research on strategic choices and their performance implications.

BACKGROUND LITERATURE AND RESEARCH QUESTIONS

Vertical governance and technological performance

Although the strategy literature has paid a great deal of attention to the factors that influence the make or buy decision, relatively little empirical work has addressed the performance implications of these decisions (e.g., D'Aveni and Ravenscraft, 1994; Masten, 1993; Rumelt, 1974, 1982). This lack of attention is surprising given that the heterogeneity of firm performance represents one of the fundamental questions in the strategy field (Rumelt, Schendel, and Teece, 1994) and that the integration decision represents a large, difficult-to-reverse investment that directly affects firms' capabilities and property rights.

Although there exists little empirical work that has rigorously examined whether integration affects performance, numerous conceptual papers have described the potential advantages associated with outsourcing and vertical integration (see Mahoney, 1992, for a review). Outsourcing not only shifts costs from the focal firm to suppliers, but it may also create economic value within the supply chain, enabling a firm to tap into specialized capabilities (e.g., Dyer and Ouchi, 1993; Helper and Sako, 1995; Mowery, Oxley, and Silverman, 1996; Nishiguchi, 1994; Womack *et al.*, 1990) and to shorten product development cycles through concurrent development (e.g., Clark, Chew, and Fujimoto, 1987; Clark and Fujimoto, 1991). Teece (1992) argues that de-integrated corporate hierarchies avoid the decision making complexities typically associated with integrated concerns and are therefore better suited to competing in environments where the convergence of technologies from disparate sectors and time to market are of importance.

However, vertical integration may also enhance performance, chiefly because of the coordination benefits associated with internalization. While increased coordinative abilities may be due either to an improved flow of administrative information or to technological interdependencies, enhancements in electronic data exchange technologies suggest that the most significant coordinative advantages are based on technological interdependencies that exist between activity stages. This latter point is supported by research which suggests that locating manufacturing activities outside of the firm's boundaries can degrade a firm's capabilities by impairing cross-functional coordination (e.g., Allen, Lee, and Tushman, 1980; Chesbrough and Teece, 1996; Griffin and Hauser, 1992; Hatch and Mowery, 1998; Teece, 1996).

The opposing conclusions reached by these two groups of arguments are the result of different sets of underlying assumptions. Arguments emphasizing the benefits of outsourcing assume that many critical capabilities reside outside the boundaries of the firm and that appropriation problems are minimal. Such a situation is most likely in environments involving mature, autonomous technologies (Teece, 1984). In these settings, performance is directly tied to the ability to identify, isolate, and solve a specific set of independent technical problems. Under these conditions, contracting hazards are minimal, specialization is favored, and outsourcing will enhance performance by allowing specialized suppliers to benefit from scale economies by pooling demand and from learning economies by focusing on a limited number of well-defined activities. Recent empirical results are consistent with these arguments (e.g., Liebeskind *et al.*, 1996).

In contrast, arguments extolling the benefits of integration assume that the most valuable capabilities reside within the firm and that appropriation problems are significant. As a principle advantage of internal organization is the ability to provide enhanced information transfer and coordination across activities within the production system, this argument implies that performance is principally driven by second- and higher-order interactions among a set of value-chain activities. While the adoption of an architectural innovation may diminish the value of established information filters and communication channels (e.g., Henderson and Clark, 1990), integration is likely to remain a superior form of governance for managing technologies that require a great deal of coordinated adjustment within the production system (e.g., Chesbrough and Teece, 1996; Langlois, 1992; Teece, 1996).

Unobserved attributes, vertical governance, and technical performance

The preceding section suggests that the value of employing a particular organizational form is strongly dependent upon crucial assumptions regarding the nature of underlying resources, supplementary attributes of the focal firm, and other industry conditions. However, absent a theory of resource value, these arguments will be necessarily incomplete. While existing anecdotal (e.g., Kotabe and Murray, 2001) and some large-scale empirical evidence (e.g., D'Aveni and Ravenscraft, 1994) is suggestive of a direct relationship between vertical integration decisions and performance, this direct comparison is appropriate only if firms' governance choices are not influenced by other firm- or transaction-level characteristics.

The proposition that firms' governance choices are not influenced by various firm- and transaction-level characteristics, however, is untenable. Simple comparisons of integration decisions across firms facing similar environments—such as General Motors and Chrysler in the automobile industry or IBM and Compaq in the personal computer industry—suggest that firms differ dramatically in their sourcing decisions. For instance, while IBM is highly integrated, Compaq outsources a number of its component assembly processes. A number of conceptual articles (e.g., Quinn and Hilmer, 1994; Barney, 1999) and a limited body of case study research (e.g., Argyres, 1996) echo this point, arguing that relative firm-level capabilities affect firms' governance decisions.

It is further likely that many of the firm- and transaction-specific characteristics alluded to above also influence firms' technological performance. For instance, in a series of influential papers, Henderson and Clark (1990) and Henderson (1993) have argued that firms' ability to successfully adopt new technologies hinges on the match between the architectural requirements of the new technology and firm-specific information filters and communication channels. Recent evidence that differences in certain HR practices influence the speed and quality with which semiconductor production facilities are able to implement new process technologies paper suggests a similar conclusion (Macher and Mowery, 2001).

If there are unobserved attributes that influence both governance choice and performance, then a self-selection bias will be present, and normative implications drawn from these analyses may be incorrect (e.g., Heckman, 1979; Maddala, 1983; Masten, 1993; Masten, Meehan, and Snyder, 1991). The presence of unobserved heterogeneity in performance models potentially biases parameters for indicator variables used to measure the influence of a firms' governance choices since firms select governance forms based on their own maximizing analyses rather than on a random basis. Thus, in standard ordinary least squares models, the effects of these decisions can be confounded by unobserved factors that prompted the firm to choose one vertical integration approach over another in the first place. This leads us to state the following research question:

Research Question 1: Do unobserved attributes underlying firms' vertical integration decisions influence the governance-technological performance relationship?

Governance misalignment and technical performance

While any number of unobserved transaction-, firm-, or industry-level characteristics may give rise to a self-selection problem, a theoretical discussion of firms' boundary choices and their implications requires researchers to identify these factors and their causal relationships. Transaction cost economics (TCE) helps in this effort by elucidating specific transaction-level characteristics that influence the efficiency of alternative forms of governance. More specifically, a fundamental tenet of transaction cost theory, the notion of discriminating alignment, is that efficiency will be enhanced when a fit exists between the chosen governance arrangement and the underlying attributes of that transaction and the broader contracting environment (e.g., Williamson, 1985, 1991). For example, TCE holds that vertical integration is beneficial under small numbers situations in aligning the interests of exchange parties and providing administrative rules and procedures to adjudicate differences, thereby facilitating sequential adaptation (Monteverde and Teece, 1982; Walker and Weber, 1984; Williamson, 1975). However, these benefits must be weighed against the greater administrative costs associated with internal governance.

Based on this line of reasoning, it follows that the performance implications of outsourcing versus vertical integration should hinge upon the alignment between the chosen governance structure and the attributes of the transaction and its contracting environment. For example, for complex exchange relationships involving transaction-specific investments, a firm that selects a simple governance structure lacking adequate safeguards will be exposed to moral hazard and hold-up risks. By contrast, the potential consequences of adopting an excessively complex governance structure for a simpler relationship include a loss in flexibility and decision-making speed due to the imposition of bureaucratic controls (Williamson, 1985, 1991). The bureaucratic decision making processes implied by hierarchy may also dull incentives for innovation (Chesbrough and Teece, 1996; Teece, 1996). Thus, in contrast to contentions that governance has an absolute effect on performance, transaction cost theory proposes that the fit between a governance arrangement and the transaction's attributes influence the efficiency of these intra- and interorganizational exchanges. Given these theoretical arguments and the paucity of research on the alignment-performance relationship (e.g., Anderson, 1988; Silverman *et al.*, 1997), we address the following research question:

Research Question 2: Does the fit between firms' vertical governance decisions and relevant transactional attributes highlighted by TCE influence technological performance?

METHODOLOGY

Data

The empirical setting for this paper is the production of semiconductor devices. Such transactions are interesting to focus on for several reasons. First, production capabilities are of tremendous importance in this industry. The ability to produce semiconductor devices using advanced processes is directly tied to both improved end-product performance and reduced cost. As a result, there is a strong incentive for all industry participants to incorporate advanced processes into their products. Second, while many semiconductor firms remain integrated into both design and production, a number of firms have focused their operations solely on design and have outsourced their production needs. Thus, despite the importance of production activities, there is no single, universally accepted method for the organization of these transactions. Finally, even in instances where semiconductor production is internalized within the firm, production usually takes place in a location that is physically separated from design and other upstream activities. Consequently, in this industry, the potential influence of geographic proximity does not confound the effects of firms' governance decisions on performance. These considerations suggest that the semiconductor industry provides an ideal setting in which to examine the relationship between organizational governance and a set of competitively important technical capabilities.

The data for this paper were derived from a 1997 report of firm-level activity within the semiconductor industry published by the Integrated Circuits Engineering Corporation (ICE). This document summarizes the responses to surveys conducted in 1996 regarding the corporate, production, and alliance activities of 176 global integrated circuit manufacturers. The portion of the report utilized in this paper described 192 instances where these firms managed production externally through sourcing (i.e., buy) agreements and 367 instances where these firms managed production in wholly-owned production facilities. Preliminary data analysis indicated that the necessary product-market and process-technology information was available for 174 of the 192 (90.6%) sourcing agreements and 220 of the 367 (60.0%) internal production lines in the sample. Since a firm may access production for multiple products and processes through a single sourcing agreement or fabrication line, we record a separate observation for each product market-process technology combination.[1] For example, a firm may sell memory devices using 1.0 micron technology, memory devices using a 0.8 micron technology, and application-specific integrated circuits (ASICs) using a 1.0 micron technology. These product-market, process-technology combinations would be recorded in our sample as three separate observations. Upon disaggregation to the product market-process technology level of analysis, our final sample consisted of 714 production decisions.

Measures

Technological performance

The measure of technological performance used in this paper is defined as a function of the focal transactions transistor density provided by the focal transaction. Transistor density is the fundamental driver of both cost and product performance in the industry and represents a universally valued success factor (e.g., Hazewindus and Tooker, 1982; Gruber, 1994). Prior research has measured technological performance as a function of a primary indicator of transistor density, the line width, or feature size, at which information is etched onto a semiconductor circuit (e.g., Eisenhardt and Schoonhoven, 1996; West and Iansiti, 1998). Since the average feature size varies across product-market applications, we transformed the raw feature size value by standardizing it with respect to the mean and standard deviation for all other observations within the same product-market subfield. Further, since smaller feature size implies better technological performance, we take the negative of this standardized value (i.e., technological performance $= -(x - \bar{x}_j)/(s_j)$, where x is the focal transaction's feature size, $\bar{x}_j$ is the average feature size in product-market j, and s_j is the standard deviation of feature sizes in this product-market).[2]

Thus, a positive value for technological performance implies a company is using a technology that is more advanced than the average technology used by its immediate competitors. In contrast, a negative value indicates that substandard technology is being utilized. The magnitude of the measure indicates the degree to which the technology is ahead or behind the technology curve for its product-market subfield. Data on firms' technological performance in 1996 were obtained from the 1997 edition of Integrated Circuit Engineering's *Profiles of IC Manufacturers and Suppliers*.

Explanatory variables

In order to examine the proposition that technological performance differs across transactions over which different levels of managerial control are maintained, we define the variable *Governance Choice (Buy)* to be equal to one for transactions where a firm engages in a contractual relationship for its production activities, and equal to zero for transactions where a firm maintains full ownership over its production activities.

The primary argument put forth in this paper is that technological performance is influenced by the degree to which each particular transaction is appropriately governed. To test this proposition, we develop a measure, *Governance Misfit*, that captures the probability that another governance form is more appropriate given the contractual hazards surrounding the exchange. In developing this measure, we follow Anderson (1988) and Silverman *et al.* (1997) in estimating the likelihood that a particular governance form will be chosen as a function of transaction-level attributes derived from TCE. Specifically, we estimate the most likely value for *Governance Choice (Buy)* for a given transaction, using the following probit model:

$$\text{Prob}(Y_i = 1) = \Phi(\beta' X_i) \qquad (1)$$

where Y_i is the governance choice variable for the ith observation, X_i is a vector of characteristics surrounding the transaction that determine the contractual hazards, β is a vector of estimated coefficients for these characteristics, and $\Phi(\cdot)$ is the standard normal cdf. The degree of *Governance*

[1] Following industry practice, we distinguish seven product markets: analog devices, application specific integrated circuits (ASICs), discrete devices, digital signal processors (DSPs), memory devices, microprocessors, and telecommunications devices. The primary process technologies in use as of 1996 included 1 micron, 0.8 micron, 0.5 micron, 0.35 micron, and 0.25 micron technology.

[2] Although smaller semiconductor feature sizes support superior performance, to ease interpretation of our results, we reversed the sign of our technological performance measure so that positive values indicate supernormal performance, and negative values indicate substandard performance.

Misfit is then defined as $1-\Phi(\beta' X_i)$ when *Governance Choice (Buy)* is equal to one (i.e., when a transaction is outsourced) and as $\Phi(\beta' X_i)$ when *Governance Choice (Buy)* is equal to zero (i.e., when a transaction employs internal governance). Thus, governance misfit measures the probability that too much governance is employed for transactions that are internally governed and the probability that too little governance is employed for transactions that are outsourced.

The vector of characteristics used to determine the level of contracting hazards in the first stage probit model includes transaction-level measures for the *ex ante* number of suppliers, the specificity of the assets exchanged, and the uncertainty of product demand, as well as controls for firm industry tenure and firm size. The measure of *ex ante* number of suppliers accounts for the effect that shifts in bargaining power due to the number of available suppliers will have on firms' make or buy decisions (e.g., Pisano, 1990). This measure is obtained by counting the number of firms that supplied production for the relevant process technology during our sample time frame.[3] Following Levy (1985), the measure of product market demand uncertainty captures unanticipated shifts in the demand for a specific type of semiconductor device by calculating the variance surrounding a forecasted trend in demand for products within the relevant product-market subfield. Specifically, demand uncertainty is measured as the sum of squared errors from a regression of the relevant product-market's historical demand rates for the years 1990–96.

While a direct measure of asset specificity was not available for this research, we were able to utilize a previously developed industry-specific construct that captures the conditions likely to give rise to high asset specificity. The logic supporting this measure is based on cross-product-market differences in the need for coordination and mutual adaptation across value chain activities. These differences account for factors such as the number of design attributes that need to be specified, the reliability with which those attributes can be measured, and the degree to which these attributes influence the performance of other elements of the product (e.g., Chesbrough and Teece, 1996). Such adaptation is, in turn, facilitated by the development of specialized resources (e.g., reporting structures, information systems, and common language) that promote technical dialogue (Henderson and Clark, 1990; Henderson, 1993; Monteverde, 1995). In the semiconductor industry, coordination requirements are thought to be particularly heavy in analog and leading-edge memory product markets.[4] Consequently, we identify high asset specificity conditions as those where the exchange involves either analog or memory products (asset specificity = 1). Asset specificity is coded as zero for all other transactions.

Control variables

Although our interest is in developing a parsimonious model of the relationship between the governance mechanisms used to manage production activities and technological performance, other firm- and transaction-specific characteristics may influence our results. For instance, larger firms may face a degree of institutional insulation and bureaucratization that decreases their responsiveness to shifting industry conditions (Haveman, 1993). In contrast, larger organizations often have superior financial and human resource endowments that enable them to internalize production and invest in state-of-the-art production equipment. Large firms may also have greater market power or positional advantages compared to their smaller rivals that affect their incentives and abilities to adapt to changing environments (e.g., Delacroix and Swaminathan, 1991; Baum and Oliver, 1991; Baum and Mezias, 1992). Similarly, firm age may influence both a firm's performance and its choice of a particular governance form. For instance, the liability of newness concept suggests that older firms benefit from accumulated experience (e.g., Amburgey, Kelly, and Barnett, 1993; Carroll and Delacroix, 1982), but others have countered that firms become increasingly ossified as they age (e.g., Barnett, 1990; Ranger-Moore, 1997).

In order to ensure that our estimated relationships are not unduly influenced by such factors,

[3] A squared term is included to improve model fit.

[4] Although the production of both analog and digital memory products requires heavy transaction specific investment, the underlying cause of the specific investment varies between the two markets. While the production of analog devices may not require state-of-the art production technology, heavy customization requirements necessitate a high degree of adaptation across design and production activities. The production of digital memory devices, in contrast, generally requires that design and production engineers develop state-of-the-art processes (Monteverde, 1995).

all models include a series of control variables. We defined *Firm Size* as the log of the firm's average domestic sales for the period 1994 through 1996. *Firm Tenure* was measured as the number of years that had passed since the firm was first founded or, for conglomerates, the number of years since the firm first began selling semiconductors. We accounted for differences in sourcing strategy that may simultaneously influence governance form and performance through two additional controls. The variable, *Governance Duration*, is defined as the number of years that the focal sourcing agreement or production line has been active. This control captures effects due to the depth of external relationships or internal experiences. Similarly, the variable, *Buy Portfolio*, accounts for differences attributable to the breadth of a firms' sourcing network and is defined as the number of contemporaneous production sourcing transactions managed by the firm.

Descriptive statistics

Table 1 provides summary and zero-order correlation statistics for the entire sample. While the positive correlation between technological performance and the use of buy transactions is consistent with the popular contention that outsourcing provides the flexibility to access superior technical capabilities, the large and significant intercorrelations between technological performance, firms' governance choices, and a number of the control variables indicate the need for multivariate analyses to partial out the influence of each variable on technological performance.

Technological performance model specification

As noted above, a complication arises when analyzing the influence of governance form on performance in that firms are able to willfully choose the form of governance they use to manage a particular transaction. Since firms are able to self-select the observed form of governance based on their own performance maximizing analyses, it is likely that the observed level of technological performance is conditional upon unobserved factors that influence firms' governance choices. To correct for these potential biases, we employed a two-stage technique derived from work in labor economics (e.g., Heckman, 1978, 1979; Lee, 1982, 1983; Lee, Maddala, and Trost, 1980). This modeling approach can be summarized as reestimating regression coefficients by introducing an adjustment term, called the inverse Mills ratio, to the performance model.

The starting point for this procedure lies is the realization that the regression model is misspecified if a selection process underlies the sampling procedure. Heckman's original examples included the study of market wages for women (Heckman 1976, 1979). He recognized that the sample of working women is likely not random since the decision to participate in the labor force is an outcome of unobserved factors (e.g., the presence of small children in the family). In the context of semiconductor production, a similar problem arises because firms may select to outsource or vertically integration production of semiconductor devices based on unobserved characteristics associated with the firm or transaction. Failure to address this selection process leads to the possibility that performance differentials may be incorrectly attributed to firms' governance decisions, when they are in fact due to unobserved factors associated with the governance decision.

Heckman (1976, 1979) proposed a way to deal with this self-selection problem. In the present study, the intuition is that the estimates in the performance model need to be corrected by controlling for the propensity of the firm to choose a particular form of governance. Technically, one first estimates a first-stage probit model to specify a selection equation (Equation 1) and then calculates the inverse Mills ratio, which is used as a control variable in the second-stage performance model. The inverse Mills ratio, λ_{ji}, is given as $\lambda_{1i} = (\phi(\beta' X_i))/(\Phi(\beta' X_i))$ for observations that are outsourced (i.e., $j = 1$) and $\lambda_{0i} = -\phi(\beta' X_i)/([1 - \Phi(\beta' X_i)])$ for observations that are produced in-house (i.e., $j = 0$). In both cases, ϕ $(\cdot)$ is the standard normal pdf, and $\Phi(\cdot)$ is the standard normal cdf. Second-stage models that incorporate this correction provide consistent and unbiased estimates (e.g., Maddala, 1983; Greene, 1997). An existing application of this technique in the management literature can be found in Shaver (1998).

RESULTS

First-stage governance choice estimates

Table 2 presents the results from three potential first-stage governance choice models. The probit

Table 1. Descriptive statistics and correlation matrix for performance model[a]

Variable	Mean	S.D.	1	2	3	4	5	6	7	8
1. Technological Performance	0.000	0.996								
2. Firm Industry Tenure	22.941	13.720	−0.068†							
3. Firm Size	5.968	2.818	−0.008	0.398***						
4. Governance Duration	8.492	9.102	−0.243***	0.381***	0.255***					
5. Buy Portfolio	2.499	3.458	0.228***	−0.318***	−0.276***	−0.386***				
6. Asset Specificity	0.458	0.499	0.000	−0.100**	−0.045	−0.147***	0.085*			
7. Governance Choice (Buy)	0.335	0.472	0.289***	−0.403***	−0.409***	−0.475***	0.750***	0.122**		
8. Governance Misfit	0.455	0.138	0.367***	−0.348***	−0.324**	−0.457***	0.633***	0.112**	0.843***	
9. Correction for Self-Selection (λ)	1.443	0.938	−0.507***	0.627***	0.553***	0.516***	−0.481***	−0.225***	−0.604***	−0.669***

[a] $N = 714$. $^{\dagger}p < 0.10$; $^{*}p < 0.05$; $^{**}p < 0.01$; $^{***}p < 0.001$

Table 2. Probit estimates for first-stage governance choice model[a,b]

Independent variables	Model I	Model II	Model III
Intercept	1.175***	0.160	−0.236
	(0.136)	(0.441)	(0.465)
Firm Tenure	−0.035***	−0.028***	−0.031***
	(0.005)	(0.005)	(0.005)
Firm Size	−0.153***	−0.175***	−0.163***
	(0.021)	(0.024)	(0.025)
Ex Ante Number of Suppliers		0.223***	0.220***
		(0.047)	(0.047)
Ex Ante Number of Suppliers Squared		−0.010***	−0.010***
		(0.001)	(0.001)
Demand Uncertainty		0.088	0.465*
		(0.126)	(0.187)
Asset Specificity		0.271*	1.264***
		(0.134)	(0.381)
Demand Uncertainty Asset Specificity			−0.728**
			(0.261)
N	714	714	714
Log likelihood	−361.12	−279.88	−275.93
$-2[L(c) - L(\beta)]$[c]	188.08(2)***	350.56(6)***	358.46(7)***
$-2[L(\text{Model I}) - L(\beta)]$		162.48(4)***	170.38(5)***
$-2[L(\text{Model II}) - L(\text{Model III})]$			7.90(1)**

[a] Positive coefficients indicate a greater probability of external governance (i.e., buy).
[b] †$p < 0.10$; *$p < 0.05$; **$p < 0.01$; ***$p < 0.001$
[c] Log likelihood for null model was −455.161. Appropriate degrees of freedom are reported in parentheses.

model utilized in this first step is fit using all observations in the sample and differentiates firms that outsource from those that internalize production. Model I presents a baseline model that consists of an intercept term and measures of firm tenure and size. Models II and III introduce measures derived from transaction cost theory that are widely held to affect governance choice. In particular, Model II introduces measures of the *ex ante* number of available suppliers, the level of product demand uncertainty, and the level of specificity in the assets used in production. Since uncertain environments may lead to contractual renegotiation that can be hazardous in the presence of specific investments, Model III introduces an interaction term between demand uncertainty and asset specificity.

The results from the first-stage governance choice models are largely consistent with expectations and results published in the literature. The negative coefficients associated with the measures for firm tenure and firm size are compatible with observations regarding the production decisions of firms within the semiconductor industry. The advent of specialized manufacturing firms (i.e., foundries) has allowed recent entrants to the industry to utilize vertical sourcing agreements for their production needs (e.g., Macher, 2001). Given the increasing importance of foundries as well as the observation that the construction costs of semiconductor fabrication plants often reach well over $1 billion, it is not surprising to find that smaller, potentially resource-constrained, firms tend to utilize vertical sourcing relationships. Consistent with previous academic research, the significant positive coefficient associated with the *ex ante* number of suppliers indicates that firms in the semiconductor industry internalize production when small numbers bargaining problems are severe and outsource as the number of available suppliers increases (e.g., Pisano, 1990). At first glance, the positive coefficients associated with the measure for asset specificity might seem surprising. However, the negative interaction effect observed in Model III is consistent with the basic predictions of transaction cost economics (Coles and Hesterly, 1998; Mahoney, 1992). Specifically, firms internalize transactions when it is likely that they will need to renegotiate supplier contracts under conditions of high asset specificity.

The robustness of the estimated coefficients across model specifications suggests that the transaction- and firm-level effects included in our models uniquely influence firms' make or buy decisions for our sampled transactions. Given the stability of the estimated coefficients across specifications as well as the results of the log-likelihood ratio tests presented at the bottom of Table 2, the results obtained in Model III are used in the formulation of the inverse Mills ratio for the multivariate regression estimates presented for the second-stage performance model.

Second-stage performance estimates

Table 3 provides the results of our performance models. Model I presents a baseline specification that consists of an intercept term, our control variables, and a dichotomous variable for the direct effect of governance. While Model I does not control for self-selection, and is therefore likely to be misspecified based on our arguments, we present these results to demonstrate how the correction for self-selection affects the coefficient for governance choice. Model II adds the correction for self-selection (λ). However, the estimated coefficients in Model II remain restricted to be equivalent across make and buy transactions. Model III addresses this restriction by providing unrestricted estimates for each of the covariates across transactions utilizing internal production ('Make' models) and vertical outsourcing agreements ('Buy' models). Model IV introduces the measure for governance misfit and provides separate estimates for each of the different modes of governance choice. As expected, F-tests of the null hypothesis that all of the coefficients are jointly zero are strongly rejected in all models ($p < 0.001$). The estimated coefficients for lambda are also negative and strongly significant in all models ($p < 0.001$), indicating that firms have self-selected the most favorable form of governance.

The significant intercorrelations between a number of the variables in our models, and in particular between governance misfit and the self-selection correction (i.e., λ), led us to investigate potential multicollinearity problems through variance inflation factors (VIFs), conditioning indices, and variance decomposition proportions. The maximum VIF obtained in the four models is 5.32, which is substantially below the rule-of-thumb cut-off of 10 for multiple regression models (Neter, Wasserman, and Kutner, 1985: 392). The maximum conditioning indices for these models was 34.15. The variance decomposition proportions associated with this index indicated that the underlying dependencies involved the intercept term

Table 3. Estimates for second-stage technological performance models[a]

Independent variables	Model I	Model II	Model III		Model IV	
			Make	Buy	Make	Buy
Intercept	0.175	0.140	−0.149	0.413**	−0.049	1.625***
	(0.129)	(0.101)	(0.115)	(0.141)	(0.372)	(0.406)
Firm Industry Tenure	0.004	0.028***	0.028***	0.026***	0.028***	0.021***
	(0.003)	(0.003)	(0.003)	(0.006)	(0.003)	(0.006)
Firm Size	0.045**	0.131***	0.146***	0.113***	0.148***	0.088***
	(0.014)	(0.012)	(0.015)	(0.023)	(0.017)	(0.024)
Governance Duration	−0.018***	0.002	0.001	−0.010	0.001	−0.010
	(0.005)	(0.004)	(0.004)	(0.018)	(0.004)	(0.018)
Buy Portfolio	0.002	−0.014	−0.000	−0.025+	0.001	−0.028*
	(0.015)	(0.012)	(0.023)	(0.013)	(0.023)	(0.013)
Governance Choice (Buy)	0.597***	0.149				
	(0.124)	(0.100)				
Governance Misfit					−0.247	−1.908**
					(0.870)	(0.600)
Correction for Self-selection (λ)		−0.998***	−0.984***	−1.118***	−1.004***	−0.838***
		(0.048)	(0.053)	(0.143)	(0.087)	(0.166)
N	714	714	475	239	475	239
Model F	18.86***	97.67***	80.00***	12.79***	66.55***	12.75***
Adjusted R^2	0.111	0.449	0.455	0.199	0.454	0.229

[a] $p < 0.10$; * $p < 0.05$; ** $p < 0.01$; *** $p < 0.001$

and the misalignment variable. While the maximum value moderately exceeds the value of 30 used to identify underlying near dependencies with borderline tightness (Belsley, Kuh, and Welsch, 1980: 112), it is well-below the cutoff value of 100 used to identify substantial variance inflation and 'potential harm to regression estimates' (Belsley *et al.*, 1980: 153). Thus, we conclude that the regression estimates presented in Table 3 are not biased by the presence of multicollinearity.[5,6]

The highly significant and positive coefficient on governance choice presented in Model I suggests that semiconductor production transactions managed through outsourced relationships exhibit superior technological performance characteristics as compared to those managed internally by vertically integrated concerns. Model I also suggests that the performance of all transactions diminishes over time. These findings are consistent with the popular argument that outsourcing provides access to suppliers' specialized capabilities.

The results presented in Model II stand in stark contrast to those presented in Model I. Model II provides no evidence that governance choice influences technological performance. Moreover, the highly significant coefficient for the self-selection correction term (λ) in Model II indicates that unobserved characteristics underlying a decision to make or buy strongly influence the technological performance of that transaction relative to equivalent transactions where the alternative governance form was chosen. The negative parameter estimate for the self-selection term further indicates that the greater the firm's propensity to buy based on its unobserved characteristics, the higher its technological performance (see Dolton and Makepeace, 1987, for details on the interpretation of the self-selection coefficient term). These results suggest that the influence of governance choice on technological performance observed in Model I is largely driven by a self-selection process and that firms' governance choices for production in this industry are appropriately treated as endogenous.

While the specification presented in Model II facilitates comparison with Model I, it is limited to the extent that the estimated coefficients are restricted to be equivalent across the two governance choices. Model III relaxes this constraint by separately estimating models for 'make' and 'buy' transactions in the presence of the self-selection correction.

Model III also provides the information necessary to assess whether, on average, firms that chose to 'make' perform better than if they had chosen to 'buy', and whether firms that chose to 'buy' perform better than if they had chosen to 'make.' The average expected technological performance for each of these four contingencies is obtained by multiplying the coefficient estimates in Model III by the vector of firm attributes for the relevant sub-sample of make or buy transactions. The results of these calculations indicate that the expected technological performance for the 475 transactions that were conducted within a firm is -0.204 and the technological performance that one would predict if these same transactions were outsourced is -0.299. Similarly, while the expected technological performance for the 239 transactions that were outsourced is $+0.405$, the technological performance that one would predict if these transactions were internalized would be only $+0.281$. These results indicate that firms in this sample that internalized their production decisions performed better than they would have been expected to if they outsourced these transactions, and that firms in our sample that chose to outsource their production activities perform better than if they had chosen to internalize these activities.

An alternative, but related, means of testing the discriminating alignment proposition is to

[5] Even if the correlations between the variables in our models included an underlying near dependency, this would not necessarily imply that our estimated parameters are biased. As stated by Belsley *et al.* (1980: 115) 'Near dependencies among the variables may degrade parameter estimates (e.g., inflate standard errors) without necessarily harming them for the purposes of hypothesis testing.' Thus, while the resulting confidence intervals obtained under these conditions may be wider, the parameter estimates are not biased. Given that the examined effect is statistically significant and in the hypothesized direction across models that contain relevant combinations of the collinear variables, we conclude that our estimates have not been biased by excessive collinearity.

[6] A second potential complication in the analysis is the nonindependence of the observations. The models reported in this paper assume that each of the observations in our sample are equally weighted and independent. However, a few firms appeared multiple times in our sample. While the transactions employed by these firms were certainly not identical, we suspected that there might have been common elements across the transactions involving each firm. In the regression analyses we addressed this concern by calculating heteroskedastic-corrected standard errors. Such an approach provides a robust estimator for clustered data where observations are assumed to be independent across firms, but not independent within firms. The results using these alternative methods are consistent both in magnitude and significance with those reported in this paper.

examine whether the degree of governance misfit directly influences firms' technological performance. Incorporating a measure of governance misfit directly into the analysis provides a test regarding the performance of firms that 'bought', but had good reasons to 'make' and firms that 'made', but had good reasons to 'buy.' The results in Model IV indicate that governance misfit does have a negative impact on technological performance for buy transactions ($p < 0.01$); however, the corresponding parameter estimate does not reach significance in the 'make' performance model. These findings indicate that firms' technological performance is reduced by governance underfit—when the firm fails to implement adequate safeguards to offset hazards in the contracting environment. While the excessive bureaucracy associated with governance overfit may penalize other dimensions of performance (e.g., excessive administrative costs), our models provide no evidence that excessive governance reduces technological performance.

Finally, the findings associated with our control variables deserve comment. For instance, after controlling for governance selection, both firm industry tenure and firm size are associated with greater technological performance of both make and buy transactions (both $p < 0.001$). No evidence is found to suggest that the expected level of technological performance varies with the duration of either type of transaction. There is evidence of a weak negative relationship between the size of a firm's sourcing portfolio and technological performance, suggesting that technological performance is enhanced when firms limit the breadth of their sourcing transactions.

DISCUSSION

The field of strategic management is interested in describing why firms differ in their investment choices and subsequent performance (e.g., Rumelt *et al.*, 1994). Indeed, much of the work in the field can be categorized into studies that have examined factors that influence one element of strategic choice, organizational governance, and those that have examined how firms' choices affect performance. Much of the research in the first group estimates governance choice models rather than examining performance directly. Normative implications are often drawn from such models under the implicit assumption of the presence of a selection environment that ensures that observed governance choices are efficient. While the second group of studies examines the performance implications of strategic decisions, a complication commonly arises when such decisions are closely linked to underlying firm, transaction, and environmental attributes, and these decisions are made based on firms' performance expectations. In such situations, correlation or ordinary regression analyses are likely to produce misleading empirical results.

In order to obtain unbiased results in these settings, empirical models must simultaneously address firms' governance choices as well as their drivers and consequences. We believe that the two-stage modeling approach used in this study has wide applicability in the field of strategy research. This is nicely illustrated in a recent study on modes of foreign market entry conducted by Shaver (1998), which demonstrated how self-selection bias could influence normative implications drawn from studies on the performance implications of firms' foreign market entry decisions. More generally, the procedure has the benefit of allowing researchers to build integrated models of firms' strategic choices as well as the drivers and performance implications of those choices (cf. March and Sutton, 1997). It therefore has potential applications in any research design that examines the implications of managerial choice. At the level of business strategy, notable topics include the influence of first-mover effects, competitive engagement, or strategic group decisions. At the corporate level, this technique could be applied to diverse topics such as market entry decisions, diversification strategies, or the design of governance structures.

This study examines the relationship between governance choice and technological performance while controlling for selection biases. Two main findings are presented. First, in contrast to popular arguments suggesting that vertical integration or outsourcing will lead to superior technological performance, this study shows that governance decisions per se do not significantly influence technological performance directly. Rather, observed differences in the performance of transactions governed by different organizational forms are driven by factors underlying governance choice. While the increasing rapidity of technological change and the increasing dispersion of knowledge suggest an increased role for outsourcing in the economy (e.g., Teece, 1992), the relationship between governance choice and performance is dependent on the distribution of relevant capabilities and the degree to which performance is driven by autonomous or systemic innovation (Chesbrough and Teece, 1996; Teece, 1996). The results presented in this study caution against drawing universalistic normative implications for firms deciding on whether or not to internalize or outsource their value-chain activities and point to the value of contingency-based theoretical approaches. They also suggest that, absent a fundamental change in transactional characteristics in this industry, outsourcing may well continue but will not completely displace vertical integration.

Second, following TCE's fundamental tenet of discriminating alignment, we find that deviation from the optimal form of governance, as dictated by transactional attributes associated with various contracting hazards, may have a detrimental effect on performance. Specifically, we find that one element of performance, technological performance, is diminished when contractual safeguards are inadequate for the hazards present in a given exchange. These hazards can arise due to hold-up problems in market-mediated exchange (e.g., Klein, Crawford, and Alchian, 1978) as well as the need for coordination across value-chain activities when innovation is driven by system-level needs rather than performance at the component level (Chesbrough and Teece, 1996; Monteverde, 1995; Teece, 1996). While other elements of performance, such as administrative efficiency, may be adversely affected by the presence of excessive governance, this finding contests the notion that governance under-fit and governance over-fit symmetrically affect performance.

The paper also offers some interesting implications for future research. While the significance of our measure of governance misfit supports TCE-based arguments regarding the influence of transactional characteristics on both governance choice and performance, the continued significance of the self-selection term suggests that other factors that influence governance choice are also influencing technological performance. This is in line with Williamson's recent arguments which suggest that the optimal form of governance is likely to be contingent on both attributes of the transaction and the preexisting strengths and weaknesses of the focal firm (Williamson, 1998: 48; see also Williamson, 1999). Thus, future work may contribute to this stream of research by identifying some of these unobserved transaction- and firm-level attributes, possibly employing insights from other theoretical perspectives such as the resource-based view of the firm or real option theory. For instance, it would be useful to explore how the use of specific information filters and communication channels (e.g., Henderson, 1993) or HR practices (Macher and Mowery, 2001) influence firms' governance decisions and subsequent technological performance. Similarly, it would be productive to examine whether and how firm-level differences in absorptive capacity influence firms' vertical integration decisions (Cohen and Levinthal, 1990). Research along these lines might also inform the relative importance of firm- and transaction-level factors as they relate to firms' performance.

Another attractive opportunity for research would be to extend this framework by analyzing whether and how other dimensions of governance influence technological performance. Although markets and hierarchies share some common features, they exhibit many distinct strengths and weaknesses that may differentially affect performance (Williamson, 1985). Thus, it would be productive to specifically examine the relationships between degree of ownership, various dimensions of coordination, and performance. For instance, it has been suggested that geographic agglomerations such as those present in the Silicon Valley provide a degree of coordination that improves a firm's ability to successfully introduce new products (Eisenhardt and Schoonhoven, 1996). More generally, Teece (1996) argues that various organizational archetypes (i.e., the virtual, alliance, multiproduct, and Silicon Valley types) may be more or less suited to systemic and autonomous innovations. As firms employ a wide range of different types of vertical sourcing arrangements, future research might differentiate these alternative forms of governance and investigate their performance implications (e.g., Dyer, 1996, 1997; Oxley, 1997).

Finally, the study's findings indicate how firms' vertical governance decisions influence one specific dimension of performance. Extensions might consider how firms' boundary decisions influence other performance dimensions such as overall firm profitability, excess cash flow, or risk. Efforts to isolate and catalogue the drivers of firms' governance choices, to explore other forms of governance, and to examine other dimensions of performance will provide the foundation for more integrative research on the implications of organizational governance and the roles played by theories emphasizing different levels of analysis (e.g., transaction cost theory, the resource-based view, and real options). Such work may also offer a basis for more sound normative guidelines for practitioners. We suspect that research in directions such as these will continue to be of great importance as corporate de-integration remains a central concern for strategy research.

ACKNOWLEDGEMENTS

The authors thank Erin Anderson, Nile Hatch, Jeffrey Macher, Tammy Madsen, Doug Miller, Kirk Monteverde, Rachelle Sampson, and two anonymous SMJ reviewers for helpful comments on earlier versions of this manuscript. We also wish to acknowledge the helpful comments provided by seminar participants at the 2001 Wharton Technology Mini Conference and the 2000 Academy of Management Conference. Financial support for this research was provided by the Ohio State University Center for International Business Education and Research (CIBER), the Purdue University Center for Manufacturing Management (CMME), and INSEAD's R&D Department. All errors remain the responsibility of the authors.

REFERENCES

Allen TJ, Lee DMS, Tushman ML. 1980. R&D performance as a function of internal communication, project management, and the nature of work. *IEEE Transactions on Engineering Management* **27**: 2–12.

Amburgey TL, Kelly D, Barnett WP. 1993. Resetting the clock: the dynamics of organizational change and failure. *Administrative Science Quarterly* **38**: 51–73.

Anderson E. 1988. Strategic implications of Darwinian economics for selling efficiency and choice of integrated or independent sales forces. *Management Science* **34**: 599–618.

Anderson MC. 1997. A primer in measuring outsourcing results. *National Productivity Review* **17**: 33–41.

Argyres N. 1996. Evidence on the role of firm capabilities in vertical integration decisions. *Strategic Management Journal* **17**(2): 129–151.

Barnett WP. 1990. The organizational ecology of a technological system. *Administrative Science Quarterly* **35**: 31–60.

Barney JB. 1999. How a firm's capabilities affect boundary decisions. *Sloan Management Review* **40**: 137–145.

Baum JAC, Mezias SJ. 1992. Localized competition and organizational failure in the Manhattan hotel industry. *Administrative Science Quarterly* **37**: 580–604.

Baum JAC, Oliver C. 1991. Institutional linkages and organizational mortality. *Administrative Science Quarterly* **36**: 187–218.

Belsley DA, Kuh E, Welsch RE. 1980. *Regression Diagnostics*. John Wiley: New York.

Bettis RA, Bradley SP, Hamel G. 1992. Outsourcing and industrial decline. *Academy of Management Executive* **6**: 7–22.

Carroll GR, Delacroix J. 1982. Organizational mortality in the newspaper industries of Argentina and Ireland: an ecological approach. *Administrative Science Quarterly* **27**: 169–199.

Chesbrough HW, Teece DJ. 1996. When is virtual virtuous? Organizing for innovation. *Harvard Business Review* **74**(1): 65–73.

Clark KB, Fujimoto T. 1991. *Product Development Performance*. Harvard Business School Press: Boston, MA.

Clark KB, Chew WB, Fujimoto T. 1987. Product development in the world auto industry. *Brookings Papers on Economic Activity* **3**: 729–781.

Cohen WM, Levinthal DA. 1990. Absorptive capacity: a new perspective on learning and innovation. *Administrative Science Quarterly* **35**: 128–153.

Coles JW, Hesterly WS. 1998. The impact of firm-specific assets and the interaction of uncertainty: an examination of make or buy decisions in public and private hospitals. *Journal of Economic Behavior and Organization* **36**: 383–409.

D'Aveni RA, Illinitch AY. 1992. Complex patterns of vertical integration in the forest products industry: systemic and bankruptcy risks. *Academy of Management Journal* **35**: 596–626.

D'Aveni RA, Ravenscraft DJ. 1994. Economies of integration versus bureaucracy costs: Does vertical integration improve performance? *Academy of Management Journal* **37**: 1167–1206.

Delacroix J, Swaminathan A. 1991. Cosmetic, speculative, and adaptive organizational change in the wine industry: a longitudinal study. *Administrative Science Quarterly* **36**: 631–661.

Dierickx I, Cool K. 1989. Asset stock accumulation and sustainability of competitive advantage. *Management Science* **35**: 1504–1511.

Dolton PJ, Makepeace GH. 1987. Interpreting sample selection effects. *Economic Letters* **24**: 373–379.

Dyer JH. 1996. Specialized supplier networks as a source of competitive advantage: evidence from the auto industry. *Strategic Management Journal* **17**(4): 271–291.

Dyer JH. 1997. Effective interfirm collaboration: how firms minimize transaction costs and maximize firm value. *Strategic Management Journal* **18**(7): 535–556.

Dyer JH, Ouchi W. 1993. Japanese style partnerships: giving companies a competitive edge. *Sloan Management Review* (Fall): 51–63.

Eisenhardt KM, Schoonhoven CB. 1996. Resource-based view of strategic alliance formation: strategic and social effects in entrepreneurial firms. *Organization Science* **7**: 136–150.

Greco J. 1997. Outsourcing: the new partnership. *Journal of Business Strategy* (July/August): 48–54.

Greene WH. 1997. *Econometric Analysis* (3rd edn). Prentice-Hall: Upper Saddle River, NJ.

Griffin A, Hauser JR. 1992. Patterns of communication among marketing, engineering and manufacturing—a comparison between two new product teams. *Management Science* **38**: 360–373.

Gruber H. 1994. *Learning and Strategic Product Innovation: Theory and Evidence for the Semiconductor Industry*. Elsevier Science: Amsterdam.

Harrigan KR. 1983. *Strategies for Vertical Integration*. Lexington Books: Lexington, MA.

Hatch N, Mowery D. 1998. Process innovation and learning by doing in semiconductor manufacturing. *Management Science* **44**(11): 1461–1477.

Haveman HA. 1993. Organizational size and change: diversification in the savings and loan industry after deregulation. *Administrative Science Quarterly* **38**: 20–50.

Hazewindus N, Tooker J. 1982. *The U.S. Microelectronics Industry*. Pergamon: New York.

Heckman J. 1976. The common structure of statistical models of truncation, sample selection, and limited dependent variables and a simple estimator for such models. *Annals of Economic and Social Measurement* **5**(4): 475–492.

Heckman J. 1978. Dummy endogenous variables in a simultaneous equation system. *Econometrica* **46**(6): 931–959.

Heckman J. 1979. Sample selection bias as a specification error. *Econometrica* **47**: 153–161.

Helper S, Sako M. 1995. Supplier relations in Japan and in the United States: Are they converging? *Sloan Management Review* (Spring): 77–84.

Henderson R. 1993. Underinvestment and incompetence as response to radical innovation: evidence from the photolithographic alignment equipment industry. *Rand Journal of Economics* **24**(2): 248–270.

Henderson RM, Clark KB. 1990. Architectural innovation: the reconfiguration of existing product technologies and the failure of established firms. *Administrative Science Quarterly* **35**(1): 9–30.

Integrated Circuit Engineering (ICE). 1997. *Profiles Reports: A Survey of Worldwide IC Manufacturers and Suppliers*. ICE: Scottsdale, AZ.

Kotabe M, Murray J. 2001. Outsourcing service activities. *Marketing Management* **10**(1): 40–45.

Klein B, Crawford R, Alchian A. 1978. Vertical integration: appropriable rents and the competitive contracting process. *Journal of Law and Economics* **21**: 279–326.

Langlois RN. 1992. Transaction cost economics in real time. *Industrial and Corporate Change* **1**: 99–127.

Lee L. 1982. Some approaches to the correction of selectivity bias. *Review of Economic Studies* **49**: 355–372.

Lee L. 1983. Generalized econometric models with selectivity. *Econometrica* **51**: 507–512.

Lee L, Maddala G, Trost R. 1980. Asymptotic covariance matrices of two-stage probit and two-stage tobit methods for simultaneous equation models with selectivity. *Econometrica* **48**: 491–503.

Levy DT. 1985. The transactions cost approach to vertical integration: an empirical examination. *Review of Economics and Statistics* **67**: 438–445.

Liebeskind JP, Oliver AL, Zucker L, Brewer M. 1996. Social networks, learning, and flexibility: sourcing scientific knowledge in new biotechnology firms. *Organization Science* **7**: 428–443.

Macher J. 2001. Vertical disintegration and process innovation in semiconductor manufacturing: foundries versus integrated producers. Georgetown University working paper.

Macher J, Mowery DC. 2001. Measuring competence: performance and practice in semiconductor manufacturing. Georgetown University working paper.

Maddala GS. 1983. *Limited-Dependent and Qualitative Variables in Econometrics*. Cambridge University Press: Cambridge, UK.

Mahoney J. 1992. The choice of organizational form: vertical financial ownership versus other methods of vertical integration. *Strategic Management Journal* **13**(8): 559–584.

March JG, Sutton RI. 1997. Organizational performance as a dependent variable. *Organization Science* **8**: 698–706.

Masten SE. 1993. Transaction costs, mistakes, and performance: assessing the importance of governance. *Managerial and Decision Economics* **14**: 119–129.

Masten SE, Meehan JW, Snyder EA. 1991. The costs of organization. *Journal of Law, Economics, and Organization* **7**: 1–25.

Monteverde K. 1995. Technical dialog as an incentive for vertical integration in the semiconductor industry. *Management Science* **41**(10): 1624–1638.

Monteverde K, Teece DJ. 1982. Supplier switching costs and vertical integration in the automobile industry. *Bell Journal of Economics* **13**: 206–213.

Mowery DC, Oxley JE, Silverman BS. 1996. Strategic alliances and interfirm knowledge transfer. *Strategic Management Journal*, Winter Special Issue **17**: 77–92.

Neter J, Wasserman W, Kutner MH. 1985. *Applied Linear Statistical Models* (2nd edn). Irwin: Homewood, IL.

Nishiguchi T. 1994. *Strategic Industrial Sourcing: The Japanese Advantage*. Oxford University Press: New York.

Oxley JE. 1997. Appropriability hazards and governance in strategic alliances: a transaction cost approach. *Journal of Law, Economics, and Organization* **13**: 387–409.

Pisano GP. 1990. The R&D boundaries of the firm: an empirical analysis. *Administrative Science Quarterly* **35**: 153–176.

Quinn JB, Hilmer F. 1994. Strategic outsourcing. *Sloan Management Review*, Summer: 43–55.

Ranger-Moore J. 1997. Bigger may be better, but is older wiser? Organizational age and size in the New York life insurance industry. *American Sociological Review* **62**: 903–921.

Reich R, Mankin E. 1986. Joint ventures with Japan give away our future. *Harvard Business Review* **64**(2): 78–86.

Rumelt RP. 1974. *Strategy, Structure, and Economic Performance*. Division of Research, Harvard Business School: Boston, MA.

Rumelt RP. 1982. Diversification strategy and profitability. *Strategic Management Journal* **3**(4): 359–369.

Rumelt RP, Schendel DE, Teece DJ. 1994. Fundamental issues in strategy. In *Fundamental Issues in Strategy: A Research Agenda*, Rumelt RP, Schendel DE, Teece DJ (eds). Harvard Business School Press: Boston, MA; 9–47.

Sampson RC. 2001. The cost of inappropriate governance in R&D alliances. New York University working paper.

Shaver JM. 1998. Accounting for endogeneity when assessing strategy performance: does entry mode choice affect FDI survival? *Management Science* **44**(4): 571–585.

Silverman BS, Nickerson JA, Freeman J. 1997. Profitability, transactional alignment, and organizational mortality in the U.S. trucking industry. *Strategic Management Journal*, Summer Special Issue **18**: 31–52.

Teece DJ. 1984. Economic analysis and strategic management. *California Management Review* **26**(3): 87–110.

Teece DJ. 1992. Competition, cooperation, and innovation: organizational arrangements for regimes of rapid technological progress. *Journal of Economic Behavior and Organization* **18**: 1–25.

Teece DJ. 1996. Firm organization, industrial structure, and technological innovation. *Journal of Economic Behavior and Organization* **31**: 193–224.

Walker G, Weber D. 1984. A transaction cost approach to make-buy decisions. *Administrative Science Quarterly* **29**: 373–391.

West J, Iansiti M. 1998. Experience, experimentation, and the accumulation of knowledge: the evolution of R&D in the semiconductor industry. Harvard Business School working paper.

Williamson OE. 1975. *Markets and Hierarchies*. Free Press: New York.

Williamson OE. 1985. *The Economic Institutions of Capitalism*. Free Press: New York.

Williamson OE. 1991. Comparative economic organization: the analysis of discrete structural alternatives. *Administrative Science Quarterly* **36**: 269–296.

Williamson OE. 1998. Transaction cost economics: how it works, where it is headed. *De Economist* **146**(1): 23–58.

Williamson OE. 1999. Strategy research: governance and competence perspectives. *Strategic Management Journal* **20**(12): 1087–1108.

Womack J, Jones D, Roos J. 1990. *The Machine that Changed the World*. Macmillan: New York.

[18]

Strategic Management Journal
Strat. Mgmt. J., **25**: 723–749 (2004)
Published online in Wiley InterScience (www.interscience.wiley.com). DOI: 10.1002/smj.391

THE SCOPE AND GOVERNANCE OF INTERNATIONAL R&D ALLIANCES

JOANNE E. OXLEY[1]* and RACHELLE C. SAMPSON[2]
[1] *University of Michigan Business School, Ann Arbor, Michigan, U.S.A.*
[2] *Stern School of Business, New York University, New York, U.S.A.*

Participants in research and development alliances face a difficult challenge: how to maintain sufficiently open knowledge exchange to achieve alliance objectives while controlling knowledge flows to avoid unintended leakage of valuable technology. Prior research suggests that choosing an appropriate organizational form or governance structure is an important mechanism in achieving a balance between these potentially competing concerns. This does not exhaust the set of possible mechanisms available to alliance partners, however. In this paper we explore an alternative response to hazards of R&D cooperation: reduction of the 'scope' of the alliance. We argue that when partner firms are direct competitors in end product or strategic resource markets even 'protective' governance structures such as equity joint ventures may provide insufficient protection to induce extensive knowledge sharing among alliance participants. Rather than abandoning potential gains from cooperation altogether in these circumstances, partners choose to limit the scope of alliance activities to those that can be successfully completed with limited (and carefully regulated) knowledge sharing. Our arguments are supported by empirical analysis of a sample of international R&D alliances involving electronics and telecommunications equipment companies. Copyright © 2004 John Wiley & Sons, Ltd.

INTRODUCTION

In today's fast-paced, knowledge-intensive environment, research and development (R&D) alliances have become a popular vehicle for acquiring and leveraging technological capabilities. However, such alliances also pose particularly thorny challenges related to the protection of technological knowledge, since successful completion of alliance objectives often requires a firm to put valuable knowledge at risk of appropriation by alliance partners. Firms must therefore find the right balance between maintaining open knowledge exchange to further the technological development goals of the alliance, and controlling knowledge flows to avoid unintended leakage of valuable technology. Prior research in transaction cost economics suggests that choosing an appropriate governance structure or organizational form is one mechanism that firms use to promote knowledge sharing *and* protection in an alliance (e.g., Pisano, 1989; Oxley, 1997; Kale, Singh, and Perlmutter, 2000; Sampson, 2004). However, there may be circumstances where even the most 'protective' alliance form (for example, the equity joint venture) does not reduce leakage concerns sufficiently to ensure the level of knowledge sharing required to achieve alliance objectives. Where this is the case, firms must either forego the benefits of collaborative R&D altogether or find alternative means to reduce the hazards of cooperation.

In this paper, we consider the choice of alliance 'scope' as an alternative way to control the threat

Keywords: R&D; alliances; alliance scope; governance
*Correspondence to: Joanne E. Oxley, University of Michigan Business School, 701 Tappan St., Ann Arbor, MI 48109-1234, U.S.A. E-mail: oxley@umich.edu

of knowledge leakage and protect technological assets in an R&D alliance. Establishing the scope of activities for an R&D alliance involves decisions such as whether to restrict joint activity to pre-competitive R&D only or to extend it to include manufacturing and/or marketing. Related to this is the question of whether a development project can be effectively 'modularized' and conducted in relative isolation by the partner firms, only to be brought together at the final stages (as contrasted with an arrangement that involves extensive cooperation and knowledge sharing throughout the duration of the project). These scope decisions have important implications for the extent to which alliance partners expose valuable know-how to each other. In circumstances where the costs of knowledge leakage are deemed to be particularly high, one might expect that alliance scope will be narrowed in order to limit exposure.

We develop these ideas in more detail below, and identify circumstances where we expect partner firms to make adjustments to the scope of alliance activities, and in particular to the choice between 'pure' R&D alliances and alliances involving a broader range of activities. We argue that the hazards of knowledge sharing will be most salient when partner firms are direct competitors in end product or strategic factor markets, prompting them to narrow the scope of alliance activities in this case.

We test our ideas in an empirical study utilizing a sample of R&D alliances involving companies in the electronics and telecommunications equipment industries, where profitability depends critically on firms' abilities to create and commercialize new technologies quickly and efficiently. Our empirical results suggest that alliance partners are more likely to limit their joint activities to 'pure' R&D when they are direct competitors in final product and geographic markets. We find that competitors are especially averse to adding joint marketing activities to their R&D collaborations, suggesting that the competitive consequences of market-related knowledge leakage is a particularly salient concern. Alliance scope is also driven in part by the relative absorptive capacity of partner firms, however, which influences their ability to effectively cooperate broadly, particularly in joint manufacturing: the degree of overlap in partner firms' technological assets is a strong predictor of alliance scope along this dimension.

Finally, our empirical analysis explores to what extent alliance scope and governance choices act as substitute mechanisms for protecting technological assets and other firm-specific capabilities in R&D alliances. Consistent with prior research, we find that firms choose a more protective alliance structure (i.e., an equity joint venture) when alliance scope is broad. We also find the reverse effect, where the choice of an equity joint venture encourages alliance partners to engage in joint activities that go beyond 'pure' R&D. Overall, the empirical tests provide support for our arguments that allying firms narrow the scope of their alliance activities in response to competitive threats—perhaps in recognition of the fact that protective alliance governance may be inadequate to control the risk of leakage when partners are direct competitors.

The remainder of the paper is organized as follows: In the first section we discuss how alliance scope fits within a larger dynamic system of alliance formation decisions. We then discuss the scope decision itself and motivate the definition of scope used in our study of R&D alliances. This is followed by our hypotheses regarding the alliance scope decision, and how this relates to the choice of governance structure for the alliance. Our empirical analysis follows, and we conclude with a discussion of implications, limitations, and potential extensions of the research.

ALLIANCE FORMATION DECISIONS

When a firm wishes to undertake an R&D project for which it does not currently possess all of the relevant technical knowledge (or other critical resources), forming an alliance is potentially an attractive solution. Before an alliance is established, however, managers have many important decisions to make: Which firm(s) should we ally with? What range of activities should be performed within the alliance (i.e., what should be the scope of the alliance)? How should responsibilities and authority be allocated? What control mechanisms or governance structure should be adopted?[1] All

[1] There is also of course the question of whether and why firms should form inter-firm alliances in the first place. We abstract from this issue here by focusing on situations where the decision to collaborate in R&D has already been made. See Eisenhardt and Schoonhoven (1996) for a useful review of the strategic and social explanations for alliance formation and an empirical analysis of the industry, firm, and top management team factors explaining rates of alliance formation.

of these questions are interrelated, and each choice made is likely to ripple through the other key decisions, perhaps over varying time horizons. The set of decisions involved in the establishment of an alliance thus represents a complex, dynamic, endogenous system: a system that is analytically and empirically intractable without significant simplification.

In order to render the problem tractable, prior research on alliance formation has, broadly speaking, approached this complex system of alliance decisions by adopting one of two major simplifications. Each of these simplifications has yielded important insights. The first approach has been to look only at the question of which firms are chosen as alliance partners, ignoring the related issues of alliance scope and governance. So, for example, Gulati (1995a) shows that firms choose alliance partners based on 'social structural' considerations, such as prior interactions and other network ties, as well as on considerations of strategic interdependence related to interactions in resource or final product markets. Mowery, Oxley, and Silverman (1998), on the other hand, highlight the importance of knowledge complementarities and partner-specific absorptive capacity in the partner choice decision.

The second simplification, adopted in transaction cost economic treatments of alliance formation, has been to focus exclusively on the governance choice decision, effectively abstracting from decisions regarding partner choice and alliance scope. Key findings from this literature are that certain characteristics of alliance activities are associated with the adoption of more hierarchical or protective governance structures—most notably the equity joint venture (see, Pisano, Russo, and Teece, 1988; Pisano, 1989, for early examples). Joint ventures are particularly prevalent where alliance objectives require partners to share complex and/or tacit knowledge, especially in technologically innovative projects. Transaction cost analysis suggests that this is so because firms have incentives to misappropriate the knowledge assets of alliance partners and/or to free-ride on partners' innovative efforts. Contractual governance of exchanges involving complex, tacit knowledge is particularly problematic as contracts are very difficult to specify, monitor, and enforce in these circumstances. By adopting an equity joint venture structure, the hazards of opportunism can be mitigated, because incentives are more closely aligned when ownership of the venture is shared, and monitoring is enhanced due to the increased disclosure requirements among joint venture partners (Pisano, 1989).

Prior research on alliances by transaction cost economists also provides evidence of a connection between alliance governance and the scope of activities undertaken within an alliance. Empirical associations have been found between increased alliance scope and the adoption of hierarchical governance structures, where alliance scope is defined as the number of technologies or functional activities involved (Pisano, 1989; Oxley, 1997), or as the extent to which innovative projects involve the creation of new technology rather than application of existing technology (Sampson, 2004). However, the reduced-form analysis in this prior work treats the scope of alliance activities as exogenous. Little attention has been given to alliance scope as a decision variable, despite the fact that the *Alliance Analyst* (a newsletter for participants in inter-firm alliances) suggests that determining alliance scope is 'one of the most important tasks [alliance] partners will undertake ... The partners must establish boundaries of geography, product categories, customer segments, brands, technologies, and fixed assets between the new entity and the parents. They must identify the activities in which the alliance may engage and those reserved for the parents.'[2]

In the analysis presented below we remedy this oversight by exploring the determinants of alliance scope. In particular we examine the implications of partner characteristics for the scope of activities included in the alliance *and*, consequently, for the choice of governance structure.[3] We also explore the nature of the relationship between alliance scope and governance as alternative means for firms to protect knowledge resources in R&D alliances.

---

[2] *The Alliance Analyst*, 15 July 1997, quoted in Khanna (1998).

[3] In formulating our approach to the problem in this way, we take the choice of alliance partner as exogenous. Fully endogenizing the partner choice decision is a useful direction for future research but is beyond the scope of this paper. We nonetheless checked for potential selection bias in our empirical results by also estimating a model of partner choice using a randomized sample of 'non-alliance pairs' generated from our sample firms. This robustness check is discussed at the end of the Results section.

DEFINING ALLIANCE SCOPE

Choosing the scope of activities to include in an alliance linking a particular set of firms establishes both the probability and the cost of opportunistic behavior by alliance partners. The more extensive, interdependent, complex, and uncertain are the activities performed in the alliance, the greater is the potential risk of opportunism. This is because the extent of coordination and more intimate face-to-face contact necessary to achieve success increases along these dimensions (Kogut, 1988; Kogut and Zander, 1992; Gulati and Singh, 1998) and uncertainty raises the costs of monitoring and assessing partner behavior (Pisano, 1989). Attaching and enforcing claims on technological knowledge contributed to or generated by joint activities is also more difficult when activities are complex and interdependent.

Unfortunately, as the *Alliance Analyst* quote above suggests, alliance scope is a multidimensional construct, which poses significant challenges for theoretical and empirical analysis. Furthermore, many aspects of the scope of alliance activities, such as the number of product categories or customer segments involved, or the dollar value of a joint project, are not reported by alliance participants and so are unavailable through secondary data sources such as press announcements of new alliances. This may explain the dearth of prior research on alliance scope.[4]

The most accessible dimension of alliance scope in terms of conceptual clarity and data availability is the functional or 'vertical' scope of the alliance, i.e., to what extent the partners combine multiple and sequential functions or value chain activities within the alliance, such as R&D, manufacturing and/or marketing. An increase in the vertical scope of an alliance predictably exacerbates the complexity of the collaborative challenge, all else equal (Reuer, Zollo, and Singh, 2002). The 'horizontal' scope of activities, related to the size, complexity, and uncertainty of the particular project, also undoubtedly varies within functional areas and across alliances. However, evaluation of horizontal scope is a much more subjective and challenging exercise because, in contrast to vertical scope, these project-specific features cannot be ascertained from available secondary reports of alliance events. In the following we therefore focus attention on a simple measure of vertical scope that is particularly relevant to R&D-related alliances, i.e., comparing alliances that involve R&D activities alone against those that combine R&D with other activities, specifically manufacturing and/or marketing.

When firms consider establishing an R&D alliance they usually do so with the broad technological goal of the alliance already in mind (e.g., development of 'next-generation' integrated circuits, as in the case of a Fujitsu–Analog Devices alliance, or of components for high-speed fiber optical communications in a Hewlett-Packard–IBM alliance).[5] After establishing this overarching R&D goal, partner firms must still decide how to organize the activities needed to bring the planned technologies or products through to commercial-scale production. In particular they must decide whether manufacturing and marketing activities will be reserved for the partner firms in isolation, or whether they will be performed jointly under the umbrella of the alliance. Research in the field of technology and operations management suggests that the key to reducing time-to-market and improving quality of new product introductions is to employ 'activity overlapping, information transfer in small batches, and the use of cross-functional teams' (Loch and Terwiesch, 2000). Interest in this approach to innovation—the so-called 'concurrent engineering' or 'simultaneous engineering' approach—began with the popular work of Imai, Nonaka, and Takeuchi (1985) and Takeuchi and Nonaka (1986) and has since sparked a strong body of research that has demonstrated its benefits in a variety of industrial settings (e.g., Clark and Fujimoto, 1991; Cusumano and Selby, 1995; Sabbagh, 1996). Although the benefits of overlapping activities in some uncertain and 'high-velocity' environments have been questioned (e.g., Eisenhardt and Tabrizi, 1995; Krishnan, Eppinger, and Whitney, 1997), recent empirical work has again confirmed the benefits of an overlapping activities approach to development even in uncertain

[4] Important exceptions are Khanna (1998) and Khanna, Gulati, and Nohria (1998). These papers provide the only theoretical analysis to date of alliance scope and its potential impact on the dynamics of learning alliances. However, the conceptualization of alliance scope used in this prior work is itself multidimensional and abstract and has to date defied empirical operationalization.

[5] Both of these alliances were established in the early 1990s and are reported in the SDC database described in the Data section, below.

environments, this time in the electronics industry (Terwiesch and Loch, 1999). Given the importance of intense, frequent cross-functional communication in the simultaneous engineering approach, the implications for R&D alliances are clear: if the prime concern is to bring the best product to market in the timeliest fashion, then jointly performing R&D *and* manufacturing and marketing activities within the alliance is likely the superior arrangement.

Despite this received wisdom regarding the benefits of overlapping activities and the importance of cross-functional communication, 'pure' R&D alliances are still more common than 'mixed-activity' R&D alliances.[6] What explains this apparent divergence between theory and practice? Why do firms isolate R&D activities organizationally from other aspects of the commercialization process, if there are operational benefits from combining functions such as manufacturing with R&D? We assert, and argue more carefully below, that it is the potential for costly partner opportunism that prompts firms to limit knowledge-sharing requirements by reducing alliance scope in some contexts.

Keeping abreast of technological opportunities and threats and protecting existing technological capabilities are critical activities in formulating an effective technology strategy. Companies increasingly turn to competitive intelligence experts to provide them with the information on latent opportunities and threats that is necessary for proactive strategic response (Klavans, 1993). As a consequence, competitive intelligence activities have become embedded in the fiber of business throughout the world, particularly in technological arenas (Prescott, 2001). This rise in professional competitive intelligence activities poses particular opportunities and threats to companies collaborating in R&D activities. Collaboration in R&D may provide many opportunities for firms to glean competitive intelligence from alliance partners, including: (i) hints about partner strategies and directions of technological search, or the feasibility of particular ideas; (ii) competitive benchmarking data; (iii) identification of key personnel who may be hired away; (iv) codified knowledge (in formulas, designs, and procedures); (v) deep exposure to tacit knowledge in skills and routines.[7] Some of these types of information leakage, especially (i), (ii), and (iii), are arguably unavoidable in any R&D alliance, at least with respect to technologies and processes relevant to alliance activities. Despite this, if an alliance involves only *'pure'* R&D activities, prior evidence suggests that alliance partners are often able to partition the activities such that the exposure of a firm's tacit knowledge in skills and routines to its partners is relatively small, and necessary exposure of codified knowledge can be controlled through appropriate use of contractual safeguards (Pisano *et al.*, 1988). This is particularly true for modular designs (Sanchez and Mahoney, 1996), but also appears to apply more generally. As a consequence, 'pure' R&D alliances are predominantly governed via contractual arrangements (Narula and Hagedoorn, 1999).

When the vertical scope of an R&D alliance is increased to encompass other activities (in particular, manufacturing and/or marketing) the extent of knowledge sharing and coordination inevitably rises (Reuer *et al.*, 2002). Furthermore, protection of technological knowledge becomes more challenging with increases in alliance scope, as the tacit knowledge embedded in operating routines must be exposed to alliance partners if joint operations are to proceed efficiently. To jointly bring an R&D project through to commercialization requires many more points of contact between the partner firms, with a concomitant reduction in control over information flows across the relevant organizational boundaries (Teece, 1992).[8] Furthermore, operational routines exhibit substantial inseparability, and it is likely that knowledge gained in the course of manufacturing and marketing efforts within the alliance will have important effects on other areas of partner firms' operations. As a result, it is almost impossible to effectively manage mixed activity R&D alliances without extensive sharing of tacit knowledge embedded in

[6] For example, in our sample, described below, 63 percent of the alliances in our sample are 'pure' R&D alliances, while only 37 percent involve 'mixed' activities.

[7] We thank Sid Winter for this categorization of information exposure in R&D alliances.

[8] Of course, the extent of exposure of knowledge assets also depends on the horizontal scope of the alliance activities, as discussed earlier, but we abstract from this issue here. In our empirical analysis we control for one aspect of horizontal scope for which we were able to obtain data—i.e., the innovation type. We anticipate that 'radical' innovations, which require the development of entirely new technologies, pose greater knowledge sharing and control challenges than more incremental innovations (see discussion in section on Innovation type).

operational routines, which in turn may have significant effects on the relative competitive position of partner firms.

Taken together these arguments suggest that the distinction between pure R&D alliances and alliances involving R&D in combination with other operational functions represents a useful and relevant dichotomous measure of vertical alliance scope, with pure R&D alliances being of more narrow scope than mixed activity R&D alliances. We now turn to a discussion of the determinants of the choice of alliance scope in this context.

COMPETITION, CAPABILITIES AND ALLIANCE SCOPE

Although there is only limited systematic empirical evidence of the competitive effects of alliance activity,[9] the general idea that inter-firm cooperation may change the relative competitive position of partner firms is a common theme in commentaries on the rise of international alliances during the 1980s (e.g., Reich and Mankin, 1986; Hamel, Doz, and Prahalad, 1989; Doz and Hamel, 1998). For example, Hamel *et al.* (1989: 135) suggest that when seeking collaborators for technology-related projects firms should target partners whose 'strategic goals converge while their competitive goals diverge.' The rationale behind this prescription is that if alliance partners are competitors in end-product markets (i.e., if their competitive goals 'converge') then each may be so intent on internalizing the other's knowledge—and at the same time limiting access to their own proprietary skills—that the goal of the alliance will be thwarted.

Does this problem mean that competitors cannot effectively collaborate in R&D? Clearly not: One can find examples of R&D alliances linking firms that also compete intensively in almost any industry. In autos, for example, Ford, GM, and DaimlerChrysler collaborated extensively during the 1990s to develop the so-called 'Next-Generation Passenger Vehicle' or NGPV, despite the fact that they are direct competitors. Nonetheless, as highlighted by Khanna *et al.* (1998), if partner objectives include inter-firm learning (as is the case in R&D-related alliances), we will likely observe both competitive and cooperative behavior by participating firms. Furthermore, the balance between competitive and cooperative behavior is determined by relative pay-offs; these in turn depend, among other things, on the final product market activities of the partners and in particular on the extent to which partner firms' primary product markets overlap. Where partners are direct competitors in final product markets, then the pay-off from competitive behavior (i.e., free-riding or misappropriation) within the alliance is particularly high. Furthermore, all else equal, the closer the partners' competitive domains come to complete overlap, the closer the knowledge-sharing process approximates a zero-sum game, since knowledge gained in the alliance will be applied by each partner firm in the same competitive arena (see, for example, Branstetter and Sakakibara, 2002).

Given this set of circumstances, even if the alliance activities are embedded in a 'protective' governance structure such as an equity joint venture, the residual hazards of opportunism may be high enough to deter extensive knowledge sharing among alliance participants (Inkpen, 2000). In anticipation of this problem, we hypothesize that alliance partners with substantial competitive overlap will limit the scope of alliance activities so as to allow the general alliance objectives to be achieved with more limited knowledge sharing:

Hypothesis 1: The greater the overlap among partner firms' end-product markets, the lower the probability of broad alliance scope.

End-product markets are not the only source of competitive concerns that partner firms might consider when establishing the scope of alliance activities. Another important source of concern may be competition in strategic factor or resource markets. As emphasized by research on the resource-based view of the firm, sustainable competitive advantage rests on a firm's ability to assemble valuable, rare, and inimitable resources or capabilities (Barney, 1986; Dierickx and Cool, 1989). Specialized technological capabilities are quintessential exemplars of such firm-specific resources; the value and uniqueness of a firm's technological capabilities are protected through patents and other 'appropriability mechanisms' (Levin *et al.*, 1987) and are

[9] See Mitchell and Singh (1996) for a review of the relevant literature and some useful evidence on the effects of collaboration on firm survival. Stuart (2000) and Koka and Prescott (2002) explore and elaborate on the potential benefits generated by a firm's alliance portfolio.

often misunderstood by competitors (Henderson and Clark, 1990), further enhancing inimitability.

One predictable outcome of collaborative R&D is that inter-organizational learning increases the similarity of alliance partners' resources and capabilities after alliance formation (Nakamura, Shaver, and Yeung, 1996; Dussauge, Garrette, and Mitchell, 2000). Furthermore, the greater the extent of knowledge sharing in the alliance, the more that partner–firm capabilities will come to resemble one another over time (Mowery, Oxley, and Silverman, 1996). This implies that inter-firm cooperation predictably undermines the rarity and inimitability of a firm's resources, at least *vis-à-vis* those of alliance partners. As a result, the incentives to team up in technology development projects may be different for leading and lagging firms, in a similar way that the incentives to agglomerate differ (Shaver and Flyer, 2000): market leaders are more likely to want to 'go it alone' to protect firm-specific know-how, while laggards may be willing to team up to try to leapfrog industry leaders. Similar logic suggests that the alliance scope decisions of leaders and laggards may also differ systematically. In particular, firms far behind the industry frontier in terms of technology, human capital, training programs, suppliers, and distributors or other important resources (Shaver and Flyer, 2000: 1176) may be more willing to expose competitively significant know-how to alliance partners if in doing so they can together produce a new product that will improve their competitive position *vis-à-vis* other 'coalitions' in the industry (Gomes-Casseres, 1996). This suggests the following hypothesis:

Hypothesis 2: The probability of broad alliance scope is higher when all partner firms are industry laggards.

In addition to the extent or quality of alliance partners' technological and other capabilities, the positioning of partners' respective technological portfolios in 'technology space' also may have an impact on the scope of activities undertaken within an R&D alliance. We can think of overlap in partner capabilities as lying along a continuum: at one extreme, partners have no areas of technological expertise in common; at the other extreme, partners have completely overlapping areas of technological expertise. In most cases, allying firms will fall at various points along the continuum between these two extremes. As we argue below, placement along the continuum affects both the *willingness* and the *ability* of partners to engage in alliance activities of broad scope.

When there is extensive overlap in alliance partners' areas of technological expertise, partner firms are likely to draw on the same external pools of technological knowledge and thus may perceive of each other as direct competitors in relevant resource markets (including, for example, specialized labor markets). If this is the case then the scope of alliance activities may be restricted to limit exposure of knowledge relevant to competition in strategic factor markets. As the areas of common interest diminish (i.e., as technological overlap decreases), these competitive effects are likely to be less pronounced. This argument implies a negative relationship between technology overlap and alliance scope.

There is, however, a potentially competing effect of low technology overlap on alliance scope, unrelated to leakage hazards. In order to effectively share knowledge, partner firms must possess adequate absorptive capacity, i.e., the 'ability to recognize the value of new external knowledge, assimilate it, and apply it to commercial ends' (Cohen and Levinthal, 1990: 128). When alliance partners are very 'far apart' in terms of their areas of technological expertise, deficiencies in absorptive capacity are likely to limit the objectives achievable within an alliance, so that only those requiring relatively restricted knowledge sharing are feasible (Mowery *et al.*, 1996; Park and Ungson, 1997; Lane and Lubatkin, 1998). In the context of the R&D alliances in our study this suggests that, while pure R&D projects may be achievable through modularization of the R&D tasks, the extensive knowledge sharing required to jointly take the project through R&D to commercialization will be more problematic at low levels of technology overlap.

From the arguments above, we might expect the competitive aspects of the impact of technology overlap on alliance scope to dominate when firms are very close in technological space, while the absorptive capacity effect would be more salient when partner firms are technologically distant. This implies an inverted-u shaped relationship between technology overlap and alliance scope. However, whether the net effect of technology overlap is positive or negative in the relevant operational range of actual alliances is an empirical question. We therefore simply note two distinct hypotheses, and test for nonlinearity in the empirical analysis:

Hypothesis 3a: The higher the technology overlap between partners the lower the probability of broad alliance scope (due to factor market competition).

Hypothesis 3b: The lower the technology overlap between partners the lower the probability of broad alliance scope (due to inadequate absorptive capacity).

Although our primary interest in this study is the determination of alliance scope, we are also interested in the relationship between the scope and governance decisions of alliance partners. Prior research, which treats alliance scope as exogenous, suggests that broader alliance scope increases the hazards of contractual alliances and therefore increases the need for more protective alliance governance, such as that afforded by an equity joint venture (Pisano, 1989; Oxley, 1997). Our arguments about the determination of alliance scope are also based on the premise that broad scope increases the hazards of cooperation; following previous research, we would thus also expect firms to be more likely to choose an equity joint venture rather than a contractual alliance in this case. However, in contrast to prior work, we view alliance scope as an alternative means of controlling partner opportunism. Furthermore, we argue that narrowing scope may actually be the preferred alternative when competitive overlap is particularly extensive, since the residual hazards of broad-scope alliances may be too high even when embedded in a protective governance structure. Nonetheless, we expect that more protective governance structures will make firms more confident in broadening alliance scope, all else equal. In this sense, adoption of a protective governance structure and narrowing of alliance scope are substitutes; either may be used to control partner opportunism.[10] This intuition leads to our final hypotheses:

Hypothesis 4a: The probability that partner firms select an equity joint venture is higher when alliance scope is broad.

Hypothesis 4b: The probability of broad alliance scope is higher when partner firms select an equity joint venture.

In the empirical analysis that follows, we test these hypotheses by estimating the relationship between two dichotomous outcome variables—alliance scope and governance—and a series of variables that characterize the competitive and technological positions of the alliance partner firms, as well as other relevant characteristics. By estimating these relationships first in isolation and then jointly (both as a system of seemingly unrelated regressions and in a simultaneous equations model) we are able to shed additional light on the results derived from reduced form governance equations in prior research. Further, we can identify the extent to which the two decisions are truly linked, and to what extent reductions in scope may substitute for adjustments to formal governance mechanisms (i.e., adoption of an equity joint venture structure) in response to competitive concerns.

DATA

For these tests, we use a sample of R&D alliances involving firms in the electronic and telecommunications equipment industries. Recent technological changes in these industries make them ideal for a study of R&D collaboration. In the late 1980s, the electronic and telecommunications equipment industries converged and a period of rapid growth and technological development ensued. To illustrate, a 2000 OECD report noted that 'ICT intensity' (i.e., expenditures on information and communication technology as a percentage of GDP) had reached almost 7 percent on average by 1997 and that the 'rapid consolidation in the ICT industry is having an impact on the competitive structure of this sector' (OECD, 2000: 11). Profitability and survival in this environment became critically dependent on a firm's ability to create and commercialize new technologies quickly and efficiently. As a result, firms began establishing R&D alliances at an unprecedented rate as a way to spread the risk and cost of technological development (Duysters and Hagedoorn, 1996).

[10] Of course, this does not mean that both protective governance and narrow alliance scope cannot be used simultaneously to control alliance hazards, but rather that both mechanisms perform similar functions.

Our primary source of data is the Securities Data Company (SDC) Database on Alliances and Joint Ventures. The SDC database contains information on alliances of all types, compiled from publicly available sources such as SEC filings and news reports, as well as industry and trade journals. While SDC has some information on alliances back to 1970, consistent data collection extends from 1988 onwards, and coverage of alliances after this date is more comprehensive. Coverage is still far from complete, as firms are not required to report their alliance activities. Nevertheless, this database currently represents one of the most comprehensive sources of information on alliances.

Our sample consists of all R&D alliances involving one or more firms in the telecommunications equipment and/or electronics industry commencing in 1996, as reported in the SDC database. Compilation of the sample involved assembling the names of all firms in Compustat listing either SIC 366 (telecommunications equipment) or SIC 367 (electronics) as their primary area of activity in 1996, and searching the SDC database for all R&D alliances involving one or more of these firms. As such, our sample is limited to those involving at least one public company. These criteria led to a sample of 208 R&D alliances (i.e., alliances involving collaborative R&D activities exclusively or in combination with manufacturing and/or marketing activities) representing firms headquartered in over 20 countries. Approximately 85 percent of the sample alliances include at least one U.S. firm; the next most represented nation is Japan, with almost 20 percent of the sample alliances involving one or more Japanese firms. The sample includes both 'domestic' alliances (58%), where all partner firms are headquartered in the same country, and international alliances (42%), involving partners from different nations. The high incidence of international alliances and the concentration of allying firms in the United States and Japan is consistent both with prior observations of alliance activity (e.g., Hagedoorn, 1993; Hergert and Morris, 1988) and with the distribution of firms in the information and communications technology sector.

Information on firms' technological portfolios is taken from the Micropatent database, which contains information recorded on the front page of every U.S. patent granted since 1975, including inventor and assignee names as well as technological classifications. Since firms do not always assign patents to the subsidiary where the technological expertise was created, a corporate-level patent portfolio must be created. To achieve this, we first identified all of the subsidiaries of each firm in the sample via the *Directory of Corporate Affiliations*. Patents assigned to the firm and any of its subsidiaries were then identified for the purposes of constructing our measure of technology overlap, described in detail in the following section.

MEASURES

Dependent variables

Alliance scope

Using information from the SDC database, we create a dummy variable to capture the vertical scope of alliance activities. SCOPE is set to one when alliance activities include manufacturing, and/or marketing *in addition to* collaborative R&D. As argued above, such alliances are broader in scope than alliances involving R&D activities exclusively.

To explore the multidimensionality of alliance scope in supplementary analyses we further categorized the alliances into R&D only, R&D plus manufacturing, R&D plus marketing, and R&D plus both (i.e., R&D plus manufacturing *and* marketing).

Alliance governance

Also using information from the SDC database, we create a dummy variable to capture alliance governance mode. GOVERNANCE equals 0 when the alliance is organized by contract, 1 when organized by equity joint venture.

Independent variables

To capture the extent to which partners may be direct competitors, we use two complementary measures of competitive overlap: product market and geographic market competition.

Product market competition

Measuring the intensity of competition among diversified firms operating in multiple business segments is highly problematic, and no well-accepted method exists. We therefore adopt the most straightforward approach and construct a simple measure capturing whether allying firms have their primary operations in the same industry; while our sample focuses on firms with primary operations in SIC 366 or 367, these firms are allied with partners in a diverse set of industries. MARKET OVERLAP is set to 1 if both partner firms have their primary operations in the same 4-digit SIC code, 0 otherwise. For multilateral alliances (i.e., alliances with more than two partner firms), MARKET OVERLAP is set to 1 if at least two of the partner firms have the same primary 4-digit SIC.

Geographic market competition

We create a dummy variable, GEOGRAPHIC OVERLAP, which is coded 1 for alliances involving firms headquartered in the same country, 0 if allying firms are headquartered in different nations. This measure is used to proxy for geographic market competition based on the assumption that firms headquartered in the same country perceive each other to be more direct competitors when compared with firms originating in different countries. We recognize that this assumption may be less valid in electronics than it is in the telecommunications equipment industry where national firms tend to be favored suppliers to telecommunications providers in that country, many of whom are public organizations. However, in the absence of adequate country-level data on total market presence we must rely on this imperfect proxy. To partially control for alternative interpretations and confounding effects we also include control variables related to the focal industry and the institutional environment of the home countries (see below).

Market leaders and laggards

Market leadership is a highly subjective and multifaceted issue. One approach is to focus on *technological* leadership using a patent-based measure, such as citation-weighted patent counts (e.g., Stuart, 2000). However, given the diversity of the firms involved in our sample alliances, and wide variance in the importance of patenting activity across industries (Levin *et al.*, 1987) interpretation of patent counts is problematic. Furthermore, even for R&D alliances, the relevant resources that a firm wishes to protect may go far beyond technological resources to include, for example, human capital, training programs, suppliers, and distributors (Shaver and Flyer, 2000). We therefore use a more qualitative, composite measure of market leadership, based on a 1997 listing of the Top 100 Technology Companies compiled by *Red Herring* magazine. Based on the magazine editors' opinion, this list represents the leading public and private technology firms, in terms of 'revenue, market share, funding, innovativeness of each company's technology and business model, the track record of its venture capitalists, the strength of its partnerships, and the significance of its market' (*Red Herring*, 1997). This description corresponds well to the association suggested in our earlier discussion, i.e., that market leadership in an industry is fundamentally linked to the possession of firm-specific resources. We create a dummy variable LAGGARDS that is equal to one if *none* of the alliance partners appear in the *Red Herring* list of top technology companies, zero otherwise.

Technology overlap

While a patent count is not a particularly useful measure of market leadership, a patent portfolio is nonetheless a useful indicator of a firm's areas of technological expertise. In order for a new technology to be patented (so allowing the inventor to claim exclusive rights over the product or process described therein) the invention must first pass the scrutiny of the patent office as to its novelty and improvement over existing technology. Extensive research has demonstrated the relationship between patents and other indicators of technological capabilities: Strong, positive relationships exist between patents and new products (Comanor and Scherer, 1969), patents and literature-based invention counts (Basberg, 1982), and non-patentable inventions (Patel and Pavitt, 1997). In addition, when a patent is granted, the underlying technology is classified according to the U.S. patent classification system. This classification system provides a means to identify each firm's area of technological expertise. From this, we can examine the extent of overlap among partner firms' technological portfolios (Jaffe, 1986).

To construct our measure of technological overlap, we first generate each partner firm's technological portfolio by measuring the distribution across patent classifications of the patents applied for in the 4 years prior to alliance formation. This distribution is captured by a multidimensional vector, $F_i = (F_i^1 \ldots F_i^s)$, where F_i^s represents the number of patents assigned to partner firm i in patent class s. The extent of the overlap among partner firms' areas of technological expertise is then:

$$\text{TECHNOLOGY OVERLAP} = \frac{F_i F_j'}{\sqrt{(F_i F_i')(F_j F_j')}}$$

for all $i \neq j$.[11] TECHNOLOGY OVERLAP varies from zero to one: a value of zero indicates no overlap in partner firms' areas of technological expertise, while a value of one indicates complete overlap. This measure normalizes the length of the within-class vectors to one and essentially captures the angle between the firm vectors. This means that the measure is not sensitive to the total number of a firm's patents within a class.

Control variables

To capture other factors that are predictably associated with risks of knowledge leakage (and hence with either reductions in alliance scope or adoption of an equity joint venture structure) we include several control variables. These are as follows.

Innovation type

R&D projects can run the gamut from those involving development of new products or processes based on incremental modifications of existing technology, to radical, ambitious projects where firms seek to develop the 'next generation' of a particular product. Since the nature of the obstacles to cooperation may differ markedly among projects at different positions along this continuum (Sampson, 2004), we developed a measure of innovation type, based on the synopses of alliance activities provided by the SDC database.[12]

We create a dummy variable, INCREMENTAL, which takes on a value of one if the innovative goal of the alliance represents an incremental modification of existing technology. An innovation is categorized as incremental when the synopsis suggests that alliance activities are focused on development of new products or processes based on existing technologies. This would include, for example, the alliance between 3M and IBM to jointly develop 3M's 'Ecart' 2.4 GB tape cartridges for IBM tape products. The base technology exists (3M's Ecart tape cartridges) and needs only to be customized to a new user (IBM). The omitted category of alliances involves more radical innovations, primarily alliances pursuing 'next-generation' technologies. Examples here include the alliance between Hewlett-Packard and IBM to jointly develop and manufacture advanced fiber optic components designed to provide high-speed fiber optical communications between computers, or the alliance between Fujitsu and Analog Devices for the development of 'next-generation' integrated circuits.

Pisano (1989: 166) argues that projects involving existing technology raise fewer hazards for alliance partners because, 'the preexistence of a product or process technology ... enables parties to delineate property rights at the outset with far less ambiguity than when the relevant technology does not exist.' This argument suggests that incremental innovations raise fewer risks of technology leakage. As such, steps to control partner access to knowledge-based resources by either reducing the scope of the alliance or instituting stronger governance through the equity joint venture are less necessary for incremental innovations.

Multilateral alliance

Prior research suggests that having more than two partner firms in a particular alliance makes monitoring more difficult and so increases the risk of unintended knowledge leakage. Empirical evidence has consistently shown that equity joint ventures are more frequently used to organize such multilateral alliances, all else equal (Oxley, 1997; Sampson, 2004). Increased coordination problems

[11] This measure (adopted from Jaffe, 1986) calculates technology overlap between pairs of firms. For alliances involving more than two firms, we calculate the measure for every combinatorial pair of firms in the alliance and take the average of these values. Use of minimum or maximum values of technological overlap produced qualitatively identical results in the empirical analysis.

[12] This coding scheme was a simplified version of the one used in Sampson (2004), developed in concert with an R&D professional. Coding was executed independently by two coders searching the SDC alliance synopses for references to extension/modification of existing technology vs. those pursuing radical change. Concordance between the two coders exceeded 90 percent; a third coder acted as a tiebreaker in the few cases of disagreement.

associated with multilateral alliances may also prompt a reduction in the scope of activities undertaken. Firms may be less willing to undertake broad alliance activities, given the added complexity of coordinating with multiple partners. To capture these potential effects, we include a dummy variable, MULTILATERAL, which equals one if the number of partner firms in an alliance exceeds two, zero otherwise.

Previous relationship

There is some evidence (albeit mixed) that in repeat alliances between particular partner firms trust develops such that the hazards of opportunism are reduced and contractual governance is preferred over a wider range of alliance activities (Gulati, 1995b). If trust does indeed develop over repeated interactions then one might expect that firms with prior alliance ties will be more willing to engage in alliances of broad scope with extensive knowledge sharing, even if they compete to a significant degree in product or resource markets. We therefore include a dummy variable, REPEAT ALLIANCE, equal to one if there are earlier alliances (of any type) recorded in the SDC database that bring together the same set of partner firms in the 6 years prior to the current alliance, zero otherwise.

Industry controls

We add dummy variables for each of the two focal 3-digit SIC industry codes used as a starting point for the development of our sample of alliances. SIC 366 equals one if at least one partner in the alliance has its primary activities in electronics; SIC 367 equals one if at least one partner has its primary activities in telecommunications equipment. Both dummy variables may equal one if an alliance brings together a partner primarily operating in electronics with one in telecommunications equipment.[13]

[13] Note that these dummy variables are constructed from the SICs of the actual participants in the alliance, as opposed to the SICs of the parent firms. While in all of our alliances at least one partner *or its parent* must have primary operations in SIC 366 or 367, there are cases where neither *partner* in the alliance is directly involved in 366 or 367.

Institutional environment

Following prior research (e.g., Sampson, 2004) we include a control variable designed to capture the integrity of the contracting environment in the allying firms' home countries. This measure, JUDICIAL SYSTEM, developed by *Business International Corporation*, provides an assessment of the 'efficiency and integrity of the legal environment as it affects business, particularly foreign firms' (Business International Corporation). We use the minimum value of this variable (which is scaled from one to ten, with higher scores indicating higher judicial efficiency) among the countries represented in the alliance. This variable controls for the fact that firms may rely more on stronger private governance (such as the equity joint venture) or reduce alliance scope when contract law development and enforcement are inadequate.

We also include a measure of NON-TARIFF BARRIERS in the home countries of the allying firms, to control for increases in alliance scope or local equity participation that may be driven by local content requirements or other investment regulations. In constructing the measure we use UNCTAD data, adapted by the World Bank (Wacziarg, 2001), which captures the non-tariff barrier coverage ratio for the early 1990s. We use the highest value on this measure among the home countries represented in an alliance.

METHODS

In the empirical analysis reported below, we begin by estimating alliance scope as a function of the variables featured in our hypotheses, along with the relevant controls described above. Since alliance scope is a dichotomous variable we use probit analysis for this estimation. We also conduct additional analyses using the multiple alliance scope configurations (R&D, R&D plus manufacturing, R&D plus marketing, R&D plus both manufacturing and marketing) to explore for possible differential effects across categories. We use simple cross-tabulations and multinomial logit for this analysis. Next, we estimate a model of alliance governance selection similar to the dichotomous probit model of scope. Here, we also include alliance scope in the governance equation because, as discussed above, previous studies have found alliance scope to be a relevant determinant of governance choice.

Estimating the scope and governance equations separately in this way allows us to test our primary hypotheses and also to compare our results with those of prior research on alliance governance. However, in order to test our hypothesis that alliance scope and governance are alternative mechanisms to address threats of knowledge leakage, we must allow for potential interdependencies in the relationship. We do so by employing two alternative joint estimation methods. First, we estimate the scope and governance equations as a pair of seemingly unrelated regressions, using a bivariate probit model. The bivariate probit jointly determines parameter estimates for the alliance scope and governance equations based on a partially overlapping set of decision criteria. This joint estimation allows for contemporaneous correlation between the error terms of the scope and governance equations, reflecting our arguments that the two decisions are related. In this way, the bivariate probit model controls for unobservable factors that affect the choice of both alliance scope and governance. The resulting parameter estimates are more efficient, largely because of the increased information available for the estimation (i.e., via the correlation between the errors of the two equations) (Greene, 1997).

Second, we estimate the scope and governance choices via a simultaneous probit specification. This specification allows for endogeneity of one or both of the dependent variables using full information maximum likelihood (FIML) estimation. While the bivariate probit allows for unobservable factors that may affect both scope and governance, the simultaneous system approach is more appropriate if we expect the choice of scope to directly affect the choice of alliance governance, and vice versa. Thus, the key advantage of the simultaneous approach in this case is that we can estimate a non-recursive model, where scope influences governance and governance influences scope. Of course, the drawback of using such an approach is that simultaneous system estimates are very sensitive to specification errors—much more so than bivariate probit estimates (Greene, 1997). For this reason, we estimate both the non-recursive model (as a simultaneous probit) and the recursive (as a bivariate probit) to ensure the robustness of our results.

RESULTS

Table 1 presents descriptive statistics for the sample of 208 alliances included in the empirical analysis. Several interesting features of our alliance sample emerge from a reading of this table: First, with respect to our focal dependent variables, the majority (63%) of the sample alliances have narrow scope, i.e., are R&D-only alliances, while 37 percent involve manufacturing and/or marketing in combination with R&D and are therefore designated as broad scope. Also, 20 percent of the alliances are equity joint ventures, the remainder being contractual alliances. Consistent with prior research, the correlation between scope and governance is positive and significant, at 0.209. Several of the explanatory variables are also correlated but none at levels that prompt concerns about multicollinearity.

Few alliances in our sample (13%) bring together firms that have their primary activities in the same 4-digit SIC code, while the majority (58%) of alliance partners are headquartered in the same country and are industry laggards (60% of the alliances do not include any leading firms). The low mean value for technological overlap (0.079 on a 0–1 scale) suggests that on average alliance partners have quite low levels of overlap in their technological resources (as proxied by patenting activity). Most alliances are not aimed at radical or 'next-generation' innovation: the technological goals of 90 percent of alliances are classified as incremental innovation. These sample characteristics are interesting in themselves; they may also speak to some of the factors considered in the earliest stages of alliance formation, i.e., during partner selection and definition of broad alliance goals. Although consideration of these other alliance decisions goes beyond the scope of the paper, we discuss selection issues related to the robustness of our empirical results at the end of this section.

Our first estimation results are shown in Table 2a, which presents several alternative specifications of a simple probit model of alliance scope. Column 1 includes all of the variables related to Hypotheses 1–3 (i.e., hypotheses on the determination of alliance scope), as well as alliance-level control variables; column 2 adds a square term to test for nonlinearities in the effect of technology overlap on alliance scope. Columns 3 and 4 add industry and institutional environment

Table 1. Descriptive statistics

Correlations (correlation, significance)	1	2	3	4	5	6	7	8	9	10	11	12	13
1 SCOPE	1.000												
2 GOVERNANCE	0.210	1.000											
	0.002												
3 MARKET OVERLAP	−0.069	−0.058	1.000										
	0.322	0.405											
4 GEOGRAPHIC OVERLAP	−0.198	−0.181	−0.123	1.000									
	0.004	0.009	0.078										
5 LAGGARD	0.164	0.170	0.038	−0.142	1.000								
	0.018	0.014	0.590	0.041									
6 TECHNOLOGY OVERLAP	0.104	0.046	0.239	0.064	−0.118	1.000							
	0.137	0.510	0.001	0.360	0.091								
7 INCREMENTAL	0.094	−0.016	−0.229	0.016	0.095	−0.239	1.000						
	0.175	0.823	0.001	0.815	0.172	0.001							
8 MULTILATERAL	0.019	0.153	0.298	−0.123	−0.106	0.285	−0.179	1.000					
	0.791	0.028	0.000	0.078	0.128	0.000	0.010						
9 REPEAT ALLIANCE	−0.011	−0.019	−0.066	0.040	−0.041	0.092	−0.158	−0.118	1.000				
	0.879	0.786	0.342	0.571	0.561	0.187	0.023	0.091					
10 TELECOMMUNICATIONS	−0.113	0.027	−0.002	0.043	0.123	−0.062	0.037	0.032	−0.157	1.000			
	0.105	0.703	0.977	0.536	0.077	0.371	0.594	0.644	0.024				
11 MICROELECTRONICS	−0.017	0.063	0.030	−0.125	0.022	0.067	−0.081	0.030	0.057	−0.237	1.000		
	0.806	0.363	0.672	0.073	0.751	0.339	0.247	0.672	0.412	0.001			
12 JUDICIAL SYSTEM	−0.141	−0.101	−0.089	0.406	−0.057	0.074	−0.029	−0.035	−0.087	0.140	−0.119	1.000	
	0.064	0.183	0.244	0.000	0.456	0.332	0.708	0.645	0.254	0.065	0.117		
13 NON-TARIFF BARRIERS	0.086	−0.125	0.105	−0.226	0.026	−0.067	0.051	−0.118	0.068	−0.045	−0.015	−0.397	1.000
	0.260	0.099	0.165	0.003	0.729	0.376	0.499	0.120	0.372	0.555	0.848	0.000	
Mean	0.370	0.202	0.135	0.582	0.596	0.079	0.913	0.135	0.082	0.216	0.322	9.592	3.891
Median	0.00	0.00	0.00	1.00	1.00	0.00	1.00	0.00	0.00	0.00	0.00	10.00	3.60
Minimum	0.00	0.00	0.00	0.00	0.00	0.71	0.00	0.00	0.00	0.00	0.00	2.50	0.70
Maximum	1.00	1.00	1.00	1.00	1.00	1.00	1.00	1.00	1.00	1.00	1.00	10.00	10.90
S.D.	0.484	0.402	0.342	0.494	0.492	0.145	0.282	0.342	0.275	0.413	0.468	0.983	0.378
n	208	208	208	208	208	208	208	208	208	208	208	174	175

Table 2a. Alliance scope

Probit regression model.
Dependent variable is alliance scope
Positive coefficients indicate increased probability that firms select broad alliance scope.

	(1)	(2)	(3)	(4)	(5)
MARKET OVERLAP	−0.513*	−0.476	−0.527*	−0.479	−0.459
	(0.307)	(0.311)	(0.309)	(0.367)	(0.369)
GEOGRAPHIC OVERLAP	−0.553***	−0.556***	−0.547***	−0.560**	−0.546**
	(0.191)	(0.192)	(0.195)	(0.233)	(0.235)
LAGGARDS	0.439**	0.458**	0.498**	0.307	0.364*
	(0.195)	(0.196)	(0.199)	(0.214)	(0.219)
TECHNOLOGY OVERLAP	1.767***	4.573**	1.750**	1.583**	1.579**
	(0.684)	(1.951)	(0.692)	(0.760)	(0.766)
(TECHNOLOGY OVERLAP)^2		−6.290			
		(4.137)			
INCREMENTAL	0.539	0.634	0.506	0.596	0.537
	(0.383)	(0.395)	(0.385)	(0.439)	(0.437)
MULTILATERAL	0.035	−0.053	0.076	0.020	0.001
	(0.300)	(0.307)	(0.299)	(0.328)	(0.326)
REPEAT ALLIANCE	−0.037	−0.151	−0.104	−0.080	−0.141
	(0.344)	(0.356)	(0.349)	(0.382)	(0.388)
TELECOMMUNICATIONS			−0.465*		−0.452*
			(0.243)		(0.266)
MICROELECTRONICS			−0.231		−0.245
			(0.207)		(0.231)
JUDICIAL SYSTEM				−0.064	−0.057
				(0.115)	(0.116)
NON-TARIFF BARRIERS				0.043	0.037
				(0.082)	(0.082)
INTERCEPT	−0.869**	−1.002**	−0.705	−0.378	−0.211
	(0.431)	(0.447)	(0.447)	(1.292)	(1.311)
n	208	208	208	169	169
Log-likelihood	−125.83	−124.61	−123.69	−101.91	−100.21
Chi-square	22.50***	24.94***	26.79***	18.33**	21.74**
d.o.f.	7	8	9	9	11
Prob. > χ^2	0.002	0.002	0.002	0.032	0.026

Significant at * 10%, ** 5%, and *** 1% level for 2-tailed tests.
Standard errors in parentheses.

controls respectively to the specification, and column 5 includes all of these control variables together. This stepwise inclusion of different classes of controls allows us to test for their contribution to the model's explanatory power and so identify an appropriate specification for subsequent analysis.

The model in column 1 provides broad preliminary support for our hypotheses regarding the determinants of alliance scope. First, with respect to product market competition, we find that MARKET OVERLAP has a negative coefficient, as predicted, although only at the 10 percent significance level. Stronger support is found for the predicted effect of the geographic aspect of product market competition: as hypothesized, the coefficient on GEOGRAPHIC OVERLAP is negative and significant. Thus, allying firms that are headquartered in the same country are significantly less likely to engage in activities beyond pure R&D. These results support our contention that firms in the same product market or geographic arena perceive each other to be more direct competitors, making them less willing to risk leakage of competitively significant know-how through extensive knowledge sharing in their R&D-related alliances.

We also find strong support for Hypothesis 2: If the alliance partners are industry laggards the probability of broad alliance scope is increased, suggesting that when laggards team up in an R&D

alliance they are more willing to risk exposing competitively significant know-how to their partners, perhaps in the hope of leapfrogging industry leaders.

Technological overlap is also positively and significantly related to alliance scope, consistent with our absorptive capacity argument. Low levels of overlap among the partner firms' areas of technological expertise appear to hinder their ability to effectively share knowledge to the extent required in broad-scope alliances (Hypothesis 3b). There is no support for the hypothesized heightened factor market competition among firms at high levels of technological overlap (Hypothesis 3a). Furthermore, when we add a square term to the equation (column 2) to test for nonlinear effects, the coefficient on the square term is insignificant and a likelihood ratio test reveals that the inclusion of this variable does not add significantly to the explanatory power of the model ($\chi^2[1] = 2.31$). None of the other coefficients in the model are materially changed, except for a marginal drop in significance of the market overlap variable. This uniformly positive effect of technology overlap is not surprising given the characteristics of our sample, mentioned earlier: the low mean value of technology overlap suggests that the majority of our firms have few if any areas of technological expertise in common. Thus, at least in our sample, competition in technology resource markets does not appear to drive alliance scope. None of the remaining variables in columns 1 or 2 is significant: incremental R&D projects, multilateral alliances and prior alliances between partners do not appear to affect the choice of alliance scope.

When industry controls are added to the specification (column 3), the results reported above remain essentially unchanged and, although the telecommunications dummy variable is significant at the 10 percent level, a likelihood ratio test reveals that the inclusion of these controls does not add significantly to the explanatory power of the model ($\chi^2[2] = 4.29$). Similarly, addition of institutional environment variables, whether alone (column 4; $\chi^2[2] = 0.82$) or in combination with industry controls (column 5; $\chi^2[4] = 4.23$), adds no significant explanatory power. Subsequent estimations of the scope decision therefore use only the variables in column 1.[14]

To further explore the impact of the variables featured in our hypotheses on alliance scope, Table 2b presents a series of sub-sample means for four categories of alliances: R&D only, R&D plus manufacturing, R&D plus marketing, and R&D

[14] Given the fact that the industry control variables are almost significant at the 10 percent level (Prob > $\chi^2[2] = 0.12$), we checked the robustness of the results of all subsequent models with the industry controls included. There were no significant differences in our results with these variables included.

Table 2b. Multiple scope categories: subsample means
Mean of independent variables by scope category

	R&D only $n = 131$	R&D plus manufacturing $n = 19$	R&D plus marketing $n = 37$	R&D plus both $n = 21$
MARKET OVERLAP	0.153	0.263	0.000***	0.143
	(0.032)	(0.101)	(0.000)	(0.077)
GEOGRAPHIC OVERLAP	0.656	0.368**	0.378**	0.667
	(0.042)	(0.111)	(0.080)	(0.103)
LAGGARDS	0.534	0.737*	0.622	0.810***
	(0.044)	(0.101)	(0.080)	(0.086)
TECHNOLOGY OVERLAP	0.067	0.182**	0.062	0.086
	(0.012)	(0.044)	(0.020)	(0.029)
INCREMENTAL	0.893	0.895	0.973**	0.952
	(0.027)	(0.071)	(0.027)	(0.047)
MULTILATERAL	0.130	0.158	0.135	0.143
	(0.029)	(0.084)	(0.056)	(0.077)
REPEAT ALLIANCE	0.084	0.105	0.081	0.048
	(0.024)	(0.071)	(0.045)	(0.047)

Significant at * 10%, ** 5%, and *** 1% level for adjusted Wald (R&D only is comparison group in each case).
Standard errors in parentheses.

plus both (i.e., R&D in combination with manufacturing *and* marketing). There are several points to note here. Of the 77 'broad-scope' alliances, 19 involve only manufacturing in addition to R&D, 37 have marketing only in combination with R&D, and 21 involve R&D, manufacturing, and marketing—the broadest of the categories. The difference in the market overlap variable across these categories is striking. There are *no* R&D plus marketing alliances that bring together firms in the same industry (as indicated by the 4-digit SIC code). This is in stark contrast to narrow-scope alliances (R&D only), where 15 percent of alliance partners are in the same 4-digit industry. Neither of the other two categories of broad-scope alliances (i.e., both of the categories that involve manufacturing) have levels of market overlap that are significantly different from that in narrow-scope alliances, based on adjusted Wald tests of subpopulation means. Geographic overlap, on the other hand, is significantly lower in R&D plus manufacturing alliances and R&D plus marketing alliances as compared with R&D-only alliances, suggesting that firms headquartered in the same country are less likely to engage in these broader alliances, perhaps because of competitive concerns.

Interestingly, very broad alliances, involving both manufacturing *and* marketing in conjunction with R&D, are more likely to involve same-country partners; however, they are also significantly more likely to involve lagging firms. It thus appears that partners in these alliances are less concerned with the impact on inter-partner competition in final product markets, but are focused instead on quickly catching up with industry leaders. This 'catch-up' is likely best achieved through alliances that marry joint R&D with joint manufacturing and marketing, to promote activity overlapping and cross-functional communication, and so compress product development cycles (Loch and Terwiesch, 2000). Finally, technological overlap appears to be a highly significant driver of the decision to add manufacturing activities to a 'pure' R&D agreement, but not for marketing. This reinforces our inference that technological overlap is most important in terms of absorptive capacity, and relates in particular to the ability of firms to effectively engage in joint manufacturing.

Table 2c reports estimation results of two multinomial logit models, both of which have R&D-only alliances as the omitted category, to see whether the patterns observed in our analysis of sub-sample means remain when the coefficients are estimated jointly. The results of the first model, in columns A1 and A2, are indeed very consistent with the simple analysis of means. We cannot estimate the effect of market overlap here because of the lack of variation on this variable in the R&D plus marketing category, but all of the other major regularities discussed above are still evident. Furthermore, if we combine the last two scope categories, so that broad-scope alliances are categorized into R&D plus manufacturing only and R&D plus marketing (whether alone or in combination with manufacturing), we can add market overlap back into the specification (columns B1–B2). Again with R&D only as the omitted category, we see that the results are very consistent with our previous observations.

We next turn our attention to a 'baseline' analysis of the governance decision, in preparation for joint estimation of the scope and governance decisions and testing of Hypotheses 4a and 4b. Our theoretical arguments suggest that the scope and governance decisions are linked in the sense that both may attenuate leakage problems. As such, these two decisions have partially overlapping decision criteria. Applying the logic and findings of previous research nonetheless suggests some departures from the specification and results of the scope model. For example, while prior work argues for the retention of the geographic overlap variable in the governance equation, we remain agnostic about the sign of the effect of this variable on governance choice: For alliances involving firms headquartered in the same country (GEOGRAPHIC OVERLAP = 1), the elevated hazards associated with greater competition may be partially or completely offset by the enhanced ability to enforce contracts in the home nation (an effect that has been observed in previous research on international alliances, e.g., Oxley, 1999) even holding constant the efficiency of the judicial system. We did not hypothesize a parallel effect on alliance scope of enhanced access to the courts because we expect that court ordering tends to be generally ineffective for broad-scope alliances due to the extent of tacit knowledge sharing involved.

In addition, we omit LAGGARDS and TECHNOLOGY OVERLAP from the governance equation. The rationale for these exclusions is as follows: for LAGGARDS, although we argue that lagging firms will be more willing to engage in broad-scope alliances to leapfrog leaders, there is

Table 2c. Multiple scope categories: multinomial regression analysis

Multinomial Logit regression model.
Columns A1, A2, and A3 use a four-outcome dependent variable:
(Scope = 0 if R&D only, = 1 if R&D plus manufacturing only, = 2 if R&D plus marketing only, = 3 if R&D plus manufacturing and marketing)
Columns B1 and B2 use a three-outcome dependent variable:
(Scope = 0 if R&D only, = 1 if R&D plus manufacturing only, = 2 if R&D plus marketing, with or without manufacturing)

	R&D plus Manufacturing (A1)	R&D plus Marketing (A2)	R&D plus Both (A3)	R&D plus Manufacturing (B1)	R&D plus Mktg or Both (B2)
MARKET OVERLAP				0.028	-1.418^{**}
				(0.682)	(0.679)
GEOGRAPHIC OVERLAP	-1.402^{**}	-1.171^{***}	0.177	-1.416^{**}	-0.751^{**}
	(0.565)	(0.400)	(0.518)	(0.568)	(0.341)
LAGGARDS	0.992	0.143	1.389^{**}	0.999^{*}	0.645^{*}
	(0.604)	(0.406)	(0.598)	(0.605)	(0.352)
TECHNOLOGY OVERLAP	5.561^{***}	0.756	1.922	5.452^{***}	1.686
	(1.588)	(1.631)	(1.746)	(1.611)	(1.346)
INCREMENTAL	0.674	1.640	0.989	0.704	0.998
	(0.947)	(1.099)	(1.123)	(0.969)	(0.819)
MULTILATERAL	−0.416	−0.009	0.210	−0.336	0.234
	(0.807)	(0.605)	(0.736)	(0.817)	(0.535)
REPEAT ALLIANCE	0.219	0.254	−0.490	0.234	−0.100
	(0.897)	(0.729)	(1.099)	(0.893)	(0.644)
INTERCEPT	-3.076^{***}	-2.351^{**}	-3.957^{***}	-3.109^{***}	-1.715^{*}
	(1.134)	(1.142)	(1.308)	(1.157)	(0.884)
n		208			208
Log-likelihood		−199.33			−162.93
Chi-square		37.48^{***}			34.36^{***}
d.o.f.		18			14
Prob. > χ^2		0.005			0.002

Significant at * 10%, ** 5%, and *** 1% for 2-tailed tests.
Standard errors in parentheses.

no reason to expect that they would employ a non-standard calculus with respect to the governance structure adopted, contingent on this scope decision. For TECHNOLOGY OVERLAP, our argument rests primarily on the feasibility of achieving alliance objectives that require broad knowledge sharing, rather than on appropriability considerations *per se*. Consequently, we include TECHNOLOGY OVERLAP in the scope but not the governance model specification.[15]

Table 3 presents results of our estimation of the governance choice model. Column 1 includes our composite SCOPE measure as an independent variable, and column 2 instead uses two dummy variables: one for the addition of manufacturing activities (MANUFACTURING) and one for addition of marketing activities to the alliance (MARKETING). Constrained models with industry and institutional environment variables omitted were also estimated and compared with the unconstrained models shown, to see if inclusion of the controls is warranted. In each case, likelihood ratio tests indicated that both sets of control variables should be included in the governance specification.[16]

These governance models produce results that are largely consistent with relevant findings from

[15] As a robustness check, we re-estimated the governance model with the addition of these currently omitted variables and found substantially similar results. These results are available from the authors on request.

[16] $\chi^2[4] = 57.74$ and 56.53 for LR tests comparing fully constrained models (with no industry or institutional environment controls) against the models shown in columns 1 and 2, respectively.

Table 3. Governance choice

Probit regression model
Dependent variable is alliance governance form.
Positive coefficients indicate increased probability that firms select an equity joint venture

	(1)	(2)
MARKET OVERLAP	−1.243**	−1.575**
	(0.588)	(0.638)
GEOGRAPHIC OVERLAP	−0.454	−0.664**
	(0.282)	(0.295)
INCREMENTAL	−0.703	−0.669
	(0.466)	(0.483)
MULTILATERAL	0.766**	0.884**
	(0.360)	(0.374)
REPEAT ALLIANCE	0.322	0.346
	(0.460)	(0.471)
SCOPE	0.680***	
	(0.259)	
MANUFACTURING		0.985***
		(0.326)
MARKETING		0.138
		(0.296)
TELECOMMUNICATIONS	0.512*	0.517*
	(0.304)	(0.313)
MICROELECTRONICS	0.030	0.081
	(0.281)	(0.287)
JUDICIAL SYSTEM	−0.145	−0.094
	(0.141)	(0.143)
NON-TARIFF BARRIERS	−0.224**	−0.227**
	(0.112)	(0.110)
INTERCEPT	1.723	1.313
	(1.550)	(1.589)
n	169	169
Log-likelihood	−66.18	−63.65
χ^2	28.73***	33.79***
d.o.f.	10	11
Prob. > χ^2	0.001	0.000

Significant at * 10%, ** 5%, and *** 1% level for 2-tailed tests.
Standard errors in parentheses.

previous studies (e.g., Pisano, 1989; Oxley, 1997). In particular, the coefficient on scope is positive and significant in column 1. Thus, consistent with previous estimation of reduced-form governance models, we find that equity joint ventures are preferred over contractual alliances when alliance activities are broad, i.e., when they include manufacturing and/or marketing in addition to R&D. Interestingly, this result seems to be driven primarily by the addition of manufacturing activities rather than marketing activities. In column 2, the coefficient on MANUFACTURING is positive and highly significant, while for MARKETING the coefficient is not significant.

Also consistent with prior studies, both specifications indicate that equity joint ventures are more likely to be selected when there are more than two partners linked in the alliance (the coefficient on MULTILATERAL is positive and significant). However, in contrast to prior studies, the breadth of innovation sought in the alliance (as proxied by our dummy variable, INCREMENTAL) does not appear to have any effect on alliance governance choice. The presence of prior alliances linking the partner firms also has no discernible effect on alliance governance in either specification.[17]

With respect to market competition variables, there are some interesting and counter-intuitive features of the results. In both specifications, MARKET OVERLAP has a negative and significant effect, suggesting that firms in the same industry are more likely to organize an alliance via contract than to establish an equity joint venture. This runs counter to our argument that greater product market competition increases the competitive impact of knowledge leakage, since in this case we would expect alliance activities involving partners in the same industry to be *more* likely to be embedded in the protective governance structure afforded by an equity joint venture (holding the scope of activities and other relevant alliance characteristics constant). GEOGRAPHIC OVERLAP also has a negative coefficient, but this is only significant in column 2. In this case, we anticipated the possibility of a negative effect because of firms' greater ability to enforce contracts with other domestic companies, relative to foreign-owned entities. This effect might therefore counteract the impact of increased geographic competition and the associated desire to organize the alliance as an equity joint venture (all else equal). We return to this issue and consider alternative explanations for the market overlap results in the concluding section.

Finally, some of the coefficients on the institutional environment variables are also quite counter-intuitive. The efficiency of the judicial system does not appear to affect the likelihood that an alliance will be organized as an equity joint venture, in contrast to findings of previous research (e.g., Sampson, 2004). Even more puzzling, there is a negative association between the extent of non-tariff barriers and the likelihood of an equity joint venture. We would expect a positive effect in this

[17] Repeating the analysis with a count of previous alliances yields similarly insignificant results on REPEAT ALLIANCE and does not change any of the other coefficients in the model.

case since non-tariff barriers may include requirements for local equity participation. Absent any compelling explanation for these results (which are quite robust across the different model specifications and estimation methods reported below) we merely note them as interesting observations for future investigation.

Together with the results of the scope analysis, these findings establish that there are significant areas of overlap in the determinants of alliance scope and governance for the alliances in our sample. This in turn suggests that joint estimation of the two equations may be a more appropriate empirical approach, and one that may provide additional insights over the reduced-form governance equations adopted in prior research. Table 4 displays the results of the first of our alternative methods of joint estimation: the bivariate probit model. This estimation method controls for omitted variables that may affect the scope and governance decisions, but does not allow for non-recursive endogeneity between scope and governance.

The results in columns A1–A2 in Table 4 are generally consistent with those in the simple probit models and in particular reinforce our arguments regarding the determinants of alliance scope: all of the variables that were significant in the simple probit model of alliance scope remain significant when the equation is estimated jointly with the governance equation, with the exception of MARKET OVERLAP, which is no longer significant. Interestingly, few of the variables in the governance equation are significant with bivariate probit estimation, except for alliance scope, which remains positive and highly significant. INCREMENTAL has a negative coefficient (as suggested in previous research such as Pisano, 1989), but is only marginally significant at the 10 percent level.

To tease out the different effects of hazards related to joint manufacturing and joint marketing that were suggested by the earlier analysis of multiple scope categories, Table 4 also presents the results of two additional specifications for alliance scope and governance. The first, in columns B1–B2, replaces the dependent variable in the scope equation with MANUFACTURING, the dummy variable indicating that the alliance includes manufacturing activities in addition to R&D (irrespective of the presence of marketing activities). Similarly, in columns C1–C2 we replace the dependent variable in the scope equation with MARKETING, the dummy variable for marketing activities in the alliance. These alternative specifications yield very interesting results that add significant nuance to our interpretation of the relationship between alliance scope and governance.

Comparing the two modified scope equations (columns B1 and C1) reconfirms the findings from our earlier consideration of multiple scope categories, i.e., that the addition of manufacturing is primarily a capabilities issue: technological overlap is positively and significantly associated with the addition of manufacturing (column B1), but does not have any effect on the decision to engage in joint marketing (column C1). Laggards are also more likely to engage in manufacturing, but there is no discernible effect on the decision to add marketing to an R&D alliance (although we should note that our results may be masking the effect observed in Table 3b, where we saw that alliances involving both manufacturing *and* marketing were the ones most likely to involve laggards.)

Also quite evident when comparing the two modified scope equations is the different effect of market overlap and geographic overlap—both of these variables are negative and significant when predicting the addition of marketing, but have no significant effect on the decision to include manufacturing in the alliance. None of the control variables have any significant effect in either model.

Consideration of the governance equations in columns B2 and C2 adds to this picture of different responses to hazards raised by joint manufacturing and marketing. While the coefficient on MANUFACTURING is positive and highly significant in column B2, the coefficient on MARKETING in C2 does not even approach conventional levels of significance. Indeed, very few of the variables are significant at all in the governance equation in C2, whereas in B2 we see a negative and significant effect for MARKET OVERLAP and a positive and significant effect for MULTILATERAL. These results together suggest that firms decide whether or not to engage in joint manufacturing based on the needs of the project and the capabilities of the partners, and then mitigate the hazards posed by joint manufacturing through choice of an appropriate governance structure. On the other hand, potential hazards raised by joint marketing are primarily mitigated by a reduction of alliance scope—i.e., partners simply avoid joint marketing when they foresee problems—and this decision

Table 4. Alliance scope and governance: bivariate probit models

Bivariate probit (seemingly unrelated) model.
Columns A1 and A2: dependent variables are SCOPE (= 0 for R&D only, = 1 for R&D plus manufacturing and/or marketing) and GOVERNANCE (= 0 for contract, = 1 for equity joint venture)
Columns B1 and B2: dependent variables are MANUFACTURING (= 0 for R&D only, = 1 for R&D plus manufacturing, with or without marketing) and GOVERNANCE
Columns C1 and C2: dependent variables are MARKETING (= 0 for R&D only, = 1 for R&D plus marketing with or without manufacturing) and GOVERNANCE. Positive coefficients indicate increased probability that firms select either broad alliance scope or equity joint venture.

	SCOPE (A1)	GOVERNANCE (A2)	MANF (B1)	GOVERNANCE (B2)	MKTG (C1)	GOVERNANCE (C2)
MARKET OVERLAP	−0.344	−0.512	0.289	−1.327***	−0.828*	−1.048
	(0.329)	(0.343)	(0.348)	(0.448)	(0.491)	(1.160)
GEOGRAPHIC OVERLAP	−0.571***	0.083	0.078	−0.411	−0.399*	−0.405
	(0.199)	(0.223)	(0.232)	(0.261)	(0.220)	(0.607)
LAGGARD	0.400***		0.610***		0.258	
	(0.131)		(0.221)		(0.393)	
TECHNOLOGY OVERLAP	1.213***		1.663**		0.245	
	(0.437)		(0.653)		(1.106)	
INCREMENTAL	0.575	−0.627*	0.297	−0.389	0.707	−0.728
	(0.393)	(0.376)	(0.441)	(0.408)	(0.552)	(0.558)
MULTILATERAL	0.128	0.406	−0.118	0.777**	0.338	0.699
	(0.310)	(0.288)	(0.337)	(0.339)	(0.338)	(0.571)
REPEAT ALLIANCE	−0.270	0.183	−0.135	0.250	−0.155	0.278
	(0.343)	(0.342)	(0.420)	(0.402)	(0.457)	(0.467)
SCOPE		2.142***				
		(0.175)				
MANUFACTURING				2.504***		
				(0.205)		
MARKETING						0.782
						(2.767)
TELECOMMUNICATIONS		0.321*		0.432*		0.455
		(0.186)		(0.239)		(0.300)
MICROELECTRONICS		0.020		0.136		0.012
		(0.107)		(0.200)		(0.277)
JUDICIAL SYSTEM		−0.128		−0.142		−0.175
		(0.108)		(0.131)		(0.148)
NON-TARIFF BARRIERS		−0.166**		−0.198**		−0.216*
		(0.076)		(0.079)		(0.130)
INTERCEPT	−0.805*	0.773	−1.733***	1.225	−1.128*	2.046
	(0.433)	(1.138)	(0.509)	(1.323)	(0.586)	(1.936)
n	169		169		169	
Log-likelihood	−166.58		−138.16		−162.14	
χ^2	160.87***		186.36***		34.97***	
d.o.f	17		17		17	
LR test of rho = 0(χ^2)	3.84*		3.51*		0.01	

Significant at * 10% ** 5%, and *** 1% level for 2-tailed tests.
Standard errors in parentheses.

is essentially independent of the choice of governance structure.[18]

Our last set of analyses, reported in Table 5, explores the relationship between the different

[18] The fact that rho, the correlation between the scope and governance error terms, is significant for the estimation shown in columns B1–B2 (i.e., broad scope = R&D plus manufacturing) but not for the C1–C2 estimation (i.e., broad scope = R&D plus marketing) supports this conclusion. That is, the decision to enter into joint manufacturing and the alliance governance choice are linked, while the joint marketing and alliance governance choices do not appear to be so.

Table 5. Alliance scope and governance: simultaneous probit models

Columns A1 and A2. dependent variables are SCOPE (= 0 for R&D plus manufacturing and/or marketing) and GOVERNANCE (= 0 for contract, = 1 for equity joint venture)
Columns B1 and B2. dependent variables are MANUFACTURING (= 0 for R&D only, = 1 for R&D plus manufacturing, with or without marketing) and GOVERNANCE
Columns C1 and C2 dependent variables are MARKETING (= 0 for R&D only, = 1 for R&D plus marketing with or without manufacturing) and GOVERNANCE. Positive coefficients indicate increased probability that firms select either broad alliance scope or equity joint venture

	SCOPE (A1)	GOVERNANCE (A2)	MANF (B1)	GOVERNANCE (B2)	MKTG (C1)	GOVERNANCE (C2)
MARKET OVERLAP	−0.999	−1.078	−0.098	−1.660***	−1.393*	0.090
	(0.745)	(0.584)	(0.761)	(0.624)	0.816	1.458
GEOGRAPHIC OVERLAP	−0.823**	−0.102	−0.198	−0.707**	−0.547*	0.465
	(0.323)	(0.396)	(0.334)	(0.313)	(0.317)	(1.085)
LAGGARD	0.489		0.721*		0.394	
	(0.345)		(0.388)		(0.328)	
TECHNOLOGY OVERLAP	1.892*		2.482**		0.405	
	(1.012)		(1.051)		(0.981)	
INCREMENTAL	0.414	−1.026*	0.133	−0.642	0.541	−1.798
	(0.602)	(0.537)	(0.603)	(0.466)	(0.687)	(1.528)
MULTILATERAL	0.313	0.685	−0.196	0.926**	0.607	0.303
	(0.545)	(0.443)	(0.595)	(0.370)	(0.532)	(0.863)
REPEAT ALLIANCE	0.006	0.350	0.015	0.338	−0.089	0.422
	(0.404)	(0.631)	(0.428)	(0.499)	(0.432)	(1.010)
GOVERNANCE = EJV	0.632**		0.848***		0.413	
	(0.285)		(0.301)		(0.288)	
SCOPE		0.613**				
		(0.263)				
MANUFACTURING				0.940***		
				(0.318)		
MARKETING						0.397
						(0.272)
TELECOMMUNICATIONS		0.162*		0.572*		1.130
		−(0.092)		(0.330)		(0.884)
MICROELECTRONICS		0.024		0.150		0.492
		−(0.074)		(0.305)		(0.718)
JUDICIAL SYSTEM		−0.053		−0.032		−0.455
		−(0.037)		(0.166)		(0.425)
NON-TARIFF BARRIERS		−0.053**		−0.221*		−0.163
		−(0.025)		(0.119)		(0.204)
INTERCEPT	0.352	0.797*	−2.109***	0.964	−1.503**	5.691
	−(0.224)	−(0.449)	(0.684)	(1.705)	(0.750)	(5.170)
n		169		169		169
Log-likelihood		−99.26		−72.28		−92.39
d.o.f.	8	10	8	10	8	10

Significant at * 10%, ** 5%, and *** 1% level for 2-tailed tests.
Standard errors in parentheses.

aspects of alliance scope and governance further and, in particular, allows us to test Hypotheses 4a and 4b, that broad scope increases the likelihood that firms select an equity joint venture and, conversely, firms selecting an equity joint venture are more willing to undertake broad alliance activities. Here we estimate non-recursive simultaneous equation models that allow us to fully endogenize scope and governance, including scope in the governance equation and vice versa.[19]

[19] Recursive models that omit governance from the scope equations, similar to the bivariate probit, yield results that are materially identical to those in Table 4.

Notice first that the results of these analyses are quite consistent with those of the bivariate probit estimation, with some slight changes in significance for some of the variables. Furthermore, the results in columns A1 and A2 suggest that scope and governance do indeed act as substitutes: GOVERNANCE = EJV is positive and significant in the scope equation *and* the coefficient on scope is positive and significant in the governance equation. Thus, when alliance scope is narrowed the need to embed the alliance activities in a protective governance structure, such as the equity joint venture, is reduced. Conversely, when the alliance is organized as an equity joint venture, the partners are more willing to engage in activities of broad scope. These results thus support Hypotheses 4a and 4b. Estimation of the models with modifications to the scope equation once again adds nuance to this interpretation of the relationship between scope and governance and reinforces the earlier results. The substitution effect between scope and governance is very clear in the model in columns B1–B2, where we see that the adoption of an equity joint venture structure significantly increases the probability of joint manufacturing, and vice versa. In contrast, there is no significant link between joint marketing activities and governance choice, but the effect of market and geographic competition, reducing partner firms' willingness to engage in joint marketing, is still readily apparent.

Before we turn to a broader discussion of the implications of these findings, it is worth noting one remaining issue regarding the robustness of the empirical results. In this study we have taken a significant step beyond the focus on a single alliance decision common to previous research, and have analyzed the joint determination of alliance scope and governance decisions. We nonetheless were forced to ignore other potentially endogenous decisions in order to render the problem analytically and empirically tractable. In particular, it is quite possible that partner selection decisions are based, at least in part, on some of the same considerations that we have identified as affecting the scope and governance of alliance activities. If firms select partners that are most compatible in terms of technological overlap and competitive position and presence, for example, then ignoring this selection process may introduce bias into our analysis of subsequent alliance decisions (Shaver, 1998). Furthermore, the characteristics of our alliance sample do point towards the possibility of such selection bias—for example, we see that relatively few alliances in our sample involve leading firms, or partners with their primary activities in the same industry, or partners with very high levels of technological overlap—all conditions that might be expected to raise the hazards of cooperation, according to our arguments. Fully incorporating the partner choice decision into our analysis clearly goes beyond the scope of this paper, but endogenizing this decision represents an interesting avenue for future research.

In the meantime, as a robustness check to test whether partner selection issues significantly bias our current findings we generated a sample of 'non-allying' firm pairs, randomly drawn from the pool of firms in our sample and then filtering for the condition that the pair did not have any joint alliances reported in the SDC database. We then added this non-allying sample to our sample of alliances and estimated a probit model that predicted alliance formation based on partner characteristics. The results of this analysis suggest that MARKET OVERLAP and LAGGARD are indeed significant predictors of alliance pairings. Then, to explore the extent of any self-selection bias in our results, we estimated a two-stage model of alliance scope, with the partner selection decision as a first stage. Our results on the choice of alliance scope are robust even with this correction.[20]

DISCUSSION AND CONCLUSION

The findings reported above provide support for our contention that the alliance scope decision is an important aspect of alliance management. Our results also provide support for the argument that decisions to restrict alliance scope are made, at least in part, as a response to the elevated leakage concerns associated with knowledge sharing in particular competitive contexts, and that restrictions to alliance scope may substitute for protective governance structures in such circumstances. However, we also showed that protecting technological capabilities is not the complete story, as issues of absorptive capacity and feasibility are implicated in the alliance scope decision, particularly with respect to joint manufacturing. Below, we

[20] These results are available on request from the authors.

discuss the possibility of an alternative explanation for these results, and explore the implications of our findings for alliance research and for studies of firm boundaries and governance more broadly. We also speculate on some useful directions for future research.

One potential concern about our findings is the possibility of alternative explanations. In particular, it is possible that the scope and governance decisions may be driven more by antitrust issues than by the private risks of collaboration highlighted in our theoretical arguments. Firms headquartered in the United States and/or operating in the same 4-digit SIC code may be subjected to greater scrutiny from antitrust authorities when they expand collaboration beyond pure R&D activities and this could explain why we observe a negative effect of product and geographic market overlap on alliance scope. However, there are two reasons to believe that antitrust concerns are not the primary driver here. First, our domain of study—R&D alliances—mitigates this concern: The National Cooperative Research Act (NCRA) of 1984, along with its 1993 extension, the National Cooperative Research and Production Act (NCRPA), encourages (or at least does not discourage) the formation of consortia and joint ventures for R&D, both alone and in combination with production, by members of the same industry.[21] Second, given that equity joint ventures in general tend to come under closer scrutiny from antitrust authorities than non-equity alliances, we would expect to see the effects of product and/or geographic market overlap showing up in the governance equation consistently more strongly than in the alliance scope equation if antitrust concerns were a significant issue. In general, this is not what we observe.

The analysis presented in this paper suggests that alliance managers can and do pay attention to the competitive implications of potential loss of control of technological assets that comes with cooperation in R&D, and that these considerations play a role in the design of the content and governance of R&D alliances. As such, the study reaffirms some of the main conclusions of previous studies of alliance organization undertaken within the transaction cost economics tradition. The study also highlights limitations of the exclusive focus on alliance governance in much of this previous research, however. Although the point has long been made that transaction cost economizing involves joint determination of the content and governance of transactions (Riordan and Williamson, 1985; Oxley, 1999), it has rarely been implemented in prior empirical work; instead, the most common empirical approach has been to 'take the transaction as given' and to test TCE predictions regarding the effect of transaction characteristics on governance. As suggested earlier, this reduced-form approach has simplified analysis and has yielded important insights, but it has also led to criticisms that the transaction cost approach pays insufficient attention to 'value creation' in alliances (Zajac and Olsen, 1993) and to the social and/or competitive context in which a transaction is embedded (Granovetter, 1985; Day and Klein, 1987). By focusing explicitly on decisions regarding the scope of alliance activities, and on the influence of the competitive context on the scope decision, we are able to confront these criticisms here. And indeed our results suggest that the reduced-form analysis of alliance governance employed in prior TCE studies may have overstated the role of governance in alliance management. When the scope and governance equations for our sample of R&D alliances are estimated jointly, we find that some of the factors previously associated with differences in alliance governance appear to operate more directly on the scope of alliance activities, particularly when the hazards of cooperation are related to joint marketing activities as opposed to joint manufacturing. Of course our analysis is far from definitive in this regard, but it suggests an important avenue for further research on alliance management (and in the study of governance more generally); i.e., to further explore the joint determinants of the content and governance of these relationships.

Our study also points to important complementarities between theoretical approaches that emphasize hazard mitigation in alliances (such as TCE) and those that place more emphasis on crafting

[21] The NCRA altered antitrust treatment of research consortia in two significant ways: first, it ensured that cooperative research ventures were subjected to rule of reason analysis rather than *per se* rules (i.e., their pro-competitive effects must be weighed against potential anti-competitive concerns). Second, it limited the liability of registered consortia participants to single rather than treble damages. These safeguards were extended to joint ventures involving both production and R&D with the 1993 NCRPA. See Jorde and Teece (1993) for a more thorough analysis of the NCRA and the NCRPA.

inter-firm alliances to promote effective knowledge sharing, such as the information-processing perspective (Sobrero and Roberts, 2001) or knowledge management perspective (Inkpen, 2000). While research in this latter tradition tends to take more of an organizational process approach, it does direct attention to the role of cognitive limitations in restricting the objectives achievable within a particular alliance. Our study suggests that these cognitive limitations (as manifested in the importance of adequate absorptive capacity to support broad-scope alliances involving joint manufacturing) are an important consideration, in addition to—though not to the exclusion of—concerns with mitigating the hazards of potential opportunism. In particular, our results suggest that the decision to bring manufacturing within the scope of the alliance is driven largely by feasibility issues related to overlapping capabilities and absorptive capacity; in situations where such joint manufacturing raises the fear of leakage and other competitive concerns (such as when multiple partners are involved and so monitoring of alliance activities and delineation of property rights is more complex) partners choose to embed their activities in a protective alliance structure such as an equity joint venture. The relationship between these issues of feasibility and hazard mitigation deserves further attention in future work. Extension of our own study in this direction is hampered by the crudeness of our measures (for both our dependent and independent variables), highlighting an important limitation of the analysis. Future work would benefit from more fine-grained and multidimensional measures, particularly of alliance scope, partner capabilities, and competitive and social context. This is a challenging proposition for large-scale analysis and would undoubtedly require access to primary data sources (for example, through a survey of alliance participants), but it is an area ripe for research.

In sum, the analysis presented here indicates that when firms design R&D alliances with an eye to acquiring, protecting, and leveraging technological resources, the choice of an appropriate governance structure is not the only critical decision to be made; delineating the scope of alliance activities is also an important and complementary decision. Although preliminary, our analysis suggests that this is a line of enquiry with potentially important implications for the theory and management of inter-firm alliances. In addition, the work contributes to an emergent research stream aimed at understanding the relationship between firm strategy and efficient economic organization. In common with other recent work in this area (e.g., Nickerson, Hamilton, and Wada, 2001) our research suggests that a complete understanding of firm boundary decisions requires the adoption of a combined positioning–economizing perspective (Nickerson, 1997) that takes into account the interdependence of market position, firm resources, and governance.

ACKNOWLEDGEMENTS

Many people contributed to the development of the ideas and analysis in this paper. In particular, we would like to thank the following people: Christina Ahmadjian, Jay Barney, Nicolai Foss, Bill Greene, Scott Masten, Anita McGahan, David Mowery, and Sid Winter for valuable feedback and suggestions; Francisco Polidoro and Puay Khoon Toh for stellar research assistance; and the *SMJ* Special Issue co-editors, reviewers, and fellow contributors for constructive comments, suggestions, and encouragement. Remaining errors are naturally our responsibility.

REFERENCES

Barney JB. 1986. Strategic factor markets: expectations, luck, and business strategy. *Management Science* **32**(10): 1231–1241.

Basberg BL. 1982. Technological change in the Norwegian whaling industry: a case study in the use of patent statistics as a technology indicator, *Research Policy* **11**: 163–171.

Branstetter LG., Sakakibara M. 2002. When do research consortia work well and why? Evidence from Japanese panel data. *American Economic Review* **92**(1): 143–159.

Clark KB, Fujimoto T. 1991. *Product Development Performance: Strategy, Organization and Management in the World Auto Industry*. Harvard Business School Press: Boston, MA.

Cohen WM, Levinthal DA. 1990. Absorptive capacity: a new perspective on learning and innovation. *Administrative Science Quarterly* **35**: 128–152.

Comanor WS, Scherer FM. 1969. Patent statistics as a measure of technical change, *Journal of Political Economy* **77**: 329–398.

Cusumano MA, Selby RW. 1995. *Microsoft Secrets: How the World's Most Powerful Software Company Creates Technology, Shapes Markets, and Manages People*. Free Press: New York.

Day GS, Klein S. 1987. Cooperative behavior in vertical markets: the influence of transaction costs and competitive strategies. In *Review of Marketing*, Houston MJ (ed). American Marketing Association: Chicago, IL; 39–66.

Dierickx I, Cool K. 1989. Asset stock accumulation and sustainability of competitive advantage. *Management Science* **35**(12): 1504–1514.

Doz YL, Hamel G. 1998. *Alliance Advantage*. Harvard Business School Press: Boston, MA.

Dussauge P, Garrette B, Mitchell W. 2000. Learning from competing partners: outcomes and durations of scale and link alliances in Europe, North America and Asia. *Strategic Management Journal* **21**(2): 99–126.

Duysters G, Hagedoorn J. 1996. Internationalization of corporate technology through strategic partnering: an empirical investigation. *Research Policy* **25**: 1–12.

Eisenhardt KM, Schoonhoven CB. 1996. Resource-based view of strategic alliance formation: strategic and social effects in entrepreneurial firms. *Organization Science* **7**(2): 136–150.

Eisenhardt KM, Tabrizi BN. 1995. Accelerating adaptive processes: product innovation in the global computer industry. *Administrative Science Quarterly* **40**(1): 84–110.

Gomes-Casseres B. 1996. *The Alliance Revolution*. Harvard University Press: Cambridge, MA.

Granovetter M. 1985. Economic action and social structure: the problem of embeddedness. *American Journal of Sociology* **91**(November): 481–501.

Greene WH. 1997. *Econometric Analysis* (3rd edn). Prentice-Hall: Upper Saddle River, NJ.

Gulati R. 1995a. Social structure and alliance formation patterns: a longitudinal analysis. *Administrative Science Quarterly* **40**: 619–652.

Gulati R. 1995b. Does familiarity breed trust? The implications of repeated ties for contractual choice in alliances. *Academy of Management Journal* **38**(1): 85–112.

Gulati R, Singh H. 1998. The architecture of cooperation: managing coordination costs and appropriation concerns in strategic alliances. *Administrative Science Quarterly* **43**: 781–814.

Hagedoorn J. 1993. Understanding the rationale of strategic technology partnering: interorganizational modes of cooperation and sectoral differences. *Strategic Management Journal* **14**(5): 371–385.

Hamel G, Doz YL, Prahalad CK. 1989. Collaborate with your competitors—and win. *Harvard Business Review* **67**(1): 133–139.

Henderson RM, Clark KB. 1990. Architectural innovation: the reconfiguration of existing product technologies and the failure of established firms. *Administrative Science Quarterly* **35**(March): 9–30.

Hergert M, Morris D. 1988. Trends in international collaborative arrangements. In *Cooperative Strategies in International Business*, Contractor FJ, Lorange P (eds). Lexington Books: Lexington, MA; 99–109.

Imai K, Nonaka I, Takeuchi H. 1985. Managing the new product development process: how the Japanese companies learn and unlearn. In *The Uneasy Alliance*, Clark KB, Hayes RH, Lorenz C (eds). Harvard Business School Press: Boston, MA; 337–377.

Inkpen AC. 2000. Learning through joint ventures: a framework of knowledge acquisition. *Journal of Management Studies* **37**(7): 1019–1045.

Jaffe AB. 1986. Technological opportunity and spillovers of R&D: evidence from firms' patents, profits and market value. *American Economic Review* **76**: 984–1001.

Jorde TM, Teece DJ. 1993. Rule of reason analysis of horizontal arrangements: agreements designed to advance innovation and commercialize technology. *Antitrust Law Journal* **61**: 579–619.

Kale P, Singh H, Perlmutter H. 2000. Learning and protection of proprietary assets in strategic alliances: building relational capital. *Strategic Management Journal* **21**(3): 217–237.

Khanna T. 1998. The scope of alliances. *Organization Science* **9**(3): 340–355.

Khanna T, Gulati R, Nohria N. 1998. The dynamics of learning alliances: competition, cooperation and relative scope. *Strategic Management Journal* **19**(3): 193–210.

Klavans R. 1993. Technology strategy and competitive intelligence. In *Global Perspectives on Competitive Intelligence*, Prescott JE, Gibbons PT (eds). Society of Competitive Intelligence Professionals: Alexandria, VA; 129–136.

Kogut B. 1988. Joint ventures: theoretical and empirical perspectives. *Strategic Management Journal* **9**(4): 319–332.

Kogut B, Zander U. 1992. Knowledge of the firm, combinative capabilities and the replication of technology. *Organization Science* **3**: 383–397.

Koka BR, Prescott JE. 2002. Strategic alliances as social capital: a multidimensional view. *Strategic Management Journal* **23**(9): 795–816.

Krishnan V, Eppinger SD, Whitney DE. 1997. A model-based framework to overlap product development activities. *Management Science* **43**(4): 437–451.

Lane PJ, Lubatkin M. 1998. Relative absorptive capacity and interorganizational learning. *Strategic Management Journal* **19**(5): 461–477.

Levin RC, Klevorick AK, Nelson RR, Winter SG. 1987. Appropriating the returns from industrial research and development. *Brookings Papers on Economic Activity* **3**: 783–833.

Loch CH, Terwiesch C. 2000. Product development and concurrent engineering. In *Innovations in Competitive Manufacturing*, Swamidass PM (ed). Kluwer: Dordrecht; 263–274.

Mitchell W, Singh K. 1996. Survival of businesses using collaborative relationships to commercialize complex goods. *Strategic Management Journal* **17**(3): 169–195.

Mowery DC, Oxley JE, Silverman BS. 1996. Strategic alliances and inter-firm knowledge transfer. *Strategic Management Journal*, Winter Special Issue **17**: 77–91.

Mowery DC, Oxley JE, Silverman BS. 1998. Technological overlap and inter-firm cooperation: implications

for the resource-based view of the firm. *Research Policy* **27**: 507–523.
Nakamura M, Shaver JM, Yeung B. 1996. An empirical investigation of joint venture dynamics: evidence from US–Japan joint ventures. *International Journal of Industrial Organization* **14**: 521–541.
Narula R, Hagedoorn J. 1999. Innovating through strategic alliances: moving towards international partnerships and contractual agreements. *Technovation* **19**(5): 283–294.
Nickerson JA. 1997. Toward an economizing theory of strategy: The choice of strategic position, assets, and organizational form. Doctoral dissertation, University of California, Berkeley.
Nickerson JA, Hamilton BH, Wada T. 2001. Market position, resource profile, and governance: linking Porter and Williamson in the context of international courier and small package services in Japan. *Strategic Management Journal* **22**(3): 251–274.
OECD (Organization for Economic Cooperation and Development). 2000. *OECD Information Technology Outlook 2000: ICTs, E-Commerce and the Information Economy*. OECD: Paris.
Oxley JE. 1997. Appropriability hazards and governance in strategic alliances: a transaction cost approach. *Journal of Law, Economics and Organization*, **13**(2): 387–409.
Oxley JE. 1999. Institutional environment and the mechanisms of governance: the impact of intellectual property protection on the structure of inter-firm alliances. *Journal of Economic Behavior and Organization* **38**(3): 283–309.
Park SH, Ungson GR. 1997. The effect of national culture, organizational complementarity and economic motivation on joint venture dissolution. *Academy of Management Journal* **40**(2): 279–307.
Patel P, Pavitt K. 1997. The technological complexities of the world's largest firms: complex and path-dependent, but not much variety. *Research Policy* **26**: 141–156.
Pisano G. 1989. Using equity participation to support exchange: evidence from the biotechnology industry. *Journal of Law, Economics and Organization* **5**(1): 109–126.
Pisano G, Russo M, Teece D. 1988. Joint ventures and collaborative arrangements in the telecommunications equipment industry. In *International Collaborative Ventures in U.S. Manufacturing*, Mowery D (ed). Ballinger: Cambridge, MA; 23–70.
Prescott JE. 2001. Competitive intelligence: lessons from the trenches. In *Proven Strategies in Competitive Intelligence*, Prescott JE, Miller SH (eds). Wiley: New York; 1–22.
Red Herring. 1997. The Herring 100: The best private and public technology companies of 1997. *Red Herring* 1 September.
Reich RB, Mankin ED. 1986. Joint ventures with Japan give away our future. *Harvard Business Review* **64**(2): 78–86.
Reuer JR, Zollo M, Singh H. 2002. Post-formation dynamics in strategic alliances. *Strategic Management Journal* **23**(2): 135–151.
Riordan MH, Williamson OE. 1985. Asset specificity and economic organization. *Industrial Journal of Industrial Organization* **3**: 365–378.
Sabbagh K. 1996. *Twenty-First Century Jet*. Scribner: New York.
Sampson RC. 2004. The cost of misaligned governance in R&D alliances. *Journal of Law Economics and Organization* (forthcoming).
Sanchez R, Mahoney JT. 1996. Modularity, flexibility, and knowledge management in product and organization design. *Strategic Management Journal*, Winter Special Issue **17**: 63–76.
Shaver JM. 1998. Accounting for endogeneity when assessing strategy performance: does entry mode choice affect FDI survival? *Management Science* **44**(4): 571–585.
Shaver JM, Flyer F. 2000. Agglomeration economies, firm heterogeneity and foreign direct investment in the United States. *Strategic Management Journal* **21**(12): 1175–1193.
Sobrero M, Roberts EB. 2001. The trade-off between efficiency and learning in interorganizational relationships for product development. *Management Science* **47**(4): 493–511.
Stuart T. 2000. Interorganizational alliances and the performance of firms: a study of growth and innovation rates in a high-technology industry. *Strategic Management Journal* **21**(8): 791–812.
Takeuchi H, Nonaka I. 1986. The new product development game. *Harvard Business Review* **64**(1): 1–11.
Terwiesch C, Loch CH. 1999. Measuring the effectiveness of overlapping development activities. *Management Science* **45**(4): 455–465.
Teece DJ. 1992. Competition, cooperation, and innovation: organizational arrangements for regimes of rapid technological-progress. *Journal of Economic Behavior and Organization* **18**(1): 1–25.
Wacziarg R. 2001. Measuring the dynamic gains from trade. *World Bank Economic Review* **15**(3): 393–429.
Zajac EJ, Olsen CP. 1993. From transaction cost to transaction value analysis: implications for the study of interorganizational strategies. *Journal of Management Studies* **30**(1): 131–145.

[19]

Strategic Management Journal
Strat. Mgmt. J., **22**: 197–220 (2001)

TECHNOLOGICAL ACQUISITIONS AND THE INNOVATION PERFORMANCE OF ACQUIRING FIRMS: A LONGITUDINAL STUDY

GAUTAM AHUJA[1]* and RIITTA KATILA[2]
[1]*College of Business Administration, University of Texas at Austin, Austin, Texas, U.S.A.*
[2]*Robert H. Smith School of Business, University of Maryland, College Park, Maryland, U.S.A.*

This paper examines the impact of acquisitions on the subsequent innovation performance of acquiring firms in the chemicals industry. We distinguish between technological acquisitions, acquisitions in which technology is a component of the acquired firm's assets, and nontechnological acquisitions: acquisitions that do not involve a technological component. We develop a framework relating acquisitions to firm innovation performance and develop a set of measures for quantifying the technological inputs a firm obtains through acquisitions. We find that within technological acquisitions absolute size of the acquired knowledge base enhances innovation performance, while relative size of the acquired knowledge base reduces innovation output. The relatedness of acquired and acquiring knowledge bases has a nonlinear impact on innovation output. Nontechnological acquisitions do not have a significant effect on subsequent innovation output. Copyright © 2001 John Wiley & Sons, Ltd.

In this paper we examine the impact of acquisitions on the subsequent innovation performance of acquiring firms. Studying the impact of acquisitions on postacquisition innovation performance is important from at least three perspectives. First, this evaluation is important from the perspective of organizational learning and innovation, and helps us understand how organizations absorb and use external knowledge. Firm-level theories of technical change suggest that a firm's innovativeness is an outcome of increases in its knowledge base (Griliches, 1984, 1990; Pakes and Griliches, 1984; Henderson and Cockburn, 1996). While a firm's knowledge base can grow through a series of knowledge-enhancing investments by the company over time, firms can also grow their knowledge through acquiring or 'grafting' of external knowledge bases (Cohen and Levinthal, 1989; Huber, 1991). Interestingly, while the relationship between firms' investments in knowledge and their innovation output has been studied extensively (Hall, Griliches and Hausman, 1986; Griliches, 1990), relatively little research has focused on the role of acquisitions in growing the firm's knowledge base (Granstrand and Sjolander, 1990; Huber, 1991; Gerpott, 1995). This lacuna is all the more surprising given findings which indicate that obtaining technological know-how and developing technical capabilities are increasingly important motives for acquisitions (Link, 1988; Granstrand *et al.*, 1992; Chakrabarti, Hauschildt, and Suverkrup, 1994; Wysocki, 1997a, 1997b).

Scholars studying the market for corporate control have also examined the relationship between acquisitions and innovation output (Hitt *et al.*, 1991, 1996; Hoskisson, Hitt, and Ireland, 1994). However, in contrast to the findings of the innovation literature, studies in the corporate control tradition have generally found that acquisitions have a negative impact on the postacquisition innovation output of acquiring firms. Agency problems, reduction in managerial commitment to

Key words: innovation; acquisitions; knowledge
*Correspondence to: Gautam Ahuja, College of Business Administration, University of Texas at Austin, Austin, TX 78712-1174, U.S.A.

Received 27 August 1998
Final revision received 12 October 2000

innovation, and the absorption of managerial energy in the acquisition integration process at the expense of routine management have been posited as possible explanations for these results (Hitt *et al.*, 1991, 1996). This conclusion, however, is puzzling since acquisitions continue to be a popular strategy for corporate growth. In recent years more dollars have been invested in acquisition activity than in any other equivalent time period in history (Curry, 1997). Evaluating the postacquisition innovation output of acquiring firms provides one indicator, albeit an indirect one, of the returns to corporate investments in acquisition activity.

A third reason for studying the impact of acquisitions on the postacquisition performance of acquiring firms comes from the growing literature on the resource-based view of the firm. According to this perspective acquisitions are an important part of the business process of redeploying resources into more productive uses (Anand and Singh, 1997; Capron, Dussauge, and Mitchell, 1998; Capron, Mitchell, and Swaminathan, 1998). Through acquisitions firm-specific assets housed within one organization are merged with assets in another organization to improve the productivity of the combined assets (Haspeslagh and Jemison, 1991; Anand and Singh, 1997). Evaluating the postacquisition performance of firms provides evidence on the efficiency of this asset-matching and combining process.

In this paper we draw upon theories of technological innovation, learning, and the resource-based view to develop a theoretical model and predictions relating acquisition characteristics to the innovation performance of acquiring firms (Cohen and Levinthal, 1989; Grant, 1996). Innovation performance can be measured in terms of innovative inputs such as R&D expenditures, or innovation outputs such as patenting frequency (Griliches, 1984; Henderson and Cockburn, 1996). Acquisitions can affect both innovative inputs and innovative outputs (Hitt *et al.*, 1991). For instance, a firm's R&D expenditures can decrease after it conducts an acquisition as the firm eliminates certain streams of research or as managers become more risk averse (Hitt *et al.*, 1991). Yet, even while research efforts decrease, the productivity of those efforts can increase as the two hitherto separate research teams combine their skills and knowledge. In this research we focus on the impact of acquisitions on innovation *outputs* as measured by the patenting frequency of the acquiring firm. Accordingly, we adopt an innovation production function framework and model patenting frequency as the output of a production function (Griliches, 1984). The hypotheses of this study are statements about the relationship between this patented knowledge output and the firm's stocks of owned and acquired knowledge. In estimating the impact of acquisitions on the innovative output of a firm we statistically control for the levels of innovative inputs such as R&D expenditures, but leave their substantive examination to future work.

HYPOTHESES

In this research we focus on two contingencies that may be critical to clarifying the relationship between acquisitions and the postacquisition innovation performance of acquiring firms. First, we draw attention to the fact that technological reasons do not motivate all acquisitions. For example, acquisitions can be motivated by the desire to obtain access to distribution channels, to gain entry into new markets, or to obtain financial synergies or market power (Lubatkin, 1983; Balakrishnan, 1988; Chatterjee, 1991; Haspeslagh and Jemison, 1991; Capron, Dussauge, and Mitchell, 1998). Such acquisitions may provide no technological inputs to the acquiring firm and therefore cannot be expected to improve its innovation output. Second, we note that even among technologically motivated acquisitions the impact of acquisitions may depend on the characteristics of the relationship between the knowledge of the acquired and acquiring firms (Lubatkin, 1983; Singh and Montgomery, 1987; Lane and Lubatkin, 1998).

In summary, we argue that the impact of acquisitions on the acquiring firm's innovation output can be understood in the context of the technological inputs provided by the acquisition. We argue that acquisitions that provide no technological inputs cannot be expected to have a positive impact on firm innovation output (Hypothesis 1). Second, within acquisitions that do provide technological inputs, we predict that the impact of an acquisition on the postacquisition innovation output of the acquiring firms is likely to vary positively with the absolute size of the acquired firm's knowledge base, negatively with the relative size of the acquired and acquiring knowledge bases, and curvilinearly with the relatedness of the acquired and acquiring knowledge bases (Hypotheses 2, 3 and 4).

Technological vs. nontechnological acquisitions

Acquisitions can affect the acquirer's subsequent innovation output through two possible mechanisms. First, an acquisition of another firm can be viewed as an absorption of the acquired firm's knowledge base into the acquiring firm's knowledge base. Such a union can potentially expand the acquirer's knowledge base and increase its innovation output by providing economies of scale and scope in research and by enhancing the acquirer's potential for inventive recombination (Henderson and Cockburn, 1996; Fleming, 1999). Since nontechnological acquisitions add less to the knowledge base of the acquirer they are less likely to lead to such innovation output-enhancing effects.[1] However, an acquisition can also disrupt the established routines of the acquiring firm and those of its newly acquired component, and thereby reduce productivity (Jemison and Sitkin, 1986; Haspeslagh and Jemison, 1991). Prior research suggests that acquisition integration entails far-reaching disruption, and involves significant managerial attention and transactions costs (Pritchett, 1985; Haspeslagh and Jemison, 1991; Hitt *et al.*, 1991, 1996; Hoskisson *et al.*, 1994). However, whether this disruption related to nontechnological acquisitions will produce a negative impact on innovative productivity is not clear, *a priori*. On one hand it is possible that as management focuses more on acquisitions and their integration, decision making on routine technological matters can be delayed, activities such as product championing can suffer, and a crisis mentality on the management of the acquisition can lead to only residual energies being supplied to day-to-day operations even in the technological core of the company (Thompson, 1967; Pritchett, 1985; Hitt *et al.*, 1996). Alternately, it is possible that since such acquisitions do not involve a technological component by definition, they may not affect the technological subsystem (Thompson, 1967) and innovation routines of the firm, and therefore have no impact on innovation output at all. The final impact of an acquisition depends on the degree to which these effects come into play.

Accordingly, we present the following base hypothesis:

Hypothesis 1: Nontechnological acquisitions will affect the postacquisition innovation output of acquiring firms either negatively or nonsignificantly.

Technological acquisitions and the absolute size, relative size, and relatedness of knowledge bases

Technological acquisitions are acquisitions that provide technological inputs to the acquiring firm. Thus, they potentially expand the acquirer's knowledge base and provide scale, scope, and recombination benefits (Henderson and Cockburn, 1996; Fleming, 1999). However, technological acquisitions can also entail a disruption in organizational routines. Further, this disruption is most likely in the set of routines that are closest to the innovation arena, the technological subsystem of the firm. Thus, technological acquisitions can also have a negative impact on the innovation output of the acquiring firm. On balance, assessing whether technological acquisitions will have a positive or negative impact on postacquisition innovation output is likely to depend upon the quantity and nature of knowledge elements that they bring to the acquiring firm. To evaluate whether the scale, scope, and inventive recombination benefits of acquisitions outweigh their negative effects on organizational routines we compare the knowledge bases of acquired and acquiring firms along three key characteristics that have also been prominently used in prior acquisition research: absolute size, relative size, and relatedness (Lubatkin, 1983). In the following sections we examine and develop distinct hypotheses for each of these characteristics.

Absolute size of acquired knowledge base

The absolute size of an acquired knowledge base can affect the post acquisition innovation output of the acquiring firm through at least two mechanisms, both indicating a positive effect of larger

[1] As suggested by one of the reviewers, in some cases nontechnological acquisitions can increase innovation output, for instance by providing contacts with new customers.

knowledge bases. First, the integration of the two hitherto separate knowledge bases may enable enhanced economies of scale and scope (Henderson and Cockburn, 1996). For instance, a long tradition of research in technology suggests that new innovative outputs are often the result of recombining existing elements of knowledge into new syntheses (Schumpeter, 1934; Henderson and Clark, 1990; Kogut and Zander, 1992; Tushman and Rosenkopf, 1992; Utterback, 1994; Fleming, 1999). From this combinatorial perspective, the number of direct combinations that a firm can create from its own knowledge elements increases with the size of the acquired knowledge base. While a firm with five units of knowledge can generate 10 combinations using two elements at a time, acquiring another firm with three units of knowledge increases the number of combinations that become available to 28. Similarly, the merger of two knowledge bases can also provide scale or scope economies by reducing duplication in research efforts or by providing a larger research base to defray costs.

Second, acquiring a larger knowledge base may enhance a firm's absorptive capacity. Prior research indicates that as a firm expands its internal knowledge base and technological capability, it also enhances its ability to absorb and utilize external knowledge (Cohen and Levinthal, 1989; Cohen and Levinthal, 1990). Thus, when a firm acquires a knowledge base it obtains access not only to the acquired firm's internally created knowledge but also to a larger external domain of knowledge that is understood and used by the acquired firm. Thus, acquisitions increase the number of elements of both internal and external knowledge that are available to the acquiring firm and hence increase its potential for inventive recombination. Hence, we hypothesize:

Hypothesis 2: The greater the absolute size of the acquired knowledge base, the greater the subsequent innovation output of the acquiring firm.

Relative size of acquired knowledge base

The arguments above focused on the increased scale, scope, and recombination benefits possible through the acquisition of a knowledge base. Yet, several steps must be completed before newly acquired knowledge can improve the acquirer's performance. The acquirer needs to recognize the value and content of the acquired knowledge, assimilate it, and apply it (Cohen and Levinthal, 1990). The degree to which these tasks can be successfully accomplished is likely to vary with the relative size of the acquiring and acquired knowledge bases. The larger the *relative* size of the knowledge base to be integrated, the more difficult these stages are likely to be, and the more negative the impact on postacquisition innovation output.

Knowledge is primarily transferred through interactions between the acquired and acquiring units, and entails both teaching and learning on both sides (Haspeslagh and Jemison, 1991). Integration teams, meetings within and between the two R&D units, and extensive face-to-face communication are integral parts of the process by which the merging units learn about each other's technology and processes (Gerpott, 1995). Since every communication needs both a sender and a receiver, the relative size of the knowledge bases in the merger becomes relevant. If the merged knowledge bases are relatively equal in size, most of the knowledge resources of the combined firm will be devoted to the task of integrating the two knowledge bases. As the two approximately equal groups educate each other, fewer resources will be available for conducting the actual business of innovation. On the other hand, if the two knowledge bases are relatively dissimilar in size, the absorption and assimilation activity will occupy only a part of the larger group's resources even if it entails the preoccupation of the smaller of the two groups.

A second dimension of the relative size effect is the disruption of existing organizational routines (Haspeslagh and Jemison, 1991; Singh and Zollo, 1997). For successful assimilation and application of the newly acquired knowledge, many changes have to be introduced into the functioning of the organization. Pathways of communication, routing of work and authority, and formal and informal organizational structures all have to be adapted to incorporate the acquired unit's knowledge (Gerpott, 1995). If the acquired firm's knowledge base is small relative to the acquirer the modifications required are likely to be minor, and therefore not be very disruptive. However, if the acquired firm's knowledge base is large relative to the acquiring firm, fairly major changes would have to be made in the acquiring firm, leading to a significant disruption of existing processes. Accordingly, we predict:

Hypothesis 3: The greater the relative size of the acquired knowledge base, the less the subsequent innovation output of the acquiring firm.

Relatedness of acquired and acquiring knowledge bases

The third critical dimension in the unification of two knowledge bases is their relatedness (Lubatkin, 1983; Singh and Montgomery, 1987; Lane and Lubatkin, 1998). While the previous arguments concerned the *magnitude* of the acquired and acquiring knowledge bases the relatedness argument concerns the *content* of these knowledge bases. We predict that relatedness between the acquired and acquiring knowledge bases is likely to have a nonmonotonic influence on the subsequent innovation performance of acquiring firms. Innovation output will increase with increasing relatedness, but beyond some optimum innovation output will decrease with increasing relatedness.

The absorptive capacity argument suggests that the ability to use new information to solve problems is enhanced when the new knowledge is related to what is already known (Cohen and Levinthal, 1990). Elements of similar knowledge facilitate the integration of the acquired and acquiring knowledge bases (Kogut and Zander, 1992; Grant, 1996). Common skills, shared languages, and similar cognitive structures enable technical communication and learning (Cohen and Levinthal, 1989; Lane and Lubatkin, 1998). Further, the recipes for conducting research, or the innovation routines of the acquired and acquiring firms, are also likely to be different if the firms come from distant realms of technology (Kogut and Zander, 1992; Spender, 1989). In such circumstances the integration of knowledge bases can be resource consuming, or even counterproductive as routines inappropriate to either or both knowledge bases can be adopted (Haspeslagh and Jemison, 1991; Singh and Zollo, 1997).

On the other hand, an acquired knowledge base that is too similar to the acquiring knowledge base may also contribute little to subsequent innovation performance. From an absorptive capacity perspective acquired knowledge can help improve performance through two effects. First, acquired knowledge can provide a cross-fertilization effect as old problems can be addressed through new approaches, or by a combination of old and new approaches (Cohen and Levinthal, 1990). Second, new acquired knowledge can serve as the basis for absorbing additional stimuli and information from the external environment. If an acquisition brings in knowledge that is too closely related to the existing knowledge base of the acquiring firm, both these benefits might be limited.

Knowledge bases with moderate degrees of relatedness provide the benefits of enhancing the variety of possible combinations that the firm can use, while maintaining the elements of commonality that facilitate interaction between the acquired and acquiring knowledge bases. Based on the above arguments we suggest that acquisitions that are characterized by a moderate degree of overlap in the knowledge bases are likely to provide the most significant positive impact on the acquiring firm's subsequent innovation output. Accordingly, we predict:

Hypothesis 4: The relatedness of the acquired knowledge base will be curvilinearly (inverted U) related to the subsequent innovation output of the acquiring firm.

DATA AND METHODS

Organizational knowledge base

In the previous section we argued that an acquisition can be viewed as the union of two knowledge bases. Measuring an organizational knowledge base is then the key operational issue in testing the hypotheses. Specifically, we need to identify empirical measures that capture the absolute size, relative size, and relatedness of knowledge bases. In the arguments below we suggest that an organization's patent portfolio provides a means for capturing these dimensions and mapping an organization's knowledge.

A patent, by definition, represents a unique and novel element of knowledge. A set of patents then represents a collection of discrete, distinct units of knowledge. Identifying a set of patents that a firm has demonstrated familiarity with, or mastery of, can be a basis for identifying the 'revealed' knowledge base of a firm, the distinct elements of knowledge with which the firm has revealed a relationship (Kim and Kogut, 1996).

The patents owned by a firm represent the knowledge that the firm is acknowledged as having created (Jaffe, Trajtenberg and Henderson, 1993). Such patents are naturally elements of the firm's knowledge base. However, the firm's patents also build on prior patents, the knowledge created by the same firm in the past or by other firms which have preceded it in that line of inquiry. These prior patents are cited in the patents as recognition of their contribution to the knowledge embodied in the focal patent. By creating a patent that builds on these prior patents, the firm provides evidence that the knowledge contained in those past patents is a part of the firm's knowledge set. Thus, the patents cited by the firm should also be included in its knowledge base. The Appendix describes in more detail the logic underlying this approach to measuring organizational knowledge bases.

Using patent data to measure organizational knowledge bases corresponds closely to the conceptual abstraction of a firm's knowledge base as a set of knowledge elements (Grant, 1996). The number of cited and obtained patents provides a measure of the size of the knowledge base. The individual patents in the firm's knowledge base provide the basis for comparing the firm's knowledge base with other knowledge bases. Since each patent number uniquely identifies a distinct component of knowledge, the higher the number of patents that are common across two knowledge bases, the higher the relatedness between those knowledge bases (see also Stuart and Podolny, 1996; Mowery, Oxley and Silverman, 1998).

Methods

A panel data design was used to test the hypotheses. We selected a sample of firms from the global chemicals industry independent of their acquisition behavior, and traced the acquisition behavior of these firms over a 12-year period, from 1980 to 1991 (more details on the sample are provided later). Several of the sample firms were very active in the acquisitions arena, while others conducted few or no acquisitions at all. We attempted to collect data on every acquisition conducted by these firms. The acquisition characteristics of these firms were modeled as time-varying influences on the subsequent innovation performance of these firms. A distributed lag analysis was used to identify the effects of an acquisition on innovation performance in the 4 years succeeding the acquisition.

This approach addresses several methodological problems that arise in evaluating the impact of acquisitions on the postacquisition performance of acquiring firms. First, examining a single industry over a common period controls for industry and period effects that have been cited as a problem with prior acquisition research (Fowler and Schmidt, 1988). Second, this research design naturally includes both firms that are active in acquisitions and those that are inactive. In the absence of the latter group of firms it is difficult to refute the argument that good or bad performance by acquiring firms was not shared by similar nonacquiring firms (Fowler and Schmidt, 1988). Third, this approach resolves the problem of handling firms with multiple acquisitions. To reduce the problem of confounding caused by firms making several acquisitions over the study period, some researchers have omitted such firms from analyses (Fowler and Schmidt, 1988). However, omitting such firms may lead to biased findings. With our research design such firms can be retained in the analysis by including acquisition and firm characteristics as time-varying covariates in a panel regression. Finally, our panel data set includes distributed lag effects. The distributed lag technique enables us to use multiple-period lagged values of the independent variables as additional regressors in the estimated equation (Judge *et al.*, 1988). By using this approach, we can, in principle, trace the effects of the acquisition on the performance of the acquiring firm for several periods after the acquisition.

Model specification and econometric issues

We now describe our econometric approach, followed by a discussion of the industry setting and the variables used to test the hypotheses. The dependent variable of the study, innovation output, as measured by patent counts, is a count variable and takes only non-negative integer values. The linear regression model's assumptions of homoskedastic, normally distributed errors are thus violated. A Poisson regression approach is appropriate for such data (Hausman, Hall, and Griliches, 1984; Henderson and Cockburn, 1996). Accordingly, we specified the following Poisson regression model:

$$P_{it} = \exp (X_{it-1}\gamma + A_{it-1}\beta_1 + A_{it-2}\beta_2 + A_{it-3}\beta_3 + A_{it-4}\beta_4) \quad (1)$$

where P_{it} is the number of patents obtained by firm i in year t, X_{it-1} is a vector of control variables affecting P_{it}, and $A_{it\text{-year}}$ is the lagged vector of the acquisition variables for years $t - 1$ to $t - 4$.

Intuitively, this specification implies that the number of patents obtained by any firm in any year is randomly distributed following a Poisson process, where the covariate vectors X_{it-1} and A_{it-1}, A_{it-2}, A_{it-3}, and A_{it-4} determine the mean of this process. Changes in the value of individual covariates thus influence patenting frequency by affecting the mean of the Poisson distribution from which observations are drawn, in a manner identical to an ordinary regression.

Since the impact of an acquisition is likely to be felt over a number of years, rather than entirely in any one year, we used a distributed lag model (Judge *et al.*, 1988). To capture the lag effects, the one-period, two-period, three-period, and four-period lagged values of all acquisition related variables were included as covariates in the above model. This distributed lag model tests the impact of acquisitions for up to 4 years after the year the acquisition was originally made. Thus, illustratively speaking, in these models the acquiring firm's innovation performance in 1986 is potentially influenced by acquisitions made in 1982, 1983, 1984, and 1985. In sensitivity tests (results available from the authors), instead of using four lags we also estimated the models using three and five lags and found results substantively identical to those reported here.

The use of distributed lags provides two benefits. First, it enables us to examine the time pattern of the impact of acquisitions on firm innovation output. For instance, if an acquisition contributes to improved performance for the first 2 years but thereafter leads to no further improvement in innovation output, the 1- and 2-year lagged acquisition variables will be positive and significant, while the 3- and 4-year lagged acquisition variables will be nonsignificant. Second, since the total impact of an acquisition is likely to be distributed over several periods following the acquisition and may be statistically inconsequential in any one period, the regression coefficients on the distributed lags can be summed to obtain the *total* impact of an acquisition across time (Gujarati, 1988: 507). By using the variances and covariances of the individual lag coefficients from the regression output the variance of the summed coefficient can be calculated. The summed coefficient can then be used for hypothesis testing. For instance, the hypothesis that the total impact of acquisitions summed across all years is zero can be tested by computing the following t-statistic and checking whether it is statistically significant at the desired confidence level (Greene, 1993; Gujarati, 1988).

$$t = (\beta_{t-1} + \beta_{t-2} + \beta_{t-3} + \beta_{t-4}) / \sqrt{\text{Variance } (\beta_{t-1} + \beta_{t-2} + \beta_{t-3} + \beta_{t-4})}$$

where the β_{t-i}'s are the regression coefficients for the ith period lagged acquisition variable (Judge *et al.*, 1988). The variance for the summed coefficients can be computed using the following relation (Gujarati, 1988: 507):

$$\begin{aligned}\text{Variance } (\beta_{t-1} + \beta_{t-2} + \beta_{t-3} + \beta_{t-4}) &= [\text{Var}(\beta_{t-1}) + \text{Var}(\beta_{t-2}) \\ &+ \text{Var}(\beta_{t-3}) + \text{Var}(\beta_{t-4}) \\ &+ 2\text{Cov}(\beta_{t-1}\beta_{t-2}) + 2\text{Cov}(\beta_{t-1}\beta_{t-3}) \\ &+ 2\text{Cov}(\beta_{t-1}\beta_{t-4}) + 2\text{Cov}(\beta_{t-2}\beta_{t-3}) \\ &+ 2\text{Cov}(\beta_{t-2}\beta_{t-4}) + 2\text{Cov}(\beta_{t-3}\beta_{t-4})]\end{aligned} \quad (2)$$

The conditioning vector X in Equation 1 helps us to control for alternative explanations. For instance, since an acquisition represents the absorption of one firm by another, a simple explanation for an increased postacquisition innovation output would simply be increased innovative inputs—the one postacquisition firm reflects the combined efforts of the two preacquisition firms. By including innovative inputs as time-varying covariates in the conditioning vector X, we can directly control for such effects. Other possible determinants of innovation outputs such as firm size, diversification, nationality, and time are also controlled for through this vector.

This specification does not account for unobserved heterogeneity, or the possibility that observationally equivalent firms may differ on unmeasured characteristics. For instance, firms may enter the sample with inherently different innovation generating capabilities. Such unobserved heterogeneity, if present and not controlled for, can cause estimation problems. First, it can lead to overdispersion in the data. For the Poisson distribution the variance is restricted to equal the mean. If this restriction is false and the data are overdispersed in that the variance exceeds the mean, the computed standard errors in a Poisson regression are understated. Second, unobserved firm effects can lead to serial correlation among the residuals of observations from the same firm. Under overdispersion or serial correlation, computed regression coefficients remain consistently estimated; however, the standard errors are inaccurate. Thus, hypothesis testing and inference can be invalidated. To address the possibility of unobserved heterogeneity we used the Presample Panel Poisson approach (Blundell, Griffith, and Van Reenen, 1995).

In the Presample approach, unobserved heterogeneity is modeled as an additional covariate in the basic Poisson model. The values of the dependent variable in the periods immediately preceding the study period are used to construct an instrumental variable. This instrumental variable serves as a 'fixed-effect' for the firms in the panel and helps to partial out unobservable differences across firms. Thus, presample information on the firms provides the basis for controlling for unobserved heterogeneity. In the context of the current research the Presample variable can be interpreted as a measure of the unobserved differences in knowledge stocks between the sample firms.

Although the presample approach can help in the reduction of overdispersion and serial correlation, a more complete treatment of these potential problems would be to use an estimation approach that accounts for any remaining overdispersion and serial correlation even after the inclusion of a presample variable. The Generalized Estimating Equations (GEE) methodology provides a direct approach to modeling longitudinal Poisson data with serial correlation in a regression context (Liang and Zeger, 1986). The GEE estimation procedure involves two stages. In the first stage of the procedure beta coefficients are estimated with the assumption of independence across observations. This process yields consistent estimates of the beta parameters and the residuals from this regression provide an estimate of the 'working correlations' between the errors of different observations. This working correlation matrix is then used as an input in a second regression. The beta coefficients and standard errors from this second regression provide consistent estimates of the underlying parameters while accounting for the observed correlation between observations. To ensure that all residual correlation was accounted for, we used the GEE procedure to estimate all models. Even for the GEE approach instead of using the model-based standard errors we used the more conservative (i.e., larger) empirical standard errors. This ensures that other potential misspecifications of the variance function, such as any residual overdispersion, are also accounted for.

Sample and data

We tested the hypotheses on a longitudinal data set comprising the acquisition and patenting activities of 72 leading firms from the global chemicals industry. Focusing on the largest firms of the industry was necessary to ensure the availability and reliability of data. Obtaining information on the key variables for smaller firms is extremely difficult. This focus on large firms is also consistent with prior research on acquisitions (Hitt *et al.*, 1991, 1996). We identified the 100 leading players in the chemicals industry from lists of the largest chemicals firms that are published annually by trade journals such as *Chemical Week* and *C&E News*. To avoid survivor bias, the selected sample was drawn from the lists at the beginning of the study period. In these published lists subsidiaries were often listed separately from parent firms. After combining subsidiaries with parent firms, a sample of 82 independent firms was identified for inclusion in the sample. However, for 10 firms data could not be reliably obtained and they were subsequently dropped from the analysis. The remaining firms include the key firms in the industry over the study period and comprise of 30 European, 26 American, and 16 Japanese firms. The panel is unbalanced as some of the firms were acquired by other firms or restructured in a fashion that made comparison difficult beyond a particular year. Even though the sample was focused on the largest firms in the chemicals industry the inclusion of 72 firms provides significant depth to the sample and ensures that there is considerable variety within the sample on the key variables of this study. For instance, the number of employees for firms in the sample varies from a minimum of 2300 to a maximum of more than 181,000. Similarly, the number of patents obtained yearly varies from 0 to 760. Financial figures and personnel data on these firms were obtained from Compustat, Worldscope, Japan Company Handbooks, Daiwa Institute Research Guides, and trade publications and company annual reports. For all firms, financial data were converted to constant (1985) U.S. dollars to ensure standardization within the sample. A full list of the sample firms is available from the authors.

The chemicals industry is an appropriate setting for several reasons. First, technology-based acquisitions have been a significant feature of this industry (*Chemical Week*, 1983; Gibson, 1985). Second, patents are generally regarded to be effective, and used widely and consistently in the chemicals industry (Levin *et al.*, 1987; Arundel and Kabla, 1998). For the firms in the sample we obtained yearly patent counts from 1975 to 1992 and acquisition and firm attribute data for the years 1980 to 1991. The need to use lagged relationships between patent counts and the other variables and to construct a control for unobserved heterogeneity reduced the final panel for regression analysis to 9 years. We describe the data collection and coding procedures for the three sets of variables in some detail below.

We used U.S. patent data for all firms, including the foreign firms in the sample. This was necessary to maintain consistency, reliability, and comparability, as patenting systems across nations differ in their application of standards, system of granting patents, and value of protection granted. The United States represents one of the largest markets for chemicals, and firms desirous of commercializing their inventions typically patent in the United States if they patent anywhere. Studies by Dosi, Pavitt and Suete (1990) and Basberg (1983) show empirically that U.S. patent data provide a good measure of foreign firms' innovativeness. Prior research using patent data on international samples has also followed this strategy of using U.S. patent data for international firms (e.g., Stuart and Podolny, 1996; Patel and Pavitt, 1997).

Data on acquisitions were obtained through detailed archival research on the chemicals sector. Three main types of data sources were used to identify acquisition activity and to collect data on acquisition transactions: (1) commercial data bases, including general business news media such as the Dow Jones News Retrieval Text Index, and chemicals sector-specific data bases such as Metadex, (2) general business print media such as the Frost and Sullivan Predicasts Index (United States, International, and Europe), *Mergers and Acquisitions Journal* and Moody's Manuals, and industry-specific publications such as *Chemical Week* and *Plastics Technology*, and, (3) government publications and consultant reports for the chemicals industry. We were able to identify 1287 acquisition announcements for the sample firms over the period of 1980 to 1991.

To identify the technological acquisitions within this larger sample two approaches were used. First, we obtained detailed news stories associated with each acquisition announcement. We were able to obtain news stories providing details for 516 of the 1287 acquisitions identified for the sample firms. For the remaining acquisitions no further details, or inadequate details, were available. Second, we searched the U.S. Patents Data Base to determine if the acquired firms had obtained any patents in the 5 years preceding the acquisition. 165 of the acquired firms had obtained patents during the 5 years preceding their acquisition. For all subsequent analyses we retained only those acquisitions for which we were able to find either corroborating news stories or patenting activity (534 acquisitions in total). Acquisitions for which we were able to obtain no further information were omitted. The paucity of information on these acquisitions suggests that these acquisitions were very small or unimportant. Therefore, our analysis is based on the 534 acquisitions for which we have relatively complete information.

We used two criteria to distinguish technological acquisitions from all other acquisitions. First, we examined the news stories to establish if the acquiring firm reported technology as a motivating factor for the acquisition or if technology was a part of the transferred assets. We classified the acquisition as technological if either of these conditions was met. Second, we classified the acquisition as technological if the acquired firm had any patenting activity in the 5 years preceding the acquisition. Of the 534 acquisitions on which we had information, 283 met at least one of the two above criteria and were classified as technological acquisitions. The remaining 251 acquisitions were classified as nontechnological.

This classification scheme reflects a fairly inclusive definition of technological acquisitions. Firms need not have patented to be classified as technology acquisitions. Further, acquired firms obtaining even a single patent in the 5 years prior to the acquisition are classified as technological acquisitions.

This classification scheme provides two benefits. First, it uses two indicators to identify technological acquisitions and therefore enables more complete identification of technological acquisitions than either indicator by itself would. For instance, our patent-based measures cannot be computed if only a part of a firm is acquired, rather than the complete entity, or if the technology has not been patented at all. In such circumstances the news story could provide an indication of whether the transaction entailed technology or not. Similarly, if the acquisition was motivated by multiple factors of which technology was not necessarily the one mentioned in the news story, the patents measure provides an indicator of whether technology was involved. Thus, using the input from the news stories supplements our patent-based measures.

Second, using this scheme of classification makes our statistical tests more conservative. With this scheme we are likely to capture any acquisition that includes a technological component. Further, we are more likely to err in the direction of defining acquisitions as technological even when they have a relatively small technology component. If such misclassifications occur, we reduce our likelihood of finding a positive impact of technological acquisitions on the innovation performance of acquiring firms.

Finally, we need to select a time period or window for measuring a firm's knowledge base. At one extreme only the current year's patents could be considered relevant. At another extreme any patent obtained by the firm in the past could be included in computing its current knowledge base. Prior research suggests that knowledge capital depreciates sharply, losing significant value within 5 years (Griliches, 1979). Although the depreciation rate for knowledge capital is likely to vary across industries, a boundary of 5 or 6 years seems reasonable and has been used by other researchers (Podolny and Stuart, 1995; Henderson and Cockburn, 1996). In this research we use the patents obtained by a firm in the preceding 5 years to compute our knowledge base measures. We also computed some of the measures using patents for the preceding 6 years. These measures were highly correlated with the 5-year measures, suggesting that the construct is not unduly influenced by changes in the time period used to compute it.

Variable definitions and operationalization

Dependent variable

Patents Innovation output, the dependent variable, is measured through the patenting frequency of firms, that is, the number of successful patent applications by a firm in a given year. Patents have both significant strengths and weaknesses as measures of innovation output. First, patents are directly related to inventiveness: they are granted only for 'nonobvious' improvements or solutions with discernible utility (Walker, 1995). Second, they represent an externally validated measure of *technological* novelty (Griliches, 1990). Third, they confer property rights upon the assignee and therefore have *economic* significance (Kamien and Schwartz, 1982: 49; Scherer and Ross, 1990: 621).

Patents also correlate well with other measures of innovative output. Empirical studies find that patents are closely related to measures such as new products (Comanor and Scherer, 1969), innovation and invention counts (Achilladelis, Schwarzkopf and Cines, 1987), and sales growth (Scherer, 1965). Expert ratings of corporate technological strength are also highly correlated with the number of patents held by corporations (Narin, Noma, and Perry, 1987). Further, surveys of patent holders indicate that the rate of utilization of patents is reasonably high, with estimates indicating that between 41 percent and 55 percent of all patents granted are put to commercial use for at least a limited time (Griliches, 1990). Similarly, about 50 percent of all patents granted are still being renewed and a renewal fee is being paid 10 years after the patents had originally been applied for (Griliches, 1990; Schankerman and Pakes, 1986). Given a nonnegligible renewal fee, this indicates a significant usefulness for the majority of patents for a significant time period.

However, the use of patents as a measure of innovative output also has limitations. Some inventions are not patentable, others are not patented, and the inventions that are patented differ greatly in economic value (Cohen and Levin, 1989; Griliches, 1990; Trajtenberg, 1990). Research, and the logic of appropriability, indicate that the degree to which the first two of these factors are a problem varies significantly across industries (Cohen and Levin 1989; Levin *et al.*, 1987). Limiting the study to a single industrial sector or a few closely related sectors minimizes such problems as the factors that affect patenting propensity are likely to be stable within such a context (Basberg, 1987; Cohen and Levin, 1989; Griliches, 1990).

We measure $Patents_{it}$, as the number of *successful* patent applications, or granted patents, for the acquiring firm i in year t. The granted patent carries the date of the original application. We use this date to assign a granted patent to the particular year when it was originally applied for. This procedure permits consistency in the treatment of all patents and controls for differences in delays that may occur in granting patents after the application is filed (Trajtenberg, 1990). Note that the patent count for the dependent variable is based on the patents of the acquiring firm obtained 1–4 years *after* the acquisition in all cases. These patents are therefore different from the patents on which the independent variables are based. The independent variables described below are based on patents obtained by the acquired and acquiring firms, respectively, in the 5 years *before* the acquisition.

Independent variables

Number of nontechnological acquisitions As noted earlier, acquisitions were coded as technological acquisitions if either of the following two criteria are met: first, if the acquiring firm reported technology as a motivating factor for the acquisition or if technology was reported as a part of the transferred assets of the acquired firm; second, if the acquired firm had any patenting activity in the 5 years preceding the acquisition. Acquisitions that did not meet either of the above criteria were coded as nontechnological acquisitions. We used four lagged versions of this and all other acquisition related variables listed below.

Absolute size of acquired knowledge base To obtain this variable we used the following procedure: for each acquiring firm for each year, we prepared a list of the patents its acquisitions had obtained in the preceding 5 years. These patents were then combined with the patents cited by them. Thereafter, all duplicates were removed to ensure that a patent number appeared only once on this list. The acquired knowledge base was then computed as the number of patents (i.e., knowledge elements) on this list.

Relative size of acquired knowledge base This variable was obtained by dividing *Absolute size of acquired knowledge base* by the size of the acquiring firm's knowledge base. The size of the acquiring firm's knowledge base was computed using the same procedure as the size of the acquired firm's knowledge base (see above). In a few cases the acquired knowledge base was larger than the acquiring knowledge base. In these cases we used the larger number as the denominator. Since our theoretical mechanism is concerned with the relative *proportion* of the merged firms' resources that are likely to be occupied with integrative rather than inventive activity, a number greater than 1 is not meaningful.

Relatedness of acquired knowledge base To measure the relatedness of the acquired knowledge base, the following procedure was followed. First, the list of patent numbers that appeared in both the acquired firm's knowledge base and in the acquiring firm's knowledge base was prepared. Then, the number of elements on this list was divided by *Absolute size of acquired knowledge base*.

Number of technological acquisitions where patents unavailable This is the number of acquisitions for which news stories indicated that technology was a component of the transferred assets but where no patents could be identified with the acquisition. This would occur if either the acquired unit had obtained no patents or if the acquired unit was a part of a larger parent firm and patents were not assigned separately to the acquired unit and therefore could not be separated from the parent firm's patents.

Controls

We included several control variables in the models. These control variables include yearly R&D expenditures (*R&D*), firm size as measured by natural log of number of employees (*Logemployees*), firm diversification as measured by entropy (*Diversification*), and a measure of national cultural distance between the acquired and acquiring firms (*Foreign acquisitions*). The following formula was used to calculate the *Diversification* measure: Entropy $= \Sigma P_j \times \ln(1/P_j)$, where P_j is defined as the percentage of firm sales in segment j and $\ln(1/P_j)$ is the weight for each segment j (Palepu, 1985). *Foreign acquisitions* was computed as

$$\text{Foreign\_acquisitions} = \sum_{i=1}^{4} \{(I_{ij}-I_{iu})^2/V_i\}/4 \qquad (3)$$

where I_i stands for the index of the ith cultural dimension, j and u are subscripts indicating the countries j and u, and V_i is the variance of the index of the ith dimension (Hofstede, 1980). The index thus indicates the cultural distance between the acquirer's country (j) and the acquired firm's country (u) (Kogut and Singh, 1988). We anticipate that foreign acquisitions might be more difficult to integrate than domestic acquisitions. Finally, in all models we included the firm heterogeneity control variable *Presample patents* (the sum of patents obtained by a firm in the 3 years prior to the firm's entry into the sample) and dummy variables for acquirer nationality and calendar year.

RESULTS

Table 1 provides descriptive statistics and correlations. The table indicates the diversity of firms included in the sample. Even though the sample involves the prominent players in the industry, there is considerable variance on all the key variables such as *Patents*, *R&D*, *Logemployees*, and the acquisition variables. The variables reflecting the hypothesized effects are not very highly correlated among themselves or with the control variables. However, the correlations between some of the control variables are high, notably the correlation between *R&D* and *Logemployees* (0.72), and the correlation of the firm fixed-effect variable *Presample patents* with *R&D* (0.89) and *Logemployees* (0.72). Robustness tests (reported at the end of this section), however, indicate that the results on the hypothesized effects were strong and unaffected by these high correlations between some of the control variables.

Table 2 provides results for all models using GEE Poisson estimators (reported with empirical standard errors). The variables reflecting the hypothesized effects were entered into the regression individually and likelihood ratio tests are reported for all models. Model 1 in Table 2 presents the base model with firm- and acquisition-related control variables. Models 2–6 include the *Number of nontechnological acquisitions*, *Absolute size of acquired knowledge base*, *Relative size of acquired knowledge base*, *Relatedness of acquired knowledge base* and *Relatedness of acquired knowledge base*^{2} variables entered successively. We use the full model to discuss the results of the hypothesis tests. The summed coefficients and the associated standard errors for the lagged acquisition variables for these models are given at the bottom of the table.

In Hypothesis 1 we predicted a negative or nonsignificant relationship between the number of nontechnological acquisitions conducted in a year and the subsequent innovation output of the acquiring firm. The coefficients of *Number of non-technological acquisitions* are nonsignificant for all four periods individually. The summed coefficient, which represents the total effect of acquisitions across the 4 years, is not statistically significantly different from zero. Thus, we do not find an appreciable impact of acquisitions without a technological component on the innovation output of the acquirer for the four periods following the acquisition.

The coefficients of the four *Absolute size of acquired knowledge base* variables in Model 6 in Table 2 are positive and significant, supporting Hypothesis 2. The summed coefficient reflecting the total impact of the acquisition is also positive and significant. However, as the summed coefficient indicates (0.0004), the absolute magnitude of this effect is low. A one-unit increase in the knowledge base of the acquired firm leads to a 0.04 percent increase in the acquirer's innovation

Table 1. Means, standard deviations, minimum and maximum values and bivariate correlations for all variables

Variable	Mean	S.D.	Min.	Max.	1	2	3	4	5	6	7	8	9	10	11	12	13	14	15	16	17	18	19	20	21	22	23	24	25	26	27	28	29	30	31	32	33	34
1 *Patients$_t$*	93.90	137.65	0.00	760.00																																		
2 *No. of nontechnological acquisitions$_{t-1}$*	0.35	0.79	0.00	6.00	0.12																																	
3 *No. of nontechnological acquisitions$_{t-2}$*	0.34	0.78	0.00	6.00	0.12	0.26																																
4 *No. of nontechnological acquisitions$_{t-3}$*	0.29	0.68	0.00	5.00	0.11	0.29	0.23																															
5 *No. of nontechnological acquisitions$_{t-4}$*	0.25	0.69	0.00	7.00	0.12	0.21	0.29	0.30																														
6 *Absolute size of acquired knowledge base$_{t-1}$*	26.08	275.94	0.00	6227.00	0.13	0.07	−0.03	0.01	−0.02																													
7 *Absolute size of acquired knowledge base$_{t-2}$*	34.59	332.94	0.00	6227.00	0.15	−0.03	0.05	−0.03	0.00	0.02																												
8 *Absolute size of acquired knowledge base$_{t-3}$*	34.83	338.80	0.00	6227.00	0.18	−0.02	−0.03	0.01	−0.03	0.03	0.01																											
9 *Absolute size of acquired knowledge base$_{t-4}$*	33.20	337.48	0.00	6227.00	0.19	0.00	−0.02	−0.03	0.02	0.51	0.03	0.01																										
10 *Relative size of acquired knowledge base$_{t-1}$*	0.02	0.08	0.00	0.90	−0.04	0.12	0.00	0.00	−0.05	0.44	−0.01	0.00	0.16																									
11 *Relative size of acquired knowledge base$_{t-2}$*	0.02	0.09	0.00	0.90	−0.01	−0.01	0.12	−0.01	−0.01	0.00	0.56	−0.01	0.00	−0.02																								
12 *Relative size of acquired knowledge base$_{t-3}$*	0.02	0.09	0.00	0.90	0.01	0.04	0.00	0.09	0.00	0.04	0.00	0.55	−0.01	0.05	−0.02																							
13 *Relative size of acquired knowledge base$_{t-4}$*	0.02	0.08	0.00	0.90	0.02	−0.02	0.03	0.03	0.05	0.37	0.03	0.00	0.56	0.12	0.05	0.00																						
14 *Relatedness of acquired knowledge base$_{t-1}$*	0.01	0.06	0.00	1.00	0.03	−0.02	0.02	−0.03	−0.03	0.04	0.00	−0.01	0.01	0.09	−0.01	−0.02	0.27																					
15 *Relatedness of acquired knowledge base$_{t-2}$*	0.01	0.06	0.00	1.00	0.04	0.00	−0.01	0.02	−0.03	−0.01	0.03	0.00	−0.01	0.00	0.08	0.00	−0.01	−0.02																				
16 *Relatedness of acquired knowledge base$_{t-3}$*	0.01	0.06	0.00	1.00	0.05	0.03	0.00	−0.01	0.03	0.07	−0.01	0.03	0.00	0.07	0.00	0.07	0.00	−0.02	−0.02																			
17 *Relatedness of acquired knowledge base$_{t-4}$*	0.01	0.06	0.00	1.00	0.06	0.02	0.03	0.00	−0.01	0.09	0.06	−0.01	0.03	0.02	0.06	0.00	0.07	0.04	−0.01	−0.01																		
18 *Relatedness of acquired knowledge base$_{t-1}^2$*	0.00	0.05	0.00	1.00	−0.01	−0.03	0.02	−0.03	−0.03	0.01	−0.01	0.00	0.01	0.04	−0.01	−0.01	0.36	0.92	−0.01	−0.01	0.00																	
19 *Relatedness of acquired knowledge base$_{t-2}^2$*	0.00	0.05	0.00	1.00	0.00	−0.02	−0.03	0.02	−0.03	0.00	0.00	−0.01	0.00	0.01	0.03	−0.01	0.00	−0.01	0.92	−0.01	−0.01	−0.01																

Table 1. Continued

Variable	Mean	S.D.	Min.	Max.	1	2	3	4	5	6	7	8	9	10	11	12	13	14	15	16	17
20 *Relatedness of acquired knowledge base*$_{t-3}^2$	0.00	0.05	0.00	1.00	0.00	0.00	−0.02	−0.03	0.03	0.02	0.00	0.00	−0.01	0.03	0.01	0.03	−0.01	−0.01	−0.01	0.92	−0.01
21 *Relatedness of acquired knowledge base*$_{t-4}^2$	0.00	0.05	0.00	1.00	0.01	−0.01	0.00	−0.02	−0.03	0.05	0.02	0.00	0.00	0.01	0.03	0.01	0.04	0.02	−0.01	−0.01	0.93
22 *No. of technological acquisitions where patents unavailable*$_{t-1}$	0.17	0.47	0.00	3.00	0.19	0.22	0.16	0.17	0.20	0.03	−0.03	−0.02	0.02	0.06	−0.04	0.00	−0.02	0.01	0.01	0.03	0.00
23 *No. of technological* acquisitions where *patents unavailable*$_{t-2}$	0.16	0.46	0.00	3.00	0.21	0.10	0.23	0.18	0.17	−0.01	0.01	−0.03	−0.02	0.00	0.05	−0.03	0.01	−0.03	0.01	0.02	0.04
24 *No. of technological* acquisitions where *patents unavailable*$_{t-3}$	0.14	0.43	0.00	3.00	0.22	0.15	0.11	0.23	0.22	0.01	−0.01	−0.01	−0.03	0.00	−0.02	0.02	−0.02	0.05	−0.03	0.02	0.00
25 *No. of technological* acquisitions where *patents unavailable*$_{t-4}$	0.13	0.41	0.00	3.00	0.19	0.15	0.15	0.08	0.20	−0.02	0.00	−0.02	−0.02	−0.02	0.00	−0.02	−0.01	0.02	0.06	−0.03	0.00
26 *Foreign acquisitions*$_{t-1}$	0.20	0.86	0.00	14.66	0.02	0.23	0.06	0.07	0.04	0.15	−0.02	0.00	−0.01	0.33	−0.03	0.11	0.01	0.12	−0.02	0.17	−0.01
27 *Foreign acquisitions*$_{t-2}$	0.19	0.85	0.00	14.66	0.03	0.06	0.24	0.05	0.08	0.00	0.12	−0.01	0.00	0.01	0.29	−0.03	0.12	0.01	0.13	−0.02	0.19
28 *Foreign acquisitions*$_{t-3}$	0.15	0.60	0.00	6.86	0.04	0.24	0.10	0.15	0.10	0.08	0.00	0.01	−0.01	0.11	0.00	0.15	−0.02	−0.01	0.02	0.19	−0.01
29 *Foreign acquisitions*$_{t-4}$	0.15	0.69	0.00	8.97	0.03	0.10	0.21	0.07	0.11	0.01	0.06	−0.01	0.01	0.01	0.08	−0.01	0.16	−0.01	0.00	0.01	0.17
30 *R&D*$_{t-1}$	19.32	30.24	0.23	161.25	0.89	0.18	0.19	0.16	0.16	0.08	0.09	0.10	0.11	−0.04	−0.02	−0.01	−0.01	0.03	0.05	0.07	0.07
31 *Logemployees*$_{t-1}$	2.76	1.14	0.83	5.20	0.70	0.31	0.31	0.28	0.28	0.13	0.15	0.15	0.15	0.00	0.03	0.05	0.04	0.05	0.05	0.06	0.05
32 *Diversification/Entropy*$_{t-1}$	0.53	0.41	0.00	1.60	0.04	0.10	0.08	0.07	0.06	0.10	0.12	0.11	0.10	0.05	0.09	0.10	0.10	0.01	0.01	0.00	0.01
33 *U.S.A.*	0.35	0.48	0.00	1.00	0.10	−0.04	−0.04	0.00	0.02	0.04	0.07	0.09	0.09	−0.01	0.03	0.04	0.08	0.08	0.07	0.07	0.07
34 *Japan*	0.24	0.43	0.00	1.00	−0.19	−0.24	−0.23	−0.22	−0.20	−0.04	−0.05	−0.05	−0.05	−0.08	−0.09	−0.09	−0.09	−0.07	−0.07	−0.07	−0.06
35 *Presample patents*$_{t-1}$	2.28	3.26	0.01	15.80	0.94	0.11	0.12	0.09	0.10	0.07	0.08	0.11	0.11	−0.07	−0.05	−0.03	−0.03	0.05	0.05	0.05	0.05

Variable	18	19	20	21	22	23	24	25	26	27	28	29	30	31	32	33	34
20 *Relatedness of acquired knowledge base*$_{t-3}^2$	−0.01	−0.01															
21 *Relatedness of acquired knowledge base*$_{t-4}^2$	0.00	−0.01	−0.01														
22 *No. of technological acquisitions where patents unavailable*$_{t-1}$	−0.01	−0.02	0.00	−0.02													
23 *No. of technological* acquisitions where *patents unavailable*$_{t-2}$	−0.03	−0.01	−0.01	0.00	0.28												
24 *No. of technological* acquisitions where *patents unavailable*$_{t-3}$	0.03	−0.02	−0.01	−0.02	0.19	0.31											
25 *No. of technological* acquisitions where *patents unavailable*$_{t-4}$	−0.01	0.03	−0.02	−0.01	0.14	0.21	0.31										
26 *Foreign acquisitions*$_{t-1}$	0.08	−0.02	0.09	−0.01	0.39	0.14	0.12	0.04									
27 *Foreign acquisitions*$_{t-2}$	0.00	0.08	−0.01	0.09	0.10	0.40	0.16	0.14	0.15								
28 *Foreign acquisitions*$_{t-3}$	−0.02	0.01	0.12	−0.02	0.18	0.17	0.41	0.22	0.42	0.23							
29 *Foreign acquisitions*$_{t-4}$	−0.01	−0.01	0.00	0.10	0.05	0.15	0.14	0.43	0.15	0.37	0.25						
30 *R&D*$_{t-1}$	0.00	0.00	0.01	0.02	0.24	0.30	0.30	0.26	0.07	0.11	0.15	0.12					
31 *Logemployees*$_{t-1}$	0.01	0.01	0.01	0.00	0.29	0.28	0.28	0.24	0.13	0.14	0.18	0.17	0.72				
32 *Diversification/Entropy*$_{t-1}$	0.02	0.02	0.01	0.02	−0.03	−0.07	−0.08	−0.12	0.01	0.00	−0.06	−0.08	−0.05	0.14			
33 *U.S.A.*	0.07	0.07	0.07	0.07	0.05	0.04	0.07	0.08	−0.13	−0.12	−0.12	−0.11	−0.02	0.08	0.13		
34 *Japan*	−0.05	−0.04	−0.04	−0.04	−0.17	−0.17	−0.15	−0.16	−0.05	−0.07	−0.07	−0.07	−0.21	−0.53	−0.10	−0.42	
35 *Presample patents*$_{t-1}$	0.01	0.01	0.01	0.01	0.20	0.21	0.21	0.18	0.00	0.02	0.05	0.04	0.89	0.72	0.02	0.10	−0.20

Table 2. GEE presample Poisson regression with distributed lag analysis predicting patents, variable

Variable	1	2	3	4	5	6
Intercept	1.906*** [0.333]	1.916*** [0.334]	1.994*** [0.296]	2.055*** [0.293]	2.041*** [0.295]	2.048*** [0.292]
No. of nontechnological acquisitions$_{t-1}$		0.004 [0.017]	0.009 [0.017]	0.009 [0.016]	0.008 [0.016]	0.004 [0.017]
No. of nontechnological acquisitions$_{t-2}$		−0.002 [0.018]	−0.0003 [0.017]	0.004 [0.016]	0.005 [0.016]	0.004 [0.016]
No. of nontechnological acquisitions$_{t-3}$		−0.001 [0.023]	−0.006 [0.023]	−0.002 [0.023]	−0.003 [0.023]	−0.002 [0.024]
No. of nontechnological acquisitions$_{t-4}$		−0.010 [0.014]	−0.012 [0.014]	−0.008 [0.013]	−0.010 [0.013]	−0.008 [0.012]
Absolute size of acquired knowledge base$_{t-1}$			0.0001** [0.00004]	0.0001*** [0.00002]	0.0001*** [0.00002]	0.0001*** [0.00003]
Absolute size of acquired knowledge base$_{t-2}$			0.0001*** [0.00003]	0.0001*** [0.00002]	0.0001*** [0.00002]	0.0001*** [0.00002]
Absolute size of acquired knowledge base$_{t-3}$			0.0001*** [0.00002]	0.0001*** [0.00002]	0.0001*** [0.00002]	0.0002*** [0.00002]
Absolute size of acquired knowledge base$_{t-4}$			0.0000 [0.0000]	0.0001** [0.00004]	0.0001** [0.00004]	0.0000* [0.0001]
Relative size of acquired knowledge base$_{t-1}$				−0.393** [0.165]	−0.387** [0.167]	−0.348* [0.156]
Relative size of acquired knowledge base$_{t-2}$				−0.447*** [0.144]	−0.442*** [0.142]	−0.455*** [0.144]
Relative size of acquired knowledge base$_{t-3}$				−0.388*** [0.125]	−0.405*** [0.125]	−0.488*** [0.121]
Relative size of acquired knowledge base$_{t-4}$				−0.479** [0.196]	−0.430* [0.198]	−0.369* [0.209]
Relatedness of acquired knowledge base$_{t-1}$					−0.207 [0.176]	0.221 [0.558]
Relatedness of acquired knowledge base$_{t-2}$					0.002 [0.112]	0.821* [0.384]
Relatedness of acquired knowledge base$_{t-3}$					0.158 [0.111]	0.847* [0.371]
Relatedness of acquired knowledge base$_{t-4}$					0.151 [0.121]	0.836** [0.335]
Relatedness of acquired knowledge base$^2_{t-1}$						−0.820 [0.911]

Table 2. Continued

Variable	1	2	3	4	5	6
Relatedness of acquired knowledge base$^2_{t-2}$						−1.580** [0.572]
Relatedness of acquired knowledge base$^2_{t-3}$						−1.301* [0.572]
Relatedness of acquired knowledge base$^2_{t-4}$						−1.251** [0.501]
No. of technological acquisitions where patents unavailable$_{t-1}$		0.023 [0.026]	0.034 [0.024]	0.016 [0.023]	0.018 [0.023]	0.009 [0.024]
No. of technological acquisitions where patents unavailable$_{t-2}$		0.011 [0.024]	0.025 [0.022]	0.016 [0.021]	0.016 [0.021]	0.014 [0.022]
No. of technological acquisitions where patents unavailable$_{t-3}$		0.039 [0.030]	0.053* [0.027]	0.043 [0.026]	0.047 [0.025]	0.045 [0.025]
No. of technological acquisitions where patents unavailable$_{t-4}$		0.063* [0.020]	0.068* [0.027]	0.066* [0.026]	0.073** [0.026]	0.077** [0.027]
Foreign acquisitions$_{t-1}$	0.034** [0.012]	0.031 [0.016]	0.026 [0.017]	0.035 [0.019]	0.038* [0.017]	0.040* [0.017]
Foreign acquisitions$_{t-2}$	0.020* [0.010]	0.019 [0.013]	0.014 [0.012]	0.023 [0.015]	0.022 [0.015]	0.024 [0.015]
Foreign acquisitions$_{t-3}$	−0.020 [0.013]	−0.042* [0.017]	−0.051** [0.016]	−0.042* [0.017]	−0.046** [0.016]	−0.043* [0.017]
Foreign acquisitions$_{t-4}$	0.009 [0.013]	−0.008 [0.015]	−0.010 [0.015]	−0.004 [0.017]	−0.008 [0.017]	−0.011 [0.017]
R&D$_{t-1}$	−0.0004 [0.001]	−0.001 [0.002]	−0.001 [0.002]	−0.001 [0.002]	−0.001 [0.001]	−0.001 [0.001]
Logemployees$_{t-1}$	0.554*** [0.105]	0.553*** [0.109]	0.534*** [0.099]	0.517*** [0.097]	0.522*** [0.097]	0.516*** [0.096]
Diversification/ Entropy$_{t-1}$	−0.176* [0.076]	−0.187* [0.083]	−0.217* [0.085]	−0.180* [0.080]	−0.183* [0.080]	−0.174* [0.076]
US firm	0.553*** [0.128]	0.549*** [0.130]	0.526*** [0.120]	0.509*** [0.118]	0.507*** [0.117]	0.514*** [0.114]
Japanese firm	0.997*** [0.233]	1.006*** [0.229]	0.991*** [0.219]	0.942*** [0.217]	0.950*** [0.217]	0.951*** [0.216]
Presample patents$_{t-1}$	0.135*** [0.022]	0.137*** [0.024]	0.145*** [0.024]	0.142*** [0.023]	0.143*** [0.023]	0.142*** [0.022]

output. If we consider a hypothetical acquisition in which the acquired firm has a knowledge base with 50 elements (50 elements corresponds approximately to a firm with 10 owned patents plus 40 cited patents) then the above coefficient suggests that this acquisition should lead to a 2 percent increase in innovation output (0.0004 × 50 = 0.02) for the acquiring firm after the acquisition, other things being equal.

In Hypothesis 3, a negative relationship was proposed between the relative size of an acquisition and the subsequent innovation performance of the acquiring firm. This prediction was also borne out. Specifically, the individual and

Table 2. Continued

Variable	1	2	3	4	5	6
N	598	598	598	598	598	598
Pearson chi sq./d.f.	10570/579	10491/571	9923/567	9454/563	9448/559	9280/555
−2 log-likelihood vis à vis the preceding model		158***	1136***	938***	12***	336***
Summed coefficients (Model 6)						
No. of nontechnological acquisitions						−0.001 [0.054]
Absolute size of acquired knowledge base						0.0004** [0.000]
Relative size of acquired knowledge base						−1.660** [0.661]
Relatedness of acquired knowledge base						2.725** [1.332]
Relatedness of acquired knowledge base^2						−4.951** [2.020]
No. of technological acquisitions where patents unavailable						0.145* [0.081]
Foreign acquisitions						0.011 [0.049]

*$p < 0.05$; **$p < 0.01$; ***$p < 0.001$ (one-tailed tests for hypothesized variables, two-tailed tests for controls)
The table gives parameter estimates; standard errors are in brackets. Year dummies are included, but not shown.

summed coefficients for *Relative size of acquired knowledge base* are negative, indicating that acquiring firms that are large relative to the acquirer leads to a decline in postacquisition innovation output for the acquirer. We also find support for Hypothesis 4. As shown in Model 6 of Table 2, the coefficients for *Relatedness of acquired knowledge base* are positive, and that for its squared term, *Relatedness of acquired knowledge base*^{2}, are negative. These findings support the argument that the relatedness of acquisitions has a curvilinear impact on the subsequent innovation output of acquiring firms. The summed coefficients at the bottom of Table 2 support the same results.

Among the control variables, the coefficients on *Number of technological acquisitions where patents unavailable* for the 1-year, 2-year, and 3-year lags are positive but not significant. However, the 4-year lag is positive and significant. The summed coefficient is positive and significant. Thus, technological acquisitions for which we were unable to identify patents also improve postacquisition innovation output.

Foreign acquisitions, which represents the cultural distance between acquired and acquiring firms, has a nonsignificant impact on innovation output. Although the 1-year and 3-year lagged variables are marginally significant but in opposite directions (positive and negative, respectively), the summed coefficient is positive but not statistically significant (Table 2). Although these results are not conclusive it appears that, on balance, foreign acquisitions neither help nor hurt innovation output. This finding is consistent with recent research on international acquisitions, which finds that acquisitions in which the acquirer and the acquired are from different countries do not result in greater postacquisition conflict, and, in fact, often lead to superior postacquisition performance (Weber, Shenkar, and Raveh, 1996; Very *et al.*, 1997). Nevertheless, further research is required to understand this issue more completely.

Logemployees, the control measuring the size of the acquirer, was found to be positively associated with patenting frequency. *Diversification* is negatively associated with patenting frequency. Prior results on the impact of diversification on innovative activity have been mixed; studies find that diversification has both a positive and a negative impact on innovation (Cohen and Levin, 1989). Diversification can either encourage innovation by providing a stimulus of multiple knowledge bases within a single firm and by leading to cross-fertilization of ideas, but it can also imply a loss of focus in a given technological area as research efforts are spread in multiple directions. The results of this research support the latter interpretation.

The *Presample patents* variable and several of the year dummies (not reported) were also significant in all the models. This indicates that it was important to control for unobserved firm effects as well as period effects in these data. The Year dummies for 1983–86 were negative and significant, while all the other year dummies were nonsignificant, relative to the omitted category—1991. These results indicate that patenting had significantly increased for this set of firms in the later years (1987–91), relative to the earlier years (1983–86). The nation dummy variables reflecting acquirer nationality, Japanese and U.S., were also positive and significant, indicating that Japanese and U.S. firms were likely to obtain more patents than European firms.

Sensitivity analyses

We also ran several sensitivity tests to check the robustness of the results. We reconstructed all the knowledge base measures after separating the patents obtained by a firm (Own Patents) from the patents cited by the firm (Cited Patents), and ran two distinct sets of models. In the first set we used knowledge-based measures for acquired and acquiring firms based on only the Own Patents of the firm (i.e., we excluded Cited Patents). In the second set we computed all knowledge-based measures for acquired and acquiring firms based on only the Cited Patents of the firms (i.e., we excluded Own Patents). The results of these analyses are presented in Table 3. In the interest of brevity, we only report the summed coefficients and associated *t*-tests for the models (the full regression table is available from the authors).

Table 3. GEE presample Poisson regression predicting patents$_t$. Own-cited patents measures. Summary of the summed coefficient results for all lagged variables

Variable	Own patents 1	Cited patents 2
No. of nontechnological acquisitions	−0.007 [0.057]	0.0006 [0.054]
Absolute size of acquired knowledge base (Own patents)	0.002*** [0.001]	
Relative size of acquired knowledge base (Own patents)	−1.632* [0.708]	
Relatedness of acquired knowledge base (Own patents)	2.411 [3.562]	
Relatedness of acquired knowledge base^2 (Own patents)	−3.568 [7.533]	
Absolute size of acquired knowledge base (Cited patents)		0.0004*** [0.000]
Relative size of acquired knowledge base (Cited patents)		−1.653*** [0.465]
Relatedness of acquired knowledge base (Cited patents)		2.752* [1.266]
Relatedness of acquired knowledge base^2 (Cited patents)		−4.759** [1.865]
No. of technological acquisitions where patents unavailable	0.196* [0.083]	0.141 [0.081]
Foreign acquisitions	−0.003 [0.003]	0.011 [0.048]

Reported coefficient estimates are the sum of the four lagged coefficients (β_{t-1}, β_{t-2}, β_{t-3}, β_{t-4}) for each variable.
*$p < 0.05$; **$p < 0.01$; ***$p < 0.001$ (one-tailed tests for hypothesized variables, two-tailed tests for controls)

The summed coefficients in Table 3 indicate that with knowledge base measures based only on Own Patents (Model 1), the hypotheses (2 and 3) on *absolute size* of acquired firm and *relative size* of acquired firm were strongly supported. However, the coefficients for the *relatedness* hypothesis (4) carried the right signs (+ for relatedness and − for its square term), but they were statistically nonsignificant. In the analysis based on Cited Patents only (Model 2) all hypotheses (2, 3, and 4) were strongly supported. Thus, using only patents, or only citations to measure knowledge bases, provides rather similar results to using both patents and citations for the absolute size and relative size effects. However, in terms of measuring relatedness, the patents-only measures do not appear to capture relatedness as well as the citations-based measures. A comparison of the results in Table 3 (based on Own Patents or Cited Patents taken individually) with the results in Table 2 (based on combining Own and Cited patents to provide a single measure of knowledge elements), however, suggests that the relatedness measure based on both own and cited patents collectively is more predictive than the measures based on either Own Patents or Cited Patents only.

In the results discussed above we used the Presample Patents variable as a measure of unobserved differences in the knowledge bases of firms. The patent production function literature provides several alternate approaches to construct indices, reflecting differences in firms' knowledge stocks (Griliches, 1984; Hall *et al.*, 1988; Henderson and Cockburn, 1996). We constructed several such indices using different assumptions and input data. First, we used capitalized historical R&D expenditures to construct a knowledge stock index (Hall *et al.*, 1988).[2] Second, we constructed an index based on the moving average sum of R&D expenditures for the previous n years. We used a depreciation rate of 0.20 (consistent with Henderson and Cockburn, 1996, and other studies in this tradition) for the first approach and n = 5 years for the second approach, and estimated models using these indices in lieu of the R&D variable. We also computed the knowledge stock index using cumulative lagged depreciated patent stocks rather than cumulated lagged depreciated R&D, and used this as a control variable in our regression models in place of the Presample Patents measure. In additional estimation we used the one period lagged values of the dependent variable in lieu of the presample variable. The results for all these approaches (available from the authors) are very similar to the earlier reported results.

As noted earlier, the correlations between the three control variables, *R&D*, *Logemployees* and *Presample patents* were high. Although this high collinearity does not bias coefficient estimates, it can affect the stability of the estimated coefficients. Consequently, omitting even a few observations can change the sign or the significance of the affected variables (Greene, 1993). To ensure that our results were robust, we drew 50 random samples of 90 percent of the observations and estimated the full model for each of these samples. The results of this sensitivity analysis indicated that the results of our hypothesis testing were robust.

DISCUSSION

The results of the study indicate support for our theoretical predictions. Specifically, separating nontechnological acquisitions from technological acquisitions and distinguishing between technological acquisitions on the basis of the absolute size, relative size, and relatedness of the knowledge bases of the acquired firm helps in predicting postacquisition innovation output. We do not find any statistically significant impact of nontechnological acquisitions on subsequent innovation output. Within technological acquisitions, absolute size of the acquired knowledge base has a positive impact on innovation output, while relative size of the acquired knowledge base reduces innovation output. The relatedness of acquired and acquiring knowledge bases has a nonlinear impact on innovation output, with acquisition of firms with high levels of relatedness and unrelatedness both proving inferior to acquiring firms with moderate levels of relatedness. These findings have implications for theory, research, and practice. We discuss these below.

Implications for theory and research

Recent research has highlighted the role of acquisitions as a mechanism for the redeployment of resources that are subject to market failure (Anand and Singh, 1997; Capron, Dussauge, and Mitchell, 1998). In this study we investigated the effectiveness of such redeployment in the case of one set of resources that are very susceptible to market failure, knowledge-based or technological resources. Our results provide both good and bad news.

[2] Specifically, for the ith firm and tth period, knowledge stock (K_{it}) was computed using the formula $K_{it} = 1 - \delta(K_{it-1}) + R_{it}$, where δ stands for depreciation rate for knowledge capital, and R_{it} represents the current-period knowledge flow addition. The initial stock of capital was computed by dividing the first observed year's flow by $\delta + g$, where g is the firm's historic rate of growth of real expenditures on R&D (Henderson and Cockburn, 1996).

The positive outcome of this evaluation is that, *when the characteristics of acquisitions are accounted for*, acquisitions improve the technological performance of the acquiring firm. Although prior research concluded that acquisitions reduce innovative outputs, this study suggests that under the appropriate circumstances, even after controlling for innovative inputs such as R&D, acquisitions can introduce a positive shock onto innovation output. This finding is also consistent with the sheer volume of acquisition activity in the high-technology sector that suggests that managers also view acquisitions as a mechanism for accessing technology.

The results of this study are also important from a broader economic perspective. The rapid growth of technical knowledge in the past few decades has meant that building and maintaining expertise in multiple technologies is difficult for even the largest corporations (Granstrand and Sjolander, 1990; Arora and Gambardella, 1994). Yet, bringing together different streams of knowledge is becoming an important precondition to successful innovation in many industries (Grant, 1996; Powell, Koput, and Smith-Doerr, 1996). This need for increased differentiation in the development of knowledge and increased integration in its application has given new importance to technology markets (Demsetz, 1991; Grant, 1996). Our results indicate that the process of obtaining technological assets from external sources and matching them with internally developed assets to enhance their productivity can work, at least insofar as the frequency of innovation output is concerned.

The above arguments highlighted the brighter aspects of the results. Another perspective on these results, however, draws attention to their less positive facets. Prior research has identified three kinds of deficiencies in the context of organizational learning: hubris, or underestimating the likelihood of failure or the difficulty of the task at hand; overexploration, or venturing into domains of completely new knowledge; and, overexploitation, or focusing only on the immediate neighborhood of the well known and understood and overlooking more distant options (Levinthal and March, 1993). Our data and results indicated evidence of all three kinds of errors. To the extent that relative size serves as a measure of the relative difficulty of integration, it appears that underestimating the magnitude of the integration task is not uncommon. Further, the variation on relatedness, and the curvilinear relationship that we identified between acquisition relatedness and innovation output, suggests that in selecting acquisitions managers err in both directions, acquiring both businesses that are only distantly related or unrelated to the existing business, as well as those that are too closely related to the current business. The problems of acquiring unrelated businesses are well documented and our results further reinforce those lessons (Singh and Montgomery, 1987). Additionally, our findings help to highlight errors in the opposite direction. Managers do make mistakes in picking acquisitions that are too closely related to their extant domains, and these mistakes are penalized with poorer performance. Thus, we find some empirical evidence of 'competency traps' (Levinthal and March, 1993).

Implications for measurement

Our findings also have implications for the literature on acquisitions in general. Managerial hubris and agency problems (Roll, 1986; Hayward and Hambrick, 1997), imitation effects (Haunschild, 1993), inappropriate application of learning (Haleblian and Finkelstein, 1999), and underestimating the process impediments to postacquisition integration (Jemison and Sitkin, 1986; Hitt *et al.*, 1996; Singh and Zollo, 1997) are all valid and complementary explanations for why acquisitions consistently fail to help the acquiring firm. The findings of this paper suggest an additional explanation, one that is perhaps more sympathetic to managerial motivations and decisions. Evaluating all acquisitions on the same performance metric, for instance financial performance, may not be appropriate. Acquisitions motivated by different objectives may differ in their timing and mode of impact on firm performance.

The recent literature has provided several exciting ways of using patent data to measure the construct of knowledge (Mowery, Oxley, and Silverman, 1998; Jaffe *et al.*, 1993; Stuart and Podolny, 1996). We build on these studies by presenting an additional set of measures and applying them to the context of acquisitions. The patent-based approach to knowledge measurement used by this study, and similar approaches used in prior studies (e.g., Stuart and Podolny, 1996), have several strengths. A primary strength of such approaches is that by using information on individual elements of knowledge these approaches make possible a fine-grained assessment of a firm's knowledge base. Research has argued that a firm's knowledge base should ideally be reflected as an asset on its balance sheet (Granstrand and Sjolander, 1990). Although that remains unattainable at present, these kinds of measures do provide some basis for quantifying this asset and enabling further analysis. A second strength of these approaches is that they enable continuous, and highly informative and detailed, measurement of difficult constructs such as technological relatedness. Finally, another virtue of these approaches is that they permit quantification of knowledge bases from publicly available data, in a longitudinal context.

The use of distributed lags to study the dynamics of postacquisition performance represents another potential contribution of this study. Although the distributed lag technique has been used by economists, relatively few researchers in the strategy area have used this methodology. This approach could be extremely useful in a variety of strategic contexts that entail studying the impact of an event or strategy on the performance of a unit across several periods.

Implications for managers

The findings of this study draw managers' attention to several paradoxical aspects of the acquisition selection and integration process. This study indicates that larger absolute size and smaller relative size of acquisitions are associated with superior postacquisition performance for the acquiring firm. Further, a moderate level of relatedness is preferable to acquiring a firm that is very closely related, or distant to the acquiring firm. Thus, from the acquisition selection standpoint this research suggests that a balance on both size and relatedness of acquisitions is favored.

Limitations and future research

Several limitations of this study are worth noting. The restriction of the sample to a single industrial context reinforces the need for conducting the study in other industries. Additionally, the relevance and utility of the patent-based measures of knowledge are likely to be limited to industries in which patents are themselves meaningful indicators of innovation. In addition to the chemicals sector some of the other contexts in which these measures could be applied include biotechnology, semiconductors, industrial machinery, and advanced materials.

Another limitation of this study that suggests an area of future research is the use of patents to measure innovative output. Although patents are reasonably good indicators of innovative output, they are best regarded as intermediate outcomes between acquisitions and value creation. While our results indicate that acquisitions accelerate the creation of technologically new combinations and innovations, they do not allow us to directly measure the value generated by these innovations. Examining the economic value of postacquisition innovations would be a natural extension of the work in this study and would enable a more complete assessment of the contribution of technological acquisitions to technological development.

Finally, in this paper we used only a simple count measure for nontechnological acquisitions. However, such acquisitions could themselves vary on many dimensions and reflect many different objectives. The development of a schema to measure the various dimensions of nontechnological acquisitions and relating them to different types of firm outcomes would be another fruitful direction of further research.

ACKNOWLEDGEMENTS

We thank Jim Fredrickson, George Huber, Jim Westphal, the editor, and two anonymous reviewers for comments that helped us to refine our thinking. We especially thank Risto Miikkulainen for assistance and advice on the computer programs used to compute the patent citation variables. We gratefully acknowledge the financial support of the University of Texas at Austin. The first author also acknowledges the support of the Center for International Business Education and Research at the University of Michigan.

REFERENCES

Achilladelis B, Schwarzkopf A, Cines M. 1987. A

study of innovation in the pesticide industry: analysis of the innovation record of an industrial sector. *Research Policy* **16**(2–4): 175–212.

Anand J, Singh H. 1997. Asset redeployment, acquisitions and corporate strategy in declining industries. *Strategic Management Journal*, Summer Special Issue **18**: 99–118.

Arora A, Gambardella A. 1994. Evaluating technological information and utilizing it: scientific knowledge, technological capability, and external linkages in biotechnology. *Journal of Economic Behavior & Organization* **24**(1): 91–114.

Arundel A, Kabla I. 1998. What percentage of innovations are patented? Empirical estimates for European firms. *Research Policy* **27**: 127–141.

Balakrishnan S. 1988. The prognostics of diversifying acquisitions. *Strategic Management Journal* **9**(2): 185–196.

Basberg B. 1983. Foreign patenting in the U.S. as a technology indicator: the case of Norway. *Research Policy* **12**: 227–237.

Basberg B. 1987. Patents and the measurement of technological change: a survey of the literature. *Research Policy* **16**(2–4): 131–141.

Blundell R, Griffith R, Van Reenen J. 1995. Dynamic count data models of technological innovation. *Economic Journal* **105**(429): 333–344.

Capron L, Dussauge P, Mitchell W. 1998. Resource redeployment following horizontal acquisitions in Europe and North America, 1988–1992. *Strategic Management Journal* **19**(7): 631–661.

Capron L, Mitchell W, Swaminathan A. 1998. Resource redeployment following horizontal acquisitions in Europe and North America, 1988–1992. *Strategic Management Journal* **19**(7): 631–661.

Chakrabarti A, Hauschildt J, Suverkrup C. 1994. Does it pay to acquire technological firms? *R&D Management* **24**(1): 47–56.

Chatterjee S. 1991. Gains in vertical acquisitions and market power: theory and evidence. *Academy of Management Journal* **34**(2): 436–448.

Chemical Week. 1983. The push to buy technology. **134**(Sept. 14): 44.

Cohen W, Levin R. 1989. Empirical studies of innovation and market structure. In *Handbook of Industrial Organization (II)*. Schmalensee R, Willig RD (eds). Elsevier: Amsterdam; 1059–1107.

Cohen WM, Levinthal D. 1989. Innovation and learning: the two faces of R&D. *Economic Journal* **99**(397): 569–596.

Cohen W, Levinthal D. 1990. Absorptive capacity: a new perspective on learning and innovation. *Administrative Science Quarterly* **35**: 128–152.

Comanor W, Scherer F. 1969. Patent statistics as a measure of technical change. *Journal of Political Economy* **77**(3): 392–398.

Curry S. 1997. The year of the very big deals. *Fortune* Nov. 24: 36–37.

Demsetz H. 1991. The theory of the firm revisited. In *The Nature of the Firm*, Williamson O, Winter S (eds). Oxford University Press: New York; 159–178.

Dosi G, Pavitt K, Suete L. 1990. *The Economics of Technical Change and International Trade*. New York University Press: New York.

Fleming L. 1999. Recombinant uncertainty in technological search. working paper, Harvard Business School.

Fowler KL, Schmidt DR. 1988. Tender offers, acquisitions, and subsequent performance in manufacturing firms. *Academy of Management Journal* **31**(4): 962–974.

Gerpott T. 1995. Successful integration of R&D functions after acquisitions: an exploratory empirical study. *R&D Management* **25**(2): 161–178.

Gibson WD. 1985. Tracking technology around the globe. *Chemical Week* **136**(24): 34.

Granstrand O, Bohlin E, Oskarsson C, Sjoberg N. 1992. External technology acquisitions in large multi-technology corporations. *R&D Management* **22**(2): 111–133.

Granstrand O, Sjolander S. 1990. The acquisition of technology and small firms by large firms. *Journal of Economic Behavior & Organization* **13**(3): 367–386.

Grant RM. 1996. Prospering in dynamically-competitive environments: organizational capability as knowledge integration. *Organization Science* **7**(4): 375–387.

Greene W. 1993. *Econometric Analysis*. Macmillan: New York.

Griliches Z. 1979. Issues in assessing the contribution of research and development to productivity growth. *Bell Journal of Economics* **10**(Spring): 92–116.

Griliches Z. 1984. Introduction. In *R&D, Patents and Productivity*, Griliches Z (ed). University of Chicago Press: Chicago, IL; 1–20.

Griliches Z. 1990. Patent statistics as economic indicators: a survey. *Journal of Economic Literature* **28**(4): 1661–1707.

Gujarati D. 1988. *Basic Econometrics*. 2nd edn, McGraw-Hill: New York.

Haleblian J, Finkelstein S. 1999. The influence of organization acquisition experience on acquisition performance. *Administrative Science Quarterly* **44**(1): 29–56.

Hall B, Griliches Z, Hausman J. 1986. Patents and RandD: is there a lag? *International Economic Review* **27**(2): 265–283.

Hall B, Cummins C, Laderman E, Mundy J. 1988. The R&D master file documentation. NBER Technical Working Paper Series, no. 72.

Haspeslagh PC, Jemison DB. 1991. *Managing Acquisitions: Creating Value Through Corporate Renewal*. Free Press: New York.

Haunschild PR. 1993. Interorganizational imitation: the impact of interlocks on corporate acquisition activity. *Administrative Science Quarterly* **38**(4): 564–592.

Hausman J, Hall B, Griliches Z. 1984. Econometric models for count data with an application to the patents–R&D relationship. *Econometrica* **52**: 909–938.

Hayward MLA, Hambrick DC. 1997. Explaining the premium paid for large acquisitions: evidence of CEO hubris. *Administrative Science Quarterly* **42**(1): 103–127.

Henderson RM, Clark KB. 1990. Architectural innovation: the reconfiguration of existing product technologies and the failure of established firms. *Administrative Science Quarterly* **35**(1): 9–30.

Henderson R, Cockburn I. 1996. Scale, scope, and

spillovers: the determinants of research productivity in drug discovery. *RAND Journal of Economics* **27**(1): 32–59.
Hitt MA, Hoskisson RE, Ireland RD, Harrison JS. 1991. Effects of acquisitions on R&D inputs and outputs. *Academy of Management Journal* **34**(3): 693–706.
Hitt MA, Hoskisson RE, Johnson RA, Moesel DD. 1996. The market for corporate control and firm innovation. *Academy of Management Journal* **39**(5): 1084–1119.
Hofstede G. 1980. *Culture's Consequences: International Differences in Work-Related Values*. Sage: Beverly Hills, CA.
Hoskisson RA, Hitt MA, Ireland D. 1994. The effects of acquisitions and restructuring (strategic refocusing) strategies on innovation. In *The Management of Corporate Acquisitions: International Perspectives*. Von Krogh G, Sinatra A, Singh H (eds). Macmillan: Houndmills; 144–169.
Huber G. 1991. Organizational learning: the contributing processes and a review of the literatures. *Organization Science* **2**(1): 88–117.
Jaffe A, Trajtenberg M, Henderson R. 1993. Geographic localization of knowledge spillovers as evidenced by patent citations. *Quarterly Journal of Economics* **108**: 577–598.
Jemison DB, Sitkin SB. 1986. Corporate acquisitions: a process perspective. *Academy of Management Review* **11**(1): 145–163.
Judge GG, Hill RC, Griffiths WE, Lütkepohl HH, Lee T. 1988. *Introduction to the Theory and Practice of Econometrics*. 2nd edn, Wiley: New York.
Kamien M, Schwartz N. 1982. *Market Structure and Innovation*. Cambridge University Press: New York.
Kim D-J, Kogut B. 1996. Technological platforms and diversification. *Organization Science* **7**(3): 283–301.
Kogut B, Singh H. 1988. The effect of national culture on the choice of entry mode. *Journal of International Business Studies* **19**(Fall): 411–432.
Kogut B, Zander U. 1992. Knowledge of the firm, combinative capabilities, and the replication of technology. *Organization Science* **3**(3): 383–397.
Lane PJ, Lubatkin M. 1998. Relative absorptive capacity and interorganizational learning. *Strategic Management Journal* **19**(5): 461–477.
Levin RC, Klevorick AK, Nelson RR, Winter SG, Gilbert R, Griliches Z. 1987. Appropriating the returns from industrial research and development: comments and discussion. *Brookings Papers on Economic Activity* **3**: 783–831.
Levinthal D, March J. 1993. The myopia of learning. *Strategic Management Journal*, Winter Special Issue **14**: 95–112.
Liang K-Y, Zeger S. 1986. Longitudinal data analysis using generalized linear models. *Biometrika* **73**(1): 13–22.
Link A. 1988. Acquisitions as sources of technological innovation. *Mergers & Acquisitions* **23**(November/December): 36–39.
Lubatkin M. 1983. Mergers and the performance of the acquiring firm. *Academy of Management Review* **8**(2): 218–225.
Mowery D, Oxley J, Silverman B. 1998. Technological overlap and interfirm cooperation: implications for the resource-based view of the firm. *Research Policy* **27**(5): 507–523.
Narin F, Noma E, Perry R. 1987. Patents as indicators of corporate technological strength. *Research Policy* **16**(2–4): 143–155.
Pakes A, Griliches Z. 1984. Patents and R&D at the firm level: a first look. In *R&D, Patents and Productivity*, Griliches Z (ed). University of Chicago Press: Chicago, IL.
Palepu K. 1985. Diversification strategy, profit performance and the entropy measure. *Strategic Management Journal* **6**(3): 239–255.
Patel P, Pavitt K. 1997. The technological competencies of the world's largest firms: complex and path-dependent, but not much variety. *Research Policy* **26**: 141–156.
Podolny JM, Stuart TE. 1995. A role-based ecology of technological change. *American Journal of Sociology* **100**(5): 1224–1260.
Powell WW, Koput KW, Smith-Doerr L. 1996. Interorganizational collaboration and the locus of innovation: networks of learning in biotechnology. *Administrative Science Quarterly* **41**(1): 116–145.
Pritchett P. 1985. Managing people problems in deals that don't close. *Journal of Buyouts & Acquisitions* **3**(7): 35–38.
Roll R. 1986. The hubris hypothesis of corporate takeovers. *Journal of Business* **59**(2): 197–216.
Schankerman M, Pakes A. 1986. Estimates of the value of patent rights in European countries during the post-1950 period. *Economic Journal* **96**: 1052–1077.
Scherer F. 1965. Corporate inventive output, profitability and sales growth. *Journal of Political Economy* **73**(3): 290–297.
Scherer F, Ross D. 1990. *Industrial Market Structure and Economic Performance*. Houghton Mifflin: Boston, MA.
Schumpeter J. 1934. *The Theory of Economic Development*. Harvard University Press: Cambridge, MA.
Singh H, Montgomery CA. 1987. Corporate acquisition strategies and economic performance. *Strategic Management Journal* **8**(4): 377–386.
Singh H, Zollo M. 1997. Knowledge accumulation mechanisms and the evolution of post-acquisition integration strategies. Working paper, University of Pennsylvania.
Spender J-C. 1989. *Industry Recipes: An Inquiry into the Nature and Sources of Managerial Judgement*. Blackwell: New York.
Stuart TE, Podolny JM. 1996. Local search and the evolution of technological capabilities. *Strategic Management Journal*, Summer Special Issue **17**: 21–38.
Thompson J. 1967. *Organizations in Action*. McGraw-Hill: New York.
Trajtenberg M. 1990. *Economic Analysis of Product Innovation: The Case of CT Scanners*. Harvard University Press: Cambridge, MA.
Tushman M, Rosenkopf L. 1992. Organizational determinants of technological change: toward a sociology of technological evolution. *Research in Organizational Behavior* **14**: 311–347.

Utterback J. 1994. *Mastering the Dynamics of Innovation: How Companies can Seize Opportunities in the Face of Technological Change*. Harvard Business School Press: Boston, MA.
Very P, Lubatkin M, Calori R, Veiga J. 1997. Relative standing and the performance of recently acquired European firms. *Strategic Management Journal* **18**(8): 593–614.
Walker R. 1995. *Patents as Scientific and Technical Literature*. Scarecrow Press: Lanham, MD.
Weber Y, Shenkar O, Raveh A. 1996. National and corporate cultural fit in mergers/acquisitions: an exploratory study. *Management Science* **42**(8): 1215–1227.
Wysocki B. 1997a. Why an acquisition? Often, it's the people. *Wall Street Journal* October 6: 1.
Wysocki B. 1997b. Many mergers driven by search for fresh talent. *Chicago Tribune* December 28: 67.

APPENDIX

The list of patents obtained by a firm, and the patents cited by the firm in the patents obtained by it, is used as the measure of the firm's knowledge base. Since a firm's own patents represent knowledge created by the firm, their use as components of the firm's knowledge base is natural. However, the use of patents *cited* by the firm as a component of the firm's own knowledge base may require further explanation.

Patent citations represent an acknowledgement of the contributions of prior work in the development of the current invention. As Jaffe *et al.* (1993: 580) note:

> The granting of the patent is a legal statement that the idea embodied in the patent represents a novel or useful contribution over and above the previous state of knowledge, as represented by the citations. Thus, in principle, a citation of Patent X by Patent Y means that X represents a piece of previously existing knowledge upon which Y builds.

A citation thus represents a restriction on an owner's rights to exploit her own patent, and is therefore not likely to be made if the cited patent does not truly represent knowledge underlying the current patent.

The legal aspects of the patenting process also ensure that patents that *do* truly represent preexisting knowledge are cited with accuracy in patents that build upon them. First, all applicants are required by law to acknowledge all 'prior art' through citations. Second, to oversee this process, each patent and its citations must be cleared by a patent examiner before the patent is actually granted. The patent examiner, who is usually an expert in the technological domain under which the patent falls, conducts a search of the prior art in the area and adds any citations that the patent applicant's own search may have missed. Since citations circumscribe the exploitable domain of the applicant's patent, the patent applicant is likely to contest these citations if they do not accurately represent prior related knowledge. Together, this system represents a set of checks and balances that ensures that patent citations are relatively robust indicators of knowledge flows (Trajtenberg, 1990; Jaffe *et al.*, 1993). Thus, the assignee of a patent is likely to know the contents of the patents cited by her, at least to a level of familiarity, and perhaps to mastery.

Part V
Real Options

[20]

© *Academy of Management Review*
1997, Vol. 22, No. 4, 974–996.

A REAL OPTIONS LOGIC FOR INITIATING TECHNOLOGY POSITIONING INVESTMENTS

RITA GUNTHER MCGRATH
Columbia University

In this article I extend real options theory to technology positioning projects and specify how the relationship between boundary conditions and uncertainty influences the value of a technology option, as well as the appropriate timing of its exercise. I also take a strategic perspective on uncertainty itself, concluding that option value can be amplified by investments to shift boundaries, ideally in ways that are idiosyncratic to the firm.

Conventional valuation and budgeting techniques show their limitations when one applies them to investments requiring substantial commitment under conditions of significant uncertainty, such as investments in technology (Dixit & Pindyck, 1994). Although some scholars suggest real options theory as an alternative conceptual route of attack, its application to strategic technology positioning problems merits further development, which is my objective in this article (Bettis & Hitt, 1995; Bowman & Hurry, 1993; Mitchell & Hamilton, 1988; Trigeorgis, 1993).

In this article I extend real options concepts in two ways. First, I develop theory specifying how the relationship between boundary conditions and uncertainty influences the value of technology options, as well as the appropriate timing of their exercise. Second, I take a strategic perspective on uncertainty itself, arguing that value can be amplified by investments to shift boundaries, ideally in ways that reduce uncertainty idiosyncratically.

IDIOSYNCRATIC REDUCTION OF UNCERTAINTY AND RENTS

In strategic management capturing "rents" is a key performance objective, where the term represents excess profits that do not immediately induce a competitive response (see Alchian, 1991). One obvious way in which technologies generate rents for a firm is through monopoly, as with patented drugs. Even though competitors can observe and understand

Four anonymous *Academy of Management Review* reviewers offered valuable commentary. I much appreciate suggestions by Arnold C. Cooper and Ian C. MacMillan and comments by Neil Doherty, and I gratefully acknowledge the financial support of the Strategic Technology Assessment Research (STAR) Program of the Sol C. Snider Entrepreneurial Center at the Wharton School.

why the firm is successful, they are prevented by formal appropriability regimes from being able to mount an effective attack.

Somewhat more subtle are situations in which perfect competition is thwarted by the use of firm-specific routines (see Nelson & Winter, 1982, and Peteraf, 1993). Productive routines in technology are sometimes lucky accidents (Barney, 1986), but they also can reflect learning, in which those in a firm gain greater understanding of how to influence action-outcome relationships (Cheng & Van de Ven, 1996). In either case, knowledge and insight asymmetries can emerge by virtue of the experiential and path-dependent nature of the development process (Dierickx & Cool, 1989), resulting in different firms having different uncertainty profiles. Investments in development are attractive, by implication, when results idiosyncratically improve the relative uncertainty a firm faces in ways that can be kept firm specific (Amit & Schoemaker, 1993). My argument builds on these ideas to suggest constructs that can idiosyncratically influence both uncertainty and, consequently, the value of a technology option.

TECHNOLOGY POSITIONING PROJECTS GENERATE REAL OPTIONS

Real options theory scholars seek to understand classes of investment decisions having a structure similar to financial options contracts but for which the assumptions made in valuing financial options do not hold. The distinguishing characteristic of an options approach lies in firms making investments that confer the ability to select an outcome only if it is favorable.

Price and Value of a Technology Option

Financial options models rely on several core assumptions to arrive at an option price. The following are among the most important: the underlying asset on which the option is written must be priced, this price must be known, and the asset must be continuously tradable (Black & Scholes, 1973). The more volatile the price of the underlying asset, the more valuable the option becomes, as potential positive returns are greater, although potential losses are limited to the price of the option. These models cannot be applied to pricing options created through investment in technology positioning projects, for their core assumptions do not hold. For instance, the price of an underlying technology asset is not known, and it may not be continuously tradable.

In other respects, however, technology options have important similarities to financial options. The "price" of a technology option is the cost of development (which may occur in sequential stages). Completing the development of the technology creates an asset, consisting of the underlying right to commercialize the technology. "Exercise" of such an option involves the further investments needed to commercialize the technology (which may also occur sequentially). The option also may be "traded"—for instance, by being licensed out or spun off. In this regard the option is like an American call: it can be exercised or traded any time after technology development is completed, or not at all. Completing commercialization creates another asset, which is the underlying right of the firm to extract returns from the exploitation of the now-commercialized technology (Kester, 1981).

Just as an investor may elect not to strike a financial option and lose only the amount of the options contract, a firm may elect to stop investing and limit losses to the sunk costs associated with the discontinued project. And, just as the value of a financial option increases as the price volatility of the underlying asset increases, so the value of a real technology option increases as the variance of the expected value of net revenues minus commercialization cost increases. The greater the expected variance in net operating revenues less commercialization costs, the higher the possible value of the underlying asset, although possible losses are limited.

Factors influencing price volatility therefore can be seen to influence the value of a financial option, and factors influencing variances similarly influence the value of a real technology option. The greater the uncertainty with respect to net revenues and costs, the more likely it is that large variances will be anticipated. This brings us to the critical role that the nature of uncertainty plays in establishing the value of a real technology option.

Technical and External Uncertainty, Timing, and Option Value

In their 1994 book, *Investment Under Uncertainty,* Dixit and Pindyck distinguish between two forms of uncertainty: "technical" uncertainty and "input cost" uncertainty. Technical uncertainty relates to the likely costs and probabilities of accomplishing technical success; a firm reduces this type of uncertainty only through investment. Technical uncertainty creates pressure on the firm to invest immediately. Delays, at best, incur a discounting penalty and, at worst, expose the firm to the risk of competitive preemption. Positive feedback effects to scale, path dependence, and network externalities may simply lock out late entrants (see Arthur, 1994). Later entry also can subject the firm to what Dierickx and Cool (1989) term "time compression diseconomies," relative to early movers. Early, aggressive entry in the face of technical uncertainty can be critical. However, moving early can subject a firm to a punishing array of difficulties, including the bearing of the costs of infrastructure, the appropriation of rents by owners of needed cospecialized assets, and other aggravations (Lieberman & Montgomery, 1988).

These timing dilemmas, when a firm faces technical uncertainty, significantly increase the benefit to the firm of adopting an options approach for two reasons. First, although the volatility of new markets is greater than that of established markets, options have greater value in such markets. Second, viewing the technological challenge as a series of sequentially exercised options accommodates a milestone-oriented, iterative management process. This simultaneously permits project redirection, advances learning, and permits investment to be discontinued at the earliest possible time, conserving the firm's precious resources (Bowman & Hurry, 1993; Eisenhardt & Tabrizi, 1995).

A second form of uncertainty, which Dixit and Pindyck (1994) term "input cost" uncertainty, relates to factors exogenous to the firm. No amount of investment makes a difference in this form of uncertainty, which creates pressure on the firm to delay investment until information is revealed with the passage of time.

Yet a third form of uncertainty exists, lying between the two forms Dixit and Pindyck identify. It is present when sources of uncertainty are largely "external" to the firm (in other words, not technical in nature) but can be influenced by strategic action. If it is possible for the firm to act as an agent of endogenous change, it may seek to shape contingencies in its favor. This gets to the heart of the strategic component of the use of a real options framework. The form uncertainty takes shapes the boundary conditions within which option value may be estimated and strategic action taken.

External Uncertainty and the Value of the Option to Wait

In determining whether a firm can take strategic action, one must first assess temporal truncation—where the passage of time alone brings uncertainty to an end. For example, let us consider a hypothetical investment by a firm in technologies for pollution control. Suppose that adoption will occur only if environmental control regulations contain certain provisions. Developing the technology will cost the firm an estimated $10 million; it is equally probable that the net present value of commercialization costs will be either $100 million or $50 million. The net present value of revenues under favorable regulation (a 50 percent probability) will likely be $180 million. Absent favorable rules (a 50 percent probability), revenues are expected to be $0. The decision tree for this investment under risk decision[1] is presented in Figure 1. The expected value of the project is:

$$0.5(-100 + 0.5 \times 180 + 0.5 \times 0) + 0.5(-50 + 0.5 \times 180 + 0.5 \times 0) - 10 = \$5 \text{ million}$$

or an expected profit of $5 million. The firm should immediately invest under conventional net present value rules.

When we recall, however, that the options approach allows a firm to select an outcome only if it is favorable, a rather different picture emerges. What we have here is a situation in which uncertainty with

[1] Since this model utilizes probability estimates, this is a decision in a risk situation rather than a decision under uncertainty.

FIGURE 1
Technology Positioning Option Decision Tree

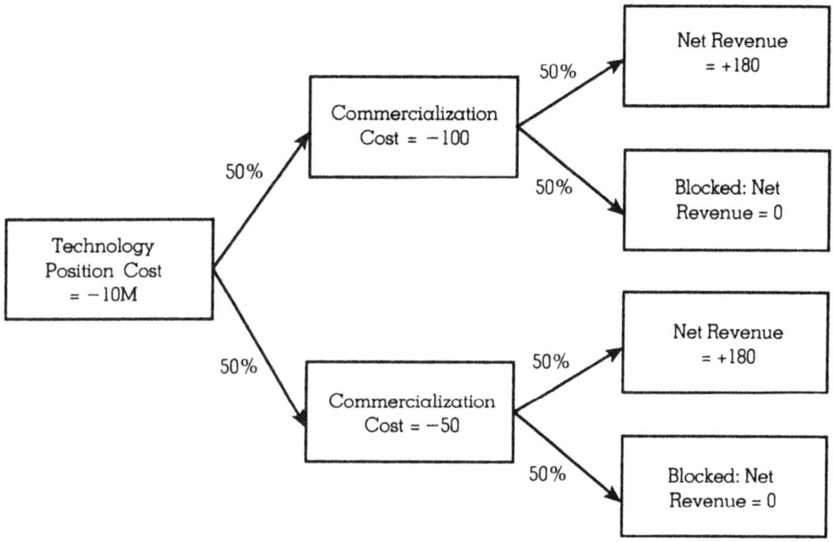

respect to the pending regulation will be truncated by the passage of time. Eventually, open issues will be resolved. Let us say the firm recognizes this and elects to hold off on further investments until the regulatory uncertainty is temporally truncated. If, for simplicity, we assume that the firm's discount rate is 10 percent and that the regulatory issue will be decided in 1 year, the expected value of such a "wait-and-see" strategy is:

$$0.9(0.5[0.5\{180 - 100\} + 0.5\{180 - 50\} - 10] + 0.5(-10) = \$38.25 \text{ million}$$

The presence of temporally truncated uncertainty increases the value of holding the option, as opposed to exercising it, since it decreases the value of immediate investment. The firm should simply wait and see. If we apply the concept of external uncertainty to this problem, however, another, less obvious approach emerges. Let us suppose that the firm can conceive of actions that will help shape emergent rules in its favor. In such a situation, it may elect to hold its *technology* option but instead may make strategic investments on *other* fronts. Such alternative investments, if successful, amplify the value of the underlying claim to commercialize, prior to commencing commercialization.

Amplifying Preinvestments

In the hypothetical example given above, uncertainty both constrains the positive potential of the project and establishes an expiration date for options on the technology. The option will only be "in the money" if new regulations are adopted, which will happen at some future point. Let us suppose, instead, that the firm pays a group of lobbyists to help push through favorable regulation (MacMillan, 1978). If the lobbyists could guarantee the immediate adoption of favorable rules, the firm would capture 180 – (0.5 × 50 + 0.5 × 100) – 10 = \$95 million in expected value, versus the \$38.25 million expected previously. By implication, the firm could spend up to \$56.75 million on a lobbying strategy having a 100 percent chance of immediate success, which theoretically justifies an expenditure of \$5 million or so with a strategy having only a 10 percent chance of success.

As a thriving lobbying industry suggests, it may behoove firms to maximize the value of their options by influencing key legal boundary conditions rather than by investing in technology per se. When it is possible for firms to make amplifying preinvestments, these preinvestments create a context in which the technology can flourish, which has the effect of increasing the value of the underlying technology asset and the value of the option.

Effects of Boundary Conditions on Variance of Net Returns and Option Value

Variance for technology options behaves in a way similar to volatility in financial options. The greater the variance in net revenues that might be accessed by commercializing the technology, the greater the option value. By implication, to the extent that variance of expected returns is constrained, so too is the potential for gain. This depresses the value of the option. When we relax the constraints, however, as with an amplifying preinvestment, we increase the value of the option.

If the variance of the returns is **var π**, the variance of net revenues from operations is **var r**, and the variance of commercialization costs is **var c**, then var π = var(r – c) = var r + var c. If E[r] is expected net revenues and E[c] is expected commercialization costs, this leads to two distinct cases:

1. **Case 1: E[r] > E[c]; var r > var c.** Here, the upper limit of var π is determined by var r. Amplifying strategies that relax any boundary conditions suppressing the variance of net revenues will increase the value of the option.
2. **Case 2: E[r] > E[c]; var c $\geqslant$ var r.** Here, the upper limit of var π is determined by boundary conditions that significantly obstruct lower limits on commercialization costs. In other words, the boundary conditions cause commercialization costs to be inordinately high. In this case, amplifying strategies that reduce the variance of commercialization costs, var c, to below the variance of net revenues, var r, will increase the value of the real technology option.

This suggests an important implication for the application of real options theory to technology investments—namely, that valuation of a technology option cannot be conducted in isolation of potential amplification opportunities. Amplifying strategies, in turn, cannot be pursued without taking into account the relations among revenues and costs. The value of a particular technology option to a particular firm, therefore, is deeply embedded in the strategic context of that firm and cannot be considered apart from it. Practically, this suggests that pursuing technology decisions without considering other facets of the overall strategy could lead to significant suboptimality, particularly if the boundary conditions shaping uncertainty lend the situation to amplifying preinvestment opportunities.

This brings us to a discussion of the nature of key boundary conditions that influence the uncertainty profile of a technology option and the associated amplifying preinvestments that could release these bounds to enhance the value of the option.

BOUNDARY CONDITIONS AFFECTING THE UNDERLYING CLAIM TO CUMULATIVE RETURNS FROM OPERATIONS

Here, I first take up case 1 situations, in which the objective is to determine whether boundary conditions set limits on the positive potential magnitude of var r. This is of particular concern for firms facing "hypercompetitive" (D'Aveni, 1994; MacMillan, 1988) environments, in which any rents flowing from successful commercialization will come under severe competitive pressures. Under such circumstances, to paraphrase Hobbes, conditions are brutal, exchanges nasty, and the duration of any competitive advantage short.

In order to justify taking out an option on a technology, a firm's scale of cumulative rent streams from deployment of the asset must be large enough to recover initial investment in the face of attempts at appropriation. This recovery often is done in a compressed time frame. Two factors determine the scale of cumulative returns—the size of the rent stream and the length of time it can be sustained—each of which poses its own set of uncertainties. These factors and their determinants, as well as possible amplifying preinvestments, are summarized in Table 1. The relationships depicted in Table 1 represent proposed effects of uncertainty, strategies, idiosyncratic resources, and the value of the option.

Boundary Conditions and Uncertainty Regarding Size of Rent Streams

Uncertainty regarding demand. The idea that successful innovations should "satisfy some want" (Schmookler, 1966) long has been recognized as a core premise in the technology and innovation management literature (Kamien & Schwartz, 1975; Myers & Marquis, 1969; Mowery & Rosenberg, 1982; Scherer, 1979), for, absent want, there will be no demand. As Mowery and Rosenberg (1982) observe, however, ex ante clarity on

TABLE 1
Key Contingencies Influencing Uncertainty and Strategic Opportunities for Rent

Sources of Uncertainty	Contingent Factors that Bound Value of Option	Effect on Option Value	Opportunity for Amplifying Preinvestment	Opportunity for Deployment of Idiosyncratic Endowments
Total cumulative net revenues	Convexity and monotonicity of demand	Plus	Aggressive early moves that generate positive returns to scale	Reputation and track record that offer access to capital and other resources
	Adoption speed uncertainty	Minus	Reduce obstacles to adoption, increase perceived attractiveness	Idiosyncratic customer linkages
	Likelihood of blocking	Minus	Counterblocking	Institutional/ collaborative relationships
	Likelihood of expropriation	Minus	Negotiated environment	Idiosyncratic negotiating "stakes"
Sustainability	Likelihood of matching	Minus	Standard setting	Collaborative relationships
	Likelihood of imitation	Minus	Development of appropriability regimes, collaborative arrangements	Collaborative relationships or membership in standard-setting bodies
Commercialization cost	Infrastructure requirements	Minus	Parallel or joint development	Exclusive control over critical contingencies or resources
	Parallel technology development	Minus	Parallel or joint development	Preemptive alliance with leading partner
	Cospecialized assets requirements	Minus	Parallel or joint development	Firm-specific, cospecialized assets or preemptive alliance with asset owner
Technology development cost	Spillover potential	Plus	Internal cross-fertilization	Competences and customer relations
	Life-cycle effects	Minus	Membership in community technical organization	Preexisting standard-setting relationships

market demand is difficult to establish, particularly for very new technologies. Potential customers find it difficult to assess the value of a technology with which they have no experience. Under such circumstances, conventional market research often is not only inaccurate but actually can be misleading (von Hippel, 1988).

Although it may not be possible to estimate demand accurately, it may be possible to identify demand structures that are more or less attractive (recalling a philosophy for evaluating technologies proposed by Goodman & Lawless, 1994). If, for instance, potential applications of the technology have an underlying demand structure with properties of monotonicity and convexity, the technology asset's value should be greater and so, too, should the value of the option.

For example, let us contrast the demand structure of a technology that would dramatically *ameliorate* the symptoms of a widespread, frequently recurring ailment, such as the common cold, with one that would *cure* a disease that afflicts 1 person in 100,000. In the first case, usage is repetitive and incidence widespread, suggesting both a greater potential mean and greater potential variance in demand than in the second case, in which there is but a single dosage required for a highly limited population. This suggests that one may be able to assess whether the potential variance of demand opportunities is vast or limited. Examples of factors signaling how attractive the structure of demand is include the following: whether there are many or few problems the technology might solve; whether problems that the technology might solve are widespread; and, if the problem is widespread, whether it is also growing, whether it recurs rather than is resolved, whether recurrence is frequent, and so on.

The more the demand structure suggests that demand over time will be monotonic and convex, rather than concave, the greater the potential mean demand, and, more important, the greater the demand's potential variance, which increases potential returns and thus the value of the option.

When the firm can recognize and capitalize on advantages created by positive returns to scale, it may preinvest by setting the stage for aggressive early moves (Lieberman & Montgomery, 1988). Examples of such moves might include preemptively investing to secure network position or network commitments in industries with significant network externality effects (Conner, 1995). Similarly, an established firm may use such a firm-specific asset as a solid reputation with the financial community to idiosyncratically gain access to those resources (such as equity capital) that allow it to engage in an aggressive launch.

Uncertainty regarding speed of adoption. Slow adoption rates can limit cumulative returns significantly by delaying revenue inflows and reducing the time before alert competitors can compete away rent streams. The importance of speed of adoption has prompted a considerable number of scholars to assess the conditions under which diffusion is likely to be rapid or slow (see Rogers, 1995). Slow appears to be the norm. Typically, both the innovating firm and its customers must establish a base of experience before widespread adoption occurs (Levinthal & March, 1993: 106). Factors that tend to delay adoption include whether and how much education or training is required to use the technology; whether adoption implies a significant change in usage patterns; whether there is a need for coevolved infrastructure; whether customers have significant sunk costs in older technologies; and whether customers perceive adoption as risky. Factors tending to increase confidence in rapid adoption include low risk to the potential customer, alignment between beneficiaries and buyers, and the likelihood of the firm establishing an early dominant design (Dewar & Dutton, 1986; Katz & Shapiro, 1985; Lieberman & Montgomery, 1988).

Slow adoption affects the scale of potential revenues in two ways. First, it delays the onset of revenue streams, with a concomitant discounting penalty. This reduces both the mean and variance of potential cumulative revenue streams.[2] Second, slow adoption creates a period in which new competitors may enter in a competition for the dominant design, prompting expensive introduction of new product features (Anderson & Tushman, 1990). Alternatively, firms using established technologies have more time to react (Cooper & Smith, 1992). Slow adoption, in short, reduces the variance of cumulative returns and, consequently, the value of the option.

The adoption rate, however, may be amenable to an amplifying preinvestment strategy. To the extent that a firm can find ways to reduce the risks or costs to customers, or to demonstrate clear superiority over existing procedures, adoption can be accelerated. Capturing the support of what von Hippel (1988) refers to as "lead users," achieving critical mass with a core group of distributors or with opinion leaders, and helping customers ease the transition (e.g., by providing backward compatibility) are all likely strategies for a firm.

Similarly, a firm may use firm-specific resources to execute these strategies, which offers the additional potential for firm-specific reduction in uncertainty. A firm's past successful relationships, for instance, can create routine specificity between the firm and its groups of customers and suppliers—what Venkataraman (1990) terms its "transaction set." Existing customers, other things being equal, more likely will adopt new technologies from existing suppliers than from unknown suppliers. As a history of trusting interactions reduces the perceived risk of purchasing the new technology, adoption can be accelerated (Block & MacMillan, 1993).

Uncertainty regarding blocking. "Blocking" refers to situations in which a firm is denied access to inputs, applications, customers, markets,

[2] var $x_t = \sum(x - x_{t,ave})^2/n$. If $x_{t+1} = ax_1$, where $1 - a$ is the discount rate, then var $x_{t+1} = a^2$ var x_t, so var $x_{t+1} <$ var x_t if a is less than 1.

or other business essentials by competitive or regulatory action. Since the variance of a sum is equal to the sum of individual variances, the total variance of the cumulative revenue streams is reduced by the variance of each individual blocked revenue stream.[3] Thus, any point of blocking reduces the value of both the mean and the variance of revenues, therefore reducing the value of the option.

The dynamics of blocking have been discussed with respect to barriers to entry (Porter, 1980), preemptive behavior (Fudenberg & Tirole, 1985; Gilbert & Newberry, 1982), market power (Boulding & Staelin, 1990; Montgomery, 1985), foreign direct investments (Caves, 1971), pricing rivalry (Chen & MacMillan, 1992), and competitive signaling (Chen, Venkataraman, MacMillan, & Black, 1991). A common theme in these studies is that, consciously and with intent, a party external to the firm creates obstacles.

Studying this literature yields some interesting insights regarding the sources of likely blocking behavior. For example, those whose control over key influencers is threatened by the technology are likely to use this control to block adoption (MacMillan, 1978; MacMillan & Jones, 1986). Rivals with monopolistic or quasi-monopolistic market control further may block entry by interfering with access to suppliers, distributors, or key customers (Teece, 1986). Advocacy groups with an interest in preventing the adoption of the technology may impose explicit or subtle legislative, regulatory, or public opinion prohibitions with a blocking effect (Teece, 1986). Finally, those with a vested interest in preventing open competition may block access to geographic markets or may impose expensive conditions upon obtaining that access (Mowery & Teece, 1993).

If the danger of blocking is high, the value of the option can be amplified by the firm to the extent that it successfully can preinvest in counterblocking strategies prior to committing resources to the technology. The environmental technology example I referred to earlier reflects such a situation. Relevant idiosyncratic resources, as Table 1 suggests, are those institutional or collaborative relationships allowing a firm to uniquely eliminate the threat of blocking, ideally while preserving a blocked position for competitors.

Uncertainty regarding expropriation. Expropriation is related to blocking, in that the firm faces the threat of difficulty from powerful external players, but differs in its outcome. When blocked, a firm simply is denied access to critical resources and markets. Under expropriation, however, the firm is required to cede a portion of its rent streams, and, as the net revenues are reduced, so too is their variance.[4]

---

[3] Let x and y be segments with revenue variances of var x and var y, respectively. If the variance of the combined segments is var(x + y), then var(x + y) = var x + var y. This means that if segment x is blocked, the resulting variance of revenues is var(x + y) − var x, which is less than the original total variance.

[4] $\text{var } x = \Sigma(x - x_{ave})^2/n$. If $y = \alpha x$, where α is the proportion of revenue streams not expropriated from the original revenue x, then $\text{var } y = \alpha^2 \text{ var } x$, and $\text{var } y < \text{var } x$, if $\alpha < 1$.

The threat of expropriation may come from either governmental or interorganizational sources. Governments, for reasons of national interest, commonly place rent-sharing obligations on firms. These may take the form of requirements that technologies be made available to other firms, that local employees receive employment, and that the firm support infrastructure development or training programs, or they may involve outright commandeering of the firm's assets (Kobrin, 1982; Teece, 1986).

Interorganizational dynamics also can exacerbate expropriation. Discriminatory regulation or exclusionary business networks can inhibit innovators' access to markets or can force them to disclose the technology (Cusumano & Takeishi, 1991). Entities other than the innovator can control some crucial aspect of the value chain and extract rents that might otherwise go to the innovating firm. Similarly, organized labor can extract a portion of rents.

With expropriation threats, the firm's creation of a negotiated environment is an important amplifying strategy. To the extent that the firm can make favorable deals with potential expropriators prior to launching the technology, it may be able to amplify the value of its option. In fact, there is a real danger in not prenegotiating: the firm will find itself in a much weaker bargaining position after having made an irreversible technology investment. But when the firm is in possession of unique stakes desired by parties to the negotiation, it can potentially achieve an idiosyncratically superior arrangement.

Boundary Conditions and Uncertainty Regarding Sustainability of the Rent Stream

Uncertainty regarding matching. The problem of matching is one of equifinality. Rent streams are appropriated under matching—not through imitation but through substitution. Parallel efforts by both direct and indirect competitors may address the same needs and problems as the proposed technology. They may find a different path to a similar end. This has the effect of reducing the net revenues, relative to circumstances without the threat of matching, and it also reduces the variance of potential revenues. Probability of matching reduces the value of the option.[5]

Although scholars have not paid much attention to it in the resource-based literature, matching raises questions regarding the value of idiosyncratic endowments. Unless others cannot mobilize their own idiosyncractic endowments to create equivalent value for customers, rents may *still* be competed away. The diffusion of graphical user interfaces is an example. Originated by the Xerox Corporation, these interfaces later were adopted by Apple Computer and subsequently embedded in the "Windows" products of the Microsoft Corporation. Although the hardware plat-

[5] If x is the expected stream of revenues without matching, and $y = ax$ is the reduced revenue stream with matching, then, as above, var $y <$ var x.

forms for each variant of the interface differed significantly, their utility, from the perspective of the end user, was similar.

The threat and likelihood of matching are influenced by three additional factors: (1) motivation and capacity of rivals, (2) customer immobilities, and (3) order of entry and lock-out effects. To the extent that capable competitors are threatened, motivation to match increases (Chen & MacMillan, 1992). Immobilities can be created when early-adopting customers must make investments to use the innovator's products (imposing a cost to switch to another supplier), when transaction-specific learning creates reluctance to change to another way of doing things (Wernerfelt, 1985), when firms establish long-term contracts, or when customers are offered some form of incentive (such as a favorable discount) to remain with the innovating firm. Buyer immobility also can result from psychological costs. Firm reputation (Fombrun & Shanley, 1990), name recognition (Lieberman & Montgomery, 1988), and well-established relationships can create inertia as well. When positive feedback or network effects yield increasing returns to early movers, those making later attempts to match may never be able to overcome established, embedded technologies (David & Bunn, 1988; Katz & Shapiro, 1985).

As Rosenkopf and Tushman (1994) argue, one strategy for amplification consists of collaborative arrangements that set standards favoring the firm's technology over competing technologies. To the extent that a firm is party to standard setting when others are not, this form of access can create an idiosyncratic advantage.

Uncertainty regarding imitation. Unlike blocking, expropriation, or matching, imitation truncates revenue streams by increasing supply of a product or service, thereby decreasing its rarity and, thence, price (Barney, 1991). Imitators doubly disadvantage the innovating firm in that their cost to imitate is often a fraction of the cost to originate (MacMillan, McCaffery, & Van Wijk, 1985; Mansfield, 1988). As in the case of matching, this has the effect of reducing both the mean and the variance of potential revenues and, concomitantly, reducing the value of the option.

One solution to the problem of imitation is for a firm to take advantage of "tight" appropriability regimes for such organizational assets as patents, trademarks, and copyrights (Ghemawat, 1984; Rumelt, 1987). It is important to note that not only the existence of the regime, but its enforcement as well, is necessary for effective prevention of imitation. Collaborative preinvestments in reinforcing appropriability regimes thus can amplify later technological breakthroughs. Firms engaging in network strategies, strategic alliances, joint ventures, and collaborative research programs may find that these relationships, too, can increase rent potential by virture of quasi-monopolistic collaborative strategies (Koh & Venkatraman, 1991; Shrader, Lincoln, & Hoffman, 1989). Despite the fact that rents are shared, collaborative strategies are one of the few mechanisms, in an increasingly global competitive environment, limiting competition and delaying the appropriation of rent streams.

A firm's uncertainty regarding the size of cumulative net revenue streams affects the streams' expected variance and, thus, option value in case 1 situations, where var r is greater than var c. Let us turn next to uncertainty with respect to commercialization costs, which influence the value of the option for the case 2 situation, where var c is greater than var r.

BOUNDARY CONDITIONS AND COMMERCIALIZATION COSTS

Scholars writing about commercialization of technology identify three broad categories of costs: (1) investments in production assets, (2) infrastructure development investments, and (3) costs to develop parallel technologies and cospecialized or complementary assets (Mansfield, 1988). They address extensively the estimation of production asset costs, concluding generally that the more ambitious or asset intensive the project, the more expensive it is likely to be (Greer & Moses, 1992). Less obvious considerations apply to the second and third categories, because they often either pose unforeseen costs or promise unforeseen benefits to the firm. The higher the anticipated variance is of commercialization costs, the lower the value of the option will be; also, more preinvestment relaxing these boundary conditions will enhance the value of the option.

Uncertainty regarding accessibility to necessary infrastructure. Firms may incur significant and unexpected costs to create infrastructures and coevolved systems supporting the delivery, sale, and servicing of their outputs (Itami, 1987). Necessary but often unanticipated usurpers of corporate resources, infrastructure costs can include creation of physical infrastructure, such as distribution, service, communication, and transportation systems; creation of social infrastructures, including commonly accepted business practices or standards (Rosenkopf & Tushman, 1994); and creation of a human capital infrastructure (Becker, 1964). Clearly, to the extent that the availability of such infrastructures is uncertain now or in the future, the expected costs of commercialization and their variance will be greater, setting a lower boundary on the value of the option.

To the extent that infrastructure development is either not necessary or can be completed collaboratively prior to the completion of technology development, the variance in commercialization cost, var c, will be reduced and the value of the option amplified. There is also a role for firm-specific resources. If the firm can capture an exclusively advantaged position with respect to critical contingencies or infrastructure (e.g., by obtaining exclusive rights of way), it may shut out competitors' access to the infrastructure and preserve a position of privilege.

Uncertainty regarding availability of parallel technologies. A second broad area of investment necessary in commercializing a technology has to do with the uncertainty regarding the development of parallel technologies that may be needed to create demand or to use the target technology effectively. Software for personal computers and content for home video and game systems illustrate situations in which the sale of one technology is dependent upon the parallel development of another. Creating such complementary assets can be both costly and uncertain, increasing var c.

Alliances have a role to play here, as do such technology-sharing arranagements as research and development consortia. To the extent that the firm itself possesses firm-specific parallel technologies and cospecialized assets, it can expect to uniquely reap the benefits of combining them with the new technology. Therefore, the greater the potential is to deploy existing parallel technologies or uniquely owned cospecialized assets in conjunction with the proposed technology, the greater the value of the technology option will be. We must note, however, that this argument can apply in reverse: to the extent that existing technologies are unlikely to be valuable in future commercialization, firms may find themselves facing competence-destroying innovations (Tushman & Anderson, 1986), competence traps (Levitt & March, 1988), and disruptive technologies (Christensen & Rosenbloom, 1995).

Figure 2 illustrates how the relationship between the underlying claim to cumulative returns from operations and the cost of commercialization influences the value of the underlying claim to commercialize. Investment makes sense only when this value exceeds the "price" the firm must pay to create the option. Price, here, is considered to be the cost to develop the technology.

FIGURE 2
Factors Influencing the Value of a Technology Positioning Option

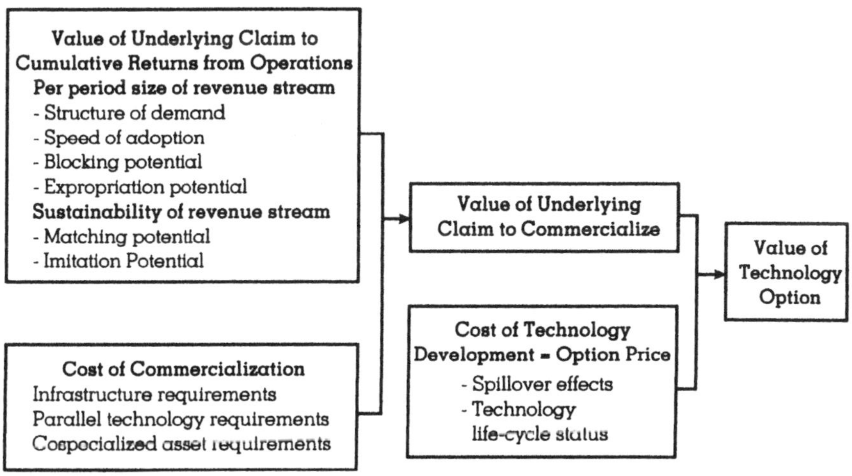

FACTORS INFLUENCING PRICE OF THE OPTION: DEVELOPMENT COST

In this model the cost of the technology, other than the direct cost of development, stems from two important components. The first is the extent to which developing a technology offers a firm positive "spillovers" (Itami, 1987: 118). The second is the extent to which the technology is in the early stages of the technology cycle.

Spillover effects. As scholars have argued in the learning literature (Cohen & Levinthal, 1990), investments intended to create new knowledge for the firm often have important spillover effects, which, to some extent, explains why firms engage in such practices as "informal know-how-trading" (von Hippel, 1987) or scientific collaboration (Hagedoorn, 1993). A firm often begins such activities with the intention of not only acquiring new knowledge relevant to developing technologies but also of improving the base business. Itami (1987) discusses this dynamic in depth. If, he argues, the new insight and new capabilities the firm develops in the pursuit of new technologies can be used to enhance the attractiveness or lower the cost of existing products, the firm's cost of development may be defrayed and rent streams, as a whole, enhanced.

A firm amplifies spillover benefits in one way when it utilizes the developing technology to enter modest markets with the intention of using these as a springboard to target other markets when the technology is sufficiently well developed. The strategy the firm uses may emphasize incremental product introductions to capture "adjacent" markets as stepping stones to a more ambitious strategy, or it may emphasize the disruption of previous technology barriers to open entirely new opportunity spaces (Itami, 1987; McGrath, MacMillan, & Tushman, 1992). A common thread in both approaches is that the firm deliberately uses entry into specialized subfields or modest new markets to garner experience. Thus, the greater the potential for early entry into commercial markets, the more certain one can be that costs can be recovered en route—hence, the lower the "price" of the technology option. The extent to which the firm exploits existing idiosyncratic competences and current idiosyncratic relations with customers significantly enhances the potential for early spillover benefits and creates idiosyncratic opportunities to reduce the cost of the real option.

A second form of spillover can occur when a firm can envisage selling or trading the option prior to commercialization. In this case, the firm may be able to generate returns when others perceive the option as valuable, without incurring the costs of exercise.

Technology life-cycle status. Several scholars have shown that technology cycles have significant effects on technology strategies in that they influence the basis of competition and the nature of the uncertainties faced (Anderson & Tushman, 1990; McGrath et al., 1992; Tushman & Anderson, 1986). During periods of incremental change, learning is "exploited"

(March, 1991), and positions of advantage tend to be reinforced. During periods of ferment and disruption, investments in "exploration" must be made, and incumbent positions are at risk.

In incremental, evolutionary technology development, firms often innovate within the context of a dominant design (Tushman & Anderson, 1986), meaning that key dimensions of merit in the marketplace have been established, the firms' customers have been educated, and their coexisting technologies share a common architecture. Each of these acts to reduce development costs. When disruptive technologies are introduced, however, firms often need to make investments to identify what attributes of the technology will be crucial, to acquire or train people in the new field, to communicate the benefits of the technology, and to develop parallel or supporting technologies.

Interestingly, the uncertainty of eras of ferment makes the use of technology options more attractive, as hedges against placing the wrong technology bets. Paradoxically, however, creating the option (by developing the technology) is apt to be more expensive. Because dimensions of merit are unclear, a firm may anticipate expensive design variations, experimentation, investment in assets of questionable future use, and investments to create or prompt development of parallel or complementary technologies, as well as considerable community-level coordination costs (Rosenkopf & Tushman, 1994). Therefore, earlier in the technology cycle, costs are greater and more uncertain than they are at later points. At the same time, possible gains motivate firms to take out these options.

It may pay the firm handsomely to invest in participation in what Rosenkopf and Tushman term "community technological organizations" (1994: 410), which help shape the standards and specifications for an emerging technology, as well as in the technology itself.

MODELING THE VALUE OF OPTIONS ON TECHNOLOGY

As Figure 2 shows, the value of an option on a technology can be described in terms of three contructs: (1) the cost to develop the technology, which is equivalent to the price of taking out the technology positioning option; (2) the cost of commercialization; and (3) the value of the underlying claim to cumulative returns from operations. As in financial options, greater uncertainty tends to increase the attractiveness of making an options investment, provided the firm can contain possible losses.

An option is also increasingly attractive if the firm can approach it through sequential development so that incurring the cost of development for one stage does not irreversibly commit the firm to the next stage of development. Even when the amount of potential losses on technology development cannot be predicted, expenditures to reduce technical uncertainty (which only investment will resolve) may still be warranted if sequencing is possible.

Uncertainty also suggests a role for both strategy and for idiosyncratic resource endowments. I treat the uncertainty profiles of firms here as heterogeneous, suggesting that a strategic objective for investments in technology is to reduce uncertainty for oneself, without similarly reducing it for competitors. To the extent that investments in a technology positioning project show promise of drawing from firm-specific skills, assets, and routines, the firm may further be able to idiosyncratically reduce uncertainty in ways that are specific to it and inappropriable by competitors.

In situations where the nature of uncertainty the firm faces for a project is such that no amount of investment can reduce it, but the uncertainty will be truncated temporally, the firm may opt to postpone the option. Where there is "external" uncertainty, however, the firm may take strategic actions to amplify the value of the option. I identify and summarize opportunities for deploying both amplifying preinvestments and idiosyncratic resource endowments in Table 1.

DISCUSSION

One of my objectives in this article was to develop a parsimonious, yet comprehensive, model of the forces influencing the value of a technology option, while recognizing that the assumptions used to value financial options cannot be made for real options. Although not exhaustive, the framework I present in Figure 2 integrates previous research on technology strategy and uncertainty to articulate whether, and when, initiating an option shows the potential to gain the firm access to future rents.

My treatment here of uncertainty and firm-specific uncertainty reduction specifically and unambiguously integrates real options models, technology assessment, and strategic management theory. It makes clear, for example, that existing resource endowments are apt to be important for technology positioning when they allow a firm to idiosyncratically reduce critical uncertainties in ways that are not obvious to competitors. By implication, it also suggests that endowments that either systematically create an uncertainty disadvantage for the firm or that are no longer differentiated with respect to competition have little strategic value in the technology positioning decision. A legacy computer system, for instance, which certainly qualifies as a part of a firm's heritage, may prove more of a hindrance than an asset if it is incapable of helping the firm resolve critical uncertainties in a timely way.

The model I present also suggests that appropriate strategic action with respect to a given technology actually may have less to do with the technology than with other factors "external" to it and often to the firm. Not only is a firm sometimes better served by waiting than by proceeding with an investment, but it may extract maximum value by proceeding along a different strategic front. The argument I summarize in Table 1 also suggests specifically which forms of strategic action may be most appropriate under different boundary conditions.

Hence, this view of technologies as options links the technology strategy of a firm directly with its broader business, political, and social strategies. For example, by arguing that, in some cases, a firm's technology development should pause while contextual and regulatory actions are taken to create inappropriability, I directly connect in this framework such institutional-level issues as legitimacy with competitive strategy. This motivates a view of technology that does not exist in isolation but, rather, is one element in an overall strategy to capture rents, clearly embedded in a competitive and institutional context.

Scholars have not yet introduced the concept of amplifying preinvestments to create an appropriate context in the real options literature. The framework I present here both introduces this idea and suggests when and how such amplifying preinvestments can be of strategic use to a firm. Further, I suggest approaches to estimating the value, and therefore the acceptable level of investment, in such amplifying strategies.

IMPLICATIONS FOR FURTHER RESEARCH

Operationalization and subsequent empirical testing clearly are indicated in terms of further research. Moving toward the development of both direct and indirect measures is the next logical step. Direct measures might include such variables as demographics to anticipate growth in demand, interest group communications and legislative/regulatory agendas to measure blocking potential, and relative standing on industry key ratios to assess comparative levels of resource efficiency. Indirect measures could well be developed from managerial or functional experts' perceptions, either by questionnaire or through such techniques as the Delphi (Dalkey & Helmer, 1963).

The concepts I have developed here also have much to offer, both to the further development of resource-based theory and to the strategic application of real options theory. The concept of idiosyncratic reduction of uncertainty begins to move us as a field toward an unambiguous definition of when resources are likely to offer strategic value and when they are not. With respect to the real options literature, this article helps clarify the nature of options (as costs incurred to create underlying claims on future rents), as well as provides a coherent logic suggesting when those options should be exercised, when they should be held, and when they should be allowed to expire.

REFERENCES

Alchian, A. A. 1991. Rent. In J. Eatwell, M. Milgate, & P. Newman (Eds.), ***The World of Economics:*** 591–597. New York: W. W. Norton.

Amit, R., & Schoemaker, P. 1993. Strategic assets and organizational rent. ***Strategic Management Journal,*** 14: 33–46.

Anderson, P., & Tushman, M. L. 1990. Technological discontinuities and dominant designs: A cyclical model of technological change. ***Administrative Science Quarterly,*** 35: 604–633.

Arthur, W. B. 1994. ***Increasing returns and path dependence in the economy.*** Ann Arbor, MI: University of Michigan Press.

Barney, J. B. 1986. Strategic factor markets: Expectations, luck, and business strategy. ***Management Science,*** 32: 1231–1241.

Barney, J. B. 1991. Firm resources and sustained competitive advantage. ***Journal of Management,*** 17: 99–120.

Becker, G. S. 1964. ***Human capital.*** New York: National Bureau of Economic Research.

Bettis, R., & Hitt, M. 1995. The new competitive landscape. ***Strategic Management Journal,*** 16: 7–19.

Black, F., & Scholes, M. 1973. The pricing of options and corporate liabilities. ***The Journal of Political Economy,*** 81: 637–654.

Block, Z., & MacMillan, I. 1993. ***Corporate venturing: Creating new businesses within the firm.*** Cambridge, MA: Harvard Business School Press.

Boulding, W., & Staelin, R. 1990. Environment, market share, and market power. ***Management Science,*** 36: 1160–1177.

Bowman, E. H., & Hurry, D. 1993. Strategy through the option lens: An integrated view of resource investments and the incremental-choice process. ***Academy of Management Review,*** 18: 760–782.

Caves, R. E. 1971. International corporations: The industrial economics of foreign investment. ***Economica,*** 38: 1–27.

Chen, M.-J., & MacMillan, I. C. 1992. Nonresponse and delayed response to competitive moves: The roles of competitor dependence and action irreversibility. ***Academy of Management Journal,*** 35: 359–370.

Chen, M.-J., Venkataraman, S., MacMillan, I. C., & Black, S. S. 1991. ***The role of irreversibilities in competitive exchange: Behavioral considerations from organization theory and game theory.*** Working paper, Snider Entrepreneurial Center, The Wharton School, University of Pennsylvania, Philadelphia, PA.

Cheng, Y., & Van de Ven, A. H. 1996. Learning the innovation journey: Order out of chaos? ***Organization Science,*** 7: 593–614.

Christensen, C. M., & Rosenbloom, R. S. 1995. Explaining the attacker's advantage: Technological paradigms, organizational dynamics and the value network. ***Research Policy,*** 24: 233–257.

Cohen, W. M., & Levinthal, D. A. 1990. Absorptive capacity: A new perspective on learning and innovation. ***Administrative Science Quarterly,*** 35: 128–152.

Conner, K. R. 1995. Obtaining strategic advantage from being imitated: When can encouraging "clones" pay? ***Management Science,*** 41: 209–225.

Cooper, A. C., & Smith, C. G. 1992. How established firms respond to threatening technologies. ***Academy of Management Executive,*** 6: 55–70.

Cusumano, M., & Takeishi, A. 1991. Supplier relations and management: A survey of Japanese, Japanese-transplant and U.S. auto plants. ***Strategic Management Journal,*** 12: 563–588.

D'Aveni, R. 1994. ***Hypercompetition: The dynamics of strategic maneuvering.*** New York: Free Press.

Dalkey, N. C., & Helmer, O. 1963. An experimental application of the Delphi methods to the use of experts. ***Management Science,*** 9: 458–457.

David, P., & Bunn, J. 1988. The economics of gateway technologies and network evolution. ***Information Economics and Policy,*** 3: 165–202.

Dewar, R. D., & Dutton, J. E. 1986. The adoption of radical and incremental innovations: An empirical analysis. ***Management Science,*** 32: 1422–1433.

Dierickx, I., & Cool, K. 1989. Asset stock accumulation and sustainability of competitive advantage. ***Management Science,*** 35: 1504–1513.

Dixit, A., & Pindyck, R. 1994. ***Investment under uncertainty.*** Princeton, NJ: Princeton University Press.

Eisenhardt, K., & Tabrizi, B. 1995. Accelerating adaptive processes: Product innovation in the global computer industry. ***Administrative Science Quarterly,*** 40: 84–110.

Fombrun, C., & Shanley, M. 1990. What's in a name? Reputation building and corporate strategy. ***Academy of Management Journal,*** 33: 233–258.

Fudenberg, D., & Tirole, J. 1985. Preemption and rent equalization in the adoption of new technology. ***Review of Economic Studies,*** 70: 383–401.

Ghemawat, P. 1984. Capacity expansion in the titanium dixoide industry. ***Journal of Industrial Economics,*** 33: 145–163.

Gilbert, R. J., & Newberry, D. M. G. 1982. Preemptive patenting and the persistence of monopoly. ***American Economic Review,*** 72: 514–526.

Goodman, R. A., & Lawless, M. W. 1994. ***Technology and strategy.*** New York: Oxford University Press.

Greer, W. R., & Moses, O. D. 1992. Extending technology: The estimation and control of costs. ***Journal of Accounting and Public Policy,*** 11: 43–66.

Hagedoorn, J. 1993. Understanding the rationale of strategic technology partnering: Interorganizational modes of cooperation and sectoral differences. ***Strategic Management Journal,*** 14: 371–385.

Itami, H. 1987. ***Mobilizing invisible assets.*** Cambridge, MA: Harvard University Press.

Kamien, M., & Schwartz, N. 1975. Market structure and innovation: A survey. ***Journal of Economic Literature,*** 13: 1–37.

Katz, M. L., & Shapiro, C. 1985. Network externalities, competition, and compatibility. ***The American Economic Review,*** 75: 424–440.

Kester, W. C. 1981. ***Growth options and investment: A dynamic perspective on the firm's allocation of resources.*** Unpublished doctoral dissertation, Harvard University, Cambridge, MA.

Kobrin, S. 1982. ***Managing political risk assessment: Strategic response to environmental change.*** Los Angeles: University of California Press.

Koh, J., & Venkatraman, N. 1991. Joint venture formations and stock market reactions: An assessment in the information technology sector. ***Academy of Management Journal,*** 34: 869–892.

Levinthal, D., & March, J. G. 1993. The myopia of learning. ***Strategic Management Journal,*** 14: 95–112.

Levitt, B., & March, J. G. 1988. Organizational learning. ***Annual Review of Sociology,*** 14: 319–340.

Lieberman, M. B., & Montgomery, D. B. 1988. First mover advantages. ***Strategic Management Journal,*** 9: 41–58.

MacMillan, I. C. 1978. ***Strategy formulation: Political concepts.*** St. Paul, MN: West.

MacMillan, I. C. 1988. Controlling competitive dynamics by taking strategic initiative. ***Academy of Management Executive,*** 2: 111–118.

MacMillan, I., & Jones, P. 1986. ***Strategy formulation: Power and politics.*** St. Paul, MN: West.

MacMillan, I. C., McCaffery, M. L., & Van Wijk, G. 1985. Competitors' responses to easily imitated new products—exploring commercial banking product introductions. ***Strategic Management Journal,*** 6: 75–86.

Mansfield, E. 1988. The speed and cost of industrial innovation in Japan and the United States: External vs. internal technology. ***Management Science,*** 34: 1157–1168.

March, J. G. 1991. Exploration and exploitation in organizational learning. ***Organization Science,*** 2: 71–87.

McGrath, R. G., MacMillan, I. C., & Tushman, M. L. 1992. The role of executive team actions in shaping dominant designs: Towards the strategic shaping of technological progress. ***Strategic Management Journal,*** 13: 137–161.

Mitchell, G. R., & Hamilton, W. F. 1988. Managing R&D as a strategic option. ***Research-Technology Management,*** 27: 15–22.

Montgomery, C. A. 1985. Product-market diversification and market power. ***Academy of Management Journal,*** 28: 789–798.

Mowery, D. C., & Rosenberg, N. 1982. The influence of market demand upon innovation: A critical review of some recent empirical studies. In N. Rosenberg (Ed.), ***Inside the black box: Technology and economics.*** Cambridge, England: Cambridge University Press.

Mowery, D. C., & Teece, D. J. 1993. Japan's growing capabilities in industrial technology: Implications for U.S. managers and policymakers. ***California Management Review.*** 35: 9–33.

Myers, S., & Marquis, D. G. 1969. ***Successful industrial innovations: A study of factors underlying innovation in selected firms.*** NSF 69-17. Washington, DC: National Science Foundation.

Nelson, R. R., & Winter, S. J. 1982. ***An evolutionary theory of economic change.*** Cambridge, MA: Belknap and Harvard University Press.

Peteraf, M. A. 1993. The cornerstones of competitive advantage: A resource-based view. ***Strategic Management Journal,*** 14: 179–192.

Porter, M. E. 1980. ***Competitive strategy.*** New York: Free Press.

Rogers, E. M. 1995. ***The diffusion of innovations*** (4th ed.). New York: Free Press.

Rosenkopf, L., & Tushman, M. L. 1994. The coevolution of technology and organization. In J. Baum & J. Singh (Eds.), ***Evolutionary dynamics of organizations:*** 403–424. New York: Oxford University Press.

Rumelt, R. P. 1987. Theory, strategy and entrepreneurship. In D. J. Teece (Ed.), ***The competitive challenge: Strategies for industrial innovation and renewal.*** New York: Harper & Row.

Scherer, F. M. 1979. ***Industrial market structure and economic performance.*** Chicago: Rand McNally.

Schmookler, J. 1966. ***Investion and economic growth.*** Cambridge, MA: Harvard University Press.

Shrader, C. B., Lincoln, J. R., & Hoffman, A. N. 1989. The network structures of organizations: Effects of task contingencies and distributional form. ***Human Relations,*** 42: 43–66.

Teece, D. J. 1986. Capturing value from technological innovation: Integration, strategic partnering and licensing decisions. ***Research Policy,*** 15: 285–305.

Trigeorgis, L. 1993. Real options and interactions with financial flexibility. ***Financial Management,*** Autumn: 202–224.

Tushman, M., Anderson, P. 1986. Technological discontinuities and organizational environments. ***Administrative Science Quarterly,*** 31: 439–465.

Venkataraman, S. 1990. ***Transaction sets, architecture of complexity and new venture growth and stagnation.*** Working paper, Sol C. Snider Entrepreneurial Center, The Wharton School, University of Pennsylvania, Philadelphia, PA.

von Hippel, E. 1987. Cooperation between rivals: Informal know-how trading. ***Research Policy,*** 16: 291–302.

von Hippel, E. 1988. ***The sources of innovation.*** New York: Oxford University Press.

Wernerfelt, B. 1984. A resource-based view of the firm. ***Strategic Management Journal,*** 5: 171–180.

Wernerfelt, B. 1985. Brand loyalty and user skills. ***Journal of Economic Behavior and Organization,*** 6: 381–385.

Rita Gunther McGrath is an assistant professor in the Management of Organizations Division, Columbia Business School, and Research Director of the Snider Entrepreneurial Center, The Wharton School. She received her Ph.D. from The Wharton School, University of Pennsylvania. Her current research focus is on the strategic management of highly uncertain situations.

[21]

Capabilities as Real Options

Bruce Kogut • Nalin Kulatilaka

Wharton School, University of Pennsylvania, Philadelphia, Pennsylvania 19104

School of Management, 595 Commonwealth Avenue, Boston University, Boston, Massachusetts 02215

kogut@wharton.upenn.edu • nalink@bu.edu

Abstract

Strategy research consists of a balance between positive and normative theory. Normative theories suggest particular heuristics, or cognitive representations, to find appropriate solutions. Heuristics permit faster solutions to real-time problems; they also suffer from the potential of negative transfer to inappropriate applications.

The theory of real options provides the appropriate heuristic framing of competencies and exploratory search. A real options approach marries the theory of financial options to foundational ideas in strategy, organizational theory, and complex systems. We join these approaches to identify three pairs of concepts: scarce factor and the underlying asset in option theory, inertia and irreversibility, and the ruggedness of landscape and option values. Strategic theories of resources largely define a core competence as unique and nonimmutable. In doing so, this definition has wrongly forgotten Barney's initial insight into scarce factor markets as determining the valuation of a competitive asset. Financial theory of real options derives its heuristics of investing in exploratory search by inferring future value of today's investments from market prices.

We apply the three conceptual pairs to the evaluation of capabilities as real options through a formal descriptive model. The valuation of core capabilities is derived from observing the price dynamics of correlated strategic factors in the market. Because of inertia, managers cannot easily adjust the wrong set of organizational capabilities to the emergence of market opportunities. However, firms that have made investments in capabilities appropriate to these opportunities are able to respond. From this description, we define core competence as the choice of capabilities that permits the firm to make the best response to market opportunities. The heuristic framing of capabilities as real options guides the normative evaluation of the balance between exploitation and exploration.

(*Strategy; Options; Capabilities; Heuristics; Irreversibility; Landscapes*)

Strategizing is the application of heuristic frames to analyze the world and to generate normative evaluations of potential avenues of implementation. Strategy research reflects competing ideas about how the world looks, or what the world needs. However, like their counterparts in engineering or architecture, strategy researchers distinguish themselves from practitioners by their attention to an articulation of theory and evidence. It is this serious concern for the design of practice by an investigation of the appropriate theoretical framing that is the mark of applied sciences in professional schools (Simon 1969).

There is currently debate in strategy research between the importance for a firm to "position" itself in the market or to own unique resources. Ned Bowman (1995) made the distinction between strategies that look in the mirror and those that look through the looking glass. In the parlance of contemporary strategy research, resource and knowledge theories of the firm are inward looking, whereas market positioning and industry analysis are outward looking.

It is not surprising that during a time of restructuring and reengineering, strategy researchers should shift the emphasis from industry analysis to the internal sources of competitive advantage. The current emphasis on looking in the mirror begs the question of how to choose among alternatives. The resource-based and knowledge theories view the unique capabilities of the firm as the cornerstone of sustainable rents. These approaches share the common insight that a chosen strategy presumes the capability of implementing the vision. A heuristic appropriate to identifying capabilities required to support a strategy is provided by the notion of core competence, as proposed by Hamel and Prahalad (1994). And yet, it is often overlooked that Hamel and Prahalad (1994) essentially invert the framing of a resource-based view of strategy by arguing for the analysis of white spaces in the market topography of existing businesses to identify valuable avenues of exploration.

The ideas of core competence and white spaces share strong parallels with the heuristic application of real option theory to strategy, whereby investments in exploration create capabilities to address future opportunities (Bowman and Hurry 1993). In this regard, a real option is strongly reminiscent of the distinction between exploitation and exploration (Hedlund and Rolander 1990,

Organization Science, © 2001 INFORMS
Vol. 12, No. 6, November–December 2001, pp. 744–758

1047-7039/01/1206/0744/$05.00
1526-5455 electronic ISSN

March 1991). A real option is the investment in physical and human assets that provides the opportunity to respond to future contingent events. A good example of a real option is the investment in a sales operation in a developing country that loses money but provides learning of the environment that will provide the future capability to expand if the country should grow. Firms as adaptive systems strike a balance between refinements of existing processes and explorations of variations on new techniques and new markets. The static analysis of deciding to allocate effort to each of these activities is especially complicated, because current efforts result in short-term efficiencies that can overwhelm long-term efforts of exploration (March 1991). Thus, the dynamics by which capabilities interact and are learned pose a complex combinatorial problem.

Organizational theory has been slow to embrace the idea that organizations can proactively exploit risk rather than just absorb it. Yet if firms and their environment are engaged in a coevolutionary dynamic Lewin et al. (1998), then it is useful for a firm to ponder the match between its future capabilities and future environments. We propose that the theory of real options provides an appropriate theoretical foundation for the heuristic frames to identify and value capabilities and exploratory activities. Because capabilities are platforms that create a generic set of resources, they represent investments in future opportunities (Kogut and Kulatilaka 1994a). In effect, real option valuation marries the resource-based view with industry positioning by disciplining the analysis of the value of capabilities by a market test.

We proceed by first characterizing the value and limitations of heuristics. We then develop the use of real options as a heuristic and show how it is grounded in, and throws light on, three lines of thought of strategy: resource-based view of the firm, traditional organizational theory, and complex adaptive systems. A central concept in organizational theory is inertia. Its allied notion in financial economics of "irreversibility" is critical to the definition of a real option. The argument that we develop is that capabilities reflect irreversible investments because of the costliness of rapidly transforming the organizational knowledge in a firm. This knowledge is composed of the set of technological and organizational complements, very much in the spirit of the sociotechnical tradition. The difficulty facing a firm is that improvements in complements provide a competitive advantage, while also generating a high level of inertia that inhibits the firm from radical change.

We examine these ideas through a stylized mathematical *description* of the problem of adopting radical change. Our formalization clarifies that the benefit of a real options heuristic is the imposition of a market test to derive the valuation of capabilities. It also allows for a precise definition of a core competence as derived from the valuation of inert resources given the environment.

Strategy as Heuristic

Following distinctions made in cognitive science, we separate a heuristic into its cognitive frame and the rules of search. A cognitive (or heuristic) frame refers to the "representation" of the problem and solution space. The heuristic rules of search are the algorithms by which solutions are found in the represented solution space.[1]

Simon (1969) introduced both of these elements by defining heuristics as procedural search in distinction to the substantive rationality of economics and operations research. Simon noted that the solutions to many problems are not computable; that is, the search algorithm cannot in finite time determine the optimal answer. The problem of computation is classically illustrated by the traveling salesman problem in which the objective is to minimize the travel costs of a salesperson who has to visit 50 cities. The 50! calculation is computable, but not within any reasonable horizon. For this reason, many heuristic algorithms have been proposed to provide "satisfactory," but not clearly optimal, solutions to this problem.

Strategy often has this level of complexity and also often lacks a method of determining the optimal choice. In order to know whether a firm should enter into a particular business, it is important to understand the costs and quality of the product or service that can be delivered. It is also important, consequently, to understand the response of competitors to the competitive entry of an innovation and its embedded bundle of attributes (e.g., price, quality). There are, then, two embedded decisions—one determining the capabilities, the other the market strategy. The dimensional problem of identifying all these elements and understanding their interactions quickly defies a declarative analysis (i.e., net present value) or an exhaustive procedural search across all combinations.

Heuristics have the advantage of countering some cognitive biases, but at a cost. In some cases, the heuristic will lead to nonoptimal decisions. In a study on plant scheduling, Bowman (1963) found that managers would do better if they used linear estimates from their experience rather than tried to optimize in response to each situation. The implication is that managers would do better to rely upon experiential heuristics than seek to optimize each situation. In real time, the search for optimal strategies can be too costly or liable to be influenced by recency effects (e.g., the arrival of new information). Kunreuther (1969) modified these findings that rules cued to selective environmental information improve actual decisionmaking. This finding implies that there may be certain metaheuristics that identify the environment and thereby indicate the application of less robust, but more appropriate, decision rules.

The merit of a heuristic is its real-time utility. One of the cornerstones of evolutionary organizational theory is March and Simon's (1958) notion of routines, which is an organizational enactment of heuristic problem solving. (See Nelson and Winter 1982.) Studies on innovation show remarkable trade-offs between costs and time for innovations (Scherer 1967, Mansfield 1988; see Midler's 1993 discussion). Consequently, because the development and use of heuristics are situated in particular contexts that are only partially understood, they can be misapplied. The Cohen and Bacdayan (1994) experiments illustrate these trade-offs between routines as heuristics and misapplications. Their study of the emergent rules in a simple card game showed that simple heuristics routines guided the behavior of play. Better routines, of course, were associated with better outcomes. When the game changed, the players tended to engage in "negative transfer," that is, they transferred the acquired heuristics to the new game even though inappropriate.

Strategizing is, then, the application of imperfect heuristics to problem solving and implementation. Unlike the implicit heuristics that represent what Argyris and Schon (1978) call "theories in use," formal strategizing applies explicit schemas to search for appropriate decisions. Nevertheless, the underlying theories to these heuristics are often only implicit. Even explicit heuristics are liable to be applied to the inappropriate setting if hidden assumptions prove to be wrong. Because complex interactions mark organizational choice, the potential for the cognitive misrepresentation of the problem is large. Thus, an important feature of any framework is a process of discovery and experimentation.

It is this process that the heuristic of real options attempts to impose and evaluate. If strategy is seen as the choice of capabilities that provide the appropriate flexibility for a stochastically changing landscape, then it is critical to try to infer the value of design and investment that provides the flexibility to respond to opportunities. An organization cannot, however, expend all of its resources on search, nor can it simply ignore the importance of change and evolution. A real option heuristic is a way to discern the value of particular paths of exploration in evolving environments. Moreover, as we develop below, it is a heuristic that is grounded in theories of strategy, organizational ecology, and complex systems.

Strategy and Real Options

The core competence concept arose in the late 1980s during the height of reengineering, propelled by acquisitions and new information technologies. It is a direct response to the reputed financial pressures from financial markets dominated for the first time by institutional investors. The formulation by Hamel and Prahalad (1994) suggests that the initial data are in the spirit of understanding the intended strategy of the firm, which should be grounded in a distinctive competence, an idea that dates back to Selznick (1957).

The theoretical foundations to this view are several, from the reasoning about why knowledge is hard to imitate to the evolutionary theories of firm growth. From a decision-theoretic perspective, the core competence framing readily lends itself to a real option interpretation. A real option is technically defined by an investment decision that is characterized by uncertainty, the provision of future managerial discretion to exercise at the appropriate time, and irreversibility.

These three elements are jointly required for the application of a real options heuristic. An option has value only if there is uncertainty, though defining the relevant source of the uncertainty is not trivial. An operationally important element of design is the provision of discretion, such as the staging of an R&D project to correspond to discrete points of go-no go decisions.

Irreversibility is an easily overlooked feature and signifies the inability to costlessly revisit an investment or decision. Irreversibility is a subtle idea that carries the notion of the arrow of time.[2] For example, the decision to make an investment today bears the risk that the invested assets can only be sold later at a discount. In this context, irreversibility is the inability to recover the investment costs already expended for the product division. Irreversibility is accentuated if the divesting of an investment also engages costs attached to the unbundling of integrated and coupled assets.

The concept of irreversibility is critical to why inertia of organizational capabilities is the source of the value of real options. Irreversibility does not mean that firms cannot change, or that transformation is not possible. Clearly, if a firm had no competitive pressure, it could gradually implement the transformation process. It is because the benefits of transformation decline as competitors themselves adopt new capabilities that a firm faces a problem of irreversibility in this case. A delay may lower the costs, but transformation comes too late.

Because time has an arrow, the decision to delay making investments in knowledge needed to enter a new market has a time subscript. The value of this decision will be different if it is considered next year; other firms may have entered and the early rents are now dissipated. Because strategy entails a decision to invest in capabilities to sustain a market strategy, foregoing this investment means that the firm does not have the option to launch the strategy if the market becomes favorable. However, not all capabilities are irreversible. The failure of a firm to invest in information technologies at one point of time does not preclude contracting out for such services in the future. More likely, the opportunity to invest diminishes over time as others come to acquire the necessary capabilities. There is, therefore, a time dimension between making a decision to invest and its actual implementation during which the value of the investment will change.

Irreversibility implies that the asset should be "scarce" and difficult to replicate in a timely way in order to support a strategy at a particular time. If, through imitation and substitution, this factor will be more abundant in the future and its value will be less, the option value is only realized through the current investment to exploit transient opportunities.[3] To Barney (1986), the creation of entrepreneurial rents is fortuitous. If managers understood the value creation process, the knowledge through imitation would lead to the erosion of these rents.

Consequently, a core competence is a scarce factor as Barney (1986) defines it, that embeds complex options on future opportunities. The important difference between this early statement of the resource-based view of the firm and core competence is the latter's insistence on the value of a resource as derived from its future but uncertain use. In the sense that Barney relies on market valuations to back into his identification of unique assets, he is consistent with the view that the market values the use of these assets in reference to their potential use by firms bidding for their ownership. Dierickx and Cool (1989) note that Barney makes the incomplete inference by ignoring irreversibility. Firms must have differential information by factor of the arrow of time (or what they call "time compression"). Because it takes time to build and absorb capabilities, a firm cannot spontaneously replicate scarce assets. Consequently, some firms will discover profitable projects where the "excess rents" are earned through their organizational complementarities, not through superior information.

This conclusion emphasizes that the scarcity of a core competence should be reflected in equilibrium financial prices, even if a firm cannot earn excess rents by buying this competence in the market. This point is exactly Barney's contention that a market for scarce factors forces external prices to reflect the present and future value of the internal assets. The market attempts to value the scarcity of these assets for generating current and future cash flows given a firm's *position* in the market. In other words, scarcity itself does not determine the value of a competence. It is a necessary but not sufficient condition. Rather, scarcity is interesting if the competence permits a firm to achieve a competitive position in the market place.

Some writers fail to make this observation altogether. For example, Teece et al. (1997, p. 516) write that "core competences must accordingly be derived by looking across the range of a firm's (and its competitors) products and services." This statement is, however, seriously incomplete. The missing element in this analysis is, of course, the scarce factor market. A firm may be well endowed with patents making it difficult for competitors to imitate. However, the important question is whether these endowments, which we might also call more generically the knowledge of the firm, is useful not only to current, but also to future applications. This question is not answered by a notion of dynamic capabilities, or of combinative capabilities, unless the normative criterion is the identification and investment in core competence in reference to potential uses. This criterion, by definition, must consider the market and the position of the firm. This objection is not petty, for it is easy to imagine that without market discipline on the analysis, the potential candidates for core competence quickly multiply.

There is another way to think about this problem, suggested by Winter (1987), as a broader formulation along the lines of optimal control. Winter (1987, p. 180–181) states, "From evolutionary theory comes the idea that a state description may include organizational behavioral patterns or routines that are not amenable to rapid change, as well as . . . more conventionally defined assets. It is by this route that a variety of considerations that fall under the rubrics *knowledge* and *competence* may enter the strategic state description."[4] Conventionally, optimal control describes the state characteristics of the environment and allows the decision maker discretion over a few control variables, e.g., technologies or output. Winter's suggestion is to capture the constraints and opportunities of capabilities through a richer description of the state variables. It is this insight that we use below in our formal treatment.

The Organizational Ecology of Irreversible Investments

The concept of core capabilities as embodying exploratory investments to hedge the future runs counter to some streams of thought on organizational design. Organizational theory historically has viewed uncertainty as threatening the stability of the technical core of an organization. By proposing the idea of "uncertainty avoidance," March and Simon (1958) suggested that an organization's design serves the function of eliminating variance. This idea also appears in Thompson's (1967) landmark book that analyzes the many ways that firms buffer themselves from uncertainty. Similarly, Pfeffer and Salancik (1978) motivate the theory of resource dependency as the creation of organizational mechanisms to reduce uncertainty.

The contribution of organizational ecology is to formulate more explicitly the relationship between environmental uncertainty and organizational strategies in a dynamic setting. In particular, Hannan and Freeman (1977) distinguish between environments that change smoothly from those whose change is granular, i.e., more abrupt. This distinction between two kinds of uncertainty—a distinction that we exploit mathematically below—implies that in granular and uncertain environments, generalists will do better than specialists. This notion of generalists is defined operationally in their empirical work as organizations whose competence corresponds to a broad array of possible environmental outcomes.

Given its stress on inertia, it is no doubt surprising that organizational ecology implies option thinking. In the framing of options, generalists are organizations whose competencies are robust across many future states of the world. The formulation by Tuma and Hannan (1984) makes the analogy between options and survival strategies more explicit. They note that a hazard model is the probability of hitting a lower boundary in a stochastic diffusion process which governs the growth of the organization. Indeed, the conditional probability of exercising the option is also the hazard of hitting the boundary of a diffusion process. The difference between the two approaches, of course, is that a real option model typically considers the upper boundary, that is, the probability of increasing growth by exercise rather than the implied strategy in organizational ecology of minimizing the hazard of death.[5] The inertial qualities of an organization are consequently central to understanding the value of a firm's assets for future deployment given the uncertainty and graininess of the environment. Indeed, it is exactly because of McKelvey's (1999) contention that the environment changes more rapidly than organizations that there is value in investing assets to respond to future changes.

The boundary problem in organizational ecology is based on the same presumption as for real options, namely, that an organization consists of a reservoir of inert and irreversible resources. Organization ecology, like organizational theory in general, has sought to separate out features of the organization that can be easily changed and hence *peripheral* from those that are inert and hence *core*. Following Thompson, the core has frequently been regarded as technical, a point of view also adopted by Scott (1995).

Core and periphery imply a dimension of distance. This notion of distance in core and periphery is one of the hardest concepts in organizational science, and yet it is also found in the fundamental organizational concepts of local versus exploratory learning and radical or incremental innovation. These concepts share the idea that firms can be mapped onto a multidimensional space representing different combinations of technological and organizational practices. (In the next section, we turn to understanding this space as rugged.) To simplify these dimensions, consider a two-dimensional space with the coordinates representing a combination of a technology and organizational practices. The notion of inertia poses the question of whether it is easier to move along the technological or organizational dimensions if one wants to change.

The organizational literature on innovation has implied that the technological dimension is especially problematic by emphasizing the difference between incremental and radical innovation. The dimension of organization appears as independent from this consideration. Tushman and Anderson (1986) have offered the insight that innovations can be characterized as radical or incremental, depending upon whether they destroy or enhance a firm's competence. (See also Henderson 1993.) This reasoning ultimately leads to the consideration that the radicalness of an innovation has less to do with the novelty of the technology than its conformity with existing knowledge of the firm, i.e., the ways by which work is organized and power is distributed. Because the way work is organized will vary by firms, then the radicalness of a technological innovation cannot be determined independent of a particular organizational context. Switching, or adoption, costs are strongly contingent on the current organization of work.

One of the most perplexing questions in organizational behavior is the failure to identify clear matches between technologies and organizational structures. (See the review given in Dosi and Kogut 1993, and the summary of the work comparing U.S. and Japanese organizations in Lincoln 1993.) Dosi and Kogut (1993) proposed that the failure to find robust relationships has been due to the tendency to theorize element-to-element correspondence, such as high volume production with vertical hierarchy.[6] The empirical results do not show that these are complements when other factors are controlled. Alternatively, Dosi and Kogut suggest that the correspondence might be set to set, where a set of organizational practices maps onto a set of technologies. The data might not reveal that

A and B exist as complements; all we observe is A and C and D and B. Complementarities need not be unique between any given technology or organization, but they still should be relationally bounded. The recent findings by MacDuffie (1996) on "bundles" of human resource practices in auto plants indicate that there is a logic which relates organizing practices to each other and to technologies. The experience of General Motors and other car manufacturers is that adopting the new capabilities of flexibility and speed requires changes in automation and organization. Between these two sets, there are many functionally equivalent complements, but there are no unique element-to-element correspondences.

This description also captures Dosi and Kogut's idea of coevolution of technology and organization through two key features. First, technology and organization do not represent random assignments, nor is their coupling simply at the discretion of managers. Rather, the matches of a technology and organizing principle are constrained to reasonable set-to-set correspondence. However, within these "developmental" constraints, improvements in technology and organization are correlated through experiential learning. For example, the introduction of mechanical equipment to move the chassis from one line to the next required the organizational innovation to increase the "tightness" of the coupling of serial work processes in the factory. In other words, technology and organization are dynamically coupled in their evolution.

The costs of altering tightly coupled components of technology and organization imply that firms will persist in their old ways beyond the recommendation of the net present value. This persistence defines a range of inertia, or what is called a hysteresis band. Because organizational change is disruptive and hence discontinuous, managers hesitate to radically change their organizations, hoping perhaps that future states of the world would provide more appealing environments. Thus, contrary to the normative value in responding flexibly, inertia is rationally encouraged in highly volatile environments if change is costly and the environment is granular.

Figure 1 provides a simple illustration of this point. A firm can choose between two complementary systems, called low and high variety. The important issue is whether the relative value of gaining the capability of variety is enough to offset the costs of discontinuous change. The choice of capabilities is, as we depict it, derived from the market price placed on variety. Because of uncertainty over the evolution of the value of variety and the costs of adoption, managers rationally might choose to persist with inferior techniques before they are confident of future developments. Inertia, then, is not simply a property of stickiness, but reflects expectations regarding the value and costs of change. Inertia increases with uncertainty, because managers are rationally hesitant to incur the cost of change to capabilities that may become easily worthless if the environment reverts to its previous state. (Clearly, inertia can also stem from considerations of loss aversion or status quo bias of managers.)

In a normative vein, it stands to reason from the point of view of an organizational ecology that a firm should experiment in activities that promote its future survival (Lewin and Volderba 1999). In this sense, organizational ecology offers an escape from the inward-looking bias of the resource-based view of strategy. For enhancing future survival, a firm should invest in platforms that correspond to expectations regarding the evolution of the external environment. Platforms are technological and organizational investments that permit a firm to enter into a wide menu of future markets. Firms that build general platforms are more likely to survive and grow (Kim and Kogut 1996). It is exactly the evaluation of this correspondence between exploration of new capabilities and the evolution of the market environment that is provided by the application of a real options heuristic.

Complex Adaptive Systems and Option Theory[7]

A hallmark feature of complex systems is the recognition that environmental change is marked by sharp nonlinearity (Lewin and Volderba 1999). An option is defined, of course, by returns that are nonlinear and contingent on the stochastic state of the world. Hence, it is not surprising that option valuation is appropriate in complex and nonlinear environments.

To avoid confusion, we distinguish between experimentation and market exploration. Search is the application of products and services to new markets and landscapes. Experimentation is the learning of new techniques and combinations of technical and organizational elements. In practice, the market search and experimentation are likely to be linked, and, consequently, it is confusing to insist too strongly on their separation. It is nevertheless useful to remember that new ways of doing things and addressing new markets are not the same. To understand what innovation stream to choose and what kinds of flexibility to develop requires an evaluation of the implications of experiments and market exploration for the contours of the innovative landscape. For this, we need to match more closely the stochasticity of the environment to experiments of new combinations of organizational and technological elements.

A landscape is the performance contour generated by

Figure 1 The Implications of Hysteresis on the Choice of New Techniques

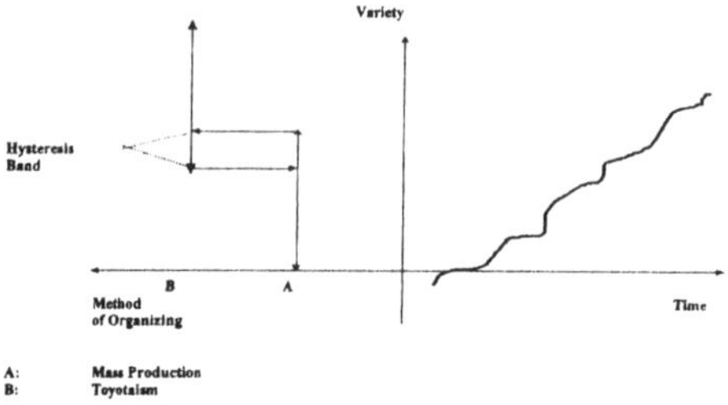

the aggregation of each firm's position in a multidimensional space given by its configuration of resources. Without imposing a core and periphery, it represents distance among organizational and technical features in a multidimensional space. (If features are identified as "there" or "not there," then the measure of proximity is the Euclidean hamming distance in a multidimensional space.) The ruggedness of the landscape implies, *much like the theory of strategic groups*, that firms compete around discrete combinations of resources which correspond to different market niches, or which provide functionally equivalent methods of production. Moving from one to the other is problematic because competencies are scarce (that is, difficult to imitate) *and* also because a landscape coevolves through competitive interactions.

A useful heuristic in this kind of representation is to know the value of directional change in the landscape. The value of changing resources and hence changing position in this landscape requires an evaluation of the cost of change against the future unknown reward. This is what option theory does; it puts a value on the investment in the capability to change position in the landscape contingent on the environmental outcome. It does this by inferring from today's market valuations the expected value of changing position in the future. Unlike the fitness landscapes found in organizational ecology and biology, real options looks at the *value* of a position, where contours correspond to different valuations placed on the assets of an organization.

Complex adaptive system thinking has found it difficult to give heuristic advice other than the importance of creating a process by which to respond to uncertainty. In his thoughtful conclusion to a special issue on complexity, Cohen (1999, p. 375) concludes that aim of recent "efforts is not so much a theory that predicts what a given complex system will do, but rather a framework." However, ultimately, a framework needs to provide decision heuristics to evaluate choices.

The normative literature in organizational theory (e.g., Tushman and O'Reilly 1997, Brown and Eisenhardt 1998) has in recent years recognized the value of flexible responses to radical uncertainty as an organizational capability in complex environments. Brown and Eisenhardt (1998, p. 151) explicitly connect probes with options thinking. If the future evolution of opportunities was obvious, then a firm could focus the efforts of change in one direction. However, because the direction of change is unknown, it pays to invest in probing. But clearly, a heuristic to choose the probes is required.

It is useful to marry the perspective of probes as options with the idea of a rugged landscape, an idea that originates in biology but that has useful applications to understanding the performance implications of complex interactions among complements. (See Kauffman 1993, Levinthal 1997, McKelvey 1999.) Consider a firm that is located near a local peak. Holding its environment constant, probing consists of learning about superior ways to combine its technological and organizational elements. By learning, we imply a notion of knowledge that is incomplete and not entirely explicitly understood. It is incomplete, as suggested by the observation that firms innovate incrementally to improve existing practices. It is partly tacit; or else firms would easily converge to best practice. To return to the set-to-set discussion in Dosi and Kogut (1993), local learning is the discovery of the correspondence between new elements in the existing set of techniques.

The case of local learning is different than moving from one peak to another. Moving to a new peak implies an architectural change in the language of Henderson and Clark (1990). Such change is no longer experimentation around individual modules, but requires a complex redesign around new sets of technical and organizational elements. Normally, radical architectural change is viewed as incurring catastrophic risks because of, for example, a competitor's introduction of a product that requires a major change in a firm's capabilities. But in fact, such change can be orchestrated through exploration and experimentation that effectively builds ridges between peaks.

Baldwin and Clark (2000) propose a real options approach for understanding the choice of modules when performance outcomes are unknown. The explicit valuation of the activity of experimentation raises an important issue of the design of the firm. It would seem, arguably, that the best design for exploration is based on modularity, whereby a firm can pick and choose the best components. Since modules might be viewed as independent experiments, a reasonable inference is that the firm, like a market, should be designed around independent teams. This argument is, in fact, congruous with Simon's (1969) argument of the social decomposition of organizations into relatively independent units.

The heuristic framework of Baldwin and Clark (2000) cannot, however, be applied to the case of radical architectural change. This is no longer a mix-and-match problem of modules, but represents the switch from one conjoint set of technological and organizational correspondences to a new set. In this environment, a firm cannot exploit its path dependence. It rather must evaluate the distant contours of the value landscape and decide the optimal timing to switch new capabilities. As we will show, this problem is, in fact, less severe than commonly portrayed because exploration permits the building of ridges between value peaks.

A Formal Description

To ground a heuristic of real options in positive theory, we have indicated the commonalities in strategy, organizational ecology, and complex systems by pairing three concepts of scarce factors and the underlying asset, irreversibility and inertia, and the landscape ruggedness and option values. To make these pairings explicit, we present below a formal description that grounds the heuristics of real options in organizational theory. We make use of the following three principal ideas. First, the stickiness of organizational and technological combinations requires a notion of time, that is, of irreversibility as indicated by the dynamic market valuation of organizational assets. (There are, of course, start-up and transaction costs, which we incorporate into switching costs.) Second, that distance in discrete combinations results in a radical difference between learning and recombination of modules within a family of organization and technical elements, as opposed to between families. Last, uncertainty can be decomposed into continuous and granular measures of change.

We seek below to offer the theoretical underpinnings to understanding capabilities as an option, not to value explicitly a real option. (Dixit and Pindyck 1994 and Amram and Kulatilaka 1999 provide a thorough collection of such applications.) For many applications, there exist reasonable avenues of valuation. The formal description has the important advantage of clearly defining a core competence in reference to a market valuation, as well as providing a clear statement of the trade-offs between learning and exploitation, on the one hand, and experimentation and exploration on the other. The graphs, which are drawn based on linear profit functions and particular parameter values, illustrates these trade-offs and switching points.

Looking Outside the Firm: Market Pricing

How should we value an investment in new capabilities? Clearly, the value of a capability depends not only on the internal assets, but also on how those assets are deployed, and on the external market conditions. Thus, the price of a correlated asset in the relevant "scarce factor" market represents the initial point of departure. The value of a capability is then inferred from the observed price dynamics which replicate the payoff to the real option. This replication is the device through which market discipline is imposed on the identification and selection of core capabilities. It is not the static comparison of the capability and strategic factor that matters, but rather the information that is gleaned in the changes in prices over time.

To elucidate the intuition, consider again the framing of a real options problem. The organizational assets of a firm provide an option to spend a fixed amount to procure a new capability by purchasing a physical asset at the end of one year. If the option is exercised, then the resulting project value has the risk characteristics of an existing traded firm. For example, a pharmaceutical firm is considering an entry into biotechnology. It currently has a strong capability in conventional drug development that provides an option to enter into biotechnology at an estimated cost. This cost is idiosyncratic to this firm. However, once it enters into the market, its new business carries a market risk similar to other biotechnology firms. This example illustrates why the price of other firms does not give the value of the core capability, because the cost of entry is idiosyncratic to each firm. However, the price dynamics of other firms provide information on the factors (e.g., risk) that drive the value of the option to enter in this market.

The value of a financial option depends on the current share price. Because the source of this exogenous uncertainty is the market price of a frequently traded market financial security (the share of stock), financial options can be dynamically replicated with a portfolio of stocks and risk-free investments. As a result, derivatives can be valued without knowledge of the expected return earned by the underlying financial asset.

For a special but important case, Black and Scholes (1973) derived this value through an option pricing formula. The simple but critical innovation was their eventual recognition that by composing a replicating portfolio, the value of the option could be perfectly tracked by a levered position in the traded stock. It is, however, unlikely that real options can be perfectly replicated with traded assets. The replication may require the use of product or factor prices. Even when widely traded, the prices of such real assets need not appreciate at a rate equal to its equilibrium risk-adjusted return. Instead, owners of real assets will reap various convenience benefits and incur carrying costs that affect the total returns. In such cases, valuation requires knowledge of the actual price dynamics of the factor price *and* the equilibrium risk-adjusted return.

As an initial proposal, we suggest that the theoretically most interesting way to identify the appropriate correlated asset is to decompose the market price into a bundle of attributes that pierces the revenue veil of the firm to see the underlying assets. Whereas this analysis is unquestionably hard, it should be recalled that it is consistent both with the financial market pressure to understand (i.e., strip) assets and the growth of derivatives to hedge specific components of a firm's risk. From this angle, the value of the capability depends on its contribution to the price of product or factor prices whose risk is spanned by traded assets in the economy. The value of the capability is thus obtained by explicitly specifying the profit function using these prices as an argument.

To identify and value a core competence, we must specify the evolution of the quality-adjusted price that we call θ. A simple example is a microprocessor, whereby a quality-adjusted price can be expressed as the ratio of price to the processing speed (or "mips" for millions of instructions per second). Since we cannot directly observe this price, we can choose the stock price of a firm specializing in microprocessors to give the estimates for the volatility. However, because θ is not a pure security but is the observed price of a scarce factor, its price characteristics need not necessarily evolve according to its equilibrium risk characteristics. Local supply-and-demand conditions and technological innovation determine the evolution of θ. We want to sort out smoothly evolving uncertainty from discrete granular shocks. We assume θ to be exogenously determined and characterize its evolution by stochastic process

$$\Delta\theta_t = \underbrace{\mu(\theta_t, t)\,\Delta t}_{\text{Deterministic Growth}} + \underbrace{\sigma(\theta_t, t)\,\Delta Z_t}_{\text{Smoothly evolving uncertainty}} + \underbrace{\kappa dq}_{\text{Discrete innovations}}$$

where μ is the expected growth rate of θ, σ is its instantaneous volatility, ΔZ_t is standard Normal distributed, dq is a Poisson process with intensity parameter λ, and κ is the random percentage jump amplitude conditional on the Poisson event occurring (Merton 1976).

Changes in the quality-adjusted price may reflect unpredictable shifts in consumer preferences or incremental technical change. For example, an increase in oil prices would lead consumers to prefer cars which save in fuel consumption. As long as these changes are fairly smooth, it seems reasonable to capture this uncertainty in volatility. This term represents the uncertainty in the environment around the local peak.

Other changes may be more radical and appear as discontinuous Poisson jumps, such as the arrival of new organizational innovations, and is a measure of the granularity of the environment. These changes would appear as a sudden jump in price to a firm.

The quality adjusted price approach to building a valuation model faces potential problems. The argument hinges on the premise that the risk profile of the value of the innovation is spanned by quality-adjusted prices. However, the quality-adjusted price is derived from a model of the industry pricing behavior and can suffer from "modeling error." Furthermore, the quality-adjusted price may not perfectly track the value of the innovation and may introduce a "tracking error." This error is akin to basis risk in commodity markets where the price of a commodity is specific to its location. Finally, not being a security price, the quality-adjusted price can embed a convenience value that is not easily observed or estimated. For the arbitrage-based valuation approach to work, the error components must be independent of each other and have no systematic risk. Hence, expert opinion may provide a superior method to form probability distributions of possible future market conditions for the new business in radically new landscapes.

Looking Inside the Firm: Capability Sets

Even if two firms are competing in the same industry and market, movement in prices of the strategic asset influences differently their value because of the relationship between the capabilities of the firm and the profit opportunities. To describe this formally, we make use of the notion of distance—which we showed to be a common assumption in strategy, organizational ecology, and complex systems—between discrete combinations of technology and organizational elements that define a capability. We first develop the notion of a capability set and then define the profit function of a firm in relation to its set of organizational and technological practices.

For convenience, we consider the case of an automobile producer choosing to stay within the current set of mass production or switching to new high-performance combinations called lean production (See MacDuffie 1996). A firm has the set of capabilities c, where $c \in C$ is the set of all feasible capabilities. In our case, C contains "mass" and "lean" production families with their associated organizational structures. Each family of production techniques can contain many distinct technologies. They are, however, coupled with the same organizational structure. Hence, a technology family refers to all technologies that can be operated within a single organization.

Suppose the firm is currently employing technology in the "mass" production family, i.e., $c^j_m \in c_m \in C$. The firm's problem is to decide what capabilities it should use in the current period. Specifically, its choices are (a) continue using c^j_m, (b) continue in the same family but make incremental technological improvements by employing a better mass production technique, c^j_m, or (c) make a discontinuous organizational switch and employ lean production technique, c^k_l. Choices a and b reflect "as is" evaluations; only c involves a "could be" alternative. We capture the idea of inertia through the reorganizing costs incurred by switching from one capability to another, be it from mass into lean, or conventional pharmaceutics to biotechnology. We denote these large organizational costs of switching as Δ_{ij}. For example, the cost of switching from c_n (mass production) to c_l (lean production) can be denoted as Δ_{ml}. In practice, switching between capabilities will consume time, as Dierickx and Cool (1989) note. Our model assumes that there is an inverse monotonic relationshp between cost of switching and the time taken to switch. Hence, a switch that takes more time will be represented by a higher switching costs.

Within an organizational capability, switching costs between modules are small, but not insignificant. At the same time, continuing within the same family enables the firm to capitalize on local learning effects. If the firm continues in c^i_m or moves to a better mass technique c^j_m, then it will subsequently learn by doing. However, switching from the ith to the jth technology may still incur technological costs. We define the local learning benefits in mass production as $-\delta_{mm}$ and technological switching costs δ_{ii}.

To summarize the magnitude of switching costs between all combinations of capabilities and technologies, we denote the cost of switching from c^i_m capability to c^j_l as:

$$\delta^{ij}_{ml} = \underbrace{T_{ij}}_{\text{Technological change}} + \underbrace{\Omega_{ml}}_{\text{Organizational learning}},$$

where

$$T_{ij} = \begin{cases} \overbrace{\underbrace{\begin{matrix}\delta_{ij} \\ -\delta_{ii}\end{matrix}}_{\text{technological learning}}}^{\text{technological cost}} & \begin{matrix}\text{if } i \neq j \\ \text{if } i = j\end{matrix} \end{cases}$$

$$\Omega_{ij} = \begin{cases} \overbrace{\underbrace{\begin{matrix}o_{ml} \\ -o_{mm}\end{matrix}}_{\text{organizational learning}}}^{\text{organizational cost}} & \begin{matrix}\text{if } m \neq l \\ \text{if } m = l.\end{matrix} \end{cases}$$

We can now write down the firm's objective. Each set of capabilities c^i_m has an accompanying profit function which is obtained by solving the usual profit-maximization problem:

$$\Pi(\theta, c^i_m) = \max_{y \in c^i_m} \theta.y$$

where θ is a vector of quality-adjusted input and output prices and y is the vector of input and output levels that are determined by the capability set. This simple expression indicates that the firm's ability to choose the best strategy is contingent on its organizational resources.

Dynamic Valuation of the Critical Capability Set

When future values of θ evolve stochastically, the current decision influences all future decisions as well. The decision by a mass producer of cars to invest in flexible manufacturing using lean production runs the risk that the American market suddenly decides to buy large recreational vehicles made best by standard mass production techniques. But now they face the problem that they are invested in lean manufacturing and cannot easily switch back. The tight coupling of organization and technology is essential to understanding why capabilities radically change the understanding of strategy as not only the choice of entering markets, but also as the selection of competence.

The way to fully analyze the implications of inertia is to write out explicitly the problem over time. To do this, we no longer work directly with profit functions, but instead with a value function. While technically this problem is often hard to solve, its formulation is both intuitive and insightful. At a point in time t, this formulation treats the present value of all future benefits given optimal future behavior, as represented by the value function

$V(\theta_t, c^i_m)$. The value function is the solution of the well-known Bellman equation:

$$V(\theta_t, c^i_m) = \max_{c^j_l} [(\Pi(\theta_t, c^j_l) - \delta^{ij}_{ml}) + \rho E_t[V(\theta_{t+1}, c^j_{1l})],$$

where c^i_m is the current capability c^j_1 chosen from the set of feasible technologies and organizations at time $t + 1$. This formulation is Winter's optimal control suggestion which includes organizational knowledge, or capabilities, as a state variable.

The Bellman equation has an intuitively appealing formulation, for it directly evaluates the exploitation of the choice of current capabilities (the first term in brackets) with the value of persisting or switching in the future (the second term). This equation indicates that in each period the producer contemplates switching into a new capability. If it chooses capability c^j_l, it realizes benefits of $\Pi(\theta_t, c^j_l)$, but pays switching costs of δ^{ij}_{ml}, and then arrives at the following period with value function $V(\theta_{t+1}, c^j_m)$. This value depends on the capability chosen, c^j_l, as well as on the value of the state variable next period, θ_{t+1}. Because θ_{t+1} was still unknown at time t, we take expectations; we also discount at rate ρ. (See Pindyck 1991 and Kulatilaka and Marcus 1994 for a more explicit treatment of systematic risk.)

In each period the producer chooses the capability c^j_l, that maximizes the value of the project. This choice can be interpreted as defining the *dynamic capability* as

$$c^{**} = \underset{c^j_l}{\operatorname{argmax}} [(\Pi(\theta_t, c^j_l) - \delta^{ij}_{ml}) + \rho E_t[V(\theta_{t+1}, c^j_l)].$$

In the absence of switching costs, the solution to this optimization problem is simple: Choose in each period the capability c^j_l that maximizes $\Pi(\theta_t, c^j_l)$ in that period. This is the *static critical capability* discussed earlier. However, the presence of switching costs makes a forward-looking analysis necessary. In the case of costly reorganization, the probability distribution of future prices affects the *current* choice of technology and organization.

This definition of a dynamic capability defines our reinterpretation of a "core competence." Core competence is the capability set (i.e., combination of organization and technology elements) that permits the firm to dynamically choose the optimal strategy for a given price realization of the strategic factor.

Hysteresis and Inertia

With the above concepts, we can now analyze more fully the hysteresis band first given in Figure 1. Hysteresis is a critical concept because it shows that inertia need not be the consequence of myopia but is itself sensitive to environmental turbulence and to a firm's competence. A more competent firm is, ironically, more subject to inertia.

Figure 2 Static and Dynamic Hysteresis

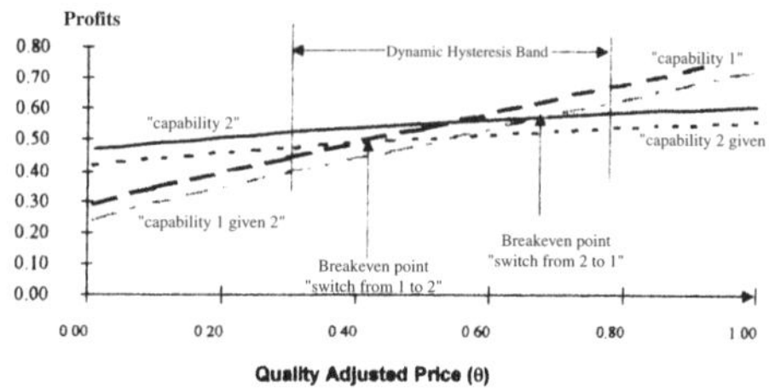

The band between the switching costs is underestimated by comparing single-period breakeven profits, net of switching costs, under the two capabilities. Because of the possibility that θ may revert back to previous values (e.g., because of a sudden drop in oil prices favoring gas-guzzling cars), the firm persists in its current mode and waits to see how prices evolve in the future. At some point, however, θ takes on values that justify not only the one-time switching costs but also the probability-weighted costs attached to switching back. If a firm is unable to choose the optimal response, these conditions lead to a competency trap that is expressed by a hysteresis band. In Figure 2, the profit functions for two capability sets and the resulting hysteresis band is graphed.

Competency Traps and Learning to Learn

Because of the benefits of learning by doing, simply exploiting current capabilities leads to cumulative and incremental improvement. In effect, the profit function can be described as shifting outward over time. By staying in its current activities, the firm becomes increasingly more competent. Techniques of mass production are expressed in well-understood routines that couple technology and people through known organizing principles of work.

The danger remains, of course, that θ will suddenly jump to a range or cross a critical threshold in which the firm's competence is no longer profitable. In a sense, its accumulated learning in the old techniques is a *"competency trap."* (See the discussion in March 1991.) Yet, as a consequence, by improving in mass production, it is less attractive to change organizational capabilities. Hence, a firm might rationally preserve its way of doing things because it has become so good at doing the (now) wrong thing. Dougherty (1995) has labeled this "core incompetence." Exploitation of current knowledge drives learning by doing; the pitfall is that this learning increases the rigidity of the firm.

To speed its transition to new techniques, the firm may decide proactively to allocate funding to exploration by experimenting with new techniques. This diversion of resources slows down its accumulation of learning with the current technology. At the same time, it increases the value of the option to switch to new capabilities by lowering the costs of switching. To characterize this wider menu of choices, Figure 3 depicts the decision of a firm that has accumulated a particular breadth of knowledge in the current production techniques, as well as in learning derived from experiments with new methods. (We can think of these experiments as "joint ventures," such as the Nummi venture between General Motors and Toyota.)

The net effects of learning are ambiguous and depend upon the rate by which new knowledge is gained through learning by doing relative to experimentation. In Figure 4, this comparison is graphed by showing the upward change in profit functions over time due to these two learning effects. By construction, we show the gains to experimentation dominating learning by doing. Thus, while it is a truism that firms need to balance exploitation and exploration, the possibility for local learning may drive out distant search (Levinthal and March 1993).

There is a more important insight provided by the investments in exploration, namely, that the literature on innovation overemphasizes the difficulty posed by discontinuous change. In more contemporary parlance, the prospects of successful radical change are viewed as poor because the chance of jumping from one performance peak to a distant peak is considered improbable. But exploratory investments permit the building of ridges between peaks. By exploring the current assets that can be

Figure 3 **Expanded Capability Sets**

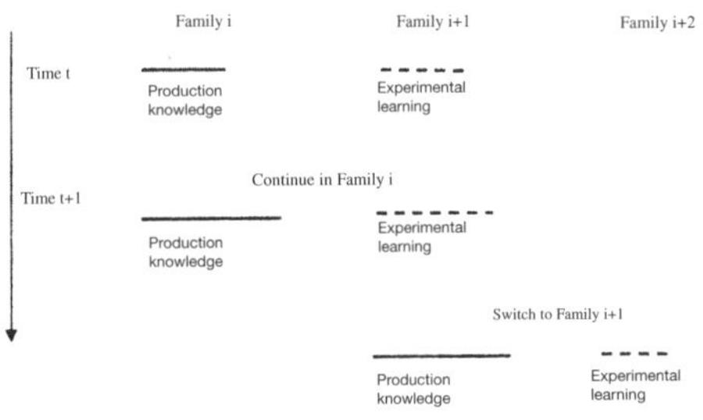

Figure 4 **Effects of Learning**

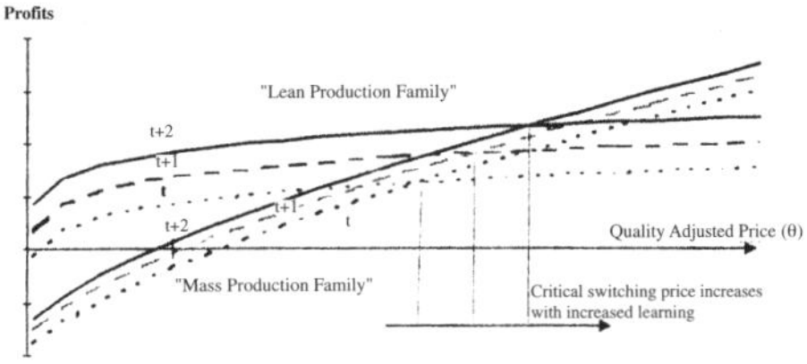

recombined and coupled with new ones, a firm is able to reduce the risks of falsely choosing new capabilities. Through recombination, exploration reduces the organizational costs to successfully adopt radical change in its capabilities.

Discussion: How Good a Heuristic?

Real option theory provides complex heuristics to apply, though there are reasonable ways to simplify the application (Bowman and Moskowitz 1997). Equally important is that the real options heuristic carries the danger of negative transfer to inappropriate situations. A particularly troubling complication for real options analysis is the explicit consideration of competitive interactions and positioning. Competitive interactions endogenize the dynamics of the external market price. The valuation of a strategic option requires an identification of a market price by which to derive the replicate of the underlying asset. In financial markets, this price is easily given by stock or future prices. An important and reasonable assumption is that exercising the option does not influence the value of the replicating portfolio.

This assumption does not always hold for real options for two reasons. First, by exercising an option to enter a market, a firm often influences prices through increasing supply. Second, by entering (or exiting) a market, competitors will alter their behavior. As a result, the market price is endogenous to the decision whether to exercise the option.

This problem is partly resolved by recognizing that the value of theta reflects the assessment on entry. But this assumption hardly provides insight into the identity of possible entrants and their strategic behavior. A structural approach is explicit regarding the nature of future competition. Kulatilaka and Perotti (1998) follow this approach by evaluating the decision to launch a new technology in the context of different conjectures about market structure. This solution marries the industry structure analysis to core competence, but through the stipulation that the analysis is forwardlooking rather than focused on current market structure.

Conclusions

Real option analysis provides the theoretical foundations to the use of heuristics for deriving capabilities. Through conditioning an understanding of competence in relation to a market test (e.g., Barney's notion of a strategic factor market), it identifies the coupling of organization and technology as the leading explanation for the irreversibility of investments in capabilities. In a narrow sense, it faults discounted cash flow analysis as the principal tool of understanding the value of a firm. But more profoundly, the recognition that the coupling of people and technology is a source of option value challenges simplistic notions of firms as "pure asset plays." In this respect, it strengthens the argument that a firm's most enduring advantage lies in its human resources (Pfeffer 1994).

Ironically, then, the derivation of the option value from the embedded knowledge in organizational assets deflects a purely financial evaluation of the firm. Because organizations consist of coupled systems, the value of the firm is not reflected in the present value of its constituent parts, but in the combinative potential of deploying these capabilities for innovation in existing markets or for addressing new markets. It suggests that firms are dynamic systems consisting of the complex coupling of technology and people through organizational design. The paradoxical conclusion to the sustained application of financial modeling to firms is that, in the end, the fundamental basis of the value of the firm is its organizational capability to exploit current assets and explore future opportunities.

The recent efforts in organizational theory to embrace complex systems analysis underscores the importance of understanding distance in capabilities in conjunction with the volatility of the environment. An appropriate heuristic arising out of complex systems is to understand landscape contours as representing market expectations of the value of discrete organizational capabilities. An options approach indicates that firms construct exploratory ridges between peaks to hedge against adverse changes in the landscape. But to understand the direction of exploration, it is necessary not only to know distances and locations of peaks, but also to know their heights. It is logical that the positive theory of organizational inertia should imply that the most valuable heuristic is one that identifies the best options attached to the direction and timing of organizational changes.

Acknowledgment

The authors would like to thank Ned Bowman, Carliss Baldwin, Howard Kunreuther, and Arie Lewin for many helpful discussions, the anonymous referees for their comments, and the managers of Lucent for a reality check on some ideas.

Endnotes

[1]See Minsky (1985, pp. 243–253) for an example.

[2]Georgescu-Roegen (1971, p. 196) provides an explicit and early discussion of irreversibility, hysteresis (as discussed later), and the arrow of time. The arrow of time and evolution are central themes in Prigogine and Stengers (1984). A volume of essays on irreversibility (with little closure on definition) is edited by Boyer et al. (1991).

[3]For a discussion, see Kogut and Zander (1992) on knowledge and combinative capabilities as options, and Zander and Kogut (1995) and Szulanski (1995) for empirical studies that measure inimitable as tacitness.

[4]Winter (1987) suggests net present value as a measure which is appropriate for the case without uncertainty. Most surveys on the use of capital budgeting techniques show that almost all large corporate firms use net present value calculations for investment decisions. See Kogut and Kulatilaka (1992) and Baldwin and Clark (1992, 1994) for a discussion of why investment in capabilities is not a net present value of cash flows, but a real option valuati.

[5]Howard Kunreuther points out to us that a strategy of minimizing death results in suboptimal behavior, as shown in an extant literature in the decision sciences. Still, there is a technical correspondence between evaluating the value of the firm at either boundary. As Black and Scholes (1973) noted early in the development of option pricing, the boundary defined by bankruptcy can be used to value the stock. Note also Tuma and Hannan's (1984) acknowledgement of the nonlinear and stochastic process that governs firm growth in their Chapters 12 and 15 discussion of stochastic calculus.

[6]This point is implicit in the lattice formulation of Milgrom and Roberts (1990), where a firm's choice is constrained by technical complementarities. See Levinthal (1997) for complementarities as a combinatorial problem in organizational space.

[7]We thank Arie Lewin for encouraging the development of the ideas below.

References

Amram, Martha, Nalin Kulatilaka. 1999. *Real Options: Managing Strategic Investments in an Uncertain World.* Harvard Business School Press, Harvard, Cambridge, MA.

Argyris, Chris, Donald Schoen. 1978. *Organizational Learning.* Addison-Wesley, Reading, MA.

Baldwin, Carliss, Kim Clark. 1992. Capabilities and capital investment: New perspectives on capital budgeting. *J. Applied Corporate Finance* **5** 67–82.

——, ——. Capital budgeting systems and capabilities investments in U.S. companies after World War II. *Bus. History Rev.* **68**(1) 73–109.

——, ——. 2001. *Design Rules: The Power of Modularity.* MIT Press, Forthcoming.

Barney, Jay. 1986. Strategic factor markets: Expectations, luck, and business strategy. *Management Sci.* **32**(10) 1231–1241.

Black, F., M. Scholes. 1973. The pricing of options and corporate liabilities. *J. Political Economy* **81** 637–659.

Bowman, Edward. 1963. Consistency and optimality in managerial decision making. *Management Sci.* **9** 310–321.

——. 1995. Strategy history: Through different mirrors. *Adv. in Strategic Management* **11** 25–45.

——, D. Hurry. 1993. Strategy through the option lens: An integrated view of resource investments and the incremental choice process. *Acad. Management Rev.* **18**(4) 760–782.

——, Gary Moskowitz. 1997. The use of options analysis in strategic decision making. Working Paper, Reginald H. Jones Center, Wharton School, University of Pennsylvania.

Boyer, Robert, Bernard Chavance, Olivier Godard, eds. 1991. *Les Figures de l'irréversibilité en Économie*. Paris: EHESS.

Brown, Shona, Kathleen Eisenhardt. 1998. *Competing on the Edge: Strategies as Structured Chaos*. Harvard Business School Press, Boston, MA.

Cohen, Michael. 1999. Commentary on the *Organization Science* special issue on complexity. *Organ. Sci.* **10** 373–376.

——, Paul Bacdayan. 1994. Organizational routines are stored as procedural memory: Evidence from laboratory study. *Organ. Sci.* **5** 554–568.

Dierickx, Ingemar, Karel Cool. 1989. Asset stock accumulation and the sustainability of competitive advantage. *Management Sci.* **35** 1504–1511.

Dixit, Avinash, Robert Pindyck. 1994. *Investment Under Uncertainty*. Princeton University Press, Princeton, NJ.

Dosi, Giovanni, Bruce Kogut. 1993. National specificities and the context of change: The coevolution of organization and technology. B. Kogut, ed. *Country Competitiveness: Technology and the Organizing of Work*. Oxford University Press, New York.

Dougherty, Deborah. 1995. Managing your core incompetencies for innovation. *Entrepreneurship, Theory, Practice* **19** 113–135.

Georgescu-Roegen, Nicholas. 1971. *The Entropy Laws and Economic Progress*. Harvard University, Cambridge, MA.

Hamel, Gary, C. K. Prahalad. 1994. *Competing for the Future: Breakthrough Strategies*. Harvard Business School Press, Boston, MA.

Hannan, Michael, John Freeman. 1977. The population ecology of organizations. *Amer. J. Soc.* **82** 929–964.

——, ——. 1984. Structural inertia and organizational change. *Amer. Soc. Rev.* **49** 149–164.

Hedlund, Gunnar, Dag Rolander. 1990. Action in hierarchies: New approaches to managing the MNC. Christopher A. Bartlett, Yves Doz, Gunnar Hedlund, eds. *Managing the Global Firm*. Routledge, London and New York.

Henderson, Rebecca. 1993. Underinvestment and incompetence as responses to radical innovation: Evidence from the photolithographic alignment equipment industry. *Rand J.* **24** 248–270.

——, Kim Clark. 1990. Architectural innovation: The reconfiguration of existing product technologies and the failure of established firms. *Admin. Sci. Quart.* **35** 9–30.

Hull, John C. 1997. *Options, Futures, and Other Derivatives*, 3rd Edition. Prentice-Hall, Englewood Cliffs, New Jersey.

Kauffman, Stuart. 1993. *The Origins of Order: Self Organization and Selection in Evolution*. Oxford University Press, New York.

Kim, Dong Jae, Bruce Kogut. 1996. Technological platforms and diversification. *Organ. Sci.* **7** 283–301.

Kogut, Bruce, Nalin Kulatilaka. 1992. What is a critical capability? Working paper, Reginald H. Jones Center, Wharton School, University of Pennsylvania, Philadelphia, PA.

——, ——. 1994a. Options thinking and platform investments: Investing in opportunity. *California Management Rev.* **36** 52–71.

——, ——. 1994b. Operating flexibility, global manufacturing, and the option value of multinationality. *Management Sci.* **40** 123–139.

Kulatilaka, Nalin, Alan J. Marcus. 1994. Hedging foreign project risk. *J. Internat. Financial Management Accounting* **5**(2) 142–156.

——, Enrico Perotti. 1998. Strategic growth options. *Management Sci.* 1021–1031.

Kunreuther, Howard. 1969. Extensions of Bowman's theory on managerial decision-making. *Management Sci.* **16** B415–B439.

Levinthal, Daniel. 1997. Adaptation on rugged landscapes. *Management Sci.* **43** 934–950.

——, James March. 1993. The myopia of learning. *Strategic Management J.* **14** 95–112.

Lewin, Arie, T. Carroll, C. Long. 1998. The co-evolution of new organizational forms. *Organ. Sci.* **10** 535–550.

——, Henk Volderba. 1999. Prolegomena on co-evolution: A framework for on research on strategy and new organizational forms. *Organ. Sci.* **10** 519–534.

Lincoln, James. 1993. Comparison of Japanese and US organizational structures. B. Kogut, ed. *Country Competitiveness: Technology and the Organizing of Work*. Oxford University Press, New York.

MacDuffie, John Paul. 1996. Human resource bundles and manufacturing performance: Flexible production systems in the world auto industry. *Indust. Labor Relations Rev.* 97–221.

Mansfield, Edward. 1988. Speed and cost of industrial innovation in Japan and United States. *Management Sci.* **34** 1157–1168.

March, James. 1991. Exploration and exploitation in organizational learning. *Organ. Sci.* **2** 71–87.

——, Herbert Simon. 1958. *Organizations*. Wiley, New York.

McDonald, Robert, Daniel Siegel. 1984. Option pricing when the underlying asset earns a below-equilibrium rate of return: A note. *J. Finance.* **39** 261–265.

McKelvey, Bill. 1999. Avoiding complexity catastrophe in coevolutionary pockets: Strategies for rugged landscapes. *Organ. Sci.* **10** 294–321.

Merton, Robert C. 1976. Option pricing when underlying stock returns are discontinuous. *J. Financial Econom.* **3** 125–144.

Midler, Christopher. 1993. *L'auto qui n'existe pas. Management des projets et transformation de l'enterprise.* InterEditions, Paris, France.

Milgrom, Paul, John Roberts. 1990. The economics of modern manufacturing: Technology, strategy, and organization. *Amer. Econom. Rev.* **80** 511–528.

Minsky, Marvin. 1985. *The Society of Mind*. Simon and Schuster, New York.

Nelson, Richard R., Sidney G. Winter. 1982. *An Evolutionary Theory of Economic Change*. Harvard University Press, Cambridge, MA.

Pfeffer, Jeffrey. 1994. *Competitive Advantage through People*. Harvard Business School Press, Boston, MA.

——, Gerald Salancik. 1978. *The External Control of Organizations: A Resource Dependence Perspective*. Harper and Row, New York.

Pindyck, Robert S. 1991. Irreversibility, uncertainty, and investment. *J. Econom. Literature.* **29**(3) 1110–1148.

Prigogine, Ilya, Isabelle Stengers. 1984. *Order Out of Chaos. Man's New Dialogue with Nature*. Bantam Book, New York.

Scherer, F. M. 1967. Research and development resource allocation under rivalry. *Quart. J. Econom.* **81** 359–394.
Scott, W. Richard. 1995. *Institutions and Organizations*. Sage, Thousand Oaks, CA.
Selznick, Philip. 1957. *Leadership in Administration, A Sociological Interpretation*. Row, Peterson, Evanston, IL.
Simon, Herbert. 1969. *The Sciences of the Artificial*. MIT Press, Cambridge, MA.
Szulanski, Gabriel. 1995. Unpacking stickiness: An empirical investigation of the barriers to transfer best practice inside the firm. *Acad. Management Best Paper Proceedings* 437–441.
Teece, David, Gary Pisano, Amy Shuen. 1997. Dynamic capabilities and strategic management. *Strategic Management J.* **18** 509–534.
Thompson, James D. 1967. *Organizations in Action: Social Science Bases of Administrative Theory*. McGraw Hill, New York.

Accepted by Arie Lewin.

Tuma, Nancy Brandon, Michael T. Hannan. 1984. *Social Dynamics: Models and Methods*. Academic Press, New York.
Tushman, Michael, Philip Anderson. 1986. Technological discontinuities and organizational environments. *Admin. Sci. Quart.* **31** 439–465.
——, Charles O'Reilly. 1997. *Winning Through Innovation: A Practical Guide to Leading Organizational Change and Renewal*. Harvard Business School Press, Boston, MA.
Winter, Sidney. 1987. Knowledge and competence as strategic assets. D. Teece, ed. *The Competitive Challenge—Strategies for Industrial Innovation and Renewal*. Ballinger, Cambridge, MA. 159–184.
Zander, Udo, Bruce Kogut. 1995. Knowledge and the speed of the transfer and imitation of organizational capabilities: An empirical test. *Organ. Sci.* **6** 76–92.

[22]

Strategic Management Journal
Strat. Mgmt. J., **25**: 121–138 (2004)
Published online 4 November 2003 in Wiley InterScience (www.interscience.wiley.com). DOI: 10.1002/smj.368

ENTRY IN THE PRESENCE OF DUELING OPTIONS

TIMOTHY B. FOLTA[1]* and JONATHAN P. O'BRIEN[2]
[1] *Krannert Graduate School of Management, Purdue University, West Lafayette, Indiana, U.S.A.*
[2] *Mendoza College of Business, University of Notre Dame, Notre Dame, Indiana, U.S.A.*

This paper investigates the influence of industry uncertainty on the decision by established firms to enter a new industry. Specifically, we examine the tension between the option to defer, which discourages entry in the presence of uncertainty, and the option to grow, which may encourage entry in the presence of uncertainty when there are early mover advantages. Empirical analysis on data from a broad array of industries revealed that the effect of uncertainty on entry is not monotonic. Our findings are the first to find support for the nonmonotonic effect of uncertainty that has only recently emerged in theoretical treatments of real options theory, and amplify the importance of considering both the option to defer and the option to grow when contemplating entry. Furthermore, we found evidence that the relationship between uncertainty and entry is moderated by: (a) irreversibility, which influences the value of the option to defer; (b) the total value of growth opportunities; and (c) early mover advantages, which magnify the value of growth options.

This paper explores how exogenous industry uncertainty influences the decision by established firms to enter a new industry. Although prior studies have implicitly assumed that uncertainty is not pertinent to the entry decision by established firms (Montgomery and Hariharan, 1991; Silverman, 1999), the real options perspective argues that uncertainty will be a critical factor whenever an investment decision is characterized by some degree of irreversibility and there is potential for the future exercise of managerial discretion (Kogut and Kulatilaka, 2001). However, even within the real options literature, debate continues about the influence of uncertainty on the entry/investment decision. The conflict between the *option to defer* and the *option to grow* plays a central role in this debate.

The option to defer obtains its value from the potential for a manager to defer the entry decision, which allows a firm to 'keep its options open' and avoid the opportunity costs associated with making an irreversible investment (McDonald and Siegel, 1986). Factors that increase the value of the option to defer, such as greater sunk costs associated with the investment, make entry less likely. Alternatively, the option to grow gains its value from the possibility that early investment will help the firm to develop a 'capability' that will allow it to take better advantage of future growth opportunities in the industry (Kulatilaka and Perotti, 1998). Accordingly, more valuable growth options encourage investment and make entry more likely.

Since these two types of real options have opposite effects on entry, and both increase in value with increasing uncertainty, predicting the

Key words: uncertainty; growth option; real option; entry; early mover advantages
*Correspondence to: Timothy B. Folta, Krannert Graduate School of Management, Purdue University, West Lafayette, IN 47907-1310, U.S.A. E-mail: foltat@mgmt.purdue.edu

Received 15 February 2002
Final revision received 22 July 2003

net impact of uncertainty on entry is problematic. However, recent theoretical work by Kulatilaka and Perotti (1998) suggests that different options will dominate the entry decision at different levels of uncertainty, thus leading to a nonmonotonic effect of uncertainty on entry. In investigating whether insights from real options theory enlighten firm entry decisions, we seek to make two important contributions to the literature. First, we test whether uncertainty has a nonmonotonic effect across a broad sample of industries. Second, we theoretically and empirically examine how the influence of uncertainty on entry is moderated by factors that influence the value of the growth and deferment options.

BACKGROUND

Since a real option is the right, but not the obligation, to take some action in the future (Amram and Kulatilaka, 1999), the option will be valuable whenever an investment decision entails significant uncertainty and cannot be costlessly reversed.[1] The traditional approach to examining firm investment behavior, neoclassical investment theory, is based on the reasoning that investments should be made when the simple net present value (NPV_S) of the opportunity equals or exceeds zero, the investment threshold. In the context of market entry, this suggests the following decision rule:

$$\text{Enter if}: NPV_S \geq 0 \qquad (1)$$

However, this reasoning does not accurately describe the influence of uncertainty on entry because it fails to consider the two crucial factors that are tenets of real options theory. First, most investments are at least partially irreversible since they cannot be fully recovered and costlessly redeployed in the event of a negative shock. Second, managers can adapt and revise their strategies in response to unexpected market developments that cause cash flows to deviate from their original expectations. Real options theory provides a theoretical basis for considering why firms set investment thresholds at any point other than that suggested by neoclassical investment theory (i.e., at NPV_S equals zero).

Even though it is rare for real managers to explicitly employ real options models when making investment decisions (Copeland and Keenan, 1998), evidence suggests that managers are often willing to overrule traditional investment criteria in the interest of strategic flexibility (Hayes and Garvin, 1982; Donaldson and Lorsch, 1983). One such form of strategic flexibility is the flexibility to defer an investment decision when substantial uncertainty exists about the future conditions in the targeted industry. The 'option to defer' is intended to describe just this type of flexibility. The basic intuition is that a firm sacrifices its option to defer when it makes an irreversible investment, losing the potential to make a different decision should new information arrive that affects the desirability or timing of the investment. Thus, when a firm makes an irreversible investment, it incurs an opportunity cost, D, that is equal in value to the option to defer. A focus on the option to defer suggests that the investment threshold should be raised (i.e., should be increased above zero) by the value of these opportunity costs. The key insight from this perspective is that entry is discouraged with greater uncertainty because D escalates with uncertainty. In previous research, evidence that uncertainty has a depressive effect on investment has been interpreted as support for the explanatory power of real options theory.[2]

To date, most of the empirical research on real options theory has focused on this option to defer. Although there is relatively strong support for the prediction that macroeconomic uncertainty discourages economy-wide investment levels (e.g., Pindyck, 1986; Episcopos, 1995), research examining firm investment levels has produced mixed results.[3] The latter is curious since real options approaches to investment specifically attend to firm-level thresholds, and thus it is at this level of analysis that the influence of uncertainty should be most evident (Carruth *et al.*, 2000). Research that examines whether or not investment occurs, rather than investment levels, may be more amenable to

[1] Consistent with Amram and Kulatilaka (1999), we define uncertainty as the randomness of the external environment that cannot be altered by the actions of individual firms or managers.

[2] See Carruth, Dickerson, and Henley (2000) for a review of these arguments and the associated empirical work. A focus on *total* uncertainty distinguishes real options analysis from traditional investment theory, which argues that NPV is influenced only by the systematic component of uncertainty.

[3] Several studies report the expected negative relationship between uncertainty and firm investment levels (Huizinga, 1993; Guiso and Parigi, 1999), while others report weakly negative or no relationship (Campa and Goldberg, 1995; Driver, Yip, and Dakhil, 1996; Leahy and Whited, 1996).

real option analysis because, as Dixit and Pindyck (1994) note, the theory does not determine the level of investment per se, but rather it identifies factors that may affect the *threshold* at which investment should occur. Accordingly, it is noteworthy that studies focusing on the probability of investment have linked increased uncertainty with both a decreased likelihood of foreign entry (Campa, 1993), and a decreased likelihood of acquiring an alliance partner (Folta and Miller, 2002).

Although previous studies have indicated that real options analysis is helpful in describing investment patterns, there remain significant opportunities for further study. For example, the empirical focus has been almost exclusively on how uncertainty *deters* investment, thus ignoring that investment may provide valuable growth options. Since the value of growth options increases with uncertainty, higher uncertainty may lower investment thresholds for projects with significant growth options, thus inducing a positive relationship between uncertainty and investment. Kester (1984) was perhaps the first to recognize that traditional investment thresholds are too high when growth opportunities are embedded in investments. In recent years, growth options have been used by management scholars to justify investment in a number of contexts, such as the acquisition of new firm capabilities (Kogut and Kulatilaka, 2001), and patterns of sequential investment in new ventures (Hurry, Miller, and Bowman, 1992), industry segments (Kim and Kogut, 1996), international markets (Chang, 1995), and technology (McGrath, 1997). Although this work has served as a useful bridge between strategy and finance, it has failed to empirically consider how uncertainty influences the choice between waiting and immediate entry.

The choice between waiting and immediate entry has strong implications for business- and corporate-level strategy. Entering earlier may better position the firm to build a technological advantage, develop brand recognition, or accrue any number of resources that may lower costs or increase revenues relative to later movers (Lieberman and Montgomery, 1988). Whether entry should be undertaken immediately to take better advantage of growth opportunities, or delayed until the environment becomes less uncertain, depends on the nature and the size of the deferment and growth options, which may depend critically on the total uncertainty in the target industry. The next section elaborates on the relative impact of the options underlying these alternatives.

THEORY AND HYPOTHESES

Both the option to defer and growth options are potentially valuable whenever there is uncertainty about the underlying asset. In the context of new market entry, we are concerned with uncertainty as it relates to the potential cash flows associated with entering the target industry. Below, we explain how the option to grow is embedded in entry decisions, how it may conflict with the option to defer, and highlight the factors that influence the relative value of both options.

Option influences on the entry decision

No prior investment is necessary to initiate the option to defer since any firm can choose to 'not invest' and every firm incurs opportunity costs, D, when it makes an irreversible commitment (Trigeorgis, 2000). As mentioned earlier, this logic alters the decision rule (1), since the simple net present value of immediate entry must not only be greater than zero, but also greater than D. However, since D will always be greater than or equal to zero, the modified decision rule can be simply stated as

$$\text{Enter if}: NPV_S + CF \geq D \qquad (2)$$

where NPV_S represents the simple net present value of the opportunity with deferred entry, and CF represents the cash flows forgone if entry is delayed one period. This equation implies that a decision to 'not enter' may be attributed to either an elevated entry threshold due to a valuable option to defer (D), or to the possibility the firm is not a viable candidate for entry because the present value of all future cash flows ($NPV_S + CF$) is less than zero.

Unlike the option to defer, which all firms have, growth options must be initiated at some extra cost from the outset (Trigeorgis, 2000: 4). For example, entry gives a firm the right, but not the obligation, to expand operations in the future if industry conditions turn out favorable. Since managers have discretion over the exercise of growth options, their value, G, escalates with uncertainty due to the asymmetry in their pay-off distribution: if unexercised, their lowest value is zero, while their upper value is virtually unbounded. Thus, even if the simple NPV for entry is negative, it may be worthwhile to enter the industry if there is a significant potential for the industry conditions to far exceed the expected values that were used in calculating NPV_S.

It is important to recognize that the incremental cash flows, CF, may be more valuable if entry occurs earlier. This is the primary focus of the strategic management research on entry timing. However, the value of growth options, G, may also be enhanced by some multiplier, α, if moving early gives the firm an increased future ability to expand beyond initial expectations (Lieberman and Montgomery, 1988), where $G \geq 0$ and $\alpha \geq 1$. Therefore, in competitive markets, the entry decision should reflect the potential to enhance the value of the growth option with immediate entry compared to deferred entry:

$$\text{Enter if}: (NPV_S + CF + \alpha G) \geq (D + G) \quad (3)$$

There are several key insights that can be gleaned from this model with respect to the firm's probability of entering a new market:

(a) Entry is more likely when α is larger (i.e., when there are competitive advantages to moving early). If $\alpha = 1$, there is no competitive advantage to early entry, so the entry decision is reduced to decision rule (2).
(b) Entry is more likely when CF is large. CF may be a function of early mover advantages, or it may represent forgone cash flow under monopoly conditions.
(c) Entry is more likely as the option premium grows, where the option premium is defined as the net value of the options (i.e., $\alpha G - (D + G) = [\alpha - 1]G - D$).

These first two points are consistent with prior research. The last point, (c), is relatively unexplored and is the focus of this paper because it illuminates the relationship between uncertainty and entry (and investment more generally). The implication is that, in the absence of competition (i.e., $\alpha = 1$) uncertainty should negatively influence the entry decision. However, when there are competitive advantages to earlier entry (i.e., $\alpha > 1$) the total effect of uncertainty becomes ambiguous because both G and D increase with uncertainty. Predicting how uncertainty impacts entry requires assumptions about the relative magnitudes of each option and how their valuations grow with respect to uncertainty.

The conflict between growth and deferment options

Figure 1 illustrates how G and D may vary with uncertainty, and how their interplay may affect the incentive to enter immediately, holding NPV_S constant. In this figure, G, D, and α combine to create the option premium, which determines the total effect of uncertainty on entry. Panel A demonstrates that, with $\alpha = 2$, increases in uncertainty have a negative effect on entry when the slope of D exceeds the slope of G, and a positive effect on entry when the slope of G exceeds the slope of D.[4] Although the figure is stylized, the basic intuition remains: when there are competitive advantages to early entry, increasing uncertainty may not categorically dissuade entry through a monotonic decrease in the option premium (Amram and Kulatilaka, 1999; Kulatilaka and Perotti, 1998).

There are two reasons why the value of an option to grow may be more sensitive to uncertainty than the option to defer. Relative to delayed entry, immediate entry may result in a larger market share and higher profits because the firm benefits from first mover advantages, creating greater upside potential (Kulatilaka and Perotti, 1998). The second reason is that the value of the option to defer is bounded, while the value of the growth option is not. Option pricing theory informs us that the maximum value of a call option is bounded by the current price of the underlying stock. Intuitively, the maximum value of a call option is the maximum amount that an investor could lose by simply purchasing the underlying asset now. In the case of entry, the value of the option to defer, D, cannot exceed the total irreversible commitment required to enter the industry. Thus, although D is monotonically increasing in uncertainty, it is also asymptotic to the total sunk costs required to enter the industry, which occurs only at very high levels of uncertainty. In contrast, growth options have no such upper bound on their value. As the uncertainty about future industry conditions increases,

[4] More generally, increasing uncertainty, U, will positively influence entry when $(\alpha - 1)[dG/dU] > dD/dU$.

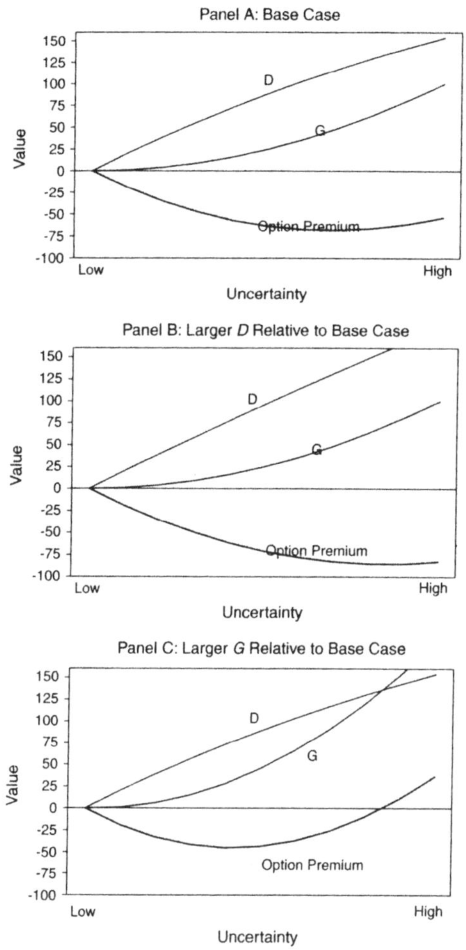

Figure 1. The value of the option to defer (D) and the option to grow, (G) with respect to uncertainty

so does the potential economic value of gaining a competitive advantage in that industry. Since the upside of the growth option is not bounded, the value of the growth option may be substantially greater than the value of the deferment option at extreme levels of uncertainty.

To the best of our knowledge, Campa (1993) is the only study to test for a nonmonotonic effect of uncertainty. His results imply that the effect of exchange rate uncertainty on entry into wholesale markets is negative and monotonic. It should be noted, however, that his industry context may lack growth opportunities, and exchange rate uncertainty represents only one form of potential uncertainty. Further work is needed to verify whether uncertainty has a nonmonotonic effect on entry. If growth options are present in a reasonable number of industries, we expect that the overall net impact of uncertainty on the probability of entry will be nonmonotonic.

Hypothesis 1a: The impact of uncertainty on entry is nonmonotonic.

Hypothesis 1b: Uncertainty will negatively influence entry when uncertainty is low and positively influence entry when uncertainty is high.

Factors influencing the relative value of growth and deferment options

The remaining hypotheses relate to factors that may systematically alter the relative value of growth options and deferment options, and thereby shift the critical point at which the relationship between uncertainty and entry changes from negative to positive. Relative to the base case (Panel A) provided in Figure 1, Panel B alters the slope of D to illustrate how more valuable deferment options influence the option premium. We see that the uncertainty has a negative effect on the option premium over a larger proportion of its range. If D were steep enough (i.e., valuable deferment options), uncertainty would have a monotonically decreasing effect on the option premium.

The value of the option to defer is influenced by the degree of irreversibility of the investment required for entry. If the investments required for entry are completely reversible, then the deferment option has no value and there is no point in delaying entry, since those investments can be fully recouped in the event that the firm decides to exit the industry at some point in the future. However, as those investments become more irreversible, there is a higher opportunity cost associated with entering. Thus, higher irreversibility should be associated with more valuable deferment options, making entry less likely.

Hypothesis 2: More valuable options to defer (i.e., greater irreversibility) will attenuate the U-shaped effect of uncertainty on entry.

The curvilinear effect described in Hypotheses 1a and 1b should also be affected by the size of growth options. Relative to the base case (Panel A) provided in Figure 1, Panel C alters the slope of G to illustrate how the option premium varies in uncertainty with larger growth options. It shows that, relative to the base case, the option premium becomes positively related to uncertainty at lower levels of uncertainty. This suggests that the effect of uncertainty on entry will switch from negative to positive at lower levels of uncertainty. Similarly, if G were flat enough (suggesting the growth option is small), uncertainty would have a monotonically decreasing effect on the option premium.

Hypothesis 3: More valuable growth options will accentuate the U-shaped effect of uncertainty on entry.

The same result illustrated in Panel C of Figure 1 can be obtained if we hold the value of G constant relative to the base case (assuming $G > 0$), but increase the importance of early mover advantages, such that $\alpha > 2$. Thus, when early entry conveys an increased ability to benefit from future growth opportunities (perhaps due to learning or experience curve advantages), there are added incentives to move early in highly uncertain environments. In some industries, growth opportunities may be secured through competitive preemption. If there is only space for n firms in an industry, then entry by firm $n + 1$ will not be economical because it will reduce profits below the threshold level (Schmalensee, 1978).[5] This may lead to an incentive for early entry to secure growth options before the competitive space is filled. Therefore, in contexts where early mover advantages are more pronounced, the value of growth options will be magnified and the effect of uncertainty will switch from negative to positive at lower levels of uncertainty.

Hypothesis 4: When early entry enhances the value of growth options, the U-shaped effect of uncertainty on entry is accentuated.

METHODS

Data sources

Most of the variables used in this study were derived from the Compustat Industrial and Business Segments databases. Compustat I contains detailed financial information at the level of the firm, and is the source for most of our firm-level variables. Compustat II, which provides financial data for each of the firm's lines of business, was used to develop most of the industry-level variables and to detect instances of entry into new industries. Since Compustat II was not available for years prior to 1980, our analysis encompassed the 17,897 firms and 144,947 firm/year observations that were listed in Compustat II between 1980 and 1999. The only variable not constructed from Compustat data was our measure of uncertainty, which was derived from GDP data taken from the Bureau of Economic Analysis at the U.S. Department of Commerce.

Dependent variable: entry

In Compustat II, firms may report up to 10 individual business segments, with each segment having up to two primary segment SIC codes and four product SIC codes.[6] Thus, for any given year, a firm may report up to 60 SIC codes. We defined *Entry* as activity by an existing firm in an industry in which the firm had not reported involvement in the previous 2 years. To detect entry, we looked for instances where a firm listed a primary segment SIC code that did not match any of the possible 60 SIC codes reported by the firm in each of the 2 previous years. If such activity in a new industry was found, and the firm was in Compustat II in the prior year, that new primary segment SIC code was coded as an entry. Using this approach, we identified 19,354 instances of entry between 1981 and 1999. In some instances, firms have multiple entries in a given year.[7]

The primary advantage of using the Compustat data to identify entry by established firms into new industries concerns generalizability. Unlike alternative databases, such as Trinet or the FTC dataset,

[5] Of course, firms may sometimes make dramatic errors in forecasting either the market size or the number of other entrants (Sahlman and Stevenson, 1985).

[6] Approximately 46 percent of observations report a single segment, and less than 0.5 percent report 10 segments.

[7] The data do not distinguish between direct entry and entry by acquisition. However, the motives for entry outlined above should apply to both modes of entry.

Compustat data cover a significant period of time and several phases of the business cycle.[8] Furthermore, the Compustat data capture a large portion of activity in all sectors of the American economy, whereas Trinet and FTC data focus on the manufacturing sector. Thus, the Compustat data are not only more generalizable across industries and across time, but also help ensure that our sample will contain considerable variation in uncertainty and in the value of deferment and growth options.

There are also some disadvantages to using Compustat data to identify entry into new industries. Probably the greatest concern is that industry segment SIC codes may be inconsistently assigned. In order to attenuate this concern, all SIC codes in Compustat were recoded from 4-digit level of analysis to the 3-digit level.[9] An additional concern is that firms have some discretion in how they group business activities in a single segment, leading to the potential agglomeration of diverse business activities into the same segment. However, Davis and Duhaime (1992) argue that the problem should not be substantive because of FASB reporting requirements. Again, by moving to a higher level of aggregation (3-digit SIC), we attenuate this possibility.

Independent variables

All of the independent variables were computed for each year and lagged 1 year to avoid potential problems of endogeneity with the instances of entry. Descriptive statistics and correlations are illustrated in Table 1.[10]

Uncertainty

There are at least two key challenges to measuring the uncertainty that is pertinent to the decision to enter a new industry. The first challenge involves choosing among the multiplicity of sources that account for the randomness in the external environment. We focus on the randomness of demand because demand impacts prices, which in turn largely determines profitability. Moreover, the strategic advantage of growth options should be particularly valuable in high states of demand, where profits per unit of output are higher (Kulatilaka and Perotti, 1998). Finally, there is strong precedent for operationalizing uncertainty using demand (Episcopos, 1995; Price, 1995), including Kulatilaka and Perotti (1998), the study from which this study builds most directly.[11]

Testing our hypotheses necessitates the development of time-varying measures of industry-specific uncertainty. All prior real options studies have measured macroeconomic uncertainty, except Kogut (1991), Folta (1998), and Folta and Miller (2002), who measured industry-specific uncertainty. However, all three of these studies calculated the variance of some output or indicator (e.g., stock price, GDP, sales) over time, an approach that has two critical shortcomings. First, it fails to account for the trends in the data, which will increase the measured variance although they may not constitute an element of uncertainty if they are predictable. Second, this approach does not allow for the possibility that the variance may be heteroskedastic (i.e., not constant over time), a characteristic that is typical of many economic time series (Campa, 1993).

To address both of these concerns, we measure uncertainty with the conditional variance generated from generalized autoregressive conditional heteroskedasticity (GARCH) models (Bollerslev, 1986). To approximate unique time-varying estimates of uncertainty for each industry requires time series data that correspond to (a) our sample period (1981–99), (b) our full vector of industries, and (c) an adequate number of data points prior to our sample period. Although we could not satisfy these requirements at the 3-digit level, we obtained data from the Bureau of Economic Analysis at the U.S. Department of Commerce on Gross Domestic Product (GDP) for 51 industries that largely overlap the full array of 3-digit SICs in Compustat.

[8] Our data cover 19 years, while FTC and Trinet cover only 4. Also, the FTC dataset covers an unusual period in economic history immediately subsequent to the first oil shock and Nixon's wage and price controls.

[9] For example, while a segment may be coded as manufacturing either disk drives (SIC 3684) or computer peripheral parts (SIC 3688), it will likely always be classified inside of the more general 3-digit SIC of 368.

[10] Despite high correlations for several variables, analysis revealed that multicollinearity should not be a problem since (a) all tolerances were well above the most conservative commonly used cut-off of 0.01 (Neter *et al.*, 1996); and (b) reduced models excluding variables that were highly correlated with our variables of interest produced substantively similar results.

[11] Other ways in which uncertainty has been represented in empirical studies include volatility in stock market index returns (Pindyck, 1986; Episcopos, 1995; Folta and Miller, 2002), exchange rates (Campa, 1993), inflation (Huizinga, 1993), and output prices (Huizinga, 1993).

Table 1. Descriptive statistics and Pearson correlations

Variable	Mean	S.D.	Min.	Max.	1	2	3	4	5	6	7	8	9	10	11	12	13	14	15	16	17	18	19
Relatedness	0.03	0.07	0.00	0.97																			
Diversification	0.14	0.23	0.00	0.90	0.21																		
F-Size	4.28	2.48	−0.69	13.09	0.13	0.40																	
F-Capital Intensity	7.93	335.04	0.00	85,352.00	0.00	−0.01	−0.01																
F-R&D	0.05	0.33	0.00	40.00	−0.01	−0.06	−0.12	0.00															
F-Sell	0.02	0.24	0.00	43.95	−0.01	−0.01	−0.02	0.00	0.00														
Beta	1.12	0.35	−2.58	2.49	0.03	0.02	−0.02	0.00	−0.01	0.00													
Concentration	0.19	0.17	0.01	1.00	−0.20	−0.03	0.00	0.00	0.01	0.00	−0.04												
I-Size	8.90	2.11	0.39	16.18	0.20	−0.02	0.04	0.00	0.01	−0.01	−0.08	−0.40											
I-ROS	0.06	2.86	−230.00	0.98	0.00	0.00	0.00	0.00	0.00	0.00	−0.02	−0.02	0.05										
I-Capital Intensity	16.41	1041.00	0.19	85,352.00	0.00	0.01	0.00	0.00	0.00	0.00	0.02	0.01	−0.03	−0.99									
I-R&D	0.01	0.04	0.00	0.79	0.13	−0.01	0.00	0.00	0.02	0.00	0.05	−0.10	0.18	−0.01	0.00								
I-Advertising	0.01	0.02	0.00	0.92	−0.05	0.00	−0.02	0.00	0.00	0.00	0.04	0.02	−0.16	−0.02	0.00	−0.04							
I-GDP	758.57	199.78	279.18	1629.00	−0.09	−0.05	0.03	0.00	0.01	0.00	−0.15	0.06	−0.03	−0.06	0.05	−0.02	−0.05						
Intangibles	0.02	0.04	0.00	0.44	−0.09	−0.04	0.03	0.00	0.01	0.00	−0.09	0.10	−0.07	0.01	−0.01	−0.06	0.01	0.06					
Inverse Leverage	0.83	0.12	0.22	1.00	0.09	0.00	−0.01	0.00	0.01	0.00	0.05	−0.03	−0.04	−0.02	0.00	0.20	0.04	−0.07	−0.11				
Market-to-Book	1.04	0.80	0.04	46.20	0.03	−0.02	0.01	0.00	0.01	0.00	−0.01	0.07	−0.09	−0.02	0.01	0.20	0.13	−0.01	0.10	0.12			
Growth	0.03	0.12	−0.45	0.88	−0.03	0.04	−0.02	0.00	−0.02	0.00	0.00	0.02	−0.09	0.03	−0.03	−0.05	0.03	−0.22	−0.07	0.00	−0.07		
Scale Advantages	0.06	0.15	0.00	1.00	−0.12	−0.01	0.00	0.00	0.00	0.00	−0.03	0.43	−0.49	−0.04	0.02	−0.05	0.20	0.08	0.12	−0.13	0.09	0.06	
Uncertainty	86.84	75.43	11.28	543.34	0.03	−0.01	0.00	0.00	0.00	0.00	−0.02	0.01	0.00	−0.07	0.07	−0.06	−0.06	0.41	−0.01	0.02	0.00	−0.09	0.10

With $n = 74,915$, correlations with an absolute value of 0.01 or more are significant at $p < 0.05$.

GDP by industry, which represents an industry's contribution to overall GDP, provides an excellent measure of total industry demand. See the Appendix for a description of these 51 industries.

After obtaining annual measures of GDP for 51 broad industries for the period 1947–2000, we fit GARCH models to each of the individual time series. This enabled us to approximate unique time-varying estimates of demand uncertainty for 51 broad industries. The GARCH model produces an estimate of the conditional variance, which captures the uncertainty that is not predictable about any trend that might exist for each period in the time series. Specifically, we employ the GARCH-in-mean, or GARCH-M model, which is parameterized by two values that specify the number of lags for the squared error terms and the number of past variances to be included in the computation of the current variance. In general, a one-period lag on both parameters (i.e., a GARCH-M[1,1] model) provides excellent fit for modeling a wide variety of asset returns (Solnik, 1996). Diagnostic checks of our data indicate that the parsimonious GARCH-M(1,1) model provides excellent fit.[12] We used the square root of the annual conditional variance generated from this model as our estimate of industry-specific *Uncertainty*.

Variables representing the irreversibility of the entry decision

While it is impossible to measure irreversibility directly, theory suggests several approximations. For example, Arrow (1968) suggests that intangible assets may have very little use outside their current application because they are likely to suffer from market failure, making trade on the open market difficult relative to physical assets (Long and Malitz, 1985). Thus, the irreversibility of an investment decision is negatively related to the tangibility of the assets in which a firm invests. Following precedent, including Friend and Lang (1988), Titman and Wessels (1988), Gompers (1995), and Rajan and Zingales (1995), we measure *Intangible* for each industry by the median value of the ratio of intangible assets to total assets for all firms in that industry in the previous year. We generate a second measure of irreversibility because, while intangible assets are expected to be highly irreversible, the salvage value of tangible assets may also vary across industries. Tangible or intangible assets that have high salvage value (i.e., are less irreversible) can support a high level of debt, while assets that have low salvage value will have to rely on equity financing (Williamson, 1988). Thus, the investments required to enter high-leverage industries should be more reversible than the investments required to enter low-leverage industries (Titman and Wessels, 1988; Shleifer and Vishny, 1992; Gompers, 1995). The variable *Inverse Leverage* is computed by taking, for all firms competing in an industry, the median value of one minus the ratio of long-term debt to total book value of assets.

Variables representing growth options or early mover advantages

One broad proxy for the presence of growth options is the market-to-book ratio. Myers (1977) argued that firms with a high market-to-book ratio should be associated with a higher proportion of unexercised growth opportunities. Thus, industries with higher average market-to-book ratios should generally have more valuable growth options. The variable Market-to-Book is the median industry market-to-book ratio, and is intended to capture expected future industry growth opportunities.

Industry growth rates should also increase the importance of growth options relative to deferment options. When an industry grows quickly, a firm must invest in order to exercise the option to grow with the market. Having a strategic advantage is particularly valuable in states of high demand when profits per unit of output are higher (Kulatilaka and Perotti, 1998). Thus, firms targeting higher growth industries should have more valuable growth options. The variable *Growth* is intended to capture expectations about the future growth rate of the industry. The GARCH model that was used to produce our measure of uncertainty also produces an estimate of the predicted value of industry GDP for each year and each of the industries where uncertainty was estimated. To capture expected industry growth, we computed *Growth* as the predicted value of industry GDP for the year that entry occurred, less the actual value of industry GDP in the previous year, all divided by the actual value of industry GDP in the previous year.

[12] We evaluated model fit based on evidence that the residuals were distributed as white noise, the statistical significance of the model's hyperparameters, and comparison of the model with alternative lag structures.

Preemption is a way in which firms can secure competitive advantages when the industry is only able to support a limited number of players. If firms delay entry too long, they may get locked out and find it infeasible to enter at a later date because they cannot reach minimum efficient scale. Access to scale economies can provide firms an enduring advantage (Lieberman and Montgomery, 1988). Thus, target industries defined by larger scale economies should be stronger candidates for early entry. An estimate of each industry's minimum efficient scale (MES) was constructed by the median value of assets for all business segments that compete in each industry.[13] Since it is the MES relative to total industry size that influences the advantages of early entry, we develop a measure of relative scale, *Scale Advantages*, by dividing the estimated MES by total industry assets.

Control variables

In addition to the variables specified above, proper specification of the theoretical model requires the inclusion of variables correlated with the expected NPV of the entry opportunity. Several industry-level variables should influence the attractiveness of entering a given industry. *I-Profit* is defined as the industry median operating profit to sales ratio. *Concentration* is the industry's Herfindahl concentration index. *I-Size* is the natural log of the total assets of all business segments competing in each industry. *Beta* controls for the systematic risk of each industry, and is calculated as the covariance between the returns on each industry's stock index and the market return over the previous 60 months. The intensity of investment in capital, *I-Capital Intensity*, is measured by the industry median ratio of assets to total sales. The intensity of R&D and advertising is approximated by *I-R&D* and *I-Advertising*, respectively, and is measured by the median industry ratios of R&D and advertising, respectively, to assets.[14]

Several firm-level factors could also impact the static NPV of the entry decision. The degree of relatedness between the industry entered and the portfolio of industries in which the firm already competes should greatly influence the expected value from entry. Thus, we include a measure of industry relatedness similar to that proposed by Teece *et al.* (1994), which measures the likelihood that a firm operating in industry j will also operate in industry m, corrected for the expected degree of relatedness under the null hypothesis that diversification is random. This measure was calculated at the 3-digit industry level and allowed to vary over time. *Relatedness* was the distance between the target industry and nearest industry that was already in the firm's portfolio.

Diversification controls for how diversified the firm was prior to the new entry by measuring the sum of squared shares of each of the firm's business segments. *F-Size* controls for the size of each firm by taking the natural log of total firm assets. The variables firm-level capital intensity (*F-Capital Intensity*), firm-level advertising intensity (*F-Advertising*), and firm-level R&D intensity (*F-R&D*) were computed in a similar fashion as their industry-level counterparts, but for each individual firm.

Analysis

Similar to Montgomery and Hariharan (1991), we use state-based sampling (Manski and McFadden, 1981) to model the dichotomous entry decision with a series of multivariate binomial logit models that compare instances of entry with a random sample of all the nonentries.[15] The sample of nonentries was created by randomly selecting (with replacement) 60,000 firm–year observations, and then randomly assigning to each observation an industry in which the firm had not competed in at least the previous 2 years. After eliminating observations with missing data, the final sample includes 14,915 instances of entry and 60,000 instances of nonentry.

Since our hypotheses predict that *Uncertainty* may have a nonmonotonic (i.e., a U-shaped) effect on the probability of entry, we include both *Uncertainty* and *Uncertainty*^{2} as predictor variables in our model. We expect that the sign of the coefficient on *Uncertainty* will be negative and the coefficient on *Uncertainty*^{2} will be positive, which

[13] Results are substantively identical if the minimum value of assets, the 10th percentile, or the 25th percentile is used instead of the median.

[14] *I-Advertising* was calculated based on firm-level data, while *I-Capital Intensity* and *I-R&D* were based on segment-level data.

[15] State-based sampling yields unbiased and consistent coefficients for all variables except the constant term, which can be corrected by subtracting from it the log(proportion of all entries in sample/proportion of all nonentries in the sample), where the numerator is 1, and the denominator is [60,000/(144,947 * 51–313,247)].

is consistent with a relationship that is generally negative but turns positive at high levels of uncertainty. As stated in Hypotheses 2–4, this relationship will be moderated by the presence of deferment or growth options. Decisions to enter into target industries with larger deferment options will lead to *Uncertainty* having a negative effect over a larger range of the sample. Alternatively, decisions to enter target industries with larger growth options and strong first mover advantages will lead to *Uncertainty* having a positive effect over a larger proportion of its range. Accordingly, we test these hypotheses by interacting the variables approximating deferment options, growth options, and first mover advantages with both *Uncertainty* and *Uncertainty*^{2}. These interactions represent the unique contribution of real options theory relative to other theories pertinent to entry (such as industrial organization) which emphasize only main effects.

RESULTS

The logit models presented in Table 2 test whether the independent variables significantly improve our ability to explain the choice between entry and nonentry. Model 1 presents the base model, Model 2 includes *Uncertainty*^{2}, and Models 3–7 include the interactions for the moderating effects on uncertainty. Firm- and industry-level variables are included to control for the NPV_S of the entry opportunity. Also included in each of these models, although not reported, are year fixed effects. The significance of individual coefficients is interpreted using two-tailed Wald chi-square tests. However, since our hypotheses are concerned with interaction effects, the relevant test is a likelihood ratio test comparing a model including interaction terms with a base model excluding interactions.

The effect of uncertainty

Consistent with much of the prior research on the relationship between uncertainty and investment, Model 1 indicates that *Uncertainty* has a significant negative effect ($p < 0.001$) on the probability of entry. Our point of departure from these previous studies, however, is that we hypothesize a curvilinear (i.e., a U-shaped) effect for uncertainty. A likelihood ratio test indicates that the addition of *Uncertainty*^{2} in Model 2 significantly improves model fit relative to Model 1 ($p < 0.001$). Moreover, the coefficient for *Uncertainty*^{2} is positive, while the coefficient for *Uncertainty* is negative, indicating that the effect of uncertainty is nonmonotonic—greater uncertainty decreases the likelihood of entry, but after some critical level it increases the likelihood of entry. While this finding is consistent with the expectations stated in Hypothesis 1, it is important to determine whether the effect of uncertainty is truly nonmonotonic within the sample's range of uncertainty. By taking the first derivative of the logistic regression equation with respect to *Uncertainty*, we determined that the critical point where the relationship switches from negative to positive (henceforth, the turning point) occurs at about the 94th percentile of *Uncertainty* ($\approx$199).[16] Thus, the effect of *Uncertainty* is nonmonotonic and U-shaped, as predicted in Hypotheses 1a and 1b.

Hypotheses 2–4 suggest that if target industries differ in the magnitude of deferment and growth options, the effect of uncertainty will shift to reflect these differences. Models 3–7 introduce interaction effects to test our hypotheses regarding the moderating role of deferment and growth options. As indicated in the table, likelihood ratio tests confirm that all of the interaction terms improve fit ($p < 0.001$) relative to Model 2. Since the interpretation of multiple moderating effects is difficult, we provide illustrations of each of these effects in Figure 2, which demonstrates the effect of *Uncertainty* over the moderating variable's range at three different levels: the 5th percentile of the moderating variable's range; the median; and the 95th percentile.

The moderating effect of irreversibility on uncertainty

Models 3 and 4 incrementally introduce interaction terms involving our two measures of irreversibility: *Intangible* and *Inverse Leverage*. Since the option to defer is expected to have greater relative importance when *Intangible* and *Inverse Leverage* are higher, we expect that *Uncertainty* will have a negative effect on entry over a larger portion of its range when either of these two variables is large. Panel A, which is derived from Model 3,

[16] Approximately the same turning point was found when conducting the analysis where all industries were defined at the 2-digit level.

Table 2. Maximum likelihood estimates for the determinants of entry[a]

Variables	1	2	3	4	5	6	7
Intercept	−2.176***	−1.835***	−1.802***	−3.181***	−1.885***	−1.743***	−1.769***
Relatedness	20.974***	20.820***	20.807***	20.787***	20.749***	20.810***	20.782***
Diversification	0.974***	0.982***	0.982***	0.984***	0.986***	0.986***	0.982***
F-Size	−0.050***	−0.049***	−0.049***	−0.048***	−0.048***	−0.049***	−0.049***
F-Capital Intensity	0.000	0.000	0.000	0.000	0.000	0.000	0.000
F-R&D	−0.261***	−0.258***	−0.260***	−0.266***	−0.260***	−0.265***	−0.262***
F-Sell	−0.365	−0.366	−0.366	−0.367	−0.371	−0.360	−0.368
Beta	0.146***	0.151***	0.153***	0.122***	0.169***	0.158***	0.150***
Concentration	−0.602***	−0.618***	−0.629***	−0.636***	−0.623***	−0.620***	−0.600***
I-Size	−0.012	−0.010	−0.010	−0.003	0.007	−0.011	−0.014
I-ROS	0.304**	0.302**	0.294**	0.221*	0.164	0.275**	0.376***
I-Capital Intensity	0.001**	0.001**	0.001**	0.001*	0.000	0.001*	0.001***
I-R&D	−0.102	−0.020	−0.065	−0.239	−0.546	−0.052	−0.095
I-Advertising	−9.782***	−9.980***	−10.029***	−10.072***	−10.267***	−9.674***	−10.241***
I-GDP	−0.001***	−0.002***	−0.002***	−0.002***	−0.002***	−0.002***	−0.002***
Intangibles	−3.332***	−3.249***	−2.858**	−2.904***	−3.331***	−3.234***	−3.362***
Inverse Leverage	1.474***	1.455***	1.448***	3.104***	1.433***	1.447***	1.464***
Market-to-Book	0.127***	0.119***	0.121***	0.114***	−0.031	0.121***	0.142***
Growth	−1.022***	−0.918***	−0.940***	−0.909***	−0.837***	−3.468***	−0.899***
Scale Advantages	−3.150***	−2.972***	−3.078***	−2.957***	−2.775***	−2.986***	−4.529***
Uncertainty	−7.20E-04***	−4.16E-03***	−4.38E-03***	1.66E-02***	−7.60E-03***	−5.16E-03***	−4.67E-03***
$Uncertainty^2$		1.00E-05***	1.00E-05***	−2.00E-05*	1.50E-05***	1.30E-05***	1.00E-05***
Uncertainty × Intangibles			−3.05E-02				
$Uncertainty^2$ × Intangibles			2.30E-04***				
Uncertainty × Inverse Leverage				−2.45E-02***			
$Uncertainty^2$ × Inverse Leverage				3.90E-05***			
Uncertainty × Market-to-Book					3.53E-03***		
$Uncertainty^2$ × Market-to-Book					−5.58E-06***		
Uncertainty × Growth						2.35E-02***	
$Uncertainty^2$ × Growth						−3.00E-05***	
Uncertainty × Scale Advantages							9.69E-03***
$Uncertainty^2$ × Scale Advantages							1.60E-05***
Likelihood Ratio	22,702.611	22,768.665	22,794.526	22,818.496	22,844.191	22,824.101	22,860.250
Likelihood Ratio Test (vs. Model 2)			51.723***	99.662***	151.052***	110.873***	183.171***

* $p < 0.05$; ** $p < 0.01$; *** $p < 0.001$
[a] Year fixed effects included in models, but not reported.

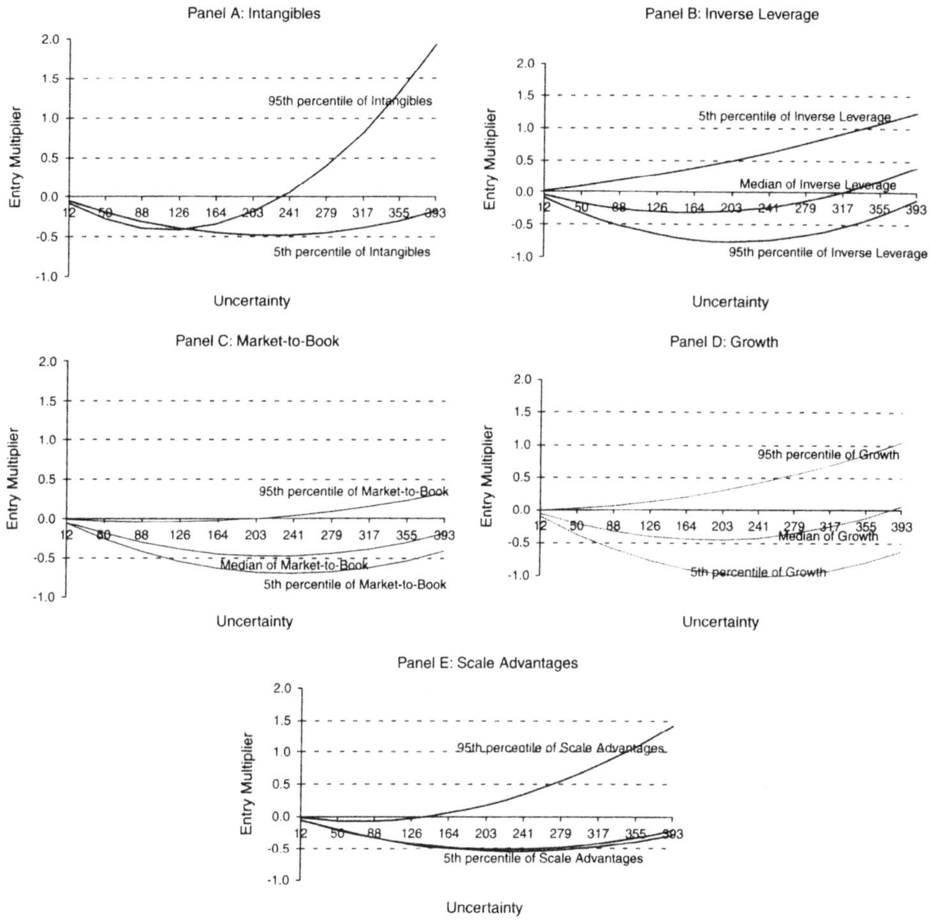

Figure 2. The effect of uncertainty at different percentiles of the moderating variables. Each figure was produced by taking the derivative of the relevant logistic regression equation (with respect to *Uncertainty*) at each of three different levels (i.e., 5th, median, and 95th percentiles) of the relevant moderating variable. The vertical axes in these figures represent the computed multiplier of entry, while the horizontal axes depict *Uncertainty* plotted from the 1st to the 99th percentile. In Panel A both the 5th percentile and median of *Intangibles* equal zero

reveals that the moderating effect of *Intangible* on *Uncertainty* is not consistent with Hypothesis 2. The figure suggests that when firms target industries requiring more investment in intangible assets, *Uncertainty* has a negative effect on entry over a smaller portion of its range. Panel B depicts the predictions derived from Model 4, which introduces the interaction terms involving *Inverse Leverage*. In contrast to Panel A, Panel B is consistent with Hypothesis 2. Interestingly, *Uncertainty* has a monotonically positive effect on entry into industries that require investments that are highly reversible (i.e., *Inverse Leverage* at 5th percentile). Conversely, in industries where investments are more irreversible, (i.e., *Inverse Leverage* at 95th percentile), *Uncertainty* has a negative effect on entry over 94 percent of its range. In sum, we found mixed results for the expectation that uncertainty will have a negative effect over a larger portion of its range when

Table 3. Economic significance of uncertainty on entry

Quantile of *Uncertainty*	Value of *Uncertainty*	Probability of entry[a]	% of 3-digit Industries with *Uncertainty* exceeding quantile	Market-to-Book		Growth		Scale Advantages	
				5th quantile	95th quantile	5th quantile	95th quantile	5th quantile	95th quantile
10th	23.7	1.22%	97.4%	1.15%	1.28%	1.95%	0.68%	1.52%	0.48%
20th	34.0	1.18%	97.0%	1.08%	1.27%	1.81%	0.69%	1.46%	0.47%
30th	42.7	1.15%	90.6%	1.02%	1.27%	1.71%	0.69%	1.41%	0.47%
40th	55.4	1.10%	88.0%	0.95%	1.26%	1.57%	0.70%	1.34%	0.46%
50th	67.2	1.06%	87.1%	0.89%	1.26%	1.46%	0.71%	1.29%	0.46%
60th	82.1	1.02%	57.7%	0.82%	1.25%	1.34%	0.72%	1.23%	0.46%
70th	99.4	0.98%	44.0%	0.75%	1.25%	1.23%	0.74%	1.17%	0.47%
80th	114.5	0.95%	29.9%	0.71%	1.25%	1.15%	0.75%	1.13%	0.48%
90th	163.3	0.89%	15.4%	0.60%	1.27%	0.97%	0.83%	1.03%	0.52%
95th	268.8	0.91%	6.0%	0.54%	1.41%	0.88%	1.12%	1.00%	0.81%
99th	392.6	1.22%	2.1%	0.71%	1.80%	1.26%	1.89%	1.26%	2.01%

[a] Calculated at the mean values of all variables except *Uncertainty*.

firms target industries requiring investments that are more irreversible.

The moderating effects of growth options and early mover advantages

Models 5–7 incrementally introduce interaction terms involving the three measures of the total value of growth options: *Market-to-Book, Growth*, and *Scale Advantages*. The interactions involving *Market-to-Book* are entered in Model 5, and Panel C depicts the moderating effect of this variable on uncertainty. This figure illustrates that when growth options are valuable (i.e., *Market-to-Book* is at the 95th percentile), the effect of *Uncertainty* on entry turns positive at the 67th percentile of *Uncertainty*. However, when growth options are less valuable (i.e., *Market-to-Book* at 5th percentile), the effect of uncertainty on entry does not turn positive until about the 94th percentile of *Uncertainty*. Thus, Model 5 indicates that when firms target industries with larger market-to-book ratios, entry is encouraged over a wider range of *Uncertainty*.

Model 6 introduces the interactions involving *Growth*. Panel D in Figure 2 demonstrates that for high-growth target industries (i.e., *Growth* at 95th percentile, expected growth of 23% or higher), the effect of *Uncertainty* on entry is monotonically positive. In contrast, sharply declining industries (i.e., *Growth* at 5th percentile, expected decline of at least 13%), the effect of *Uncertainty* does not become positive until its 96th percentile. Overall, the results for the variables *Market-to-Book* and *Growth* indicate strong support for Hypothesis 3.

Finally, Model 7 tests whether early mover advantages enhance the value of growth options. Consistent with expectations, Panel E illustrates that in contexts where there are potentially strong early mover advantages (i.e., *Scale Advantages* at 95th percentile), the effect of *Uncertainty* on entry turns positive at about the 54th percentile of *Uncertainty*. In contrast, the effect of *Uncertainty* does not turn positive until around its 94th percentile for most industries (as illustrated by both the median and 5th percentile of *Scale Advantages*).

Economic significance of uncertainty

To interpret the economic significance of uncertainty and the variables approximating growth potential, it is necessary to convert the estimated log odds ratios to actual probabilities of entry.[17] Table 3 illustrates how changes in *Uncertainty* affect the probability of entry. Holding all other variables constant at their mean values, the probability of entry declines from 1.22 percent to 0.89 percent as *Uncertainty* increases from its 10th percentile (23.7) to its 90th percentile (163.3). This represents a 27 percent decrease in the probability of entry due to changes in *Uncertainty*. The

[17] We used Model 2 in Table 2 to estimate the probabilities. The estimated probabilities are calculated after having adjusted the constant term for state-based sampling.

table also illustrates the U-shaped influence of *Uncertainty* on the probability of entry, because as *Uncertainty* increases beyond its 90th percentile to the 99th percentile (392.6), the probability of entry increases from 0.89 back up to 1.22 percent. These findings clearly suggest that changes in *Uncertainty* have an economically significant impact on entry.

Table 3 also reveals that the moderating influence of growth options is strong and economically significant. For example, when *Uncertainty* moves from its 10th percentile to the 90th, the probability of entry declines by 48 percent, 50 percent, and 32 percent, respectively, when targeting industries where growth options are small (*Market-to-Book, Growth*, and *Scale Advantages* at their 5th percentile, respectively). However, when focusing on target industries where growth options are large (*Market-to-Book, Growth*, and *Scale Advantages* each at their 95th percentile) the probability of entry actually increases by 11 percent, 35 percent, and 56 percent, respectively, as *Uncertainty* moves from its 90th percentile to its 95th percentile. These findings suggest that properly accounting for growth options can lead to substantial improvement in our ability to predict entry.

Control variables

Since our sample constitutes a wide range of industries, results for control variables may not be directly comparable to studies that focused only on manufacturing industries, such as Montgomery and Hariharan (1991) and Silverman (1999).[18] Nevertheless, the control variables generally have the expected effect. Among the variables unique to this study, *Relatedness* had a positive effect, which suggests that firms are more likely to enter industries that are highly related to an industry in which they already compete. This finding is consistent with predictions from the resource-based view.

DISCUSSION

Assumptions about a negative relationship between uncertainty and investment have dominated empirical attempts to validate the explanatory power of real options theory in the last 10 years. Such a focus assumes that firm decision-making is dominated by the option to defer investment in the face of uncertainty and ignores the possibility that growth opportunities are also enhanced with greater uncertainty. We empirically demonstrate that the relationship of uncertainty to entry is not monotonic, as previous research had theorized. We find that throughout 93 percent of the range of uncertainty, uncertainty has a negative effect on entry, implying that the option to defer dominates growth options in most contexts. However, the value of growth options outweighs the value of deferment options at high levels of uncertainty (i.e., beyond the 93rd percentile) and induces a positive effect of uncertainty on entry. Furthermore, the turning points may be much lower when firms target industries offering larger growth options. Our findings are the first to find support for the nonmonotonic effect of uncertainty that has only recently emerged in theoretical treatments of real options theory, and amplify the importance of considering both the option to defer and the option to grow when contemplating entry.

Some discussion of our findings pertaining to intangible assets is warranted. While we postulated that an industry's ratio of intangible assets to total assets approximated irreversibility, we found that more intangible assets accentuate the U-shaped relationship of uncertainty on entry. One possible explanation is that greater levels of investment in intangible assets may actually be associated with greater growth options, in addition to greater irreversibility. Myers (1977) actually postulated that the proportion of intangible assets to tangible assets represents the proportion of growth options to assets in-place. It may also be that early entry may be more important when competing on R&D or brand equity. Thus, while there is some empirical and theoretical precedent for using intangibles to approximate irreversibility, our results suggest that its effect on growth options dominates concerns about irreversibility.

Our findings should be particularly interesting to scholars of strategic management, who have long recognized that certain projects should be undertaken because they have 'strategic' value. Although managers seldom use real options theory in capital budgeting, our empirical evidence suggests that they recognize subtle factors that alter the value of growth opportunities. This is encouraging for scholars who believe that real

[18] An unreported model concentrating only on the manufacturing industry produced results for our control variables that correspond closely to these previous findings.

options theory is an economic explanation for certain aspects of managerial intuition. Our evidence suggests that real options theory enriches the study of entry from the perspective of industrial organization. While the structural conditions of industries clearly influence expected performance in those industries, our results also suggest that managers are constantly factoring uncertainty into their entry decisions. Moreover, many of the factors that have been traditionally related to industry structure (and hence expected performance) also seem to have important threshold effects, as evidenced by their significant interaction with uncertainty. It is possible that industrial organization research has overemphasized the performance effects of structural attributes. Our real options approach suggests that much of the influence that such attributes have on entry may be due to their threshold effects. Researchers should consider the threshold effects induced by structural attributes, to the extent that they correlate with irreversibility or growth opportunities.

We have also demonstrated that real options theory enriches the study of entry from the perspective of resource-based theory. Uncertainty has a potent effect on entry even after controlling for firm resource profiles, including the relatedness to the target industry. Previous applications of the resource-based view have focused on performance effects and ignored the possibility that firm-specific thresholds for investment may either induce or restrict entry. There remain many fruitful opportunities for further work in this realm. For example, while we have controlled for firm-level factors, our work emphasizes how industry-level factors influence option value. Resource-based and knowledge-based views will help in deducing how firm-level differences influence option value. For example, firms with higher absorptive capacity may have lower investment thresholds because they are better suited to capitalize on growth options. We can also conjecture that differences in firm reputation or managerial capability may also affect investment thresholds due to their impact on growth opportunities. Thus, firm-level factors may help to explain why some firms enter and others do not in the face of similar industry characteristics, an issue that has not been satisfactorily resolved in either the real options literature (Carruth *et al.*, 2000) or in work on early mover advantages (Lieberman and Montgomery, 1998).

We believe that this research provides a generalizable statement on the importance of growth options for entry decisions by diversified firms. The need for such an approach should not be underemphasized. Nevertheless, we may be underemphasizing the importance of growth options because we lack data that may allow for a more precise evaluation of the various factors that influence the value of growth options. For example, if we were better able to discern the precise submarket that firms were entering, we would have a better sense for the structural forces that might enable a first mover advantage for growth opportunities. Another factor that may inhibit our ability to observe growth option effects is that we are focusing on existing industries with incumbents. It is likely that the effect of uncertainty turns positive at lower levels of uncertainty in new industries, where first mover advantages are more likely to be pertinent. Future research should examine our hypotheses in the context of new industries.

Despite the noted shortcomings and the broad research agenda ahead, we feel that this research provides an important first step in empirically examining the dueling options present in entry decisions. In general, these results suggest that, consciously or not, managers consider the value of real options and generally recognize the factors that influence their value. We have demonstrated that the effect of uncertainty is not monotonic, on average, and that turning points are influenced by factors that should theoretically influence options to grow and options to defer. Since growth options may lie at the heart of strategic investment decisions, this first step is significant.

ACKNOWLEDGEMENTS

We are grateful to Tim Cason, Marvin Lieberman, Joe Mahoney, Bill Robinson, Derek Ruth, Jennifer Ryan, Brian Silverman, and seminar participants at the University of Toronto and Purdue University for helpful comments on earlier drafts of this manuscript, and to Doug Johnson for his contributions on an earlier paper related to this project.

REFERENCES

Amram M, Kulatilaka N. 1999. *Real Options: Managing Strategic Investment in an Uncertain World*. Harvard Business School Press: Boston, MA.

Arrow KJ. 1968. Optimal capital policy with irreversible investment. In *Value, Capital and Growth: Essays in Honor of Sir John Hicks*, Wolfe JN (ed). Edinburgh University Press: Edinburgh.

Bollerslev T. 1986. Generalized autoregressive conditional heteroskedasticity. *Journal of Econometrics* **31**: 307–327.

Campa J, Goldberg LS. 1995. Investment in manufacturing, exchange-rates and external exposure. *Journal of International Economics* **38**(3–4): 297–320.

Campa JM. 1993. Entry by foreign firms in the United States under exchange-rate uncertainty. *Review of Economics and Statistics* **75**(4): 614–622.

Carruth A, Dickerson A, Henley A. 2000. What do we knout about investment under uncertainty? *Journal of Economic Surveys* **14**(2): 119–153.

Chang SJ. 1995. International expansion strategy of Japanese firms: capability building through sequential entry. *Academy of Management Journal* **38**(2): 383–407.

Copeland TE, Keenan PT. 1998. How much is flexibility worth? *McKinsey Quarterly* **2**: 38–49.

Davis R, Duhaime IM. 1992. Diversification, vertical integration, and industry analysis: new perspectives and measurement. *Strategic Management Journal* **13**(7): 511–524.

Dixit AK, Pindyck RS. 1994. *Investment under Uncertainty*. Princeton University Press: Princeton, NJ.

Donaldson G, Lorsch JW. 1983. *Decision Making at the Top: The Shaping of Strategic Direction*. Basic Books: New York.

Driver C, Yip P, Dakhil N. 1996. Large company capital formation and effects of market share turbulence: micro-data evidence from the PIMS database. *Applied Economics* **28**(6): 641–651.

Episcopos A. 1995. Evidence on the relationship between uncertainty and irreversible investment. *Quarterly Review of Economics and Finance* **35**(1): 41–52.

Folta TB. 1998. Governance and uncertainty: the tradeoff between administrative control and commitment. *Strategic Management Journal* **19**(11): 1007–1028.

Folta TB, Miller KD. 2002. Real options in equity partnerships. *Strategic Management Journal* **23**(1): 77–88.

Friend I, Lang LHP. 1988. An empirical-test of the impact of managerial self-interest on corporate capital structure. *Journal of Finance* **43**(2): 271–281.

Gompers PA. 1995. Optimal investment, monitoring, and the staging of venture capital. *Journal of Finance* **50**(5): 1461–1489.

Guiso L, Parigi G. 1999. Investment and demand uncertainty. *Quarterly Journal of Economics* **114**(1): 185–227.

Hayes RH, Garvin DA. 1982. Managing as if tomorrow mattered. *Harvard Business Review* **60**(3): 70–79.

Huizinga J. 1993. Inflation uncertainty, relative price uncertainty, and investment in United States manufacturing. *Journal of Money Credit and Banking* **25**(3): 521–549.

Hurry D, Miller AT, Bowman EH. 1992. Calls on high-technology: Japanese exploration of venture capital investments in the United States. *Strategic Management Journal* **13**(2): 85–101.

Kester WC. 1984. Today's options for tomorrow's growth. *Harvard Business Review* **62**(2): 153–160.

Kim DJ, Kogut B. 1996. Technological platforms and diversification. *Organization Science* **7**(3): 283–301.

Kogut B. 1991. Joint ventures and the option to expand and acquire. *Management Science* **37**(1): 19–33.

Kogut B, Kulatilaka N. 2001. Capabilities as real options. *Organization Science* **12**(6): 744–758.

Kulatilaka N, Perotti EC. 1998. Strategic growth options. *Management Science* **44**(8): 1021–1031.

Leahy JV, Whited TM. 1996. The effect of uncertainty on investment: some stylized facts. *Journal of Money Credit and Banking* **28**(1): 64–83.

Lieberman MB, Montgomery DB. 1988. First-mover advantages. *Strategic Management Journal*, Summer Special Issue **9**: 41–58.

Lieberman MB, Montgomery DB. 1998. First-mover (dis)advantages: retrospective and link with the resource-based view. *Strategic Management Journal* **19**(12): 1111–1125.

Long M, Malitz I. 1985. The investment-financing nexus: some empirical evidence. *Midland Corporate Finance Journal* **3**(3): 53–59.

Manski CF, McFadden D. 1981. Alternative estimators and sample designs for discrete choice analysis. In *Structural Analysis of Discrete Data with Econometric Applications*, Manski CF, McFadden D (eds). MIT Press: Cambridge, MA; 2–50.

McDonald R, Siegel D. 1986. The value of waiting to invest. *Quarterly Journal of Economics* **101**(4): 707–727.

McGrath RG. 1997. A real options logic for initiating technology positioning investments. *Academy of Management Review* **22**(4): 974–996.

Montgomery CA, Hariharan S. 1991. Diversified expansion by large established firms. *Journal of Economic Behavior & Organization* **15**(1): 71–89.

Myers SC. 1977. Determinants of corporate borrowing. *Journal of Financial Economics* **5**(2): 147–175.

Neter J, Kutner MH, Nachtsheim CJ, Wasserman W. 1996. *Applied Linear Regression Models* (3rd edn). Irwin: Chicago, IL.

Pindyck RS. 1986. Capital risk and models of investment behavior. Working paper, Sloan School of Management.

Price S. 1995. Aggregate uncertainty, capacity utilization and manufacturing investment. *Applied Economics* **27**(2): 147–154.

Rajan RG, Zingales L. 1995. What do we know about capital structure: some evidence from international data. *Journal of Finance* **50**(5): 1421–1460.

Sahlman WA, Stevenson HH. 1985. Capital market myopia. *Journal of Business Venturing* **1**: 7–30.

Schmalensee R. 1978. Entry deterrence in ready-to-eat breakfast cereal industry. *Bell Journal of Economics* **9**(2): 305–327.

Shleifer A, Vishny RW. 1992. Liquidation values and debt capacity: a market equilibrium approach. *Journal of Finance* **47**(4): 1343–1366.

Silverman BS. 1999. Technological resources and the direction of corporate diversification: toward an integration of the resource-based view and transaction cost economics. *Management Science* **45**(8): 1109–1124.
Solnik B. 1996. *International Investments*, Vol. 3. Addison-Wesley: Reading, MA; Ch. 3, Appendix.
Teece DJ, Rumelt RP, Dosi G, Winter SG. 1994. Understanding corporate coherence: theory and evidence. *Journal of Economic Behavior & Organization* **23**(1): 1–30.
Titman S, Wessels R. 1988. The determinants of capital structure choice. *Journal of Finance* **43**(1): 1–19.
Trigeorgis L. 2000. *Real Options: Managerial Flexibility and Strategy in Resource Allocation*. MIT Press: Cambridge, MA.
Williamson OE. 1988. Corporate finance and corporate governance. *Journal of Finance* **43**: 567–591.

APPENDIX: INDUSTRY CODING SCHEME

Industry	Name	SICs	Industry	Name	SICs
1	Agriculture, Forestry & Fishing	1–10	27	Local Passenger Transit	41
2	Metal and Coal Mining	10–12	28	Trucking/Warehousing	42
3	Oil & Gas Extraction	13	29	Water Transport	44
4	Other Mining	14	30	Air Transport	45
5	Construction	15–17	31	Pipelines	46
6	Lumber & Wood	24	32	Transportation Services	47
7	Furniture & Fixtures	25	33	Communications	48
8	Stone, Clay & Glass Prod.	32	34	Electric, Gas, & Sanitary Serv.	49
9	Primary Metal Indus.	33	35	Wholesale	50–51
10	Fabricated Metal Prod.	34	36	Retail	52–59
11	Machinery	35	37	Banking & Credit Agencies	60, 61
12	Electric, Electronic & Instruments	36, 38	38	Security & Commodity Brokers	62
13	Autos & Equipment	371	39	Insurance Carriers	63
14	Other Transp. Equip.	372–379	40	Insurance Agents & Brokers	64
15	Misc. Manufacturing Indus.	39	41	Real Estate & Holding	65–67
16	Food Products	20	42	Hotels/Lodging	70
17	Tobacco	21	43	Personal Services	72
18	Textile	22	44	Business & Misc Professional Serv.	73, 84, 87
19	Apparel	23	45	Auto Repair, Services & Parking	75
20	Paper	26	46	Misc. Repair Services	76
21	Printing	27	47	Motion Pictures	78
22	Chemicals	28	48	Amusement/Recreation	79
23	Petroleum & Coal Products	29	49	Health Services	80
24	Rubber & Plastic Products	30	50	Educational Services	82
25	Leather Products	31	51	Social Services	83
26	Railroad Transportation	40			

[23]

DEFERRAL AND GROWTH OPTIONS UNDER SEQUENTIAL INNOVATION

Michael J. Leiblein and Arvids A. Ziedonis

ABSTRACT

This paper examines the application of real option theory to sequential investment decision-making. In an effort to contribute to the development of criteria that discriminate between investments that confer growth options from those that confer deferral options, we introduce a conceptual model that explains technological adoption as a sequence of embedded options. Upon the introduction of each successive technological generation, a firm may either defer investment and wait for the arrival of a future generation or invest immediately to obtain experience that provides a claim on adoption of subsequent generations. We propose that deferral and growth option value is dependent on the magnitude, frequency, and uncertainty of inter-generational change, and the nature of rivalry.

INTRODUCTION

Real option theory has shown promise as a useful framework for evaluating investment decisions under uncertainty (Dixit & Pindyck, 1994; Trigeorgis,

Real Options Theory
Advances in Strategic Management, Volume 24, 225–245

ISSN: 0742-3322/doi:10.1016/S0742-3322(07)24008-2

1998; Amram & Kulatilaka, 1999; Copeland & Antikarov, 2001). In contrast to traditional discounted cash flow techniques, the real option approach recognizes that managers are capable of reacting flexibly to changing market conditions and that risk may vary over different investment stages. When investments are at least partially irreversible, traditional decision rules based on discounted cash flows may fail to account for the benefits of sequential decision-making in response to the arrival of new information. Real option logic provides a method for evaluating the benefit associated with the opportunity to react flexibly. Numerous types of real options have been identified in the literature, including the option to defer, the option to stage investment, the option to alter operating scale, the option to abandon operations, the option to switch inputs or outputs, and the option to grow (Trigeorgis, 1998, pp. 2–3).

While the early real options literature emphasizes the value associated with deferring investment in the face of uncertainty (e.g., McDonald & Siegel, 1986), more recent work focuses on growth options that provide the ability to expand in the future (e.g., Pindyck, 1988; Kulatilaka & Perotti, 1998). The logic underpinning both deferral and growth options infers that traditional (discounted cash flow) investment theory fails to account for the value of flexibility under uncertainty, yet the two types of options recognize different sources for this value. Deferral option logic highlights the benefits of waiting for new information to arrive (McDonald & Siegel, 1986). In contrast, growth option logic stresses the value associated with early investments that provide platforms for future, follow-on investment. This ambiguity makes it possible to apply the theory to develop fully contradictory recommendations. For example, when considering entry into a new market, option theory can be used to support a decision to delay investment until uncertainty is resolved (McDonald & Siegel, 1986) or to enter early and create future growth options (Kogut, 1991; McGrath, 1997; Pindyck, 1988; Kulatilaka & Perotti, 1998).

It is our contention that the difficulty in distinguishing, ex ante, between different sources of option value has limited the contribution of real option theory to strategic management. A fundamental issue in the field of strategic management is the relationship between firm conduct and competitive advantage (Rumelt, Schendel, & Teece, 1994). Existing theory suggests that competitive advantage is most likely to be realized in market sectors where competition is constrained (Porter, 1980) or in "thin" factor markets where information asymmetry or other ex ante barriers to competition provide opportunities to access or develop resources below their market value (Barney, 1986; Peteraf, 1993). Existing work also describes how both early

(Lieberman & Montgomery, 1988) and late (Lieberman & Montgomery, 1998) investment may affect the ability to pursue these sources of competitive advantage. If real option theory can be sufficiently developed to provide insight in securing competitive advantage based on the timing (early or late) or nature (reversible or irreversible) of investment, it can contribute significantly to ongoing conversations within the strategy domain – possibly by suggesting linkages with existing theories of competitive advantage.

The objective of this chapter is to propose a set of criteria to distinguish between situations that are more or less likely to impart growth or deferral option value. We do so both conceptually and through an examination of technology adoption decisions in environments characterized by continual technical progress. In such situations, managers are often confronted with serial introductions of new technologies that either extend or replace previous versions. By adopting today, the firm faces the risk that a better technology may become available tomorrow and possibly render the original technology obsolete. While waiting would avoid this risk, it does so at the opportunity cost of utilizing the technology in the interim. The adoption decision therefore involves consideration not only of the technology available immediately, but also expectations regarding the timing and attributes of a future innovation. For example, Sayers (1950) documented the influence of such expectations on the adoption of competing engine technologies in ship building in the early 20th century (as quoted by Rosenberg, 1976, p. 530):

> There were times, between the wars, when marine engineering was changing in such a rapid yet uncertain way that firms in the highly competitive shipping industry delayed investment in the replacement of old high-cost engines by the new low-cost engines. In the middle 'twenties progress was rapid in all three propulsion methods – the reciprocating steam-engine, the geared steam-turbine and the diesel motor. Minor variations are said to have brought the number of possible combination types up to nearly a hundred. For some classes of ships there was momentarily very little to choose between several of these combinations, and shipowners were inclined to postpone placing orders until a little more experience and perhaps further invention had shown which types would be holding the field over the next ten years. Put in economic terms, the shipowners' position was that, though total costs of new engines might already be less than running costs of old engines, the profit on engines of 1923 build might be wiped out by the appearance in 1924 of even lower-cost engines, the purchase of which would allow a competitor (who had postponed the decision) to cut freights further. Also there was uncertainty as to which of the two types of 1923 engine would prove to work at lower cost. If shipowner A installed engine X, and shipowner B installed engine Y, whose costs in 1923 appeared to equal those of X, a year's experience might show that in fact Y costs were much lower than X costs, in which event shipowner A would have done better to wait until 1924 before installing new engines.

The following passage by Adams and Dirlam (1966) also cited by Rosenberg in his 1976 article (p. 531) illustrates a similar dilemma faced by managers of U.S. Steel Corporation in deciding whether to adopt a new steel-making process:

> U.S. Steel conceded that "some form of oxygen steel-making will undoubtedly become an important feature in steelmaking in this country," but it declined to say when or whether to commit itself to introducing this innovation. Indeed, three years later, *Fortune* still pictured the Corporation as confronted by "painfully difficult choices between competing alternatives – for example, whether to spend large sums for cost reduction *now* [1960], in effect committing the company to *present* technology, or to stall for time in order to capitalize on a new and perhaps far superior technology that may be available in a few years."

The more recent Nucor case (Ghemawat, 1997; Christensen, 1997) illustrates how prior investments and uncertainty regarding the value of future products or technologies affect the attractiveness of subsequent opportunities. During the late 1980s, Nucor, a small steel manufacturer, decided to invest in a new and uncertain compact strip production process at a new "mini-mill" steel plant to be constructed in Crawfordsville, Indiana. Reports indicate that there were a number of severe start-up problems at this new plant and that initial financial returns on this investment were modest. Subsequent improvements to the process and the mini-mill concept created opportunities for dramatically improved performance, however. Firms such as Nucor that invested early in mini-mill technology gained experience that enabled them to quickly expand their stake in the new technology (for example, Nucor opened its second mini-mill plant in Hickman Arkansas, at a much lower cost than that required to open the Crawfordsville plant). In contrast, companies that deferred investment in the new process quickly fell behind and struggled to apply the rapidly improving technology in their own operations.

Similar investment decisions are increasingly common in contemporary high-technology markets such as semiconductors, software, and pharmaceuticals. When there are expectations of continued technological advance, managers often must decide between "keeping their options open" by deferring investment and "securing a stake" by committing at least some funds to explore a promising new technical trajectory. Because real option theory addresses both sides of this dilemma, technology adoption under sequential innovation provides the opportunity to identify situations where growth or deferral options dominate.

In the next section we briefly review the technology adoption literature and comment on the implications of real option logic for decision-making under sequential innovation. In section theory development, we relate growth and deferral option logic to technology adoption. In section empirical context, we outline an empirical strategy for identifying deferral and growth options in technology migration choices. A concluding section discusses the implications of our propositions and identifies avenues for future research. This discussion also serves to illustrate the broader applicability of our approach in research on strategic choice.

GROWTH AND DEFERRAL OPTIONS

A real option is analogous to a financial "call" option (Black & Scholes, 1973), which gives an investor the right, but not the obligation, to purchase a valuable asset (such as a share of stock) on a future (expiration) date at a certain (exercise) price. This "option" feature creates an asymmetry in the distribution of returns – by purchasing an option, the owner gains access to greater upside potential than downside exposure. At a future date, management may terminate the investment or take steps to fully exploit future positive outcomes (i.e., exercise the option). As in financial option theory, the value of the flexibility provided by the ability to selectively act on an initial investment depends critically on the value of the underlying asset, the cost to exercise the option, uncertainty regarding the future value of the underlying asset, the duration of time until the right to selectively act expires, and the expected cost of capital.

Growth and deferral options illustrate two principal means through which flexibility may affect investment decisions and their timing. Growth options characterize expenditures that create follow-on opportunities that may or may not subsequently be exploited. Such options are particularly relevant in high technology and R&D based industries where multiple opportunities arise from the development of a single infrastructure or platform (e.g., Myers, 1977; Kester, 1984). In such environments, small initial investments in exploratory research, pilot production, or early generations of a new product or technology may furnish a preferential claim or "platform" for future favorable investments (Kim & Kogut, 1996). This preferential claim is often based on the ability to transfer experiential learning from early investments into later projects or to preempt competition, possibly by developing brand loyalty with influential customer segments. Since the value of subsequent investment is contingent on the earlier investment decision, early investment can be considered as the entry price for the opportunity to participate in the expected sequence of subsequent related projects. A growth option differs from a first mover advantage conceptually due to the emphases by the former on the opportunity to stage investment over time and on the opportunity to reduce risk by reacting flexibly to changing environmental conditions.

A number of recent papers in the strategic management literature describe how uncertainty affects the growth option value associated with early investments. For instance, Kogut (1991) and Kogut and Kulatilaka (1994, 2001) explore the option value associated with investing in "platform" capabilities from which firms may respond more readily to future changes in their external environment. Similarly, McGrath and Nerkar (2004) find that pharmaceutical firms are more likely to create an option (by taking out a patent) in a new area of exploration when the "upside-potential" is large. The conceptual link between early investments, uncertainty, and growth option value has also been used to justify investment in international markets (Chang, 1995) and international joint ventures (Reuer & Tong, 2005) and establishing standards in industries exhibiting network effects (Lin & Kulatilaka, 2007).

Option theory has also been used to describe how risk may be mitigated by delaying or deferring investment. In industries such as resource extraction, deferral options are often associated with an initial payment such as a lease or license that grants the right to delay development of a plot of land or oil field. Development would be postponed until favorable environmental conditions such as high demand or low interest rates emerge (Kemna, 1993). Ordinarily a deferral option requires no expenditure to initiate, however, but simply bears upon the optimal timing of an irreversible investment. If the firm was to make an irreversible commitment under uncertainty, it would forfeit the possibility that it would not make the investment if unfavorable information would later arrive – therein lies the deferral option. The cost of the deferral option is comprised of the foregone marginal benefits that the firm would realize had it made the initial investment. Costs to delay foregone cash flows or possible entry by rival firms must be weighed against the benefits of receiving new information, therefore. Given estimates that in established industries such as chemicals, pharmaceuticals, petroleum, and electronics as few as 20% of R&D projects result in economically successful products or processes (Mansfield, Rapoport, Schnee, Wagner, & Hamburger, 1971), the option to defer or stage investment is likely to be a valuable source of flexibility (Dixit & Pindyck, 1994; McDonald & Siegel, 1986; Pindyck, 1991). Consistent with this view, several studies citing the option to defer have reported results linking higher levels of uncertainty with lower levels of investment (Campa & Goldberg, 1995) and a reduced probability of foreign market entry (Campa, 1993).

Theoretical models have been developed to analyze the effect of growth and deferral options on investment decision-making. Pindyck (1988) analyzes production capacity choice and expansion decisions when investments are irreversible and market demand is uncertain. In Pindyck's model, the cost of an immediate investment includes the lost option value of forestalling or averting the investment should adverse information emerge. The model also recognizes that under uncertain demand, plant expansion provides a growth option to add additional capacity later should positive market conditions occur. Considering these opposing options and modeling the firm's investment capacity choice as a series of incremental investment decisions, Pindyck shows that the greater the demand uncertainty, the higher the threshold demand must be to justify an irreversible investment. Greater uncertainty is also associated with a higher optimum level of investment, suggesting that growth option value associated with capacity investment is increasing in uncertainty. Based on these model outcomes, Pindyck concludes that irreversibility and unpredictable demand serves as a deterrent in initial plant investment.

Kulatilaka and Perotti (1998) conduct a second analysis of the effect of growth and deferral option value on investment. Their work expands Pindyck's approach by considering the strategic (preemptive) effect of investment under imperfect competition. They show that when possible preemption is taken into account, investment in an initial growth option can result in greater ex post profits relative to investment deferral. Even though the value of the deferral option is shown to be increasing in uncertainty, this increase is smaller than that of the growth option when there are significant preemptive effects associated with the initial growth option. When preemptive effects are small however, deferral option value increases at a higher rate than growth option value under increasing uncertainty. Kulatilaka and Perotti's model thus implies that the effect of uncertainty on the relative values of growth and deferral options is ambiguous.

Lin and Kulatilaka (2007) consider the preemptive implications of network effects on the relative values of growth and deferral options in industries where the creation of standards is important. In their model, early investment generates network externalities that shapes the expectations of consumers and encourages them to adopt the initial investor's technology. Under these conditions, Lin and Kulatilaka find that growth option value is magnified relative to deferral option value. When network effects are strong, growth options dominate deferral options under high uncertainty.

Despite the modeling efforts outlined above, empirical studies of real options attempting to reconcile conflicting prescriptions of real options theory regarding the timing of adoption or investment are rare. Folta and O'Brien (2004) is a notable exception. Their study examines the conflict between the effect of deferral and growth option logic on market entry decisions. Following Kulatilaka and Perotti (1998), they argue that growth options are more sensitive to uncertainty because (a) first mover advantages associated with market entry create greater upside potential, and (b) deferral options are bounded by the total amount of the irreversible investment required to enter (whereas growth options are not subject to an upper bound). Deferral option logic dominates when uncertainty is "low" and growth option logic dominates when uncertainty is "high," thereby generating a non-monotonic relationship between uncertainty and the entry decision. Folta and O'Brien also consider several factors and their effect on the uncertainty and entry relationship, such as the degree of "irreversibility" of the initial entry investment, the level of the initial investment, and the extent to which entry produces an early-mover competitive advantage.

Consistent with their hypothesis that the value of deferral options typically exceeds the value of growth options, Folta and O'Brien find that uncertainty negatively affects the probability of entry throughout most of the range of measured uncertainty in a pooled cross-sectional analysis of 17,897 firms observed from 1980 to 1999. Under very high levels of uncertainty (above the 93rd percentile in their study) however, the value of growth options overshadows the value of deferral options. In industries where early-mover advantage is substantial, the authors find that growth options become valuable at significantly lower levels of uncertainty. Support for their hypothesis regarding the effect of irreversibility on the uncertainty-entry relationship is nonetheless mixed.

While Folta and O'Brien's results are illuminating, their empirical analysis is limited to determining whether the degree of uncertainty or irreversibility (as proxied by the degree of investment in intangibles or inverse financial leverage) affects the decision to enter a new market. Their data do not allow them to observe whether a firm actually deferred investment from one period to the next or made early-stage option investments that were exercised with follow-on investments in a subsequent stage, issues that are central in the adoption of new technologies.

THEORY DEVELOPMENT

The preceding section describes the potentially conflicting prescriptions offered by real option theory and suggests that the comparative importance of growth and deferral options is likely to vary with contextual factors. In this and the following section we develop criteria that may be used to distinguish between situations that favor growth or deferral option value and outline a methodology to empirically test these predictions in environments characterized by sequential investment. Our model incorporates two technological generations, an existing technology currently available for adoption ("current technology") and a potentially more advanced technology expected to arrive sometime in the future ("future technology"). The model also considers two decisions, (a) whether and when to adopt the current technology, and (b) whether to adopt the future technology. We strive to contribute to the discussion regarding criteria that affect the relative significance of growth and deferral options by examining how differences in expectations regarding the value of the future technology, uncertainty in the values of the current and future technologies, the arrival time of the future technology, and the relatedness between the current and future technology are likely to affect investment at the two decision points.

The first criterion for discriminating between situations that offer greater growth or deferral option value highlights the importance of identifying the underlying source of uncertainty. A well-known insight from real option theory is that option value is positively correlated with increasing uncertainty – higher uncertainty increases the "upside" potential of the focal investment with no corresponding increase in "downside" risk. Deferral and growth options differ, however, on the object of the uncertainty. The "uncertainty" of interest in a deferral option is the variance in the value of the current technological opportunity. A deferral option's value is equal to the opportunity cost of irreversibly investing in a current technology of uncertain value today. This opportunity cost increases with the likelihood of outcomes that negatively influence the value of the current technology. In contrast, the objective of a growth option investment is to maximize the potential value captured from the future technology. As long as a mechanism to maintain a preferential claim on follow-on investments exists, the value of a growth option increases with the upside potential or uncertainty in the future technology's value.

Uncertainty in the values of the current and future technologies may or may not be related. If the current and future technologies are strong substitutes then changes in the value of the future technology will directly influence the value of the current technology. In this case, variations in the values of the two technologies are co-determined and it is appropriate to use a single measure of uncertainty when considering the two investment decisions. In terms of option values, higher uncertainty in the value of the future technology lowers the option value of deferring investment in the current technology, while increasing its growth option value. On the other hand, if the future technology represents only a weak substitute (e.g., when it provides an incremental advance over the current technology), then changes in the value of the future technology have a limited effect on the value of the current technology. Under this condition, there is little relation between the distribution of potential values of the two technologies and the proper application of deferral or growth option logic will require an assessment of different notions of "uncertainty."

A second condition that may distinguish between deferral and growth options involves the mechanism used to secure a claim on the opportunity to act flexibly in the future. Deferral option value is determined by the opportunity to make a different decision should new information arrive that affects the relative value of the focal investment compared to the set of alternative investment opportunities. This requires trading off potential first mover advantages and immediate cash flows associated with early investment against the advantages of a potentially lower purchase price or the arrival of potential alternative investments. The mechanisms that provide this opportunity are those that mitigate the loss of potential first mover advantages (e.g., ownership of complementary assets) and contextual factors (e.g., a limited opportunity environment) that affect the number and attractiveness of alternative investments.

Growth option value, in contrast, is dependent on the establishment of a preferential claim to invest in a future opportunity. In the basic setup for a growth option, (a) a firm identifies a new investment trajectory and expends some relatively fixed and presumably small price for pursuing that investment trajectory (i.e., an option purchase), (b) information arrives over time changing the level of uncertainty regarding the value of the new investment trajectory, and (c) in response to this information, the firm either exercises its option by investing fully or discontinues the project. A critical mechanism in this model is the action that insures that any advantage vis-à-vis the follow-on opportunity is not competed away when uncertainty is favorably resolved. It is often argued that this preferential claim is based on the ability to access proprietary information or develop deep knowledge regarding the follow-on opportunity. For instance, firms may use minority equity or joint venture investments to privately ascertain the management quality and growth prospects of a potential acquisition target (Folta, 1998; Kogut, 1991).

The different means of value creation emphasized in growth and deferral option logic suggests that it is possible to identify situations that favor one type of option over its alternative. As mentioned above, the expected value of a deferral option is determined by the opportunity cost of immediately investing in the current technology. This opportunity cost is driven by the supply and demand for critical resources that would be exhausted through immediate investment in the current technology. As these critical resources become scarce or the number of potential alternative investment projects increases, the opportunity cost of committing and the deferral option value also increase. The ability to capture value through a growth option requires the existence of a preferential claim on value created through the favorable resolution of uncertainty regarding the future technology. The strength of this claim is significantly influenced by the nature of the two technologies. If the future technology is considered a competence enhancing innovation, then the knowledge gained via experience with the current technology will be applicable to the use of the future technology, providing investors in the current technology with a preferential claim on investment in the future technology.

The expected arrival time of the future opportunity renders a third method to distinguish between deferral and growth option value. For both financial and real options, the time to expiration affects investors' cumulative exposure to uncertainty and the expected opportunity cost of investing capital in alternative interest bearing investments. The influence of time on the value of real options is more complex, however. As discussed above, for an investment to operate as a real growth option, the value derived from the future follow-on investment must be contingent on making the initial investment and a mechanism must exist that protects the advantage created by the initial investment. In financial options this mechanism is the legal contract associated with the option purchase. In a real growth option, maintaining a preferential claim on the value derived from a future technology is likely to require ongoing investment. For instance, it may be necessary to invest in R&D or monitoring programs to facilitate "absorption" of the future technology, survey the relative viability of alternatives, or assess the willingness to pay by customers for output generated by the future technology. The accumulation of such ongoing costs directly reduces the value of the growth option – its expected value is therefore likely to decrease the more distant the forecasted arrival time of the future technology.

EMPIRICAL CONTEXT

The preceding discussion describes the role of expectations regarding the source of uncertainty, the claim on upside value, and duration in growth and deferral option value. In this section we outline a methodology to empirically test these predictions based on observable behavior in environments characterized by sequential technological investment. We utilize the classification of adoption strategies developed by Grenadier and Weiss (1997). Their model incorporates two technological generations, an existing technology currently available for adoption ("current technology"), and a more advanced technology expected to arrive sometime in the future ("future technology"). A firm immediately adopting both the current and the future technology is termed a "compulsive" adopter. A firm that immediately adopts the current technology, but does not adopt the second generation, is considered to be following a "buy-and-hold" strategy. A firm that waits until the arrival of the future technology prior to adopting the current technology is considered to be adopting a "laggard" strategy. A firm that declines to adopt the current generation of technology, but adopts the future generation is considered to be enacting a "leapfrog" strategy. A firm that does not adopt either technology is termed a "bystander" (we do not consider this final possibility in our analysis) Fig. 1 depicts the adoption strategies classified by Grenadier and Weiss (1997).

The logic underpinning our analysis begins by first considering the adoption decision for the currently available technology. If the expected benefits exceed expected costs, the firm may adopt this technology. Upon the emergence of a new technology, the firm may upgrade to the new technology or remain with the current technology. The firm's decision to adopt the current technology, therefore, turns not only on the direct value of the current technology, but also on the value of the embedded option to upgrade to the future technology when it arrives, and any strategic benefit it would obtain from a first-mover advantage. The claim on this upgrade option value may be generated by knowledge gained by learning from using the current technology that is applicable to the new technology. For example, firms with experience in the current generation may be able to leverage trouble-shooting and problem-solving skills, expertise regarding the technical feasibility of fabricating products with different fabrication and assembly techniques, or informal cross-functional and cross-facility contacts (Clark & Fujimoto, 1991; Leonard-Barton, 1995) into the future generation. Experience in the current technology may therefore lower the cost of adopting the future technology through the establishment and refinement of appropriate information filters, communications channels, routines, and procedures (e.g., Cyert & March, 1963; Arrow, 1974; Nelson & Winter, 1982; Henderson & Clark, 1990). The embedded upgrade option associated with adoption of the current technology can therefore be viewed as a future growth option.

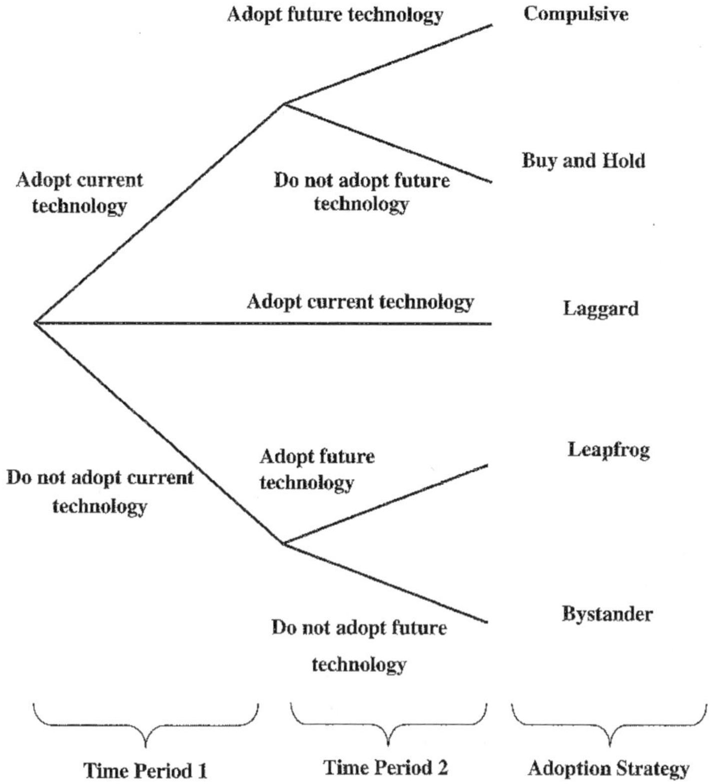

Fig. 1. Potential Timing of Investments in Current or Future Technology.

The firm may alternatively elect to defer adoption until the future technology arrives. At that time, the firm has three choices: (1) adopt the future technology outright, (2) adopt the old (formerly current) technology, or (3) do neither. The option to defer has value if there is a positive probability that the future technology would represent a drastic innovation that would threaten the value of the current technology. To the extent that the firm could adopt the current technology at a discounted cost upon arrival of the future technology, the incentive to delay adoption would also be increased. Potential costs of switching to the current technology may also deter adoption. For example, Kogut and Kulatilaka (2001) argue that an organization's difficulty in costlessly revisiting or reversing an adoption decision may deter application of the technology in the first place. Other switching costs such as network effects may also make it difficult for a firm to shift to a different technology in some industries (Lin & Kulatilaka, 2007). Lastly, any preemptive effects from prior adoption by rivals may discourage or delay adoption.

We collapse the relationships described immediately above and in the prior section into dimensions that capture the expected value of the flexibility to adopt potential follow-on technologies and the expected value associated with deferring investment. The strategies are presented in two panels that comprise Table 1. The top panel describes situations where subsequent generations of a technology are expected to provide drastic improvements in price and performance over the current technology and leverage capabilities that are utilized in the current technology. Such technological advances have been termed "radical" (e.g., Henderson, 1993, p. 250) and represent significant improvements along a single technical trajectory (Dosi, 1982). The lower panel depicts instances where improvements in the future technology are sufficiently moderate so that the current technology is expected to be competitive with the future technology. Henderson (1993, p. 250) refers to such innovations as "incremental." Following Henderson (1993), we acknowledge that the incentive to invest in innovation is affected by both market power and organizational capabilities.

To capture both time and variance components in classifying uncertainty, we follow the taxonomy developed by Luehrman (1998). At one extreme, when a new generation of technology is expected to arrive quickly and/or the technology's value is relatively certain, the environment is characterized as exhibiting low "cumulative volatility" (Luehrman, 1998). The left column in both the upper and lower panels of Table 1 represents such situations. In the right column, a greater period of time is expected to pass prior to the arrival of the subsequent generation and/or the value of the future technology is highly uncertain, thus corresponding to high cumulative volatility. The two rows in

Table 1. Technology Adoption Strategy Predictions.

	Low Cumulative Volatility	High Cumulative Volatility
Technologies are substitutes		
Competence enhancing	Leapfrog	Compulsive
Competence destroying		
Technologies are not substitutes		
Competence enhancing		
Competence destroying	Laggard	Buy and hold

both the upper and lower panel portray the expected value of the future technology relative to the current technology and depict whether capabilities necessary to utilize the current technology are applicable to effectively employing the future technology (i.e., the future technology is competency enhancing or competency destroying). The contents of each cell depict the technology migration strategy that is predicted to occur based on an evaluation of the growth and deferral option value expected in each situation.

The cell in the top left corner of the upper panel in the table describes circumstances where future innovations are (a) economically drastic, (b) rapidly, but predictably, introduced, and (c) draw on similar knowledge bases. Since innovations in the upper panel are substitutes for the current technology, the future technology is likely to drastically reduce the value of the current technology. The lack of variability in the value and arrival time of the future technology limits the value of the option to upgrade to the new technology. Moreover, the short expected time until the arrival of the future technology reduces the opportunity for a firm to recoup its investment in the current technology. While the use of similar skills to those that would be employed if a firm adopted the prior generation implies the existence of a strong claim on any potential growth option value, low cumulative volatility suggests that any latent option value is limited. Thus, firms positioned in the top-left cell are less likely to adopt the current technology and more likely to adopt the future technology – they are expected to follow a leapfrogging strategy. The option to defer adoption of the current technology dominates the growth option in this situation.

The lower panel describes situations where the future technology is not expected to be a strong substitute for the current technology. The expectation of relatively minor cost or quality improvements suggests that the current technology will remain competitive with the future technology, thus reducing the likelihood that the future technology will be adopted. The lower left cell in this panel describes environments in which the future technology is rapidly and predictably introduced and requires different skills to use. The high degree of confidence regarding the value of the future technology as well as the expectation that the future technology will arrive in the near future implies that there is little growth option value associated with the current technology. Moreover, firms that delay adoption of the current technology until arrival of the future technology may be able to take advantage of reductions in the cost of adopting the current technology. In addition, the expectation that use of the future technology will draw upon a new knowledge base severely limits inter-generational knowledge spillovers. In this situation therefore, a laggard strategy is implied. As with the leapfrogging strategy, deferral option value exceeds growth option value of adoption of the current technology in this case.

The cell in the top right corner of the upper panel in the table represents settings in which the likelihood and timing of introduction of a future technology is unpredictable over an extended period of time. Moreover, effective adoption of such a technology requires skills consistent with those employed along the current technical trajectory and the technology is expected to provide a significant performance improvement over the preceding generation. Since the future technology represents a drastic improvement over the current technology, it is likely that it will be adopted upon its introduction. The net effect of the expectation of the future technology on investment in the current generation, however, requires one to tradeoff the cost of cannibalization implied by the eventual introduction of a drastically superior technology against the opportunity to develop skills that may eventually be applied to a future technology of uncertain value. To the extent that the slow change implied by the expected long duration of time until the arrival of the future technology mitigates the expected costs of cannibalization, managers facing this scenario are more likely adopt the current technology. A compulsive adoption strategy is therefore predicted. In contrast to the two strategies above, a compulsive strategy is associated with higher growth than deferral option value for current technology adoption.

The cell located in the lower right-hand corner of the lower panel in Table 1 illustrates situations where the future technology is expected to provide a marginal, but highly uncertain, performance benefit. The anticipated marginal performance benefit indicates that the threat of cannibalization is low, increasing the value of the current technology while reducing the probable value of the future technology. The high level of uncertainty regarding this expectation, nevertheless, suggests the existence of potential growth option value. This potential growth option value is mitigated by the anticipated need for new skills in the future technology. The lack of a common skill base across the current and future technologies suggest there will be few opportunities to lever competencies across generations and that adoption of the current technology will not provide a strong, preferential claim on the future technology. While the ceiling imposed on the expected value of the future technology limits a firm's ability to capture growth option value via investment in the current technology, the relatively long duration of time that the current technology is able to generate returns encourages its immediate adoption, thus also limiting deferral option value. Since the values of both deferral and growth options are likely to be modest, firms in this situation are likely to employ a buy and hold technology strategy.

The final four cells in Table 1 illustrate circumstances that are not amenable to real option analysis. The two empty cells in the upper panel portray situations reminiscent of competence destroying technological discontinuities in the airline, cement, and mini-computer industries discussed by Tushman and Anderson (1986). In these settings, the future innovation is expected to provide a significant improvement over the existing technology using very different skills. Given the lack of learning spillovers or other factors that may provide a preferential claim on the new technology, the investment implications for the current technology are unclear. The empty cells in the lower panel illustrate circumstances where the future technology is expected to utilize skills similar to those employed in the current technology but to provide only a marginal improvement in functionality. While these situations may support some growth option value, the modest magnitude of this value indicates that the influence of any option value will be dominated by the traditional net present value of the current technology.

DISCUSSION

Motivated by the conflicting prescriptions of real option analysis with respect to growth and deferral option value, this chapter describes how expectations regarding future technological developments are likely to affect a firm's technology migration strategy. We consider the effect of the expected economic significance of future innovations, the frequency with which these future innovations arrive, the amount of inter-generational learning that can be transferred from an existing to a new technology, and the degree of uncertainty surrounding these innovations on the decisions to adopt current and future technologies.

The conceptual framework presented in this chapter contributes to the recent discussion regarding distinctions between growth and deferral options. Since the values of both options rise with uncertainty, and because the two options have opposing effects on the incentive to invest, the theoretical literature suggests that the effect of uncertainty is ambiguous (Abel, Dixit, Eberly, & Pindyck, 1996).

We submit that by jointly considering the duration of time until the next decision period, the degree of uncertainty regarding the comparative values of the future and current technologies, and the extent to which knowledge may spill over across technological generations, it is possible to assess the relative magnitude of growth and deferral option value. By illuminating the tradeoff between growth and deferral option value we hope to address a major issue that has limited the development of real option theory and its dissemination into practice.

In addition to addressing the real option literature, our work also contributes to the technology strategy and inter-generational learning literature. The introduction of real options analysis to studies of technology adoption provides a general theoretical framework with which to analyze the factors that influence whether and when firms should adopt a given technology. In so doing, we highlight two types of expectations that are likely to affect firms' current technology postures in the presence of expectations regarding future technologies – uncertainty and frequency of change. Finally, our model has implications for an ongoing debate in the economic literature that has shown that as the significance of the future technology increases, or becomes "drastic" in the sense used by Arrow (1962), the likelihood of adoption of the current technology may be mitigated (e.g., Arrow, 1962; Gilbert & Newberry, 1982; Reinganum, 1983). The model provides a conceptual explanation for analytical work that has shown that the "Arrow" effect depends on the degree to which innovation destroys existing market power and the uncertainty surrounding the timing of the introduction of the innovation (e.g., Gilbert & Newberry, 1984; Reinganum, 1984).

This paper also helps to clarify the role of inter-generational learning. While the notion of inter-generational learning spillovers is well established in the technology literature (Irwin & Klenow, 1994), its use as a measure of the strength of the preferential claim on subsequent generations of technology has received little attention. The mapping of inter-generational learning onto an option-based model of technology adoption allows an explicit comparison of the economic benefits provided by this learning vis-à-vis alternative investments. Backward induction reveals that the value of inter-generational spillover is directly tied to the attractiveness of the future technology. This value is a function of the expected economic significance of the future technology. Real option analysis, however, demonstrates that uncertainty in the value of the future technology and the duration of time until the arrival of the future technology provide an additional upside to the future technology value.

This paper is not without limitations. First, we ignore portfolio effects of real options. To the extent that there is unobserved heterogeneity across firms in their portfolio of sunk investments, some firms will be able to productively adopt technologies that other firms cannot without experiencing greater risk (e.g., Dixit & Pindyck, 1994; Trigeorgis, 1998). Moreover, firms' innovative capabilities are not explicitly considered. We also do not account for the effects of competitive interaction in adoption decisions.

Future research should address some of the limitations discussed above, as well as empirically test the arguments presented.

This exercise suggests a number of opportunities for management scholars to study the similarities and differences between stand-alone and dynamic approaches to new technology adoption. As the complexity and rapidity of technological change increases, these research directions will likely take on greater importance.

ACKNOWLEDGMENTS

The authors thank Janet Bercovitz, Jennifer Kuan, Joanne Oxley, Rosemarie Ziedonis, seminar participants at the Conference on Real Options in Entrepreneurship and Strategy at the University of North Carolina, seminar participants at the University of Michigan, and the editors for thoughtful comments and suggestions. Financial support for Ziedonis from the Mack Center for Technological Innovation at the Wharton School of the University of Pennsylvania is gratefully acknowledged.

REFERENCES

Abel, A. B., Dixit, A. K., Eberly, J. C., & Pindyck, R. S. (1996). Options, the value of capital, and investment. *Quarterly Journal of Economics, 111*(3), 753–777.

Adams, W., & Dirlam, J. B. (1966). Big steel, invention, and innovation. *Quarterly Journal of Economics, 80*(2), 167–189.

Amram, M., & Kulatilaka, N. (1999). *Real options: Managing strategic investment in an uncertain world.* Boston, MA: Harvard Business School Press.

Arrow, K. (1974). *The limits of organization.* New York: Norton Press.

Arrow, K. J. (1962). *The rate and direction of inventive activity.* Princeton, NJ: National Bureau of Economic Research.

Barney, J. (1986). Strategic factor markets: Expectations, luck, and business strategy. *Management Science, 32*, 1231–1241.

Black, F., & Scholes, M. (1973). The pricing of options and corporate liabilities. *Journal of Political Economy, 81*(3), 637–654.

Campa, J. M. (1993). Entry by foreign firms in the United-States under exchange-rate uncertainty. *Review of Economics and Statistics, 75*(4), 614–622.

Campa, J. M., & Goldberg, L. S. (1995). Investment in manufacturing, exchange rates and external exposure. *Journal of International Economics, 38*, 297–320.

Chang, S. J. (1995). International expansion strategy of Japanese firms: Capability building through sequential entry. *Academy of Management Journal, 38*(2), 383–407.

Christensen, C. M. (1997). *The innovator's dilemma.* Boston, MA: Harvard Business School Press.

Clark, K. B., & Fujimoto, T. (1991). *Product development performance: Strategy, organization, and management in the world auto industry*. Boston, MA: Harvard Business School Press.

Copeland, T., & Antikarov, V. (2001). *Real options: A practitioner's guide*. New York: Texere.

Cyert, R. M., & March, J. G. (1963). *A behavioral theory of the firm*. Englewood Cliffs, NJ: Prentice-Hall.

Dixit, A. K., & Pindyck, R. S. (1994). *Investment under uncertainty*. Princeton, NJ: Princeton University Press.

Dosi, G. (1982). Technological paradigms and technological trajectories: A suggested interpretation of the determinants and directions of technical change. *Research Policy*, *11*, 147–162.

Folta, T. B. (1998). Governance and uncertainty: The tradeoff between administrative control and commitment. *Strategic Management Journal*, *19*, 1007–1028.

Folta, T. B., & O'Brien, J. P. (2004). Entry in the presence of dueling options. *Strategic Management Journal*, *25*, 121–138.

Ghemawat, P. (1997). *Games businesses play*. Cambridge, MA: The MIT Press.

Gilbert, R. J., & Newbery, D. M. G. (1982). Preemptive patenting and the persistence of monopoly. *American Economic Review*, *72*, 514–526.

Gilbert, R. J., & Newbery, D. M. G. (1984). Uncertain innovation and the persistence of monopoly: Comment. *American Economic Review*, *74*(1), 238–242.

Grenadier, S., & Weiss, A. (1997). Investment in technological innovations: An option pricing approach. *Journal of Financial Economics*, *44*, 397–416.

Henderson, R. M. (1993). Under-investment and incompetence as responses to radical innovation: Evidence from the photolithographic alignment equipment industry. *RAND Journal of Economics*, *24*(2), 248–270.

Henderson, R. M., & Clark, K. B. (1990). Architectural innovation: The reconfiguration of existing product technologies and the failure of established firms. *Administrative Science Quarterly*, *35*, 9–30.

Irwin, D., & Klenow, P. (1994). Learning-by-doing spillovers in the semiconductor industry. *Journal of Political Economy*, *102*(6), 1200–1227.

Kemna, A. G. Z. (1993). Case studies on real options. *Financial Management*, *22*(3), 259–270.

Kester, W. C. (1984). Today's options for tomorrow's growth. *Harvard Business Review*, *62*(2), 153–160.

Kim, D. J., & Kogut, B. (1996). Technological platforms and diversification. *Organization Science*, *7*(3), 283–301.

Kogut, B. (1991). Joint ventures and the option to expand and acquire. *Management Science*, *37*(1), 19–33.

Kogut, B., & Kulatilaka, N. (1994). Operating flexibility, global manufacturing, and the option value of a multinational network. *Management Science*, *40*, 123–139.

Kogut, B., & Kulatilaka, N. (2001). Capabilities as real options. *Organization Science*, *12*(6), 744–758.

Kulatilaka, N., & Perotti, E. C. (1998). Strategic growth options. *Management Science*, *44*(8), 1021–1031.

Leonard-Barton, D. (1995). *Wellsprings of knowledge: Building and sustaining the sources of innovation*. Boston, MA: Harvard Business School Press.

Lieberman, M., & Montgomery, D. (1988). First-mover advantages. *Strategic Management Journal*, *9*(Summer), 41–58.

Lieberman, M., & Montgomery, D. (1998). First-mover (dis)advantages: Retrospective and link with resource-based view. *Strategic Management Journal, 19*(12), 1111–1125.

Lin, L., & Kulatilaka, N. (2007). Strategic growth options in network industries. *Advances in Strategic Management, 24*, 177–198.

Luehrman, T. A. (1998). Investment opportunities as real options. *Harvard Business Review, 76*(4), 51–67.

Mansfield, E., Rapoport, J., Schnee, J., Wagner, S., & Hamburger, M. (1971). *Research and innovation in the modern corporation*. New York: W. W. Norton and Company, Inc.

McDonald, R., & Siegel, D. (1986). The value of waiting to invest. *Quarterly Journal of Economics, 101*, 707–727.

McGrath, R. G. (1997). A real options logic for initiating technology positioning investments. *Academy of Management Review, 22*(4), 974–996.

McGrath, R. G., & Nerkar, A. (2004). Real options reasoning and a new look at the R&D investment strategies of pharmaceutical firms. *Strategic Management Journal, 25*, 1–21.

Myers, S. C. (1977). Determinants of corporate borrowing. *Journal of Financial Economics, 5*, 147–175.

Nelson, R. R., & Winter, S. G. (1982). *An evolutionary theory of economic behavior and capabilities*. Cambridge, MA: Harvard University Press.

Peteraf, M. A. (1993). The cornerstones of competitive advantage: A resource-based view. *Strategic Management Journal, 14*, 179–191.

Pindyck, R. S. (1988). Irreversible investment, capacity choice, and the value of the firm. *The American Economic Review, 78*(5), 969–985.

Pindyck, R. S. (1991). Irreversibility, uncertainty, and investment. *Journal of Economic Literature, 29*(September), 1110–1152.

Porter, M. E. (1980). *Competitive strategy*. New York: Free Press.

Reinganum, J. F. (1983). Uncertain innovation and the persistence of monopoly. *American Economic Review, 73*, 741–748.

Reinganum, J. F. (1984). Practical implications of game theoretic models of R&D. *American Economic Review*, 61–66.

Reuer, J. J., & Tong, T. W. (2005). Real options in international joint ventures. *Journal of Management, 31*(3), 403–423.

Rosenberg, N. (1976). On technological expectations. *The Economic Journal, 86*, 523–535.

Rumelt, R. P., Schendel, D. E., Teece, D. J. (Eds). (1994). *Fundamental issues in strategy: A research agenda*. Boston, MA: Harvard Business School Press.

Sayers, R. S. (1950). The springs of technical progress in Britain, 1919–39. *The Economic Journal, 60*, 289–290.

Trigeorgis, L. (1998). *Real options: Managerial flexibility and strategy in resource allocation*. Cambridge, MA: MIT Press.

Tushman, M., & Anderson, P. (1986). Technological discontinuities and organizational environments. *Administrative Science Quarterly, 31*, 439–465.

[24]

MANAGEMENT SCIENCE
Vol. 53, No. 10, October 2007, pp. 1618–1633
ISSN 0025-1909 | EISSN 1526-5501 | 07 | 5310 | 1618

informs
DOI 10.1287/mnsc.1070.0705

Real Options in Technology Licensing

Arvids A. Ziedonis
Stephen M. Ross School of Business, University of Michigan, Ann Arbor, Michigan 48109,
azied@umich.edu

This paper examines the use of options contracts by firms acquiring rights to commercialize university technologies. By combining information about the sequence of licensing decisions with characteristics of the firms and technologies involved, I explore factors that shape decisions to purchase and exercise option contracts for these technologies. Decisions by firms that considered but did not purchase an option or a license are included in the sample. Consistent with the basic premise of real-options theory, I find that firms are more likely to purchase option contracts for more uncertain technologies. Also in line with theoretical predictions, I find that firms that are better able to evaluate an external technology are less likely to purchase options before licensing. The results also highlight more complex motives for exercising options in technology licensing. On the one hand, firms appear to benefit from their ability to learn about the technology during the option period. On the other hand, firms that are better able to "absorb" the technology during the contract period may have reduced incentives to subsequently license the invention.

Key words: real options; licensing; university-industry linkages; knowledge transfer
History: Accepted by Scott Shane, technological innovation, product development, and entrepreneurship; received May 14, 2004. This paper was with the author 1 year and 8 months for 2 revisions. Published online in *Articles in Advance* August 3, 2007.

1. Introduction

A lively literature has emerged applying real-options thinking to R&D decision making and investment (Mitchell and Hamilton 1988, Trigeorgis 1996, McGrath 1997, Folta 1998, MacMillan and McGrath 1998, Hamilton 2000), the development of organizational capabilities (Kogut and Kulatilaka 2001), expansion decisions (Kogut 1991, Kogut and Kulatilaka 1994, Kulatilaka and Perotti 1998), and entry decisions (Miller and Folta 2002). Yet, empirical tests of real-options theory in the strategy literature remain inconclusive due to difficulties in identifying ex ante what is and is not a real option and in falsifying hypotheses about the motivations of firms to undertake and exercise such options (Adner and Levinthal 2004).

This paper examines the use of option contracts by firms acquiring rights to commercialize university technologies. Under an option agreement, the firm purchases the right, but not the obligation, to license the technology at a future date at some cost. By combining information about the sequence of licensing decisions with characteristics of the firms and technologies involved, I explore factors that shape decisions to purchase and exercise option contracts for early-stage technologies. The sample encompasses decisions by firms considering licensing inventions from the University of California (UC) between 1979 and 1998. An important feature of the data is that it contains "prelicensing" agreements (e.g., secrecy and option agreements) that allowed me to observe sequential steps in the licensing transaction. Moreover, secrecy agreements identify the pool of firms that considered licensing a particular invention from UC, including firms that chose not to license the technology.

Consistent with the basic premise of real-options theory, I find that firms are more likely to purchase option contracts for inventions characterized by greater technological uncertainty. I also find that firms that are better able to evaluate a university technology are less likely to purchase options before licensing. The results also highlight more complex motives for exercising options in university technology licensing. On the one hand, firms appear to benefit from their ability to learn more about the technology during the option period. On the other hand, firms that are better able to "absorb" the technology during the contract period may have reduced incentives to subsequently license the invention.

Section 2 provides a brief overview of the real-options literature. Section 3 develops three hypotheses on how the use of options contracts in the licensing decision depends on technological and firm-specific knowledge. Section 4 outlines the technology licensing process at UC. Section 5 describes the sample, explains the methodology, and presents the variables. Section 6 reports the empirical findings. Finally, §7 discusses several implications that stem from these results and identifies areas for future research.

2. What Is a Real Option?

A real option is analogous to a financial call option (Black and Scholes 1973), which gives an investor the right, but not the obligation, to buy a valuable asset (such as a share of stock) at a future (expiration) date at a certain (exercise) price. Real-options theory expands the notion of option value to encompass nonfinancial or so-called "real" investments (Dixit and Pindyck 1994). In this context, the "option" corresponds to a small initial investment (e.g., a pilot project, limited joint venture, or exploratory R&D) that creates the opportunity to make a larger subsequent investment (e.g., a full-scale plant, merger, or commercialization effort) in the event findings generated by the initial investment are favorable. The price of the option corresponds to the initial expenditure, and the exercise price corresponds to the cost of the follow-on investment.

A real-option investment has value because of uncertainty regarding future payoffs associated with full-scale investment. A central prediction of real-options theory is that the greater ex ante uncertainty or variation of expected future payoffs, the greater the value of the initial option investment. Greater variation in expected payoffs generates greater upside potential of the option with no increase in downside risk because the option would not be exercised if the outlook were not favorable at the end of the option period. Option value is thus realized through the resolution of uncertainty.

The real-options literature has emphasized two forms of uncertainty—exogenous and endogenous—that can be resolved during the option period (Roberts and Weitzman 1981, Dixit and Pindyck 1994, Folta 1998). Exogenous uncertainty is deciphered through waiting, and resolution is independent of actions of the asset holder. For example, consider a firm weighing an investment decision in the face of pending legislation that it is unable to influence, but that will affect the returns to its investment. In this circumstance, uncertainty regarding the legislation is unaffected by whether the firm invests or not, and therefore is exogenous to the investment decision (McDonald and Siegel 1986, Majd and Pindyck 1987). Prior studies have focused on exogenous uncertainty resolution and the exercise of such a "deferral" option. For example, Folta and Miller (2002) examine acquisition patterns by pharmaceutical firms in the biotechnology industry using an options framework. Established companies that hold minority stakes in new biotechnology firms are said to hold options on future acquisition. The authors find that holders of these stakes are more likely to acquire the biotechnology firm (i.e., exercise the option) when exogenous uncertainty in the industry subfield of the target firm is reduced (as measured by volatility in subfield stock-market returns).

Endogenous uncertainty, in contrast, is resolved by active involvement by the firm and encourages early and sequential investment (Dixit and Pindyck 1994). For example, Kogut (1991), Kogut and Kulatilaka (1994, 2001), and Kogut and Kim (1996) explore the creation of "growth" option value by developing organizational capabilities; capability development creates "platforms" from which firms may respond more readily to future changes in their external environment. Absent such active investment, the firm would be less favorably positioned to capitalize on changes in market conditions.

As discussed in more detail in §4, the licensing process at UC provides a unique empirical setting to examine the application of growth options. Inventions marketed by the university cover embryonic inventions requiring significant development before being marketable, and are thus characterized by significant technological and commercial uncertainty. Option contracts offer the potential licensee the opportunity to stage its investment in a university technology with the ability to discontinue investment (by not exercising the option to license) in the event that unfavorable information is uncovered during the option period. Contractual requirements that the firm evaluate the technology in-house are designed to facilitate the resolution of endogenous uncertainty—uncertainty that firms can reduce through their own efforts. Finally, the short time period of the option contract (one to two years) limits the opportunity to use the option to resolve exogenous uncertainty for these early-stage inventions.

The empirical context of this study also enables me to overcome several obstacles faced by prior studies in real options. As Adner and Levinthal (2004) note, virtually any investment has real-options properties, so identifying what is *not* a real option is challenging. Moreover, contracts granting one party the explicit right to exercise an option are rarely observed. For example, real-options studies of joint ventures by Kogut (1991) and Reuer and Tong (2005) were unable to discern whether the joint venture agreements contained language granting one of the parties the option to assume the joint venture. Similarly, in their study of equity partnerships, Folta and Miller (2002) analyze 285 minority equity partnerships in which one partner may obtain a majority stake, but are unable to identify whether these partnership agreements included the explicit right to acquire a majority stake ex ante. Lastly, because the selection and exercise of a real option may be unobservable, it can be difficult to falsify predictions arising from the theory (Adner and Levinthal 2004). The licensing data outlined above enabled me to address each of these challenges. First,

I am able to differentiate among firms that purchased or did not purchase an option. Moreover, I observe instances where firms purchased option contracts that explicitly confer the right but not the obligation to license a technology at the end of the contract term. The data also enabled me to identify circumstances under which firms do (and do not) exercise option contracts.

3. Hypothesis Development

Inventions offered for licensing from research universities typically are by-products of basic research, and therefore are often characterized as "early-stage" technologies. The embryonic nature of these technologies can make it difficult to assess their commercial potential and fit with a firm's existing product lines or development efforts. For example, in a survey of 62 university technology transfer offices by Jensen and Thursby (2001), respondents reported that only 12% of licensed inventions were ready for commercial use, and that manufacturing feasibility was known for only 8% of inventions. Prototypes were available for 29% of the licensed inventions, but 48% were termed "proof of concept only" at the time of licensing (i.e., they lacked a working prototype). Therefore, for many of these licensed inventions, technical, manufacturing, and commercial feasibility are all subject to varying degrees of uncertainty. As discussed in §2, a main prediction of real-options theory is that the probability of purchasing an initial option is increasing in the degree of uncertainty that can be resolved regarding future payoffs. This prediction leads to Hypothesis 1:

Hypothesis 1. *The greater the uncertainty embodied by an invention, the more likely a firm will purchase an option contract prior to making a licensing decision, all else equal.*

A firm's scientific or technological capabilities also may influence the decision-making process. An extensive theoretical and empirical literature has demonstrated that possession of related knowledge is often necessary to digest and utilize external information. For example, Cohen and Levinthal (1990) and Rosenberg (1990) show that firms undertake R&D to improve their capability to monitor, assimilate, and exploit external technologies. Arora and Gambardella (1994) distinguish between two components of a firm's "absorptive capacity": (1) the ability to *evaluate* external information, and (2) the ability to *utilize* external information. They argue that both of these dimensions affect the propensity of the firm to enter into technological collaborations, albeit in different ways. In a study of collaborative agreements between pharmaceutical firms and specialized biotechnology firms and universities, these scholars show that firms with higher in-house technological capabilities related to development and commercialization are more likely to enter into external collaborations. Conversely, firms that are better able to evaluate external projects are more discerning and select fewer projects on average. Similarly, Zahra and George (2002) distinguish between "potential" absorptive capacity, the capability to assess broad external technological information, and "realized" absorptive capacity, the ability to utilize that information in a specific application domain to generate new innovations.

Uncertainty, therefore, may not only be a characteristic of an invention, but may also differ among firms. Cohen and Levinthal (1990, p. 136) note, for example, that an organization's related prior knowledge "permits the firm to predict more accurately the nature and commercial potential of technological advances" and that "possession of related expertise will permit the firm to better understand and therefore evaluate the import of intermediate technological advances that provide signals to the eventual merit of a new technological development." Prior knowledge, consequently, should also influence whether the firm would purchase an option prior to making the licensing decision. A lack of technological knowledge results in a high level of uncertainty regarding viability of the technology, but the purchase of an option can reduce that uncertainty. Conversely, high levels of prior related technological knowledge by a potential licensee suggests that less uncertainty would be resolved under an option agreement. In these instances, the option to delay would be less valuable, thus less likely to be purchased by the firm. This result generates the second hypothesis:

Hypothesis 2. *Firms better able to evaluate an invention will be less likely to purchase an option contract prior to making a licensing decision, all else equal.*

Prior research suggests that knowledge resources have commercial value in specific contexts, and that these resources are most valuable in closely related applications. For example, Montgomery and Wernerfelt (1988) show that firms achieve the highest returns when diversifying into closely related industries. Focusing on technological resources, Arora and Gambardella (1994) find that prior patenting in biotechnology classes by pharmaceutical firms positively influenced the likelihood that they would enter in strategic alliances with new biotechnology firms to commercialize a new technology. Similarly, Silverman (1999) reports that firms are most likely to diversify into industries where the firm's existing patent portfolio is most applicable. In the licensing context, a firm's related technological knowledge supports the firm's ability to derive economic value from the license (i.e.,

Ziedonis: *Real Options in Technology Licensing*
Management Science 53(10), pp. 1618–1633, ©2007 INFORMS 1621

commercially exploit the technology). Related technological knowledge, therefore, would be expected to influence the firm's decision to commit to a license (or not). This leads to a third prediction:

HYPOTHESIS 3. *Firms with a higher level of technological knowledge related to an invention will be more likely to license the invention, all else equal.*

4. Institutional Context

The nine-campus UC system serves as the empirical setting for this study.[1] This university has been one of the leading recipients of technology licensing income among U.S. universities during the past three decades; in 1996, it earned $63.2 million in licensing revenues, more than any other U.S. university (Association for University Technology Managers 1998). Licensing by two other major university licensors has been examined by Shane (2002) (Massachusetts Institute of Technology) and Elfenbein (2004) (Harvard University), but neither study has considered the use of option agreements in the licensing process.

Several aspects of the licensing process at UC are useful for a study of real options. First, all firms interested in licensing the technology must sign a secrecy agreement and agree to "nondisclosure" provisions as a precursor to further steps in the licensing process. Once a secrecy agreement has been signed, the Office of Technology Transfer (OTT) at UC reveals limited confidential information about the invention to the potential licensee, including the identity of the inventor. In exchange, the firm agrees not to utilize the technology for purposes other than evaluation, and not to disclose information about the invention to others. Secrecy agreements are not exclusive (i.e., secrecy agreements with multiple firms may be in effect simultaneously). The university does not, however, disclose the existence of other secrecy agreements or interest in the invention by other firms.

The sequential licensing process at UC allows firms to pursue the technology further through either an exclusive option or an exclusive license contract. An option contract at UC grants a potential licensee an exclusive right to use information disclosed by the university to determine the technological and commercial viability of the invention. In exchange, the firm typically pays the filing fee and other costs of obtaining the patent, as well as an option fee. (Option fees are significantly lower than fees for exclusive licenses.) An additional cost of the option is the requirement that the firm exercise "due diligence" and make "good-faith efforts" toward evaluating the technology in-house. These activities are monitored by OTT through periodic progress reports submitted by the firm. During the option period, which typically lasts one to two years, the firm is contractually barred from marketing the invention or incorporating the technology into its own production processes. Upon expiration, the firm may either negotiate a license or abandon further efforts and terminate the transaction. The firm does not recover the option fee or patent application costs, however, and title to any resulting patent remains with the university.[2]

The ultimate step in the process is the license agreement. A license agreement grants the licensee rights to fully develop the invention covered by the patent, if issued, or patent application, pending issue. Moreover, the license, which may be subject to field-of-use or geographic restrictions, commits the firm to undertake efforts to commercialize the technology. Payments to the university frequently include up-front fees, royalty payments, and minimum royalties. The license also contains language regarding due diligence and reporting requirements, milestones, problem resolution rules, and duration of contract (usually equal to the life of the patent). UC licenses do not include know-how provisions, so the firm's possession of related technological knowledge is even more critical in these transactions: It not only improves the ability of the firm to assess the value of a university invention, but also enhances the firm's ability to commercialize the technology.

To summarize, once a firm has agreed to secrecy provisions, it may license the patent outright or purchase an exclusive option prior to licensing (or choose to do neither). Those firms choosing an option may also choose to license the technology at the end of the option period, but they are not obligated to do so. For any patent in the sample, there may be several potential licensees signing secrecy agreements, but only one firm taking an option or ultimately licensing the patent.

5. Data and Methodology

The primary data for this study come from unpublished archival records of technology licensing at UC maintained by OTT. In managing the licensing of university technologies, OTT tracks inventions disclosed by all UC faculty, students, and research staff, including (1) whether a patent application was filed in the United States or other countries; (2) whether a patent was issued; (2) whether the patent was licensed to a private firm, the identity of the licensee, and the terms of the license (e.g., exclusive/nonexclusive); (3) the amount, if any, of license fees and royalties

[1] A tenth campus, UC-Merced, was established too late (2005) to be included in the sample.

[2] Option agreements typically spell out ex ante a range of terms subject to negotiation at the end of the option period (e.g., possible geographic or field-of-use restrictions).

for each license; (4) patenting and licensing costs and reimbursements; and (5) magnitude and pattern of sales resulting from the licensed technology.

Virtually all licenses and option contracts cover patented inventions. Potential licensees were not required to sign secrecy agreements at UC until 1979. To identify all firms receiving confidential information regarding available technologies, I therefore consider only utility patents licensed after 1978. I also restrict the sample to 309 patents that were licensed on an exclusive basis between 1979 and 1998.[3] The grant of exclusivity serves as an incentive for investment in further development by the licensee. The sample period ends with patents issued in 1992.

For each licensed patent, a potential licensee could make one of the following four mutually exclusive (and collectively exhaustive) choices in the licensing process: (1) secrecy agreement only, (2) secrecy agreement + exclusive license, (3) secrecy agreement + option, or (4) secrecy agreement + option + exclusive license. Each choice among these four alternatives represents one observation. In total, the overall sample includes 669 firm-choice observations on 309 patents, with 258 firms signing at least one agreement.

I supplement the UC OTT data with publicly available information from the U.S. Patent and Trademark Office (USPTO) on all patents issued to UC and link these patents to invention disclosures and licenses. Using these data, I construct measures of a firm's technological knowledge assets (described more fully in §5.2). Unfortunately, patent-based measures are unavailable or poorly measured for two types of firms, "start-up" firms with limited or nonexistent patent histories,[4] and firms not listed as assignee on any patent issued within the previous six years.[5] I therefore drop choices made by these firms from the analysis. Of the 258 firms making licensing decisions, start-ups account for 36, and existing firms with no patent history account for 86. Excluding these firms from the analysis leaves 136 firms, or approximately 53% of the original 258 firms. These 136 firms account for 457 of the original 669 licensing decisions, or 68%. This sample forms the basis for the empirical analysis.

5.1. Specification

The discrete nature of the dependent variable suggests a qualitative choice model such as a logit for estimating the relationship between licensing choice and the attributes of the firm and characteristics of the technology under consideration. The logit can be inappropriate for many choice situations when the independence of irrelevant alternatives (IIA) property is violated, however. Another limitation of the logit choice model is that estimation rests on the assumption that error terms among all of the alternatives are uncorrelated. In the licensing decision process depicted in §4, this assumption is violated in the following sense. If the firm chooses to take an option prior to making the licensing decision, then the firm has greater information regarding the technology on which to make its licensing decision than if it had not chosen an option first. The components of the error terms of these two licensing alternatives (license or reject) that pertain to the option choice are therefore hypothesized to be jointly distributed (and likewise for the licensing alternatives given that option was not chosen). The nested multinomial logit model (NMNL) (McFadden 1981) overcomes some of the IIA and error correlation limitations of the logit model.[6] I conduct a Hausman-McFadden Specification Test (Hausman and McFadden 1984) to test the assumption that the IIA condition does not hold among the four licensing choices, and consequently the applicability of the NMNL model in describing the licensing choice set faced by the firm. The χ^2 statistic from the test is 10.34, which allows me to reject at the 5% significance level the null hypothesis that the IIA condition holds, suggesting that structural relations among the alternatives exists and that the alternative specification of four nonnested licensing choices should be rejected. I describe the Hausman test in greater detail in the online appendix (provided in the e-companion).[7]

In this paper, I employ the NMNL model and assume that error terms have generalized extreme

[3] Nonexclusive licenses typically do not involve due diligence or milestone requirements, entail lower fees, and involve technologies (e.g., research tools) that do not require substantial investment in further development.

[4] As Elfenbein (2004) and Lowe and Ziedonis (2006) discuss, the licensing decision process works very differently for start-ups, which are typically founded with the intent to commercialize the technology. Unlike more established firms, for example, start-ups often seek option agreements to delay commercial development while sufficient capital is secured.

[5] For firms with no patent history, I am unable to account for variation in their technological make-up, so substantial unobserved heterogeneity would be introduced into the specification for these firms if they were to be included. If the no-patent history firms are systematically different than the firms with patents, including the no-patent history firms would generate biased coefficients. Therefore, the results reported in §5 are conditional on the firm having patented actively prior to the licensing decision.

[6] The NMNL model is appropriate when the set of alternatives faced by the decision maker can be partitioned into subsets so that the ratio of probabilities (odds ratio) for any two alternatives within the subset is independent of the characteristics of alternatives not in the subset (the IIA condition), while this condition does not hold between choices *not* in the same subset.

[7] An electronic companion to this paper is available as part of the online version that can be found at http://mansci.journal.informs.org/.

Figure 1 Partitioning of Firm Option and Licensing Choices in Nested Logit Specification

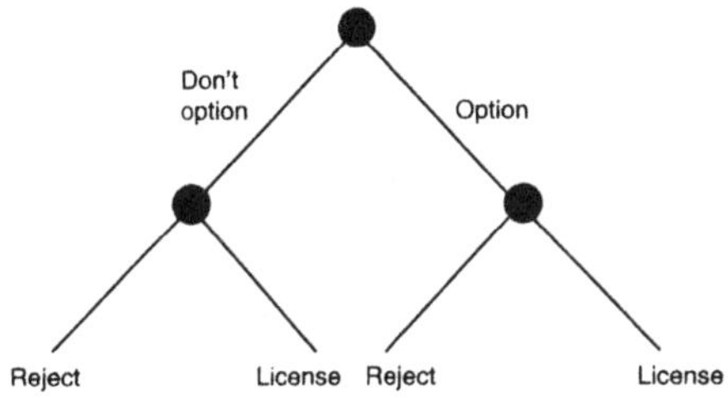

value (GEV) distributions.[8] The specification is as follows. The licensing decision faced by the potential licensee is divided into two partitions, as depicted in Figure 1. The upper subset (conditional on a firm signing a secrecy agreement) represents the decision to take an option prior to making the licensing decision or not. If the firm forgoes an option, it then either licenses or rejects (i.e., not licenses) the technology. If the firm purchases an option, then at the end of the option period it is faced with the choice to license or reject the technology.

It is important to note that the sequential process depicted in Figure 1 does not necessarily correspond to the decision process undertaken by the firm when considering whether to license the university technology. Rather, the partitions are constructed to avoid the IIA problem introduced when all four possible alternatives are commingled. Notably, this construction also enables me to separate the licensing decision from the option decision empirically, while preserving the correlation between these two choices.[9]

[8] The generalized extreme value distribution is compatible with the assumption of stochastic utility maximization (McFadden 1981, Boersch-Supan 1990).

[9] An alternative specification is the ordered probit (or logit). In this model, firms are assumed to rank order the licensing and option choices by their expected value—choosing to license immediately for high expected value inventions, purchasing an option for inventions with medium expected value, and rejecting a license for low-valued inventions. Such a specification requires the strong assumption that every decision maker in the sample would unambiguously order the choices in the same way, E(license) > E(option) > E(reject) (Borooah 2002). For some firms reviewing some technologies however, the value of waiting to resolve the underlying uncertainty may make the option more attractive relative to licensing immediately, therefore violating this assumption. The ordered model also assumes "parallel regressions" (Long 1997), i.e., that each probability curve corresponding to the ordered choices is parallel and differs only in being shifted right or left. I conducted the Brant (1990) test of the null hypothesis that the coefficients are equal across each binary equation. This test produces a χ^2 statistic of 12.32, which has a 1.5% level of significance. The parallel regression assumption is therefore violated, suggesting that a more general discrete choice model that does not impose this restriction is more appropriate.

Each of the four possible licensing alternatives to firm n can be characterized by its utility, $U_{O,L,n}$, where $i = 1, 2$ refers to the alternatives when an option is not taken and $i = 3, 4$ refers to the alternatives when an option is taken. Assume that this utility can be decomposed into components representing the observable portion of utility, and components representing unobservable components of utility (error terms) as follows (dropping the subscript n for convenience):

$$U_{O,L} = V_O + V_L + V_{O,L} + \varepsilon_O + \varepsilon_L, \tag{1}$$

where V_O is the observable component of utility that is obtained from the option choice and V_L is the observable component of utility that is obtained from the licensing choice. ε_O and ε_L are the random components of utility from the respective choices. The observable components of utility are defined as

$$V_O = \beta'_O X_O \quad \text{and} \quad V_L = \beta'_L X_L, \tag{2}$$

where X_O and X_L are vectors of variables corresponding only to attributes of the option and licensing choices, respectively. β'_O and β'_L are vectors of parameters to be estimated. The advantage of disaggregating the firm's utility equation in this fashion is that, following McFadden (1981), the joint choice probabilities can now be expressed as a product of marginal and conditional probabilities

$$P_{O,L} = P(L \mid O) \cdot P(O), \tag{3}$$

in which each term on the right side of the equation can be expressed in the form of a logit model.

As stated above, the random components are assumed to conform to a generalized extreme value distribution. Because the random components of the utilities from the alternative choices may be correlated, I specify a GEV joint distribution function implied by the nested logit (McFadden 1981) that accounts for correlation among these error components as follows. I assume that the two licensing choices available (if the firm forgoes the option) are correlated because each licensing choice would be made with the same available information. Similarly, alternatives when an option is taken also are correlated because each choice would benefit from the information gleaned during the option period. This form of correlation is parameterized in the error distribution function by a "correlation coefficient" λ that represents correlation among alternatives.

The conditional probability that a license will be chosen or rejected given that the firm does not choose an option first is

$$P(\textit{License Choice} \mid \textit{No Option}) = \frac{e^{(V_L)_i}}{\sum_{i=1,2} e^{(V_L)_i}}. \tag{4}$$

Similarly, the conditional probability that a license will be chosen or rejected given that an option was chosen first is

$$P(License\ Choice \mid Option) = \frac{e^{(V_L)_i}}{\sum_{i=3,4} e^{(V_L)_i}}. \quad (5)$$

The marginal probability of a choice to option or not option, *P(Option Choice)*, can be expressed in an analogous fashion:

$$P(Option\ Choice) = \frac{e^{V_O+(1-\lambda_j)I_j}}{\sum_{j=No\ Option,\ Option} e^{V_O+(1-\lambda_j)I_j}}, \quad (6)$$

where $I_{No\ Option} = \ln[\sum_{i=1,2} e^{(V_L)_i}]$ is the natural log of the denominator in Equation (4) and $I_{Option} = \ln[\sum_{i=3,4} e^{(V_L)_i}]$ is the natural log of the denominator in Equation (5).

The likelihood function corresponding to the probability expression in (1) is

$$L = \prod_{n=1}^{N} \prod_{i=1}^{4} [P_n(L \mid O)P_n(O)]^{y_i}, \quad (7)$$

where $y_i = 1$ if observation n chooses alternative i, and zero otherwise. The coefficients are estimated using the full information maximum likelihood (FIML) method.

5.2. Independent Variables

There are two primary classes of independent variables—those influencing the option decision and those influencing the licensing choice. These variables are described below.

5.2.1. Technological Uncertainty. Uncertainty regarding the commercial potential of the patent affects the option choice and is likely to be higher for technologies that are more "basic" or more "distant" from commercialization. Ideally, one would measure such uncertainty by determining the stage of development of each invention at the time of disclosure, as was done by Jensen and Thursby (2001) in their survey of technology transfer offices. Such a measure is not available at the level of invention for the technologies examined in this study, unfortunately. Following Lanjouw and Schankerman (2001), I therefore measure technological "maturity" using the number of backward citations contained in the UC patent under consideration. In less developed technological areas (with less prior art to be cited), it is likely that there would be greater uncertainty regarding the commercial potential of patented inventions. In contrast, a greater number of prior art citations would indicate a more derivative invention and less uncertainty regarding the invention's potential.[10]

5.2.2. Technological Knowledge. To estimate the firm's stock of technological knowledge, I use measures based on U.S. patent data.[11] For the 136 potential licensees in the sample, I identify approximately 203,000 patents awarded to the firm, its parent and/or subsidiaries between 1975 and 1998. I then weight each patent by the number of forward citations it receives. Following the discussion in Hall et al. (2000), I control for "right-side" truncation bias by normalizing each patent in the firm's portfolio by the average number of citations to all patents issued in the same year and in the same primary (3-digit) USPTO technological class. The resulting citation weight is a measure of importance relative to other patents in its cohort and is comparable across years. I use this method to construct citation-based weights for each patent in the sample firms' portfolios.[12] The citation weight $\tilde{C}_i$ applied to patent i is

$$\tilde{C}_i = \frac{FC_i}{(\sum_{n=1}^{N_{c,t}} FC_{n,c,t})/N_{c,t}}, \quad (8)$$

where FC_i is the total number of citations received by patent i from future patents by the end of 1998 ("forward" citations), and $(\sum_{n=1}^{N_{c,t}} FC_{n,c,t})/N_{c,t}$ is the mean number of forward citations received by all patents issued in the same year t and in the same 3-digit technology class c as patent i.

Because not all knowledge represented by a firm's patent portfolio is equally relevant to the licensing of a patent in a particular area, I also account for the "relatedness" of the firm's knowledge base to

[10] The initial and exercise costs of the option also could influence the firm's choices. For example, if the option is priced too high, a firm may forgo the option although a technology's prospects are highly uncertain. Similarly, a firm may perceive the OTT's licensing fees as excessive and forgo a license. Although I am able to observe the option and licensing fees paid, I do not include cost as an explanatory variable in the analysis for three reasons. First, I am not able to observe the price of an option or license that was not chosen. Second, as described above, option contracts typically specify that if the option is exercised, the optionee and the university agree to "negotiate in good faith" to arrive at mutually agreeable and reasonable license terms. Because a primary objective of the OTT is to encourage commercialization of university technologies, this suggests that the OTT would be unlikely to seek an excessive initial or exercise price that may deter an interested party, but rather agree to a fee deemed reasonable by the licensee. Third, and most important, option and licensing fees represent a small portion of the overall cost of the option or license. A substantial proportion of the licensing or exercise cost hinges on the firm's subsequent commercialization strategy. Because the firm may choose alternative levels and modes of investment under the license, it is not possible to calculate an exercise price ex ante (Bowman and Moskowitz 2001, Adner and Levinthal 2004).

[11] See Jaffe and Trajtenberg (2002) for validation of patent-based measures of technological knowledge despite inherent weaknesses in patent data.

[12] An additional feature of this weighting method is that it also controls for differences in the propensity to patent across technological fields.

the patent being licensed.[13] The dimension of related knowledge that enables the firm to realize value from the licensed patent (Arora and Gambardella 1994) is measured by the share of the firm's patent portfolio at the time of the licensing decision that is assigned to the same primary class as the class of the licensed patent. A high share of patenting activity by the firm in a particular primary patent class indicates that a substantial proportion of the firm's innovative activity has occurred in the area represented by that patent class. For university inventions within this "focus" area, firms may be better able to utilize the university technology and therefore anticipate a higher expected payoff than from a more technologically distant invention.

The variable used to measure the influence of the related technological knowledge for firm j on its ability to realize a return from a licensed university patent in primary class c at the time of the licensing decision t is expressed as follows:

$$Focus_{j,c,t} = \left[\left(\sum_{t-6}^{t}\sum_{j}\tilde{C}_i \cdot p_i\right)_c \Big/ \left(\sum_{t-6}^{t}\sum_{j}\tilde{C}_i \cdot p_i\right)\right], \qquad (9)$$

where $(\sum_{t-6}^{t}\sum_j \tilde{C}_i \cdot p_i)_c$ is the citation-weighted sum of firm i's issued patents that were applied for within six years at the time of the licensing decision t and are in the same primary patent class as the class c of the university patent under consideration, and $\sum_{t-6}^{t}\sum_j \tilde{C}_i \cdot p_i$ is the sum of all citation-weighted patents issued to the firm that were applied for by date t.

The dimension of a firm's technological knowledge that enables it to assess the range of potential payoffs of the patent under consideration is related to the firm's option choice. One proxy of this expertise is the "quality" of the output generated by the firm's research activities in a certain area. In a study of 17 pharmaceutical firms, Narin et al. (1987) found the number of citations to a firm's patent portfolio to be significantly correlated with the National Institutes of Health's expert advisory panel evaluations of the "creativity and innovativeness," "overall contribution," and "success at pursuing basic research" of the firm's R&D efforts. We would expect these characteristics to also be correlated with the firm's capability in evaluating scientific developments (Arora and Gambardella 1994). I therefore estimate the firm's expertise in the technological area related to the university patent by calculating the mean citation weight of the firm's patents that are in the same patent class as the university patent. This instrument reflects the extent to which patents in patent class c held by firm j are cited, relative to the citation of other patents in the same cohort:

$$Expertise_{j,c,t} = \left(\sum_{t-6}^{t}\sum_{i}\tilde{C}_i\right)_c \left(\frac{1}{N_{c,t}}\right). \qquad (10)$$

Mowery et al. (1998) find "decreasing returns to similarity" in their study of the effect of technological overlap on the formation and performance of R&D alliances. To control for possible decreasing returns in the influence of knowledge assets on the option and licensing choices, I include squared terms for *Focus* and *Expertise* in the option and licensing submodels.[14]

5.3. Control Variables

Most of the research generating inventions in this sample was funded by federal agencies such as the National Institutes of Health or the National Science Foundation. In some cases, however, the research was performed under corporate sponsorship. Although UC does not give preference to corporate sponsors in the licensing process, sponsorship does suggest a particular interest in the research underlying the patented invention. I therefore control for corporate sponsorship in the option submodel with the indicator variable *Sponsor* that equals one if the firm has sponsored research that has led to the patent under consideration, and equals zero otherwise. Similarly, firms that have previously licensed a UC invention may be better able to evaluate a new technology in related areas. I therefore create the variable *Previous Licensee*, which equals one if the firm has previously licensed a patent in the same class from the university, and equals zero otherwise.

In a study of licensing of Massachusetts Institute of Technology (MIT) patents, Shane (2002) found that when markets for technology are ineffective, inventor-founded start-up firms are more likely than established firms to be the licensee. Although start-up

[13] Technological relatedness measures using patents are typically based on the distribution of the firm's patents among primary patent classes (see, for example, Jaffe 1989). Such measures attempt to gauge the degree of congruence between pairs of patent portfolios. Firms with a larger share of patents in the same or closely related patent classes as the patent being licensed should have more related technological knowledge (both codified and uncodified) bearing on that invention.

[14] Although I theoretically distinguish between evaluative and commercialization knowledge capabilities, they are likely correlated (Arora and Gambardella 1994). We would therefore expect that a firm that is better able to evaluate scientific developments may also be better able to derive economic value from a licensed technology. In the specification, *Expertise* is modeled as directly affecting only the option choice and *Focus* directly affecting only the licensing choice; the nested logit specification accounts for correlation between these two dimensions of absorptive capacity and their joint effects on the licensing decision, however. The joint probability of the option and licensing choices in the specification is a product of the conditional logit depicting the licensing choice and the marginal logit representing the option choice of the model depicted in Equation (3) (McFadden 1981).

firms are excluded from my sample, inventor-related firms may have a higher expected value of commercialization for some inventions. Moreover, because a significant portion of technological knowledge is tacit, one avenue by which the firm may obtain such knowledge could be through a relationship with the inventor. Inventor-related firms also may have "inside" knowledge regarding the number and identity of other firms signing secrecy agreements for the licensed patent, which could affect the inventor firm's licensing decision. To control for such relationships between the inventor and potential licensees, I include an indicator variable, *Inventor Firm*, which equals one if the university inventor has a fiduciary, other executive, or scientific advisory board position with the firm and zero otherwise.[15]

Competition. Competition from innovative activity by others may have an impact on the firm's licensing and option decisions. I measure competition in two ways. First, I control for variation in the size of the pool of firms making licensing decisions for each patent. The variable *Competitors* is simply the count of firms signing secrecy agreements for each licensed patent.[16] Second, I control for the influence of technological activity by a firm's most likely competitors on the option decision. When technological activity is high, the possibility that other firms may be able to "get to market" with a competing technology may be increased. Under such conditions, a firm may not have the luxury to delay the commercialization activities that would be required under a license but prohibited under an option contract. I measure the changes in patenting activity in the primary patent class of the sample patent by all other firms signing secrecy agreements. For each firm j signing a secrecy agreement, *Technological Competition* equals the difference in the share of patents in the same primary patent class as the licensed university patent of all issued patents applied for by the other firms signing secrecy agreements for a particular patent between the year of the earliest secrecy agreement and six years earlier. A positive value for *Technological Competition* indicates increasing inventive activity in the university patent's primary patent class over the previous six years by other potential licensees, whereas a negative value indicates decreasing inventive activity by those firms over the six-year period. A value of zero suggests either no change in the share of patenting activity in patent class u in the previous six years or that no other firms k signed secrecy agreements (where $k \neq j$).

$$\begin{aligned}\textit{Technological Competition}_{j,u,t} &= \left[\left(\sum_{j\neq k}\sum_{i}\tilde{C}_i\cdot p_i\right)_u \Big/ \left(\sum_{j\neq k}\sum_{i}\tilde{C}_i\cdot p_i\right)\right]_t \\ &\quad - \left[\left(\sum_{j\neq k}\sum_{i}\tilde{C}_i\cdot p_i\right)_u \Big/ \left(\sum_{j\neq k}\sum_{i}\tilde{C}_i\cdot p_i\right)\right]_{t-6}, \qquad (11)\end{aligned}$$

where the first term is the fraction of citation-weighted patents i in university patent class u applied for by all firms k in the year of the earliest secrecy agreement, t, and the second term is the fraction of citation-weighted patents in patent university patent class u applied for by all firms k in the year six years prior to the year of the earliest secrecy agreement.

Invention Quality. Invention technological and commercial "quality" is likely to influence the ultimate licensing decision. Following Trajtenberg et al. (1997), I measure invention "quality" or "importance" by the number of citations it receives from other patents (*Importance*). As discussed earlier, a patent's technological importance is likely to be correlated with the number of patent citations it receives. To control for right-truncation bias, I count only those citations within the first six years of the issue date of the UC patent.[17]

The dependent and independent variables described above are summarized in Table 1.

Table 2 provides descriptive statistics based on the sample of 457 licensing decisions. The number of potential licensees for each university patent ranges from one to 19, with a mean slightly below five (*Competitors* = 4.78). The mean proportion of weighted patents by potential licensees held in the patent class of the university invention is approximately 15% (*Focus* = 0.15). Roughly 15% of the decisions were made by firms that sponsored the research leading

[15] Firms also could partially acquire know-how related to the invention through consulting relationships with the inventor or by hiring graduate students involved with the research producing the technology under consideration for licensing. Unfortunately, information on such activity is not available for the firms in this study.

[16] Care must be taken in interpretation of results for this variable. The university does not divulge the identity of other firms signing secrecy agreements for a particular invention, nor does it divulge the existence of other secrecy agreements. It is therefore unlikely that a firm would explicitly incorporate this information into its licensing decision. The coefficient for *Competitors* would more likely be an artifact of the construction of the empirical model; the expectation that any one firm will be the licensee of a patent declines as the number of firms signing secrecy agreements increases. For example, if $p(L \mid N) \sim 1/N$, then $p(L)$ is decreasing in N.

[17] Potential endogeneity between the number of patent citations and the licensing decision poses a more difficult problem with this measure. One reason for differences in the level of citation of licensed and unlicensed patents could be that licensed patents are likely to be more intensively developed than unlicensed patents. This greater attention may generate more cumulative invention, resulting in a higher level of citation for licensed patents. The licensee itself is likely to be responsible for much of this endogenous citation. To partially control for endogeneity, therefore, I omit self-citations.

Ziedonis: *Real Options in Technology Licensing*
Management Science 53(10), pp. 1618–1633, ©2007 INFORMS 1627

Table 1 Summary of Variable Descriptions

	Description	Choice
Dependent variable		
Decision	Discrete variable that equals 1, 2, 3, or 4, corresponding to the licensing alternative chosen by the firm	
Independent variable		
Invention characteristics		
Technological uncertainty		
Maturity	Number of "backward" citations to prior patents by licensed university patent	Option
Competition		
Competitors	Number of firms signing secrecy agreements for each licensed university patent	License
Technological Competition	For those firms signing secrecy agreements, change in share of firm patents in the same patent class as university patent under consideration between year of earliest secrecy agreement and six years earlier	Option
Importance		
Importance	Number of citations within six years of patent issue date, excluding self-cites	License
Firm characteristics		
Technological knowledge		
Focus	Citation-weighted share of firm's patents in the same primary patent class as the patent under consideration	License
Expertise	Mean citation weight of firm's patents in the same primary patent class as the patent under consideration	Option
Other firm characteristics		
Sponsor	Indicator variable that equals one if firm-sponsored research leads to licensed patent, zero otherwise	Option
Previous Licensee	Indicator variable that equals one if firm has previously licensed a patent in the same class, zero otherwise	Option
Inventor Firm	Indicator variable that equals one if university inventor has a fiduciary or other executive relationship with firm, zero otherwise	License
Compustat	Indicator variable that equals one if firm is listed in Standard & Poor's Compustat Database (1998)	—

to the licensed invention (*Sponsor* = 0.16) or previously licensed a UC patent in the same class (*Previous Licensee* = 0.16). Only 1% of the licensing decisions were made by inventor firms, however (*Inventor Firm* = 0.01). Each UC patent cited was cited by more than five patents within six years on average (*Importance* = 5.61), although the number of citations varied from zero to 50. In turn, the average UC patent cited slightly more than four patents as prior art (*Maturity* = 4.36), and this value varied between zero and 21.

Table 2 Descriptive Statistics for Independent Variables ($N = 457$)

Variable	Mean	Standard deviation	Minimum	Maximum
Focus	0.15	0.23	0	1
Expertise	0.83	0.93	0	6.4
Sponsor	0.16	0.37	0	1
Previous Licensee	0.15	0.36	0	1
Inventor Firm	0.01	0.08	0	1
Maturity	4.36	4.49	0	21
Importance	5.61	6.54	0	50
Competitors	4.78	3.84	1	19
Technological Competition	0.01	0.14	−1	1

6. Results

Table 3 reports results for the nested logit model. Panel A of the table contains results for the 457 option outcomes and includes the inclusive value coefficient. Panel B reports results for 400 licensing choices that were made when an option was not taken, and Panel C presents results for the 57 licensing decisions made after an option was taken. As described in §4, all three sub-models are estimated simultaneously using the full information maximum likelihood method.

6.1. Option Choice

Turning first to Panel A of Table 3, *Maturity* is negative and significant (−0.29). This result indicates that firms are less likely to purchase an option contract prior to making a licensing decision when the technology being considered is characterized by less uncertainty. This finding is consistent with the basic premise of real-option theory, that options are more valuable under conditions of greater uncertainty, and supports Hypothesis 1.

The second set of variables in this panel measures the firm's ability to evaluate external technologies, whether it has sponsored research underlying

Table 3 Nested Logit Results for Option and Licensing Decisions

Panel A—Option choice: Dependent variable = *P*(*Option*)		
Invention characteristics		
Maturity	−0.29***	(0.04)
Firm characteristics		
Expertise	−0.88***	(0.23)
Expertise^2	0.16***	(0.05)
Sponsor	−0.74*	(0.42)
Previous Licensee	0.28	(0.35)
Competition variable		
Technological Competition	0.38	(1.01)
Inclusive value parameter		
λ	0.01	(0.07)
Panel B—License choice conditional on option not chosen: Dependent variable = *P*(*License* | *No Option*)		
Firm characteristics		
Focus	4.61***	(1.60)
Focus^2	−4.25**	(2.09)
Inventor Firm	345.16	(1566.27)
Competition variable		
Competitors	−0.49***	(0.06)
Invention characteristics		
Importance	0.11***	(0.03)
Panel C—License choice conditional on option chosen: Dependent variable = *P*(*License* | *Option*)		
Firm characteristics		
Focus	−31.49**	(15.06)
Focus^2	32.20**	(14.89)
Inventor Firm	—	—
Competition variable		
Competitors	−0.49**	(0.21)
Invention characteristics		
Importance	0.18**	(0.08)
N	457†	
Log likelihood	−396.91	
$\chi^2(\beta'X = 0)$	473.25***	

Note. Standard errors are in parentheses.
*$p < 0.10$, **$p < 0.05$, ***$p < 0.01$.
†$N = 457$ for option decision, $N = 400$ for licensing decision conditional on option not taken, $N = 57$ for licensing decision conditional on option taken.

the licensed invention, and whether it has previously licensed a UC technology in the same patent class as the licensed patent. The variables *Expertise* and *Expertise*^{2} are related to the firm's ability to evaluate external technologies. The negative and significant coefficient for *Expertise* (−0.88) indicates that the better the firm is in evaluating external technologies, the less likely it is to purchase an option to obtain additional information prior to making a licensing decision, consistent with Hypothesis 2. This finding suggests that firm evaluation capabilities play an important role in the value of an option, and a portion of option value can be firm specific. Stated differently, options appear to be less valuable for firms with greater "absorptive capacity" because these firms are better positioned to evaluate the technology outright. The positive and significant coefficient for the squared term (0.16) is consistent with diminishing marginal returns to this "absorptive capacity" in the option decision. The coefficient for *Sponsor* is negative and weakly significant, suggesting that firms that sponsor research at UC are less likely to purchase an option prior to making a licensing decision.[18] The coefficient for *Previous Licensee* is not significant, implying that previous licensing experience from UC has no influence on the firm's decision of whether to purchase an option for the university invention.

The coefficient for the competition variable *Technological Competition* is not significant, suggesting that the OTT policy of not divulging the identity or existence of other signatories of secrecy agreements may be effective. The result also suggests that potential licensees may not have good information on the patenting activities of other potential competitors. In either case, it appears that firms do not factor in the patenting activities of other signatories in making their option decisions.

The inclusive value parameter coefficient, $(1 - \lambda)$, is estimated to be close to one, indicating a lack of cross-elasticity across submodels. This estimate suggests that separating the licensing choices according to the option choice is valid (Hausman and McFadden 1984).

6.2. Licensing Decision When Option Is Not Chosen

Panel B of Table 3 reports coefficients for the variables thought to influence the licensing decision when firms forgo the option contract. The coefficient on *Focus* (4.61) is positive and significant. Consistent with Hypothesis 3, this result suggests that firms with higher levels of related technological knowledge are more likely to license without additional review beyond that provided by the secrecy agreement.

Of the control variable coefficients, the coefficient for *Inventor Firm*, signifying whether the university inventor has a fiduciary relationship with the potential licensee, is positive but indistinguishable from zero. This finding is consistent with the UC licensing policy that does not give preference to inventors in licensing. The coefficient for the competition variable *Competitors* is negative and significant, indicating that as the number of firms signing secrecy agreements climbs, the probability that a particular firm would license the university patent declines. Finally, the "importance" of the invention, measured by the number of citations it receives, is positively and significantly correlated with the probability of licensing

[18] Although UC does not formally give preference to sponsors in licensing technology, a sponsoring firm's licensing decision may be influenced above and beyond what is captured in the *Focus* variable. I therefore conducted a robustness check including a dummy variable for *Sponsor* in the licensing submodels (not reported). The coefficient for *Sponsor* is positive and significant for firms that made a licensing decision when an option is not chosen. The main results are unaffected, however.

Ziedonis: *Real Options in Technology Licensing*
Management Science 53(10), pp. 1618–1633, © 2007 INFORMS 1629

Table 4 Differences of Means Tests for Firm Characteristics (*t*-Test): Licensing Choices Conditional on Option, No Option (Firms Choosing License Only)

	No option		Option		
Variable	Number of observations	Mean	Number of observations	Mean	*t*-test statistic
Focus	148	0.13 (0.02)	10	0.22 (0.13)	−0.68
Expertise	148	0.83 (0.01)	10	0.79 (0.25)	0.18
Inventor Firm	148	0.02 (0.01)	10	0.00 (0.00)	1.74*
Compustat	148	0.79 (0.03)	10	0.60 (0.16)	1.14

Note. Standard errors are in parentheses.
*$p < 0.10$, **$p < 0.05$, ***$p < 0.01$. Unequal variances are assumed.

(*Importance* = 0.11). Again, this result is not surprising in that more technologically important inventions also are more likely to be commercially desirable (Harhoff et al. 1998, Lanjouw and Schankerman 2001, Sampat and Ziedonis 2004).

6.3. The Option Exercise Decision

Panel C of Table 3 reports the results for firms that made licensing decisions after taking an option. This subset consists of 57 firm decisions.

In contrast to the result for firms that forgo options, the coefficient for related technological assets, *Focus*, is negative and significant (the squared term is positive and significant).[19] This result implies a negative relationship between *Focus* and the probability of licensing and therefore is inconsistent with Hypothesis 3. This finding also suggests a different licensing decision calculus for firms that purchase options prior to making a licensing decision.

Why might we observe different behavior by firms licensing patents directly and by those taking licenses at the end of an option period? As a first step toward understanding this finding, I conduct "difference of means" tests on firm characteristics for the two groups of licensees—those firms that license without taking an option and those who license after taking an option first. The objective of this exercise is to determine whether there are systematic firm-specific differences between these two groups that may explain the observed difference in behavior. The results are reported in Table 4.

Table 4 reports that 148 licenses were taken when an option was not purchased (out of 400 licensing decisions). When an option was purchased first, 10 licenses were ultimately taken (out of 57 licensing decisions). These results provide little evidence that differences in the firm characteristics measured in this study underlie conflicting licensing behavior when options are and are not chosen (although a small number of observations for firms taking options undercuts the power of these tests). A difference in the fraction of inventor firms (represented by *Inventor Firm*) between the two groups is the only characteristic exhibiting (weak) significance.

To investigate whether technology field effects explain the differences in firm licensing behavior depending on whether or not an option is chosen, I disaggregate the sample of all firm decisions into those regarding broad biomedical and nonbiomedical technology areas using the USPTO Technology Assessment and Forecast Report designations. See Ziedonis (2001, Appendix C, pp. 189–191) for details on how these reports were used to identify biomedical technology areas. Ten biomedical U.S. patent classes (128, 424, 435, 436, 514, 530, 536, 604, 606, and 623) are represented in the UC sample, accounting for 70% of all firm license and option decisions. The most heavily represented biomedical patent class is 514 (drug, bio-affecting, and body treating compositions), accounting for 26% of all decisions in the overall sample and 37% of the decisions on biomedical patents. The remaining 30% of decisions concern nonbiomedical patents.

The results, reported in Table 5, provide little evidence that differences between these technology fields explain the differences in firm licensing behavior between firms that chose options and those that did not. In column 2 of Table 5, we see that 72% of the 400 licensing choices when an option was not chosen were for biomedical technologies, compared to 61% of the 57 licensing choices when an option was chosen. The *z*-statistic for the difference in these proportions is 1.35 and is not significant. (The proportions for nonbiomedical technologies in column 3 of Table 5 show similar results.)

Table 5 Two-Sample Test of Proportions for Licensing Decisions for Biomedical/Nonbiomedical U.S. Patent Classes by Option Choice

	Technology area	
	Biomedical	Nonbiomedical
Licensing choice conditional on option not taken		
Number of decisions	286	114
Proportion of decisions	0.72	0.28
Licensing choice conditional on option taken		
Number of decisions	35	22
Proportion of decisions	0.61	0.39
Difference in proportions	0.09	−0.11
z-statistic	1.35	−1.03

[19] Less important, the coefficient for *Competitors* is negative, as expected. There were no inventor-related firms choosing options, therefore a coefficient for *Inventor Firm* does not appear in this panel.

7. Discussion and Conclusion

A growing number of studies have used real-options thinking to examine a wide range of strategic decisions. Despite a well-developed theoretical literature, there has been very little empirical work utilizing actual option contracts to verify the basic tenets of this theory. This paper has undertaken such an objective, using data on licensing of university technologies to understand how managers use real options. Because much of the empirical literature on the use of "options" in technology acquisition and management relies on "virtual" or "implied" options, the exploitation of actual option agreements in an empirical study represents an important contribution to the literature.

This paper also contributes to the empirical literature on technology licensing. Previous empirical studies (e.g., Taylor and Silberston 1973, Caves et al. 1983, Anand and Khanna 2000) have examined the incidence of licensing and the structure of license contracts. Because these studies have selected observed licenses as the unit of analysis, they have been unable to examine factors that influence the decision whether to license a particular technology. I overcome this limitation by distinguishing between firms that signed exclusive licenses for patented inventions from UC and firms that considered but ultimately decided not to license the same set of patented inventions. The detailed data employed in this study also enabled me to examine the licensing process more closely by disaggregating the licensing transaction into licensing and prelicensing agreements.

I find that when deciding whether to license a university technology, firms in my sample are more likely to purchase options for inventions characterized by greater technological and commercial uncertainty. This finding is consistent with one of the principal predictions of real-options theory, that options are more valuable when there is substantial uncertainty regarding future payoffs from an investment in new technology.

As discussed in §2, uncertainty can be reduced due to actions taken by firms or by exogenous events that occur during the option period (Folta 1998). Given the embryonic nature of university inventions and the provisions of option contracts at UC (that require "good-faith" efforts and due diligence in evaluating the commercial potential of a new technology), I argued that the value of sequencing the licensing process in this context primarily hinges on the resolution of endogenous uncertainty. Put differently, these arguments primarily represent "growth" rather than "deferral" options.

As a supplemental test to check whether differences in exogenous uncertainty across broad technology areas may explain the exercise of an option, I disaggregated the university patents for which an option was taken into "biomedical" and "nonbiomedical" technologies using the USPTO Technology Assessment and Forecast Report designations discussed earlier. These broad technology areas are likely to exhibit different patterns of exogenous uncertainty change. Biomedical technologies are often subject to lengthy regulatory approvals. In contrast, exogenous events such as new product announcements by rivals or the emergence of substitute technologies affecting the commercial promise of nonbiomedical technologies such as electronics and computer software are more likely to be revealed during a one- to two-year option period. If the primary value of pursuing an option contract is due to the resolution of exogenous uncertainty, we should expect a higher share of options to be exercised for nonbiomedical technologies relative to biomedical inventions. Inconsistent with this view, however, I find that roughly 40% of the options were exercised in each technology area and that the difference in these proportions was not significant. The outcome of this simple test suggests that the decision to exercise an option within the context of this study is primarily driven by the uncertainty resolved through firms working with and evaluating new technologies.

The results of this study also suggest that option value can be firm-specific; firms in the sample that are better able to evaluate an external technology are less likely to purchase an option. This finding adds empirical support to arguments advanced by Cohen and Levinthal (1990), Rosenberg (1990), and Arora and Gambardella (1994) that knowledge investments in R&D by firms have benefits that transcend the traditional view of R&D as primarily undertaken to generate inventions internally. These authors argue that a firm's internal knowledge base enables it to evaluate and exploit technologies developed outside its boundaries. In this study, firms appear to benefit from such "absorptive capacity" by making more streamlined decisions when identifying and developing external technologies.

Finally, the results also show that licensing behavior is more complex than may be apparent if we examined licensing contracts only, and depends on whether the firm has chosen to sign an option agreement first. For decisions made by firms forgoing option agreements (the majority of decisions in my sample), possession of technological knowledge related to commercialization is generally positively related to the probability of licensing, as expected, and is consistent with the findings in Arora and Gambardella (1994). More surprising, however, the relationship between technological knowledge and the likelihood of licensing tends to be quite different for firms that first sign an option agreement. In this case, firms with patents concentrated in areas related to the university invention are less likely to subsequently license the university patent.

Several explanations are consistent with this last finding. First, recall that the firm determines whether to take an option to gain additional information before making its licensing decision. The availability of an option may increase the probability that less attractive inventions will not languish. During the option period, firms whose patenting activity is more heavily concentrated in areas related to the university invention (e.g., high-focus firms) may be better able to "weed out" inventions that lack commercial viability. Ultimately, those firms may be more likely to reject a license than firms whose patenting activity is less concentrated in the university patent class.

Firms also may differ in their motivations for licensing optioned university inventions. For firms taking options in areas close to their own patenting focus, the licensing decision may be driven largely by a desire to exploit the firm's existing knowledge base. If this is true, the rejection rate by these firms may be relatively high if the commercial potential of an optioned invention is low. In contrast, for firms that take options on inventions more technologically distant from their patenting focus, their option and licensing decisions may be driven more by reasons of "exploration" or entry into new technological areas. These firms may be more willing to commit to a license.

A final explanation stems from the "Arrow paradox of information" (Arrow 1962). The Arrow paradox states that to sell an information good, the seller must reveal information to the buyer for the buyer to evaluate the information good for sale. Once the information is revealed, however, the buyer has little incentive to pay for it. By revealing information about the invention in stages (under secrecy, option, and license agreements) while simultaneously increasing the obligations and requirements of the potential and eventual licensee under these agreements, the university licensing process mitigates this problem to some degree. This procedure may not, however, resolve the Arrow paradox completely. The option "due diligence in evaluation" requirement is intended to prevent firms from simply purchasing the option to prevent access by other firms. Once a firm has exercised due diligence in evaluation during the period of the option, it undoubtedly has a better understanding of the technology and its own ability to commercialize it. If a firm can appropriate returns to knowledge gained under the option agreement by "inventing around" the university invention or utilizing the knowledge internally, its incentives to license the patent at the end of the option period may be lowered. Firms whose patenting activities are more concentrated in areas related to the university invention may be better able to "learn" from the information disclosed during the option period, and therefore be less likely to subsequently license that invention. In this case, the firm could benefit from the university invention without paying royalties to the university to license the patent.

Whether such an outcome of the licensing process is problematic depends upon one's point of view. From the perspective of a university attempting to generate licensing income, foregone revenues from unconsummated licensing agreements following options may be a problem requiring improved monitoring of firm behavior, increased diligence on potential patent infringement, or avoidance of option agreements with firms with strong patent portfolios in the field of the university invention.

Additional research could differentiate between the alternative explanations for the post-option licensing behavior articulated above. We might expect the "Arrow paradox" to be less problematic when the patent appropriability regime of the university patent is strong, which may make it more difficult for a firm to appropriate the returns to the patent without licensing commercialization rights from the university. Using available data, it is straightforward to identify differences in the appropriability regime between optioned patents that are licensed and those not licensed. However, because firms also would be more likely to license a patent when appropriability is high because it would be more difficult for other firms to invent around a licensed patent (Shane 2002), we would still be unable to differentiate between these motivations. This identification problem points to the importance of future research in conducting interviews of both managers at firms and university licensing professionals to augment archival data in future research.

Additional research could also investigate the role of complementary assets other than related technological knowledge on the university technology licensing decision. Teece (1986) identifies possession of complementary assets such as production, distribution, and marketing capabilities as an important factor in commercializing new inventions. Shane (2001) shows that the importance of access to complementary assets influenced the whether a university technology licensed from MIT would be commercialized by a start-up firm. Using data from UC similar to that used in this study, however, Lowe and Ziedonis (2006) found that difficulty in commercializing university technology by start-ups due to a lack of access to complementary assets may be mitigated by later acquisition established firms. Using survey data of R&D lab managers, Arora and Ceccagnoli (2006, p. 297) find the effect of complementary assets held by the licensee to be "small or negligible." None of these studies specifically consider the effect of complementary assets on university technology option and license choices by established firms, suggesting a need for future research in this area.

In addition to implications for managers, the findings in this study may also be pertinent for university technology transfer policy. By making options available, the university encourages firms to consider inventions they might otherwise reject at the initial licensing phase. Some of these inventions may ultimately be licensed, potentially benefiting both the licensee and the university. These findings also suggest, however, that firms may benefit from the availability of options in other ways that are less attractive from the university's perspective. If a firm can utilize the option period to understand and subsequently apply knowledge about the invention to its own technological base, its incentive to license the invention may be diminished. To the extent that licensing revenues are a principal means by which university technology transfer offices appropriate returns from an exercised option, this interpretation suggests a less favorable outcome for the university.[20]

These divergent conclusions have contrasting implications for university technology managers. The first class of explanations suggests that options should be tailored to account for differences in firms' abilities and motivations. For example, a firm that purchases an option to determine if a university invention can be exploited by its related knowledge base may be interested in a different agreement structure for the option and subsequent license agreement than a firm pursuing an option as a means to consider entry into a new technological area. By accounting for a firm's capabilities and motivation in tailoring option agreements, the technology transfer office may be able to further facilitate the licensing process. In contrast, to the extent the "Arrow paradox" is pervasive in university technology licensing, the simple granting of options may be insufficient to protect the interest of the university technology transfer office. Strengthening due diligence provisions, increasing monitoring of firm behavior (both during the option period and after its expiration), or avoiding granting options to firms with strong patent portfolios in the field of the university invention may be necessary.[21] This prescription would suggest that the technology transfer office embrace a narrow, primarily pecuniary, view of technology transfer. However, the ultimate effect of reduced licensing income for the university may be to spur related inventive effort by the optionee, an outcome that would support a broader mission of technology transfer. From a more inclusive social welfare perspective, commercialization and the use of university inventions that would otherwise have been neglected by the private sector represents a success story, regardless of who captures the value.

[20] Universities may offer options for other reasons, such as to secure corporate research support in exchange for the right of first refusal to license future research outcomes. Such options may have contrasting implications for university and firm incentives and are different than those studied here, which are options to license previously invented technology.

[21] This introduces a possible dilemma for OTT: Firms with strong related patent portfolios may be the only firms willing to option an invention with high commercial uncertainty.

8. Electronic Companion

An electronic companion to this paper is available as part of the online version that can be found at http://mansci.journal.informs.org/.

Acknowledgments

The author gratefully acknowledges support from the Alfred P. Sloan Foundation, the Andrew Mellon Foundation, and the Mack Center for Technological Innovation at the Wharton School of the University of Pennsylvania. The author thanks Janet Bercovitz, Fred Feinberg, John Freeman, Bronwyn Hall, Rebecca Henderson, Jennifer Kuan, David Mowery, Richard Nelson, Joanne Oxley, Gary Pisano, Don Siegel, Kenneth Train, Brian Wright, Rosemarie Ziedonis, the associate editor, four anonymous reviewers, and seminar participants at the University of California, Berkeley, the University of Pennsylvania, the University of Michigan, the University of Illinois, the University of Southern California, the 2004 Harvard Business School Strategy Conference, and the 2004 Harvard Business School Conference on Entrepreneurship and Innovation for comments and suggestions. Lynn Fissell, Bronwyn Hall, Suzanne Majewski, and Suzanne Quick generously provided data used in this study.

References

Adner, R., D. A. Levinthal. 2004. What is *Not* a real option: Considering boundaries for the application of real options to business strategy. *Acad. Management Rev.* **29**(1) 74–85.

Amram, M., N. Kalutilaka. 1999. *Real Options: Managing Strategic Investment in an Uncertain World.* Harvard Business School Press, Boston, MA.

Anand, B. N., T. Khanna. 2000. The structure of licensing contracts. *J. Indust. Econom.* **48**(March) 103–135.

Arora, A., M. Ceccagnoli. 2006. Patent protection, complementary assets, and firms' incentives for technology licensing. *Management Sci.* **52**(2) 293–308.

Arora, A., A. Gambardella. 1994. Evaluating technological information and utilizing it: Scientific knowledge, technological capability, and external linkages in biotechnology. *J. Econom. Behav. Organ.* **24** 91–114.

Arrow, K. J. 1962. Economic welfare and the allocation of resources of invention. R. R. Nelson, ed. *The Rate and Direction of Inventive Activity: Economics and Social Factors.* Princeton University Press, Princeton, NJ, 609–625.

Association for University Technology Managers. 1998. AUTM licensing survey, fiscal year 1997. Report, Norwalk, CT.

Black, F., M. Scholes. 1973. The pricing of options and corporate liabilities. *J. Political Econom.* **81**(May–June) 637–659.

Boersch-Supan, A. 1990. On the compatibility of nested logit models with utility maximization. *J. Econometrics* **43** 373–388.

Borooah, V. K. 2002. *Logit and Probit: Ordered and Multinomial Models.* M. S. Lewis-Beck, series ed. Sage University Papers Series on Quantitative Applications in the Social Sciences, 07-138. Sage Publications, Thousand Oaks, CA.

Bowman, E. H., G. T. Moskowitz. 2001. Real options analysis and strategic decision making. *Organ. Sci.* **12**(6) 772–777.

Brant, R. 1990. Assessing proportionality in the proportional odds model for ordinal logistic regression. *Biometrics* **46** 1171–1178.

Caves, R. E., H. Crookell, P. J. Killing. 1983. The imperfect market for technology licenses. *Oxford Bull. Econom. Statist.* **45**(3) 249–267.

Cohen, W. M., D. A. Levinthal. 1990. Absorptive capacity: A new perspective on learning and innovation. *Admin. Sci. Quart.* **35** 128–152.

Dixit, A. K., R. S. Pindyck. 1994. *Investment Under Uncertainty.* Princeton University Press, Princeton, NJ.

Elfenbein, D. 2004. Contract structure and performance of technology transfer agreements: Evidence from university licenses. Working paper, University of California, Berkeley, CA.

Folta, T. B. 1998. Governance and uncertainty: The trade-off between administrative control and commitment. *Strategic Management J.* **19** 1007–1028.

Folta, T. B., K. Miller. 2002. Real options in equity partnerships. *Strategic Management J.* **23** 77–88.

Hall, B. H., A. Jaffe, M. Trajtenberg. 2000. Market value and patent citations: A first look. NBER Working Paper 7741, National Bureau of Economic Research, Cambridge, MA.

Hamilton, W. F. 2000. Managing real options. G. S. Day, P. J. H. Schoemaker, R. E. Gunther, eds. *Wharton on Managing Emerging Technologies.* John Wiley & Sons, New York, 271–288.

Harhoff, D., F. M. Scherer, K. Vopel. 1998. Citations frequency and the value of patented inventions. *Rev. Econom. Statist.* **81**(3) 511–515.

Hausman, J., D. McFadden. 1984. Specification tests for the multinomial logit model. *Econometrica* **52**(5) 1219–1240.

Jaffe, A. B. 1989. Characterizing the "technological position" of firms, with application to quantifying technological opportunity and research spillovers. *Res. Policy* **18** 87–97.

Jaffe, A. B., M. Trajtenberg. 2002. *Patents, Citations, & Innovations: A Window on the Knowledge Economy.* MIT Press, Cambridge, MA.

Jensen, R., M. Thursby. 2001. Proofs and prototypes for sale: The tale of university licensing. *Amer. Econom. Rev.* **91** 240–259.

Kogut, B. 1991. Joint ventures and the option to expand and acquire. *Management Sci.* **37**(1) 19–33.

Kogut, B., D. Kim. 1996. Technological platforms and diversification. *Organ. Sci.* **7**(3) 283–301.

Kogut, B., N. Kulatilaka. 1994. Operating flexibility, global manufacturing, and the option value of a multinational network. *Management Sci.* **40**(1) 123–139.

Kogut, B., N. Kulatilaka. 2001. Capabilities as real options. *Organ. Sci.* **12**(6) 744–758.

Kulatilaka, N., E. C. Perotti. 1998. Strategic growth options. *Management Sci.* **44**(8) 1021–1031.

Lanjouw, J. O., M. Schankerman. 2001. Characteristics of patent litigation: A window on competition. *RAND J. Econom.* **32**(1) 129–151.

Long, J. S. 1997. *Regression Models for Categorical Dependent Variables.* Sage Publications, Thousand Oaks, CA.

Lowe, R. A., A. A. Ziedonis. 2006. Overoptimism and the performance of entrepreneurial firms. *Management Sci.* **52**(2) 173–186.

MacMillan, I. C., R. G. McGrath. 1998. The value, assessment, and enhancement of industry investments in fundamental research. Working paper, The Wharton School, University of Pennsylvania, PA.

Majd, S., R. S. Pindyck. 1987. Time to build, option value, and investment decisions. *J. Financial Econom.* **18**(1) 7–27.

McDonald, R., D. Siegel. 1986. The value of waiting to invest. *Quart. J. Econom.* **101**(4) 707–728.

McFadden, D. 1981. Econometric models of probabilistic choice. D. McFadden, ed. *Structural Analysis of Discrete Data with Econometric Applications.* MIT Press, Cambridge, MA, 198–272.

McGrath, R. G. 1997. A real options logic for initiating technology positioning investments. *Acad. Management Rev.* **22**(4) 974–996.

Miller, K. D., T. B. Folta. 2002. Option value and entry timing. *Strategic Management J.* **23**(7) 655–665.

Mitchell, G. R., W. F. Hamilton. 1988. Managing R&D as a strategic option. *Res.-Tech. Management* **31**(3) 15–22.

Montgomery, C. A., B. Wernerfelt. 1988. Diversification, Ricardian rents, and Tobin's *q*. *RAND J. Econom.* **19**(4) 623–632.

Mowery, D. C., J. E. Oxley, B. S. Silverman. 1998. Technological overlap and interfirm cooperation: Implications for the resource-based view of the firm. *Res. Policy* **27** 507–523.

Narin, F., E. Noma, R. Perry. 1987. Patents as indicators of corporate technological strength. *Res. Policy* **16**(2–4) 143–155.

Reuer, J. J., T. W. Tong. 2005. Real options in international joint ventures. *J. Management* **31**(3) 403–423.

Roberts, K., M. L. Weitzman. 1981. Funding criteria for research, development, and exploration projects. *Econometrica* **49** 1261–1288.

Rosenberg, N. 1990. Why do firms do basic research with their own money? *Res. Policy* **19** 165–174.

Sampat, B. N., A. A. Ziedonis. 2004. Patent citations and the private value of patents. U. Schmoch, ed. *Kluwer Handbook of Science and Technology Indicators.* Kluwer Academic Publishers, Boston, MA, 277–298.

Shane, S. 2001. Technology regimes and new firm formation. *Management Sci.* **47**(9) 1173–1190.

Shane, S. 2002. Selling university technology. *Management Sci.* **48**(1) 122–137.

Silverman, B. S. 1999. Technological resources and the direction of corporate diversification: Toward an integration of the resource-based view and transaction cost economics. *Management Sci.* **45**(8) 1109–1124.

Standard & Poor's Compustat Database. 1998. McGraw-Hill Company, New York.

Taylor, C. A., Z. A. Silberston. 1973. *The Economic Impact of the Patent System: A Study of the British Experience.* Cambridge University Press, Cambridge, UK.

Teece, D. J. 1986. Profiting from technological innovation: Implications for integration, collaboration, licensing and public policy. *Res. Policy* **15** 285–305.

Train, K. 1986. *Qualitative Choice Analysis: Theory, Econometrics, and an Application to Automobile Demand.* MIT Press, Cambridge, MA.

Trajtenberg, M., R. Henderson, A. B. Jaffe. 1997. University vs. corporate patents: A window on the basicness of invention. *Econom. Innovation New Tech.* **5** 19–50.

Trigeorgis, L. 1996. *Managerial Flexibility and Strategy in Resource Allocation.* MIT Press, Cambridge, MA.

Zahra, S. A., G. George. 2002. Absorptive capacity: A review, reconceptualization, and extension. *Acad. Management Rev.* **27**(2) 185–203.

Ziedonis, A. A. 2001. The commercialization of university technology: Implications for firm strategy and public policy. Unpublished dissertation, University of California, Berkeley, CA.

[25]

© Academy of Management Review
2004, Vol. 29, No. 1, 74–85.

WHAT IS *NOT* A REAL OPTION: CONSIDERING BOUNDARIES FOR THE APPLICATION OF REAL OPTIONS TO BUSINESS STRATEGY

RON ADNER
INSEAD

DANIEL A. LEVINTHAL
The Wharton School, University of Pennsylvania

We argue that the greater the extent to which choice sets evolve as a consequence of firms' exploration activities, the less structured the firms' abandonment decisions become and, in turn, the less distinguishable a real option is from more generic notions of path dependence—a sequential stream of investment in and of itself does not constitute a real option. While organizational adaptations can extend the applicability of real options, they impose tradeoffs that may lead to the underutilization of discoveries made in the course of exploration.

The concept of real options has created considerable excitement in the management literature in recent years (Kogut & Kulatilaka, 2001; McGrath, 1997). The appeal of real options is quite natural. Firms face uncertain futures, and the investment opportunities they face are, to an important degree, a function of their prior investment commitments. Thus, at a surface level, the real options framework appears to precisely fit firms' strategic challenges by linking current actions to uncertain futures. In addition, relative to other justifications for exploration, such as rationales for slack search (March & Simon, 1958), real options provide a rhetoric that comfortably fits the standard language of corporate finance. Less obvious than the benefits offered by the real options framework but equally important are its inherent limitations in guiding organizations' decision making under uncertainty.

The underlying logic of the real options framework is based on the realization that future investment opportunities are contingent on prior investment commitments. Thus, in contrast to net present value analysis, real options analysis accounts for the sequential nature of choice processes. Such an observation is not alien to researchers working within the tradition of the behavioral theory of the firm (Cyert & March, 1963) or evolutionary economics (Dosi, 1982; Nelson & Winter, 1982; Penrose, 1959) who have long been sensitive to path dependence. What is distinctive about the use of real options as both an analytic tool (Dixit & Pindyck, 1994; Trigeorgis, 1997) and a strategic heuristic (Luehrman, 1998; McGrath, 1997) is that the real options logic offers the prospect of assigning actual values (cardinal in the former case or ordinal in the latter case) to stage-setting investments.

However, while the logic can always be used to generate values and orderings, the validity of these results depends on some key assumptions. The technical violations of strict real options assumptions and the methods for their accommodation are, by now, well known (e.g., Copeland & Antikarnov, 2001). More subtle, however, and more limiting to the validity of real options as a tool in organizational decision making are violations that stem from organizational processes in the face of different modes of uncertainty resolution.

The real options framework is intended to exploit the flexibility inherent in sequential investments. We argue that this flexibility stems from the possibility of abandoning investment initiatives, rather than from the simple substitution of a stream of smaller payments for a larger lump sum payment. This implies that the effective management of real options requires a high degree of rigidity in the specification of the agenda of the initiatives and the criteria for

We thank Bert Cannella, Javier Gimeno, Bruce Kogut, Christoph Loch, Anand Swaminathan, Peter Zemsky, participants at the 2001 Strategy Research Forum and the Jones Center Workshop at Wharton, and our three reviewers for their valuable comments on prior drafts of this manuscript.

their success. We consider the implications, positive and negative, of such constraints and compare them to more generic path-dependent approaches to managing investment under uncertainty.

In the next section we briefly lay out the formal structure of real options. We then examine some of the critical assumptions that underlie the application of options to firms' investment decisions. When uncertainty resolution emerges as an outcome of firm action, the sharp temporal demarcation made in the options literature between "Stage 1" and "Stage 2" investments is called into question. In such settings, beyond learning about a specific initiative, the flexibility associated with later investment decisions often stems from the possibility of discovering a wide variety of related opportunities, even in the face of unfavorable initial outcomes. We show that the greater the extent to which initiatives are open ended, the more problematic the application of the real options framework is. Flexibility in search can undermine the flexibility associated with abandonment. Abandonment is essential for limiting downside risk, a key virtue attributed to real option investments. We consider how organizational processes, such as the allocation of decision rights that limit the range of possible action and the specification of well-defined temporal boundaries, can extend the applicability of real options reasoning but force strategic tradeoffs of their own. We conclude by considering differences between real options approaches and more generic path-dependent processes, and we suggest how they might be distinguished empirically.

REAL OPTIONS

Real options investments are characterized by sequential, irreversible investments made under conditions of uncertainty (Dixit & Pindyck, 1994). The framework suggests that purchasing a real option on a strategically important opportunity allows firms to postpone commitment until a substantial portion of the uncertainty about the opportunity has been resolved. After making an initial investment, management is then to turn its attention to other matters and wait for a signal as to whether or not it is appropriate to harvest or cultivate the initial investment.

Consider the events that transpire prior to the exercise of financial options, on which the real options model is based (Figure 1). First, an investor purchases an option (Stage 1). Then, during the course of the holding period, the value of the option changes in response to external events. Throughout the holding period, the financial markets provide a clear signal as to the current value of the option. Finally, events transpire so that the investor chooses to exercise the option, or, alternatively, the expiration date specified in the option contract is reached and the option expires (Stage 2). Investments with this structure are optionlike in that Stage 2 investments are not a necessary consequence of

FIGURE 1
The Structure of Real Options

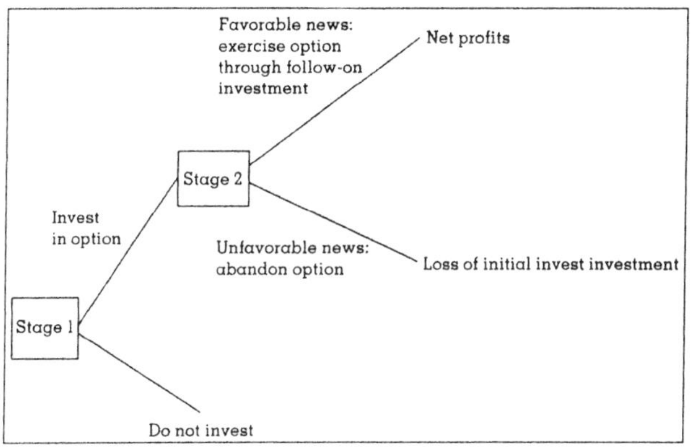

having made an initial Stage 1 investment but, rather, can be conditioned on the realization of interim information.

Note two critical features of this process: (1) the value of the option (and the underlying asset) is exogenous to the investor's activity—the investor cannot take steps to make the intrinsic characteristics of the asset more attractive; (2) the market signal of option value is readily observable and is independent of the investor's behavior. If these properties carry over to an investment of a nonfinancial sort (such as in plant and equipment, or in technology licenses), then the logic of options can directly carry over. The greater the extent to which these properties are violated, the more problematic the application of an options framework is.

The "wait and see" setting of financial options represents the extreme case of investment environments, but one for which the methodology is ideally suited. Because much attention in the management literature is focused on the ways in which the firm can affect outcomes and variances (e.g., McGrath, 1997), it is important to examine what happens to the applicability of options logic as we move away from a world of wait and see to a world of "act and see," in which uncertainty resolution is endogenous to firm activity.

FIGURE 2a
Boundaries of Applicability for Net Present Value and Real Options

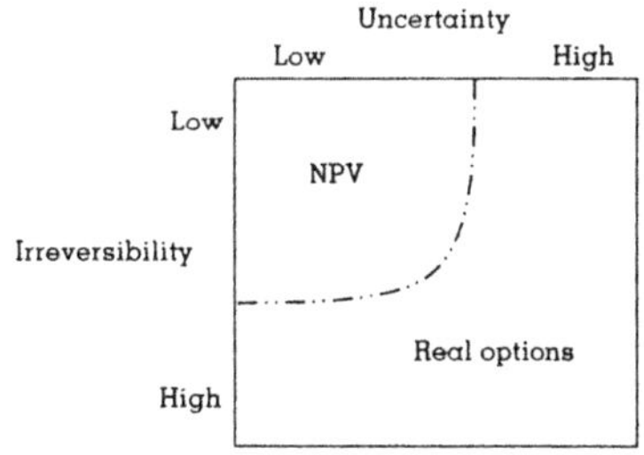

FIGURE 2b
Boundaries of Applicability for Real Options and Path-Dependent Opportunities

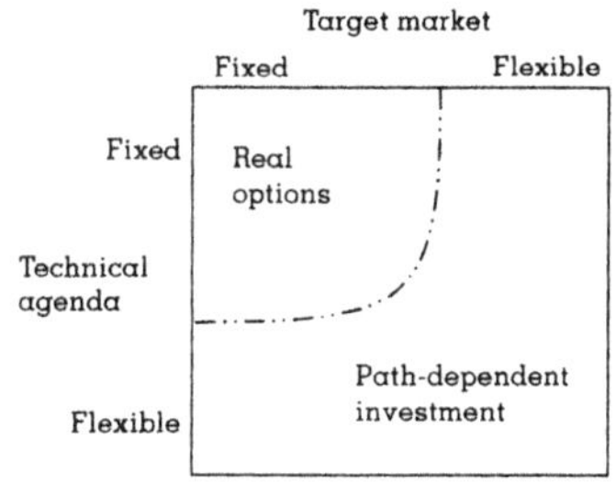

BOUNDARIES OF REAL OPTIONS

The boundaries of the logic of real options are often considered in relation to the breakdown of net present value analysis: to the degree that investment choices have the property of high uncertainty and irreversibility, a real options valuation provides a better characterization of the investment's true value than does a net present value calculation because of the latter's inability to account for the value of delaying commitments (see Figure 2a).

Less examined but no less important are the boundaries along which real options logic is itself strained. As we move from a world of real options on tradable assets, in which the firm has no hand in resolving uncertainty and the set of possible actions in response to this uncertainty resolution can be specified at the time of the initial investment, to real options on strategic opportunities whose outcomes are intimately linked to firm action, the clean demarcations between investment stages begin to blur and the application of real options becomes more challenging analytically and, focal to this paper, organizationally.

When target markets and technical agendas are flexible (see Figure 2b), the discrete investment logic of real options is eroded, and activities may be characterized more appropriately as more generic path-dependent processes that fall under such labels as *probe and learn* (Lynn, Morone, & Paulson, 1996), *incremental search* (March & Simon, 1958; Nelson & Winter, 1982), or *innovation journeys* (Van de Ven, Pollery, Garud, & Venkataraman, 1999). Alternatively, if the scope of the option investment is fixed a priori—that is, if the opportunities on which one is taking an option can be clearly specified at the inception of the option—then the decision to abandon an initiative can be clearly articulated and the flexibility associated with an option investment can be readily maintained.

In the context of financial options, one can clearly state a priori when a given option will be "in the money" and worth exercising. However, in the case of real options on strategic opportunities in which new possibilities are identified as a consequence of a firm's actions, such a prior specification may not be possible or even desirable. Experiments, even unsuccessful ones, not only provide information about intended investment paths but also provide information about other possibilities—possibilities that may not even have been envisioned at the time of the initial investments. Outcomes that are negative vis-à-vis the initial proposed initiative may still suggest or engender other possible actions. For instance, the failure of a technology development effort to reach a given technical hurdle in a specified time horizon does not preclude additional efforts or different approaches; similarly, if a new product fails to win acceptance in a given target market, it may still be successful in other possible target markets.[1]

At a basic logical level, the boundless set of possibilities associated with a strategic investment initiative presents a problem analogous to that posed by Popper (1959) in the context of hypothesis testing. Popper's proposition is that a hypothesis is never proved; rather, it is not disproved in any given test. Conversely, in a given attempt to realize an opportunity, we can only prove that an initiative can be successful in a particular setting; we cannot use negative outcomes to preclude the possibility that it will be successful in other target markets or with further technical refinements.

This "impossibility of proving failure" is an inherent feature of firm initiatives under uncertainty and poses a fundamental challenge to the applicability of the real options framework. Further, because actors at different levels of an organization have different perspectives on the attractiveness of a given opportunity, they will disagree as to the proper framing of termination decisions. Thus, the open-ended nature of the search for success raises organizational challenges to abandoning options that can deter firms from exercising the very flexibility that made the real options approach attractive in the first place.

The degree to which an initiative is circumscribed is itself dependent on organizational choices concerning project scope. If an initiative is highly circumscribed in terms of what appropriate target markets are and what temporal and technical boundaries must be respected, then the degree of flexibility in the potential directions of the initiative is limited, but at a price. Imposing rigid criteria for abandonment may result in the underutilization of discoveries made in the context of initiatives that are failures with respect to their initial agenda but that introduce promising possibilities not previously imagined.

This highlights the need for proponents of real options logic to consider a more nuanced organizational perspective that incorporates the different views that exist within an organization. The firm cannot be regarded as a unitary actor. Managers charged with pursuing an opportunity and executives charged with evaluating a portfolio of opportunities will differ in their beliefs as to when an option ceases to be attractive. Such differences are a function of their incentives and opportunity structures, and they serve to compound the psychological biases regarding sunk costs and escalating commitment that can act to deter strategic redirection.

These considerations are largely overlooked in the theoretical (e.g., Dixit & Pindyck, 1994; Trigeorgis, 1997) and managerial (e.g., Amram & Kulatilaka, 1998; Luehrman, 1998) work on real options in strategy, in which researchers tend to treat the firm as a monolithic actor. They are also largely overlooked in empirical work in which scholars tend to look at investment decisions that are not integrated within the firm's organizational activities (e.g., investments in acquisitions and joint ventures). The importance of organizational factors in determining the applicability of options logic increases as real options theory is extended from the evaluation of investments in physical assets, for which the resolution of uncertainty is exogenous to firm action and the scope of possible firm response to this uncertainty is relatively constrained (e.g., making use of a plant in a foreign country to potentially take advantage of swings in relative wages), to the evaluation of investments in stra-

[1] Knight's (1921) distinction between risk and uncertainty, which lies at the heart of notions of entrepreneurship (Kirzner, 1997), is closely related to the distinction we are making here. Real options are well suited to incorporate risk, an uncertain realization from a well-specified probability distribution. In contrast, the inherent unknowability that characterizes Knightian uncertainty poses a significant challenge to characterizing means and variances of key option analysis parameters.

tegic opportunities, for which the resolution of uncertainty is largely endogenous to firm action, and the scope for possible modifications in the initial initiative is vast.

CHALLENGE OF ABANDONMENT

Ambiguity of Stages in Real Options

An investment's flexibility is revealed in the project abandonment decision. Options are flexible not because they substitute a stream of smaller payment for a larger lump sum payment but because the payment stream can be abandoned in light of negative outcomes. The less well-defined the abandonment decision, the less valid the perception of flexibility and the less appropriate the nominal application of real options logic. In the absence of proofs of failure and strict, structured abandonment deadlines, firms face a difficult organizational challenge to exploit the flexibility offered by the sequential nature of optionlike investments. As noted earlier, there is an important asymmetry between positive and negative signals that is analogous to the asymmetry of proving and disproving a hypothesis in empirical research. Popper (1959) argues that it is logically impossible to prove that a hypothesis is true; we can only show that it is not false, up to a specified probability level. In contrast, a managerial initiative can only be shown to be successful in that some market or technical criteria are met; an initiative, such as the development of a new technology, cannot conclusively be shown to be incapable of succeeding.

Consider what we learn when a new technology is "tried." We may learn that, in a given time period and with a given level of investment, researchers were not able to meet certain technical milestones. Similarly, we may learn that a particular market context was not receptive to the technology in its current form. However, these results are not impossibility theorems. They do not demonstrate that, with more time and more resources, these technical milestones could not be met. Nor do such results demonstrate that other market settings might not respond positively to the technology in its current form or that enhancements to the technology or

FIGURE 3
Malleability of Technical and Market Focus to Firm Activity

	Target market fixed	Target market flexible
Technical agenda fixed	Oil production capacity Multinational production for currency arbitrage	Patented materials (technical properties fixed, but different markets may yield different outcomes—e.g., Viagra, which failed to reduce angina but was found efficacious elsewhere)
Technical agenda flexible	Artificial heart (application is fixed but different development approaches may yield different outcomes)	Wireless internet Solar panels Artificial intelligence

a reduction in cost will not lead to success in the current market setting.

Thus, the significance of any specific market response is constrained in market space to a specific segment, and constrained in time to a specific development state. Figure 3 maps the potential malleability of a project to interim feedback, in terms of the opportunity to affect the technical agenda of the project and the markets in which the project will be evaluated, after an option has been purchased. Again, what is critical is that these new directions need not have been identified at the time of the inception of the initial option investment. They were not part of the justification for the initiation of the option, but they became critical for the continuation of the investment effort. The greater the potential for ex post discovery of possible new directions for the original initiative, the greater the difficulty in deciding when to abandon the opportunity.

The problem of abandonment is compounded by the fact that most options on strategic opportunities do not have an explicit, exogenous expiration date. How long the firm can keep the option open, whether in persisting in development efforts in nominally promising pharmaceuticals or in market creation efforts for emerging technologies, is often difficult to specify ex ante.[2] When there is no explicit time limit on efforts to enhance the value of the initial initiative, time to expiration becomes an endogenous choice. From a valuation perspective, longer option durations lead to higher option values and increase the risk of overvaluation. From an organizational perspective, flexibility in duration leads to negotiations over termination criteria, which increase the risk that firm investments will be governed by influence processes and idiosyncratic justification, rather than by a coherent portfolio strategy.

A fundamental difficulty raised by broad flexibility in response to interim signals is that while real options valuation techniques are well tailored to offering go/no-go guidance on project initiations, they are relatively equivocal regarding what to do after an initial "go" recommendation.

Clearly, the real options valuation framework can be reapplied to the evaluation of a given initiative at a future time, when some uncertainty is resolved (e.g., compound real options). Structured reevaluation, however, becomes increasingly difficult as uncertainty gets resolved in increasingly incremental steps that are a function of firm activity.[3] Incremental discovery leads to difficulty in drawing precise distinctions among a series of independent options, a single compound option, and more generic path-dependent search activities. The division of time into discrete decision points is typically artificial. Does the pharmaceutical firm that started a research project on Monday make a conscious choice on Tuesday to continue with the project? Ultimately, the structure of the discovery activity

---

[2] The fact that a real option may lack an explicit expiration date need not imply that the option persists indefinitely. In particular, competitive forces may result in the effective expiration of an option. However, even in cases where the competitive environment drives option expiration (e.g., McGrath, 1997; Trigeorgis, 1997), the timing of expiration cannot be specified ex ante. In addition, when the opportunity can be exploited in multiple markets, a rival's entry into one market does not necessarily lead to the expiration of the option in other markets, so even in these settings ex post endogeneity is present. Strategic interactions also present another dimension of endogeneity to the problem of evaluating options that leads to an additional set of challenges. As opposed to "standard" real options settings, where expected payoffs are exogenously determined, "competitive" real options analysis requires an accounting for the ways in which the firm's actions affect rivals' responses. The benefits of delay and exploration that are derived from sequential investments must be weighted against the potential benefits offered by commitment and preemption (Ghemawat, 1991).

Although still in infancy, theoretical attempts to marry real options with game theory (e.g., Grenadier, 2001; Kulatilaka & Perotti, 1998) suggest that analytic results can be obtained. These results, however, are very sensitive to underlying assumptions about parameter values — values that are impossible to assess with accuracy for many strategically important opportunities. Furthermore, for analytic tractability, these models necessarily ignore many important factors, such as the complexity and duration of competitive responses, identified in recent work on competitive dynamics (e.g., Ferrier, Smith, & Grimm, 1999).

[3] Trigeorgis (1997), for example, notes that a rich set of compound options poses analytic and computational challenges to the actual evaluation of real options. Consider that of the six factors that determine option value (value of underlying risky asset, exercise price, time to expiration, volatility of the value of the underlying asset, foregone cash flows, discount rate), the first four are especially difficult to specify with confidence for the case of strategic opportunities. See Bowman and Moskowitz (2001) for a reexamination of the effectiveness of the application of option analysis at Merck, and Lander and Pinches (1998) for a discussion of the modeling challenges to making real options analysis "practical."

FIGURE 4
"Option Traps" That Hinder the Abandonment of Opportunities

	Target market fixed	Target market flexible
Technical agenda fixed	Option trap: in the absence of expiration, the firm can maintain the option indefinitely until conditions improve "Things will get better"	Option trap: negative market signals may lead to a search for new potential markets or market interventions rather than abandonment "We can try it somewhere else"
Technical agenda flexible	Option trap: further development efforts always hold the potential for overcoming any negative market signal "We can try harder"	Option trap: too many degrees of freedom for ruling out success "We can make this work"

and the scope of the authority of the project team define these distinctions and thereby draw the boundary of applicability for options logic.

Organizational Factors

Because holding options open entails both organizational and financial maintenance costs (Garud & Nayyar, 1994), managing abandonment has direct implications for firms' success in managing real options. The "option traps" presented in Figure 4 highlight the forces that undermine timely project abandonment in different uncertainty resolution regimes. The challenge of project abandonment is complicated by the nature of organizational resource allocation (Bower, 1970) and the different perspectives and incentives of stakeholders at different levels in the organization.

This contrast is most succinctly characterized as the difference between "holding the option" and "being the option." For executives at high levels in the organization within whose purview lie a number of distinct initiatives, an individual initiative may have an optionlike quality in that the abandonment of the particular project may not entail significant consequences. However, the managers focused on that particular project may see greater potential in its pursuit, both because they are not aware of the larger set of alternative investments available to the firm and because of the career consequences associated with its abandonment. These considerations, in turn, act to increase these managers' dedication to achieving success with respect to a particular initiative. This dedication, while positive in the sense of increasing the likelihood of success, clearly hampers their ability to abandon initiatives when the learning outcomes are negative (Garud & Van de Ven, 1992). For this reason, the way in which the selection and resource allocation mechanisms are manifested throughout the organizational hierarchy is fundamental to the challenge of exploiting the flexibility inherent in real options.

These organizational drivers are compounded by psychological deterrents to abandonment. Given the difficulty firms have in incorporating the logic of sunk costs (Russo & Schoemaker, 1989), their tendency toward escalating commitments (Staw, 1981) and overconfidence (Camerer & Lovallo, 1999), the political impetus not to show failure (McGrath, 1999; Sitkin, 1992), and the natural desire to succeed, the challenge of abandonment—giving up on an opportunity that has a chance for success—is a large one.[4] This, essentially, is the dark side of managing projects with product champions (Maidique, 1980) and skunk works (Kanter, 1988), where the systems and support mechanisms put in place to create an impetus for starting innovations act directly against their objective reassessment and termination. Indeed, the challenges associated with abandoning projects can be

[4] Investment in joint ventures creates an organizational boundary that should make it easier to monitor, separate, and abandon an option investment (e.g., Kogut, 1991). Even in this ideal setting, however, the evidence of firms' effectiveness in managing exit and minimizing downside risk is mixed (Reuer & Leiblein, 2000).

greater than those associated with initiating them (Brunsson, 1982; Garud & Van de Ven, 1992).[5]

Real options logic makes a fundamental contribution to the structuring of risk. When deployed in real organizations, however, this logic must be complemented with appropriate controls (Block & MacMillan, 1993; McGrath & MacMillan, 2000). These controls need to reflect the evaluation traps created by flexible sets of possible responses to interim signals and the way in which initiatives may be usefully redirected. Given the vast set of possible paths that an investment effort may take, the "flexibility" of abandonment that is central to the evaluation of real options calls for an off-setting rigidity in specifying the set of allowable courses of action at the time of the initial investment.

The degree of flexibility in response to interim signals is likely to vary with the unit of analysis being considered and where the boundaries of the decision-making entity lie. At the level of the firm, it is difficult to presume that the firm can commit itself to a narrow range of response to information that may emerge for an initial "option" investment. Indeed, such a commitment may not be sequentially rational (Selten, 1975). However, at lower levels of the organization, subunits may be constrained in the range of initiatives that they pursue (Galunic & Eisenhardt, 1996), whether limited in the range of target markets that can be addressed or the set of technical approaches that can be explored. It is at these lower levels that real options may reside. Disciplined project management can foster organizational environments that facilitate real options strategies. Projects can be artificially constrained through milestones, strict deadlines, and regimented market focus to overcome the challenge of the impossibility of proving that all possible avenues of opportunity are unpromising and, in turn, that the abandonment of the initiative is appropriate. Such practices, however, run the risk of underexploiting the lessons learned from unsuccessful or only partially successful initiatives (e.g., Sitkin, 1992).

Our aim is not to dispute the importance of continued search activity. Indeed, we note that it is at the heart of well-admired processes such as skunk works, probe and learn, and so forth. Rather, our aim is to highlight the difficulty such processes raise for conceptualizing and managing real options. Specifically, the more freedom afforded to an option manager (i.e., the more active his or her role is in resolving uncertainty), the more likely the option is to take on a life of its own, independent of the requirement specified by the logic governing the firm's portfolio of options. When the selection and evaluation criteria differ across levels of the organization, which is increasingly the case as managers have more control and are more vested in their projects, freedom of action at the level of the project is not consonant with the consistent approach to portfolio management at the level of the firm. Making investment decisions under the assumption of clear-cut abandonment points, when organizational conditions are such that abandonment is unlikely to occur or unlikely to be timely, is likely to render an initial real options valuation to be quite misleading. Since real options framework is a theory of allocation and control, this raises a question as to the merits of its application under such conditions.

In sum, firms invest in the active development of technology, and not in *passive* bets as to winners and losers. Given the impossibility of proving failure and the absence of formal expiration dates for an option investment, a firm's internal selection regime dictates its ability to manage options. A well-managed real options strategy must necessarily guard against the natural momentum that builds up around hopeful activity.

Another way of characterizing our argument about the importance of abandonment is to recognize that, from the perspective of the option initiative, the key is to avoid false negatives (type II errors), in which valuable opportunities are foregone. However, from the perspective of managing the option portfolio, the key is not to forget about avoiding false positives (type I errors), which tie up resources that are better used elsewhere.

[5] In our own conversations with managers in high-technology firms known for the depth and breadth of their innovative capacity, they consistently spoke of the challenge of shutting down projects. At one firm, "death row" was used to describe a set of projects that the organization had formally recognized as in need of termination but that, for various reasons, kept getting stays of execution and surviving.

ALTERNATIVE SEARCH PROCESSES

While the notion of real options as a way to frame decision makers' choices under uncertainty is relatively new to the management literature, the problem of organizations confronting uncharted worlds is well established. A long-standing strand of the literature has highlighted the role and importance of search processes. Search may be local and problem driven, but it also may comprise more opportunistic, and less local, exploration. This latter notion of slack search (March & Simon, 1958) has been picked up and interpreted in terms of ideas of autonomous innovation (Burgelman, 1983) and skunk works (Kanter, 1988)—ideas that have not only been offered to capture existing behavior but also held out as normative suggestions for how organizations may cope with uncertain futures.

These images of search efforts, relatively loosely controlled but modest in financial commitments, lack the procedural rationality of real options decision making. No explicit decision is made as to what constitutes the appropriate amount of slack search at a given time; rather, the organization is viewed as having, or is encouraged to develop, a heuristic or norm regarding the appropriate level of resources to be allocated to such efforts.

Another clear distinction between this broad category of path-dependent principles and the real options framework is that specific initiatives are not endorsed or necessarily examined by higher-level actors within the organization. Slack is introduced into the resource allocation system such that modest initiatives can occur without direct corporate oversight. Of course, if such initiatives identify opportunities that appear to be worth pursuing and worth commercial development, it is likely that greater resource commitments will be necessary, and, as a result, higher-level organizational actors will need to be convinced of the initiative's merits. This commitment of more substantial resources and the organizational approval of such commitments is akin to the "striking" of a option. Again, though, it is important to note that the initial initiative was never chosen by the organization; rather, the organization chose to allocate sufficient slack such that such initiatives might emerge.

A natural question to ask is when might such less specified discovery processes be preferred to the more explicit decision calculus of real options? One complexity in considering this contrast is what one is to take as a characterization of a real options process. In our view, as a behavioral matter, the use of real options may not vary so greatly from the behavior that stems from slack search. Indeed, one of the merits of the real options framework is that it provides a procedurally rational justification for processes of slack search. In that sense, real options are a powerful tool for the "technology of foolishness" (March, 1988).[6]

Alternatively, we can make the reference point decision processes that more closely adhere to the precepts of real options decision making. One virtue of the real options approach is that since the initial stage-setting investments are more explicit and, in turn, more visible to higher-level actors within the organization, the organization will have a better sense of its portfolio of initiatives. An organization may feel that it wants some initial exploration in a variety of technologies or markets and would have a better sense of its overall exposure under the real options approach than under a system of slack search

The tradeoff to this benefit is, as we have argued above, that to make exploration initiatives conform to the structure of a real option, the boundaries of these initiatives should be tightly specified ex ante. If an initiative to explore the value of a particular technology or particular market application wanders in response to feedback from initial efforts or shifts in markets and technologies, the discipline required by real options is lost. However, such discovery processes are often heralded as valuable means by which innovation occurs (Adner & Levinthal, 2002; Day, 1990; Lynn et al., 1996). While these more generic path-dependent processes of exploration do not address the abandonment challenges we have highlighted, unlike real option approaches, they do not make implicit assumptions about termination that are critical to their validity.

[6] By "technology of foolishness," March (1988) refers to mechanisms that legitimate the allocation of resources for search and discovery processes.

Clearly, to the extent that the potential paths a stage-setting investment may take are inherently limited, the real options framework is more readily applied. In addition, in such settings the potential downside of its application in terms of foregone opportunities that are discovered ex post is sharply diminished. Such a perspective suggests that real options may be better suited to well-specified investments, such as overseas production facilities and innovation licenses, than to less structured opportunities, such as the development of new product technologies in the face of a wide set of possible technical solutions and market applications.

CONCLUSIONS

In some sense, our argument simply places the notion of real options in the broader context of sequential decision making. Actions are taken, beliefs are revised, and subsequent choices are made. One is then left with the empirical question of how to distinguish real options from the broader class of sequential decision-making processes. Demonstrating that firms' investment paths involve a sequential process of scaling up is, in itself, insufficient. There are a wide variety of processes that would generate such a pattern, such as adjustment costs and time compression diseconomies (Dierickx & Cool, 1989), and incremental learning (Cyert & March, 1963), that are not directly linked to ideas of real options.

As a result, we must move beyond measures of the performance characteristics of firms' project portfolios, whether average project losses at real options firms are smaller (because they have limited downside risk) or average project gains relative to losses are larger at real options firms (because losses are limited but upside risk is unbounded). Rather, we should examine the way in which the portfolio is adjusted over time and, in particular, firms' approaches to abandonment.

Thus, in terms of observed behavior, we would expect real options firms, relative to their traditional counterparts, to

1. abandon projects earlier and
2. have higher project abandonment rates.

Similarly, in terms of organizational processes, we would expect real options firms to have

1. stricter action mandates for business units and project teams;
2. formalized milestones and go/no-go procedures;
3. incentive systems, organizational cultures, and allocation mechanisms that are more tolerant of failure; and
4. review procedures that are more sensitive to the presence of different incentives at different levels of the organization.

More broadly, in terms of empirical research on the use of real options in organizations, we would hope to see demonstrations of patterns of systematic and structured decision making that demonstrate divergence from *both* net present value analysis *and* unstructured path dependence, as well as data that consider project termination, not just the initiation and pacing of investment.

Firms face complex and highly uncertain investment environments. The real options framework usefully highlights the links between current actions and the set of future possibilities. However, to the extent that exit criteria are not well posed in a world of action and open-ended search, this framing can be overly seductive. While real options logic may justify investments that would be rejected under the calculus of net present value, these "justified" investments may well destroy value when implicit assumptions about abandonment flexibility are wrong. In settings where the range of responses to the resolution of technical and market uncertainty is largely unconstrained, the utility of applying options logic is unclear. For such strategic investments, in contrast to financial options, exit criteria are not self-evident. Rather, in such circumstances it may be more useful to identify the possible sequence of experiments that will test the most promising market and technical paths available to the firm.

To be clear, our purpose is not to question the internal logic of real options. We do feel, however, that grouping all path-dependent activity under the real options label overextends real options logic as tool, framework, and even as metaphor, and undermines its effective application. We believe that understanding a theory's boundaries serves to make it more powerful and more precise. We argue that the cause of options thinking is best served in considering the boundaries of the domain of applicability of this logic for business strategy, and in defining its place within the broader set of tools that are available to address decision making under uncertainty. In this regard, we suggest that the answer to the question "What is (and is not) a real option?" has as much to do with the organization as it does with the opportunity.

REFERENCES

Adner, R., & Levinthal, D. 2002. The emergence of emerging technologies. ***California Management Review***, 45(1): 50–66.

Amram, M., & Kulatilaka, N. 1998. ***Real options: Managing strategic investment in an uncertain world***. Boston: Harvard Business School Press.

Block, Z., & MacMillan, I. 1993. ***Corporate venturing: Creating new businesses within the firm***. Boston: Harvard Business School Press.

Bower, J. L. 1970. ***Managing the resource allocation process***. Boston: Harvard Business School Press.

Bowman, E. H., & Moskowitz, G. T. 2001. Real options analysis and strategic decision making. ***Organization Science***, 12: 772–777.

Brunsson, N. 1982. The irrationality of action and action rationality: Decisions, ideologies, and organizational actions. ***Journal of Management Studies***, 19: 29–34.

Burgelman, R. A. 1983. A process model of internal corporate venturing in the diversified major firm. ***Administrative Science Quarterly***, 28: 223–244.

Camerer, C., & Lovallo, D. 1999. Overconfidence and excess entry: An experimental approach. ***American Economic Review***, 89: 306–319.

Copeland, T., & Antikarnov, V. 2001. ***Real options: A practitioner's guide***. New York: Texere.

Cyert, R. M., & March, J. G. 1963. ***A behavioral theory of the firm***. Englewoods Cliffs, NJ: Prentice-Hall.

Day, G. S. 1990. ***Market driven strategy: Processes for creating value***. New York: Free Press.

Dierickx, I., & Cool, K. 1989. Asset stock accumulation and the sustainability of competitive advanatage. ***Management Science***, 35: 1504–1511.

Dixit, A. K., & Pindyck, R. S. 1994. ***Investment under uncertainty***. Princeton, NJ: Princeton University Press.

Dosi, G. 1982. Technological paradigms and technological trajectories: A suggested interpretation of the determinants and directions of technical change. ***Research Policy***, 11: 147–162.

Ferrier, W. J., Smith, K. G., & Grimm, C. 1999. The role of competitive action in market share erosion and industry dethronement: A study of industry leaders and challengers. ***Academy of Management Journal***, 42: 372–388.

Galunic, D. C., & Eisenhardt, K. M. 1996. The evolution of intracorporate domains: Divisional charter losses in high-technology, multidivisional corporations. ***Organization Science***, 7: 255–282.

Garud, R., & Nayyar, P. R. 1994. Transformative capacity: Continual structuring by intertermporal technology transfer. ***Strategic Management Journal***, 15: 365–385.

Garud, R., & Van de Ven, A. H. 1992. An empirical evaluation of the internal corporate venturing process. ***Strategic Management Journal***, 13: 93–109.

Ghemawat, P. 1991. ***Commitment***. New York: Free Press.

Grenadier, S. 2001. ***Game choices: The intersection of real options and game theory***. London: Risk Books.

Kanter, R. M. 1988. When a thousand flowers bloom: Structural, collective, and social conditions for innovation in organization. ***Research in Organizational Behavior***, 10: 169–211.

Kirzner, I. M. 1997. Entrepreneurial discovery and the competitive market process: An Austrian approach. ***Journal of Economic Literature***, 35: 60–85.

Knight, F. H. 1921. ***Risk, uncertainty and profit***. Boston: Houghton Mifflin.

Kogut, B. 1991. Joint ventures and the option to expand and acquire. ***Management Science***, 37: 19–33.

Kogut, B., & Kulatilaka, N. 2001. Capabilities as real options. ***Organization Science***, 12: 744–758.

Kulatilaka, N., & Perotti, A. 1998. Strategic growth options. ***Management Science***, 44: 1021–1031.

Lander, D. M., & Pinches, G. E. 1998. Challenges to the practical implementation of modeling and valuing real options. ***Quarterly Review of Economics and Finance***, 38: 537–567.

Luehrman, T. A. 1998. Strategy as a portfolio of real options. ***Harvard Business Review***, 76(5): 89–99.

Lynn, G. S., Morone, J. G., & Paulson, A. S. 1996. Marketing and discontinuous innovation: The probe and learn process. ***California Management Review***, 38(3): 8–37.

Maidique, M. 1980. Entrepreneurs, champions, and technological innovation. ***Sloan Management Review***, 21(2): 59–76.

March, J. G. 1988. ***Decisions and organizations***. Oxford: Blackwell.

March, J. G., & Simon, H. A. 1958. ***Organizations***. New York: Wiley.

McGrath, R. G. 1997. A real options logic for initiating technology positioning investments. ***Academy of Management Review***, 22: 974–996.

McGrath, R. G. 1999. Falling forward: Real options reasoning and entrepreneurial failure. ***Academy of Management Review***, 24: 13–30.

McGrath, R. G., & MacMillan, I. C. 2000. ***The entrepreneurial mindset***. Cambridge, MA: Harvard Business School Press.

Miller, K. D., & Reuer, J. J. 1998. Firm strategy and economic exposure to foreign exchange rate movements. ***Journal of International Business Studies***, 29: 493–514.

Nelson, R., & Winter, S. 1982. ***An evolutionary theory of economic change***. Cambridge, MA: Harvard University Press.

Penrose, E. 1959. ***The theory of the growth of the firm***. New York: Oxford University Press.

Popper, K. R. 1959. ***The logic of scientific discovery.*** New York: Harper Torchbook.

Reuer, J. J., & Leiblein, M. J. 2000. Downside risk implications of multinationality and international joint ventures. ***Academy of Management Journal,*** 43: 203–214.

Russo, E., & Schoemaker, P. J. 1989. ***Decision traps: Ten barriers to brilliant decision-making and how to overcome them.*** New York: Doubleday.

Selten, R. 1975. Reexamination of the perfectness concept for equilibrium points in extension games. ***International Journal of Game Theory,*** 4: 25–55.

Sitkin, S. B 1992. Learning through failure: The strategy of small losses. ***Research in Organizational Behavior,*** 14: 231-266.

Staw, B. M. 1981. The escalation of commitment to a chosen course of action. ***Academy of Management Review,*** 6: 577–587.

Trigeorgis, L. 1997. ***Real options: Managerial flexibility and strategy in resource allocation.*** Cambridge, MA: MIT Press.

Van de Ven, A. H., Polley, D. E., Garud, R., & Venkataraman, S. 1999. ***The innovation journey.*** New York: Oxford University Press.

Ron Adner is an associate professor of strategy and management at INSEAD. He received his Ph.D. from The Wharton School, University of Pennsylvania. His research examines the impact of the demand environment on technology evolution and competition, as well as the organizational challenges that confront firms competing in heterogeneous demand environments.

Daniel A. Levinthal is the Julian Aresty Professor of Management and Economics at The Wharton School, University of Pennsylvania. He received his Ph.D. from the Graduate School of Business, Stanford University. His research interests focus on issues of organizational adaptation and industry evolution, particularly in the context of technological change.